CAMBRIDGE LIBRARY COLLECTION

Books of enduring scholarly value

Botany and Horticulture

Until the nineteenth century, the investigation of natural phenomena, plants and animals was considered either the preserve of elite scholars or a pastime for the leisured upper classes. As increasing academic rigour and systematisation was brought to the study of 'natural history', its subdisciplines were adopted into university curricula, and learned societies (such as the Royal Horticultural Society, founded in 1804) were established to support research in these areas. A related development was strong enthusiasm for exotic garden plants, which resulted in plant collecting expeditions to every corner of the globe, sometimes with tragic consequences. This series includes accounts of some of those expeditions, detailed reference works on the flora of different regions, and practical advice for amateur and professional gardeners.

A Manual of Materia Medica and Therapeutics

Having served as a military surgeon in India, where he also pursued botanical research and investigated the efficacy of Hindu medicines, John Forbes Royle (1798–1858) went on to become a professor of materia medica at King's College, London. Acknowledging the need for a thorough yet manageable textbook on the subject, he published in 1847 this manual containing entries on the medicinal substances derived from minerals, plants and animals that were used in Britain at that time. The terminology, operations and aims of pharmaceutical practice are also addressed, and the differing preparations of the London, Edinburgh and Dublin pharmacopoeias are taken into account for the benefit of students. Furthermore, the work provides information on recently discovered medicines, 'as may be seen among the Preparations of Iron and of Gold, as well as in Matico, Indian Hemp, Bebeerine &c'.

Cambridge University Press has long been a pioneer in the reissuing of out-of-print titles from its own backlist, producing digital reprints of books that are still sought after by scholars and students but could not be reprinted economically using traditional technology. The Cambridge Library Collection extends this activity to a wider range of books which are still of importance to researchers and professionals, either for the source material they contain, or as landmarks in the history of their academic discipline.

Drawing from the world-renowned collections in the Cambridge University Library and other partner libraries, and guided by the advice of experts in each subject area, Cambridge University Press is using state-of-the-art scanning machines in its own Printing House to capture the content of each book selected for inclusion. The files are processed to give a consistently clear, crisp image, and the books finished to the high quality standard for which the Press is recognised around the world. The latest print-on-demand technology ensures that the books will remain available indefinitely, and that orders for single or multiple copies can quickly be supplied.

The Cambridge Library Collection brings back to life books of enduring scholarly value (including out-of-copyright works originally issued by other publishers) across a wide range of disciplines in the humanities and social sciences and in science and technology.

A Manual of Materia Medica and Therapeutics

Including the Preparations of the Pharmacopoeias of London, Edinburgh, and Dublin, with Many New Medicines

John Forbes Royle

CAMBRIDGE
UNIVERSITY PRESS

University Printing House, Cambridge, CB2 8BS, United Kingdom

Published in the United States of America by Cambridge University Press, New York

Cambridge University Press is part of the University of Cambridge.
It furthers the University's mission by disseminating knowledge in the pursuit of education, learning and research at the highest international levels of excellence.

www.cambridge.org
Information on this title: www.cambridge.org/9781108069298

This edition first published 1847
This digitally printed version 2014

ISBN 978-1-108-06929-8 Paperback

A MANUAL

OF

MATERIA MEDICA AND THERAPEUTICS;

INCLUDING

THE PREPARATIONS OF THE PHARMACOPŒIAS

OF

LONDON, EDINBURGH, AND DUBLIN,

WITH MANY NEW MEDICINES.

BY

J. FORBES ROYLE, M.D. F.R.S.

LATE OF THE MEDICAL STAFF OF THE BENGAL ARMY;
MEMBER OF THE MEDICAL AND CHIRURGICAL SOCIETY OF LONDON;
OF THE MEDICAL AND PHYSICAL SOCIETY OF CALCUTTA;
AND OF THE ROYAL MEDICAL SOCIETY OF EDINBURGH, ETC.

PROFESSOR OF MATERIA MEDICA AND THERAPEUTICS,
KING'S COLLEGE, LONDON.

LONDON:
JOHN CHURCHILL, PRINCES STREET, SOHO.
M.DCCC.XLVII.

PREFACE.

If it should be asked, whether another work on Materia Medica was required in addition to the numbers which already exist, it must be replied, that this was undertaken at the repeated request of its intelligent Publisher, who may be supposed to be well acquainted with the wants of the Profession. This alone, however, would not have been sufficient to induce the Author to undertake the work, had he not also been aware from the complaints of Pupils, and convinced from his experience as a Teacher, that the Student of Materia Medica required something systematic to study, which brought up to the present time, should be sufficiently full for information, and yet as short and condensed, as was compatible with the avoidance of being superficial. The Author has been unable to satisfy himself on either of these points, chiefly from the difficulty of treating briefly, of so great a number of distinct substances in the three Pharmacopœias, with their several preparations, each requiring a number of distinct facts to be related respecting it, and without curtailment.

The difficulty of a brief treatment of the subject is further increased by Students attending their course of Materia

Medica in the first year of their attendance on Lectures, that is, before they have become acquainted with Chemistry, are ignorant of Botany, and have just begun their attendance on Physiology, preparatory to a study of the nature and treatment of Disease. This difficulty the Author has already alluded to in his Essay on Medical Education, p. 27, where he has stated, "that a portion of the difficulties at present experienced might be partially obviated, if Students were to commence their attendance at Lectures in the Summer, instead of the Winter Session. They might then pay attention to short preliminary Courses of Natural Philosophy and Chemistry, as well as to the ordinary Courses of Botany, of Zoology, or of Comparative Anatomy. By these means they would be gradually introduced to the objects, the principles, and the nomenclature of some of their principal studies, and thus be better enabled to commence attendance during the ensuing Winter Session, on the complete Courses of Anatomy and of Physiology, besides Chemistry and Materia Medica with Therapeutics." Until some such plan is adopted, or some of these Sciences become objects of general as well as of merely Professional Education, or until Materia Medica is removed to the second year's studies, difficulties will continue to be experienced in studying the subject. The Author has therefore introduced a notice of some of the laws and nomenclature of Chemistry, under the head of Pharmaceutical Chemistry, and some account of the Parts and Products of Plants, with their Physiology, Classification, and Medical Properties as connected with Structure.

The objects of Materia Medica and Therapeutics, and the Sciences connected with them, being treated of in the first pages of this Manual, the Author will here refer only to the Natural History arrangement which he has thought it ad visable to adopt. This he has continued to employ since the Session 1836-37, finding it, upon the whole, the most convenient for teaching, as he had previously found it the best suited for investigating a new subject, the Materia Medica of India.

The Author has explained, in his "Essay on the Antiquity of Hindoo Medicine," that his attention was first particularly directed to this subject at the request of the Medical Board of Bengal, in order to ascertain whether the Public Service might not be rendered less dependent upon the supplies from Europe, either by substituting articles indigenous in the country, or by cultivating exotics in the most suitable climates of the plains and mountains of Northern India. He made collections of all the drugs procurable in the Indian bazaars, tracing them as much as possible to the plants, animals, and countries whence they were derived. These were arranged under the three heads of the Animal, Vegetable, and Mineral Kingdoms, and catalogues were made of the whole, with their synonymes in various languages, and a notice of the medical properties ascribed to each. The articles amount to about 1000 in number, and though the Work still remains in Manuscript, some of the Botanical results were made use of in the Author's "Illustrations of Himalayan Botany." So the information respecting Animal and Mineral products was referred to, when proving the early periods at which their properties must have been investigated by the Hindoos. These investigations have been his authority for many of the historical determinations.

In a Natural History arrangement for Inorganic substances or the Mineral department, that adopted in Chemical works is decidedly the best, as so many Chemical preparations require to be treated of. In the Vegetable part, the

Author has followed the arrangement of De Candolle, as he had previously done in his large Work, and in his Syllabus of 1837. He has given in many instances the English names of the Natural Families which have been lately published by his friend, Dr. Lindley; but care must be taken to use them in the same precise manner that the Latin ones are employed. For the few Animals which require to be treated of, he has followed the arrangement of his friend, Dr. R. E. Grant, as in the above Syllabus.

The foregoing Sciences form the basis for an accurate knowledge of Materia Medica. For this purpose we require not only to know the kind and quality of the Drug to be employed, but also the sources whence it is obtained. The Pharmacopœias do not always determine these points; in fact, they sometimes give names which only suggest so many problems to be solved; for instance, some of the plants mentioned, yield no products at all, or none that are used, while others do not yield those of the best quality, which are alone supposed to be employed. In determining some of these questions, a good deal of space has been required.

The Author fears that some obscurity may at first appear in the directions for making the Preparations of the three Colleges, from the attempt to condense them as much as possible. But the points of coincidence have been made the basis of the directions, while the differences of the E. and D. Colleges have been included within brackets, with the initials of the College. He cannot but lament, as others do, that there should be so many points of what appear to be unnecessary differences: for the Students who have studied according to the directions of three Pharmacopœias become Practitioners in England, in the Army, the Navy, the East India Company's Service, and in the Colonies, where the formulæ of the London

Pharmacopœia are alone employed. Four-fifths of the differences might easily be abolished, if it was not found possible to observe uniformity in the whole, and some inert substances might with benefit be excluded. But the Author is far from being an advocate for limiting the recognised Materia Medica to too brief a list; as great advantages are very often derived from the power of changing one Remedy for another of nearly similar properties, and there are few that are exactly alike. It must also be remembered, that the British Schools of Medicine provide Practitioners for nearly all parts of the World, who, if they do not, while Students, acquire some general ideas respecting the resources of Materia Medica, will seldom do so afterwards. They usually remain content to depend entirely upon home supplies, instead of making use of the valuable resources often within their reach. To this cause chiefly we must ascribe the little addition to our knowledge respecting the History and Properties of new Drugs, and the inaccurate information which we have respecting many of those in constant use. The Author has not confined himself to the contents of the three Pharmacopœias, but has noticed many of the new medicines, as may be seen among the Preparations of Iron and of Gold, as well as in Matico, Indian Hemp, Bebeerine, &c.

The ultimate object of the previous accurate study being the acquirement of a knowledge of the modes of Action and Uses of these several Drugs as Medicines, this is necessarily the most important part of the whole study. The Author has usually towards the end of his Course grouped together all those Remedies which may be used for the same Therapeutical purposes. This he has also done in this Work in a tabular form, with general observations appended to each group. The Student will become aware of the many instances in which Medicines are relative agents from these observations. They ought to be read in connexion with the notices on the Action and Uses of Drugs at the end of each article. The Author has been obliged to compress his materials within the compress of a Manual, though the subject would require a volume to itself, to be treated in connexion with Modern Physiology, Pathology, and Practice of Physic.

Numerous Works have necessarily been consulted, and to save space they are sometimes referred to by initials. Among the older Chemical works, Murray's "System" and Aikin's "Dictionary," and among the more recent, the works of Professors Brande, Turner, 5th and 7th edit., Graham, Fownes, and Gregory, the "Handbuch der Chemie" of Gmelin, "Introd. to the Study of Chemical Philosophy" of the late Professor Daniell, and the "Rural Chemistry" of Mr. E. Solly, Thomson's "Organic Chemistry," Simon's "Animal Chemistry," and, for the chemical preparations of the London Pharmacopœia, the excellent observations of Mr. R. Phillips have been often quoted. (*Ph.*)

In the Botanical part, the Author has made use of the characters of the Natural Families of Plants as given by Jussieu, Brown, and Koch, as well as the excellent abstract of Dr. Lindley in his "Elements of Botany," those of Endlicher and of Bartling, also the elegant "Enchiridion Botanicum" of the former. For the descriptions of Plants, he has in many instances referred to the original describer, frequently to the "Encyclopédie Botanique" of Lamarck and Poiret, with De Candolle's (*D. C.* and *Dec.*) "Prodromus," and Walper's "Repertorium." For the British and European plants, usually to Koch's "Flora Germanica," and Smith's "British Flora," "Engl. Bot." (*E. B.*), and the "Bot. Magazine," (*B. M.*), with Babington's "Manual of British Botany," and on many occasions to Lindley's "Flora Medica." The Author has endeavoured to unite the full character of the Genus in that of the officinal species which belongs to it. Having previously paid attention to the identification of officinal plants, he has on the present occasion done so with as many as was in his power. Of the Botanical drawings used, some are original, others taken from the "Dict. des Sciences Naturelles," the "Genera Floræ Germanicæ" of Th. Nees v. Esenbeck, the "Flore Medicale," the "Medical Botany" of Churchill and Stephenson (*Ch.* and *St.*), and the Plates of the "Handbuch der Medicinisch Pharmaceutischen Botanik" of Fred. Nees v. Esenbeck and H. Ebermaier. (*Nees v. E.* and *Eberm.*) The "Pharmaceutische Waarenkunde" of Goebel and Kunze (*G.* and *K.*) is often referred to for representations of drugs. The Author takes this opportunity for thanking Dr. Falconer for various information, as also for his valuable description of the Assafœtida Plant.

Of works on Materia Medica, he must first mention those which he found most useful in drawing up his Course of Lectures for King's College in 1836, as he may have occasionally quoted unawares some peculiar fact or opinion, without referring it to its source. First, the very valuable "Dict. de Matière Médicale" of Merat and De Lens (*M.* and *De L.*), 1834, and the excellent "Dispensatory" of Wood and Bache (*W.* and *B.*), 2nd ed. 1834, in which not only are all the Preparations of the British Pharmacopœias given in full, but also those peculiar to the United States, and the Pharmacological information of Europe brought up to the time at which the work was written. He at the same time made use of Fée, "Cours d'Histoire Naturelle Pharmaceutique," the "Manuel de Matière Médicale" of Milne Edwards and Vavasseur (*E.* and *V.*), Guibourt's "Hist. Abregée des Drogues Simples," the excellent Dispensatories of Dr. A. T. Thomson (*t.*) and of the late Dr. Duncan; also the "Grundriss der Pharmakognosie des Pflanzenreiches," and at a later period, the "Lehrbuch der Pharmaceutischen Zoologie," both of T. Martius, and the "Medizinische Zoologie" of Brandt and Ratzeburg. From this work the zoological representations have been taken.

On the present occasion he has not only referred to many of these writers, but has added a more particular examination of the Dublin Pharmacopœia of 1826, that of London, 1836, and the revised Edinburgh, 1842, being the last of their respective Colleges. He has consulted Dr. Christison's "Dispensatory" (*c.*), 1842, for the preparations of the E. P., and Dr. M. Nelligan's "Materia Medica," (*n.*) 1844, for those which are peculiar to the D. P. The divisions of the old wine measure still used by the D. College are given at p. 5, where it is inadvertently stated that they still order the weighing of liquids. Their *libra*, when treating of liquids, means the old wine *pint.* Soubeiran's "Traité de Pharmacie" has also afforded much valuable information; and though among the last mentioned, not the least valuable, is Dr. Pereira's (*p.*) "Elements of Materia Medica and Therapeutics," 1st and 2nd ed. 1840 and 1842, which forms so extensive and accurate a Magazine of everything connected with Materia Medica. The Pharmaceutical Society, which promises to contribute so much improvement and accuracy to our Pharmacy, is often quoted by the title of its Journal, P. J., 1842-1846, as well as other Journals, Medical and Scientific.

Among the authors principally consulted on the Therapeutical mode of arrangement may be mentioned, Alibert, Barbier, Guersent, Trousseau and Piddoux, Bayle, "Travaux Therapeutiques," Murray's and Dr. A. T. Thomson's "Elements of

Materia Medica," and Dr. Paris's "Pharmacologia," 7th edit., which he would especially recommend to the Student's attention, as embracing a number of important points for his consideration.

The Practical Works which have been most frequently referred to, though not always quoted, are the "Cyclopædia of Practical Medicine," Dr. Copland's "Dictionary of Practical Medicine," the systematic Works of Cullen, Good, Craigie, &c., Dr. Prout on Stomach and Urinary Diseases, Dr. Elliotson's "Principles and Practice of Medicine," and the elegant and invaluable "Lectures" of his late colleague, Dr. Watson.

In conclusion, the Author must thank the Publisher, as well as the Printers, for the elegant and clear typography, and Mr. Bagg, for the skill and taste he has displayed in the wood-cuts.

TABLE OF CONTENTS.

MANUAL

OF

MATERIA MEDICA AND THERAPEUTICS.

"MATERIA MEDICA AND THERAPEUTICS," being the title of one of the courses of lectures attended by students of medicine, is adopted as that of this Manual, because it treats of the same subjects; that is, it gives an account of the substances and agents which are employed as remedies for the relief or cure of disease. These consist either of material substances or of the general powers of nature. Advantage is also taken, in some measure, and in particular cases, of mental affections and passions.

The subjects divide themselves naturally into two distinct branches.

1. MATERIA MEDICA; meaning, correctly, the material substances employed as medicines; but it usually includes all the other means which are employed with the same object. It treats of their natural characters, sensible properties, chemical qualities, and mode of action as medicines.

2. THERAPEUTICA, from θεραπεύω, *to take care of* the sick, to heal. Therapeutics, in the most comprehensive sense, includes the application of remedies for the prevention or cure of disease. As connected with Materia Medica, it treats of the modes of action and of the effects of medicines as employed for the restoration of healthy action, and the consequent removal of disease. Since a variety of substances are capable of producing the same general effects, though they may differ from each other in minute particulars, they are frequently grouped together, so as to form classes of Medicines. These are convenient for practical purposes, because what is best suited to the peculiarities of a constitution, or to the different stages of any particular case, may be more easily selected; or one remedy may be substituted for another, when the first has begun to lose its effect.

"Materia Medica and Therapeutics hold a middle place between the purely scientific and the strictly practical branches of professional study. Of the former, Natural Philosophy and Chemistry treat of the properties and intimate relations of all natural substances, as well as of the powers of nature. They form necessary preliminaries for fully understanding any of the natural sciences, and thus become a part of the studies of other professions as well as of medicine. Botany and Comparative Anatomy, treating of the structure and physiology of plants and of animals, as well as of their classification, geographical distribution, and uses to man, are equally essential objects of study to those who wish only to attend to agricultural or to horticultural pursuits, or to improving the breeds of animals. Even Human Anatomy and Physiology are often studied by those who desire only to become acquainted with the internal structure and functions of the body to perfect themselves as painters or as sculptors, or who desire to study the beauty and design displayed in the works of the Creator.

"The strictly practical branches of professional study—of which the principles, however, require also to be scientifically studied—are embraced in the departments of Surgery, Medicine, and Midwifery, which elucidate the nature and treatment of diseases, both external and internal, as well as those peculiar to women and children, and of the process of parturition. To these is superadded Forensic Medicine, a complicated branch of study, as it is connected with every other, and requires a knowledge both of healthy and of diseased structure, as well as of the effects of deleterious agents on the constitution.

"The practical sciences require for their study and practice a complete knowledge not only of their own particular subjects and of the above-mentioned preliminary sciences, but also of the agents and substances, whatever be their nature, which are called Remedies, and which are employed to alleviate or to remove all departures from the healthy state, or from what constitutes disease." *

A complete knowledge of Medicines consists in an acquaintance with their physical and chemical characters, their physiological action, and therapeutical effects. Medicines have been defined to be, all substances which have the power of modifying the actual state of one or more of our organs, and which possess this property independent of their nutritive qualities. Hence they are administered in disease, for the purpose of curing or relieving. M. Barbier gives as a distinctive character of remedies, the property of not being decomposable, nor of being easily transformed into chyle by the action of the stomach, but of being capable of modifying the state of this organ. Alimentary substances, on the contrary, are digested, and transformed into chyle. This, however, will not apply to all; for some of the vegetable salts, as the acetates, citrates, and tartrates, are converted into carbonates in passing from the stomach into the excretions; while there are no means of proving whether other substances strictly nutritive, such as fibrin and albumen, are decomposed or not. It is clear that aliments are

* Medical Education,—a Lecture delivered at King's College by the Author. Session 1844-45.

assimilated to our organs, and become an integral part of our bodies; whilst remedies do not contribute in a direct manner to nutrition. It is equally difficult to define remedies as distinct from poisons; for, in fact, many of the same substances act either as remedies or as poisons, according to the quantities in which they are applied to our organs.

Before knowing how and when to prescribe a medicine, its nature ought to be thoroughly understood. A knowledge of medicines, therefore, comprehends an acquaintance with their external character, their sensible properties, and their chemical nature, as well as their modes of preparation. With these, should be included some knowledge of the Natural History of the Animals, Plants, or Minerals which yield them. In order to prescribe them as efficient agents for producing changes in different organs and functions, we must be acquainted with their mode of action on the several tissues and organs, as well in a state of health as in disease,—that is, with both their Physiological action and their Therapeutical effects. With all this must be combined a knowledge of the forms in which they may most fitly be prescribed, the substances with which they may be combined, or with which they are incompatible, the doses in which they must be given, and the cautions which peculiar circumstances may render necessary: all which is embraced in the theory and art of prescribing, together with the rules for the diet and treatment of the sick and convalescent.

It is hardly necessary to state, that for the purposes of Materia Medica and Therapeutics, sufficient only of the sciences connected with the above subjects requires to be known so as to enable the mode of action and effects of substances, when employed as remedies, to be appreciated. These cannot be fully discussed in this Manual, for they include, in fact, a large portion of the objects of scientific study. Thus, the affections of the mind and the passions are treated of in systems of Mental and of Moral Philosophy. The general powers or forces of nature, such as Heat, Light, Electricity, and Magnetism, as well as the nature and constitution of the Atmosphere and of Water, as also the subject of Climate, are discussed both by the Natural and by the Chemical philosopher.

The material substances commonly called Medicines being necessarily obtained from the Mineral, the Vegetable, and Animal Kingdoms, might be expected to be treated of with the minerals, plants, or animals which yield them, in works of Natural History, that is, of Mineralogy, of Botany, and of Zoology. This, however ,is seldom the case except so far as their external characters, which require to be observed for the purposes of classification. But the internal structure of the plants and animals which yield them are examined by the vegetable and the comparative anatomist. The composition of *organic* beings and of their products, as well as of mineral substances, is ascertained by the chemist. The mineral kingdom is sometimes distinguished, by the name of *inorganic*, from the vegetable and animal, which differ very conspicuously, in the different parts of each, however constituted or howsoever composed, being *organised*. They are thus calculated to perform particular functions, which are controlled by the vital powers, and form the objects of study of both the Vegetable and Animal Physiologist.

Though for the study of Materia Medica it is not requisite to master these various sciences, yet as the object is to acquire an accurate knowledge of the nature of medicines, the modes of classification pursued in these several sciences may be adopted as the best method of attaining the object. Indeed, this mode of studying animal, vegetable, and mineral substances may with advantage be considered as abridged views of their respective sciences, in which the principles and classification are briefly treated of, and the details exemplified by medicinal substances.

The products of the mineral kingdom being inorganic, are also more simple in composition, and therefore desirable to study first; so that we may proceed from simply observing the external characters and chemical composition of minerals, to the complicated structure and functions of organized bodies. Medicines from the mineral kingdom may be arranged either according to their external characters, as in some systems of Mineralogy, or according to their chemical composition, as is the case in others, as well as in systems of Chemistry. The latter affords numerous advantages, besides enabling the substances which are produced in nature to be arranged together with those which are the result of pharmaceutical operations, as many of these are of a chemical nature.

Before proceeding to treat specifically of individual substances, it might be expected that something should be said of the modes of distinguishing Minerals by their external characters, or of the primitive forms of Crystals and of the laws of Crystallization, or of the several formations of rocks, whether primary, secondary, tertiary, or alluvial, from which different Medicinal substances are obtained.

It might also be expected that notice should be taken of the General Properties of Matter, as of the Attraction of Cohesion, which attracts the particles of matter to each other; or of such subjects as the Solidity, Hardness, Specific Gravity, powers of Electricity and Refraction of bodies, because these require to be noticed in the descriptions of each individual substance. But these subjects are fully treated of in their respective Sciences; and for the purposes of Materia Medica it is only required to make use of the correct Nomenclature, scientific Classification, and other information which these afford.

As it is more natural to attend first to what may be seen and touched, so in the following descriptions, the External, Physical, and Sensible Characters of each substance will be first noticed, and this before proceeding to the Chemical Composition, as this requires the *destruction* of a substance before its Analysis can be accomplished.

OPERATIONS OF PHARMACY.

Medicinal substances, as produced by nature, not being usually fit for exhibition as Medicines, require a number of preliminary processes, which are called Operations of Pharmacy.

These relate—1. To the choosing, collection, and preservation of Drugs. 2. Their preparation, to fit them for exhibition as Medicines, including their mixture and combination. For the sake of uniformity, these require to be ordered by authority, and are so in the several Pharmacopœias of London, Edinburgh, and Dublin.

The choosing of Drugs necessarily implies a knowledge of the whole subject, as all the characters both of genuine and of adulterated drugs should be well known before they are selected or purchased, or an attempt should be made to prescribe them. For this purpose all the external characters, as colour, smell, taste, form, consistence, comparative weight, fracture, degree of solubility, point of fusion, &c., also their chemical nature and composition, must be noticed, and their purity be ascertained by means of *Chemical Tests.*

The collection of Drugs requires, in addition, that the influence of different Physiological states, both in plants and animals, should be attended to. Hence season, situation, aspect, age, habit, being wild or cultivated, are all of importance.

The Preservation of Drugs requires attention to the best methods of drying both mineral and organic substances, as well as to the necessity of protecting most of them from the influence of air, of moisture, and of light. To protect them from the latter, black or green bottles are ordered for particular preparations.

Drugs require to be weighed, both for Pharmaceutical preparations and in dispensing. Troy, called also Apothecaries' Weights, are employed for these purposes, and are divided as below, with each weight denoted by its sign.

One Pound,	℔ i.	= 12 ounces,	℥ xij	= 5760	grs.
„ Ounce,	℥ i.	= 8 Drachms,	ʒ viij	= 480	„
„ Drachm,	ʒ i.	= 3 Scruples,	℈ iij	= 60	„
„ Scruple,	℈ i.	= 20 Grains,	gr. xx	= 20	„
„ Grain,	gr. i.				

Liquids used formerly to be weighed, as they still are in the Dublin Pharmacopœia. But as it is much more convenient to measure them, this method therefore is now generally adopted. The Wine Measure was formerly employed; but the Imperial Gallon is now used, and is thus divided and distinguished :—

IMPERIAL MEASURE ADOPTED IN THE PHARMACOPŒIAS, L. AND E.

				Minims.	Wine Measure in former Pharmacopœia.
One Gallon,	Ci. =	8 Pints,	O viij. =	76800	61440 = O viij.
„ Pint, .	O i. =	20 Fluidounces,	f℥ xx. =	9600	7680 = f℥ xvj.
„ Fluidounce,	f℥ i. =	8 Fluidrachms,	fʒ viij. =	480	480 = fʒ viij.
„ Fluidrachm,	fʒ i. =	60 Minims,	♏ lx. =	60	60 = ♏ lx.
„ Minim,	♏				

The Imperial Gallon and Pint evidently contain much more than the Wine Measure, in the proportion of about 5 to 4; but the College, by dividing the Pint into f℥xx, instead of into f℥xvi, obtain nearly the same quantities for the ounce, drachm, and minim. (v. Phillip's Transl. of the London Pharmacopœia.)

Bodies occupying the same space are, however, well known to differ much in weight, as, for instance, Lead and Cork, in consequence o the former containing more matter in the same space than the latter; or, in other words, Lead has a greater *density* than Cork. Bodies are, therefore, weighed under two points of view: first with respect to their absolute weights, and second with reference to that which is peculiar to each *species*, and is hence called *Specific Gravity*. This refers to the comparative weights of different bodies occupying the same space and referred to a common standard. As the quantity of matter within the same space differs very much according as it is more or less expanded by heat, so the Sp. Gr. of bodies is always referred to one temperature, that is, 60°; but the London College mentions 62° of Fahr. The quantity of a medicinal substance, as of an acid, an alkali, or a spirit, may, moreover, vary very much in the same bulk according as it is more or less diluted with water. The strength, therefore, which is the same thing as the Sp. Gr., requires to be ascertained both for Pharmaceutic and for Medicinal purposes. Water is, for convenience, taken as the standard to which the comparative weights of other bodies are referred; but its Sp. Gr., though usually reckoned as 1, is by some taken at 1000, to avoid fractional parts.

The Sp. Gr. of a liquid may easily be ascertained by weighing it in a bottle which holds exactly 1000 grains of water at 60°. Solids are weighed first in air, and then, when suspended by a hair, in water. In this case, they displace a quantity of water equal to their own bulk, and weigh less than in air, because they are supported by the surrounding water with a force equal to the weight of water which has been displaced. Rule: find the difference between the weight of the body in air, and when weighed in water; take this difference to divide the weight of the body in air, and the quotient will be the specific gravity. The Sp. Gr. of aëriform bodies is ascertained by weighing certain measured quantities when passed into a vessel exhausted of air, and of which the weight has been previously ascertained. The different gases vary very much in their Sp. Gr., but they are all referred to Atmospheric air as a standard. By careful experiment, it has been found that 100 cubic inches of air weigh 31·0177 grains at 60° of Temperature and 30 inches of Barometrical pressure.

Professor Daniell has given the following table of the Sp. Gr. of the lightest gas, of air, of steam, and of water:

Cubic inches.	Weights, Grains.	Sp. Gr. Air 1.	Sp. Gr. Water 1.
Hydrogen, 100	2·136	0·0694	0·0000846
Air, 100	31·000	1·0000	0·0012277
Steam, 100	19·220	0·6240	0·0007611
Water, 100	25250·000	814·0000	1·0000000

MECHANICAL OPERATIONS OF PHARMACY.

Before Drugs can be exhibited as Medicines, they require to undergo a variety of processes, some of a Mechanical, others of a Chemical nature. Of these some are intended merely to effect Mechanical Division. This is useful in two ways, first in assisting the chemical action of bodies upon each other, and secondly, in rendering them more easy of administration as Medicines. Some substances must be subjected to preliminary operations, as Cleaning, Cutting, Bruising, Grating, Rasping, Filing ; or they may be powdered in mortars of wood, iron, wedgwood, glass, &c. ; or ground in mills or between rollers. The different modes are sometimes distinguished by distinct names:

Pulverization by Contusion, as Pounding of tough substances.
" Trituration, Rubbing to a fine powder.
" Grinding, as in Mills and between Rollers.
" Friction, as with a Grater, file, or rubbing.
" Porphyrization, on a Slab with a muller.

Mediate Pulverization, when substances are added to assist the process, and are afterwards washed out.

Levigation . . . Fine Trituration with water, or any fluid in which the solid is not soluble.

Granulation . . . when melted metal is agitated till it cools, or is shaken in a box, or poured from a height into cold water.

As the finest powders prepared by the above means always contain some coarse particles, so methods are adopted by which bodies may be mechanically separated. Thus, Pharmaceutists adopt—

To separate solids from solids,

Elutriation, or washing. The fine particles, being suspended, are poured off from the coarser, and are then allowed to settle.

Sifting, as with Sieves.	Sieves may be simple or compound, of wire, perforated zinc, hair, or gauze; or the fine particles may be dusted through bags.

To separate fluids from solids,

Decanting: Deposition,	when performed for the sake of the solid.
" Defæcation	" " " of the fluid.

Or the fluid may be sucked off with a sucking-tube, or removed with a syphon.

Filtration . . .	with Funnels and Filters; woollen or paper filters, or powdered glass, or Charcoal and sand, may be used.
Expression, . . .	for separating vegetable juices, or pulp of fruits, oils, &c., from ligneous fibre, &c.
Clarification, or Despumation,	by adding different substances, as Albumen, or the white of egg, or isinglass, when a scum rises to the surface, or falls to the bottom, carrying impurities with it.

Fluids may be separated from fluids, when there is no affinity between them, and they are of different specific gravities; as by decanting, skimming-off, the use of the Separatory, or by a Syringe, &c.

Several preparations fitted for exhibition as Medicines are prepared by the above mechanical processes with the addition of mixing, such as Powders, Pills, Confections, Electuaries, Mellita, Mixtures, Cerates, Ointments, Plasters.

Operations of Pharmacy dependent on Heat. Some of those ncluded under this head are not usually considered instances of true Chemical Action, but as dependent on the ordinary laws of Nature. Here, however, it is immaterial in what light they are viewed, as the processes are the same, and depend chiefly on the effects of Heat in diminishing or increasing the cohesion of different bodies.

Temperature is measured by Fahrenheit's Thermometer, of which the Freezing Point is 32° and the Boiling Point 212°. A gentle Heat means any degree between 90° and 100°. A Water-Bath signifies an apparatus for heating by boiling water or by steam; a Sand-Bath, one in which a vessel is placed and warmed by the gradual heating of sand.

Fusion, or melting produced by increasing temperature, is practised with metals, wax, &c. Bodies first dilate, and then, if not decomposed, they melt, and each always at the same temperature. No further increase of temperature then takes place, as all subsequent additions of caloric become latent. Fusion may be effected in metallic vessels, in the open air, as in furnaces, or in crucibles: the London College mentions Hessian and Cornish ones. In some cases, fluxes are required; but these are not mentioned in the L. Pharmacopœia.

SOLUTION—When a fluid, as water, alcohol, ether, oil, &c., overcomes the cohesion of a body, and incorporates its particles within itself, without decomposing them, itself remaining transparent. When unable to dissolve more, it is said to be saturated. Heat usually, but not always, increases the solvent power of a liquid. Solution is favoured by the quantity of the solvent, by division, and by agitation, also by pressure. Cold is produced during solution, from a portion of the caloric of the liquid being required to enable the solid to pass into the same state. Hence freezing mixtures are produced by dissolving salts in water, or by mixing them with ice or snow.

Solutions of an Homogeneous solid. 1. In pure Water, as of Acids, Alkalis, and of Salts, or of certain vegetable and animal principles, as Gum, Sugar, Starch, or of Gelatine and of Albumen. 2. In Alcohol and Proof Spirit. Vegetable Alkalis are very soluble in alcohol, so is Iodine. 3. In Ether: this solvent is little employed in the Pharmacopœia, but is capable of dissolving many of the same substances as Alcohol, and more of some resins and fats. 4. In Oil. This dissolves Camphor and some of the acrid and narcotic principles of plants, whence the French employ several Oils.

Solutions of an Heterogeneous solid are differently named, according to the temperature employed. 1. Maceration, is an operation performed at ordinary temperatures, as from 60° to 80°, and which is continued for some time, as from 12 hours to a few days. *a.* With pure water as a solvent, forming cold infusions, which are useful when we wish to prevent aroma being dissipated, or to obtain a light infusion, free from principles which would be taken up if heat was employed. *b.* In Rectified or Proof Spirit, forming Tinctures, the former for the resinous, and the latter for the more gummy products of plants. The Tinctures may be either *Simple* or *Compound*, that is, when one, or when more than one, substance is acted upon by the solvent. Sometimes Ammonia is added, forming Ammoniated Tinctures. *c.* Ether is occasionally employed in making what are then called Ethereal Tinctures. *d.* Wine, used as a solvent, forms Medicated Wines, and is preferred for taking up some of the soluble principles: as in Vinum Colchici, V. Ipecacuanhæ, V. Opii, V. Veratri. *e.* Vinegar is used as a menstruum and preferred for some vegetable principles; but in most cases it is probable that some chemical change is also produced, as in Acetum Opii, A. Colchici, A. Scillæ.

2. Percolation is a kind of Maceration, but superior to it in its power of exhausting a body of its active soluble principles. This is effected by more minute subdivision of the solid, and by passing the liquid through it, and thus bringing all its particles in contact with the whole

of the menstruum. The solvent can then be used upon fresh portions of solid, which may be similarly exhausted until the liquid becomes of the desired degree of strength.

3. Digestion is similar to Maceration, but the action is promoted by a heat of from 90° to 100°.

4. Infusion—When boiling water, that is, at 212°, is poured on the leaves, barks, or roots, &c. of plants, and allowed to cool down. Many substances require first to be bruised or cut, so as to be permeable to the water. Polished metallic vessels, as they cool more slowly from radiating heat less freely, are usually preferred to make infusions in.

5. Decoction—When boiling water is used, and kept boiling for a shorter or longer period, all the principles soluble in water become dissolved, and some others are suspended with them. Decoctions are preferred when the full effects of some medicines are required. In other cases the aromatic principles are dissipated.

VAPORIZATION—Those operations in which liquid bodies become converted into vapours or gases. This may take place only at the surface of the liquid, when it is called Evaporation; or, if the vapour is formed by the addition of caloric throughout the whole mass, Ebullition.

Evaporation is adopted when the volatile liquid is allowed to escape, and the solid residue retained for use; as in the case of Extracts, Inspissated juices, in crystallizing salts, as in obtaining sea-salt. Evaporation when taking place at ordinary temperatures is called *spontaneous* evaporation; and as it takes place at the surface, shallow vessels with broad surfaces are necessary. During evaporation, cooling ensues, in consequence of the quantity of caloric required for the liquid to exist in the gaseous state. Hence the coolness produced by evaporating lotions, that of porous vessels, &c. The rate of evaporation increases as temperature is increased or pressure removed; hence it occurs at a much lower temperature in vacuo. *Inspissated Juices* are reduced to comparative dryness by this process. Where the active principles of a vegetable or animal have been obtained in solution in alcohol or water, and are similarly reduced to a proper consistence, we obtain *aqueous or alcoholic extracts.* As these are apt to be injured by heat, so it is of advantage to prepare them at as low a temperature as possible. Hence the superiority of those prepared in vacuo.

Ebullition taking place by additions of heat to the mass of liquid, a large quantity of caloric becomes latent (or hidden, as in evaporation,) to enable the liquid to exist in a gaseous state. The Boiling point varies in different liquids. Ether boils at 100°, Alcohol at 173°·5, Water at 212°, Oil of Turpentine at 316°, Mercury at 656°. It is also influenced by other circumstances, but especially by pressure, as fluids boil at 140° lower temperature in vacuo than in the open air. When pressure is increased, the boiling point is raised, and often also the solvent powers of the fluid. Boiling is employed in making Decoctions, &c., and also in the process of Distillation.

Distillation is employed in separating a volatile liquid from other substances which are either fixed or less volatile than itself. It consists of two processes; first, the application of heat to convert the volatile substance into vapour, and then the condensation of this vapour, in a separate vessel, into a liquid. The operation may be performed in a retort or still, to either of which a receiver kept cool or a refrigeratory must be fitted. The vessels may be of metal, of glass, or of earthenware, and the heat applied either directly or through the medium of a sand, water, or steam bath.

Distillation of Water. Distilled water is required to be employed in all the Pharmacopœia Preparations.

„ Distilled Waters. These contain a little of the volatile principles of plants, and may be distilled either off the plants, or by distilling some Essential Oil with water.

„ Essential Oils. Volatile Oil being diffused through the different parts of various plants, these are, if necessary, coarsely divided, and soaked in water. The oil is then distilled over with the water, from which it is afterwards separated as it floats upon the surface.

„ Acids. As of Acetic, Nitric, Hydrochloric, Hydrocyanic acids, &c., of Vinegar for purification.

„ Alcohol. This is first obtained in the form of Raw Spirit. It is rectified for the purpose of purification, and also for concentration, and has then a Sp. Gr. of ·835, Proof Spirit a Sp. Gr. of ·920. Rectified Spirit may be further strengthened by distilling it off Carbonate of Potash or dry Chloride of Calcium, which retain the water, while the volatile Alcohol is distilled off.

Distilled Spirits. These are colourless solutions in Alcohol of the volatile principles of plants, and are obtained by distillation in the same way as the distilled waters, but with rectified or proof spirit as the solvent.

Sublimation is distinguished from Distillation in the volatilized matter assuming a solid form on condensation, as in Sulphur, Sal Ammoniac, Iodine, &c.

Condensation. In the Processes of Distillation and of Sublimation bodies are first converted into vapour, and then reduced to the liquid or solid state by the simple reduction of temperature. The condensation of gases, or those which are permanent at ordinary temperatures and pressure, may be effected by increasing the pressure or by the application of great cold. Prof. Faraday, by combining the condensing powers of mechanical compression with that of very considerable depressions of temperature, has obtained a pressure equal to 50 atmospheres, and a cold equal to $-166°$ of Fahrenheit's scale, and has thus liquefied many previously uncondensed gases. Condensation of a gas may also be effected mediately, that is, by passing it through a liquid for which it has some affinity, or through which it may become permeated: *ex.* Solution of Ammonia, Liquid Hydrochloric acid, Carbonic acid water. The terms Congelation or Solidification are employed when a body assumes the solid form from the mere reduction of temperature.

Precipitation is the process when a body passes to the solid state so rapidly as to prevent the particles arranging themselves in any regular form, and therefore the *precipitate* falls as a more or less fine powder.

Crystallization. When bodies, in passing from the liquid or gaseous state, assume regular geometrical forms, the process is called *crystallization*, and the solid bodies, *crystals*. This may be effected by gradually cooling down any melted mass, as Sulphur, the Metals; or a vapour, as Sal Ammoniac or Corrosive Sublimate; or by slowly evaporating the liquid in which a solid may have been dissolved. The process then, depending on evaporation, requires to be performed in broad and shallow vessels. It is favoured by the presence of foreign bodies, &c., and by agitation; but the crystals then produced are small, as also when the process is quickly performed.

The majority of crystals, when deposited from their watery solutions, carry with them a certain portion of water, which they render solid, but much of which they also part with by the mere application of heat or by exposure to the air. This is called *water of crystallization*, and it exists in crystals in its equivalent proportion, or some multiple of it. Professor Graham considers a portion of this water, when it replaces a base, as essential to the *constitution* of the salt. This is then sometimes described as *water chemically combined.* A salt or body *combined* with water is called a *hydrate*, one without any, *anhydrous.* Salts which contain much water of crystallization, when heated, undergo aqueous fusion, and, the water being dissipated, they are left as dry salts: *e. g.* burnt Alum. Some, when exposed to the air, lose this water, and are said to *effloresce;* others absorb water, and are said to *deliquesce;* while those which undergo no change in the air are called *permanent.*

Crystals are also studied with respect to their forms and the nature of their formation. Crystallography is now an extensive and independent science, the principles only of which we need notice, as they are treated of in separate works, as well as in systems of Chemistry. As many substances have forms peculiar to themselves, it is necessary to be acquainted with those which characterize different pharmaceutical salts, as they may thus be distinguished even when very minute. Some forms, however, are common to several distinct minerals, and these are therefore sometimes grouped together according to their external characters. The forms of these crystals have been distinguished into those which are *Primary*, and others which, from being considered as modifications of them, are called *Secondary*. That the *external forms* are connected with a certain regularity of *internal structure*, is evident from the well-known facts of the *cleavage* of crystals, also their refraction of light and different degrees of expansion by heat. The primitive forms are, — 1. The Cube. 2. The Tetrahedron, contained under four equilateral triangles. 3. The Octohedron, contained under eight equilateral triangles. 4. The Hexangular Prism. 5. The Rhombic Dodecahedron, limited by twelve rhombic faces; and 6. The Dodecahedron with isosceles triangular faces. The secondary forms may be produced by modifications of the above, as by decrements of particles taking place on their *edges* and *angles*, which would produce a great variety of forms.

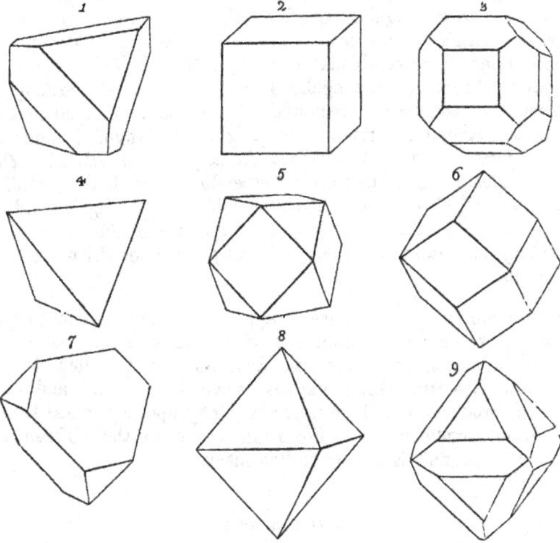

Though there can be no doubt that the internal structure of crystals must be as regular as their external forms, because it seems evident from the unequal expansion and contraction of certain crystals by changes of temperature, and also from their different modes of refracting light, yet, to use Mr. Daniell's words, crystallographers now confine themselves to the experimental determinations, and the geometrical relations of their exterior forms. Crystallography is now founded upon our ideas of the regular and exactly similar distribution of two, three, four, or any number of parts in symmetrical order; and that if one of the primary planes or axes of a crystal be modified in any manner, all the symmetrical planes and axes must be modified in the same manner.

The introduction of a systematic arrangement of crystalline forms according to their degrees of symmetry, is due to the labours of Weiss and Mohs, and its principles are explained in Prof. Daniell's Introd. to Chemical Philosophy. The classes and the forms which are referred to these will only be enumerated here, as many of them are mentioned in the descriptions of crystals in this work.

I. *Cubic, Octohedral, or Regular System*, having 3 rectangular axes all equal; contains the Cube, Regular Octohedron, Tetrahedron, Rhombic Dodecahedron, and Trapezohedron. II. *Right Square Prismatic System:* 3 rectangular axes, 2 equal, contains Square Prism and Octohedron with a Square base. III. *Rhombohedric:* 3 equal axes, not rectangular; 1 perpendicular to the 3; Rhombohedrons, Bipyramidal Dodecahedrons, Hexangular Prisms. IV. *Right Rectangular or Rhombic Prismatic:* 3 rectangular axes, no two equal; Right Rectangular Rhombic Prism, and Right Rectangular, and also Rhombic Octohedron. V. *Oblique Rectangular or Rhombic Prismatic:* 2 axes oblique, the third perpendicular to both. Oblique Rectangular and also Rhombic Prisms; Oblique Rectangular and also Rhombic Octohedrons. VI. *Double Oblique Prismatic:* 3 axes, all intersecting each other obliquely; Doubly Oblique Prism and also Octohedron.

By these various processes are obtained the Pharmaceutical preparations which are known by the names of Solutions, Tinctures, Wines, Vinegars, Infusions, Decoctions, and Extracts; also Distilled Waters; Spirits and Essential Oils; likewise some Precipitates and Salts, though the production of these depends chiefly upon Chemical Decompositions and Combinations. For a full account of the different processes consult Kane's Elements of Pharmacy.

PHARMACEUTICAL CHEMISTRY.

Substances which are throughout identical in nature, are subject only to the ordinary laws of Physics; but when substances different in nature, and minutely subdivided, come into contact, or are placed at insensible distances from each other, they become subject to a series of changes consequent on Chemical Attraction or Affinity. The result of this attraction is to unite two or more bodies together into one, which has properties usually very different from the bodies of which it is composed. One great object of Chemistry is to determine what bodies are compound, and what, from the inability of chemists to separate them into more simple bodies, are considered to be *Elements*. These are at present about 55 in number; a majority of the most important of them form objects of study in Materia Medica, either in their simple form, or as constituents of compound bodies.

As most bodies, when judged of by external characters, appear to be homogeneous, or composed of only one substance, the object of the Chemist is to ascertain whether this is actually the case, or only apparent. This he does either by the Action of Heat, or by presenting to the compound body some other substance which has a greater affinity for one of its constituents. This is called *Analysis*, or the separation of a Compound Body into its *constituent* parts, the quantities of each being ascertained. As a Compound Body is capable of combining with other bodies, which may themselves be either simple or compound, it is clear that by the process of *Chemical Analysis* a body may be reduced either into the substances from the immediate union of which it has been formed, and which are called its *Proximate Principles*, or into the *elementary* substances of which the latter consist, and which are then called its *Ultimate* Principles. When the constituent principles of a body can be so reunited as to reproduce the substance which has been analysed, the process is called *Chemical Synthesis*, and is the most certain proof of the correctness of an analysis.

These Decompositions and Recompositions, or Combinations, form the chief occupation of the Chemist, are always going on in the great operations of nature whether of growth or of decay, and require to be studied especially with regard to the laws which govern their combinations, which are called those of Chemical Affinity. It has been ascertained that though some substances, as Alcohol and Water, unite in any proportion, and others, as Salt and Water, in any proportion up to a certain extent, yet that the majority of substances (or, according to some, all substances) which form *true* Chemical combinations, unite only in *one* or in a *few fixed* and *definite* proportions. This forms the basis of what is called the doctrine of Chemical Equivalents, Definite Proportions, or of the Atomic Theory, of which the facts have been ascertained by numerous experiments, though the explanations are theoretical.

The combination of Bodies is much influenced by the different states in which they are brought into apparent contact, and much favoured by bodies being brought into the state of liquids, and also by heat, as is fully detailed in works on Chemistry. It is requisite to allude to the points requiring to be studied, in order to understand even the ordinary Pharmaceutical preparations. Bodies, it has been stated, combine with one another, not only in one, but often in several *definite* proportions; and it has further been ascertained that the quantity of one of them in the different combinations is found to be exactly double, triple, or some multiple of the other, and in the ratio either of 1 to 1, 2, 3, or 4, &c., or of 1 to 3, 5, 7. This is indicated by a peculiar *Nomenclature*. The *weights* of these equivalents have been ascertained by experiment to differ from each other. Hence their relative differences are pointed out by *Numbers*, which are in most instances peculiar to each of the *elementary* bodies. For the convenience also of a brief mode of stating all the facts respecting the composition of a Chemical substance, particular letters are adopted as the *Symbols* of the different Elements, as also *Formulæ*, which express the sum and differences of the substances employed in chemical decompositions and their results. Hence attention must be paid to the Equivalents, which are also the atomic weights of Chemical Elements, the *Symbols* by which these are distinguished, and to the *Nomenclature* by which the composition of a Body is at once known. The great advantage of attending to these points is the facility which they give for understanding complicated changes, especially in Organic Chemistry. They also show the exact quantities which are sufficient to produce particular changes, and also the weights of the products which are obtained.

To indicate the composition of Bodies as well as the proportions in which they are combined, a peculiar nomenclature is employed by chemists. Thus, Binary compounds which are not acid, of the non-metallic elements, as Oxygen, Chlorine, Iodine, Bromine, have their names terminating in *ide*, as Oxide, Chloride, Iodide, Bromide; also of the compound body Cyanogen, as Cyanide; of other substances, as Sulphur, Phosphorus, in *uret*, as Sulphuret, Phosphuret.

The number of Equivalents or Combining weights in a compound is shown—Of the first-mentioned element by a Latin numeral, as Bis, Ter, *e.g.* Binoxide of Mercury, indicating that 2 equivalents of Oxygen are combined with 1 of Mercury. If 2 equivalents of one body combine with 3 of another, that is, in the proportion of 1 to 1½, this is indicated by the word *sesqui*, as Sesquioxide, Sesquichloride. The first oxide is sometimes distinguished as the Protoxide. The affix of *Per* to an acid or an oxide indicates the highest proportion of Oxygen, &c., as Peroxide, Perchloride, Percyanide. The numbers of the second-mentioned element are indicated by Greek numerals, as Dis, Tris, *e.g.* Trisnitrate of Bismuth, where 3 equivalents of Oxide of Bismuth are combined with 1 of Nitric acid.

The nature of the acidifying principle of acids is indicated by the term *Oxyacids* being applied to those containing Oxygen, and ***Hydracids*** to those acidified by Hydrogen, with the word *Hydro* prefixed; as in Hydrochloric. Acids containing the largest proportion of Oxygen have their names terminating in *ic*, as Sulphuric acid, and that of their salts in *ate*, as Sulphate, Nitrate. A smaller quantity of Oxygen in the acid is indicated by the termination *ous*, as Sulphurous; their salts terminate in *ite*; while the addition of *Hypo* indicates a smaller quantity of Oxygen than in the compounds to which it is prefixed, as Hyposulphuric, Hyposulphurous. The excess of acid in a salt is indicated by calling it an *acid* or a *super-salt*, and the deficiency of the acid, by calling it a *sub-salt;* or, to show the excess of base, or of that which combines with and masks the properties of the acid, by calling it a *basic salt.* Double and Triple salts have their composition pointed out by the names of their Principles being all mentioned, as Tartrate of Potash and Soda or Potassio-Tartrate of Soda. A triple compound of the elementary bodies, as Oxygen, Hydrogen, and Carbon, as in Creosote, is called an Oxyhydro-carburet.

Substances combined with water are called *hydrates*, as the Hydrate of Lime; and as it seems in some cases to act the part of a base, it is then called *basic* water. Compound bodies, such as Cyanogen, which unite with elements as if they were themselves simple substances, are called *radicles.* There are many among organic compounds. These groups of a few elements in infinitely varied proportions are capable of combining with elementary substances and with one another, and of being substituted the one for the other, according to the laws of definite, multiple, and equivalent proportions. (Daniell, Chem. Phil. p. 604.)

The Symbols by which Elementary substances are distinguished are the first letter of their names in Latin, sometimes with the addition of a small second letter. The following elements comprise all those which are most generally diffused, and of about half of which the greatest portion of material substances are composed. All of these, with the exception of Silicon, form components of Materia Medica articles. The Numbers of the Equivalents are those adopted by Professors Brande and Daniell, and by Mr. Phillips in his Translation of the Pharmacopœia. In these, as generally in this country, Hydrogen is taken as unity, and being found combined in the proportion of 1 with 8 equivalents of Oxygen, the number of the latter is taken as 8. As these numbers are arbitrary, so others may be assumed: hence Oxygen, on the Continent, is made 100; but, as the others must all be proportional, Hydrogen will then become 12·5.

NON-METALLIC ELEMENTS.

Oxygenium	O	8	Brominium	Br	78
Hydrogenium	H	1	Sulphur	S	16
Nitrogenium	N	14	Phosphorus	P	16
Carbon	C	6	Boron	B	20
Chlorinium	Cl	36	Silicon	Si	8
Iodinium	I	126			

METALLIC ELEMENTS.

*Kaligenous Metals.**

Kalium (Potassium) K or Ka 40 Natrium (Sodium) Na or N 24

* Ammonium is an hypothetical metal, and might be enumerated here. But it will be mentioned under the head of Ammonia.

Terrigenous Metals.

Alkaline Earths.			*Earths Proper.*		
Barium	Ba	69	Aluminium	Al	14
Calcium	Ca	20			
Magnesium	Mg	12			

Metals Proper.

Manganesium	Mn	28	Stannum	St	59
Ferrum (Iron)	Fe	28	Stibium (Antimony)	Sb	65
Zincum	Zn	32	Arsenicum	As	38
Cuprum	Cu	32	Hydrargyrum	Hg	202
Bismuthum	Bi	72	Argentum	Ag	108
Plumbum	Pb	104	Aurum	Au	200

As some names occur very frequently, especially under the head of Tests, it is sometimes found convenient to have abbreviations independent of the Symbols, and without reference to the composition of a body. Also for Pharmacopœia, the letter P., and L. E. D. for London, Edinburgh, and Dublin.

Thus for, Water	Aq.	Citric Acid	Cit′
Distilled Water	Aq. dest.	Potassa	P. or Pot.
Sulphuric Acid .	S′ or Sul′	Soda	So.
Carbonic „ .	C′ or Carb′	Chloride of Barium . .	Cl. Ba.
Phosphoric „ .	P′ or Phosp′	Nitrate of Silver . . .	Nitr. Arg.
Oxalic „ .	O′ or Ox′	Sulphuretted Hydrogen .	Sulph. Hydr.
Acetic „ .	A′ or Acet′	Ammonia	Am.
Nitric „ .	N′ or Nitr′	Cyanogen	Cyan.
Muriatic „ .	M′ or Mur′	Hydrocyanic Acid . .	Hydrocy′
Tartaric „ .	T′ or Tar′	Ferrocyanide of Potassium	Ferrocy. Pot.

The above Symbols not only indicate their respective Elementary substances, but when alone, always stand for 1 Eq. of that Element. To indicate more than 1 Eq., numerals are added to the Symbols, as 2C, 3C, or O_2, O_3, O_4, &c. As these elements combine together, and form compounds always constant in nature, the composition of these is indicated by the juxtaposition of Symbols, or by placing the + sign between them, as HO or H+O, indicating 1 Eq. of Hydrogen combined with 1 Eq. of Oxygen, as in water. Numerals are added if more than 1 Eq. be present, as C 2O or CO_2, indicating that 1 Eq. of Carbon is combined with 2 of Oxygen, as in Carbonic Acid. Each compound has its own Eq. number, which is made up of the sum of those of its components. Thus $HO=1+8=9$, or the sum of the Eq. numbers of Hydrogen and Oxygen. So $CO_2=22$, because 1 Eq. of Carbon, 6, is added to 2 Eq. of Oxygen, $8\times2=16$. Here it may be seen that the Eq. number of Oxygen is the same in both cases, as it is indeed in all others, showing that these elements always combine together in the same relative proportions. This is the case also with the compounds, of which the Eq. number is always the same, and they are subject to the same law of definite proportions as the elements. Their composition is expressed in the same way. Thus, $HO+SO_3$, or HO, SO_3, means, in either case, 1 Eq. of Water combined with 1 Eq. of Sulphuric acid, a compound formed of 3 Eq. of Oxygen with 1 of Sulphur, having as its Eq. No. $9+(8\times3+16)=49$. When a large figure is printed before a symbol, it multiplies every symbol to the next comma, or to the next + sign, or all placed within brackets. When the Equivalent proportions of a compound are unknown, or when it is wished to state the per-centage of the components of a known body, the following method is adopted: thus

Wax is stated to be composed of		And Carbonate of Magnesia of	
Carbon	81·874	Magnesia	41·6
Hydrogen	12·672	Carbonic Acid	36·0
Oxygen	5·454	Water	22·4
	100·		100·

Besides combining in equivalent weights, substances have, when in a gaseous state, a certain relation to each other, and combine in certain proportions, that is, one measure or Volume with one or more Volumes of another gas. The resulting measure of the compound gas is either equal to the sums of the volumes of its components, or, in consequence of chemical union, it is condensed into a smaller compass, which however bears to the former a certain ratio. This, therefore, requires to be noticed in an account of a gas.

By taking advantage of the tendency of bodies to combine, and of the power of others to decompose them, are obtained the different Chemical preparations of Pharmacy. Thus by the mere process of Torrefaction or Roasting, some of the volatile parts of a body are expelled, while others undergo a change, or oxygen is absorbed. This process, therefore closely resembles Oxygenation, where oxygen combines with other substances, either by exposing them to the influence of pure Oxygen, or to that in the atmosphere, or by acting on them with a body, as Nitric acid, containing much Oxygen (cohobation). Deflagration is where a metallic body is ignited with a Nitrate, when the Nitric' yields its Oxygen, and the metal is oxidized. This process is often the converse of Reduction, where whatever is combined with a metal, as Sulphur or Oxygen, is driven off by heat, sometimes assisted by the presence of a flux. This, therefore, is often identical with Calcination, where the volatile components of a body are driven off by heat. By different methods of Decomposition, various acids are obtained from their salts. They are distinguished into Oxygen Acids, as the Nitric, Acetic, &c., and Hydrogen Acids, as Hydrochloric and Hydrocyanic. So the alkalis and other bases are obtained by the decomposition of their salts, as *e. g.* solution of Potash. The various salts are formed by bringing together the acids and bases, whether alkaline, earthy, or metallic oxides, in the requisite proportions, and ascertaining the saturation of neutral salts by means of litmus or turmeric paper. Some of the elementary substances, when in the same manner combined, are, from resemblance in nature to common salt (ἅλες), called *haloid* salts. For forming these, the law of definite proportions is of great value, as it informs us of the exact quantities of each substance to be employed, because the equivalent numbers indicate not only the proportion in which one body combines with another, but also that in which it will combine with every other. So if two neutral salts are mixed together, and mutually decompose each other, the results will also be two neutral salts, &c. Finally, the processes of Fermentation yield us our different Spirits, Wines, and Vinegars, as will be explained under their respective heads, and, under Etherification, the effects of acids on Alcohol.

MANUAL OF MATERIA MEDICA.

PART FIRST.

INORGANIC KINGDOM.

MINERAL AND CHEMICAL SUBSTANCES.

SEEING that Bodies are either Simple or Compound, and that the latter may be analysed, or the former united, it is obvious that the study of natural substances may be commenced either with the simple elementary bodies, proceeding thence to the compounds; or some one of the latter with which we are well acquainted may be taken and separated into its ultimate principles, and these studied before proceeding to others. There are two bodies, with the appearance of which all are well acquainted, and which, though apparently elementary, are now well known to be compound bodies. These are Atmospheric Air and Water, the nature and composition of both of which require to be studied before many of the processes for preparing articles of the Materia Medica can be understood.

ATMOSPHERIC AIR.

F. Air Atmosphérique. *G.* Atmosphärische Luft.

The Atmosphere, which everywhere surrounds the globe, extends to a height of 45 miles. It is an invisible gaseous body, devoid of odour and of taste, compressible, easily expanded by heat. Its Sp. Gr., according to the experiments of Sir G. Shuckburgh, is 001208 at 60° Fahr., and the Barometrical pressure of 30°. But being usually taken as the standard of comparison for gases, it is then commonly reckoned =1. 100 Cubic Inches weigh 31·0117 grains, and its pressure at the level of the Sea is equal to 15 pounds upon each square inch of surface, or a column of air one inch square, and extending to the limits of the atmosphere, weighs about 15 pounds, or the same as a column of Mercury, also one inch square, but only 30 inches high, which it is thus able to balance and support by the pressure of its weight. This weight must necessarily vary at great depths, as well as at great heights, as a greater or less mass of air will be superimposed. Hence the Barometer is employed for measuring heights, a diminution of one inch being found equal to about 1000 (922) feet. From being compressible, its density necessarily varies at different heights, the inferior strata being dense, and the upper ones rarefied. The temperature also diminishes as we ascend into the atmosphere, at the rate of 1° F. for every 100 yards, or, more correctly, for every 352 feet.

Though apparently simple in composition, it is actually composed of two very distinct gaseous bodies, Oxygen and Nitrogen, and a small portion of Carbonic Acid gas. The proportions in which these exist are:

Nitrogen gas	77·5 by measure,	75·55 by weight.
Oxygen gas	21 „	23·32 „
Aqueous vapour	1·42 „	1·03 „
Carbonic Acid gas	·08 „	·10 „

Omitting the aqueous vapour, which is variable in quantity, sometimes amounting, in hot countries, to as much as 2 per cent., the proportions of the permanent gases are stated by Humboldt, from experiments by himself and Gay Lussac, to be:

Nitrogen gas . .	0·787 by measure.
Oxygen gas . . .	0·210 „
Carbonic Acid gas . .	0·003 „

or, omitting the Carbonic Acid gas, 77 by weight of Nitrogen, with 23 of Oxygen, or by volume 79·2 of the former and 20·8 of the latter, in 100 parts.

The presence of Ammonia, in small quantities, has been detected by Liebig; and some Nitric′ is found also after thunder-storms. It has been imagined that some Hydrogen gas may exist in atmospheric air; and Muriatic Acid, it is said, has been detected in it at the sea-shore.

Dr. Murray has observed, that the Atmosphere may be regarded as a collection of all those substances which are capable of existing at natural temperatures in the aërial form, and which are disengaged by the processes carrying on at the surface of the earth. These, with other substances, as watery vapour, the effluvia from animals and vegetables, independent of Heat, Light, and the Electric fluid, form a vast mixture, the composition of which it is apparently impossible to determine.

Chemical Analysis has, however, proved that the various substances which may be mixed with the atmosphere quickly disappear, and are not to be detected by Chemical Tests, and that the composition of the air is everywhere uniform. It must, however, be admitted, that in very crowded assemblies of people, where there is a want of circulation, the quantity of Carbonic Acid gas is a good deal increased. It was at one time supposed that the constituents of the Atmosphere were retained by Chemical Attraction. Dalton, however, promulgated the opinion, that they were mechanically mixed, that the particles of the *same* gases repel, but that the particles of *different* gases do not repel, one another; and that thus one gas acting as a vacuum to another, and each being repelled by its own particles, they become diffused. Professor Graham has ascertained that each gas has a diffusive power, or *Diffusiveness*, peculiar to itself, which is inversely proportional to the square root of its density.

The properties of Atmospheric air are a mean of those of its constituents, and its chemical actions are due to the oxygen; by this it is enabled to support combustion; also the respiration of animals, a portion of it being by them converted into Carbonic Acid. Fishes also depend upon its presence in the water, though this dissolves only a small portion of air, but more of its Oxygen than of its Nitrogen. The processes of vegetation are also dependent upon the atmosphere, as it conveys water and also Carbonic acid to the leaves of plants, where the Carbon becomes fixed, and the Oxygen again set free; and thus plants contribute to purify the air which might become deteriorated by the respiration of animals. As the atmosphere varies in the quantity of moisture it contains, so it assists in the distribution of water over the surface of the globe, and is, by its mobility, the principal agent by which the extremes of temperature are moderated. According to the temperature and moisture of the atmosphere, so is the rate of evaporation, and consequently of perspiration. Hence not only many pharmaceutical preparations require a knowledge of the constitution of the atmosphere, but some of the functions of the body must be influenced by its different states, and the operation also of some classes of Medicines are modified by its different degrees of density and of dryness.

Oxygen.

Oxygen. *Vital Air.* *F.* Oxygène. *G.* Sauerstoff.

Oxygen (Symb. O. Eq. N. 8), named by the French chemists from οξυς, *acid*, and γεννaω, *I generate*, having been supposed by Lavoisier to be the only generator of acids. It was not discovered until 1774 by Priestley, though it is the most extensively diffused body in nature. It forms one-fifth by weight of the atmosphere, eight-ninths by weight of water, and probably not less than one-third of the solid crust of the globe; for Silica, Alumina, and Carbonate of Lime contain nearly one-half of their weight of Oxygen. It forms, moreover, one of the constituents of both animal and vegetable bodies.

Prop. Oxygen is a permanent colourless gas, devoid of odour and of taste. It is somewhat heavier than common air, as, at a temperature of 60° F. and a Barometrical pressure of 30 inches, 100 Cubic Inches weigh 34·25 grains; it is sixteen times heavier than an equal bulk of Hydrogen gas. Its Sp. Gr. is 1·111. 100 volumes of water dissolve 3·5 of the gas, but by pressure water may be made to take up much more of the gas. (Aqua Oxygenii or Oxygen Water.) Oxygen has the most extensive affinities, combining with every other known elementary body, except Fluorine. The bodies which are thus formed vary much in their properties. Some of them are called oxides, others alkalis, both of which, however, combine with a third set, possessed of very different properties, and which are called acids, or oxygen acids. Some bodies combine slowly with Oxygen, others with great vehemence, and with the evolution of light and heat, as in the combustion of bodies in the air, but of which the brilliancy is much increased if taking place in Oxygen gas. The respiration of animals is, in fact, a kind of combustion; the Oxygen of the air combines with the carbon of the blood, and is expelled in the form of Carbonic′ gas. But vehemence of action is moderated by its being diluted with four-fifths of Nitrogen gas in the air. The properties of Oxygen, therefore, require to be well understood, both with reference to the functions of Life and the mode of action and preparation of many Medicines.

Prep. Heat powdered and dried *black oxide of Manganese* in an iron gas bottle, till it is red hot; collect the gas. 1℔ should yield from 40 to 50 pint measures. Or mix *black ox. Mangan.* and *Sul.′* to the consistence of cream, and distil in a glass retort. Or the *red oxide of Mercury* or *Nitre* may be heated to dull redness to obtain this gas. 100 grs. *Chlorate of Potash,* heated in a retort or tube, yield 100 C. I. of very pure Oxygen.

Tests. A rough test of the purity of this gas is, introducing into it a *glowing* taper: if the gas be pure, the taper will immediately burst into a flame.

Action. Uses. Oxygen gas is stimulant when inhaled, and has hence been used, diluted with common air, in asphyxia, &c. Oxygen Water is a moderate stimulant, and may be given to the extent of a bottle or two daily.

Nitrogen.

Nitrogen. *Azote.* *F.* Azote. Nitrogène. *G.* Stickstoff.

Nitrogen (N = 14), the other constituent of the Atmosphere, was so called from being considered the producer of Nitre or of Nitrates. It was discovered in 1772 by Rutherford. Its properties may be considered the reverse of those of oxygen, as it will not support combustion nor the respiration of animals. In fact, it is fatal to them, but chiefly on account of the absence of oxygen. It is often called Azote, from α, *privative*, and ζωή, *life*. It is abundantly diffused, as it forms four-fifths of the atmosphere. Its chief use seems to be, to dilute the oxygen, though it no doubt also performs some more important functions. It exists also in small quantity in the Ammonia of the atmosphere, also in the Nitric acid which is found in it after thunder-storms. It forms a constituent of almost all animal bodies, and likewise of many vegetable products which form the food of animals.

Prop. Nitrogen in its simple state is best described by negatives, as it is devoid of colour, and is without taste or smell. It cannot support combustion, neither can it sustain respiration, and is nearly insoluble in water. It is lighter than common air. Sp. Gr. = ·975 ; 100 Cubic Inches weigh 30·15 grains. It forms, however, numerous compounds with Oxygen, &c., many of which are possessed of very active properties : *v.* Nitric acid, Ammonia, &c.

Prep. Nitrogen may be obtained by burning *Phosphorus* carefully in a jar of common air, when the whole of the Oxygen being abstracted, the Nitrogen is left comparatively pure. Or a mixture of *Sulphur* and *Iron-filings* made into a paste with water, and similarly enclosed, will slowly absorb the Oxygen.

Action. Uses. Nitrogen, being devoid of active properties, has been proposed to be employed in still further diluting common air in cases of excitement of the respiratory organs. Substances abounding in Nitrogen are suitable as food in some diseases.

Hydrogen.

F. Hydrogène. *G.* Wasserstoff.

Hydrogen (H = 1), from ὕδωρ, *water*, and γεννάω, *I generate*, does not exist free in nature, but combined is a constituent of water, of some acids, gases, and all vegetable matter. It was first correctly described in 1766 by Cavendish. It has hitherto been undecomposed, but several Chemists entertain the view that its base is analogous to a metal.

Prop. At common temperatures an invisible permanent gas, devoid of odour or taste; by exposure to intense cold, Faraday could not liquefy it. It is 14·4 times lighter than air, and its sp. gr. = ·0693. 100 C. I. weigh 2·14 grains. Water dissolves 1½ per cent. of its bulk of Hydrogen. When a lighted taper is brought in contact with the gas, it inflames, and burns with a pale yellow flame, uniting with the Oxygen of the air, and forming water. These gases, when mixed, do not unite until they are inflamed. Hydrogen, being the lightest body known, is assumed as the standard with which the equivalent numbers of other bodies are compared: its Eq. therefore is 1. But when Oxygen = 100 is taken as the standard of comparison, H = 12·5. It is interesting to us chiefly as being a constituent of Water, of Hydrocarbons, of Hydrochloric and Hydrocyanic acids, also of all vegetable and most animal substances.

Oxygen and Hydrogen.

Water. Aqua. (Distilled Water. Aqua Destillata.) *F.* Eau. *G.* Wasser.

Water (Aq. or HO=9), like the Air, is so universally diffused and well known, as not to require to be described. When pure, it is colourless and devoid of both taste and smell; but it may contain many impurities without these properties being sensibly impaired; therefore, for Chemical and Pharmaceutical purposes Distilled water should be employed.

Like Air, Water contains Oxygen, but it differs in being a strictly chemical compound. The other element is Hydrogen. One Equivalent of Oxygen, 8, is combined with one Eq. of Hydrogen, 1, making 9, the Equivalent number of water; which, therefore, or some multiple of it, is added to the equivalent number of chemical compounds when water is in combination. From its ready accessibility, water has been assumed as the standard of comparison for Specific Gravities. Its Sp. Gr. is therefore represented by 1, as in the L., or 1000, as in the E. P., as may be thought most convenient. Water, it is well known, freezes at 32° of Fahrenheit's thermometer, but attains its greatest density at 40°, and therefore ice readily floats upon water. It boils at 212°, and is then converted into steam, of which the Sp. Gr. is ·625 at 212° F., when it has the greatest density, and is composed of one volume of Oxygen combined with two volumes of Hydrogen. But water passes at all temperatures into the air by spontaneous evaporation, and causes its greater or less moisture or dryness. It enters into intimate combination with various bodies, which are then called Hydrates (from ὕδωρ, *water*), as in the cases of Lime and Potash, the Hydras Calcis and Hydras Potassæ L. P.; so also in some Liquids, as Sulphuric acid, and in Nitric acid, and in a variety of crystals; from these, however, a great proportion of water may be expelled by heat, and from some by mere exposure to the air. It forms a large proportion of most organized bodies, and dissolves a great variety of solid substances, and usually in increased proportion as its temperature is increased. It likewise dissolves many of the Gases, some, as Common Air, Oxygen, and Carbonic Acid gas, in small proportion; but others, as Ammoniacal and Hydrochloric acid gases, in immense quantities.

From the great solvent powers of water, it is seldom met with in a pure state. Rain-water even, contains some Carbonate of Ammonia or of Lime, which is floating in the atmosphere, and usually about 3½ cubic inches of common air, in 100 cubic inches of water. Spring-water generally contains Carbonate and Sulphate of Lime and Chloride of Sodium besides the usual proportion of air, and often Carbonic acid. Well-water, obtained by digging, usually contains a larger proportion of salts, and is often called *hard* water, because it curdles soap by decomposing it. River-water, though proceeding from springs, deposits upon exposure to the air many of the salts it contained, and is hence called *soft* water, because soap readily mixes with it. Some Spring-waters contain so large a proportion of impurities, as to be called Mineral waters. They will be mentioned under the heads of Acidulous, Sulphurous, Saline, Calcareous, and Chalybeate Mineral waters. Sea-water contains a still larger proportion of salts, especially common salt, with the Chloride of Magnesium and Sulphate of Magnesia.

Water necessarily commands a considerable share of attention, as it forms a portion of the aliment of both vegetables and animals; and, from its great solvent powers, is an important agent in pharmacy, as with it are formed various aqueous solutions, Distilled waters, Infusions, Decoctions, and it is employed to dilute Acids, Alkalis, and Spirits. It is useful also in some processes by becoming decomposed, when its Oxygen serves to oxidize different bodies, and its Hydrogen escapes in the form of gas. For Chemical and for most Pharmaceutical operations Distilled water (the Aqua Destillata of the Pharmacopœia) is required to be employed.

As a Therapeutical agent, also, water plays an important part, as it is often the best medium for applying either heat or cold to the body; and as it forms so large a portion of the blood, it is a chief means for increasing its fluidity, facilitating circulation, diluting secretions, and rendering them less acrid. It will act also as a solvent of many solid substances, as it passes through the system, and forms the principal part of Diluent and Demulcent Remedies.

SULPHUR.

Sulphur, E. Brimstone. *F.* Soufre. *G.* Schwefel.

Sulphur (S = 16), from *Sal*, salt, and *πυρ*, *fire;* was employed in medicine by the Greeks, Arabs, and Hindoos. It occurs in some Animal substances, as Albumen in Eggs, &c., in some plants, as in Cruciferæ, Umbelliferæ, Garlic, Fungi, &c., but chiefly in the Mineral Kingdom; also combined as in gases or salts, in some minerals and mineral waters. It is frequently found in combination with metals, as in the common ores called Pyrites, the Sulphurets of Iron, of Copper, Lead, Mercury, &c., whence it is obtained by roasting, in Germany, Sweden, and this country; the Sulphur, being volatilized, is collected in chambers. Native or Virgin Sulphur uncombined, is either a volcanic product, or occurs in beds in many parts of the world; that of commerce is brought chiefly from Italy, Sicily, and the adjacent islands, whence, in 1830, 236,338 cwt. of rough Sulphur were imported. It is afterwards purified by fusion, distillation, and sublimation, hence known under the names of Stick, Roll, Sublimed, and Flowers of Sulphur. Native Sulphur is purified by distilling it from earthen pots arranged in two rows on a large furnace. The Sulphur fuses and sublimes, and passes through a lateral tube in each pot into another placed on the outside of the furnace, which is perforated near the bottom, to allow the melted Sulphur to flow into a pail containing water, where it congeals and forms *rough* or *crude* sulphur. This being redistilled, forms refined sulphur. When fused and cast into moulds, it forms *stick* or *roll* sulphur.

Prop. An opaque brittle solid, crystallizing in acute octohedrons with an oblique base at temperatures below 232°, and at higher in the form of an oblique rhombic prism? occasionally in octohedra and tetrahedra. Fracture shining, crystalline. Sp. Gr. 1·98; when free from air-bubbles 2·080; that of the vapour is between 6·51 and 6·9. The colour in the solid state, when pure, is pale yellow; but it often varies from lemon-yellow, through green, dark yellow, and brown-yellow, according to the degree of heat to which it has been subjected. Taste insipid; odour generally none; acquires a faint and peculiar smell when rubbed. When grasped in the hand, it cracks: roll Sulphur feels greasy to the touch. Sulphur has no action on Vegetable colours. It is insoluble in water, slightly soluble in Alcohol, especially when finely divided, or the two are brought together in a state of vapour; in the same manner it is soluble in Sul. Ether, and in most fat and essential Oils, also in alkaline solutions, petroleum, &c. It is inflammable; when heated to about 300° it takes fire, and burns with a pale blue, and at higher temperatures a purple flame. It is a non-conductor of heat; when heated to about 180° it begins to volatilize, to fuse about 216°; between 226° and 280° it becomes perfectly liquid, and of a bright amber-colour; about 320° it begins to thicken, and becomes of a reddish-colour, and so viscid, that the vessel may be inverted without its running out. If in this state it is poured into water, it remains soft like wax for some time, and has been used for taking impressions of Seals, &c.; it may then be drawn into threads, which are elastic. From 482° to its boiling point, 601°, it becomes more fluid, and at 650°, if air be excluded, sublimes unchanged as an orange vapour. On cooling, it passes again through the same transitions; if slowly cooled, it forms a crystalline mass, and frequently retains its fluidity till touched by a solid body.

Sulphur is an elementary body, though it so often contains traces of Hydrogen that it was at one time thought to be a compound of that gas. It is exceedingly important as a chemical agent, forming Sulphurets with the various metals; with Oxygen it forms acids, of which the Sulphurous and Sulphuric are officinal; with Hydrogen, Hydrosulphuric Acid or Sulphuretted Hydrogen. It undergoes no alteration in the air at common temperatures. That of commerce, obtained from Pyrites, is often contaminated by metallic impurities, as Zinc Carbonate and Sulphate, Iron Oxide and Sulphuret, Arsenic, the Sulphuret, Silica, Magnesia, Alumina, and Carbonate of Lime.

Flowers of Sulphur are sometimes contaminated by the presence of a minute quantity of Sulphurous (stated by some to be Sulphuric) Acid, formed by a portion of the Sulphur during Sublimation uniting with the Oxygen in the apparatus. It should be freed from acidity by washing with hot water, when it should not affect Litmus paper.

Tests. Known by its colour, fusibility, volatility, burning with a blue flame and the evolution of the pungent vapours of Sulphurous acid gas formed during its combustion. It should completely evaporate when heated to 600°, and be perfectly soluble in boiling Oil of Turpentine. It should have no action on Litmus.

SULPHUR PRÆCIPITATUM. Precipitated Sulphur. *Lac Sulphuris*, or *Milk of Sulphur*. Is now omitted from the British Pharmacopœias, on account of its impurities, nearly two-thirds by weight of that of the shops having been Sulphate of Lime. It is very similar in most of its properties to sublimed Sulphur, but is whiter. It contains a little water. The impurities may easily be detected by heat, which will cause the Sulphur to evaporate, when the Sulphate of Lime will be left behind. One part of sublimed Sulphur was boiled with 2 parts of slaked Lime in 8 parts of water. To this solution Muriatic Acid q. s. being added, the Sulphur was precipitated. It was sometimes called *Lac Sulphuris* from holding a little water in combination ; and used to be preferred for its smoothness and want of colour, and from being easily suspended in liquids.

OLEUM SULPHURATUM. Sulphurated Oil. *Balsamum Sulphuris*, *Balsam of Sulphur*. Is now omitted from the Pharmacopœia. It was obtained by boiling Sulphur and Olive Oil, stirred together in a large iron vessel, until they gradually united. It is a dark reddish-brown viscid substance, having a very disagreeable smell.

SULPHUR (SUBLIMATUM), L. E. D. Sublimed Sulphur.

Prepared by reducing Sulphur to a coarse powder, and then subliming from a large iron retort into a sulphur-room, where the vapour is immediately condensed. Pulverulent, but when examined under a microscope, seen to be composed of minute crystals.

Prop. Its characters are the same as those of Sulphur.

Prep. E. Sublime *Sulphur* in a proper vessel, wash the powder obtained with boiling *water*, till the water ceases to have an acid taste. Dry the Sulphur with a gentle heat.

Tests. Evaporates totally at a temperature of 600°. Agitated with Aq. Dest., it has no action on Litmus. Heated with Nit', the solution diluted with water, neutralised with Carb. of Soda, and acidulated with Mur, should not yield a yellow precipitate with Sulphuretted Hydrogen, showing the absence of Arsenic. *v. Sulphur.*

SULPHUR LOTUM, D.

Prep. D. Pour hot *water* on *Sublimed Sulphur*, repeat the washing as long as Litmus indicates that the effused water is acid. Dry the Sulphur on bibulous paper.

Action. Uses. Alterative, Diaphoretic, mild Cathartic.

D. gr. v. and gr. x.—℈i. two or three times a day as an alterative; ℈i.—ʒiij. as a laxative.

UNGUENTUM SULPHURIS, L. E. D. Sulphur or Brimstone Ointment.

Prep. L. E. D. Mix thoroughly *Sulphur* ℥iij. (sublimed ℥j. E. lb. j. D.) with Hog's Lard lb. ß. (*Axunge* ℥iv. E. prepared lb. iv. D.) and *Oil of Bergamot*, ♏xx. L.

Action. Uses. Alterative, chiefly applied in Scabies and other skin diseases.

UNGUENTUM SULPHURIS COMPOSITUM, L. Compound Sulphur Ointment.

P. L. Mix *Sulphur* lb. ß., bruised *Veratrium* ℥ij., *Nitrate Potash* ʒj., *Soft Soap* lb. ß., *Lard* lb. j. ß., *Oil of Bergamot* ♏xxx.

Action. Uses. Irritating stimulant, applied in obstinate Scabies.

SULPHUR AND OXYGEN.

ACIDUM SULPHURICUM. Sulphuric Acid. *F.* Acide Sulfurique. *G.* Schwefelsäure.

Sulphuric Acid (S′ or $SO_3 = 40$) is one of the most important compounds of Chemistry. This acid is produced in small quantities in nature, as near volcanoes, in some acid springs, and exists in combination in numerous Sulphates, especially those of Lime (*Gypsum*) and of Magnesia, found as minerals, or in the water of springs. It was known to the Arabs, Persians, and Hindoos.

Sulphuric Acid appears, from its names, to have been originally made in Europe, and probably also in Persia, from the decomposition of Vitriol or Sulphate of Iron, a practice still followed at Nordhausen in Saxony. The Sulphate is first calcined, so as to expel nearly the whole of the water (of crystallization) it contains. The acid, distilled off in an earthenware retort at a red heat, comes over in vapours, which condense into a dark-coloured oily-looking liquid. This fumes when exposed to the air, and contains less than 1 Eq. of Water to 2 of Sulphuric Acid, has a Sp. Gr. of 1·9, and is known in commerce as the Nordhausen, Fuming, or Glacial Sulphuric Acid.

Prop. Sul′ may be obtained in a free or anhydrous state by carefully heating the Nordhausen Acid obtained from Sulphate of Iron in a retort, and condensing its vapours in a bottle artificially cooled. It then forms a white solid, fibrous like Asbestos, with some fine acicular crystals. At 66° it becomes liquid, and boils at 122°. Sp. Gr. 1·97. The dry acid does not redden litmus; when exposed to the air, dense white fumes are produced, from its condensing atmospheric moisture. It will combine with Water with explosive violence.

Acidum Sulphuricum, L. E. Sulphuric Acid. Acidum Sulphuricum venale, D. Oil of Vitriol. *Vitriolic Acid. Spirit of Vitriol.*

Prop. Liquid Sulphuric′ (S′, Sul′ or $S O_3 H O = 49$) is a dense oily-looking liquid, usually colourless, devoid of smell, but intensely acid, and powerfully corrosive. At first it feels oily, from destroying the cuticle, but soon acts as a caustic, charring both animal and vegetable substances by combining with the water and setting free the carbon. It freezes at from —15° to —29°, according to its density, and boils at 620° F. Its affinity for water is great, heat and condensation being produced on their union. It absorbs moisture from the atmosphere, $\frac{1}{3}$ of its weight in 24 hours, and 6 times its weight in a twelvemonth, and consequently becomes weaker the longer it is exposed. Professor Graham is of opinion that S′ combines with water in several other definite proportions. It may be diluted with it to any extent. It unites with alkalis, earths, and metallic oxides, rapidly dissolving some metals, as Iron and Zinc, when diluted with water. By the action of S′ on alcohol, Ether is produced. Several substances, as Charcoal, Phosphorus, &c., when heated with S′ decompose it, by abstracting its Oxygen, and evolving Sulphurous Acid.

Prep. S′ used to be made in this country by burning the imported Sulphur with a little Nitre. After the great increase which took place in the price of Sulphur, some manufacturers employed that obtained from Pyrites, which often contains Arsenic as an impurity. The proportion of Nitre was $\frac{1}{8}$, which was burnt with Sulphur either in the same chamber, lined with lead, and having its bottom covered with water, or in a furnace, from whence the vapours produced were conveyed into a similar chamber. The Sulphur in burning combines with 2 Eq. of Oxygen, and forms Sulphurous Acid gas, which escapes into the chamber. The N′ of the Nitrate, becoming decomposed, gives 3 Eq. of Oxygen to another Eq. of Sulphur, and thus some Sulphuric Acid is formed, which combining with the Potash of the Nitrate, forms Sulphate of Potash, which remains as a residual salt. The 2 Eq. of oxygen set free, form with the Nitrogen, Nitric oxide, which immediately takes 2 Eq. of Oxygen from the Atmosphere, and forms Nitrous acid gas, or $N O_4$; so that this, with Sulphurous Acid gas coexists in the atmosphere of the chamber, and if both are dry, no change occurs; but if moisture be present in the form of vapour, the Sulphurous Acid takes an Eq. of Oxygen from the Nitrous Acid, and becomes converted into Sulphuric Acid, while the latter becomes Hyponitrous Acid; these with a little water combining together, precipitate as a crystalline solid, forming a kind of Sulphate of Hyponitrous Acid. Immediately, however, on falling into the water, this compound becomes decomposed with effervescence. The Sulphuric acid unites with the water, and the Hyponitrous Acid escapes as Nitric oxide and Nitrous acid. These unite with the Oxygen of the atmosphere, react on the Sulphurous acid and humidity, and give rise to a second portion of the crystalline compound, which undergoes the same changes as at first. Thus the Nitric oxide is the medium for transferring Oxygen to the Sulphurous acid, to convert it into Sulphuric acid.

The mode of making Sul′ now varies from the above in many places, and has been described by Professor Graham. In this process Sulphurous Acid, from burning Sulphur, Nitric Acid vapour, and steam, are simultaneously admitted into oblong leaden chambers, so partitioned that the vapours can only advance slowly, and thus allow the whole of the Sulphuric Acid to be deposited.

When the S′ thus made has a Sp. Gr. of 1·5 (1·6 Gr.), it is drawn off, and is usually first conveyed to shallow leaden pans, where it is concentrated to a Sp. Gr. 1·70, or until it would act upon the Lead. It is then further boiled down in retorts of glass or of Platinum, until it has a Sp. Gr. of 1·84 ; when cooled, it is removed into large carboys, and forms the *Oil of Vitriol* of commerce. S′, in its most concentrated state, is a definite compound of 1 Eq. Acid and 1 Eq. of Water, which last cannot be separated by heat, as the acid and water distil over as a hydrate. The Sp. Gr. of the Acid Sulph. venale, D. = 1850.

Tests. The presence of S′, or of the soluble Sulphates, is easily ascertained by a solution of Chloride of Barium, or of the Nitrate of Barytes, as they form a white precipitate of Sulphate of Barytes, which is insoluble in either acids or alkalis. S′ should be colourless: Sp. Gr. 1·845. What remains after the acid has been distilled to dryness should not exceed $\frac{1}{400}$ part of its weight. Diluted S′ is scarcely coloured by Hydrosulphuric′. The want of colour indicates the absence of organic matter. The E. P. state its density to be 1·840, or near it: Mr. R. Philips says, he never found it under 1·844. Commercial S′ is apt to contain Nitrous Acid, or an oxide of Nitrogen, Sulphate of Lead. When diluted with its own volume of water, a scanty muddiness is produced by the deposition of Sulphate of Lead alluded to by the L. P. as amounting to $\frac{1}{400}$ part, and is pointed out by the H S′, which forms a Sulphuret of Lead. No orange fumes escape when no Nitrous acid is present, which, as well as Binoxide of Nitrogen, is indicated by a solution of the Protosulphate of Iron.

In consequence of some S′ being now made from the Sulphur obtained from Iron Pyrites, which often contains some Arsenic, this metal, in the form of Arsenious acid, is sometimes present. "Dr. G. O. Rees found 22·58 grains of this acid in f℥xx. of oil of vitriol, and Mr. Watson states, that the smallest quantity which he has detected is 35½ grains in f℥xx." (Per. p. 470.) For the detection of this impurity, the acid must be diluted, and the tests for Arsenic, q. v. applied.

ACIDUM SULPHURICUM PURUM, E. D.

Dilution with water and subsequent concentration is recommended by some, but the E. and D. P. give processes for purifying Sul′.

Prep. The D. P. directs distillation of *Sul′* in a retort of flint glass, containing a few slips of Platinum, to restrain the ebullition; the first twelfth part is to be rejected as containing too much water. Sp. Gr. 1845.

E. P. If commercial Sulphuric acid contains Nitrous acid, heat f℥ viij. of it with between 10 and 15 grains of Sugar, at a temperature not sufficient to boil the acid, till the dark colour at first produced shall have nearly or altogether disappeared. This process removes the Nitrous acid. Other impurities may be removed by distillation, as in D. P., but in a sand-bath, or with a gas-flame, and having a canopy above to keep it hot.

Dr. Christison finds that the Sulphuric acid, which is first rendered black and opake, gradually becomes pale yellow if kept for two hours near its boiling point, and that the Nitrous acid entirely disappears, without any material impregnation of Sulphurous acid. The proportion of sugar required must be first determined by an experiment on the small scale.

Tests. v. *Sul′.* Density 1845; colourless ; dilution causes no muddiness ; solution of Sulphate of iron shows no reddening at the line of contact when poured over it. E.

Inc. Many medicinal substances are incompatible with S′, as the Oxides of the Metals, some of the Earths, the Alkalis, and the Carbonates of all, also their Acetates, &c. The solutions of Acetate of Lead and Chloride of Calcium. The former it is especially necessary to remember, as it is often desirable to prescribe both Sul′ and Acetate of Lead in the same cases.

Action. Uses. Caustic, corrosive poison.

Antidotes. Chalk, Whiting, Magnesia, Soap. Dilution, Demulcents.

ACIDUM SULPHURICUM DILUTUM, L. E. D.

Prep. L. E. D. Mix gradually *Acid Sulph.* jf℥ ß. (f℥j. E. 1 pt. D.) with *Aq.* f℥xivß. (f℥xiij. E. 7 pts. D.)

Heat is evolved, condensation ensues, and a little Sulphate of Lead is precipitated. The three Diluted Acids are necessarily of different strengths. Prepared according to the L. P., it has a Sp. Gr. of 1·11, and f℥j saturates 28 grains of crystallized Carbonate of Soda. That of the E. diluted acid=1090; their comparative strengths by weight are about as 100 to 78, and by volume the difference is still greater. (R. Phillips.) (The Sp. Gr. ex D. P.=1084.)

Action. Uses. Refrigerant, Astringent, Tonic.

D. ♏x.—♏xxx. diluted with water, some bland liquor, or some bitter infusion.

Infusum Rosæ Compositum, contains about ♏ivß in each f℥.

ACIDUM SULPHURICUM AROMATICUM, E. D.

Prep. Add gradually *Sulph′* f℥iijß. (f℥vj. D.) to *Rectified Spirit* Ojß. lbs. ij. D.) Let the mixture digest at a very gentle heat for three days in a close vessel. Add *Cinnamon* bruised ℥j ß. *Ginger* do. ℥j. E. and D. moistened with a little of the acid spirit, and after twelve hours the powders may be exhausted by percolation with the rest of the spirit. Or the mixed powders may be digested for six days in the spirit, then straining the liquor.

This is intended to be a simple form of the Acid elixir of Mynsicht, and is a pleasant method for exhibiting Sulphuric Acid, as it is merely diluted with spirit instead of with water, with the addition of aromatic principles; Dr. Duncan having ascertained that the alcohol and Sul mixed in the above proportions do not react on each other, as has been sometimes supposed.

UNGUENTUM ACIDI SULPHURICI, D. Sulphuric Acid Ointment.

Prep. Mix *Sulph'.* ʒj. *Prepared Hog's Lard* ℥j. in an earthenware mortar.

Action. Uses. A powerful stimulant application in some obstinate cutaneous affections.

PHOSPHORUS.

F. Phosphore. *G.* Phosphor.

Phosphorus (P=16), from φως, *light*, and φερω, *I bear*, though so remarkable a substance, was not discovered before 1669, in the Phosphate of Soda and Ammonia of Urine, by Brandt, an alchemist of Hamburgh. Kunckel in Germany and Boyle in England had also the credit of discovering it. The Continent was for some time supplied with Phosphorus from England, prepared by Hankwitz, an apothecary of London. It is now procured almost entirely from the ashes of bones, which consist chiefly of Phosphate of Lime. It is a constituent of other animal substances, and also of some vegetables. It occurs in the mineral kingdom in the form of phosphates, which give fertility to some soils.

Prop. Phosphorus is a soft, flexible, easily cut, semi-transparent, wax-like solid, Sp. Gr.=1·77, colourless, or yellowish, devoid of taste as well as smell when pure; but immediately it is exposed to the air, garlicky fumes are evolved, and it becomes luminous in the dark, in consequence of its combination with Oxygen: this slow combustion produces heat, and sets it on fire. It is so inflammable as to take fire spontaneously in the open air; hence it is requisite to keep it under water, as the least friction excites heat enough to make it do so. When air is excluded, it melts at 108° F., and boils at 550°, passing off as a colourless vapour; at 32° it is crystalline and brittle, and from its solution in hot naphtha it may be obtained in dodecahedral crystals. Phosphorus is insoluble in water, but if kept long in it, some of the water becomes decomposed; and, if in closed vessels, according to Berzelius, the water becomes luminous if agitated. The Phosphorus also acquires a white coat, and, if much exposed to the light, becomes of a reddish tint. This has been thought to be an Oxide or a Hydrate, but by Rose to be only a peculiar mechanical state. By the aid of heat, P is soluble in Alcohol, Ether, the fixed and volatile Oils, in Naphtha, Petroleum, Sulphuret of Carbon, &c.

Prep. It is obtained on the large scale by the action of Sulphuric acid on powdered bone-ash, which consists chiefly of Phosphate of Lime. This is digested with half its weight of S′ diluted, which unites with a portion of Lime, &c. A Superphosphate of Lime is formed by combining with the Phosph set free. This is dissolved out, filtered, and evaporated. It is then mixed with one-fourth its own weight of Charcoal, and some fine quartz sand, and strongly heated. The Carbon takes the Oxygen of the free P′ becoming Carbonic Oxide, which passes over. Phosphorus is evolved, and received under water in the form of a reddish brown fusible substance. It is purified by pressing it, when melted, through chamois leather, also by melting in hot water in glass tubes, from which, on cooling, the Phosphorus separates in little sticks. This is the form it is usually met with in commerce.

Tests. Easily recognized by the above very remarkable characters.
Action. Uses. Irritant poison. Stimulant in very small doses.
Antidotes. Dilution. Demulcents. Magnesia.

Phosphorus and Oxygen.

Phosphoric Acid. *F.* Acide Phosphorique. *G.* Phosphorsäure.

Phosphoric Acid (P′ or $P_2O_5 = 72$) was first distinguished in 1760 by Marggraff. Phosphorus unites with Oxygen, and forms oxides and acids. Phosphoric′ may be obtained by the first part of the above process for obtaining Phosphorus, by decomposing the solution of Biphosph. Lime by Sesquicarb. Ammonia, separating the Lime and heating the Salt obtained, during which Am. escapes and P′ remains; or by burning P. in Oxygen or in Air; or enabling it to obtain the Oxygen from some substance which parts easily with it, such as Nitric′. This last is the method adopted by the L. P. The compounds of Phosphorus with Oxygen are among those which much interest chemists, and are fully treated of in chemical works.

Acidum Phosphoricum dilutum, L. Dil. Phosphoric Acid.

Prop. Diluted Phosphoric Acid is a colourless solution, having a Sp. Gr. of 1·064, without odour, having a powerful acid taste, and reddening Litmus. It forms salts with alkalis, earths, and metallic oxides. Even when in a concentrated state, it is not corrosive on organized structures. By evaporation it is reduced to a state of concentration and the appearance of a brown oily liquid, and is then a Hydrated Phosphoric acid. By an increase of temperature, it loses more water, and becomes Pyrophosphoric acid. When the heat is raised to redness, another equivalent of water is lost, when it becomes Metaphosphoric acid. The Diluted Acid contains 1 Eq. of acid to 3 of water, which, according to Professor Graham, acts the part of a base in the above acids. Liebig considers the three acids as different, and that Hydrogen is united in each with a different compound radical. 100 grs. of this diluted acid saturate 42 grs. of crystallized Carb. Soda, or f℈j, 24·4 grs., showing that it contains 10½ per cent. of Acid. Phosphoric Acid consists of 5 Eq. O 40 + 2 Eq. Phosph. 32=72, with 3 Eq. H O=99, or of 2½ O + 1 P=36.

Prep. Mix *Nitric'* f℥iv. with *Aq. Dest.* f℥x. add *Phosphorus* ℥j. to the diluted acid in a glass retort placed in a sand-bath, apply heat until f℥viij. pass over. Let these be returned into the retort, and again distil over f℥viij., which are to be thrown away. Evaporate the remaining liquor in a platinum capsule till only ℥ij. and ʒvj. remain. When cool, add *Aq. dest.* to make up f℥xxiv.

Here, as explained by Mr. R. Phillips, diluted acid is used, in consequence of the explosion and rapid combustion which Phosphorus causes when added to strong Nitric'. With the diluted acid the action is slow, the Phosphorus gradually melts, and combines with a portion of the Oxygen of the acid, while the remainder escapes in combination with the Nitrogen of the Nitric' in the form of Nitric oxide gas. A portion, however, of the Nitric' distils over before the whole of the Phosphorus is acidified, and therefore it is necessary to return it into the retort as directed. The Platinum capsule is required in the latter part of the process, because when the acid becomes concentrated it would act upon the glass.

Tests. Phosphoric acid with Soda forms an officinal salt (the Phosphate of Soda). Lime-water produces in it a precipitate of insoluble Phosphate of Lime. This, and the Phosphates of Barytes, Strontian, Lead, &c. are soluble in dilute Nitric Acid, also in Ammonia. The Nitrate of Silver throws down a yellow precipitate of Phosphate of Silver in a solution of a Phosphate. Arsenious Acid is also similarly affected; but it may be distinguished from it by Hydrosulphuric Acid, which causes a yellow precipitate in the Arsenious, but none in the Phosphoric Acid. The addition of Carbonate of Soda ought to cause no precipitate if the acid contains no Lime as an impurity, which it is apt to do from the mode in which it is usually made. "Chloride of Barium or Nitrate of Silver being added, whatever is thrown down is readily dissolved by Nitric';" the first indicates the absence of Sul.', the second of Hydrochloric'. "Strips of copper and silver are not at all acted upon by it," showing that no Nitric' is present; "nor is it coloured when Hydrosulphuric' is added," indicating the absence of metals.

Action. Uses. Refrigerant, Acid Tonic.

D. ♏x.—f℈i. diluted with sugar and water.

Incomp. All such substances as are incompatible with other Acids.

Boron.

F. Bore. Borium. *G.* Boron.

Boron (B=20, or 10, Berz.) was ascertained, by Sir H. Davy, in the year 1807, to be the basis of Boracic Acid. It was obtained by heating Potassium with Boracic Acid, as a dark olive-coloured powder, devoid of taste and smell, and not acted upon by the usual reagents: heated in the air or in Oxygen, it is converted into Boracic Acid.

BORACIC ACID. Acidum Boracicum. *F.* Acide Boracique. *G.* Borax Säure.

Boracic Acid ($B\,O_3$=44) is so named from Borax, a salt long known, and which is a Biborate of Soda (q. v.); from the Soda of which it may be separated by the action of Sulphuric′: a soluble Sulphate of Soda is formed, and the Boracic Acid precipitated. Boracic Acid is largely obtained in Tuscany and in the Lipari Isles, where it issues in the steam from fissures in the earth. Circular basins are dug or the fissures surrounded with cylinders of brick-work, and water is let in, which boils up, from the vapour passing into and through it. The water, having dissolved the acid, is evaporated, and, as it cools, the acid is deposited in scale-like crystals, which are then dried.

Prop. Boracic Acid is usually seen in transparent scale-like crystals, which have a feeble acid taste, and redden Litmus slightly, at the same time that they colour turmeric brown, like an Alkali. It is sparingly soluble in cold, but requires less than 3 times its weight of boiling water to dissolve it. It is very soluble in Alcohol, tinging the flame of a green colour when it is burnt. The crystals contain 3 Eq. of water of crystallization, which is expelled on their being heated: the acid then melts, and on cooling is brittle and glass-like, and may be variously coloured. As the salts of this acid promote the fusion of other bodies, Borate of Soda is much employed as a blowpipe flux.

Action. Uses. Boracic Acid is not officinal, except as a constituent of Borax, the Biborate of Soda. It was formerly supposed to have some anodyne properties, and was known by the name of *Sal sedativus.* It is now sometimes used to increase the solubility of Cream of Tartar, 1 part Bor′ with 7 parts Bitartrate of Potash forming the SOLUBLE CREAM OF TARTAR of the shops.

NITROGEN AND OXYGEN.

Nitrogen combines with Oxygen in several proportions, which it is desirable to notice, as illustrating the law of definite proportions, and because the several compounds are formed in the course of making various Pharmaceutical preparations.

Nitrous Oxide . .	$N\,O$	14 + 8 = 22.
Nitric Oxide . . .	$N\,O_2$	14 + 16 = 30.
Hyponitrous Acid .	$N\,O_3$	14 + 24 = 38.
Nitrous Acid . . .	$N\,O_4$	14 + 32 = 46.
Nitric Acid . .	$N\,O_5$	14 + 40 = 54.

ACIDUM NITRICUM, L. D. ACIDUM NITRICUM PURUM, E. Nitric Acid, *Aqua Fortis*. *Spiritus Nitri Glauberi*. *F.* Acide Nitrique, *G.* Salpetersäure.

Nitric Acid (N′ or $NO_5 = 54$) is a compound of Nitrogen acidified by Oxygen. It was known to Geber, and probably also to the Hindoos. Cavendish first clearly ascertained its composition by forming the acid by passing electric sparks through atmospheric air over a solution of Potash. Phil. Trans. vol. 75, p. 572, and vol. 78, p. 261. (1785.) It may frequently be detected in the atmosphere after thunder-storms, in consequence of the Oxygen and Nitrogen combining together with the aid of electricity. In combination with Potash, Soda, and Lime, or Ammonia, it is found, effloresced on the soil in some countries; also in some minerals; likewise in some vegetables, as in the officinal Pareira root, in the state of Nitrate of Potash.

Prop. Nitric′ is unknown anhydrous, being always seen as a liquid in combination with water. When pure, it is colourless, transparent, of a very sour corrosive taste, destroying vegetable and animal matter. It tinges the skin yellow, and causes it to peel off; gives acid properties to a large quantity of water, vividly reddening Litmus and vegetable blues. It has a strong, almost suffocating odour, fumes in the air, from condensing its moisture; fumes still more if Ammonia be present. Its affinity for water is great, absorbing it from the air, and having its Sp. Gr. and strength diminished; when mixed with it, heat is evolved. Its strength necessarily depends on the quantity of water mixed with it; the density of the L. P. acid is 1·5033 to 1·504; Mr. Phillips considers this to be the strongest procurable; but it has been brought, by distilling with a gentle heat, by Kirwan to 1·55, and by Proust to 1·62. The E. P. also employs the Nitric′ of commerce, commonly called Aqua Fortis, of Sp. Gr. 1380 to 1390. Dr. Christison considers 1500 as the densest acid which can be obtained free of Nitrous Acid, but, impregnated with this, it may be got so heavy as 1540 and upwards. The purest Nitric′, if long exposed to light, becomes yellowish from decomposition, Oxygen being given out, and Nitrous Acid formed. This may be got rid of by the action of heat, as it escapes in the form of Nitrous fumes, leaving the Nitric′ nearly colourless. Nitric′ of Sp. Gr. 1·50 will freeze at —40° and will boil at 247°, but these points vary of course with the density; if the heat be continued, the acid volatilizes, and at a red heat is decomposed. At the density of 1·5 it consists of 1 Eq. of Nitric′ and 1½ Eq. of Water, or $54+13{\cdot}5=67{\cdot}5$, and it then contains about 80 per cent. of acid.

Many vegetable substances, as Charcoal, Sugar, Alcohol, deprive Nitric′ of its Oxygen, also Phosphorus, and hence it is one of the most powerful oxidizing agents. The Metals do not in general decompose it when concentrated, but if diluted with water, this becomes decomposed as well as the acid, both yielding Oxygen to oxidize the Metal; the oxide is insoluble, or, being acted on by some of the undecomposed acid, a Nitrate is formed. The other part of the decomposed acid escapes in the form of Nitric oxide gas, which, uniting with the Oxygen of the atmosphere, ruddy fumes of Nitrous acid gas are observed. If Copper filings be used, then the solution will be of a greenish-blue colour. If Hydrochloric′ be mixed with the Nitric′, both become decomposed, and the mixture, called Nitro-Muriatic acid or *Aqua Regia*, is capable of dissolving gold.

The presence of N′ may be ascertained by the bright red colour produced when N′ touches Morphia or Brucia; saturated with Carb. of Potash, it forms Nitrate of Potash, which deflagrates.

Prep. Nitric′ may be formed by passing electric sparks through a mixture of 1 part of Nitrogen and 5 of Oxygen. The L. and E. P. direct *Nitrate of Potash* and *Sulphuric Acid* in equal quantities by weight to be mixed together in a glass retort, and the resulting Nitric′ to be distilled in a sand-bath (or by a naked gas flame, E). The D. P. employs 100 parts of salt to 97 of common Sul′.

To obtain a pure acid, the E. P. directs the Nitrate of Potash to be purified by two or three crystallizations, till Nitrate of Silver does not act on its solution in distilled water. When manufacturers make N′, they employ only half the quantity of Sul.′ and use Nitrate of Soda instead of Nitrate of Potash, because it is cheaper. The acid obtained is of a brownish colour, fumes, and is called Nitric acid of Commerce.

The quantities of Nitrate of Potash and Sulphuric′ of the Sp. Gr. 1·8433, are very nearly in the proportion of 1 Eq. of the Salt to 2 Eq. of the acid. The two acids differ from each other in their relations with water, the Nitric′ requiring 2 Eq. and the Sulphuric′ only 1 Eq. of water; hence twice as much Sulphuric′ is employed as is necessary to saturate the Potash. A less proportion is used by many manufacturers, but the advantage of the College process is, that a less heat is necessary, and a larger quantity and a purer acid, that is, more free of Nitrous Acid, is obtained. The Nitrate of Potash being decomposed by the Sulphuric′, the 2 Eq. of this combine with 1 Eq. of Potash, and form 1 Eq. of Bisulphate of Potash; the Nitric′, as set free, rises with the vapour of water, and forms aqueous Nitric′. This, when of the greatest strength, or of Sp. Gr. 1·5033 to 1·504, is composed of 2 Eq. N′+ 3 Eq. Aq. But it is more convenient, as stated by Mr. R. Phillips, to consider liquid Nitric′ (NO_5 $1\frac{1}{2}HO$) as a Sesquihydrate.

Tests. The strength of N′ may be ascertained by its Sp. Gr., which should be 1·50. 100 grains of this acid (equal to about 81 grains of dry N′) will saturate 217 grains of crystallized Carbonate of Soda. It should pass off wholly in vapour, showing the absence of any fixed salts, as of Potash, that is Nitre. The presence of Nitrous Acid is indicated by a yellow or orange colour: it may be got rid of by the gentle application of heat. Sulphuric′ is indicated by a white precipitate (Sulphate of Barytes) being formed when the acid, diluted, is tested with Sol. of Chloride of Barium, or of Nitrate of Barytes. Hydrochloric′ will be detected by a white precipitate (Chloride of Silver) being formed with solution of Nitrate of Silver.

Action. Uses. Corrosive Poison, sometimes used as a Caustic and Disinfectant.

Inc. Alkalis, Earths, Oxides of Metals, their Carbonates, and Sulphurets, all combustible bodies, Hydrochloric', Alcohol, Sulphate of Iron, Acetates of Potash and Soda, Diacetate of Lead.

Antidotes. The carbonates of earths, as Chalk, Whiting; Bicarbonates of Alkalis; Soap.

ACIDUM NITRICUM DILUTUM, L. E. D. Dil. Nitric Acid.

Prep. L. E. Mix *Nitric acid* f℥i. (pure f℥i., or Commercial Dens. 1390 f℥i. and ʒv. E.) with *Aq. dest.* f℥ix.

The Sp. Gr. of this diluted acid is 1·080. (107,7E.) 100 grs. of it saturate 31 grs. of crystallized Carb. Soda, or f3j, 18·5 grs., indicating about 10 per cent. of real acid. The D. College orders of *acid* 3 parts, *water* 4 parts. Of this the density is 1280, and it contains 38 per cent. of acid, or nearly as much as some varieties of Aqua Fortis, and therefore too much for convenience in prescribing.

Action. Uses. Refrigerant, Tonic, Alterative.

D. ♏x. — ♏xxx. diluted with water, &c. should be drunk through a quill, like the other strong acids.

IODINIUM. L. E. D.

Iodine. *F.* Iode. *G.* Iod.

Iodine (I=126), from ἰώδης, *violet*, the colour of its vapour, was obtained by M. Courtois in 1812 in the residual liquor of the process for obtaining Soda from Kelp. Though but lately discovered, its effects have long been obtained in medicine, as it is found in Sea and several Mineral waters, and in Sea-weeds, Sponge, Corals, and some Molluscous animals. In the present day, the leaf of a Sea-weed (a species of Laminaria, *Dr. Falconer*) is employed in the Himalayas, and called the *goître-leaf*, and in S. America the stems of a Sea-weed are sold by the name of *goître-sticks*, because they are chewed by the inhabitants wherever goître is prevalent.*

Prop. Iodine occurs in metallic-like scales. Its crystals may be obtained in elongated octohedrons with rhomboidal bases. It is of a bluish-grey colour, with a metallic lustre, an acrid taste, and an odour resembling that of Chlorine. Its Sp. Gr. is 4·94; but it is soft, and crumbles between the fingers, staining the skin of a yellow colour. It is rather insoluble in water, one grain requiring 7000 parts of water to dissolve it; but it is very soluble in Alcohol and Ether, forming reddish-brown solutions; it is also soluble in essential Oils. It evaporates at ordinary temperatures; but when heated, Iodine melts at 225°, and at 350° rises in a beautiful violet-coloured vapour, which is remarkable for its great weight, having a Sp. Gr., compared with that of

* Vide the Author's Illustr. of Himalayan Botany, p. 441.

air, = 8·7, but with that of hydrogen as 126 to 1. Hence its atomic number is 126. The solubility of Iodine in water is much increased by the presence of some salt in solution, and this is taken advantage of in preparing the compound solution of the London Pharmacopœia.

One of the most useful unions Iodine forms, is with Starch, as this serves as a very delicate test of its presence. If a solution of Starch be formed and allowed to cool, and then added, even though very diluted, to a solution containing Iodine, a blue precipitate of the Iodine in union with the Starch, or an Iodide of Starch or of Amidin, as it has been lately called, will be produced. It is, however, essential that the Iodine be in a free state ; hence in compound solutions it is necessary to add a few drops of Nitric′ or of Sulphuric′ to detach it from its combinations, when the blue colour of the Iodide of Starch will immediately appear. Iodine is also mixed with fatty substances, and is employed in the form of an ointment or of a liniment.

Though present in a variety of situations, Iodine is obtained for commercial purposes from the ashes of Sea-weeds, called Kelp, and is largely prepared at Glasgow. Professor Graham states, that the long elastic stems of Fucus palmatus afford most of the Iodine contained in Kelp; and as this is a deep-sea plant, it is found most on exposed sea-coasts. Dr. Traill says, that the greatest quantity is produced by Kelp made from "drift-weed," which is in a great measure composed of Fucus digitatus and F. Loreus; and that "cut weed," which consists of F. vesiculosus and F. serratus, yields much less of it. According to the experience of a manufacturer, 100 tons of Caithness kelp yield 1000 pounds of Iodine, or about a 224th part. (C.)

The most common method of obtaining Iodine is to take Kelp in powder, lixiviate with water, evaporate, and remove the Soda salts (such as the Carbonate and Sulphate) as they form; allow the liquor to cool, when crystals of Chloride of Potassium will be deposited. Decant the dark-coloured mother liquor, which contains the Iodide of Sodium with other salts; supersaturate with Sulphuric′, when an evolution takes place of Carbonic′, Sulphuretted Hydrogen, and Sulphurous acid gases. After standing for a day or so, the residuary liquor, or Iodine ley, is mixed with Binoxide of Manganese, and heat applied. Water and Iodine pass over, and are condensed in receivers.

Here the mutual action of Sulphuric′ and Binoxide of Manganese on any Chloride in the Iodine ley, will be the detachment of Chlorine (*v.* Hydrochloric Acid). This, as stated above, will decompose the Iodide, set the Iodine free; or, as usually explained, one equivalent of the Oxygen of the Binoxide, combining with the Sodium equally, sets the Iodine free, and the Sulphuric′ will combine with the Soda and the Oxide of Manganese, and thus form a Sulphate of Soda (the Oxide of Sodium), and a Sulphate of the Oxide of Manganese.

Tests. Iodine is sometimes adulterated with Charcoal, Plumbago, or Oxides of Manganese; but its chief impurity is moisture, of which it sometimes contains as much as 15 to 20 per cent., when it looks moist, and sticks to the sides of the bottle. The L. P. directs its purity to be ascertained by its little solubility in water, solubility in Alcohol; and, when heated, first melting and then subliming in violet vapours, and rendering Starch blue. The E. P., states, that when heated, it should evaporate entirely, rising in violet vapours, and that 39 grains of Iodine, with 9 grains of quicklime and f℥iij of water, when heated short of ebullition, should slowly form a perfect solution (Iodide of Calcium and Iodate of Lime), yellowish or brownish if the Iodine be pure, in consequence of a trace not being acted on, but colourless if there be above 2 per cent. of water or impurity.

For pharmaceutic preparations of fixed and uniform strength, the E. C. directs Iodine to be dried by being placed in a shallow basin of earthenware, in a small confined space of air, with 10 or 12 times its weight of fresh burnt Lime, till it scarcely adheres to the inside of a dry bottle.

Action. Uses. Stimulant of the absorbents. Alterative.

D. gr. j.—gr. v. with some mild extract.

Antidotes. Large draughts of solution of Starch, both before and after evacuating the stomach; also obviate inflammation.

IODIDE OF STARCH, proposed by Dr. Buchanan, is a mild preparation. It is directed to be made by rubbing gr. 24 of Iodine with a little water, and gradually adding ℥j. of finely powdered Starch. Dry by a gentle heat, and preserve the powder in a well-stoppered bottle.

D. ℥ß gradually increased. It has been given in very large doses.

TINCTURA IODINEI. E. D.

Prep. E. *Iodine* ℥ij. ſs. *Rectified Spirit* Oij.
D. *Iodine* ℈ij. *Rectified Spirit* by weight ℥j.

The E. prep. contains 1 gr. in 16 ♏; the D. prep. 1 gr. in 12 ♏.

TINCTURA IODINII COMPOSITUM, L.

Prep. Macerate *Iodine* ℥j. *Iodide of Potassium* ℥ij. in *Rectified Spirit* Oij. until dissolved; strain.

It contains Iodine gr. j. and Iodide of Potassium gr. ij. in ♏40.

D. This with the *Liquor Potassii Iodidi Compositus, L.* is one of the best forms for internal exhibition. It may be given with water or sherry. ♏x. gradually increased to f℈j.

UNGUENTUM IODINII COMPOSITUM, L. UNG. IODINEI, E.

Prep. Rub *Iodine* ℈ß. *Iodide of Potassium* ℈j. and *Rectified Spirit* f℈j. together, then rub up with *Lard* ℥ij.

Action. Uses. Employed as an external application in Bronchocele, scrofulous enlargements of the glands, or tumours.

The effects of Iodine as an external application, may be obtained by brushing the Tincture on the surface, or mixing it first with Soap Liniment and using it as an embrocation.

BROMINIUM, L.

Bromine. *F.* Brome. *G.* Brom.

Bromine (Br = 78), from βρωμος (*fœtor*), a strong odour, was discovered in 1826 by M. Balard, in *bittern*, the uncrystallizable residue of Sea-water. The quantity contained therein is so small, that 100 pounds of Sea-water yield only 5 grains of Bromide of Sodium or of Magnesium, of which 3·3 grains are Bromine. It exists also in rock-salt, in brine-springs, as those of Cheshire, in some mineral waters, sea-weeds, marine animals, and in the ashes of sponge.

Prop. It is a liquid of a dark red, but, in thin layers, of a hyacinth-red colour, easily volatilized, and therefore requiring to be kept well closed up, or covered by a stratum of water. The taste is acrid and unpleasant, the smell disagreeable and suffocating, something like Chlorine. The sp. gr. of liquid Bromine is 2·96, or 3·; that of its vapour, compared with air, about 5·4 ; 100 C. I. weigh gr. 168. It freezes at —4°, and congeals into a brittle crystalline solid, of a lead-grey colour and metallic lustre; and if water be present, it crystallizes in octohedra. At common temperatures, it rises into a reddish-brown vapour, like Nitrous acid, and boils at 116°. It stains the skin yellow, but not permanently, and acts as a caustic. It destroys vegetable colours, extinguishes flame, first turning the upper part red and the lower green; but some metals, as Antimony, &c., take fire in it. It is slightly soluble in water, more so in Alcohol, and very soluble in Ether; but it decomposes the Oils, Hydrobromic acid being formed.

Prep. Bromine may be detached from its combinations by passing Chlorine gas through a solution containing a Bromide, or by the action of Hydrochloric' and Binoxide of Manganese on *bittern*, or on the mother-liquor of the salt-springs of Germany. The Hydrogen of the Acid, uniting with the Oxygen of the Binoxide, forms water; one portion of the Chlorine unites with the Manganese, and another with the metallic base, whether Sodinm, Calcium, or Magnesium, with which the Bromine was combined. This being set free, easily distils over with water, and may be condensed in a receiver.

Tests. Bromine may be known by its liquidity, colour, weight, and acrid odour, by being sparingly soluble in water, more copiously in rectified spirit, best in ether ; also by its ready volatility, its action on the metals, and by forming an orange-yellow colour with starch, and a yellowish-white precipitate (the Bromide) with Nitrate of Silver.

Action. Uses. A local irritant and caustic ; increases the activity of the lymphatic system.

D. 5 or 6 drops of 1 part of Bromine in 40 of Water or of rectified Spirit. Externally as a lotion, or to moisten a poultice, a solution 4 times as strong.

Antidotes. The same as for Iodine.

Chlorinium.

Chlorine. *Dephlogisticated Marine* or *Muriatic Acid.* *F.* Chlore. *G.* Chlor.

Chlorine (Cl=36), pointed out by Scheele in 1774. The *Oxygenated Muriatic Acid* of Lavoisier, which Berthollet converted into *Oxymuriatic Acid*, as it was considered a compound of Oxygen and Muriatic′. Gay Lussac and Thenard considered it an element, and named it, from its characteristic yellowish-green (χλωρος) colour, Chlorine. It was proved to be an elementary body by Sir H. Davy in 1809. Chlorine is not found in a free state in nature, but combined with metals in great abundance in the inorganic kingdom. Chiefly as Chloride of Sodium, Rock Salt, or Common Salt, or in the waters of the ocean, and also in the organic kingdom. Dr. Prout states, that free Hydrochloric′ exists in the stomachs of animals during digestion.

Prop. Gaseous Chlorine is of a greenish-yellow colour; its Sp. Gr. is 2·5. 100 C. I. weigh about 77 grs. Under a pressure of about 4 atmospheres, or subjected to great cold, it is reduced to a bright yellow liquid. It has an astringent taste, and a suffocating, pungent odour, even when diluted with air. It is noted for its power of destroying vegetable colours; but, as this is only when water is present, and in light, it is supposed to be owing to the decomposition of the water, and the effect to be due to nascent Oxygen, which combines with colouring matter, the Hydrogen combining with the Chlorine to form Hydrochloric Acid. Cold water absorbs about twice its volume. It is also absorbed by Alcohol and organic substances; but it generally decomposes them, Hydrochlor.′ being formed. It combines with most of the simple bodies and metals, forming Chlorides; and acids with Oxygen and Hydrogen. It is incombustible, but partially supports combustion; the flame of a taper becomes red, small, and smoky in it; Phosph. and Antimony take fire in it spontaneously when in a state of division. It is a nonconductor of Electricity. Besides its bleaching, it is remarkable for its power as a disinfecting agent, and of correcting all putrid effluvia from decaying animal and vegetable matters, depending, probably, upon its great affinity for Hydrogen. It is prepared on a large scale, or may be liberated from one of the officinal compounds.

Prep. Chlorine is obtained by heating in a retort *Common Salt* 4 parts, *Binoxide of Manganese* 3 parts, and *Sulphuric Acid* 7 parts, with an equal weight of *water*, or by the mutual action of *Binoxide of Manganese* and *Hydrochloric Acid*, when the Oxygen combines with Hydrogen to form water, one equivalent of Chlorine escapes and another combines with the Manganese to form Chloride of Manganese; or we may easily obtain it by adding an *acid* to the *chlorinated Lime* or *Soda* of the Pharmacopœias. It must be collected over warm water, as it is absorbed by cold water and by Mercury, which combines with it.

Tests. Chlorine may be known by its colour, its suffocating smell, and by its bleaching properties. Nitrate of Silver produces a white curdy precipitate (Chloride of Silver) in a solution containing Chlorine.

This precipitate blackens in the light, is insoluble in Nitric′, but soluble in Ammonia, and melts into horn silver, *luna cornea*. Solution of Chlorine dissolves Gold leaf.

Action. Uses. Rubefacient. Disinfectant. Suffocating if inhaled.

Antidotes. Inhalation of Ammoniacal gas with care, and vapour of warm water. v. also *Liquor Chlorinii.*

Off. Prep. Liquor Sodæ Chlorinatæ. Calx Chlorinata.

AQUA (v. Liquor) CHLORINEI, E. D. Solution of Chlorine. *Liquid Oxymuriatic Acid. F.* Chlore liquide. *G.* Wässeriges Chlor.

A solution of Chlorine may easily be made by passing into water the gas obtained by any of the above processes. It is decomposed by light, Oxygen being evolved from the decomposition of the water, and H C′ formed by the combination of the Chlorine with the Hydrogen.

Prop. This solution has the pale greenish-yellow colour and suffocating smell of Chlorine (q. v.) with an astringent taste ; its sp. gr. = 1·008 ; at the temperature of 32° definite yellow prismatic crystals of Hydrate of Chlorine, which are decomposed by heat, are deposited. Like the gas, the solution destroys vegetable colours, as Iodide of Starch, &c., and likewise the effluvia of putrefying animal and vegetable substances. The effects may be obtained by exposing the solution to the air, or by the use of chlorinated Soda or Lime.

Prep. E. D. Take *Sul′* (commercial, E.) f℥ ij. (87 parts, D.) and *Aq.* f℥viij. (124 parts, D.) Mix them, and add them to *Muriate of Soda* gr. lx. (dried 100 parts, D.) and *Red Oxide of Lead* gr. 350, E. (*Oxide of Manganese* 30 parts, well mixed in a retort, D.) contained in a bottle with a glass stopper. Agitate till the red oxide becomes nearly white: allow the insoluble matter to subside. (Apply a gradually increasing heat and pass the gas through 200 parts of *Aq. dest.* Stop as soon as effervescence ceases in the retort. Keep the solution in well-closed glass bottles in the dark, D.)

The Dublin process yields a pure solution of Chlorine. In that of the E. P., which, according to Dr. Christison, is convenient, though requiring time, the red oxide parts with some Oxygen to oxidate the Sodium. Protoxide of Lead being formed, unites with the Sul.′, leaving white insoluble Sulphate of Lead. The Soda also combines with the Sul.′, leaving Sulph. of Soda in solution along with the liberated Chlorine, but it does not interfere with the medicinal uses.

Tests. v. Chlorine.

Action. Uses. Irritant poison. Caustic. When diluted, stimulant, either internally or as a lotion or gargle. Cautiously used, acts as an antidote in poisoning by Hydrocyanic′ and Sulphuretted Hydrogen.

D. ʒj.—ʒiv. in f℥viij. of vehicle, or as a lotion or gargle in 8 parts of fluid.

Antidotes. Magnesia, Clk., Soap, Albumen, White of Eggs, Dilution.

CHLORINE AND OXYGEN.

Like the other elementary bodies, Chlorine unites with Oxygen, and in several proportions; but as none are mentioned in the Pharmacopœias, it is unnecessary to do more than advert to these compounds. The highest oxidized of them is the Perchloric Acid. Chloric Acid consists of 1 Eq. Chlorine + 5 Eq. Oxygen; the salts are termed Chlorates, formerly *Oxymuriates:* the Chlorate of Potash, *Oxymuriate of Potash,* is officinal, *v.* POTASSIUM.

CHLORINE AND HYDROGEN.

HYDROCHLORIC ACID GAS. Muriatic Acid Gas.

Hydrochloric or Muriatic Acid gas ($H\,Cl = 37$) is a compound of Hydrogen and of Chlorine. It is sometimes found in a gaseous state in the neighbourhood of volcanoes, and in solution in the gastric juice of animals. Combined with Ammonia, it is found in the cracks and fissures of lava, and in the cool parts of brick-kilns in India.

In its pure form it exists as a colourless gas with a pungent suffocating smell, acid taste, reddening vegetable blues, and making Turmeric paper brown. It has been reduced to a liquid state by cold, and by a pressure equal to 40 atmospheres at 50° F. Heat has no effect upon it. It extinguishes all burning bodies immersed in it. It has so great an affinity for water, that it attracts it from the atmosphere, producing the appearance of fuming,—100 C. I. of the gas weigh 39·77 grs.; Sp. Gr. 1·283. Water at 40° takes up 480 times its own volume of the gas, and thus forms liquid Hydrochloric Acid. It may be obtained by the direct union of its constituents, or by any of the methods described below.

Action. Uses. Suffocating, but sometimes used as a Disinfectant.

ACIDUM MURIATICUM, E. Hydrochloric Acid. Muriatic Acid. Chlorohydric Acid. *Spiritus Salis. Spirit of Salt. Marine Acid. F.* Acide Hydrochlorique. *G.* Salzsäure, Chlorwasserstoffsäure.

This acid is a solution of the above gas in water. Geber and the Arabs were probably acquainted with it, and the Hindoos knew it by a name equivalent to *Spirit* or *sharp water of Salt.* This is the Commercial acid, and is always of a yellow colour. It commonly contains as impurities a little Sulphuric Acid, Nitrous Acid, Perchloride of Iron, Chlorine, and Bromine. It is prepared by pouring the Oil of Vitriol of commerce on Common Salt in earthen or iron vessels, especially since the extensive manufacture of Carbonate of Soda from Sulphate of Soda. Its Density, ex. E. P. = 1180. For the Properties and Tests, *v. Acidum Hydrochloricum, L.*

Acidum Hydrochloricum, L. Muriaticum (purum, E.) D.

Prop. It emits suffocating fumes. When pure, it is perfectly colourless, but is usually of a pale yellowish straw-colour, from the presence of a little Chlorine formed from the decomposition of the acid, when long kept, especially if exposed to light. It has a sour, irritating, and corrosive taste, with the odour of its gaseous acid. When heated to 112° it bubbles, from the quantity of H. C′ gas which escapes. It freezes at —60°. H. C′ combines with water in all proportions, with the evolution of heat. Sp. Gr. 1·16, when it contains about $\frac{1}{3}$ its weight of Hydrochloric Acid gas.

The action of this acid on some substances, as the Metals, Oxides, &c., requires attention, from the changes which take place in the various decompositions. Thus, when Zinc or Iron are acted on by liquid Hydrochloric′, Hydrogen is evolved in consequence of the decomposition of the acid, the Chlorine combining with the metal, forms a Chloride of Zinc or of Iron. But if an oxide of a metal be acted upon, no Hydrogen is evolved, because it combines with the Oxygen of the Oxide to form an equivalent of water, while the metal and Chlorine combine to form a metallic Chloride. Ammonia being a substance devoid of Oxygen, no decomposition of the acid takes place, and it therefore unites with the Ammonia to form Hydrochlorate of Ammonia. The same thing takes place with a vegetable Alkali, though it does contain Oxygen, but no action is considered to take place between it and the Hydrogen, and therefore the H. C′ combines with the alkali, and forms a Hydrochlorate, as that of Morphia.

A formula for preparing the pure acid is given by all the Colleges.

Prep. Take *Sul′.* ℥xx. [(pure 3 parts, E.) (of commerce 87 parts, D.)]; mix it with *Aq. dest.* f℥xij. [(*Aq.* 1 part, E.) (62 parts, D.)] Add this to *Chloride of Sodium dried* ℔ij. [*Muriate of Soda* (purified by solution in boiling Aq., concentrating, skimming off the crystals, draining, slightly washing, then well dried, 3 parts, E.) (100 parts, D.) when the mixture is cold, E. D.] Put the mixture into a glass retort: fit on a receiver containing *Aq. dest.* f℥xij. [(Aq. 2 parts, E.) (62 parts, D.)] Distil over a sand-bath (or naked gas flame with a gentle heat, E.) and let the liquid absorb the gas. Gradually increase the heat. [(Keep the receiver cool, E.) (The Sp. Gr. of this acid is 1170, E.; 1160, D.)]

In the above formula both the Chloride of Sodium and the water are decomposed. The Chlorine of the former, combining with the Hydrogen, forms Hydrochloric′, while the Oxygen of the water unites with the Sodium, to form Soda, which is seized upon by the Sulphuric acid to form Sulphate of Soda, which remains as the residual salt, and will be a Bisulphate of Soda if an excess of acid has been employed.

The E. P. directs equal parts by *weight* of the ingredients. The quantity of Sul.′ is greater than is necessary; but Dr. Christison says that less heat is required, and that the residual salt is more easily washed out with water. The salt is crystallized, to get rid of impurities, such as Nitrate of Soda, &c.

Tests. The presence of strong Hydrochloric′ is indicated by the white fumes which are produced in the neighbourhood of Ammonia; also by Nitrate of Silver producing in a solution containing it a white curdy precipitate (Chloride of Silver); this blackens in the air, is soluble in a solution of Ammonia, but is insoluble in Nitric′. Sp. Gr. 1·16. 100 grains saturate 132 of crystallized Carbonate of Soda. If pure, it is colourless. Strips of Gold are not dissolved in it, even with the assistance of heat; neither is Sulphate of Indigo decolorized; showing in both cases the absence of free Chlorine and of Bromine. It ought to be entirely dissipated by heat, without leaving any residue. Chloride of Barium gives no precipitate, if Sul.′ and Sulphates are not present, but the acid should be diluted. Neither Ammonia nor its Sesquicarbonate throw down anything, showing the absence of metals and metallic oxides, that of Iron being precipitated in form of the Red Sesquioxide. Nitrous acid or binoxide of Nitrogen (*v.* Pereira) may be recognized by the Protosulphate of Iron.

Action. Uses. Corrosive poison; Escharotic.

Antidotes. Magnesia; solution of Soap, as easily procurable; the Bicarbonates of Soda and of Potash; Milk; Demulcents. Chalk to be avoided on account of the deleterious effects of Chloride of Calcium.

ACIDUM HYDROCHLORICUM DILUTUM, L. ACIDUM MURIATICUM DILUTUM, E. D. Diluted Hydrochloric or Muriatic Acid.

Prep. L. E. *Hydrochloric′* f℥iv. *Aq. dest.* f℥xij. Mix.—D. *Mur.′* 10 parts, *Aq. dest.* 11 parts, both by measure. Mix. The Sp. Gr. of the L. E. Acid=1050. 32 grs. of crystallized Carb. of Soda saturate f℥j. The D. acid is much stronger. Its Sp. Gr.=1080.

Action. Uses. Refrigerant; Tonic; externally as a stimulant lotion and gargle.

D. ♏x.—♏xx. in some bland or sweetened fluid or bitter infusion. f℈j. with ℥j. of Honey applied with a brush in ulcerated sore throat.

Inc. Alkalis, most Earths, Oxides, and their Carbonates; Sulphuret of Potassium; Tartrate of Potash; Potassio-Tartrate of Antimony; Nitrate of Silver; Acetate of Lead.

ACIDUM NITROMURIATICUM, D.

Nitromuriatic Acid. *Acidum Nitrohydrochloricum. Aqua Regia. F.* Eau régale. *G.* Königswasser.

This acid is made by mixing Nitric′ with Muriatic′, and has probably been known since the discovery of these acids. The Arabs must have been acquainted with it, as they had a solvent for Gold.

Prop. Nitrohydrochloric′ is of a golden yellow colour, with the suffocating odour of Chlorine, and the irritant corrosive properties of the strong acids.

Prep. Mix gradually in a cooled vessel, and where the fumes can easily escape, *Nitric′* 1 part, *Muriatic′* 2 parts (both by measure). Keep the mixture in a well-closed bottle in a cool, dark place.

The resulting acid is not a mere mixture of the two acids, for both become decomposed. The Nitric′, which so readily parts with its Oxygen, gives 1 equivalent which combines with the Hydrogen of the Hydrochloric′, and some water is thus formed. Of the Chlorine set free, some escapes, the rest remains in solution with the Nitrous acid formed ; but an excess of either of the acids may remain, according to the proportions in which they are used. When exposed to light, a portion of the water becomes decomposed, and Hydrochloric acid is again formed.

Tests. It is distinguished by the property of dissolving Gold. Nitrate of Silver produces a precipitate of Chloride of Silver, which is soluble in Ammonia, but insoluble in Nitric′. When an Alkali is added, both a Chloride and a Nitrate are formed.

Action. Uses. Corrosive poison. When diluted, stimulant of the skin and of the liver.

D. ♏iij.—v. well diluted. Applied externally by sponging, or in a foot-bath of warm water, which is made of the acidity of vinegar, or in the proportion of f℥j.—f℥ij. of the acid to each gallon of water.

Carbon.

Carbonium. *F.* Charbon. *G.* Kohlenstoff.

Carbon (C = 6) is very extensively diffused in nature, as in Coal, Anthracite, Graphite, &c.; of great purity, and crystallized in the form of the regular octohedron or cube in the Diamond. It forms also a large portion of both vegetable and animal substances. Combined with Oxygen, it exists in the Atmosphere and in many mineral waters as Carbonic acid gas, and as a Carbonate in immense quantities in Chalk, Marble, Limestone, &c.

Graphite, called also Plumbago and Black Lead, is found in primary mountains, and is nearly a pure form of Carbon. It has sometimes been considered a Carburet of Iron, but the presence of this metal is not essential, as some specimens contain hardly a trace, and others as much as 5 per cent. of Iron. It is sometimes crystallized in six-sided prisms or tables, with a Sp. Gr. of nearly 2·5, opaque, steel-grey, solid, with a metallic lustre, soft to the touch, and well-known from its property of marking paper. The best specimens are obtained from Borrowdale in Cumberland.

Action. Uses. Graphite has long been employed in Medicine; internally for many of the same purposes as Charcoal, and externally as an ointment in some skin diseases.

Carbo Ligni, L. E. Wood Charcoal.

Prop. Charcoal is the form in which Carbon is usually seen. Its black colour and freedom from taste and smell are well known. It has never been decomposed, is insoluble, infusible, and unalterable. Sp. Gr. various. It is a bad conductor of heat, but an excellent one of electricity, and is remarkable for its power of counteracting putrefaction, and also for combining with and removing the odorous and colouring principles of most bodies. This is probably owing to its absorbing the odoriferous effluvia, as it does several of the gases, as Sulphuretted Hydrogen and Carbonic acid gas. Charcoal readily burns in air, emitting light and heat, combining with its Oxygen and forming Carbonic acid gas. It yields more heat than an equal quantity of wood, in consequence probably of being freed from the large proportion of water which wood contains. By the heat of coal in close vessels, Charcoal is obtained in the form of Coke.

Prep. Wood Charcoal is obtained by burning wood in covered-up heaps or in close vessels.

As the combustion takes place with only a limited supply of air, little of the Carbon of the wood is consumed, but its Oxygen, Hydrogen, and Nitrogen escape, variously combined. The charcoal which remains necessarily contains the ashes of the plant, consisting of Carbonates of Potash, Lime, &c. (q.v.) Wood yields from 14 to 23 per cent. of Charcoal. A pure Charcoal may be obtained by burning Oils and Resins with a deficient supply of Oxygen, when, the volatile matters being dissipated, the charcoal is left, and commonly called *lamp-black*. For medical purposes either kind may be ignited in a close vessel to a red heat, until all volatile matters have escaped. The Charcoal is, when cool, kept in stoppered bottles.

Action. Uses. Antiseptic and Disinfectant ; corrects the fœtor of the breath and of the stools in Dyspepsia and Dysentery.

D. Gr. x.—ʒj. internally ; ʒj. or ʒij. to Lard ℥j. as ointment.

CATAPLASMA CARBONIS LIGNI, D Mix finely powdered and fresh heated Charcoal with a simple poultice. Apply it warm to foul ulcers, to destroy the fœtor.

CARBO ANIMALIS, L. E. Animal Charcoal.

This is obtained by subjecting bones, horns, muscles, &c. to a red heat in close vessels, until vapours cease to be emitted. The residue, after being powdered, is known by the names of *bone-black* and *ivory-black*. In this state it contains 88 per cent. of Phosphate of Lime (q.v.) and Carbonate of Lime, 2 per cent. of Carburet with Siliciuret of Iron, and a trace of a Sulphuret. It is bitterish in taste, and may readily be distinguished from Vegetable Charcoal by burning a little of it on a red-hot iron. The ashes, consisting chiefly of the Phosphate of Lime, are with difficulty acted on by Sulphuric Acid ; those of Wood Charcoal, being composed of Carbonates, &c., dissolve, and form bitter solutions. Animal Charcoal is officinal on account of its attraction for the colouring matter of organic substances, a property probably owing to its extreme subdivision, and to the extent of surface which it exposes to any liquid filtered through it. The decolorizing power of

Vegetable Charcoal may be increased by mixing Chalk or pounded flint with the vegetable matter previous to its being carbonized. Animal Charcoal is extensively employed in the arts for removing the colouring matter of Syrup, and also in the preparation of Citric and of Tartaric Acid, and of the vegetable Alkalis and their salts, as Aconitina, Quina, Morphia, Veratria. The same quantity of Charcoal may be used several times, but it requires to have been first dried and subjected to a red heat. It is either mixed or boiled with the liquid to be decolorized, or the latter is allowed to filter through a layer of Charcoal. For some purposes it requires to be purified.

CARBO ANIMALIS PURIFICATUS, L. E. Purified Animal Charcoal. Purified Ivory-Black.

Prep. Take of *Animal Charcoal* ℔ j. and pour on it gradually a mixture of *Hydrochloric'* and *Aq.* of each f℥xij. Digest with a gentle heat for two days, frequently shaking. Set aside and then pour off the supernatant liquor; wash the charcoal with water till no acid is perceptible, and then dry it. The Edinburgh College, subsequent to the digestion, direct the mixture to be boiled and diluted with two pints of water. Collect the undissolved charcoal on a filter of linen or calico, and wash it with water till what passes through scarcely precipitates with solution of Carbonate of Soda. Heat to redness in a closely covered crucible.

The H C′ dissolves the Phosphate and decomposes the Carbonate of Lime, as well as any Sulphuret, with the disengagement of Carb′ gas and some Sulphuretted Hydrogen, Chloride of Calcium being left in solution. When the residuum has been thoroughly washed, it contains only a Carburet of Iron and some Silica mixed with the Charcoal.

Tests. It does not effervesce with Hydrochloric Acid (showing the absence of a Carbonate), nor is anything afterwards thrown down from this acid either by Ammonia or its Sesquicarbonate. If any Carbonate of Lime should have been present, a precipitate will be produced by the Sesquicarbonate. If Phosphate of Lime be dissolved by the HC′, it will be precipitated from the solution both by Ammonia and its Sesquicarbonate. The E. C. direct incineration with its own volume of red Oxide of Mercury, when, if pure, it will be dissipated, leaving only a 200th of spongy ash.

Uses. Animal Charcoal is officinal for Pharmaceutical purposes.

CARBON AND OXYGEN.

Carbon and Oxygen form several compounds, which are important to be known on account of their properties, though all are not officinal. CARBONIC OXIDE ($C\,O = 14$), which is interesting as a compound radical, is a colourless gas, without taste or smell, but extremely poisonous when respired. Sp. Gr. ·972.

It is formed when Carbon is burned with a limited supply of Oxygen. Though it extinguishes burning bodies, it will itself burn with a pale blue flame, as may often be seen on the surface of a coal fire. It is also formed when charcoal is slowly burned, and is necessarily deleterious in close apartments. Fresh air and its forced inhalation, or that of Oxygen, will be the best remedies for this.

OXALIC ACID is a compound of 2 Eq. Carbon with 3 Eq. Oxygen, and has been called Carbonous Acid. It might be treated of here, but being derived exclusively from organic sources, and closely resembling the other Vegetable Acids, will be treated of with them.

ACIDUM CARBONICUM.

Carbonic Acid. *F.* Acide Carbonique. *G.* Kohlen säure. *Fixed Air. Aërial Acid. Spiritus lethalis. Choke-Damp.*

Carbonic Acid (C′ or $CO_2 = 22$), so named from being a compound of Carbon and Oxygen, has long been known from its effects; but its nature was not explained until 1757 by Dr. Black. It is abundantly diffused in nature, being a constituent of the atmosphere to the extent of $\frac{1}{1000}$th part. It issues from the earth in many situations, as the Grotto del Cane in Italy, and the Valley of Poison in Java; but especially in Germany, near the Lake of Laach, where Bischoff calculates that not less than 600,000 lbs. escape daily, and in such quantities in the Brohltahl as to enable him to employ it in some chemical operations. It issues also, combined with water, from many mineral springs, giving to them their sparkling brilliancy. It is formed in large quantities in the combustion of charcoal, &c., and during fermentation. It is always being exhaled by animals in the process of respiration, and in small quantities by plants at night or in the shade. Combined with bases, it exists in large quantities in the interior of the earth, and in the mountain masses of Marble, Limestone, Chalk, &c.

Prop. Carb′, at ordinary temperatures, exists as a colourless gas, of which the solution in water has an acid taste. It is very heavy: Sp. Gr. 1·52; 100 C. I. weigh 47·25 grs. Water dissolves its own volume, but may be made by pressure to take up a much larger quantity, when it will redden vegetable blues, but not permanently, as the acid escapes when exposed to the atmosphere. When subjected to great pressure, it has been reduced to a liquid state; and, by the effect of great cold, produced by its evaporation, it has been converted into a solid at —148° F. Carb′ gas extinguishes flame and all burning bodies, except Potassium: it is also fatal to animal life.

Prep. L. P. Under the head of Potassæ Bicarbonas, Carb′ is directed to be obtained from *Chalk* rubbed to powder and mixed with water to the consistence of a syrup, upon which *Sulphuric′* is then poured, diluted with an equal weight of water. It may also be obtained from coarsely powdered Marble, or any other carbonate, by the action of the diluted hydrochloric, or any other acid. Whenever the C′ is required for Therapeutical use, it is preferable to employ the S′, as not volatile.

Tests. Carbonic Acid may easily be detected by its evanescent action in reddening Litmus paper, by rendering Lime-water turbid, by precipitating Lime and Barytes from their solutions, and by these precipitates being soluble in Acetic Acid with effervescence.

Action. Uses. Acts as a stimulant when applied externally, or taken internally, but is fatal to animal life when breathed. It may be prescribed in the forms of Liquor Potassæ effervescens, Liquor Sodæ effervescens, and in all effervescing draughts, or as Carb′ water, or ordinary bottled Soda-water, which very often contains no alkali; or in some mineral waters, which are natural solutions of C′ in water usually containing also other impurities. Death has frequently occurred from breathing this gas in descending into cellars, wells, mines, &c.; also from sleeping near brewers' vats, lime-kilns, in green-houses, or in small apartments or in cabins on board ship with a charcoal fire. It acts as a narcotic poison, and produces spasm of the glottis.

Hydrogen and Sulphur.

Sulphuretted Hydrogen. Hydro-Sulphuric Acid. *Hydrothionic Acid.* *F.* Acide Hydro-Sulfurique. *G.* Schwefelwasserstoffsäure.

Hydrosulphuric acid (H S=17) is remarkable for its offensive odour and deleterious properties.

Prop. It is a colourless gas. 100 C. I. weigh 36 grs. Sp. Gr. =177. It has been reduced to a liquid form by a pressure of 17 atmospheres. It is inflammable; Sulphurous acid and water being produced. Water absorbs about 2½ times its own bulk, acquiring the taste and smell of the gas, as well as its acid property of reddening Litmus. On exposure to the air, H S′ escapes, and the water becomes muddy from the deposition of Sulphur.

It combines with bases, and forms Hydrosulphates, as that of Ammonia; or Sulphurets, some of which, as those of Lead, Copper, Bismuth, and Silver, are blackish-coloured; that of Antimony, red; Zinc, white; Arsenic, yellow. Much useful information is obtained by employing H S′ or a soluble Hydrosulphate, as of Ammonia or of Potash, as a test, especially as it does not precipitate the Kaligenous and Terrigenous metals. Sulphuretted Hydrogen is absorbed in large quantities by Charcoal, is exhaled from putrefying animal matter, also from some vegetables, as the Cruciferæ, and likewise from decomposing vegetables generally, when a Sulphate is present, from "the decomposition of the Sulphates in water by the Carbonaceous matter of vegetables,"* and from some mineral waters.

Prep. S H may be obtained by the action of *Sul′* 7 parts diluted with *Aq.* 32 parts, poured on *Sulphuret of Iron* 5 parts.

Action. Uses. Most deleterious when respired, even when much diluted. Mineral waters, either natural (as of Harrowgate) or artificial, taken internally or used externally in the form of a bath, are stimulant, especially to the functions of the skin and of the uterine

* Professor Daniell on the Sulphuretted Hydrogen in the waters of the ocean. —Phil. Mag., July, 1841.

system. Hydrosulphate of Ammonia, D. is officinal. It is also much used as a test.

Antidotes. Inhalation of Chlorine, and Acids taken internally.

Hydrogen and Carbon.

Compounds of Hydrogen and Carbon are usually denominated Hydrocarbons. Of these, few are officinal, though we require to be acquainted with the properties of others, as they are deleterious. Thus, of those which are gaseous, the two following form the principal ingredient of Coal-gas.

Light Carburetted Hydrogen ($C\ H_2 = 8$). This gas is composed of 1 Eq. of Carbon and 2 of Hydrogen; that is, it is a Bihydruret of Carbon. It may be seen escaping in bubbles from the surface of stagnant pools, and also in stirring up fœtid mud, being formed by the decomposition of vegetable matter. It issues sometimes in immense quantities from fissures in Coal-mines, and, mixing with the Oxygen of the Air, forms the fatally explosive *Inflammable Air* or *Fire-Damp* of Miners. It may also be produced artificially from the decomposition of Acetates. It burns with a yellow flame, and may be respired; yet as Carb' and Water are the results of its combustion, a deleterious atmosphere is necessarily produced.

Olefiant Gas ($C_2\ H_2 = 14$), named from forming an oily liquid by combining with Chlorine, is a gaseous compound of 2 Eq. of Carbon and 2 of Hydrogen. Like the former, this gas is found in Coal-mines, and may be made artificially by heating strong Alcohol with 5 or 6 times its weight of Oil of Vitriol. It is not respirable, extinguishes flame, but burns with a brilliant white light.

Petroleum, L. *F.* Petrole. *G.* Steinöl.

Other Hydrocarbons are either solid or liquid: of the latter, Petroleum or Rock Oil is officinal; and Naphtha, whether obtained artificially or as a product of nature, is often used medicinally. As these natural products are considered to be produced by fossil vegetable remains, and as the artificial products are obtained by the decomposition of vegetable matter, it will be convenient to treat all of these nearly allied substances together. So also Oil of Turpentine, Tar, and the products of the distillation of Coal. (*v.* Coniferæ.)

Creasoton, L. Creosote.

This has been so named from its property of preserving meat from decay: it is a compound of Carbon, Hydrogen, and Oxygen, and is hence described as an Oxy-Hydro-Carburetum in the L. P. As it is obtained from Tar, it will most fitly be treated of with that substance.

Carbon and Nitrogen.

These two elementary substances combine together, and form a very remarkable body, called Cyanogen, from κυανος, *blue*, it being a principal ingredient of Prussian Blue. (*v. Sesquicyanide of Iron.*)

Cyanogen. *F.* Cyanogène.

Cyanogen (Cy or $C_2 N = 26$) may be obtained from Cyanide of Mercury : it is a colourless gas, of a peculiar and pungent smell, like that of peach-kernels, burning with a purplish flame, readily absorbed by water, and condensable into a colourless liquid. But it is chiefly interesting as, though being a compound body, it acts like the simple elements in combining with metals, and forms with Hydrogen an acid, the Hydrocyanic, which is of fearful importance from its rapidly deleterious effects. As Cyanogen is the type of the compound *radicles* which present themselves in Organic Chemistry, and as Hydrocyanic acid is produced naturally by some plants, it will be treated of with Laurel Water under Amygdaleæ.

Ternary Compounds.

Ternary compounds abound in the Organic Kingdom, and many of them are officinal. But such compounds of Carbon, Oxygen, and Hydrogen as Starch and Sugar, are most naturally treated of with the plants which yield them. It is also convenient to treat of the products of the Fermentation of Saccharine matter, as Wines and Alcohol, with the Grape Vine, under Ampelideæ. Also of Etherification, or the produce of the action of Acids on Alcohol, with the latter substance. Acetous Fermentation and the production of Vinegar are too closely allied to be separated from the consideration of Fermentation in general.

Vegetable Acids.

Having to notice the salts of these acids with the alkalis, earths, and metals, it would in some respects be preferable to treat of them here. But, as the account of each may be referred to, it is desirable to adhere to the plan adopted with the vegetable alkalis,—that of treating of the products with the Plants yielding them. Thus,

Citric Acid will be treated of with Citrus under Aurantiaceæ.
Tartaric Acid with Tartar under the Grape Vine in Ampelideæ.
Oxalic Acid with Oxalideæ.
Benzoic Acid with Benzoin under Styraceæ.
Succinic Acid with Amber after the Resin of Coniferæ.
Tannic Acid with Ratanhy Root under Krameriaceæ.

Nitrogen and Hydrogen.

Ammonia. Volatile Alkali. *F.* Ammoniaque. *G.* Ammoniak.

Ammonia (A or $NH_3 = 17$) was probably known to Pliny, as he mentions the strong odour evolved from the mixture of Lime and *Nitrum.** The Hindoos also were acquainted with it, and obtained it by mixing Sal Ammoniac 1 part and Chalk 2 parts. The name was derived from Sal Ammoniac, from which it was formerly obtained. The solution in water was known to the earlier Chemists, and called by them Volatile Alkali. It was first obtained as a Gas by Priestley. In 1756, Dr. Black distinguished it from its Carbonate, though Berthollet was the first to communicate precise ideas respecting its composition, which was determined by Gay Lussac. Ammonia exists at all times in small quantities in the air, and therefore in water, and is also contained in the juices of most plants, as the Birch, Beet-root, Sugar,&c. It is the chief source of the Nitrogen in plants, and is generally evolved during their decomposition. It is abundantly produced during the putrefaction of animal matter, and, in combination with Phosphoric′ and Muriatic′, exists in Urine. With heat and moisture, Urea (which is identical with Cyanate of Ammonia) is decomposed, Carbonate of Ammonia being formed. Some of its salts, as the Carbonate and Nitrate, are contained in mineral springs, as in those of Greiswolde and Kissingen; the Hydrochlorate and Sulphate (Mascagnin) are found in the neighbourhood of Volcanoes and near ignited Coal-seams. Dr. Austin ascertained that if nascent Hydrogen were presented to gaseous Nitrogen, Ammonia was formed. (Phil. Trans. vol. lxxvii. p. 379.) M. C. S. Collard has also some time since (Journ. de Chim. Méd. iii. 516) pointed out that this gas is formed during the contact of water and air wherever nascent Hydrogen and Nitrogen come in contact; and that thus it is produced daily in immense quantities, giving rise to the Nitrates which stimulate Vegetable life. (*Dict. de Mat. Med.* Merat & De Lens, i. p. 255.)

Prop. Ammonia, when pure, is a colourless transparent gas, with a pungent suffocating odour, having alkaline and caustic properties. It browns turmeric paper, and restores the blue colour of vegetables reddened by acids; but the effects are transient, from its volatility. It may be obtained by acting with caustic Lime or Potash on Hydrochlorate of Ammonia; or by heating a solution of Ammonia, and collecting the gas over Mercury. Its Sp. Gr. is 0·89. 100 C. I. weigh 18·28 grs. By a pressure of 5½ atmospheres at 50° F. it was reduced by Faraday to the state of a colourless transparent liquid, with a Sp. Gr. of 0·76. Water absorbs it with very great rapidity, and to a great extent. (*v. Solution of Ammonia.*) Alcohol and Ether also readily dissolve it.

* Probably Sal Ammoniac, as several substances were included under *Nitrum.*

Near any volatile acid, it forms a white vapour. Combining with acids, salts are formed, some of which sublime when the acid is volatile; but, when this is fixed, the Ammonia volatilizes on the application of heat. It supports neither respiration nor combustion. A mixture of 2 volumes of Ammoniacal gas and 1½ of Oxygen gas may be exploded by the electric spark. Nitrogen is produced, as well as water, proving the presence of Hydrogen. It is composed of 1 Eq. Nitrogen and 3 Eq. Hydrogen, or 1 volume of the former and 3 volumes of the latter compressed into 2 volumes.

Some chemists, however, now consider Ammonia to be a compound of a hypothetical substance, called Amide or Amidogen ($N\,H_2 = 16$) and 1 Eq. of Hydrogen, or an Amidide of Hydrogen ($N\,H_3 = 17$); and that this, by combining with 1 Eq. of H, forms another hypothetical substance, which has been called *Ammonium* ($N\,H_4 = 18$), which is supposed to act the part of a metal in various combinations.

Action. Uses. A local irritant; fatal if respired. Diluted with air, a stimulant of the nasal and bronchial passages.

Antidotes. Inhaling vapours of hot Vinegar or fumes of Mur′.

Solution of Ammonia.

This is a solution of Ammonia in water. The L. and E. P. have both a strong and a weak solution.

Prop. Colourless like the gas, with a powerful pungent odour, and acrid alkaline taste. Its density, which is less than that of water, varies with its strength, and is less as the quantity of gas dissolved is greater. Sir H. Davy ascertained that, at 50° F. and ordinary pressure, water absorbs 670 times its own bulk of Ammoniacal gas, becoming of a Sp. Gr. = 0·875, when it contained 32·5 parts or about ⅓ of gas: the lowest Sp. Gr. stated by Dalton is ·850. It freezes about —40° F. Its boiling point differs according to its density, depending chiefly on the escape of gas. Like Lime-water, Solution of Ammonia absorbs Carbonic′ gas from the atmosphere, at the same time that much Ammonia escapes. It combines with acids to form salts, and with Oil it forms Soap, in some officinal liniments. It decomposes a great many earthy and metallic salts, precipitating their oxides, and in some cases redissolving them in an excess of Ammonia, and producing a double salt, as in Ammonio-Chloride of Mercury, Ammonio-Chloride of Iron, &c.

Liquor (Aqua, E.) Ammoniæ fortior, L. Strong Solution of Ammonia.

As Solution of Ammonia is manufactured on a large scale by decomposing with caustic Lime the salts obtained from Gas liquor or from Bone spirit (*v. Ammoniæ Sesquicarbonas* and *Hydrochloras*), no formula is given for its preparation in the L. P., but it is to be of Sp. Gr. 882 (E. 880). As this is much stronger than ordinary solution of Ammonia, it may easily be reduced to the strength of Liquor Ammoniæ by adding to every f℥j. of it f℥ij. of water, by which the Sp. Gr. of the mixture will be 0·960.

The E. P. gives a formula for obtaining this and the Liquor Ammoniæ by one process: it is essentially the same as that of the L. and D. P. for obtaining the latter, differing, however, in the gas passing over into water, instead of the solution being distilled, and in being collected in two vessels: hence the solution is obtained of the two densities.

Prep. E. Slake *Quicklime* ℥xiij. with *Aq.* f℥vij. ß, triturate it quickly with finely-powdered *Mur. Ammon.*, and put in a retort; connect this with a receiver containing *Aq. dest.* ℥iv., and this with another containing *Aq. dest.* ℥viij. These must be kept cold. Heat the retort as long as gas is evolved; remove it, and heat the first receiver. Should the liquor in the last bottle not have the density of 960 (that ordered for the Aqua Ammoniæ), reduce it with that in the first, or raise it with *Aq. dest.* For details, *v.* E. P.

Action. Uses. Irritant, Vesicant and Caustic; often employed for *smelling* salts.

LIQUOR (AQUA, E.) AMMONIÆ, L. AQUA AMMONIÆ CAUSTICÆ, D.

Prep. L. D. Take of *Lime* ℥viij. (recently burnt 2 parts, D.), and slake with *Aq.* Oij. (hot 1 part, D.) Put into a retort, and add *Hydrochlorate of Ammonia* ℥x. (3 parts, D.), broken into small pieces (powdered and dissolved in *Aq.* 9 parts, D.), and the remainder of the water. Let f℥xv. of Solution of Ammonia distil over. (Distil 5 parts with a medium heat into a cold receiver, D.)

Here the Ammonia set free by the superior affinity of the Lime (Oxide of Calcium) distils over, but the Hydrochloric′ is decomposed; its Chlorine combines with the Calcium of the Lime to form Chloride of Calcium, while the Oxygen of the Lime and the Hydrogen of the Acid being set free, combine and form 1 Eq. of Water, which remains in solution with the Chloride of Calcium. The Sp. Gr. of this solution of Ammonia ought to be 0·960 (955—1000, D.), and it is composed nearly of 10 parts of Ammonia and 90 of water.

Tests. Odour, taste, and other properties, like the gas. By heat it evaporates in evanescent alkaline vapours, as shown by the transient browning of Turmeric paper. It gives no precipitate (Carbonate of Lime) with Lime-water, or with Chloride of Calcium, showing the absence of Carbonic′. It will not effervesce with dilute acids. When saturated with Nitric′, neither Sesquicarb. Ammonia nor Nitr. Silver throw down anything, proving that no earthy matter, nor H C′, nor any Chloride is present. Oxalic′ will indicate the presence of Lime.

Inc. Acids, acidulous and most metallic salts.

Action. Uses. Antacid, Rubefacient, Stimulant, Antispasmodic, Diaphoretic.

D. ♏x.—♏xxx. in water, Camphor mixture, Milk, or any demulcent liquid.

Antidotes. Vinegar, Lemon-juice, or Vegetable Acids.

Off. Prep. Hydrargyri Ammonio-Chloridum. Lin. Camphoræ Comp. Lin. Hydrarg. Comp.

TINCTURA AMMONIÆ COMPOSITA, L.

Formerly the *Spiritus Ammoniæ succinatus*, intended as a substitute for *Eau de Luce*.

Prep. Macerate Mastich ʒij. in Rectified Spirit Oij., and pour off the clear solution. Add Oil of Lavender ♏xiv., Oil of Amber ♏iv., strong solution of Ammonia Oj.

Action. Uses. Stimulant, Antispasmodic. In snake bites.

D. ♏v.—♏xx. in some bland liquid.

SPIRITUS AMMONIÆ, E.

Differs from Spiritus Ammoniæ, L. and D. in being a solution of pure Ammonia in Spirit, while the latter are solutions of the Carbonate of Ammonia in the same menstruum. They cannot, therefore, be considered as similar preparations, nor treated of together.

Prep. The E. P. obtains the Ammonia by acting on *Muriate of Ammonia* ℥viij. with *Fresh Burnt Lime* ℥xij., first slaking the latter with *Aq.* f℥vj ß., then mixing together the two salts, and heating them in a retort, to which has been adapted a tube, which passes nearly to the bottom of a bottle containing *Rectified Spirit* Oij.

Here the Ammoniacal gas, as it passes over, is dissolved in the Spirit, and a preparation very similar to the Liquor Ammoniæ, and with a strong odour of the alkali, is obtained. Sp. Gr. ·845. It does not effervesce with weak Muriatic Acid. Like the Spir. Ammoniæ, L. and D., this is employed for dissolving resinous and gummy-resinous substances, and volatile oils ; for these the Caustic Spirit is the more active solvent.

SPIRITUS AMMONIÆ AROMATICUS, E.

Prep. Take *Spir. Ammoniæ*, f℥viij., *Volatile Oil of Lemons*, f℥j., *Oil of Rosemary*, f℥jß. Dissolve the oils in the spirit by agitation.

D. ♏xv.—fʒj.

SPIRITUS AMMONIÆ FŒTIDUS, E.

Prep. Take *Spir. Ammoniæ*, f℥x ß., *Assafœtida* ℥ ß. Dissolve and distil.

D. ♏xv.—fʒj. *v.* ASSAFŒTIDA.

As an external application, Ammonia may be applied in the form of the following Liniments or Ointment.

LINIMENTUM AMMONIÆ, L. E. D.

Prep. Take of *Olive Oil* f℥ij., *Solution of Ammonia* f℥j. (f℥ij. D).

Rubefacient, Stimulant.

LINIMENTUM AMMONIÆ COMPOSITUM, E. Compound Liniment of Ammonia.

Prep. Take of Stronger Solution of *Ammonia*, D. 880, f℥v., *Tinct. of Camphor* f℥ij., *Spirit of Rosemary* f℥j. Mix them well together. This liniment may also be made weaker, for some purposes, with *Tincture of Camphor* f℥iij, and *Spirit of Rosemary* f℥ij.

Rubefacient, Vesicant, or Cauterizing.

Ammoniacal Ointment. This is formed by rubbing up Ammonia with fatty matter in proportions according to the effect required. If rubbed on the skin, and the Ammonia allowed to evaporate, rubefaction will be produced, but if confined by a compress, vesication will ensue.

CARBONATES OF AMMONIA.

From the difficulty of distinguishing in the description of old authors the several Carbonates of Ammonia, it is preferable to treat the little that is known of their history, together. We have seen that Carbonate of Ammonia is always present in the atmosphere. It is disengaged from decomposing animal remains, and is found in some springs, as well as in the juices of plants. The Hindoos would seem to be acquainted with it, as they have a formula, given by Dr. Ainslie, in which they heat together 1 part Sal Ammoniac and 2 parts of Chalk, which must produce a Carbonate of Ammonia. It was probably also, as inferred by Dr. Pereira, known to the Arabs. Raymond Lully was acquainted with the impure solution of Carbonate of Ammonia obtained from putrid Urine, and Basil Valentine mentions the Spiritus salis Urinæ.

The three Carbonates of Ammonia are all included among the preparations of the British Pharmacopœias: the simple Carbonate in a liquid form in the Spiritus Ammoniæ, the Spiritus Ammoniæ Aromaticus and Fœtidus; the Sesquicarbonate as a solid salt; and the Bicarbonate of Ammonia, D., may also often be found effloresced on the former.

CARBONATE OF AMMONIA. *F.* Carbonate d'Ammoniaque.

The Carbonate of Ammonia, which consists of 1 Eq. Ammonia, 17 +1 Eq. Carbonic Acid, 22=39, and, if hydrated, of an additional Eq. of Water, 9+39=48, may be formed by bringing C′ gas in contact with Ammoniacal gas, or by decomposing Hydrochlorate of Ammonia by the Alkaline or Earthy Carbonates in a liquid form, and distilling; or it may by evaporation be obtained in a crystalline state.

SPIRITUS AMMONIÆ, L. D.

Prep. Mix *Hydrochlorate of Ammonia* ℥x. with *Carbonate of Potash* ℥xvj., in *Rectified Spirit* and *Aq. dest.*, of each Oiij. Distil off 3 pints. L.

Here both salts are decomposed; the Carb′ combining with the Ammonia, a very volatile salt, the Carbonate of Ammonia, is formed, and distils over with the Spirit. The Chlorine of the Hydrochloric′ combining with the Potassium of the Potash, forms Chloride of Potassium, which remains in solution with 1 Eq. water, which has been formed by the union of the Hydrogen of the Acid with the Oxygen set free from the Potash. From the proportions employed, a larger quantity of Carbonate of Ammonia distils over than can be dissolved by the Spirit, and is deposited in a crystalline state.

The D. C. directs *Carb* (*i. e. Sesquicarb*) *of Ammonia* ℥iij ß., powdered, to be dissolved with a gentle heat in *Rectified Spirit* ℔iij. by weight, and filter.

This, though apparently a simple solution, is a conversion of the Sesquicarb into the Carb of Ammonia, from a portion of the Carb′ escaping during the solution. About 30 grs. are dissolved in f℥j of Spirit.

As these preparations contain only ⅔ as much Carb′ as the Sesquicarb, they are necessarily more pungent; but the activity of the Ammonia being modified by combination with the C′, they are milder than the solutions of Ammonia, but may be used for the same purposes. The Spirit of Ammonia is chiefly employed to dissolve Camphor, the volatile oils and some vegetable resins, as in the Spiritus Ammoniæ Aromaticus.

Action. Uses. Antacid, Stimulant.

D. ♏x.—♏xxx.

Spiritus Ammoniæ Aromaticus, L. D. Spirit of Sal Volatile.

Prop. It is a colourless, pleasantly fragrant, and agreeably stimulant solution of volatile oils in Ammoniated Alcohol. Sp. Gr. 0·914. It becomes brown by keeping.

Prep. L. Mix together *Hydrochlorate of Ammonia* ℥v., *Carbonate of Potash* ℥viij., *Bruised Cinnamon* ʒij, *Cloves* ʒij, *Lemon Peel* ℥iv., *Rectified Spirit* and *Aq. dest.* of each Oiv., and distil 6 pints.

D. Macerate *Spirit: Ammon:* ℔ij., by measure, *Essential Oil of Lemons* ʒij., *bruised Nutmegs* ℥ß., *bruised Cinnamon* ʒ iij., distil Oiß.

In the L. prep. the same changes take place as in the preparation of the Spir. Ammoniæ; but with the Ammonia and Spirit the essential oils of the vegetable substances distil over, which render this a more agreeable preparation.

D. ♏xv.—lx. May be prescribed with Sulphate of Magnesia.

Spiritus Ammoniæ fœtidus, L. D.

This is prepared exactly as the Spiritus Ammoniæ, with the addition of Assafœtida ℥v.

It is a solution of the Volatile Oil of Assafœtida in Ammoniated Alcohol. An efficient substitute may be made by adding Tinct. of Assafœtida to Spirit of Ammonia. The Dublin College order *Assafœtida* ℥ß to be macerated in *Spirit of Ammonia* Oij., and to distil off a pint and half of the clear liquor.

Action. Uses. Stimulant, Antispasmodic.

D. fʒß—fʒj.

Ammoniæ Sesquicarbonas, L. Ammoniæ Carbonas, E. D.

Sal Volatile. Ammonia Præparata. Ammoniæ Subcarbonas.

The Sesquicarbonate (NH_3, $1\frac{1}{2}CO_2$, $HO=59$), often called Subcarbonate, or simply Carbonate of Ammonia, has long been known by various names, as *Volatile* or *Smelling Salts*, *Salt of Hartshorn*, *Volatile Salt of Urine*, all of which indicate either its properties, or the sources from whence it was obtained. It is now obtained by the action of the Alkaline or Earthy Carbonates on Hydrochlorate of Ammonia, or sometimes on crude Sulphate of Ammonia.

Prop. Usually met with in colourless translucent cakes; fracture striated, or of a rather fibrous texture; taste sharp, alkaline, ammoniacal; odour pungent, penetrating. On exposure to the air, it loses its translucency, becomes friable, and covered with a white powder, Bicarbonate of Ammonia, which is much less pungent, and is called *Mild Carbonate of Ammonia.* This is formed from the escape of a portion of the Ammoniacal gas, or of Carbonate of Ammonia, as is indicated by the discoloration of Turmeric paper held over it. This salt is completely dissipated by heat; is soluble in less than 4 times (twice, Berz.) its weight of cold water, but in boiling water it is decomposed with the evolution both of Carb′ and of Ammonia; soluble also in proof, but sparingly so in rectified Spirit. The composition of this salt, according to Mr. Phillips, is

3 Eqs. of Carbonic Acid	22×3=66	or Carb′	55·93
2 Eqs. of Ammonia	17×2=34	„ Ammonia	28·81
2 Eqs. of Water	9×2=18	„ Water	15·26
	118		100

But it is more convenient to consider it, according to the view adopted in the Pharmacopœia, as composed of 1½ Eq. of Carb′ united to 1 Eq. of Ammonia and 1 of Aq. Thus, 1½ Eq. C′ 33+1 Eq. Ammonia 17 +1 Eq. Aq. 9=59, Hydrated Sesquicarbonate of Ammonia.

But, as observed by Dr. Pereira, "from the observations of Dalton and Scanlan, this is not a single salt or Sesquicarbonate, but a mixture or compound of the Carbonate and Bicarbonate; for, if treated with a small quantity of cold water, a solution of Carbonate of Ammonia is obtained, while a mass of Bicarbonate remains, having the form and dimensions of the Sesquicarbonate employed." From the uniformity of its composition and its crystalline structure, Dr. P. considers it to be a chemical combination of two salts. 1 Eq. Anhydrous Carb. Ammonia 39+1 Eq. Hydrated Bicarb. Ammonia 79 = 118 Hydrated Sesquicarbonate Ammonia. According to the most recent view, this salt is a compound of 3 Eq. C′+2 Eq. Oxide of Ammonium.

Prep. L.E.D. Take of *Hydrochlorate of Ammonia* ℔j. (1 part, D.), and of *Chalk* ℔j ß. (Carbonate of Soda 1 part, D.) Rub them separately into powder. Mix and sublime with a heat gradually raised.

Here mutual decomposition takes place, the Carb′ of the Chalk (Carbonate of Lime) combines with the Ammonia of the Hydrochlorate, and the desired salt is formed, and passes over in combination with 1 Eq. Aq. which has been formed by the union of the Oxygen of the Lime (Oxide of Calcium) with the Hydrogen of the H C′.

The Chlorine of this acid being set free, combines with the Calcium; a Chloride of Calcium is formed, and remains behind. Similar changes take place with the Carbonate of Soda. As 1 Eq. of Carb. Lime only is required to decompose 1 Eq. Hydrochlorate Ammonia, the D. C. directs equal parts; but the L. and E. order what is equal to 1½ Eq., as less heat is required, and more complete decomposition is secured. The peculiarity of the result obtained here, as remarked by Mr. Phillips, is, that both the Hydrochlorate Ammonia and Carb. Lime are neutral compounds, consisting each of 1 Eq. acid and 1 Eq. of base. Though a neutral salt is usually produced from the action of neutral salts, here we have a supersalt. This is explained by supposing 3 Eq. of each salt to undergo decomposition. If no loss occurred, the Carb. Ammonia would be neutral and hydrated, consisting of 3 Eq. C′ 66 + 3 Eq. Ammonia 51 + 3 Eq. Aq. 27 = 144. During sublimation, however, 1 Eq. of the Ammonia liberated with 1 Eq. of the water formed is dissipated, the Carb′ remaining undiminished. The Carbonate actually sublimed consists of 3 Eq. of Carb′ and only 2 of Ammonia, or in the proportions of a Sesquicarbonate.

Sesquicarbonate of Ammonia is sometimes made on a large scale by subliming a mixture of impure Sulph. Ammonia and Carb. Lime. The result is as above, Sulph. Lime being left. The Sulphate of Ammonia being obtained by acting with S′ or Sulph. Lime on the Carb. Ammonia of *Gas Liquor*, or that from *Bone Spirit*. It is necessarily impure, and often contaminated with tar or oily matter, and therefore requires to be refined.

Tests. The salts of Ammonia may be easily recognized by its fumes, which are exhaled when they are rubbed up with Potash. In this salt the odour and tests will at once reveal that alkali, and the Carb′ by effervescence with dilute acids. It yields a white precipitate with the Chloride of Calcium or of Barium; "the clear liquor from which the latter precipitate (Carbonate of Barytes) has subsided, yields a further precipitate on the addition of Caustic Ammonia. By this last character the Sesquicarbonate is distinguished from the neutral Carbonate." (Per.) Carb. Ammon. is not very liable to be impure when prepared by the first process. "Translucent in mass, but falls to powder in the air; entirely sublimed and soluble; changes the colour of turmeric; when saturated with Nitric′, it does not precipitate with Chlor. Barium (Nitr. Barytes, E.), or with Nitr. Silver, (L.)" If the translucency be impaired, and the salt looks white and powdery, a portion has been converted into the Bicarbonate, a less pungent salt. Anything insoluble or not sublimed is an impurity. The presence of Sulphates, as of Ammonia, will be indicated by the Barytic salts; and any Hydrochlorate of Ammonia, by a white precipitate being formed by Nitrate of Silver.

Action. Uses. Antacid, Stimulant, Antispasmodic, Diaphoretic.

D. gr. ij.—gr. x. in pills, or in solution, as below.

Inc. Acids, Acidulous Salts, Alkalis, Lime-water, Magnesia; many

Metallic salts, but not the Potassio-Tartrate of Iron, nor Sulphate of Magnesia.

LIQUOR AMMONIÆ SESQUICARBONATIS, L. AQUA AMMONIÆ CARBONATIS, E. D.

Prep. Dissolve *Sesquicarbonate of Ammonia* ℥iv. (4 parts, D.) in *Aq. dest.* Oj. (15 times its weight, D). Filter. Sp. Gr. 1090, D.

This solution has the odour and other properties of the salt, and like it is liable to change when exposed to the air, becoming less pungent. It may be employed for all the purposes of this salt, or for those of the Liquor Ammoniæ, but is less pungent, and will evolve C̸.

The Sesquicarbonate of Ammonia is sometimes employed for making effervescing draughts.

℈j. Sesquicarbonate of Ammonia saturates	f℥vj. of Lemon Juice 26 grs. of cryst. Citric acid	forming Citrate of Ammonia.
	26 grs. of cryst. Tartaric acid	forming Tartrate of Ammonia.

Action. Uses. Rubefacient, Stimulant, &c. Useful as a Test. Useful in allaying nausea and vomiting ; also slightly diaphoretic.

D. f℥ß or even f℥ij. if duly diluted.

LINIMENTUM AMMONIÆ SESQUICARBONATIS, L. Liniment of Sesquicarbonate of Ammonia.

Prep. Shake together *Sol. of Sesquicarbonate of Ammonia* f℥j., *Olive Oil* f℥iij., until they are well mixed.

The Ammonia, by combining with the Oil, forms a kind of Soap, of which the union is imperfect, in consequence of the Carbonate being employed. It is a milder preparation, and may be used for the same purposes as the Linimentum Ammoniæ. A similar preparation may be prepared by rubbing up some of the powdered salt with Lard.

AMMONIÆ BICARBONAS, D. Bicarbonate of Ammonia.

The Bicarbonate ($A\,C_2\,Aq_2$) is formed whenever the Sesquicarbonate is exposed to the atmosphere, or even when the bottle in which it is kept is frequently opened. It is sometimes called *Mild Carbonate of Ammonia*, from the Ammoniacal odour and taste being less obvious from combination with a further Eq. Carb′. It crystallizes usually in 6-sided prisms, and requires 8 parts of water for solution.

This salt is composed of 2 Eq. C′ 44+1 Eq. A. 17+2 Eq. Aq. 18 =79 ; or, per cent. C′ 55·70+A. 21·52+Aq. 22·78=100 ; or an anhydrous compound of 2 Eq. C′ and 1 of Oxide of Ammonium.

Prep. Take of *Water of Carbonate of Ammonia* any quantity, and pass through it *Carb′ gas* until the alkali is saturated. Then let crystals form; dry them without heat, and preserve in a close vessel.

Tests. "Its solution at first occasions no precipitate with Chlor. Barium or Chlor. Calcium : after a short time, however, the mixture evolves Carb′, and a white earthy Carbonate is precipitated." (*p.*) But it does not precipitate Sulphate of Magnesia.

Action. Uses. Antacid, Diaphoretic. Being milder, it may be more suitable than the Sesquicarbonate in some cases.

D. gr. v. to gr. xx. in *cold* water. For effervescing Draught. ℈j. will saturate 18 grs. of Çit′. or 19 grs. of Tar′.

LIQUOR (AQUA E. D.) ACETATIS AMMONIÆ, L. *Spirit of Mindererus.*

Acetate of Ammonia (NH_3 Ac′ + HO = 77) may be obtained in crystals in the exhausted receiver of an air-pump; but as it is a deliquescent salt, it is contained in the Pharmacopœias only in the state of a diluted solution.

Prop. This is limpid, colourless, with a faint smell, and a slight mawkish taste. If neutral, it should produce no effect on Litmus or Turmeric paper, though it is preferable to have a slight excess of acid, as it will be less irritant when used as a lotion in some cases; but the excess of alkali may sometimes be no objection when exhibited internally.

Prep. Add to *Distilled Vinegar* Oiv. [(f℥xxiv. E.) (30 parts or q. s. D.)], *Sesquicarbonate of Ammonia* ℥ivß. [(℥j. E.) (1 part, D.)], or q. s. to saturate the Vinegar.

The E. P. further directs, that if the solution have any bitterness a little distilled vinegar should be added till that taste be removed. The density of the distilled vinegar should be 1005, and that of the Aqua Acetatis Ammoniæ 1011. The quantity of the salt ordered by the L. P. is higher than requisite; but, as it is to be added to Distilled Vinegar of the Pharmacopœia strength to the point of saturation, a uniform preparation will be procured. On the Continent some of the solutions are of greater strength. D. 1030—1040. The original Spirit of Mindererus was formed by saturating strong vinegar with Spirit of Hartshorn; it was thus a solution of Acetate of Ammonia, with some Ammoniacal Soap formed by the action of the alkali on the empyreumatic oil of the spirit, which, M. Chaussier observes, added to its efficacy. Distilled Vinegar of the proper strength should alone be employed, and not, as is often the case, impure Acetic Acid diluted.

Tests. Action on Litmus or Turmeric paper will detect excess of acid or of alkali. It is not coloured by the addition of Hydrosul′., showing the absence of any metallic oxide, especially Copper or Lead; no precipitate on addition of Nitr. Silver or Chlor. Barium, the first indicating absence of Hydrochloric′, and the second of Sulph′. The water being evaporated, the residue yields Ammonia, and is dissipated by heat; any further residue is an impurity, as both the Acetic′ and the Am. are volatile. The most usual irregularity is in point of strength, and this should be ascertained by the Sp. Gr. = 1011, E. and D.; but none is given in the London Pharmacopœia.

Inc. Decomposed by the strong acids; also by Potash and Soda, and by their Carbonates; by Lime and Lime-water, Magnesia and

Sulphate of Magnesia, the Acetate and Diacetate of Lead, on account of the Carbonic′ which often remains diffused in the solution, and which then precipitates as Carb. of Lead; also several other metallic salts, as of Antimony, the Sesquichloride and Sulphate of Iron.

Action. Uses. Stimulant, Diaphoretic; Refrigerant Lotion and Collyrium.

D. f℥ij.—f℥vj. every 3 or 4 hours, with Camphor mixture, &c.

AMMONIÆ (MURIAS, E. D.) HYDROCHLORAS, L.

Hydrochlorate (Muriate) of Ammonia. *Sal Ammoniac.* *F.* Hydrochlorate d'Ammoniaque. *G.* Salmiak.

This salt ($NH_3\,HCl=54$) was known to Geber. Avicenna and Serapion mention it by the name *Noshadur.* Persian writers give *Armeena* as its Greek synonyme. The Sanscrit name is *Nuosadur;* the Author obtained it by this name in India, where it is formed in brick-kilns. (*v.* Hindoo Med. p. 40.) In Egypt it is obtained from the dung of camels. That it was known to the Romans, is evident from Pliny stating that one of the kinds of *Nitrum* gives out a strong smell when mixed with Quicklime.

Prop. It is usually in pieces of hemispherical cakes, of a white colour, without smell, having a saline acrid taste. Its texture is striated and radiated ; it is somewhat tough and ductile, opaque or crystalline and semi-transparent; nearly permanent in the air, but attracts a little moisture; some impure varieties, crystallized in conical masses are deliquescent, from containing Chloride of Calcium. Sp. Gr. 1·450. It is soluble in about its own weight of boiling water, but requires 3¼ times as much at 60°, when it produces considerable cold; hence it is one of the salts most commonly employed in freezing mixtures. It requires about 5 parts of Alcohol, but less of rectified Spirit, for solution ; when heated, it sublimes without decomposition, and is thus most frequently obtained. When its solution in boiling water is cooled down, it crystallizes in tetrahedral prisms terminated by 4 planes, octohedrons, or in plumose crystals ; the latter are formed "of rows of minute octohedrons, attached by their extremities." (*g.*)

This salt is decomposed by acids and alkalis, the S′ and N′ combining with its Ammonia and setting free the H.Cl′; while Potash, Soda, Baryta, Lime, and Magnesia, set free its Ammonia, which may be recognized by its tests, and unite with the Chlorine of the acid;—their Carbonates likewise decompose it, forming Carbonates of Ammonia. With Nitr. Silver a white precipitate (Chloride of Silver) is formed, and with Acet. Lead one of Chloride Lead, also white. With Bichloride of Platinum, a yellow precipitate (Platino-bichloride of Ammonia), which, when collected, dried, and ignited, yields spongy Platinum. (*p.*) It increases the solubility of Bichloride of Mercury, and forms an ingredient in the Liquor Hydrargyri Bichloridi.

Hydrochlorate of Ammonia is composed of Hydrochloric acid 31·48, Ammonia, 68·52=100.

By those chemists who admit the hypothetical metal Ammonium, this salt is termed Chloride or Protochloride of Ammonium. Dr. Kane considers it a Chloro-amidide of Hydrogen.

Hydrochlorate of Ammonia may readily be formed by bringing together Ammonia and H Cl′ gases; and these probably come together in volcanoes, and account for the salt being there found. In Egypt it is yielded by the soot of the dung of camels and other animals which feed on the saline plants of the desert. So in N. W. India it is obtained at the unburnt extremity of brick-kilns, where animal manure and refuse straw, &c. are employed as fuel. It is now obtained from the destructive distillation of bone, as animal Charcoal is required for the use of Sugar-refiners, the fat and marrow being first removed for the use of Soap-makers. The gelatinous and cartilaginous parts become decomposed, the Nitrogen and Hydrogen form Ammonia, and the Carbon with Oxygen some Carb.′, which unites with the Ammonia. This Carbonate of Ammonia, received and consequently dissolved in water, is called *Bone Spirit.* This salt is also obtained in the preparation of Coal Gas, which, being passed into water, forms *Ammoniacal* or *Gas Liquor.* Other salts are also formed. To these H Cl′ is sometimes added, and an impure Hydrochlorate of Ammonia obtained, which may be purified by crystallization and sublimation. By other manufacturers (*v.* Per. i. p. 316) an impure Chloride of Calcium, obtained from Salt-works, is added to the Ammoniacal liquor, when a precipitate of Carbonate of Lime is obtained, and, as before, Hydrochlorate of Ammonia in solution; in either case, it may be separated and dried by evaporation, and then purified by sublimation. Or these Carbonates may be converted into Sulphate of Ammonia, and mixed with Chloride of Sodium. On application of heat, double decomposition ensues, Hydrochlor. Ammonia is formed, and, being volatile, is obtained pure by sublimation; Sulphate of Soda remains behind.

Tests. Sal Ammoniac, as found in nature, is sometimes mixed with Chlor. Calcium, the old Muriate of Lime. This is ascertained by its greater deliquescence, and the tests for that earth. From the mode of its manufacture, it sometimes contains Iron or Lead. These would be revealed by its not being totally soluble in water, and not being sublimed by heat without residue. It should be colourless and translucent; Chloride of Barium throws down nothing, showing absence of Sulphate of Ammonia. The Lead is usually seen on the discoloured convex surface, when it has been sublimed into a leaden vessel, probably a double Chloride of Lead and Ammonia. A solution of this salt gives a black precipitate (Sulphuret of Lead) when Hydrosulphuric acid gas is passed through it. (Jackson, Med. Gaz. 1839, as quoted by Per.) Iron may be detected by Ferrocyanide of Potassium and a few drops of Nitric′.

Action. Uses. Moderately Stimulant, Irritant, Diaphoretic; Refrigerant as a lotion, from the cold produced in solution; Discutient; sometimes Anodyne in Neuralgic affections.

D. gr. v.—gr. xxx. 2 or 3 times a day, with Sugar and Aromatics. As a cold application, equal parts of Nitre and Sal Ammoniac may be employed. ℥ij. with Nitre ℥v. will reduce temperature 40°.

Inc. Strong acids; Potash, Soda, Lime, their Carb.; Acet. Lead.

METALS.

Metals may be divided into those which by union with Oxygen form 1. ALKALIS, such as Potassium and Sodium; 2. ALKALINE EARTHS, Barium, Calcium, Magnesium, Aluminum; 3. METALS commonly so called, or which form METALLIC OXIDES.

POTASSIUM.

F. Potassium. *G.* Kalium, and Kali metall.

Potassium (K or Ka = 40) is the metallic base of Potassa, or Potash, the Oxide of Potassium. In this state it exists in nature in abundance, combined with Acids and Earths, Iodine, Bromine, &c.; but it is obtained chiefly from the vegetable kingdom. It was the first of the metallic bases obtained by Sir H. Davy; this by galvanizing Caustic Potash very slightly moistened, when the metal in small globules appeared at the negative pole. It is now commonly obtained by exposing Potash to intense heat with Iron-filings or Charcoal, which take its Oxygen, and the Potassium is set free.

Prop. Potassium, at 55° F., is a soft malleable solid. It has been crystallized in cubes, is brittle at 32°, and fuses at 156°; at 60° F. its Sp. Gr. = 0·86; it is therefore light enough to float on water. It is silvery-white, but immediately tarnishes when exposed to air, from its great affinity for Oxygen, which it will take also from water, swimming and burning upon it with great brilliancy, and being converted into Potash, while the Hydrogen escapes. It is preserved in fluids, such as Petroleum and Naphtha, which contain no Oxygen, as it is one of the most powerful deoxidizing agents chemists possess.

POTASSÆ HYDRAS, L. POTASSA, E. POTASSA CAUSTICA, D. Oxide of Potassium. Potash. *Potassa fusa. Kali purum. Fixed Vegetable Alkali. F.* Potasse caustique. *G.* Kali.

Potassa (K O = 48) is a compound of Oxygen and the metal Potassium. The name Potassa was derived from the commercial name Potash, which is a Carbonate of Potassa (*v.* p. 74). Dr. Black in 1756 first clearly distinguished the Carbonates from the caustic alkali. This he called Lixivia, from the name in Pliny, but it was named *Kali* by the L. C. The ancients were no doubt acquainted with some method of depriving the alkali of its Carbonic acid, as they were acquainted with the art of making soap. (Pliny, xxviii. c. 51.)

From the affinity of Potassium for Oxygen, the Oxide or Potash is readily formed by exposing the metal to dry air or to Oxygen gas. Some of it may also be found in the gun-barrel in the process of making Potassium. But it has so great an attraction for water, that it readily absorbs it from the air, and is therefore usually seen in this, which is the officinal state, Potassæ Hydras, K O, H O = 57.

Prop. Caustic Potash, when fused and pure, is whitish, in solid, slightly crystalline masses, sometimes in tetrahedral pyramids or octohedrons; hard and brittle; Sp. Gr. 1·70; usually in cylindrical pieces or sticks, of a grayish colour, of an intensely caustic taste, with little smell. When moistened, it has a soapy feel, from dissolving the cuticle. It readily attracts moisture, and at the same time Carb from the atmosphere. It is soluble in water, producing heat when in a fused state, but some cold when crystallized; also in alcohol, with the exception of impurities. Hence it may be separated from its Carbonates, as these are insoluble in Alcohol. It liquefies ice, with the production of intense cold; is not decomposed by the most intense heat, but fuses below a red heat, and at a bright red heat evaporates in white acrid fumes. Turns green the blue colour of vegetables, but afterwards destroys them, like other organic substances, and possesses highly alkaline properties, uniting with fixed Oils and Fats to form soaps, and with acids to form salts. It combines with considerable energy with Phosphorus and with Sulphur; when fused with Siliceous Earth, it forms Glass; and when in larger proportion, a Silicate of Potash, soluble in water. With other earths it forms enamels, and even when in solution dissolves Alumina and Glucina. Its salts are soluble in water, and generally crystallizable; when added to a solution of Sulphate of Alumina, they cause the formation of Alum in crystals. Tartaric′, if added in excess, produces a precipitation of Cream of Tartar, or Bitartrate of Potash; while Chloride of Platinum throws down a reddish-yellow precipitate of Chloride of Platinum and Potassium; the salts of Potash, moreover, give a violet tinge to flame, and may by these characters be distinguished from the salts of Soda.

The common method of obtaining it is to decompose one of the most commonly obtained of the salts of Potassa, that is, the Carbonate, by means of Lime, and then evaporating the solution to dryness.

Prep. Take *Solution of Potash* cong. j. (q. s. E. D.), evaporate in a clean Iron (Silver, D.) vessel, till ebullition ceases, and only the fused Hydrate of Potash is left, L. D. (Evaporate in a clean covered Iron vessel, till an oily fluid remains, which becomes hard on cooling on a glass rod, if dipped into it, E.) Pour into proper moulds. [Pour on a bright Iron (Silver, D.) plate; and as soon as solid, cut into pieces, and keep in a well-stoppered glass bottle, E. D. (Avoid the drops spurted up during evaporation, D.)]

The solution of Potash employed should be itself pure, and so preserved as not to have attracted Carb′ from the air, while the ebullition and temperature are kept up, no Carb′ is absorbed. A clean iron vessel is sufficient, but the contact of all organic substances must be prevented.

Tests. The L. P. states that it soon deliquesces, and is entirely soluble in alcohol; but this will seldom be found to be the case. The other properties are the same as those of Liquor Potassæ.

"Boiling water commonly leaves Oxide of Iron undissolved, which should not exceed 1·25 per cent. The solution neutralized with Nitr (and it should not effervesce) gives a faint precipitate with a solution of Nitrate of Baryta (indicating a Sulphate), none with the solution of Nitrate of Silver." E. P.

Action. Uses. Escharotic, Caustic Poison, Antacid. v. *Liquor Potassæ.*

POTASSA CUM CALCE, L. E. POTASSA CAUSTICA CUM CALCE, D. Potash with Lime.

Hydrate of Potash being chiefly employed as a Caustic, and being objectionable on account of its deliquescence, this preparation is often preferred, as the presence of Lime obviates the inconvenience.

Prep. Take *Hydrate of Potash* ℥j., *Lime* ℥j.; rub together, and keep in a well-closed vessel, L. [*Aq. Potassæ*, q. s.; evaporate in a clean-covered Iron vessel to to ⅓rd (¼th, D.) its volume, add slaked *Lime* till the fluid is of the consistence of firm pulp. Preserve the product in well-covered vessels, E. D.]

Action. Uses. Caustic; made into a paste with Rectified Spirit, and applied, the neighbouring parts being defended with sticking-plaster.

LIQUOR POTASSÆ, L. POTASSÆ (CAUSTICÆ, D.) AQUA, E.

Prep. Take *Carb. Potash* ℥xv. [(dry ℥iv., E.) (of Commerce 2 parts, D.)], *Lime* ℥viij. (fresh burnt ℥ij., E.; 2 parts, D.), *Aq. dest.* boiling cong. j. (*Aq.* f℥xlv., E.; 15 parts, D.) Dissolve the Carb. Pot. in C.ß of the Aq. (in f℥xxxviij., E.) Slake the Lime with a little water in an earthen vessel, and then add the remainder of the Aq. (Slake with f℥vij., and convert it into Milk of Lime, E.) Mix the liquors in a close vessel, and agitate till they are cold. [(Add the Milk of Lime to the boiling solution of the Carb., in about 8 successive portions, boiling briskly for a few minutes after each addition, E.) (Mix the salt with the slaked Lime, and add the rest of the Aq.; when the mixture has cooled, put it into a well-closed bottle, and agitate frequently for 3 days, D.)] Set it aside for the Carb. Lime to settle, pour off the supernatant liquor, and keep it in a well-stoppered green glass bottle. [(Pour the whole into a deep, narrow glass vessel for 24 hours, then with a siphon draw off the clear liquid, which ought to be at least f℥xxxv., and of a Sp. Gr. = 1072, E.) (When the Carbonate of Lime has settled, decant the clear liquor, and keep it in well-stoppered green glass bottles. Sp. Gr. = 1080, D.)]

The Lime, having a strong affinity for Carb′, unites with that of the Carb. Pot.; the insoluble Carb. of Lime being precipitated, the free Potash remains in solution. Filters are not employed, because the Potash destroys all organic matter, and the process is tedious. The solution should be as little as possible exposed to the air, as it absorbs Carb′. The purity will depend upon the Carbonate of Potash and Lime, as well as upon that of the water employed. Dr. Christison states that decomposition of the Carb. is accelerated by ebullition.

Prop. Solution of Potash is colourless, transparent, somewhat oily-looking, without odour, but of an extremely acrid, caustic taste. The quantity of real Potash in solutions of different Sp. Gr. was ascertained by Dalton. (v. Brande, Chem. p. 541.) It feels soapy when rubbed between the fingers, is highly alkaline, rapidly absorbs Carb′ from the air, must therefore be kept in well-stoppered green glass bottles, because it acts on those made of flint glass. It forms soaps with oils and fats, and powerfully decomposes many salts, as those of Ammonia, of the Earths and Metals, throwing down their oxides, many of which it redissolves when added in excess. It corrodes both animal and vegetable textures, and precipitates from vegetable infusions any alkalis or neutral principles, while itself combines with their acids.

Tests. Sp. Gr. 1·063 (L. P.) browns Turmeric, and, like other salts of Potash, throws down a yellow precipitate with Chloride of Platinum, which is insoluble in Spirit. It should not effervesce with N′, or become milky on the addition of Lime-water, and thus show the absence of Carb′. When saturated with N′, scarcely anything should be precipitated by Carbonate of Soda (showing that no Lime or metallic impurity is present), nor with Chlor. Barium (no Sulphates), nor with Nitr. Silver (no Chlorides).

Inc. Acids, Acidulous and Ammoniacal Salts, Earthy and Metallic Salts, Chloride and Bichloride of Mercury.

Action. Uses. Antacid, Antilithic, Diuretic, Resolvent, Alterative.

Antidotes. Oil, Acids, Vinegar, Lemon-juice.

D. ℳx.—f℥j. gradually increased, with Infusion of Orange Peel, &c.

Potassii Iodidum, L. E. Potassæ Hydriodas, D.

Iodide of Potassium. *Ioduret of Potassium. Hydriodate of Potash.* *F.* Iodure de Potassium. *G.* Iod Kalium.

Iodide of Potassium (K I = 166) was first made by Courtois in 1812. Though of recent discovery, it has already had several names. It exists in Sea as well as in some Mineral waters, in Sea-weeds and Sponges, and was first employed in medicine by Coindet.

Prop. It is a colourless salt, sometimes with a slight tinge of yellow, of an acrid saline taste; often opaque, but, when carefully prepared, transparent, and crystallized in cubes or in quadrangular prisms. These contain no water of crystallization, but some is often lodged between the plates of the crystals: hence they decrepitate when heated, fuse at a low red heat, and volatilize unchanged. Permanent in dry air; soluble in ⅔ of its own weight of water, very soluble in Alcohol of Sp. Gr. ·850. It is readily decomposed by the mineral acids, Iodine being evolved, which can then be detected by the blue colour produced by the starch test. The same effect will follow if the salt be decomposed by Chlorine or a mixture of Chlorine and Nit′. It renders Iodine more soluble both in Water and in Alcohol. When dissolved, solution of Acetate of Lead produces a yellow precipitate (Iodide of Lead), and Protonitrate of Mercury a greenish Iodide of Mercury; the Pernitrate, or Bichloride of Mercury, causes a greyish-red, which soon becomes brilliant red (Biniodide of Mercury), which is redissolved by an excess of either Iodide of Potassium or of Corrosive Sublimate.

Various methods have been proposed for making this salt; but we shall restrict ourselves to those adopted by the British Pharmacopœias.

Prep. L. E. *Iodine* ℥vj. (dry ℥v. E.), *Carb. Pot.* ℥iv. (dry ℥ij. and ʒvj. E.), *Iron filings* ℥ij. (fine *Iron wire* ℥iij. E.), *Aq. dest.* Ovj. (Water Oiv. E.) Mix *Iodine* with *Aq.* Oiv., and add *Iron*, stirring frequently with a spatula for half an hour. (Boil the Iodine, Iron, and part of the water together in a glass matrass, at first gently, and then briskly, until about f℥ij. remain, E.) Apply a gentle heat, and when the liquid becomes greenish add the *Carb. Potash*, dissolved in the rest of the *water*. (While hot, add the Carb. Pot., dissolved in a few ounces of the water, and stir carefully, E.) Filter, and wash the powder with boiling *Aq. dest.*, Oij. (a little water, E.), and again filter. Evaporate the mixed liquor (at a temperature below boiling, E.) that crystals may form. (to dryness; purify this from Oxide of Iron, &c., by dissolving in less than its own weight of boiling water, or by boiling it in twice its own weight of rectified spirit, filtering the solution, and setting it aside to crystallize. More crystals may be obtained by concentrating and cooling the residual liquor, E.)—Iodide of Iron is first formed, which is decomposed by the Carb. Pot., Protocarbonate of Iron falls down, and the Iodide of Potassium, which remains in solution, is concentrated, after being filtered, and allowed to crystallize.

D. Triturate *Iodine* 1 part, with *Aq. dest.* 16 parts, and put the mixture in a glass vessel. Dilute, *Sul′* 7 parts, with *Aq. dest.* 32 parts, and pour it upon *Sulphuret of Iron* in coarse powder 5 parts, in a matrass, with a tube attached to the neck, long enough to reach to the bottom of the vessel containing the Iodine and water (to generate Sulphuretted Hydrogen gas). Pass the gas through the mixture until the Iodine disappears. (Sulphur is thrown down, and, the Iodine uniting with the nascent Hydrogen, Hydriod′ remains in solution.) Filter the liquor (to get rid of the Sulphur), concentrate by boiling to one-eighth, and filter again; (boiling is objectionable, as Iodine is given off, and C′ is absorbed, *c.*;) then add solution of *Carb. Pot.* q. s. till effervescence ceases. (Carb′ is disengaged, and Hydriodate of Potash, or Iodide of Potassium remains in solution.) Evaporate to dryness, and dissolve with the aid of heat the remaining white salt in *Rectified Spirit* 6 parts. Filter, evaporate to dryness, and preserve the residue in well-stopped vessels.

Tests. This salt is apt to be contaminated with water, Carbonate of Potash, Chloride of Potassium or of Sodium, and Iodate of Potash. L. P. S′ and Starch added together render the solution blue, in consequence of the Iodine being set free. It alters the colour of Turmeric very slightly, that of Litmus not at all (proving the absence both of acid and alkali, or of such salts). It loses no weight when subjected to heat (any Iodate of Potash will be decomposed, and Oxygen escape, and loss of weight will also occur from evaporation of water). Totally soluble in Aq. and in Alcohol (Carbonate of Potash is not soluble in the latter). The E. C. state that the solution is not affected, or is merely rendered hazy, by solution of Nitr. Bar., but will form an oily-looking mixture with the water. (If the Carbonate is present, an insoluble Carb.; if Iodate of Potash, a white precipitate of Iodate of Baryta; so with Lime-water, a white Carb. Lime will be formed.) "Gr. 10 of this salt are sufficient to decompose gr. 10·24 of Nitr. Silver. What is precipitated (Iodide Silver) is partly dissolved by Nit , and partly altered in appearance, which is not the case when Ammonia (being insoluble in it) is added." L. If it decomposes a larger proportion of Nitr. Silver than above stated, it is probably owing to the presence of Chloride of Potassium. The E. C. state that a solution of gr. v. in

Aq. dest. f℥j., precipitated by excess of sol. Nitr. Silver, and then agitated in a bottle with a little Aqua Ammoniæ, yields quickly by subsidence a clear supernatant liquid, which is not altered by an excess of Nitr′, or is rendered merely hazy. Here the Nitr. Silver will throw down any Chlorine present as a Chloride of Silver, which is soluble in Ammonia, while a very small proportion of the Iodide of Silver is taken up. "In the clear fluid, Nitr′ added to saturation of the Ammonia, or in excess, will make the Chlor. Silver reappear in the form of a white precipitate; but if there was no alkaline Chloride in the salt, the clearness of the fluid will scarcely be disturbed."

Inc. Acids, Acidulous and Metallic Salts.

Action. Uses. Irritant, Stimulant of the Absorbents, Diuretic; in Venereal nodes and Rheumatism.

Antidotes. Evacuate the stomach; give Demulcents; obviate Inflammation, and allay Irritation.

D. gr. vj.—gr. x.; ʒj. even ʒij. have been given.

UNGUENTUM POTASSÆ HYDRIODATIS, D. Ointment of (the Hydriodate) Iodide of Potassium.

Prep. Mix *Hydriodate of Potash* (i. e. Iodide of Potassium) ℈j., *Prepared Lard* ℥j.

This being a simple ointment, and devoid of colour, is preferable for some frictions, as it does not stain the skin. It may be employed much stronger than in the above preparation.

UNGUENTUM IODINII COMPOSITUM, L. UNGUENTUM IODINEI, E. Compound Ointment of Iodine. Ointment of *Ioduretted* Iodide of Potassium.

Prep. *Iodine* ʒß., *Iodide of Potassium* ʒj., *Rectified Spirit* fʒj, *Lard* ℥ij. Rub together the Iodine and Iodide of Potassium with the Spirit. Mix with the Lard (the same proportions, omitting the Spirit, E.)

Action. Uses. Promotes absorption; is therefore employed in enlarged glands, bronchocele, &c.

TINCTURA IODINII COMPOSITA, L. Compound Tincture of Iodine.

Prep. Macerate *Iodine* ℥j., *Iodide of Potassium* ℥ij., in *Rectified Spirit* Oij., till they are dissolved, then strain.

The presence of Iodide of Potassium increases the solubility of Iodine, and retains it in solution; so that it may be added to water without decomposition, or it may be given in Sherry wine.

D. ♏x.—fʒj.

LIQUOR POTASSII IODIDI COMPOSITUS, L. IODINEI LIQUOR COMPOSITUS, E. Compound Solution of Iodide of Potassium. Solution of *Ioduretted* Iodide of Potassium.

Prep. Dissolve *Iodide of Potassium* gr. x. (ʒj. E.), *Iodine* gr. v. (ʒij. E.) in *Aq. dest.* Oj. (f℥xvj. Agitate, and apply a gentle heat, E.)

The solution of the E. C. is a strong one, that of the L. P. a weak one; the doses must therefore be apportioned accordingly. It is of a reddish-brown colour, and may be given diluted with water.

D. f℥ij.—f℥iv. L.; of the E. preparation, ♏v. to ♏xv.

Potassii Bromidum, L.

Bromide of Potassium. *Hydrobromate of Potash.* *F.* Bromure de Potassium. *G.* Brom Kalium.

Bromide of Potassium (KBr = 118), discovered by Balard in 1826; introduced into the L. P. of 1836. The only officinal salt of Bromine.

Prop. It is white, without odour, of a sharp saline taste; crystallizes in transparent cubes, or four-sided flattish prisms, without any water of crystallization. Readily dissolves in Aq., less so in Alcohol. When heated, the crystals decrepitate, and may be fused without decomposition. Readily decomposed by Chlorine, which expels the Bromine; so also by the mineral acids, acidulous salts, and the metallic salts. This salt consists of 66·1 parts of Bromine with 33·9 of Potassium in 100 parts. (v. Tests for *Potassium* and *Bromine.*)

Prep. Add *Iron-filings* ℥j., and then *Bromine* f℥ij., to *Aq. dest.* Oj ß.; stir for half an hour. Apply a gentle heat till the colour becomes greenish (Bromide of Iron being formed). Then add *Carb. of Potash* ℥ij. and ʒj. dissolved in *Aq. dest.* Oj ß. Filter, wash what remains (Proto-Carbonate of Iron) in boiling *Aq. dest.* Oij. Filter again, mix the two liquors and evaporate to obtain crystals (Bromide of Potassium). In the first part of this process the Iron and Bromine, combining together, form a Bromide of Iron. On the addition of the Carb. Potash, the Oxygen of the Potash, combining with the Iron, forms the Protoxide of Iron, and this with the Carb'. forms an insoluble Carb. Iron; the Bromine and Potassium, set free, combine and form the required Bromide of Potassium.

Tests. Crystals should be colourless, totally soluble in water, and not affect Litmus or Turmeric, as it is neither acid nor alkaline. Sul' and Starch added together render it yellow (as characteristic of Bromine). Subjected to heat, they lose no weight (because no water is expelled). Chlor. Bar. throws down nothing from the solution, showing the absence of Sulphates. Gr. 10 of this salt are capable of acting upon gr. 14·28 of Nitr. Silver, and precipitating a yellowish Bromide Silver, which is dissolved by Ammonia, and but very little by Nit'. If a larger quantity of Nitr. Silver is precipitated, a Chloride is present, probably that of Potassium.

Inc. Acids, Acidulous Salts, Metallic Salts.

Action. Uses. Stimulant, Alterative, Deobstruent.

D. gr. iij.—gr. x. three times a day.

Carbonate of Potash.

Subcarbonate of Potash. Salt of Tartar. Salt of Wormwood. Kali præparatum. *F.* Carbonate de Potasse. *G.* Kohlensaures Kali.

As this salt is obtained by the burning of vegetables, it must have been known at very early times. Dioscorides describes it by the name *τεφρα κληματινης*, or *cinis sarmentorum*, ashes of vine-twigs. ("cineris lixivium." Pliny, xxxviii. c. 51.) The Arabs are usually supposed to have been the first to make known this alkali (al-*kali*); but the Hindoos, in works from which the Arabs copied, made use of the ashes of plants. Potash is found in most of the alkaline-earthy minerals, as Mica, Felspar, Leucite, Nacrite. Carbonate of Potash has been found in a few mineral springs. It is probably found in the juices of some plants. But usually Potash is combined with other acids, then forming Acetates, Malates, Oxalates, Tartrates, &c.

By incineration, the Oxygen of the vegetable acid, combining with the Carbon of the vegetable substance, forms Carb', which, combining with the disengaged Potash, forms a Carbonate of Potash. This, in its most impure state, is the Potash of Commerce, or Rough Potash of Commerce. To obtain this, land plants are burnt in countries where forests are most abundant, as N. America, Russia, Sweden, Poland. The wood is piled in heaps and burnt on the surface of the ground, in a place sheltered from the wind. "The ashes which are left consist of a soluble and insoluble portion. The *soluble* part is made up of the Carb. together with the Sulphate, Phosphate, and Silicate of Potash, and the Chlorides of Potassium and of Sodium; and the insoluble portion, of Carbonate and Subphosphate of Lime, Alumina, Silica, the Oxides of Iron and Manganese, and a little carbonaceous matter that had escaped incineration." (Wood and Bache.)

POTASSÆ CARBONAS IMPURA, L. LIXIVUS CINIS, D. Impure Carbonate of Potash. *Potashes. Pearlashes.*

This is prepared on the great scale by subjecting to the action of flame crude or Black Potash, the *black salt* of American manufacturers. Instead of fusing it to make the *Potashes* of Commerce, the alkaline mass is transferred to a large oven-shaped or reverberatory furnace, where the flame is made to play over it; and being well stirred about, the black impurities are burnt out, and the mass becomes a caustic salt of a white colour (W. and B.) with a tinge of blue, and constitutes the *Pearlash* of Commerce.

American Potash and American Pearlash (2), as ascertained by Vauquelin, contained in 1152 parts,

Caust. Hydr. of Potash	857	Sulph. of Potash.	154	Chlor. of Potassm.	20	Insol.	2	Carb' & Aq.	119	parts.
(2)	754	„	80	„	4	„	6	„	308	„

Russian Potash yields 772 parts of Caustic Hydrate; it used to be very impure, but is now more carefully prepared. For commercial purposes it is extremely necessary to have modes of ascertaining the quantity of alkali contained in any specimen of Commercial Potash. This is done by the process of Alkalimetry.

Uses. Chiefly pharmaceutical.

POTASSÆ CARBONAS, L. E. POTASSÆ CARBONAS E LIXIVO CINERE, D. Purified Potash. Pearlash.

This is Carbonate of Potash, which is not quite pure, but which is prepared by subjecting the Potash or Pearlash of Commerce to lixiviation and granulation.

Prep. L. P. Dissolve *Impure Carb. Potash* ℔ij. in *Aq. dest.* Ojß. and strain. Pour into a proper vessel, and evaporate the water. When the liquor has become thick, stir constantly with a spatula until the salt concretes. D.P. Mix and macerate *Pearlash,* in coarse powder ℔j., in *Aq.* ℔j.; filter the lixivium, and evaporate to dryness, continually stirring during the latter part of the process. The coarse powder obtained is to be preserved in close vessels. If the Potashes be not sufficiently pure before they are dissolved, let them be roasted in a crucible until they become white.

By this process the insoluble impurities, chiefly of an earthy nature, are removed, and the salt is obtained in a small granular state, white, caustic, and deliquescent. It usually contains water, some Sulphate of Potash, Chlorides of Potassium and of Calcium, and Silica.

Tests. Almost entirely dissolved by water; deliquescent; renders Turmeric brown. 100 parts lose 16 (20, E.) of water by a strong heat, and 26·3 parts of Carb′ on the addition of Sulph′. When supersaturated with Nitric′, neither Carbonate of Soda nor Chloride of Barium throw down anything (Nitrate of Baryta only a haze, E.), and Nitr. Silver but little.

POTASSÆ CARBONAS, L. POTASSÆ CARBONAS PURUM, E. POTASSÆ CARBONAS E TARTARI CRYSTALLIS, D.

Carbonate of Potash ($KO, CO_2 + 1\frac{1}{2}HO = 83{\cdot}5 + 1\frac{1}{2}HO = 88$, if crystallized) is in white roundish grains; sometimes it may be crystallized from a strong solution, by slow cooling, in opaque rhombic octohedrons. The taste acrid, alkaline, and nauseous; odour none; so deliquescent as to form a liquid, which used to be called *Oleum Tartari per deliquium;* soluble in its own weight of water, insoluble in Alcohol; alkaline in its reaction on Turmeric and the infusion of Cabbage, &c. Composed of K O 57·6, C′ 26·4, Aq. 16 = 100.

Prep. Under the article Potassæ Carbonas the L. P. directs that a more pure Carbonate of Potash may be prepared by subjecting the crystals of Bicarbonate of Potash to a red heat. This salt by losing 1 eq. of Carb′ is necessarily converted into the Carbonate. The E. P. gives the same formula; but adds another and a cheaper formula, which is nearly the same as that of the D. P. This consists in igniting purified Bitartrate of Potash (*Cream of Tartar*), when a dark-coloured powder, consisting of Carbonate of Potash and Charcoal, commonly called *black flux,* is obtained. This is roasted in a crucible without a cover, when the charcoal is burnt away, and the residue lixiviated, and the solution evaporated to dryness. The salt is granulated by brisk agitation towards the close of the operation, and then heated nearly to redness. It is a tolerably pure Carb. Potash, which must be preserved in well-stoppered bottles.

Tests. It ought to lose no weight (that is water) at a low red heat; and a solution supersaturated with pure Nit′ is precipitated either faintly, or not at all, by solution of Nitr. Baryta or Nitr. Silver, proving the absence both of Sulph. Potash and of Chloride Potassium, occasionally present from purified Pearlash being substituted for the pure Carbonate. Silica may be detected by a cloudiness or fleecy precipitate forming on N′ or H Cl′, being added to neutralization, evaporating and igniting the residue: any Silica will be insoluble in water. The C′ is readily recognized by effervescing with any of the acids, and by forming a milky solution with Lime-water. The Carbonate which is formed will effervesce and dissolve in Acetic acid. A white precipitate (Carbonate of Magnesia) is also formed when this salt is added to a solution of Sulphate of Magnesia; as this does not take place when Bicarbonate of Potash is added, this Sulphate is a useful test for distinguishing the one from the other. With Bichloride of Mercury a brick-red precipitate of Binoxide of Mercury is formed.

Inc. Acids and Acidulous Salts, Hydrochlorate and Acetate of Ammonia, Lime-water, Chloride of Calcium, Sulphate of Magnesia, Alum, and several other alkaline, earthy, and metallic salts.

Action. Uses. Corrosive, Antacid, and Poisonous like Liq. Potassæ, Diuretic, Resolvent, milder than Liquor Potassæ, Antilithic. Often employed for making effervescing Draughts. Carb. Potassæ gr. xx. = Cit′ or Tar′ gr. xviij. or f℥iv. of Lemon-juice.

D. gr. x.—ʒß.

Antidotes. Vinegar, Oil, Lemon-juice.

POTASSÆ CARBONATIS LIQUOR, L. POT. CARB. AQUA, D. Solution of Carb. of Potash. *Aqua Kali. Oleum Tartari per deliquium. Liquor Potassæ Subcarbonatis.*

Prep. Dissolve *Carb. of Potash* ℥xx. (from crystals of Tartar 1 part, D.), in *Aq. dest.* Oj. (2 parts, D.) Filter. Sp. Gr. = 1·473, L.; 1320, D.

D. ♏x.—f℥j.

POTASSÆ BICARBONAS, L. E. D.

Bicarbonate of Potash. *Potassæ Carbonas. Perfectly Saturated Carbonate of Potash. Aerated Kali. G.* Doppelt Kohlensaures Kali. *F.* Bicarbonate de Potasse.

This salt ($KO, 2CO_2+Aq=101$) was first prepared by Cartheuser in 1752, and examined by Bergmann, who devised various modes of preparing it. The older chemists obtained it by simply exposing Carbonate of Potash for some months to the air, or to an atmosphere charged with Carb′ until sufficient gas was absorbed. It may also be prepared as in the L.P. by passing a stream of Carb′ gas through a solution of Carbonate of Potash to saturation.

Prop. Bicarbonate of Potash is a colourless and transparent crystalline salt; its crystal is a modification of a right oblique angled prism. Its taste is much milder than that of the Carbonate, and it has so little alkalinity as to colour Turmeric paper only slightly. It is soluble in about 4 parts of water at 60°, and in five sixths of hot water; boiling water speedily decomposes it from the expulsion of Carb.', and it becomes a Sesquicarbonate. It is insoluble in Alcohol. Exposed to a red heat it loses 1 equivalent of Carb', likewise any water which may be deposited within its crystals, and is converted into the Carbonate of Potash. Hence this method (p. 76) is adopted to procure the pure Carbonate. The Carb' in this salt is readily detected by its abundant effervescence with acids, likewise by the insoluble precipitate formed by it in Lime or Baryta water. But a moderately diluted solution of Bicarbonate of Potash yields no precipitate with Sulphate of Magnesia or with Bichloride of Mercury; hence the former is often prescribed with it in effervescence. "The Bichloride of Mercury causes a slight white precipitate or opalescence with it." (*p.*) It is composed of per cent. K O 47·53+C' 43·56+Aq. 8·91 = 100.

Prep. L. D. Dissolve *Carb. Potash* ℔vj. (prepared from Pearlash 1 part, D.) in *Aq. dest.* cong. 1. (2 parts, D.) Pass *Carb'* (obtained by acting on white marble with diluted Mur', D.) through the solution (till it becomes turbid; filter, and again transmit the gas, D.) to saturation. Apply a gentle heat, to re-dissolve any crystals that may have formed, put the solution in a cool place to crystallize. Dry the crystals (without heat, and keep in a well-stoppered bottle, D.)

E. Mix *Carb' of Ammonia* ℥iijß. reduced to a fine powder, with *Carb. of Potash* ℥vj.; triturate them thoroughly together. Add gradually a very little water, till a smooth uniform pulp is formed. Dry this at a temperature not exceeding 140°, triturating occasionally. Continue the heat till a fine powder devoid of ammoniacal odour is obtained.

In the L. and D. processes the Carb. Pot. takes an additional Eq. of Carb', being converted into the Bicarbonate, but pressure is required for the proper absorption of the gas. In that of E. the Ammonia, and a small portion of the Carb', are expelled by the heat; the remainder unites with the Carb. Potash to form the Bicarbonate.

Tests. The usual impurities in this salt being Carbonate or Sulphate of Potash and Chloride of Potassium, the P. tests are intended to detect them. L. E. Totally dissolved by water (unless impure); the solution slightly changes the colour of Turmeric, but highly if the Carbonate be present. Sulphate of Magnesia throws down nothing from this solution, unless it be heated. A large portion of the Carbonate, Dr. C. says even 50 per cent., may be present, without Sulphate of Magnesia detecting it, when mixed with the Bicarbonate. The E. P. states that a solution in 40 parts of water does not give a brick-red precipitate with solution of corrosive sublimate; but it will do so "if the salt contains even so little as a hundredth part of Carbonate" (*c*), except when Chloride of Sodium is present. After the addition of excess of Nitric', Chloride of Barium or Nitrate of Baryta throws down nothing (unless Sulphates be present), and Nitrate of Silver very little if anything (if Chlorides be absent), by a red heat 100 parts lose 30·7 of Carb.' and of water. If the crystals be moist, the loss of water will be greater; and if Carb' be deficient, the loss will be less.

Inc. Nearly the same as with Carbonate of Potash. Acids, acidulous salts, Acetate and Hydrochlorate of Ammonia, Lime-water, Chloride of Calcium, alkaline, earthy and metallic salts.

Action. Uses. Antacid, antilithic, diuretic, resolvent.

D. ℈ß—ʒß or ʒj. For effervescing draughts, 20 grs. Bicarb. Potash = 15 grs. of Cryst. Cit′ or Tar′, or fʒiij ß of Lemon-juice.

LIQUOR POTASSÆ EFFERVESCENS. L. POTASSÆ AQUA EFFERVESCENS. E. Effervescing solution of potash.

Prep. Dissolve *Bicarb. Potash* ʒj. in *Aq. dest.* Oj. Pass through the solution *Carb′* gas under pressure, (more than sufficient for saturation. Preserve the solution in well-stoppered vessels, L.) This may be extemporaneously imitated by pouring a bottle of soda-water (*i. e. Carbonic* acid water) into a tumbler containing gr. xx. of Bicarb. of Potash.

This is a solution of Bicarb. Potash containing Carb′ gas in excess.

LEMON and KALI. A mixture of powdered *white sugar*, dried and powdered *citric acid*, and powdered *bicarbonate of potash*, employed for making extemporaneous effervescing draughts.

PULVERES EFFERVESCENTES. The E. P. orders of *Tartaric acid* ℥j, *Bicarb. Potash* ℥j. and gr. 160. Reduce both to fine powder, and divide into 16 parts. Preserve the acid and alkaline powders in separate papers of different colours.

POTASSII (POTASSÆ, D.) SULPHURETUM, L. E.

Sulphuret of Potassium. *Hepar Sulphuris. Kali Sulphuratum. F.* Sulfure de Potasse. Sulfure de Potassium Sulfaté. *G.* Schwefel Kalium.

The Sulphuret of Potassium was formerly known by the name of *Liver of Sulphur.* The solubility of Sulphur in an alkaline solution was known to Geber; but Albertus Magnus taught the method of procuring Sulphuret of Potassium by fusion. (*p.*)

Prop. When carefully prepared, it forms a hard brittle solid, of a liver-brown colour; without smell when dry, but emitting a smell of Hydrosulphuric′ when moistened; taste acrid and nauseous; its solution in water is of an orange colour, with a strong odour. When exposed to the air, it becomes moist and greenish-coloured, and then white and without odour, as a Hydrosulphate of Potash is formed from the action of water and the Oxygen, which combines both with the Sulphur and the Potassium; ultimately a Sulphate of Potash is formed from the escape of Hydrogen, while a portion of the Sulphur is deposited. It is readily decomposed by acids, as by H Cl′, as they evolve Hydrosulphuric acid, combine with the Potash, and precipitate the Sulphur; as also by most of the metallic salts, of which the metals are deposited in the form of Sulphurets. According to Mr. R. Phillips, it is composed of $3KS, KO, SO_3 = 256$; but according to others, of 2 Eq. of Pentasulphuret of Potassium with 1 of Hyposulphite of Potash, $2\,KS_{10} + KO, SO_2$.

Prep. L. E. D. Rub together *Sulphur* ℥j., *Carb. Potash* ℥iv.; heat them in a covered crucible till they melt.

When Carb. Pot. is melted with excess of Sulphur, Carb′ is expelled. The Oxygen of ¾ths of the Potash combines with 1 part of Sulphur to form Sul′, which, uniting with the undecomposed Potash, forms 1 Eq. of Sulphate of Potash. Sulphuret of Potassium is at the same time formed by the union of the Potassium with a portion of the Sulphur, more or less of it remaining in excess. This preparation, therefore, is a mixture of Per- or Pentasulphuret of Potassium with Sulphate, or rather Hyposulphite, of Potash.

Tests. Fresh broken, it exhibits a brownish-yellow colour. Dissolved in water, or in almost any acid, it exhales a smell of Hydrosulphuric acid. The aqueous solution is of a yellow colour. What is thrown down by Acetate of Lead is first red, but it afterwards becomes black. Dr. Pereira has ascertained that the alkaline monosulphurets give a black, and the polysulphurets a red precipitate with solutions of Lead. If the Sulphuret should have been long kept, and have become changed, these characteristics will not be seen.

Inc. Acids and metallic salts.

Action. Uses. Irritant, Stimulant, Diaphoretic. *Ext.* Detergent.

D. grs. iij.—x. or xv. with honey, or with soap made into pills. *Ext.* as an ointment with lard, or in a watery solution of soap, or in baths, 1 part to 1000 of water.

POTASSÆ SULPHURETI AQUA, D.

Prep. Take *Washed Sulphur* 1 part, *Solution of Potash* 11 parts. Boil for 10 minutes. Filter through paper. Keep the liquor in well-closed vessels. Its Sp. Gr. = 1117. By the mutual reaction of the ingredients, and the decomposition of the water, a solution of hyposulphite and of hydrosulphate of potash is formed, of a deep orange colour.

Action. Uses. As above. Internally and externally in cutaneous eruptions.

D. ♏x.—f℥j. two or three times a day, diluted with water.

POTASSÆ SULPHAS, L. E. D.

Sulphate of Potash. *Kali Vitriolatum. Sal Polychrestum. F.* Sulfate de Potasse. *G.* Schwefelsäures Kali.

Sulphate of Potash ($KO, SO_3 = 88$) is found near volcanoes, in a few minerals (Alum and Polyhalite), some mineral waters, in many plants, and in some animal secretions.

Fig. 10.

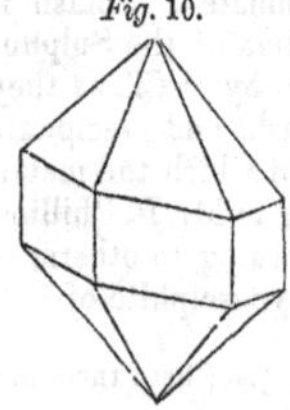

Prop. Sulphate of Potash is colourless and without odour, of a bitter saline taste; usually seen in small hard crystals formed of six-sided prisms terminated at both ends by six-sided pyramids; the prism is sometimes absent, or the angles are modified, or the crystal is double; the primary form is a right rhombic prism, or rhombic octohedron. The crystals are unalterable in the air, insoluble in Alcohol, but soluble in 16 parts of water at 60° F., and in 4 parts at 212°. They contain no water of crystallization, but a little mechanically lodged in the interstices; hence they decrepitate when heated, and melt at a red heat. If heated with Charcoal, this Salt is converted into Sulphuret of Potassium.

Prep. L.E.D. Take of the *Salt remaining after the preparation of Nitric′* (pure, E.) ℔ij. expel the excess of acid by heating the salt in a crucible, L.; boil what remains in boiling *Aq.* Cong. ij. till a pellicle forms. [(Dissolve in boiling *Aq.* C. ij. E.; in q. s. D.) (Add *White Marble* powdered q. s. till effervescence ceases, E.; add *Carb. of Potash from Pearlash* q. s. D.)] Filter and evaporate (till a pellicle forms, E.; with a very gentle heat, D.); set aside to crystallize; pour off the liquor, and dry the crystals, L.

The residual salt in the manufacture of Nitric′ is Sulphate of Potash with an excess of Sulph . This excess the L. C. directs to be driven off by heat; but the E. P. neutralizes it with Marble, and the D. P. with Carbonate of Potash.

Tests. This salt is not liable to adulteration, but the L. P. gives as its characteristics, the sparing solubility in water and insolubility in Alcohol, and that Chloride of Platinum occasions in its solution a yellow precipitate (Chloride of Platinum and Potassium), and Chloride of Barium a white one (Sulphate of Baryta), insoluble in Nitric acid. No change ought to be produced in the colour of Litmus or of Turmeric paper; no precipitate with solution of Sulphate of Silver, nor any upon the addition of Ammonia or its Sesquicarbonate.

Inc. Tartaric′, Chlorides of Barium and of Calcium, Acetate and Diacetate of Lead. Nitrate of Silver.

Action. Uses. Mild Cathartic and Deobstruent.

D. gr. x.—℥ß.

Pharm. Prep. Pulvis Ipecacuanhæ comp.

POTASSÆ BISULPHAS, L. E. D.

Bisulphate of Potash. *Potassæ Supersulphas. Sal enixum. F.* Bisulfate de Potasse. *G.* Doppelt Schwefelsaures Kali.

Bisulphate of Potash ($KO, 2\,SO_3 + 2\,HO = 146$) is obtained as the residual salt in the manufacture of Nitric′, and must have been long known; but the mode of preparing it was shown by Link towards the end of the last century.

Fig. 11.

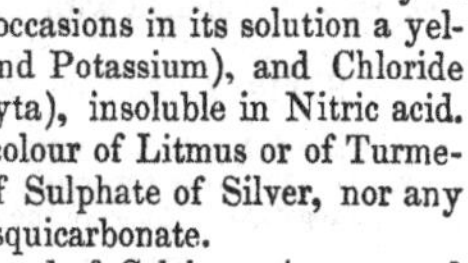

Prop. It is colourless and without odour, but has a very acid bitter taste. It crystallizes in small flat prisms belonging to the right rhombic system, when there is an excess of Sulphuric acid. Crystals very soluble in water, but insoluble in Alcohol; unalterable in dry air; moderately heated, they melt into an oily-looking fluid, and at a red heat lose their water of crystallization and one proportion of acid, and become simple Sulphate of Potash. The solution reddens vegetable blues, and "a solution in eight waters effervesces briskly with alkaline Carbonates." E.

Prep. L. E. D. Dissolve of the *Salt remaining after the distillation of Nitric'* (pure, E.) ℔ij. in boiling *Aq.* Ovj. Add *Sul'* ℔j. (commercial, f℥vij. and f℥j., E.); concentrate the solution, cool, and crystallize. The L. C. directs (so also the E. P.) an excess of Sul' to prevent the deposition of Sulphate and Sesquisulphate of Potash, in consequence of the water uniting with a portion of the Sulphuric Acid (*p*). The D. C. prepares it by saturating 1 part Sulph', diluted with 6 parts of water, with Carb. Potash obtained from Pearlash q. s.; then adding as much acid as was used in the first instance, and evaporating so that the solution may crystallize on cooling.

Tests. The Sulph' and Potash may be detected by their respective tests, and this salt may be distinguished from the Sulphate of Potash by the above acid characters.

Inc. Alkalis, Earths, and their Carbs.; many Metals, and Oxides.

Action. Uses. Purgative. Effervescing Purgative with an equal weight of cryst. Carbonate of Soda.

D. ℈j.—ʒij. diluted with water, &c.

PULVIS SALINUS COMPOSITUS, E. D. Compound Saline Powder.

Prep. Take *Sulphate Potash* ℥iij., *Sulph. Magnesia* and pure *Muriate of Soda* āā ℥iv. Dry the salts separately with a gentle heat; pulverise and triturate them well together. Preserve the compound in well-stopped vessels.

Action. Uses. This is a useful combination of several salts, in which some degree of stimulant is combined with the cathartic properties. It may be beneficially taken in costive habits.

D. ʒij.—ʒiij. dissolved in water.

POTASSÆ SULPHAS CUM SULPHURE, E. *Sal Polychrestum Glaseri. Glaser's Sal Polychrest.*

Prep. Mix *Nitrate Potash* and *Sulphur* equal parts; throw the mixture in small portions into a red-hot crucible; when the deflagration is over, and the salt cools, reduce it to powder, and preserve it in well-stopped bottles.

Here the Sulphur burns with its characteristic blue flame, taking Oxygen from the Nitric acid of the Nitrate, and a Sulphate of Potash is formed, mixed with some Sulphite, it is supposed. Nature undetermined (Christison). This salt is much more soluble than Sulphate of Potash, crystallizes in rhombic prisms, "has a sulphureous odour, as well as its solution, but Sulphuretted Hydrogen is not disengaged on the addition of a strong acid."

Action. Uses. This salt acts as a mild purgative, and may be given with an equal weight or more of Bitartrate of Potash. It was formerly much used in Dyspepsia and chronic cutaneous diseases. Dr. Duncan says that in use it agrees with the Sulphureous waters.

D. ʒß.—℥j.

Potassæ Nitras, L. E. D. Nitrum, D.

Nitrate of Potash. Nitre. *Nitrum. Sal petræ. Saltpetre. F.* Nitrate de Potasse. Nitre. *G.* Salpeter. Salpetersaures Kali.

Nitrate of Potash ($KO, NO_5 = 102$), Nitre, or Saltpetre, being a production of nature, must early have been known, especially as both the Indians and Chinese have long been acquainted with the making of fireworks, and the former have an early process for making Nitric', in which they have been followed by Geber and other Arabian authors. The names *neter* in the Old Testament, and *nitrum* in ancient authors, were applied to Carbonate of Soda, but they were also used in a generic sense.

Nitre is found effloresced on the soil in many parts of India, where there is no animal matter, and being washed out, a fresh crop is formed after a few years. The soil is sandy, with mica interspersed, which will continue to yield a supply of Potash, while the Nitric' must be furnished by the combination of the Oxygen of the atmosphere with its Nitrogen, probably, as suggested by Liebig, by the oxidation of the Ammonia which he has proved is always present in the atmosphere. Mr. Stevenson (Prinsep's Journ. ii. p. 23) has detailed the process, and shown that the saline earth contains of salts soluble in water, Sulphate of Soda, Muriate of Soda, Nitrates of Lime and of Potash. The Nitrate of Lime is easily converted into that of Potash by lixiviating the saline soil over a filter of wood-ashes, which contains Carbonate of Potash (the C' combines with the Lime, and the N' with the Potash), a Carbonate of Lime is precipitated, and the Nitrate Potash in solution is evaporated and put aside to crystallize. The salt obtained contains from 45 to 70 per cent. of pure Nitrate of Potash. It is redissolved and crystallized, but still contains impurities, which are termed so much per cent. of *refraction*. The ordinary kinds are called *rough* or *crude* Saltpetre, and the purer *East India refined*.

In Europe Nitre is prepared artificially in Nitre-beds or Nitre-walls, and in ditches covered by sheds, where urine is added to different mixtures of earth with refuse vegetables, various animal substances, and calcareous matter, &c. The whole is exposed to the action of the air. The Nitrogen, combining with the Oxygen, forms Nitrates, and the foregoing processes being adopted, similar results are obtained.

Prop. Nitrate of Potash in its purified state is colourless and semi-transparent, without odour, of a sharp and cooling, disagreeable, saline taste; crystallized usually in long, striated, six-sided prisms, terminated at each extremity with dihedral summits, or in two or six converging planes, sometimes in a dodecahedron formed of two six-sided pyramids joined base to base. The crystals are anhydrous and unalterable in the air. Sp. Gr. 1·92. Soluble in 4 parts of water at 60°, producing cold, and in an equal weight of boiling water. Insoluble

Fig. 12.

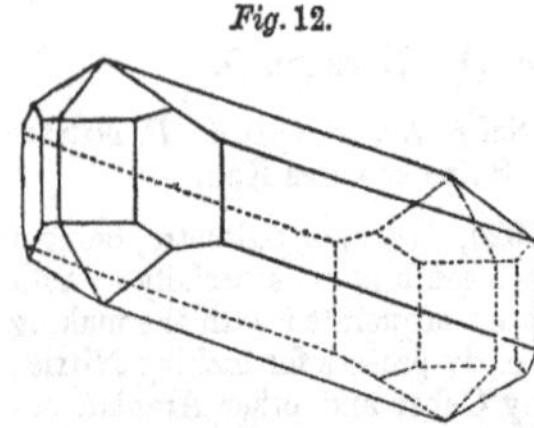

in Alcohol, and sparingly so in dilute Alcohol. Heated to about 660°, it melts into a transparent fluid ; on cooling, it forms a white semi-transparent mass, which used to be called *Sal Prunellæ* when run into small balls. By a high degree of heat, Nitre is decomposed, Oxygen gas being first given off, and afterwards mixed with Nitrogen, while Hyponitrite of Potash is left behind. Water is apt to be lodged between the plates of the crystals, particularly when these are large ; hence in gunpowder manufactories, small crystals, if equally pure, are preferred. With inflammable substances the decomposition of Nitre, when heated, is rapid, light and heat being disengaged, constituting what is called the deflagration of Nitre. This takes place also with some of the compound acids, into which Carbon enters as a constituent.

Prep. Nitre being required of the best quality for the manufacture of gunpowder, that of commerce is usually sufficiently pure for medical purposes. But it sometimes requires purification to the extent of a single solution, and re-crystallization, as in the D. formula.

POTASSÆ NITRAS PURIFICATUM, D.

Prep. Take *Nitr. Potash* one part, *boiling water* two parts, dissolve, remove the scum, filter, set aside for crystals to form.

Tests. The presence of Nitric′ and of Potash may be detected by their respective tests. Nitre should be entirely soluble in distilled water. The solution should not be affected by Chloride of Barium (no Sulphates), nor by Nitrate of Silver (no Chlorides). Calcareous salts which occur only in rough Nitre, may be detected by the Oxalate of Ammonia, throwing down a white precipitate of Oxalate of Lime. Sulphate of Potash is now seldom found in the best Nitre, and only a small proportion of Chloride of Potassium or of Sodium.

Inc. Sulph′, Alum, the Alkaline and Metallic Sulphates.

Action. Uses. Refrigerant, Diuretic. In large doses an irritant poison. *Ext.* Refrigerant and Detergent.

D. gr. v.—gr. xv. with Sugar, or in water, or in mucilaginous drinks.

Antidotes. Remove poison from Stomach, allay irritation, and subdue inflammation.

Off. Prep. Ung. Sulphuris comp. L.

POTASSÆ CHLORAS, L.

Chlorate of Potash. *Oxymuriate* or *Hyperoxymuriate of Potash. F.* Chlorate de Potasse. *Muriate Oxygéné* and *Hyperoxygéné de Potasse. G.* Chlorsaures Natron.

Chlorate of Potash ($KO, ClO_5 = 124$), though previously made, was first clearly distinguished from other salts by Berthollet. It is now largely manufactured for the preparation of detonating compounds and lucifer matches.

Prop. Chlorate of Potash is colourless, in small brilliant scales or quadrangular crystals, glittering and pearly in lustre, not unlike those of Boracic'. Taste cool, penetrating, and austere, something resembling that of Nitre. Like it, the crystals are anhydrous and unalterable in the air. Sp. Gr. 1·98. Soluble in about 30 parts of water at 32° F., in 18 parts at 60°, and at 212° F. in less than 2 parts of water; little soluble in Alcohol. The crystals crackle and become luminous in the dark when rubbed briskly. Heated, they lose about 2 per cent. of water mechanically lodged, melt at a dull red heat, and give out nearly 40 per cent. of Oxygen gas, the Acid and Alkali both being decomposed, and only Chloride of Potassium left. It deflagrates when thrown on live coal, in the same way as Nitre; but detonates violently when rubbed with combustible bodies, as Sulphur, Charcoal, Phosphorus, &c.

Prep. Chlorate of Potash is prepared by passing, to saturation, a current of *Chlorine* through a solution of 15 parts of *Carbonate of Potash* in 38 of cold *water.* The solution is then exposed to the air for a few days, agitating occasionally to allow of the escape of any free Chlorine. Scales and crystals of Chlorate of Potash are deposited. If these are separated, more may be obtained by evaporating the mother liquor. All may be purified, if washed with cold, dissolved in twice their weight of hot water, and re-crystallized.

Here effervescence ensues, chiefly from the escape of Carb' gas. The Potash becomes decomposed, its Oxygen combining with the Chlorine to form Chloric acid. This unites with some of the remaining Potash, and a Chlorate of Potash is formed: 5 equivalents of Potash are decomposed to yield the 5 Eqs. of Oxygen required to form Chloric', which combines with 1 Eq. of undecomposed Potash. The Potassium set free, combines with Chlorine to form Chloride of Potassium; and thus when Chlorate of Potash is deposited in crystals, Chloride of Potassium remains in solution with Hypochlorite of Potash, a little free Hypochloric acid, and some Chlorate of Potash.

Tests. Entirely soluble. Chloride of Potassium is the most probable impurity. This is readily detected by Nitrate of Silver, which will give a white precipitate (Chloride of Silver) if any be present; otherwise the solution will be unaffected, as stated in the L. P.

Action. Uses. Refrigerant, Diuretic, Supplier of Oxygen, Useful in Saline treatment.

D. gr. x.—gr. xv.

POTASSÆ TARTRAS, L. E. D.

Tartrate of Potash. *Tartrite of Potash. Kali Tartarizatum. Tartarum solubile. F.* Tartrate de Potasse. *G.* Einfach Weinsaures Kali.

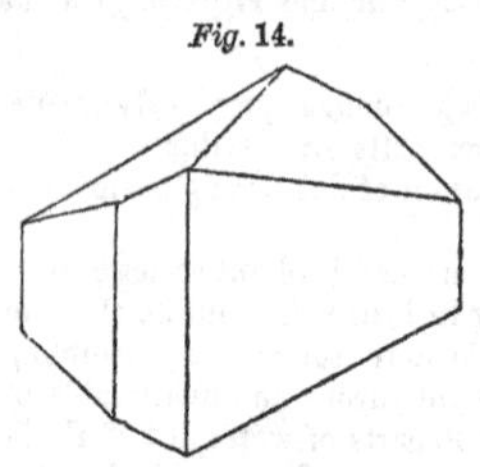
Fig. 14.

Tartrate of Potash (K O, Tar′ = 114) has not been found in nature, but has been known to chemists since the time of Lemery in the seventeenth century.

Prop. This salt is colourless, and without odour, but of a bitterish saline taste. It is usually sold in the form of a small granular powder; the evaporation during its manufacture having been carried nearly to dryness with frequent stirring. But it can be crystallized in irregular four or six-sided prisms with dihedral summits, the primary form being a right rhomboidal prism. Sp. Gr. 1·55. Deliquescent in the air, soluble in its own weight of water, and in about 240 parts of boiling Alcohol. Heated, it swells up, chars, and is converted into Carb. Potash. Its solution is readily decomposed by Sul′, or any other strong acid, as well as by several acidulous salts; crystals of the Bitart. being deposited. As in the case of the Bitart., soluble Barytic and Lime Salts, Lime-water, and Chlor. Calcium, Nitr. Silver, and the Acetates of Lead cause white precipitates of Tartrates, are soluble in N′.

Prep. Dissolve *Carb. Pot.* ℥xvj. or q. s. (from Potashes 5 parts, D.) in *boiling Aq.* Ovj. (45 parts, D.) add *Bitartrate of Potash* powdered ℔iij. (gradually 14 parts, D. till neutralized, E.) Boil, filter and boil, till a pellicle floats. Set aside to cool and crystallize. Dry the crystals, L. (The remaining liquor will yield more crystals by further concentration and cooling, L. and E.)

In this process effervescence ensues in consequence of the escape of Carb′, while the Potash of the Carbonate combines with the 2nd Eq. of the Tartaric′ in the Bitartrate, and 2 Eqs. of Tartrate of Potash are formed. This Salt is also formed in making Tartaric′ (q. v.), and may be obtained by evaporation.

Tests. If pure, this Salt is of easy solubility, neutral to Litmus and to Turmeric paper. Most acids, even the Citric, cause a deposit of crystals of the Bitartrate. The precipitate occasioned by Chlor. Barium or Acet. Lead is soluble in diluted Nitric′. L. 44 grs. in solution are not entirely precipitated by 55 grs. Nitr. Lead. E. P. Showing that there are no Sulphates, and that only the due proportion of Tartaric′ is present.

Inc. Acids (even Citric′). Acidulous Salts. Soluble Salts of Lime, and of Baryta. Acetate of Lead. Nitrate of Silver.

Action. Uses. Cathartic.

D. ʒij.—℥j.

CITRATE OF POTASH is not officinal, but it is frequently taken when effervescing draughts are prepared with either the Carbonate or

Bicarbonate of Potash and Citric Acid or Lemon-juice. Tartrate of Potash will be produced when Tartaric′ is employed.

POTASSÆ BITARTRAS, L. E. D. • TARTARUM, D.

Bitartrate of Potash. *Supertartrate and Supertartrite of Potash. Cream of Tartar. Argol. F.* Tartrate acide de Potasse. *G.* Doppelt Weinsaures Kali.

Bitartrate of Potash (K O, 2 Tar′, H O = 189) is well known by the name of Tartar, and must have been known ever since wine has been made from the grape, in the juice of which it exists. During the fermentation of wine, Sugar disappears and Alcohol is formed, and the salt not being soluble in this, is deposited on the bottom and sides of casks, as a crystalline crust, which, according to the colour of the wine, forms either *red* or *white Tartar* or *Argol.* This was known to the ancients, and is the Fæx Vini of Diosc. v. c. 13. (Hindoo Med. p. 97.) Its nature was determined by Scheele in 1769. It is largely purified both at Montpelier and at Venice.

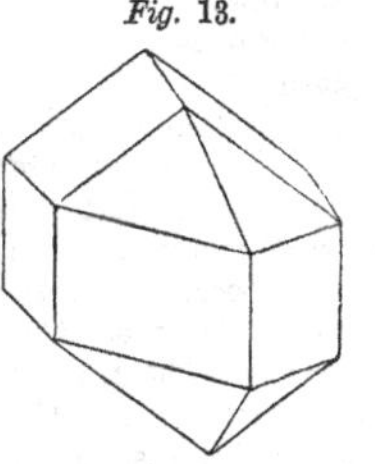

Fig. 13.

Prop. The Bitartrate of Potash of commerce is in white crystalline crusts, formed of clusters of small crystals aggregated together, which are hard and gritty under the teeth, dissolve but slowly in the mouth, and have an acid and rather pleasant taste. The crystals are semi-transparent irregular six-sided right or oblique rhombic prisms, or triangular prisms with dihedral summits. Sp. Gr. 1·95. Unaltered in the air, insoluble in Alcohol, soluble in 60 parts of water at 60°, and 18 parts at 212°. The solution is acid, reddens litmus, effervesces with alkaline Carbonates, is liable to become mouldy and decomposed. The application of heat causes the crystals first to swell up, lose 1 Eq. of water, become decomposed; gases are evolved, and a mass composed of Carbonate of Potash and of Charcoal is left, which is called *black flux.* If Tartar is deflagrated with its weight of Nitre, *white flux,* or Carbonate of Potash, is left. It is not easily decomposed by acids; its acid, *v.* Tar′, will decompose the neutral Potash Salts; Cream of Tartar being produced. Solutions of Baryta and Lime, as of Acetate of Lead, form insoluble white Tartrates of the respective substances. With salifiable bases forming soluble Tartrates, it gives rise to Double Salts, several of which are officinal, as Potassio-tartrate of Soda, Potassio-tartrate of Antimony. *v.* Soda, Iron, Antimony. Bor′ and Borax much increase the solubility of Cream of Tartar, forming BORO-TARTRATE OF POTASH, or *Soluble Cream of Tartar.* Liebig directs 47½ parts of the Bitart. to be treated with 15½ parts of crystallized Bor′ until the whole be dissolved, then evaporate to dryness.

Imp. Bitartrate of Potash contains from 2 to 6, sometimes as much as 14 per cent. of Tartrate of Lime, occasionally powdered white Quartz, or Sand. It ought to be "entirely soluble in 40 parts of boiling water; 40 grs. in solution are neutralised with 30 grs. of cryst. Carb. Soda; and when then precipitated by 70 grs. of Nitr. Lead, the liquid remains precipitable by more of the test." E. P. Alum or Bisulph. Potash in powder, may be detected by Chlor. Barium.

Inc. Strong Acids, Alkaline Carbonates, Salts of Lime and Lead.

Action. Uses. Refrigerant, Diuretic, Laxative.

D. ℨß.—ℨij. as a diuretic. ℨiv.—ℨvj. as a laxative.

Pharm. Prep. Pulv. Jalapæ Comp. Pulv. Scammonii Comp.

Potassæ Acetas, L. E. D.

Acetate of Potash. *Kali Acetatum. Sal Diureticus. Terra foliata Tartari. Digestive Salt of Sylvius. F.* Acetate de Potasse. *G.* Essigsaures Kali.

Acetate of Potash (K O, Ac′ = 99) was known to Raymond Lully in the thirteenth century, and probably earlier. It is found in the sap of many plants, as in some of those which by incineration yield Carb. Potash.

Prop. It is colourless, and with little odour, but has a sharp saline taste. It is usually seen as a shining foliated mass, made up of small pellucid scales, but by slow evaporation it may be made to crystallize in thin compressed laminæ, or in needles. It is so deliquescent as soon to become converted into an oily-looking liquid when exposed to the air. It is soluble in half its weight of water; also in Alcohol. Subjected to heat, it fuses, and then becomes decomposed. Hydrogen and Carb′ escape. Carb. Potash, as in the case of the Tartrates, remains as a residue. Acetate of Potash is decomposed by Sul′ and other strong acids, giving off an odour of Acetic Acid, and also by several Salts. Comp. Pot. 48·5 + Ac′ 51·5 = 100.

Prep. Take *Carb. Pot.* ℔j. [(dry, ℥vij. or q. s. E.) (from Tartar q. s. D.)] and add it (gradually, E. D.) to *Acetic′* f℥xxvj. and *Aq. dest.* f℥xij. [(Pyroligneous′ Oß. E.) (about 5 times its weight of Distilled Vinegar of a medium heat, D.)] till saturation takes place. Evaporate in a sand-bath, applying the heat cautiously, till the salt be dried, [(in a vapour-bath till it forms a concrete mass when cold, let it cool and crystallize in a solid cake, E.) (Evaporate to dryness, and cautiously raising the heat liquefy the salt. Dissolve in water, filter and evaporate till, on cooling, it becomes a white crystalline mass, D.) (Keep it in well-closed vessels, E. D.)]

The Acetic′ unites with the Potash, expelling the Carb′ Gas.

When distilled Vinegar is used, the solution always becomes brown. When concentrated, the D. and E. C. fuse the Salt and dissolve again, to get rid of this colour; but care is taken that the heat be not sufficient to decompose the Salt. This Salt may also be obtained by double decomposition between Acetate of Lead and Sulphate of Potash; but it then sometimes contains Lead.

Tests. Sulphuric Acid disengages Acetic vapours; a red heat converts it into Carb. Potash. Entirely soluble in water and Alcohol: should not affect either Litmus or Turmeric, though it is apt to have an alkaline reaction. Its solution is not affected by Chlor. Barium or Nitr. Silver (showing absence of Sulphates and of Chlorides). But if the solution be strong, a precipitate may be formed by the Silver test. This, however, is soluble in water or in Nitr′. If metals are present, as Lead, Hydrosul′ will give a blackish, and if Copper, Ferrocyanide of Potassium, a brownish precipitate.

Inc. Acids. Sulphates of Soda and Magnesia, and several earthy and metallic Salts.

Action. Uses. Diuretic; in large doses acts as a purgative. If continued, will render the urine alkaline as Carb. Potash does.

D. ℈ß.—ʒj. as a diuretic. ℥j.—ʒiij. as a cathartic.

Sodium.

Natrium. *F.* Sodium. *G.* Natrium.

Sodium or *Natrium* (Na=24) is the metallic base of Soda, discovered by Sir H. Davy in 1807. It is of the consistence of wax, malleable, and spreads into thin leaves. Sp. Gr. 0·972. Opaque, but with a brilliant silvery lustre; floats on water, producing a hissing effervescence, in consequence of the escape of Hydrogen, while the Oxygen of the water combining with the Sodium forms a Protoxide, or Soda, which remains in solution. Sodium fuses at 190°, and volatilizes at a white heat. It conducts both Heat and Electricity. It exists in Sea-water and Rock Salt, but in combination with Chlorine. As it oxidizes in the air, it must be preserved under Naphtha.

Soda. Protoxide of Sodium.

Mineral or Fossil Alkali *F.* Soude. *G.* Natron. *Natron* or *Nitron* of the Ancients: (but these names apply rather to the Carbonates, q.v.)

Soda (Na O=32+Aq. 9=41, the hydrate) is obtained by adding Caustic Lime to a solution of Carbonate of Soda, a precipitate takes place of Carbonate of Lime. The Soda being left in solution, may by evaporation be obtained as Hydrate of Soda. Soda is found to be a constituent of several minerals, as Sodalite, &c., and very abundant in combination with different Acids. It has been obtained in four-sided crystals, acuminated by four planes, of a greyish-white colour; acrid and caustic, soluble in both water and Alcohol; melts at a red heat, and effloresces in the open air. Its properties are very similar to those of Potash. The affinity of Soda is less for water in the air; it first becomes moist, absorbs Carbonic′, and effloresces. Glass and Soaps made with Soda are harder than those made with Potash. The Salts differ in their forms from those of Potash, and less Soda will saturate a given quantity of acid, and Soda Salts are for the most part more soluble than the corresponding ones of Potash. They do not yield crystals of Alum, when added to a solution of Sulphate of

Alumina ; nor an insoluble Bitartrate, on addition of Tartaric′ ; and Chloride of Platinum throws down no precipitate. They may be distinguished from the Alkaline Earths by no precipitate taking place on addition of Carbonate of Potash or of Soda ; while the alkaline Ferrocyanides and Hydrosulphurets, which distinguish Metallic Salts, cause no precipitates in solutions of salts of Soda.

Tests. The only positive test is, that Soda and its Salts give a rich yellow tinge to the flame of Alcohol.

Action. Uses. Similar to Potash, and not being superior, Caustic Soda is not officinal.

SODÆ CARBONAS, L. E. D.

Carbonate of Soda. *Sodæ Subcarbonas. Aerated Mineral Alkali. Fossil Alkali.* (One of the kinds of *Nitrum* of the ancients.)

Carbonate of Soda is the *neter* of the Hebrews. It was known to the early Hindoos, and is by them called *Sajji noon* (i. e. *Sajji* or Soda Salt) ; it is the *Sagimen vitri* of Geber. The Natron lakes of Egypt were known to the ancients, and it was early employed in glass-making, &c.

SODÆ CARBONAS (VENALE, D.) IMPURA, L. SIVE BARILLA, D. Impure Carbonate of Soda or Barilla. *F.* Soudes de Commerce. Soudes brutes. *G.* Kohlensaures Natron.

The substance known by the name of Soda is a Carbonate of Soda, but mixed with various impurities, according to the source whence it has been obtained ; that is, either from the different Natron lakes, from the burning of maritime plants or sea-weeds, or from the decomposition of other Salts of Soda. It is introduced into the Pharmacopœias for the purpose of obtaining from it pure Carbonate of Soda.

BARILLA is the Ash obtained by burning plants, on the shores of the Mediterranean, of the Red Sea, and Indian Ocean. These plants belong mostly to the natural family of Chenopodeæ, q. v. ; and chiefly to the genera Salsola, Salicornia, Suaeda, and Chenopodium. The quantity of Carbonate of Soda in the ash varies from 25 to 40 per cent., and is produced from the combustion of the Oxalate and other Salts of the Vegetable Acids. The Soda is no doubt obtained from the soil, for Du Hamel planted Soda-plants inland, and they yielded only Potash. Infusion of a Salsola in cold water afforded by evaporation two Salts, Carbonate of Soda and Chloride of Sodium. *Murr. Chem.* ii. p. 612. A portion of the Chloride is no doubt converted into the Carbonate during the incineration.

KELP used to be prepared on the coasts of Scotland and its islands, also on those of Ireland and Wales, and on that of Normandy in France, by burning a great variety of Algæ or Sea-weeds. The ashes when cold form *Kelp*, which is in hard cellular masses, of a bluish grey colour, and of a disagreeable alkaline taste, containing from 3 to 8·5 per cent. of Carbonate of Soda, and other salts, as in the case of Barilla, but also some Potash and Iodine (v. p. 41).

Carbonate of Soda is, however, now obtained very cheaply from Chloride of Sodium or Sea-salt. This is first converted into Sulphate of Soda by the action of Sul′, then mixed with pounded small Coal and Chalk, and heated in a reverberatory furnace and stirred. The Carbonaceous matter abstracts Oxygen both from the Sul′ and the Soda; Sulphuret of Sodium is formed, and decomposed by the Lime; Carbonate of Soda, insoluble Oxi-Sulphuret of Calcium, Caustic Soda, and Carbonaceous matter being the result. The insoluble parts are separated by lixiviation, and the Sulphur subsequently burnt away; during which the Soda is completely Carbonated. The mass now contains about 50 per cent. of Soda. Being lixiviated and evaporated, the Carbonate is obtained in large crystals. As Barilla is not sufficiently pure for medicinal purposes, the L. and D. Colleges give directions for its purification. The E. C. justly consider the Carbonate of Soda produced as above, pure enough.

SODÆ CARBONAS, L. D. Carbonate of Soda. *Natron præparatum.*

Carbonate of Soda ($NaO\,CO_2 + 10\,Aq. = 144$), prepared as above, is in large and clear colourless crystals, without odour, but having a disagreeable sub-alkaline taste, and an alkaline reaction on Turmeric. The crystals are oblique rhombic prisms, or rhomboidal octohedrons, entire or broken. In the air they effloresce, but when exposed to heat melt in their water of crystallization; as this is dissipated, the Salt becomes a white, porous, anhydrous mass, known as Dried Carbonate of Soda. Water at 60° dissolves half, and at 212° its own weight of this salt, but it is insoluble in Alcohol. It has the characteristics of a Carbonate with Acids and Earths, is distinguished from the Bicarbonate by giving a brick-red precipitate with Bichloride of Mercury, and a white one (Carbonate of Magnesia) with Sulphate of Magnesia. It is used in converting Oils and Fats into Soap. Composed per cent. of NaO 22·2 C′ 15·3 + Aq. 62·5 = 100.

Fig. 15.

Prep. L. D. Boil impure *Carb. of Soda* ℔ ij. (powdered Barilla 1 part, D.) in *Aq. dest.* O iv. (water 2 parts, for two hours, occasionally stirring, D.) Strain the liquor (while hot, and set it by, that crystals may form, L.), pound what remains of the Barilla, boil again with the same quantity of water, and repeat this a third time. Evaporate all the washings to dryness in an open iron vessel, avoiding such heat as would liquefy the salt; stir with an iron spatula till the mass becomes white. Dissolve in boiling water, and evaporate the liquid to a sp. gr. = 1220. Expose to the air at a temperature of about 32°, that crystals may form; dry these and keep in well-closed vessels. If the salt be not sufficiently pure, repeat the solution and crystallization.

Tests. Carbonate of Soda prepared from the Sulphate is usually very pure, but the Salt is apt to contain some Sulphate of Soda, and also the Chlorides of Sodium and of Potassium. "Sodæ Carbonas (crystalli). Fresh prepared, translucent, but in an open vessel, it in a short time falls to powder. It is totally soluble in water, but not at all in Alcohol. It alters the colour of Turmeric like an alkali." L. As these notes serve chiefly to distinguish this salt from others, Mr. Phillips remarks, "If pure as usual and saturated with Nitric Acid, it yields but little precipitate with the Nitrate of Silver, nor any Sulphate of Barytes with the Chloride of Barium." "A solution of 21 grains in Aq. dest. f℥j. precipitated by 19 grains of Nitrate of Baryta, remains precipitable by more of the test; and the precipitate (Carbonate of Baryta) is entirely soluble in Nitric Acid." E. P. This will leave 0·75 per cent. of the salt still in solution, if it be of due purity; so that after filtration Nitrate of Baryta will again cause a precipitate." (*c.*)

Inc. Acids, Acidulous Salts, Lime-water, Hydrochlorate of Ammonia, Earthy and Metallic Salts.

Action. Uses. Antacid; in large doses, irritant poison. Diuretic, Antilithic.

D. gr. x.—ʒß or ʒj. For effervescing draughts, gr. xx = gr. x. of Cit′ or Tar′, and fʒij.ß of Lemon-juice; used also for Seidlitz Powders.

Antidotes. Fixed Oil, Vinegar, Lemon-juice, Cream of Tartar.

SODÆ CARBONATIS AQUA, D.

Prep. Dissolve *Carb. Soda* q. s. or ℥j. in *Aq. dest.* q. s. or ℔j. so that the liquid shall have a sp. gr. = 1024.

D. fʒj.—ʒij.

SODÆ CARBONAS EXSICCATA, L. SODÆ CARBONAS SICCATUM, E. D.

Dried Carbonate of Soda ($NaO\,CO_2 = 54$) is the result of the expulsion of the whole of the water of the crystallized Salt, so that 54 grains of the Anhydrous Salt are equal to 144 grs. of the Crystallized Salt. Composed of NaO, 59·3 + C′ 40·7 = 100. It requires to be heated to redness, that it may be of uniform strength.

Prep. Heat *Carb. Soda* ℔j. (q. s. E. D.) in a proper (shallow, E. silver, D.) vessel (stirring frequently, D.) till it is dry. Heat to redness, L. E. (in a crucible, E.); rub to powder (when cold, E.) (Keep in stoppered bottles, D.)

D. gr. v. to gr. xx. in powder or in pills.

SODÆ SESQUICARBONAS.

Sesqui-Carbonate of Soda. Natron. *F. Natron. G.* Anderthalb Kohlensaures Natron.

A Sesquicarbonate of Soda ($Na\,O\,1\frac{1}{2}\,C\,O_2 + 2\,H\,O = 83$) exists in nature; for the Trona found near Tripoli in Africa, the Natron of the country to the west of the Delta of the Nile, and of the Lonar Lake described by the late Dr. Malcolmson,* have all been

* The analyses were made in the Laboratory, and the specimens are deposited in the Museum, of King's College.

proved to consist of 1½ Eq. of Carb′ to 1 of Soda; or NaO 38·55+C′ 39·76+Aq. 21·69=100. A solution of it may be obtained by heating Bicarbonate of Soda in solution to a temperature of 212°. The salt so called in the London Pharmacopœias has been proved by Mr. Everett to be a Bicarbonate, q. v.; and Dr. Pereira has stated that by the process given a Bicarbonate will always be produced.

SODÆ (SESQUICARBONAS, L.) BICARBONAS, E. D.

Sodæ Carbonas. Bicarbonate and Carbonate of Soda of the shops. *F.* Bicarbonate de Soude. *G.* Zweifach Kohlensaures Natron.

The Bicarbonate of Soda ($NaO\,2CO_2+2HO=94$) of the E. and D. P's is the same salt as the Sesquicarbonate of the L. P. That which is met with in commerce is usually a pure salt, but occasionally mixed with a small portion of the Carbonate. It exists in some mineral springs highly acidulated with Carb′, as in those of Vichy.

Prop. As usually sold, it is colourless, in powder, or in minute scale-like crystals, having a saline, slightly alkaline taste and reaction. Little changed in the air. Soluble in 13 parts of temperate water (Rose and Geiger), but in much less boiling water. Hence it crystallizes as this cools, or, indeed, as it is formed. Heated, it first loses a portion of water (some chemists consider that it contains only 1 Eq. Aq.), then half an Eq. Carb′, finally all its water, and becomes reduced to dried Carbonate of Soda. It is easily distinguished from this salt, *i.e.* the Carbonate, by its less alkaline taste, less solubility, and by not forming a white precipitate on the addition of Sulphate of Magnesia to its solution; nor is a brick-red precipitate formed with Bichloride of Mercury, but only a slight opalescence. Comp. Na O 37·05+C′ 52·26+Aq. 10·69=100. Gmelin.

Prep. L. D. Dissolve *Carb. Soda* ℔vij. (5 parts, D.) in *Aq.* cong. j. (5 parts, D), filter, pass *Carb′* gas (from the action of Mur′ on marble) through the liquid to saturation. Let the salt subside, press in folds of linen, and dry with a gentle heat.

E. Take a glass jar open at the bottom and tubulated at the top, close the bottom in such a way as to allow of the free ingress of a fluid, fill the jar with fragments of *Marble* and immerse it in a vessel containing dilute *Mur′*, then fill the apparatus with Carb′ gas; connect the tubulature closely by a bent tube with an empty bottle, and this in like manner with another filled with *Carb. of Soda*, 1 part, and *dried Carb. Soda*, 2 parts, well triturated together, and let the tube be long enough to reach the bottom of the bottle. Let the action go on till next morning, or till the salt no longer absorbs gas. Remove the damp salt which is formed, and dry it in the air, or without greater heat than 120°.

Tests. Entirely soluble in water. This solution should not be precipitated by Chlor. Platinum, or the other tests for Potash, showing all absence of this alkali. The absence of Sulphates and Chlorides is proved by Chlor. Barium and Nitr. Silver causing no precipitate in a solution saturated with Nitr′. It is most liable to contain a portion of Carb. Soda. This will, of course, give a stronger alkaline reaction and a more disagreeable taste. Its presence is detected by Sulph. Magnesia producing a white precipitate, and none in sol. of the Bicarb.

unless when heated. "A solution in 40 parts of water does not give an orange precipitate with solutions of Corrosive Sublimate," E., unless with the aid of brisk agitation, long standing, or heat, but a reddish-brown precipitate at once if so much as a hundredth part of the Carb. be present. (*c.*)

Inc. The same as the Carbonate, except Sulph. Magnesia, with which it may therefore be prescribed.

Action. Uses. Antacid, Antilithic, Diuretic.

D. gr. x. to ʒß or ʒj. For making effervescing draughts, ℈j. = 17 grs. Cit′ or 18 grs. Tar′.

TROCHISCI SODÆ BICARBONATIS, E. Soda Lozenges.

Prep. Pulverize *Bicarb. Soda* ℥j. *Pure Sugar* ℥iij. *Gum Arabic* ℥ß. beat them into a proper mass for making lozenges with mucilage.

PULVERES EFFERVESCENTES, E. Soda Powders.

Prep. Take of *Tart′* ℥j. *Bicarb. Soda* ℥j. and gr. 54, or *Bicarb. Potash* ℥j. and gr. 160. Powder the acid and either Bicarbonate finely; divide each into 16 powders. Preserve the acid and alkali powders in different coloured papers.

These are the common Soda powders, for which we may also take of either Bicarb. of Soda or Potash ℈j. and dissolve in Aq. f℥jß —f℥iij. in a tumbler, and add ½ oz. of Lemon-juice or of Tar′, or Cit′ gr. xviij. dissolved in a little water, and drink while in a state of effervescence: a Citrate or Tartrate of Soda or of Potash will be formed. By adding ʒj. or ʒij. of Rochelle Salt, or Sulphate of Magnesia, an aperient salt may be presented in the agreeable form of an effervescing draught. It must be remembered that Citrates and Tartrates become converted into Carbonates, and will, if long taken, have an alkaline reaction on the secretions, as on that of urine.

LIQUOR (AQUA, E.) SODÆ EFFERVESCENS, L. AQUA CARBONATIS SODÆ ACIDULA, D. Soda Water.

Prep. Dissolve *Bicarb.* (Sesquicarb. L.) *of Soda* ʒj. in one pint of distilled water, and pass into it, under strong pressure, a current of *Carbonic Acid* gas (obtained from Marble and Mur′ diluted with 8 waters) more than is sufficient to saturate it, and keep it in a well-stopped vessel.

Soda-water should be of this composition; but what is commonly so called is only a solution of Carb′ gas in water. By adding some Bicarb. Soda to such Soda-water, an equally efficient mixture will be formed. But the effects of the simple Carb′ gas in water and those produced when it is combined with an alkaline salt will necessarily be different. The Carb′ gas as it escapes will stimulate the stomach, while this effect will be followed by an alkaline reaction when Soda (or Potash) is present, and this may be beneficial or injurious according to the nature of the case. *v.* ANTACIDS.

Carbonic Acid gas is found in many mineral waters, to which it gives an acidulous taste and sparkling effervescence. In some it occurs with Carbonate or Bicarbonate of Soda, or with Iron, when its effects are necessarily modified according to the nature of these ingredients. *v.* MINERAL WATERS.

SODII CHLORIDUM, L. SODÆ MURIAS (PURUM, E.) E. D.

Chloride of Sodium. *Muriate of Soda.* SODÆ MURIAS PURUM, E. *Sal Fossile. Sal Marinum. Sal Gemmæ. F.* Chlorure de Sodium. *G.* Chlor Natrium.

Chloride of Sodium (Na Cl=60), or Common Salt, is abundantly diffused in nature, and, being an essential article of diet, must have been known from the earliest ages. It is found in many animal solids and fluids, and in the juices of some vegetables. It exists in large quantities in the solid form as Rock Salt, or in solution in some springs, and everywhere in sea-water. From these it is obtained by evaporation, when it crystallizes with slight variations of appearance according to differences in the process. These varieties are known by different names in commerce, as Butter, Stone, and Basket Salt, also Sea Salt, and, in large crystals, as Malden, Fishery, and Bay Salt. Most of the kinds of Salt require purification, by being again dissolved and recrystallized (Sodæ Murias purum, E.), as they are apt to contain other salts, as alkaline and earthy Sulphates and Chlorides especially of Magnesium and of Calcium.

Prop. Common Salt crystallizes in anhydrous transparent cubes; these are sometimes aggregated together, forming hollow four-sided pyramids with their sides in steps. Sp. Gr.=2·17; white, though rock salt is sometimes of a reddish hue. Taste saline, well-known; without odour; it does not affect either Litmus or Turmeric. Neither does it bleach. 1 part requires $2\frac{1}{2}$ times its weight of water to dissolve it, and its solubility is not much increased by a boiling temperature. It is insoluble in pure Alcohol, but slightly soluble in rectified Spirit, to the flame of which it gives a yellow tinge. Salt readily transmits radiant heat. When heated, it decrepitates; at a red heat, fuses and sublimes. Inalterable in the air when pure; but when impure, it deliquesces. Sul′ and Nit′ decompose Salt, as also Bor′ and Phosp′ with the assistance of heat. Nitr. Silver, the Protoxides of Lead and Mercury, Lime, Potash, and, with the aid of heat, Carb. Potash, all decompose this salt. When dissolved in water, it is supposed by some chemists to become a Hydrochlorate or Muriate of Soda, from the water being decomposed, its Oxygen combining with the Sodium and the Hydrogen with the Chlorine, then its composition will be Na O, H Cl=69. Salt is composed per cent. of Na 40+Cl 60=100.

Tests. "Scarcely any precipitate should be occasioned by Carb. Soda or Nitr. Bar." L., the first indicating the absence of earthy salts, and the latter of Carbonates. A solution is not precipitated by Sol. Carb. Am. followed by Sol. of Phosph. Soda. The former would detect Lime, and, after its action, the latter would indicate Magnesia. "9 grs. dissolved in distilled water are not entirely precipitated by a sol. of 26 grs. of Nitr. Silver," E.

Action. Uses. Stimulant, irritant externally, Emetic, Cathartic.

D. gr. x.—ʒj. as a Stimulant. ʒiv.—℥j. Cathartic. ℥jß—℥ij. with warm water, as an Emetic. lbj. of Salt to every 3 gallons of water, will make a bath of the strength of sea-water.

Pharm. Prep. Pulv. Salinus Comp. E. p. 82.

Liquor Sodæ Chlorinatæ, L.

Hypochlorite of Soda. Chloride of Soda. Labarraque's Soda Disinfecting Liquid. *F.* Chlorure de Soude. Chlorure d'Oxyde de Sodium. *G.* Chlornatron.

M. Labarraque in 1822 made known the utility of Chloruret of Soda as a disinfectant, &c., and obtained the prize of the French Society for encouraging National Industry for proposing it as a means for preventing, stopping, and destroying putrefaction.

Prop. The comp. of this substance is probably of a definite nature, for the solution, by careful evaporation, yields crystals, which, when redissolved in water, produce a solution similar to the original one. This is of a pale yellowish colour, has a slight odour of Chlorine, and a sharp but somewhat astringent taste. It has first an alkaline reaction, from the Carbonate of Soda, on Turmeric paper, and then destroys its colour, as well as that of Sulphate of Indigo. Exposed to the air, it becomes decomposed, Chlorine being evolved. Carbonate of Soda is left. Chlorine as well as Carb′ gas is evolved by acids, Chloride of Sodium being left in solution. Lime-water produces a white precipitate, indicating the presence of Carb′. It may be distinguished from Chlorinated Potash by Chloride of Platinum not producing the usual yellow precipitate, and from Chlorinated Lime by Oxalate of Ammonia causing no precipitate. By some Chemists this salt is supposed to be composed of Bicarb. Soda and Chloride of Soda, because no Carb′ is given off in the process for preparing it; but, as the composition is uncertain, the London College have adopted the name of Chlorinated Soda. The most generally received opinion, and that adopted by Dr. Pereira, is, that it is composed of 2 Eq. Bicarb. of Soda, 152 + 1 Eq. Hypochlorite of Soda, 76 + 1 Eq. Chloride of Sodium, 60 = 288.

Prep. Dissolve *Carb. Soda* ℔j. in *Aq. dest.* Oij. Through this pass *Chlorine* evolved from *Chlor. Sodium* ℥iv. and powdered *Binoxide Manganese* ℥iij. in a retort, with *Sul′* f℥iv. diluted with *Aq. dest.* f℥iij. and allowed to cool and powder them. The Chlorine is passed first through f℥v. of *Aq. dest.* and then into the solution of *Carb. Soda.*

If instead of Carbonate of Soda we employ Carb. Potash, we shall have the *Eau de Javelle* or *Chlorinated Potash*, first employed in bleaching in 1789, but little used in medicine.

Action. Uses. Disinfectant and Antiseptic. Stimulant.

Sodæ Biboras, L. Sodæ Boras, D. Borax, L. E.

Biborate of Soda. *Borate of Soda.* *F.* Borax. Borate de Soude. *G.* Boraxsaures Natron.

Fig. 16.

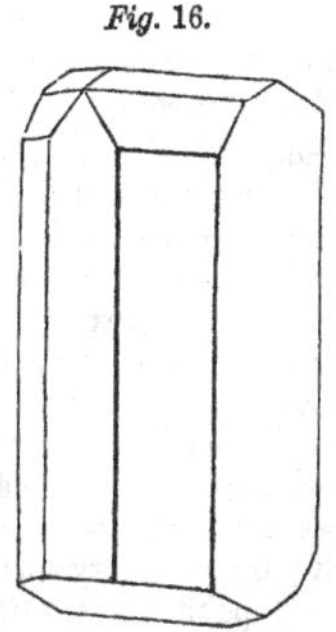

Borax or ($Na\,O, 2\,B\,O_3 + 10$ Aq.$=192$) Biborate of Soda, is supposed to have been known to the ancients, and to have been the *Chrysocolla* of Pliny. The Hindoos have long been acquainted with it; it is their *Sohaga*, Sanscrit *Tincana*, and one of the kinds of *Booruk* of the Arabs. Its nature was first ascertained by Geoffroy in 1732. It is produced by spontaneous evaporation on the shores of some lakes in Tibet, that is, in the same country with Musk and Rhubarb; brought across the Himalayan Passes into India, and imported into this country by the names of *Tincal* and *Crude Borax.* It is also obtained by saturating the Bor′ of Tuscany (p. 37) with Carb. Soda.

Prop. Crude Borax is in pale greenish pieces, covered with an earthy coating, and feels greasy to the touch. The natives of Tibet are said to cover it with some fatty matter, to prevent its destruction by efflorescence. It is purified by calcining, which destroys the fatty matter, or by washing with an alkaline ley, which converts it into a kind of soap, then dissolving and re-crystallizing. It crystallizes in irregular hexahedral prisms often terminated by 2—4 converging planes. Sp. Gr. 1·35. It is colourless, transparent, somewhat shining; taste sweetish, a little styptic, and subalkaline. It has an alkaline reaction on Turmeric. The crystals effloresce slightly in the air, are soluble in 12 parts of cold and 2 of boiling water. When heated, they lose water, swell up into a porous substance called *Borax usta v. calcinata*, and at a red heat run into a transparent glass, called *Glass of Borax*, much used as a flux. Another variety, more useful in the arts, contains only 5 Eqs. Aq., and crystallizes in octohedra, which are permanent in the air. Borax increases the solubility of Cream of Tartar, p. 86, and converts mucilage of Lichen and of Salep into a thick jelly. Comp. Na O 16·0 Bor′ 35·79 Aq. 47·37 = 100.

Tests. Not liable to adulteration. Totally soluble in water. Gives a green colour to the flame of Alcohol. Sul′ precipitates scales of Boracic′ from a concentrated solution, Sulph. Soda being left in solution. L. E.

Inc. Acids, Acid Salts, Potash, Chlorides of Lime and of Magnesia.

Action. Uses. Sub-Astringent, Detergent, Diuretic, Emmenagogue.

D. gr. v.—ʒß. ʒij. in Aq. f℥vj. as a lotion.

MEL BORACIS, L. E. D. Honey of Borax.

Prep. Mix *Borax* powdered ʒj. with *Honey* clarified ℥j.

Action. Uses. Subastringent, Detergent. Applied to Aphthæ, and to ulcers of the inside of the mouth.

SODÆ SULPHAS, L. E. D.

Sulphate of Soda. *Natrum Vitriolatum. Sal Catharticus. Sal mirabile Glauberi. F.* Sulphate de Soude. *G.* Schwefelsaures Natron.

Sulphate of Soda ($Na\,O, S\,O_3+10$ Aq.$=162$), or Glauber's Salt, is found effloresced on the soil in some countries, as in India, where it is called *kharee nimuk* or *kharee noon.* It exists in sea-water, in that of some lakes and mineral springs, also in Glauberite. It is found in the ashes of many plants, and in some animal secretions.

Prop. Fresh prepared, it is transparent and colourless, without odour, but having a nauseously bitter taste. It crystallizes in four and six-sided oblique rhombic prisms, often with dihedral summits; if much agitated, the crystals are small. An anhydrous variety crystallizes in rhombic octohedra. 3 parts of water at 60° dissolve 1 of the salt; the solubility increases to 92°, and then diminishes to 215°, at which point the salt is only as soluble as at 87°; boiling water dissolves its own weight; it is only very slightly soluble in Alcohol. In the air the crystals effloresce; if heated, they first melt in their water of crystallization, then lose half their weight, and fall into a white powder. Heated with Carbon, the salt is converted into Sulphuret of Sodium. The solution is decomposed by salts of Baryta, Lime, and Lead, insoluble Sulphates being precipitated. Comp. Na O 19·75 S′ 24·69 Aq. 55·56=100.

Sulphate of Soda is largely prepared by the action of Sul′ on Common Salt (p. 91), and is a residual Salt in several manufacturing processes (pp. 47 and 67).

Prep. L.E.D. Dissolve of the salt remaining after the distillation of *Mur′* ℔ij. (any quantity, D.) in boiling *Aq.* Oij. (Oiij. E. q. s. D.) (Add *Carb. Soda,* L. (powdered *white Marble,* q. s. E.) till effervescence ceases. (Filter, wash the insoluble matter, returning the water to the original liquid, E.)) Evaporate, filter, and crystallize. Dry the crystals, L.

In the preparation of H Cl′, by acting with Sul′ on Chloride of Sodium, Sulph. Soda is produced. But as there is always an excess of acid, it is neutralized by either of the above Carbonates, the Carb′ escaping, a further portion of Sul. Soda is produced in one case, with the expenditure of a more expensive salt; but in the more economical E. formula, some insol. Sulph. Lime is formed, from which the soluble Sulphate of Soda is easily separated.

Tests. Not liable to adulteration. Exposed to the air, it falls to powder. 100 parts lose 55·5 parts by a strong heat. Totally dissolved by water; very slightly by Alcohol (insol. *Phillips*). It does not alter the colour of Litmus or Turmeric. Nitr. Silver throws down scarcely anything (unless Chlorides be present) from a dilute solution;

Nitr. Baryta more, which is not dissolved by Nit', L.P. The presence of Iron may be detected by Ferrocyanide of Potassium, or by Tincture of Galls, and of Copper by the blue colour produced by Ammonia.

Inc. Carb. Potash, Chloride Calcium, Barytic Solutions, Acetate and Diacetate of Lead.

Action. Uses. Purgative. Often called CHELTENHAM SALTS.

D. ʒiv.—℥j. or ℥ij. ʒiij.—ʒiv. of the effloresced or anhydrous salt.

SODÆ PHOSPHAS, L. E. D.

Phosphate of Soda. Subphosphate of Soda. *Tasteless Purging Salt. Sal Mirabile perlatum.* Triphosphate of Soda and Basic water (Turner), Triphosphate of Soda and water (Graham), or Common Tribasic Phosphate of Soda. *F.* Phosphate de Soude. *G.* Phosphorsaures Natron.

This salt is remarkable for the variety of opinions entertained by Chemists respecting its nature and composition. It is used in metallurgy, and was introduced into practice by Dr. Pearson about 1800. It was found in Urine by Hellot in 1737, and analysed by Marcgraff in 1745. It is also found in the Serum of the blood, and in other animal secretions; it is obtained from bone-ashes.

Prop. Phosphate of Soda is colourless, transparent, and of a cool saline taste. It crystallizes in large oblique rhombic prisms. Sp. Gr. 1·5. Soluble in four times its weight of cold and in twice its weight of boiling water; but not in Alcohol. The crystals effloresce in the air; when heated, they undergo aqueous fusion, and then lose water, and at a red heat melt into a greenish-coloured glass, opaque when cool. Phosphate of Soda has a slight alkaline reaction. Solutions of Acetate of Lead and of Chloride of Barium produce white precipitates (Phosphates). Nitr. Silver throws down a yellow precipitate (Phosphate of Silver), unless the Phosphate has been previously heated to a red heat, and become either monobasic or bibasic Phosphate of Soda; then a white Pyrophosphate of Silver is produced. All these precipitates are soluble in Nit', and the last also in Ammonia. It is decomposed by the soluble salts of Lime, and also by those of Magnesia. In the latter case, "if Ammonia be likewise present, a very insoluble triple compound is formed, the Ammoniaco-Magnesian Phosphate, one of the varieties of urinary gravel. It is not acted on, if moderately diluted, by Ammoniacal Nitrate of Silver; which constitutes a distinction between the actions of the Silver test on this salt and on Arsenic in solution." Phosphate of Soda, according to Professor Graham, consists of 1 Eq. of Phosph', 1 of basic water, 2 of Soda, and 24 of water of crystallization, thus constituting a Trisphosphate or Tribasic Phosphate of Soda and Water, $2NaO, HO, PO_3+24$ Aq. $=361$. When subjected to a red heat, the basic water being expelled, 1 Eq. of acid remains combined with 2 of Soda, and forms a Bibasic or Diphosphate of Soda, the salt which melts into a glass.

Prep. E.D. Phosphate of Soda is only in the list of Materia Medica in the L. P. The E. and D. Colleges give formulæ. To the *acid*, or *Biphosphate of Lime* obtained by acting on *Bone-ash* with *Sul'*, as described at p. 35, and in the state of a clear liquid, which is to be heated to ebullition, add *Carb. Soda* q. s. E. (8 parts, D.) dissolved in boiling water, until the acid be completely neutralized. Put the solution aside to cool and crystallize. More crystals may be obtained by successively evaporating (adding a little *Carbonate of Soda* till the liquid exerts a feeble alkaline reaction on Litmus paper, E.). Preserve the crystals in well-closed vessels. (If the salt is not pure redissolve and crystallize, D.)

When the Carbonate of Soda is added to the solution of Superphosphate of Lime, Phosphate of Soda is formed, and remains in solution, Carb' gas escaping, and Subphos. Lime being precipitated.

Tests. If the precipitate made by Chlor. of Barium is insoluble in Nit' a Sulph. probably of Soda, is present; if that by Nitr. Silver is so, then a Chloride is present. Carb. Soda is sometimes used in excess, especially as fine crystals are then more easily obtained: its presence is readily detected by its effervescing with acids. 45 grs. dissolved in Aq. dest. f℥ij. and precipitated by a solution of 50 grs. Carbonate of Lead in f℥j. of Pyroligneous acid, will remain precipitable by solution of Acetate of Lead. E. P.

Inc. Calcareous and Magnesian Salts; many Metallic Salts, as Acetate of Lead, &c.

Action. Uses. Mild Saline Cathartic; less unpleasant than others.

D. ʒiv.—℥jß.

SOLUTIO SODÆ PHOSPHATIS, E.

Prep. Dissolve of *Phosphate of Soda* (free of efflorescence) 175 grs. in *Aq. dest.* f℥viij, and keep the solution in well-closed vessels. Employed only as a test.

SODÆ ET POTASSÆ TARTRAS, D. SODÆ POTASSIO-TARTRAS, L. POTASSÆ ET SODÆ TARTRAS, E.

Tartrate of Soda and Potash, or, of Potash and Soda. *Soda Tartarizata. Tartarized Soda. Rochelle Salt. F.* Tartrate de Potasse et de Soude. *G.* Weinsaures Natron-Kali.

Tartrate of Soda and Potash ($NaO, KO, 2Tar' + 8 Aq. = 284$) was discovered in 1672 by Seignette, an apothecary of Rochelle. Hence it is called *Sel de Seignette*, also Rochelle Salt. All the Colleges differ in naming this salt. As it has most generally been ranked as a Soda salt in medical writings, the D. name, being more convenient and as correct as that of the E. P., is preferable.

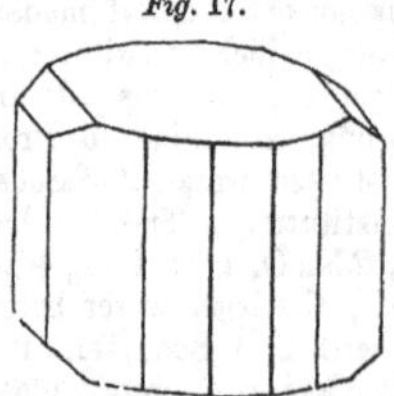

Fig. 17.

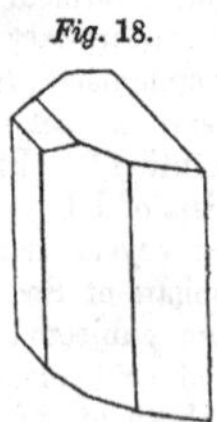

Fig. 18.

Prop. Colourless and without odour; of a mild, saline, slightly bitterish taste. Crystals transparent, often very large, in prisms with ten or twelve unequal sides; usually seen in half-crystals having six unequal sides; the primitive form is the right rhombic prism. They effloresce in a dry air, and, when heated, melt in their water of crystallization. The acid becoming decomposed, Carbonates of Potash and Soda are left with some Charcoal. This salt is soluble in 5 parts (2½ Berz.) of water at 60°, and in less boiling water. It is readily decomposed by most acids and acidulous salts, except the Bitart. Potash. The acids combine with the Soda, and precipitate Bitart. Potash. It is also decomposed by the Acet. and Diacet. Lead, and likewise by the soluble salts of Lime and of Baryta; but this is not apparent in a dilute solution. A white precipitate is also thrown down in a strong solution by Nitr. Silver (*p.*), which is soluble in an excess of water. This, therefore, like the former, is not perceptible in a dilute solution. Potash may be recognised by its peculiar tests, and when it has been precipitated, the Soda will be revealed by its tests. Composition, Tart. Pot. 40, Tart. Soda 34·5 Aq. 25·5=100.

Prep. L. E. D. Dissolve *Carb. Soda* ℥xij. (5 parts, D.) in boiling *Aq.* Oiv. (50 parts, D.) Add gradually *Bitart. Potash* ℥xvj. L. (to neutralization, E.) (7 parts, D.) Filter and concentrate till a pellicle forms; set aside to crystallize, and evaporate again for a further supply.

In this formula, the second Equivalent of Tart′ of the Bitartrate combining with the Soda, the Carb′ is expelled with effervescence, and a double salt consisting of Tartrate of Soda and of Tartrate of Potash remains in solution. As both the Carb. Soda and the Bitart. Potash are liable to vary in strength, it is better to add the latter to saturation, though it is preferable to have excess of Soda rather than of acid.

Tests. Not liable to much adulteration, from being sold in crystals; but both Bitart. Potash and Tart. Lime are sometimes present. Entirely and easily soluble in 5 parts of boiling Aq. The solution does not affect Litmus or Turmeric. S′ and Mur′ occasion a crystalline precipitate in a strong solution (Bitart. Potash). 37 grains in solution are not entirely precipitated by 43 grains of Nitr. Lead. L. and E. This will show the due proportion of Tart′, and consequently the absence of the mineral acids. The Chlor. Barium and Nitr. Silver employed by the London College require very dilute solutions (*v.* supra), as pointed out by Dr. Pereira.

Inc. Acids and Acidulous Salts, Acet. and Diacet. Lead, &c.

Action. Uses. Cathartic, Diuretic.

D. ʒij.—℥j., or in effervescence, as in Seidlitz powders.

SODÆ ACETAS, L. D.

Acetate of Soda. *Terra foliata Tartari crystallizata. Terra foliata Mineralis. F.* Acetate de Soude. *G.* Essigsaures Natron.

The Acetate of Soda (Na O, Ac′+6 Aq.=137) is considered by

Thomson to have been first described by Baron in 1747. It probably exists in many of the plants of which the ashes yield Carbonate of Soda. It is made in large quantities by the manufacturers of Pyroligneous Acid, q. v., hence it is included in the list of Materia Medica of the L. P.

Prop. When pure, a colourless salt, having a pungent, rather bitter saline taste. It crystallizes in transparent, oblique, rhombic prisms, or in striated needles, often seen in foliaceous masses. Sp. Gr. 2·1. Soluble in about 3 parts of water at 60°, and in its own weight of boiling water, and in about 24 parts of Alcohol. Exposed to dry air, it effloresces, losing about 40 per cent. of weight. Heat, cautiously applied, likewise expels its water of crystallization; but at a temperature of 600° the salt is decomposed, and at a red heat converted into the Carbonate with some Charcoal. Comp. Na O 23·36+A′ 37·22+Aq. 39·41=99·99.

Prep. Acetate of Soda is made on a large scale by saturating impure Pyroligneous acid with Chalk or Slaked Lime. The Acetate of Lime which is formed is decomposed by the requisite quantity of Sulphate of Soda. An insoluble Sulphate of Lime is precipitated, and Acetate of Soda remains in solution. This is decanted, evaporated, and crystallized. The crystals are still very impure and blackish coloured. They are purified by repeated solution, filtering, and crystallization.

The D. C. directs it to be made by saturating *distilled Vinegar* with *Carb. Soda*, filtering and evaporating the fluid till it attains the density of 1276. Crystals are deposited on cooling, which are to be dried and kept in a close vessel.

Tests. Acetate of Soda is not liable to any great adulterations as made at present; but they can be recognised by the tests of the L. P. Soluble entirely in water; not (only partially) in Alcohol. The solution does not affect Litmus or Turmeric; neither is it affected by Chlor. Barium (if no Sulph. be present), by Nitr. Silver (if no Chlorides), nor by Chlor. Platinum (if no Potash). Sul′ evolves an acetous odour. Heat converts it into Carbonate of Soda.

Inc. The strong Acids.

Action. Uses. Diuretic, Cathartic.

D. Diuretic, ℈j.—ʒij. Purgative, ʒj.—ʒiv.

BARIUM.

Baryta was discovered by Gahn and Scheele about 1774, but obtained its name from *βαρυς, heavy*. Sir H. Davy discovered that it was the Oxide of a Metal, which he named Barium.

BARIUM (Ba=69) is a brilliant silver-white metal; heavy, Sp. Gr. above 2; when heated, burning with a red light in the air, and decomposing water, combining in both cases with Oxygen, and forming an Oxide of Barium, or the earth Baryta.

BARYTA. *F.* Baryte. *G.* Schwer-erde, or Baryt-erde.

Baryta or Barytes (Ba O=77) is a porous substance, of a grayish-colour, devoid of odour, with a powerful caustic taste, alkaline reaction, corroding animal substances. Sp. Gr. 4. It combines eagerly with water, evolves heat, and becomes a Hydrate, which is not decomposed at a red heat; difficultly fused; insoluble in Alcohol, but soluble in 20 parts of cold and 3 of boiling water, forming Barytic water.

Tests. It may be detected by its alkaline reaction, the heavy white precipitates which it forms with S′ or the soluble Sulphates, and which are insoluble in water and dilute acid.

Action. Uses. Baryta is an acrid caustic, and will act as a poison on the system.

BARYTÆ CARBONAS, L. E.

Carbonate of Barytes. Terra ponderosa aerata nativa, Gm. *F.* Carbonate de Baryte. *G.* Kohlensaures Baryt.

Carbonate of Baryta (Ba O, C O_2=99) was described in 1784 by Dr. Withering, and named Witherite by Werner; it is rather common in Lancashire. It may be prepared in the form of a powder by decomposing Chlor. Barium by an alkaline Carbonate. In its native state it occurs massive with a fibrous structure, or imperfectly crystallized in a globular form, or in hexagonal prisms, or in pyramids.

Prop. It is hard, of a white or grayish colour, without odour or taste, with a vitreous lustre, and subtransparent. Sp. Gr. 4·29—4·3; nearly insoluble in water, unless there is excess of Carb′. The native is not decomposed by heat, the artificial at a white heat in contact with carbonaceous matter. Comp. Ba 77·7 + C′22·3 = 100.

Tests. Carb. Baryta, if pure, should be entirely dissolved with effervescence in H Cl′; any Sulph. Baryta present will remain undissolved. The solution of Chlor. Barium which has been formed does not give any precipitate on addition of Ammonia, showing the absence of Alumina. A brownish-yellow precipitate will indicate Iron; and H S′ throws down a black Sulphuret of Lead or Copper, or of Iron. Sul′ added in excess to the above solution, will precipitate the whole of the Baryta as a Sulphate; and if Carb. Soda is afterwards added, no precipitate should take place, and thus prove the absence of Lime. The E. P. states that 100 grs. dissolved in an excess of Nit′, are not entirely precipitated with 61 grs. of the anhydrous or 125 grs. of the crystallized Sulph. Magnesia.

Tests. Does not smell of Sulphuretted Hydrogen after ignition with Charcoal, nor effervesce with dilute N′.

Action. Uses. Carbonate of Baryta, though insoluble and tasteless, yet acts with considerable activity when introduced into the stomach of animals, probably from meeting there with acid, by which it is converted into a soluble salt. Officinal for the purpose of making the Chloride of Barium.

Barytæ Sulphas, E. D.

Sulphate of Baryta. *F.* Sulfate de Baryte. *G.* Schwefelsaures Baryt.

Sulphate of Baryta (Ba O, S O_3=117), or Heavy Spar, is more abundant as a mineral than the Carbonate. The finest specimens have been obtained from Dufton in Cumberland : the Author found it on the Himalayas, near the Convalescent depôt at Landour. (*v.* Illustr. Himal. Bot. p. xxxiii.)

Prop. Heavy Spar may be found massive or crystallized, of a foliaceous or lamellar structure; white-grey, or with a reddish hue; often translucent; heavy; Sp. Gr. 4·41 to 4·67; without odour or taste; insoluble in water. Its crystals are often bevelled tables or flat prisms of six sides, and may be divided into right rhombic prisms. This salt is formed whenever Baryta meets with S′, in whatever state of combination either the earth or the acid may previously have been; and the Sulphate of Baryta being insoluble in Nit′, this is employed to test it. Hence this earth and its soluble salts are excellent tests for S′ and the Sulphates. It is double refractive, decrepitates briskly before the blowpipe, and is with difficulty fused, but eventually melts into a hard white enamel, which is not affected by acids. Sulphate of Baryta, when heated with carbonaceous matter, has its acid decomposed, and Sulphuret of Barium is formed. From this, various salts may be formed by operating with different acids, or it may be converted into Carbonate of Baryta by heating it to a red heat with three parts of Carbonate of Potash. Comp. Ba O 66+S′ 34=100.

Action. Uses. Inert. Employed for making other salts, being cheap and usually pure.

Barii Chloridum, L. Barytæ Murias, E. D.

Chloride of Barium. *F.* Chlorure de Baryum. *G.* Chlor-Baryum.

Chloride of Barium (Ba Cl+2 Aq.=123), at one time called *Terra ponderosa Salita* and *Barytes Salita*, was discovered by Scheele in 1775. It is prepared from either the Barytic Carbonate or Sulphate.

Prop. By evaporation of its solution it may be obtained in rhombic plates, or flat or tabular quadrangular crystals with bevelled edges. Sp. Gr. 2·82. These, like the solution, are colourless and transparent, of an acrid and bitter, nauseous, disagreeable taste; efflorescent in the air when it is very dry; but in ordinary states, they are permanent. They produce no action on vegetable colours. Of these crystals 100 parts of water at 60° F. dissolve about 40 parts; but at the boiling point, 222° of a saturated solution, 78 parts are dissolved. They are slightly soluble in rectified Spirit, and in 400 parts of anhydrous Alcohol, which will then burn with a greenish-yellow flame. (Strontian salts burn red.) At a moderate heat the crystals decrepitate and lose their water of crystallization.

Prep. L. Mix H Cl′ Oß. with *Aq. dest.* Oij. gradually add *Carb. Baryta,* broken into small pieces, ℥x. Then apply heat, and on the cessation of effervescence, strain and boil down the liquor, that crystals may form.

E. D. Take of *Sulph. Baryta* ℔j. (10 parts, D.) *Charcoal* in fine powder ℥ij. (1 part, D.) pure *Mur′* q. s. Heat the Sulphate to redness; (throw into cold water and levigate, D.) pulverize it when cold, and mix it intimately with the Charcoal; subject the mixture to a low white heat, for three hours, in a covered crucible; pulverize the product, put it gradually into boiling *Aq.* Ov., and boil for a few minutes; let it rest for a little over a vapour-bath; pour off the clear liquor, and filter it if necessary, keeping it hot. Pour boiling *Aq.* Oiij. over the residuum and proceed as before. Unite the two liquids, and while they are still hot, or if cooled, after heating them again, add pure *Mur′* gradually so long as effervescence is occasioned. In this process the solutions ought to be as little exposed to the air as possible; and in the last step the disengaged gas should be discharged by a proper tube into a chimney, or the ash-pit of a furnace. Strain the liquor, concentrate it, and set it aside to crystallize.

In the L. process the Chlorine of the H Cl′ combines with the Barium of the Baryta, and forms the Chloride of Barium, while the Hydrogen, set free from the acid, combines with the Oxygen of the earth, and some water is formed.

In the E. and D. processes, the Carbon, taking the Oxygen from both the acid and the earth, escapes in the form of Carbonic Oxide, while the Sulphur, combining with the Barium, forms a Sulphuret of Barium. This, by solution in water, becomes* a Hydrosulphate of Baryta. This, on the addition of H Cl′, becomes Hydrochlorate of Baryta, which by evaporation yields crystals of Chloride of Barium, the Hydrogen and Oxygen forming water.

Tests. Sul′ and the Sulphates throw down the insoluble Sulphate of Baryta from the solution; so also do the soluble Phosphates, Carbonates, and Tartrates. Nitr. Silver also gives a white precipitate (Chlor. Silver), which is soluble in Am., but insol. in Nit′. Impurities are less apt to occur when this salt is made with the Sulphate than when made with the Carbonate. Apply the same tests.

The revised E. P. state that 100 grains in solution are not entirely precipitated by 100 grains Sulph. Magnesia.

LIQUOR BARII CHLORIDI, L. BARYTÆ MURIATIS AQUA, D.

Prep. Dissolve *Chloride Barium* ℨj. (Muriate 1 part, D.) in *Aq. dest.* f℥j. (3 parts, D.) and strain. Sp. gr. = 1230, D.

Inc. Common water, Solutions of Sulphates, Oxalates, Tartrates, Alkaline Phosphates, Borates, and Carbonates, Nitr. Silver, Acetates of Lead and Mercury, and Phosphate of Mercury.

Action. Uses. Acrid, Irritant, Stimulant, Deobstruent. Test to detect S′ and Sulphates.

BARYTÆ NITRAS, E.

Nitrate of Baryta. *F.* Nitrate de Baryte. *G.* Salpetersaures Baryt.

Nitrate of Baryta (Ba O, N O_5 = 131) finds a place in the E. P.

* Metallic sulphurets and metallic chlorides being considered to become hydrosulphates and hydrochlorates of oxides when they are dissolved in water.

on account of its utility as a pharmaceutic test in detecting some adulterations of officinal salts and acids.

Prep. This salt is to be prepared like the Muriate of Barytes (v. supra. Chloride of Barium) substituting pure Nit′ for the Mur′.

Prop. If the acid is strong, a congeries of crystals is formed; if dilute, the solution by evaporation affords crystals in octohedrons, or in small brilliant plates. The salt has a pungent and styptic taste. The crystals are anhydrous, permanent in the air; soluble in 10 or 12 parts of water at 60° and in 3 or 4 parts of boiling water. It decomposes by heat, and detonates feebly with combustible bodies.

SOLUTIO BARYTÆ NITRATIS, E. Solution of Nitrate of Baryta.

Prep. Dissolve *Nitrate Baryta* gr. xl. in *Aq. dest.* 800 grs. and keep in well-closed bottles.

Like the solution of the Chloride, gives a white precipitate with S′ and the Sulphates insol. in Nit′.

CALCIUM.

Lime, in its caustic state, was early known, being employed for making building mortar by the Egyptians, Hindoos, &c. Davy proved that it is an Oxide of a metal, which has been called Calcium, from Calx. Calcium (Ca=20) is white, brilliant, decomposes water, and, slightly heated, burns in the air, being converted into the Oxide of Calcium, or Lime.

CALX, L. E. CALX RECENS USTA, L. D.

Oxide of Calcium. Quicklime. Caustic or pure Lime. *F.* Chaux. *G.* Kalk.

Prop. Lime (Ca O=28), in its pure form, is a greyish-white earthy-looking mass, moderately hard, brittle; Sp. Gr. 2·3—3·08; having an acrid alkaline taste; corroding animal substances. When fresh burnt, it absorbs both moisture and Carb′ from the air; it will abstract water from most bodies, and is hence often employed as a drying substance. Comp. Ca 71·42+O 28·58=100.

Fresh burnt and slaked Lime, though easily procured, is seldom pure enough for medical use.

Prep. L. E. Break *Chalk* ℔j. (White Marble, E.) into small pieces and burn it in a very strong fire for 1 hour, L. (in a covered crucible, at a full red heat, for 3 hours, E.)

White Carrara Marble, Calcareous Spar, Chalk, Shells, all yield good Lime. The heat being sufficiently great, the Carb′ is expelled, and about 56 per cent. of Lime left in a caustic state, and tolerably pure; but, if shells have been employed, mixed with a little Phosphate of Lime and Oxide of Iron. It must be kept well closed up.

Tests. Water being added, Lime cracks and falls to powder; the rest as in Hydrate of Lime, L.

CALCIS HYDRAS, L. Hydrate of Lime, or Slaked Lime.

This (Ca O + H O = 37) is formed whenever water is sprinkled over caustic Lime: it is immediately absorbed with a hissing noise, the Lime splitting and crumbling into a dry, white, powdery Hydrate. It is capable of thus taking up about 31·0 of its weight of water, and at the same time disengaging so much heat, as to inflame wood. Comp. Ca O 75·68 + Aq. 24·32 = 100. It loses its water by the action of heat; but the Lime itself is very infusible, though powerful as a flux for many earths and Oxides. When heated in the Oxy-hydrogen flame, it is intensely luminous, as in the Drummond's light. Acids combine with Lime, some forming very soluble salts, as the H Cl′ the Ac′, while Ox′ and Phos′ form a nearly insoluble Oxalate and Phosphate. Its presence is readily detected in any solution by the milkiness produced by passing Carb′ gas through it, also by alkaline Carbonates and alkaline Sulphates, or by the addition of Ox′ or Oxalate of Ammonia. The Oxalate of Lime will be precipitated from a very dilute solution. S′ does not form a precipitate in a dilute solution. Salts of Lime tinge the flame of Alcohol orange. This Hydrate is soluble in water, forming Lime-water.

Tests. Soluble without effervescence in H Cl′. The solution does not precipitate with Ammonia—proving that neither Alumina nor Magnesia are present, nor Oxide of Iron, nor Silica.

Action. Uses. Used as a Masticatory in India with Betle or Pan. Caustic, Disinfectant.

LIQUOR (AQUA, E. D.) CALCIS, L. Limė Water.

Prep. Take of *Lime* (fresh burnt) ℔ß *Aq. dest.* O xij. L. (*Lime* 1 part and *Water* 20 parts, E. Slake the Lime with 1 part of hot water, and add 30 parts of cold water, D.) The Lime being slaked, is put into a bottle with about 30 times its weight of water, and then well shaken together, close the vessel and set aside, that the undissolved Lime may subside, pour off the clear liquor when it is required, and it may be replaced with fresh water, agitating briskly as before, that a fresh supply may be ready.

A simple solution of Lime in water; requiring to be kept in stoppered bottles, which are kept full, as it attracts C′ from the air, which by uniting with the Lime, forms a thin film of Carb. Lime at the surface; this afterwards precipitates in the form of white layers, but its place will be supplied by the undissolved Lime left in the bottle. Dalton ascertained that, contrary to the analogy of most bodies, *cold* water dissolved more of Lime than hot water: that is, water at 60°, $\frac{1}{778}$, and at 212°, only $\frac{1}{1270}$. Mr. R. Phillips ascertained that water near the freezing point takes up about $\frac{1}{7}$ more than water at 60°, and nearly double that of boiling water.

A pint of water at		32°	dissolves	13·25	grains of Lime.
„	„	60°	„	11·6	„
„	„	212°	„	6·7	„

Tests. Lime-water is clear and transparent, without odour, but having a disagreeable alkaline taste; changes vegetable blues to green, and forms an imperfect soap with oils; when evaporated under the vacuum of an air-pump, imperfect six-sided crystals may be obtained of the Hydrate of Lime.

Inc. Acids and Acidulous Salts, Alkaline Carbonates, Ammoniacal and Metallic Salts, Borates, and astringent Vegetable Infusions.

Action. Uses. Astringent, Antacid, Antilithic. Resolvent in glandular affections.

Pharm. Prep. Potassa cum Calce. Aqua Calcis Comp. Lime-water for Black and Yellow wash.

D. f℥ij.—f℥viij. three or four times a day.

LINIMENTUM CALCIS, E. D.

Prep. Agitate briskly together equal parts of *Lime Water* and *Linseed Oil.* (Olive Oil, D.)

Both the oils are composed of Oleic and Margaric acids and Glycerine: when mixed with Lime-water, an Oleo-Margarate of Lime (Calcareous Soap) is formed. It has long been employed as an application to burns and scalds, and employed for this purpose at the Carron Works,—hence often called Carron oil. Turpentine is sometimes added with advantage. Dr. Christison describes it as a Lime Soap with an excess of Linseed Oil; for when allowed to rest, the mixture separates into a white soap and a supernatant clear oil.

CALCIS CARBONAS.

F. Carbonate de Chaux. Craie. *G.* Kohlensaurer Kalk.

The Carbonate of Lime ($Ca\,O, C\,O_2=50$) is one of the most widely diffused of minerals, and must have been one of those most anciently employed in the arts, and likewise in medicine. Found in a great variety of forms; forming mountain masses, either crystalline, as Marble and Calc Spar, or compact, as Limestone; in both stratified and unstratified rocks, and as Chalk in great beds, as the newest of the secondary strata. It is also extensively diffused in particles throughout the soil, which by segregation become united into roundish or botryoidal masses, or it may become dissolved by water, and is found in most springs, from which it is again deposited in a stalactical form. It is found in the ashes of most plants, probably from the vegetable acids with which it was combined being converted into the Carbonic' during the process of incineration. Carbonate of Lime forms a constituent of the bones of Vertebrata, and a large part of the shells of testaceous Mollusca, of Crustacea, and of Corals. Hence Oyster-shells, Crab's-claws, Crab's-eyes, as they are called, and Corals, have all been employed in medicine, as formerly the *lapis judaicus*, which is the spine of a fossil echinus. All consist of pure Carbonate of Lime, with some animal matter intimately intermixed, sometimes a little Phosphate of Lime. Carb. Lime is found crystallized in a variety of forms, but the primitive form is an obtuse rhomboid, and that of Arragonite a rectangular prism. Many varieties are transparent, and remarkable for doubly refracting the rays of light, especially Iceland

Fig. 19. *Fig.* 20.

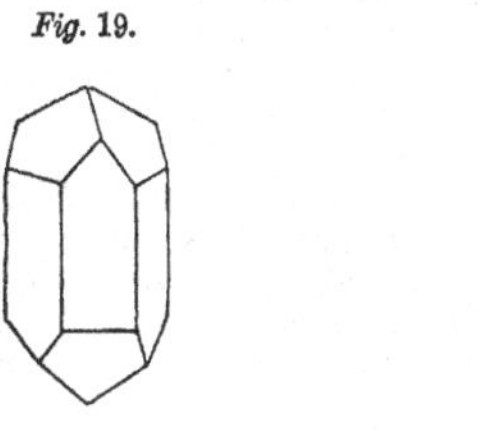

Spar. It has been artificially crystallized by Prof. Daniell in acute rhombic crystals, which contained 5 per cent. of water. It is very sparingly soluble, 1 part requiring 1600 of water. But if an excess of Carb′ be present, it is readily dissolved, and is hence found in many mineral waters, from which it is again precipitated on the escape of the C′ The solution reddens Litmus-paper, but changes the yellow colour of Turmeric-paper to brown. (*p.*) Heated in the air, Carb. Lime loses 44 per cent. of C′, Lime being left; but if heated in close and strong vessels, no change is produced, and, on cooling, artificial Marble is produced, as by Sir J. Hall. Bucholz fused it even without compression, in parts to which the access of air was prevented. Carb. Lime is readily decomposed by the acids with strong effervescence, forming soluble salts with N′ and H Cl′, and insoluble ones with S′. It may be formed artificially by adding Carb′ to Lime-water, or by decomposing any soluble salt of Lime (or the Chlor. Calcium) with the Carbonates of any of the alkalis. Comp. $Ca\,O\ 56 + C\,O_2\ 44 = 100$.

Calcis Carbonas dura. Marmor, (Album, D.) L. E. Marble.

Marble is officinal for yielding Carb′ gas (p. 52) by the action of stronger acids. It is also called Saccharine Limestone; has a glimmering appearance, from the lamellæ of its minute crystals intersecting each other in every direction. Pure white Marble is intended; that of Carrara, commonly called Statuary Marble, is the purest variety.

Tests. Marble should dissolve with effervescence in dilute H Cl′ without residue, proving the absence of Silica and some other impurities. In this solution, especially if neutralized, Ammonia ought (even after boiling) to cause no precipitate, if neither Magnesia (a constituent of some Limestones), nor Alumina, nor Oxide of Iron (the most common impurities), are present. Neither ought it to be decomposed by a solution of Sulphate of Lime in water; if any precipitate appears, it must be Sulphate of Baryta or of Strontian.

CALCIS CARBONAS FRIABILIS. CRETA, L. E. D. Chalk.

Chalk is well known as an extensive secondary formation, of a dull white earthy appearance; tasteless, but adhering to the tongue; usually friable, sometimes hard; Sp. Gr. 2·3; but either variety may be employed, though the softer is usually preferred for medical use. Its chemical characters are the same as those of Marble. In the arts it is commonly known by the name of *Whiting*, which is Chalk ground in a mill, and the grosser impurities separated by sinking in water, while the pure Chalk, being suspended, is allowed to settle, and made into small loaves. For medical use, it is similarly but more carefully prepared by the process of lævigation, and in drying is made up into small conical masses.

CRETA PRÆPARATA, L. E. D. Prepared Chalk.

Prep. Add to *Chalk* ℔j. a little water, and rub to a fine powder. Throw this into a large vessel of water, then agitate, and after a short period pour off the supernatant water still turbid, into another vessel, and set it aside for the suspended chalk to subside. Lastly, the water being poured off, dry this powder and preserve for use. The directions of E. and D. are essentially the same. Oyster shells, first freed from impurities and washed with boiling water, may be prepared in a similar manner, forming the TESTÆ PRÆPARATÆ, L. *q. v.*

CALCIS CARBONAS PRÆCIPITATUM, D. Precip. Carb. of Lime.

Prep. This precipitated Carbonate is prepared by the addition of a solution of *Carbonate of Soda* to a solution of *Muriate of Lime.* It has the advantage of minute subdivision if prepared with cold solutions.

Tests. (*v.* supra.) E. P. "A solution of 25 grs. in f℥x. of Pyroligneous acid (according to Mr. Phillips, this is capable of dissolving 4 times the above quantity of Chalk), when neutralized by Carbonate of Soda, and precipitated by 32 grs. of Oxalate Ammonia, continues precipitable after filtration by more of the test." Dr. Christison states that a considerable excess of acid is useful. A little Lime is left unprecipitated if there be 90 per cent. of pure Carbonate of Lime in the Chalk; and this slight excess of Lime is indicated by adding more Oxalate of Ammonia to the filtered liquor.

Inc. Acids and Acidulous Salts, as other Carbonates.

Action. Uses. Antacid, Absorbent, Desiccant; from allaying irritation, apparently Astringent. If long used, care must be taken that it does not accumulate in the intestines.

D. gr. x.—℈j.; but usually given in some of the following preparations.

Pharm. Prep. Hydrargyrum cum Creta. Prepared Oyster Shells also contain Carbonate of Lime. *v.* Testæ præparatæ.

MISTURA CRETÆ, L. E. D. Chalk Mixture.

Prep. Take *Prepared Chalk* ℥ß. (℥x. E.), *Sugar* ℨiij. (pure ℨv. E. ℨiij. D.)

Mixture of Acacia f℥iß. [(Mucilage f℥iij. and triturate together, E.) (Mucilage of Gum Arabic ℥j. D.)] *Cinnamon water* f℥xviij. L. [(Aq. ℔j. by measure, D.) (add gradually Aq. Oij. and Spirit of Cinnamon ℥ij. E.)] Mix.

Action. Uses. Antacid, Demulcent. Much employed in Diarrhœas arising from acidity.

D. f℥ß—f℥ij. every three or four hours.

PULVIS CRETÆ COMPOSITUS, L. E. D. Compound Powder of Chalk.

Prep. L. E. D. Reduce separately to fine powder *Prepared Chalk* ℔ß. (℥iv. E.) *Cinnamon* (bark, D.) ℥iv. (in fine powder ℥iß. E.) *Tormentil root* and *Gum Arabic* āā ℥iij. L. D. (Nutmeg in fine powder ʒj. E.) and *Long Pepper* ℥ß. L. D. Mix well.

Action. Uses. Antacid, Stimulant, and Astringent. In Diarrhœas of low states of the Constitution.

D. gr. v.—℈j.

TROCHISCUS CRETÆ, E. Chalk Lozenge.

Prep. Reduce to powder *Prepared Chalk* ℥iv. *Gum Arabic* ℥j. *Nutmeg* ʒj. *Pure Sugar* ℥vi. beat with water into a proper mass for making lozenges.

Action. Uses. Antacid. Useful in acidity of the primæ viæ.

PULVIS CRETÆ COMPOSITUS CUM OPIO, L. D. PULVIS CRETÆ OPIATUS, E.

1 grain of Opium in 40 grains of the Compound Chalk Powder, L. *v.* OPIUM.

CONFECTIO AROMATICA, L. D. Aromatic Confection.

Prep. Rub into a very fine powder *Cinnamon* and *Nutmegs* āā ℥ij. *Cloves* ℥j. *Cardamoms* ℥ß. *Saffron* (dried, D.) ℥ij. *Prepared Chalk* ℥xvi. *Sugar* (pure, D.) ℔ij. (add gradually Aq. ℔j. and rub into a pulp, D.) Keep in a close vessel, and whenever the confection is to be used add water gradually and mix until they are thoroughly incorporated. L.

Action. Uses. Antacid and Cordial. Useful in Diarrhœas, and an excellent addition to Rhubarb and Magnesia, and such powders, for children.

D. gr. v. or gr. x.—ʒj.

CALX CHLORINATA, L. E.

Chlorinated Lime. Hypochlorite of Lime. Chloride of Lime. Chlorite of Lime. *Oxymuriate of Lime.* Bleaching Powder. *F.* Chlorure de Chaux. *G.* Chlor Kalk.

This substance was first prepared by Messrs. Tennant and Mackintosh in 1798. The exact nature of the compound not having been satisfactorily determined, the present name has been assigned it in the L. and E. P., though it is very commonly called Chloride of Lime, or Bleaching Powder.

Prop. Chlorinated Lime is a dry pulverulent substance, of a greyish colour, with a hot, penetrating, bitter taste, a weak odour of Chlorine, more perceptible when the powder is shaken. When well prepared, it is very soluble in water; but commonly a considerable portion (of Lime) remains undissolved. In the atmosphere, the Carb', or any of the other acids, sets free the Chlorine, and a Carbonate of Lime is formed, with some Chloride of Calcium, which causes deliquescence. Heat also expels the Chlorine, and then Oxygen. The strongest solution has a density of 1040. (Ure.) It is of a pale yellow colour, has a slight smell of Chlorine, and holds in solution the Chlor. Lime, with a little caustic Lime, and any Chlor. Calcium which may have been formed. This solution is remarkable for its bleaching and disinfecting properties. The addition of a little acid increases its activity for such purposes.

The true composition of Chlorinated Lime is still unsettled. Dr. Ure considered it a variable preparation, and not combined in equivalent quantities, as he found the quantity of Chlorine absorbed by Lime to vary, especially according to the quantity of water which was previously added to slaked Lime. Dr. Thomson states that in Glasgow it is now so manufactured as to be almost entirely soluble in water, and that it consists of Cl Ca O$=$64. Berzelius considers it to be a Chlorite; and Balard, that a portion of the Lime gives its Oxygen to the Chlorine, and that a Hypochlorite of Lime and some Chloride of Calcium are formed. Mr. Phillips states that the Chloride of Lime appears, from the statements of Brande and Grouvelle, and also from his own experiments, to consist of Cl 36$+$2 Ca O, H O (37×2) 74 $=$110. When water is added to this, the Chloride of Lime dissolves, leaving nearly all the Lime insoluble: it is therefore probably composed of 1 Eq. Bihydrated Chloride of Lime, 18$+$36$+$28$=$82, 1 Eq. Lime, 28$=$110. If the views of Berzelius and Balard be followed, and their proportions adopted, then the composition, according to Dr. Pereira, will be 1 Eq. Trishypochlorite of Lime 128$+$1 Eq. Chloride of Calcium 56$+$4 Eq. Aq. 36$=$220; or, per cent. Cl Ca O 58·18$+$Cl Ca 25·45$+$Aq. 16·36$=$99·99. Dr. Ure, however, states that the common Chloride of Lime of commerce consists of 45·4 Lime, 40·31 Chlorine, 14·28 water$=$99·99.

Prep. Pass as much *Chlorine*, as may be sufficient to saturate *Lime* ℔j. spread out in a proper vessel. Chlorine is very readily evolved from H Cl' gently heated with *Binoxide Manganese*, v. Chlorine, p. 44.

Chloride of Lime is, however, easily obtained, being largely prepared as a Bleaching Powder. Here slaked Lime is spread out on a pile of wooden trays in a chamber built of sandstone, the joints being secured by a cement of Pitch, Resin, and dry Gypsum, into which the Chlorine is transmitted until the Lime is saturated. A larger quantity of Chlorine is absorbed when about 15 per cent. of water is previously added to the Lime, and the Lime is occasionally raked up. Here the Chlorine combines with the Hydrated Lime, or some of the

Lime gives its Oxygen to the Chlorine, some Hypochlorous Acid is produced, and Hypochlorite of Lime and Chloride of Calcium formed.

Tests. Dissolves in dilute H Cl′, emitting Chlorine. L. Greyish-white, dry: 50 grs. are nearly all soluble in Aq. f℥ij., forming a solution of the density 1027; of this, 100 measures, treated with an excess of Ox′, give off much Chlorine, and if then boiled and allowed to rest 24 hours, yield a precipitate which occupies 19 measures of the liquid. E. The precipitate is Oxalate of Lime, and the E. C. desire to have the goodness of Chloride of Lime ascertained by the amount of this precipitate, as well as by the density of the solution. But this method does not detect the amount of Chloride of Calcium, which is the most common adulteration. Dr. C. remarks that probably the simplest criterion of quality is the amount of Chlorine gas evolved by a strong acid, as originally proposed by Dr. Ure.

Action. Uses. Irritant, Stimulant, Disinfectant, Antiseptic; in solution as a lotion and gargle.

D. Internally gr. j.—gr. v. Used in Tooth-powders, Lozenges. Ointment, ʒj. with ℥j. of Lard.

Liquor Calcis Chlorinatæ. Of variable strength: ʒj.—ʒiv. in Aq. Oj. In Scabies, ℥iij. to Aq. Oj.

Calcii Chloridum, L. Calcis Murias, D. (Crystallizatum), E.

Chloride of Calcium. Hydrochlorate or Muriate of Lime. *F.* Chlorure de Calcium. Hydrochlorate de Chaux. *G.* Salzsaurer Kalk.

This salt, commonly called Muriate of Lime, was known, according to Dulk, as quoted by Dr. Pereira, in the 15th century, and called *Sal Ammoniacum fixum*, being obtained by the decomposition of Sal Ammoniac by Lime. It is found in nature in Sea-water, and in many springs and mineral waters, sometimes associated with Nitrate of Potash, but usually with Chloride of Sodium and Chloride of Magnesium. It may be readily obtained as a residuum in several of the preparations of Ammonia, as the Liquor, Spirit, and Carb., or in obtaining C′ by the action of H Cl′ on Marble.

Prop. Chloride of Calcium L. (Ca Cl=56) is known in two forms: the Anhydrous or the Murias Calcis of the D. P., which is hard, greyish coloured, and semi-translucent. Being without water, it contains nearly twice as much Chlor. Calcium as the crystallized salt. It may be fused at a red heat, and becomes phosphorescent. It is very deliquescent, as is also the crystallized salt, passing readily into the liquid state, forming what used to be called *Oleum Calcis.* It is frequently employed as a drying substance; also for attracting moisture to substances, with which it may be mixed, as the soil. It is very soluble in rectified Spirit, also in water, and is then supposed to dissolve either as Chloride of Calcium, or as Hydrochlorate of Lime (Oxide of Calcium).

Chloride of Calcium (CaCl+6Aq.=110), or Calcis Murias Crystallizatum, E., is the hydrated crystallized salt, also called Hydrochlorate of Lime. It is colourless and without smell, of an acrid and bitter taste; in striated hexagonal prisms terminated by very acute points. Water even at 32° dissolves more than its own weight, and at 60° above three or four times its weight of this salt. When heated, these crystals undergo watery fusion. Dissolved in water, they produce great cold; and hence are frequently employed as an ingredient in cold or freezing mixtures. The Chloride is used for concentrating Alcohol, from its great affinity for water.

Prep. L. Add *Chalk* ℥v. gradually to H Cl′ O ß. previously mixed with *Aq. dest.* O ß., to saturation or until the effervescence ceases. (The D. P. employs the liquor which remains after the distillation of Caustic Ammonia.) Strain, evaporate the solution until the salt is dried. Put this into a crucible, and fuse on the fire, and pour it upon a clean flat stone; when cold, break it into pieces. The *anhydrous* Chloride is produced. (But the E. P. employ Marble and diluted Mur′ to saturation, and place the filtered fluid, after having been evaporated to one-half, in a cold place for crystals to form. The *hydrated* Chloride is produced.) Both must be preserved in well closed vessels.

Here the Carb. Lime (Oxide of Calcium) is decomposed, the C′ being expelled in the state of gas; the Chlorine of the H Cl′ combines with the Calcium, and forms Chloride of Calcium; and the Hydrogen of the Acid with the Oxygen of the Lime forms 1 Eq. of water. In the subsequent part of the London process both the water used and that formed are expelled during the fusion.

Tests. The presence of Lime and of Chlorine will be revealed by their respective tests. Chloride of Calcium, L.P. should be free from colour; slightly translucent; hard and friable; totally soluble in water: the solution gives no precipitate on the addition of Ammonia (showing the absence of Magnesia), or Chlor. Barium (of Sulphates), nor, when diluted with much water, with Ferro-cyanide of Potassium (showing that it is not contaminated with Iron). E. P. "The crystallized salt is very deliquescent. A solution of 76 grs. in f℥j. Aq. precipitated by 49 grs. Oxal. Ammonia, remains precipitable by more of the test. If it contain any alkaline salt, impurity will be indicated by this method."

Calcii Chloridi Liquor, L. Calcis Muriatis (Aqua, D.) Solutio, E.

Prep. Dissolve *Chlor. Calcium* ℥iv. (3 parts, D.) in *Aq. dest.* f℥xij. (7 parts, D.) The E. P. dissolves *Muriate of Lime* (crystallized) ℥viij. in water f℥xij.

Action. Uses. Stimulant of the Lymphatics. Used for preparing Muriate of Morphia, and as a test for Oxalic Acid.

D. ♏xx.—f℥j.

Inc. Decomposed by S′ and by Sulphates and by N′; by the alkalis and their Carbonates, with the exception of pure Ammonia, which produces no change, and may therefore be prescribed with it.

Calcis Phosphas præcipitatum, D.

Precipitated Phosphate of Lime. Sub-Phosphate of Lime. Bone Phosphate of Lime. *F.* Phosphate de Chaux. *G.* Phosphorsaurer Kalk. Knochenerde.

Bone Phosphate of Lime, as its name indicates, constitutes the earthy matter of bones, teeth, and horns. Some animal excretions, as the Tartar of the teeth, the Phosphate of Lime calculus, are formed of it. It exists also in wheat, and almost all plants. It is obtained from burning bones or horns, *v.* Cornu Cervi, and is employed for obtaining Phosphorus and Phosphate of Soda. (q. v. pp. 35 and 99.)

Prop. It is a white powdery substance, insipid, and insoluble in water. At high temperatures it fuses, and is converted into an opaque enamel. Bone ashes are composed of Subphosphate with a little Carbonate of Lime. Some chemists conceive that the Bone Phosphate consists of a mixture of the two tribasic Phosphates which are analogous to those of Soda (p. 99); or, according to others, of 8 Eq. of Lime and 3 of Phosph′, or, per cent. Ca O 51·55+P′ 48·45=100.

Prep. Digest, *burnt bones* in powder 1 part, in *dil. Mur′* and *water* of each 2 parts for 12 hours. Strain the liquor. Add to it *Water of Caustic Ammonia,* q. s. to throw down the Phosphate of Lime. Wash with abundance of water and then dry it.

The Phosphate of Lime is dissolved, and any Carb. Lime decomposed by the Mur′. The Ammonia precipitates all the Phosphate of Lime, which must be carefully washed to get rid of all traces of Chlor. Calcium and of Hydrochlor. Ammonia. It has the advantage over Bone ashes and burnt Hartshorn of minute subdivision.

Action. Uses. Operation uncertain. Supposed formerly to be useful in Mollities Ossium. It is a constituent of James's Powder, and of Pulv. Antim. Comp.

D. gr. x.—ℨß.

Magnesium.

F. Magnésium. *G.* Magnium.

Magnesium (Mg=12) is a metal which has been obtained by decomposing Chloride of Magnesium by Potassium. It is of an iron-grey colour, brilliant, hard, and ductile; not acted on by water, nor by air, except at a high temperature, when it becomes oxidized, and forms Magnesia. As a Chloride, it forms a constituent of Sea-water; oxidized and combined with acids, it exists in sea-water and in numerous mineral springs, and as a Hydrate or native Magnesia in a few places. It forms a portion of Serpentine, Soapstone, Mica, Talc, and many other minerals. It exists in most plants, as in the straw of wheat; also in small quantity in the animal system, especially in the urine and in some urinary calculi. The Oxide, or Magnesia, may be obtained by burning the Carbonate, as Lime is by burning Limestone, or by adding Potash or Soda to a solution of one of its salts.

MAGNESIA, L. E. D.

Magnesia Usta, L. Calcined Magnesia. *Talc earth. F.* Magnésie. *G.* Talkerde. Bittererde.

The name Magnesia (Mg O = 20) occurs in Geber, and afterwards in alchymical works, with various meanings. The present substance was called *Magnesia alba*, and introduced as a medicine in the beginning of the 18th century. It was at first supposed to differ little from Carb. Lime. Hoffman first, and then Dr. Black (1756) clearly established the distinction between it and Lime.

Prop. Magnesia is a white and light, very finely divided, powdery substance, devoid of smell, but earthy in taste. Sp. Gr. 2·3. When moistened, it slightly greens syrup of Violets, and browns Turmeric-paper. It is hardly soluble in water, requiring 5142 parts of cold and 36000 parts of hot water. (Fyfe.) Water sprinkled on it becomes absorbed to the extent of about 18 per cent. without the evolution of heat. It is slightly soluble in Alcohol. It attracts moisture and Carb′ from the atmosphere, and becomes slowly converted into the Carbonate. It is infusible, except under the oxy-hydrogen flame, and consists per cent. of Magnesium 60+O 40 = 100.

Magnesia is remarkable for its attraction for Alumina in a humid way, so that in the analysis of some Magnesian fossils, it is found that, although Magnesia cannot be precipitated entirely from any of its salts by Ammonia, yet, if Alumina be present, its precipitation is complete. (*m.*) Acids readily unite with Magnesia, and form salts, of which those which are soluble, and especially the Sulphate, are bitter, readily distinguishing it from other earths. Caustic Potash decomposes these salts, and throws down the Magnesia, which retains about ¼ of water, forming the Hydrate, of a somewhat gelatinous consistence. The Carbonate of Potash and of Soda produce precipitates of Carb. Magnesia. The Bicarbonates give rise to no apparent decomposition, as the transparency remains unimpaired. If the Sesquicarbonate of Ammonia be added, and after this a solution of Phosphate of Soda be dropt in, a copious precipitate takes place of triple Phosphate of Magnesia and Ammonia. (*c.*) The direct addition of Phosphate of Ammonia to the solution of any salt of Magnesia will produce the same effect. Magnesia may be distinguished from Lime by Sesquicarbonate of Ammonia precipitating Lime, but not Magnesia; also by Oxalate of Ammonia, which does not precipitate Magnesia, but throws down Lime readily. Ammonia in excess throws down Magnesia, but not Lime, from neutral solutions.

Prep. L. Take *Carb. Magnesia* ℥iv.; burn it for 2 hours in a very strong fire. The E. P. directs that the heat be continued, till the powder, when suspended in water, displays no effervescence on the addition of Mur′. Both the E. and D P. direct that it be preserved in well-closed bottles.

Here, as in the case of Carb. Lime, Carb′ and water are expelled at a high temperature, to the extent of 50 or 60 per cent., and the Magnesia remains in its pure state, of which the density may be increased according as the heat is augmented.

Tests. Magnesia being prepared from the Carbonate, is apt to contain some of the impurities of the salts from which it is made, as Lime, Alumina, and Silica, and, when long kept, some of the Carbonate. H Cl′ dissolves Magnesia (50 grs. in Dil. H Cl′ f℥j. E.) without effervescence, showing the absence of Carbonate. If Silica be present, it will be left undissolved. An excess of Ammonia occasions in the solution only a scanty precipitate of Alumina, E., provided the acid, as the E. P. directs, be used in considerable excess to the Magnesia, Alumina, if present, will then be readily separated from the Muriate of Magnesia. (*c.*) No precipitate is thrown down by Bicarb. Potash (nor by Oxal. Amm. E.) added to the above solution, showing absence of Lime; and none by Chlor. Barium, showing absence of Sulph. Mag. and of Carb. Soda. Turmeric ought to be only slightly browned.

Inc. Acids, Acidulous and Metallic Salts, and Hydrochlorate of Ammonia.

Action. Uses. Antacid, Laxative. In acidity of the Stomach, when it forms soluble Magnesian Salts.

D. Antacid, gr. x.—xxx.; as a laxative, ℈j.—ʒj.; for infants, gr. ij.—gr. x.

Magnesiæ Carbonas, L. E. D.

Magnesia alba. Magnesiæ Subcarbonas. F. Carbonate de Magnésie. *G.* Kohlensaure Bittererde. Kohlensaure Talkerde.

Carbonate of Magnesia ($Mg\,O, C\,O_2=42$), at first called *Magnesia alba*, and *Comitissæ Palmæ pulvis*, was used as a medicine by the Count de Palma at Rome, whence it was also called *Pulvis albus Romanus.* It was introduced into the list of the Materia Medica by F. Hoffman. It is found in nature in some mineral waters, in some of which, however, it may exist in the form of the Bicarbonate. In an impure state it forms a constituent of Dolomitic, that is, of Magnesian Limestone, and in a comparatively pure state, a hill in the Peninsula of India, consisting of Magnesia 46, Carbonic Acid, 51, Insoluble Matter 1·5, Water 0·5, loss 1=100.

Prop. Pure Carbonate of Magnesia is sometimes found in nature in rhombohedral crystals; as usually seen, the officinal Carbonate is of a white colour, light and soft to the touch, without smell, devoid of any other than an earthy taste when properly prepared. It is unalterable in the air, and nearly insoluble in water, but more soluble in cold than in boiling water. Its solubility is much increased if C′ be present, 48 parts of water being said to be then sufficient. In fact, it is converted into the Bicarbonate of Magnesia: by spontaneous evaporation, 1 Eq. of C′ escapes, and the neutral Carbonate is deposited, being insoluble. It is decomposed by acids and by a strong heat, its Carb′ being expelled. The composition of Carbonate of Magnesia is differently viewed by different chemists. Mr. Phillips, in his last analysis, found Magnesia 40·8, Carb′ 36·0, and Water 23·2 = 100; and he considers it as a compound of 1 Eq. Bihydrated Magnesia 38 + 4 Eq. Hydrated Carb. Magnesia 204 = 242.

Prep. L. E. D. Dissolve *Sulph. Magnesia* ℔iv. (25 parts, D.) *Carb. Soda* ℔iv. ℥viij. (*Carb. of Potash* 14 parts, D.) each separately in *Aq. dest.* Cij. (400 parts, D.) and strain. Then mix and boil the liquors, stirring constantly with a spatula for a quarter of an hour; lastly, the liquor being poured off, wash the precipitated powder (collected on a filter of linen or calico, E.) with boiling distilled water, and dry it.

Here the Sulphate of Magnesia and the Carbonate of Soda (or of Potash) mutually decompose each other; the Sul′ uniting with the Soda, forms a Sulphate of Soda, which remains in solution, while the Carb′ unites with the Magnesia. The salt formed, being insoluble, is precipitated as a Hydrated Carbonate of Magnesia, but, in consequence of a portion of the Carb′ escaping, it is not strictly neutral. Carb. Potash was used in all the Pharmacopœias, as it now is in the Dublin; "but it was difficult to separate the last portions of the Sulphate of Potash from the precipitate, and the Carb. of Potash usually contains Silica, which precipitates with the Magnesia." Professor Graham states that Carb. Soda is not so suitable as Carb. Potash for precipitating Magnesia, "as a portion of it is apt to go down in combination with the Magnesian Carbonate; but it may be used, provided the quantity applied be less than is required to decompose the whole Magnesian salt in solution." Carbonate of Magnesia is sometimes pressed, when in the moist state, into the form of cubes. Considerable differences are observed in the density of Carbonate of Magnesia, according to the mode of preparation. *Dense* or *heavy* Magnesia is preferred by some, chiefly in England, and Light Magnesia by others, especially in France. Several explanations have been given of the mode of preparation, but the most distinct by Dr. Pereira, as practised in one of the most highly esteemed laboratories of this capital, where they prepare—

Heavy Magnesia by adding a cold saturated sol. of Carb. of Soda to a hot saturated sol. of Sulphate of Magnesia.

Light Magnesia, by using both solutions much diluted.

Gritty Magnesia, by mixing both of the solutions at a boiling temperature, both being concentrated.

Tests. The Carbonate, like pure Magnesia, may contain alkaline Carbonates, or Sulphate of Soda, sometimes Gypsum, Lime, and Alumina. "The water in which it is boiled should not alter the colour of Turmeric" showing the absence of any alkaline Carbonate. Chloride of Barium or Nitrate of Silver, added to the water, does not precipitate anything, the first indicating the absence both of Sulphates and of Carbonate of Soda, and the second, if insoluble in N′, of any Chloride. "When dissolved in an excess of Muriatic acid, an excess of Ammonia occasions only a scanty precipitate of Alumina, and the filtered fluid is not precipitated by Oxalate of Ammonia," E. or Bicarbonate of Potash, showing the absence of any Calcareous salt. 100 parts dissolved in dilute Sul′ lose 36·6 parts in weight of Carb′. L.

Inc. Acids and Acidulous and Metallic Salts, Hydrochlorate of Ammonia, and Lime-water.

Action. Uses. Antacid, Laxative. Very similar to Magnesia, but differs in Carb′ gas being extricated when it meets with acids in the stomach. Sometimes given in effervescence.

D. gr. v.—℈j. as an Antacid; gr. xv.—ʒj. as a Laxative, with water, milk, &c.; 14 grs.=℈j. Citric acid in effervescence.

Pharm. Prep. Hydrargyrum cum Magnesia. Pulv. Rhei Comp. Mist. Camphoræ cum Magnesia.

TROCHISCI MAGNESIÆ, E. Magnesia Lozenges.

Prep. Mix together *Carbonate of Magnesia* ℥vj. *Pure Sugar* ℥iij. *Nutmeg* ℈j., in powder, with *Tragacanth* mucilage to a mass fit for lozenges.

MAGNESIÆ BICARBONAS.

Aqua Magnesiæ Bicarbonatis. Aerated Magnesia Water. Soluble Magnesia. F. Bicarbonate de Magnesie. *G.* Zweifach Kohlensaure Bittererde.

This is found in some mineral springs in France, &c. Carbonate of Magnesia becomes soluble when a current of Carbonic acid gas is passed into a mixture of Carbonate of Magnesia and water. A preparation made in Paris in 1821 contained 6 times as much as an English preparation, which contained 36 grains in each bottle. A. Mayler mentions that Mr. Lawrence had been able to dissolve as much as 15 grains of neutral Carbonate of Magnesia in an ounce of water. The French apothecaries prepare an *Eau Magnesienne gazeuse*, which contains ʒj. of Magnesia in a bottle of 22 ounces: a Bicarb. Magnesia is, in fact, formed, with a large excess of Carb′. The second kind, called *Eau Magnesienne saturée*, is not effervescing, and contains ½ an ounce of Magnesia in water Oj., or about 9 grs. in ℥j.

Dr. Christison says a bottle which holds about ℥viij. may contain 72 grs. of Carb., and ought to hold at least a ℈ in solution. The solution prepared by Mr. Dinneford is said to contain from 17 to 19 grs. in each f℥. That examined by Dr. C. yielded what was equivalent to 8·96 of commercial Carb. in a f℥. The fluid Magnesia of Sir J. Murray, analysed by Profs. Daniell, Kane, and Davy, yielded in each f℥ 13 grs. of pure Carb. Magnesia.

A substitute, as suggested by Dr. P., may be prepared by pouring the ordinary Soda-water, that is, Carbonic acid water, over the common Carb. Magnesia contained in a tumbler; or a mixture of crystallized Sulph. Magnesia and crystallized Carb. Soda, in powder, and in atomic proportions (viz., 123 parts of the former to 144 parts of the latter salt) may be substituted for the Carbonate of Magnesia. This is something similar to the double Carbonate of Magnesia and Soda, sometimes sold as *Soluble Magnesia.*

Action. Uses. Much the same as Carb. Magnesia, but a more agreeable form for exhibition in Dyspepsia and Lithic acid Diathesis.

D. f℥iij.—f℥viij.

Magnesiæ Sulphas, L. E. D.

Sulphate of Magnesia. Epsom Salts. *Vitriolated Magnesia.* *F.* Sulphate de Magnesie. *G.* Schwefelsaure Bittererde.

This salt was first discovered in 1675 by Dr. Grew, in a spring at Epsom. It is found in many countries effloresced on the soil and on rocks which contain a Sulphate or Sulphuret. It is called *hair salt* and *bitter salt.* It exists in many mineral springs and in sea-water in the proportion of 15·5 grains in a pint. Its true nature was fully explained by Dr. Black in 1755.

Fig. 21.

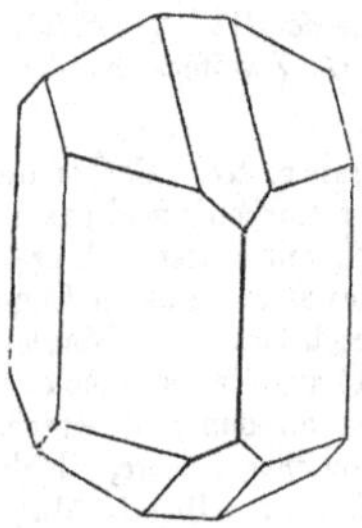

Prop. Sulphate of Magnesia ($Mg\,O, S\,O_3 + 7\,Aq. = 123$) is commonly prepared in acicular crystals, but it may be crystallized in quadrangular or hexangular prisms accuminated by two to six planes, the primary form being a right prism with a rhombic base. The salt is white or colourless, transparent, and sparkling, of a saline nauseously bitter taste. Unalterable or slightly efflorescent, according to the dryness of the air. Sometimes, but only when impure, deliquescent. Insoluble in Alcohol, soluble in its own weight of water at 60°, and in less than ¾ at 212°. Exposed to heat, the crystals melt in their water of crystallization, of which 6 Eq. are dissipated; the salt is then fused into an enamel without decomposition. If moistened when in the anhydrous state, water is reabsorbed with increase of temperature. Sulph. Magnesia is decomposed by Potash, Soda, and their Carbonates, the bases producing a precipitate of Magnesia, and their Carbs. one of the Carb. Magnesia. The Bicarbs. Potash and Soda and the Sesquicarb. Ammonia do not produce precipitates, because the Bicarb. Magnesia, which is produced, is soluble, and does not impair the transparency of the solution, unless a portion of C′ is expelled by heat. Lime, Baryta, and their soluble salts decompose it, producing a precipitate of Sulph. Lime or of Baryta. Ammonia decomposes it readily if aided by heat, otherwise partially, forming a triple Sulphate. If Sesquicarb. Ammonia be added to its solution, and then Phosph. Soda, a precipitate is obtained of Ammoniaco-Magnesian Phosphate. Comp. Mg O 32·5, S′ 32·5, Aq. 51·2=100.

Prep. The *bittern* of sea-water, after the crystallization of Common Salt, contains Sulph. Magnesia and Chlor. Magnesium. By simple evaporation the Sulph. Magnesia may be separated by crystallization. Sometimes Sul′ is added to convert the Chloride into a further quantity of Sulph. Magnesia.

At Lymington, in Hampshire, two kinds of Salt are manufactured. The Author has to thank Mr. Dyer, a late pupil at King's College, for specimens. The first, called *single*, obtained by the cooling down of a concentrated solution in wooden troughs, is moist, and contains a considerable quantity of Chloride. When redissolved and recrystallized, the second, which is called *double* Epsom Salts, is obtained; this is pure, and permanent in ordinary states of the atmosphere. Sulph. Magnesia is also prepared from Dolomitic Limestone, that is, Carbonates of Lime and Magnesia mixed together in various proportions. One method is to heat this Magnesian Limestone and to decompose it with diluted Sul. Sulphate of Lime, which is insoluble, is formed, as well as Sulph. Magnesia, which, being soluble in water, is easily separated and purified by crystallization. Or the mineral may be calcined, when the Carb′ being expelled, the caustic Lime and Magnesia are first hydrated by being moistened with water, and then the Lime converted into Chloride of Calcium, by adding only sufficient H Cl′ to effect this object. The Chloride being readily soluble in water, is by its means easily separated from the Magnesia, which is converted into the Sulphate by the addition of Sul′ or of Sulphate of Iron; or the Hydrated Lime and Magnesia may be boiled with bittern. Chlor. Calcium is formed, and remains in solution, while the Magnesia of the bittern is separated and obtained tolerably pure with the Magnesia of the Dolomitic limestone, and may as before be converted into Sulphate of Magnesia.

Tests. Apt to contain as impurities Chlor. Magnesium, Sulph. Soda, or a little Iron; but that commonly sold is sufficiently pure for medical purposes. Chlor. Magnesium may be suspected when the salt is moist. L. P. "Very readily dissolved by water. Sul′ dropped into the solution does not expel any H Cl′," showing there is little if any Chloride present, which will also be shown by the absence of a precipitate with Nitr. Silver. "100 grs. dissolved in water, and mixed with a boiling solution of Carb. Soda, yield 34 grs. of Carb. Magnesia when dried." If this quantity be obtained, the salt is unmixed with Sulph. Soda. The E. P. shows that the full proportion of Magnesia is present by another method. "10 grs. dissolved in f℥j. of water, and treated with a solution of Sesquicarb. Ammonia, are not entirely precipitated by 280 minims of solution of Phosph. Soda." Here a little Magnesia is left in solution if the salt be pure. The solution of

Phosph. Soda is 1 part in 20 of water: "Of it 280 minims are sufficient to throw down 97 per cent. of Magnesia in a pure Sulphate," *c.* Sulph. Soda used sometimes to be mixed with this salt, when it was dearer, and was made to resemble it by being rapidly crystallized with the assistance of agitation. A minute quantity of Iron is sometimes present, giving its solution a reddish tint. *v.* Tests for Iron.

Inc. Potash, Soda, and their Carbs., Lime-water, Chlorides of Calcium and of Barium, Acetate of Lead.

Action. Uses. Cathartic, Diuretic. A common constituent of a Black Dose.

D. ʒij.—℥j. or ℥ij. In Enemata, ℥j.—℥ij. in some demulcent mixture.

Pharm. Prep. Pulv. Salinus Comp. p. 80. Enema Cath. D.

CHLORIDE OF MAGNESIUM, more commonly called Muriate of Magnesia, is found in a few saline springs, and in the waters of the ocean, about 23 grains in a pint.

ALUMINUM AND ALUMINA.

The metal Aluminum or Aluminium (Al=14) was discovered by Sir H. Davy, but carefully examined by Wöhler in 1828. It is the base of its only known Oxide, ALUMINA (Al O=22), considered by some to be a Sesquioxide ($Al_2\,O_3$). In its impure state, and combined with Silica, it is abundantly diffused, being the essential constituent of all clays and likewise of many rocks. It exists in the purest form in the Sapphire, &c., less pure in Corundum and Emery, and in many minerals. It may be obtained by treating solution of Alum with an excess of Ammonia, when a copious precipitate of white gelatinous Hydrate of Alumina falls down.

Alumina is devoid of smell or taste, but adheres to the tongue; is very infusible, has a great affinity for water, attracting it from the atmosphere to the extent of $\frac{1}{3}$ of its own weight. When mixed with water, it is distinguished by its plasticity; hence, in its impure state, it has from the earliest times been employed in pottery. It has also a strong affinity for various organic substances, and, among them, for different colouring matters; salts, therefore, which contain it, have been long employed in dyeing and in calico-printing. Alumina in the state of Hydrate is soluble in caustic Potash or Soda, and likewise in dilute acids. It may be distinguished by the formation of octohedral crystals of Alum, on evaporating its solution in Dil. Sul', to which some Sulphate of Potash has been added.

ALUMEN. SULPHAS ALUMINÆ ET POTASSÆ, L. E. D.

Alum. *Argilla Vitriolata. Sulphas Aluminaris. F.* Alun. *G.* Alaun.

The name Alumen of the Romans (Pliny, xxxv. c. 15) and στυπτηρία of the Greeks (Diosc. v. c. 122) was no doubt applied to several salts of the nature of vitriols, and among them to the natural Sulphate of Iron. The Arabs also understood it as a generic term, as they include a variety of salts under the name of *Shib.* Alum, however, was probably not unknown, as Pliny (xxxv. c. 15. § 52) says, "quoniam inficiendis claro colore lanis, candidum liquidumque utilissimum est." The Egyptians and Hindoos have from very early ages been acquainted with the arts of dyeing and of calico-printing. The Hindoos are universally acquainted with the properties of Alum, and employ it for clarifying muddy water, as well as in both the above chemical arts. It may be obtained in every bazaar, and is manufactured in Cutch. The first Alum-works known to Europeans were established at Roccha, formerly called Edessa, in Syria, (whence the commercial name of Roch Alum,) then near Smyrna, &c., whence the Genoese, &c. supplied Europe. About the middle of the 15th century they were established in Italy, afterwards in Germany, Spain, and at Whitby in England in the reign of Elizabeth. (Aikin, *Dict.* i. p. 43.)

Prop. The Sulphate of Alumina and Potash ($KO, SO_3 + Al_2O_3, 3SO_3 + 24$ Aq. *Berz.*), when pure, is without odour, colourless and transparent, of a sweetish, acidulous, and powerfully astringent taste. It reddens Litmus and other vegetable colours, and English Alum strikes a green with Syrup of Violets. (*p.*) When perfectly crystallized, it is seen in the form of regular octohedrons, but often only as four-sided pyramids, or in large seemingly irregular masses; while what in commerce is called Roch Alum is in small crystalline fragments, with less transparency, and of a reddish hue. Sp. Gr. 1·71. The large masses, when immersed in water for a few days, display on their surfaces octohedral, triangular, and rectangular forms, as may be seen in the specimens in the museum of King's College, originally submitted to experiment by Professor Daniell. Alum is soluble in about 18 parts of water at 60°, but in about $\frac{3}{4}$ its own weight of boiling water. In a dry atmosphere, its crystals are slightly efflorescent; at a moderate temperature (as 92°) it melts in its own water of crystallization, boils up; and if the heat be continued, the water to the extent of 45 per cent. being evaporated, a light white spongy powder, or Burnt Alum, is left. By a stronger heat, the acid is partly expelled and partly decomposed, and the remainder, consisting of Alumina (with some Sulphate of Potash) is insoluble in water. It is also decomposed by the action of carbonaceous matter at a high temperature, forming the Pyrophorus discovered by Homburg and Lemery. Alum is decomposed by the alkalis, the alkaline earths, and by their Carbonates, which combine with its acid and precipitate its Alumina. This is soluble in an excess of the alkalis. Alum is composed of 1 Eq. of Sulphate of Alumina and 1 Eq. Sulphate of Potash, with 24 Eq. of water; or per cent. Sulph. Alum. 35·73, Sulph. Potash 18·07, Aq. 46·20 = 100.

Fig. 22.

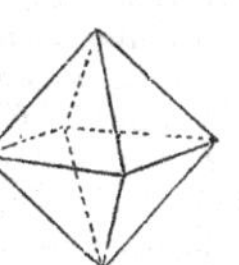

The above, which is the common kind of Alum, is sometimes distinguished by the name of *Potash Alum.* There are other kinds, which contain either an Eq. of Sulphate of Soda, hence called *Soda Alum;* or an Eq. of Sulphate of Ammonia, and then called *Ammonia Alum;* and the latter is the constitution of some of the common Alum of commerce.

Prep. Alum is manufactured in large quantities for use in the arts as well as in medicine. This is always in situations where there is some Aluminous rock; that is, one containing Alumina, and a Sulphuret, usually of Iron, sometimes a salt of Potash; by exposure to the air, either with or without heat. The Sulphur attracting Oxygen is converted into Sulphuric Acid, which combines with the Alumina and also with the oxidized iron. The Sulphate of Iron is separated, and a salt of Potash is added to the vitriolic solution of Alumina. In Cutch, Carb. of Potash is added to a solution obtained by boiling the blue clay, after it has been exposed for 5 months to the air and watered for 10 or 15 days. By due evaporation, and a repetition of the boiling and evaporation, crystals of Alum are obtained.

Tests. Alum should be colourless, and perfectly soluble in water, showing the absence of any uncombined earthy matter. From the solution, Ammonia or Potash throws down a colourless precipitate of Alumina, which is redissolved when the latter is added in excess. The freedom from colour and the solubility prove the purity of Alum. The presence of Iron may be detected by the addition of Tincture of Galls, which will produce a blueish-black colour after the Iron has been precipitated by Potash.

Inc. Alkalis and their Carbs., Lime and Lime-water, &c., Tartrate of Potash, Phosphates, Acetate of Lead, the Salts of Mercury, Gallic acid, Inf. of Galls and of Cinchona.

Action. Uses. Astringent, Styptic, both internally and as a Lotion, Collyrium or Injection.

D. gr. x.—℈j.

Alumen (Siccatum, D.) exsiccatum, L. E. Dried Alum. *Alumen ustum.*

Alum, when thoroughly heated, forms a light, spongy, opaque mass, losing its water of crystallization, but retaining its other properties.

Prep. L. E. D. Let Alum liquify in an earthen (or iron, E.) vessel; then let the fire be increased, until the ebullition has ceased. (Then reduce to powder, E. and D.) The directions of the three Colleges are essentially the same, but care must be taken that the heat is not too powerful, as then a portion of the Sul' will be driven off.

Action. Uses. Escharotic; occasionally given internally.

D. gr. v.—gr. xv.

Pulvis Aluminis Compositus, E. Compound Alum Powder.

Prep. Mix *Alum* ℥iv. *Kino* ℥j. and reduce them to fine powder.

An useful Astringent in Passive Hæmorrhages and in Chronic Diarrhœas.

D. gr. x.—ʒß.

Liquor Aluminis Compositus, L. Compound Solution of Alum.

This is a powerful astringent lotion, which used to be called Bates' Alum Water.

Prep. Dissolve *Alum* and *Sulphate of Lime* āā ℥j. in boiling Aq. Oiij. strain.

Action. Uses. Astringent, Styptic lotion. Diluted with Rose-water, used as a Collyrium and Injection.

Cataplasma Aluminis, D. Alum Cataplasm or Curd.

Prep. Agitate together *Alum* ʒj. with the *whites of two eggs*, so that they may form a coagulum.

Action. Uses. Astringent. Applied between two pieces of muslin over the eye in some kinds of Ophthalmia.

METALS PROPER.

Manganesium. *F.* Manganese. *G.* Mangan.

Manganese (Mn=28), the first of the metals proper to be treated of, is not itself officinal, and one of its Oxides is so, only on account of its Pharmacopœia use. It is hard, brittle, and of a greyish-white colour, emitting a peculiar odour when handled or in a moist atmosphere. Sp. Gr. 8. When pure, it oxidizes readily in the air, requiring to be kept under Naphtha, and is quickly dissolved by Dil. Sulphuric acid. It forms numerous combinations with Oxygen, but the Black or Peroxide is alone officinal.

Manganesii (Oxidum, E. D.) Binoxydum, L.

F. Oxyde noir de Manganese. *G.* Manganhyperoxyd.

Prop. The Binoxide of Manganese ($Mn\,O_2=44$), called also Peroxide, is that found most abundantly in nature, but is variable in appearance, sometimes crystallized in needles, often in compact masses, but most frequently is a dull earthy-looking powder, of a black or blackish-brown colour. It is usually sold in the state of a fine powder. It is devoid of both taste and smell; Sp. Gr. 4·8; insoluble in water; nearly infusible; heated, it gives out Oxygen. Treated with H Cl′, Chlorine is evolved, as also when mixed with Chloride of Sodium or Common Salt, and Sul′. Comp. Mn 63·75, O 36·25=100.

Tests. Binoxide of Manganese is seldom pure, usually containing more or less of Oxide of Iron, Carbonates of Lime and of Iron, Sulphate of Baryta, and clayey matter. "Muriatic acid, aided by heat, dissolves it almost entirely, disengaging Chlorine; heat disengages Oxygen." It also often contains much of the Sesquioxide. Its purity is judged of by the quantity of Oxygen or Chlorine evolved, or by the joint action of Hydrochloric and Oxalic acids. (*v.* Fownes, *Manual of Chemistry*, p. 271.)

Action. Uses. Officinal for aiding the evolution of Chlorine from Chloride of Sodium (p. 44), for which it is much employed; also for colouring in glass-making and pottery; seldom in medicine.

Ferrum, L. E. D.

Iron. Mars of the Alchymists. *F.* Fer. *G.* Eisen.

Iron (Fe=28) is found native, when it is supposed to be generally of meteoric origin; extensively in combination with Oxygen or Sulphur, as a Salt of various acids, as Carbonate, Sulphate, &c.; and all these in a more or less pure state, that is, mixed with earths or other metals. It also exists in vegetables, and in the blood of animals. It is extracted from Iron ores. Some of the Oxides, as Magnetic and Specular Iron ore, are heated only with Charcoal, as in Sweden, Elba, and India, when the Carbon combining with the Oxygen, the Iron is set free, and melted. The Carbonate, Iron Pyrites, Clay Iron ore, Red and Brown Hæmatites, and Spathose Iron, are first roasted, and then exposed to a fierce heat in contact with Charcoal, Coke, or small Coal, and a flux, either Lime or Clay, according as the ore is argillaceous or calcareous. These earthy matters become vitrified, and form a slag at the surface, while the heavy particles of Iron, falling down, run out by a hole at the bottom into moulds, and form *Pig* or *Cast Iron.* This is still impure, from the presence of Charcoal, Sulphur, and portions of Silicon and Aluminum. It is again twice fused in the refining and puddling furnaces, and exposed to the influence of a current of air, at a high temperature, when the whole of the Charcoal and Sulphur are burnt out, and the other impurities form a slag at the surface. The metal is taken out, beaten or pressed, and then drawn into bars, which form the *Malleable* or *Wrought Iron* of commerce.

Ferri Ramenta, L. (Limatura, E. Scobs, D.) Iron Filings. Filum, E. Fila, D. Iron Wire. *F.* Fil de Fer, Limailles de Fer. *G.* Eisendraht, Eisenfeilicht.

Iron wire and filings are ordered, because the former must be made from the most malleable, which is also the purest, Iron; while filings being finely divided, are useful for exhibition, and may be prepared from pure wire, or by means of the magnet be separated from the impurities with which they are apt to become intermixed in the process of filing.

Prop. Iron is well known for its hardness and toughness, and consequent application to an immense variety of useful purposes. It is of a whitish-grey colour, and has a styptic taste; hard, but malleable, and, though more ductile than many, exceeds all the metals in tenacity. It may be highly polished, and takes a sharp edge. Sp. Gr. 7·8. It is remarkable for its power of attracting and being attracted by the Magnet, as well as for itself and some of its compounds becoming magnetic. When heated, it becomes soft, at a white heat two pieces may be inseparably joined together, or *welded*. It melts at a bright white heat or 1587° of Daniell's pyrometer, but is not volatile. If exposed to the air when heated, or when moist, it absorbs Oxygen, and its surface becomes covered with a coat of Oxide. It burns in Oxygen gas with vivid scintillations. It combines either with 1 Eq. O, forming the Protoxide, or with $1\frac{1}{2}$, forming the Sesquioxide, often called the Peroxide. It slowly takes the Oxygen of water, Hydrogen gas escaping. When dilute Sul′ is poured upon Iron-filings, the Iron dissolves as a Protoxide, combining with the acid, and may be precipitated, on the addition of an Alkali, in combination with some water, forming a Hydrated Protoxide, of a greenish-white colour. This, when exposed to the air, absorbs more Oxygen, and is converted into the red-coloured Sesquioxide, as may be seen in several of the officinal preparations. The Protoxide may be readily converted into the Peroxide by boiling any salt containing it with a little Nitric′. The presence of the Iron may then be readily detected by testing the solution with Ferrocyanide of Potassium, or Tincture or Infusion of Galls: the former will form a blue and the latter a black precipitate.

Action. Uses. Iron, in a pure state or in filings, is inert; but, being oxidized in the stomach, acts as a tonic.

D. gr. v.—gr. xx. in Electuary with Honey or Treacle; or in Pills with some of the Bitter Extracts.

Protoxide of Iron.

Oxide of Iron (Fe O=36) is obtained when Iron is burnt in Oxygen gas. It is also obtained when Caustic Potash or Soda is added to a solution of the Sulphate of the Oxide of Iron. The white precipitate which falls becomes grey, and then of a bluish-green colour; but this is, in fact, a *Hydrated Protoxide of Iron.* It absorbs Oxygen from the atmosphere, and becomes red, being converted into the Sesquioxide. The Protoxide is a constituent of some officinal salts, as of the Sulph. Iron, and these are usually of a greenish colour, have a metallic taste, and are considered more efficacious as medicines than those containing the Sesquioxide. A similar Oxide of Iron, called *Martial Æthiops*, used to be prepared in the L. and still is in the D. P.; but it always contains a portion of the Sesquioxide.

Ferri Oxydi Squamæ, D.

Scales of the Oxide of Iron.

When Iron is heated to redness, its surface becomes oxidized to the extent of a thin coating. This is detached in thin scales in the process of hammering on the anvil.

Prop. Without taste or smell; black and brittle; attracted by the magnet. They are composed of a definite compound of the two Oxides of Iron, and are employed in making the Ferri Oxydum Nigrum. (p. 130.)

FERRI SESQUIOXYDUM, L. FERRI OXYDUM RUBRUM, E. D. FERRI CARBONAS, D.

Sesquioxide and Peroxide of Iron. *Crocus Martis. Colcothar. Chalybis Rubigo præparata. Ferri Subcarbonas. F.* Peroxide de Fer. *G.* Rothes Eisenoxyd.

The Red Oxide of Iron has been variously named, according to the mode of its preparation, though the product is essentially the same. It is abundantly found in nature, as a constituent of many red soils, crystallized in the form of rhomboids and octohedrons in Specular Iron or Iron Glance, as in Elba, and many parts of Europe, and in the Peninsula of India (Porto-Nuovo Works); in a compact state, with impurities, in Red Hæmatite or Red Iron ore. These give a reddish-brown streak, and with Borax form a green or yellow glass, are considerably magnetic, but do not, like oxydulated Iron, attract Iron-filings.

Prop. The Sesquioxide (Fe $1\frac{1}{2}$O=40) or Peroxide of Iron, artificially prepared, is a powder of a reddish-brown colour, without smell, but has a chalybeate taste, except when prepared by calcination; insoluble in water, and does not attract the magnet, unless it contain, as is sometimes the case, some of the Black Oxide. Sp. Gr. 3. It ought to dissolve in H Cl′ without effervescence; but it usually contains a little Carb′ (2 to 5, sometimes even 15, per cent., according to Mr. Phillips). It was therefore formerly called Carbonate and Subcarbonate of Iron, "which usually was, what it is now called, merely Sesquioxide of Iron." The Hydrochloric solution forms a black precipitate with Tinct. or Inf. of Galls, and a blue precipitate with Ferrocyanide of Potassium. Fe 70+O 30=100.

Different processes have at different times been adopted for making this preparation; all are effectual, in consequence of the facility with which Iron, when once oxidized, absorbs a further proportion of Oxygen. Formerly the Sulphate of Iron was calcined, when it lost its water of crystallization, and then its acid: some of this, however, yielded a portion of its Oxygen to form the reddish-coloured Sesquioxide. Or a solution of the Sulph. of Iron is decomposed by the addition of a sol. of Carb. of Potash or of Soda, when a Carb. of Oxide of Iron is precipitated. This being dried in the air, the Carb′ soon escapes almost entirely, and the Protoxide absorbs the due quantity of Oxygen from the atmosphere, and from a white and then green powder is converted into the reddish-coloured Sesquioxide. As the Ferri Carbonas D. is essentially the same, the process for making it is united with that of Ferri Sesquioxidum, L. To this preparation the E. C. applies the name Ferri Oxidum rubrum. This name is applied by the D. C. to Colcothar, which is essentially the same substance, but obtained by calcining the Sulphate of Iron.

Prep. L. E. (FERRI CARB. D.) Take boiling *Aq.* C vj. (Oß. cold Water Oiij ß. E.; Water 800 parts, D.) Dissolve *Sulph. Iron* ℔iv. (℥iv. E.; 25 parts, D.) in boiling *Aq.* Ciij. (Oß. E.; q. s. D.) then dissolve *Carb. Soda* ℔iv. ℥ij. (℥v. E.; 26 parts, D.) in boiling *Aq.* Ciij. (about thrice its weight of water, E.; q. s. D.) Mix the solutions, L. let the precipitate subside (on a calico filter, E.) Wash the precipitate well with (warm, D.) water; till it is but little affected by sol. Nit. Baryta, E. Dry it, L. D. (in hot-air or over the vapour-bath, E.)

D. (Ferri Oxydum Rubrum.) Drive off, by heat, the water of crystallization of *Sulph. Iron*, increase the heat as long as acid fumes arise. Wash the red oxide till Litmus paper is no longer reddened by the water, and dry on bibulous paper.

Tests. This preparation may be carelessly prepared, but is not likely to be adulterated. "Dissolved totally by dilute H Cl′ (aided by gentle heat, E.) with very slight effervescence, and it is precipitated by Ammonia." L. P. Earthy impurities will remain undissolved; if the presence of metals is suspected, they can be detected by their tests.

Inc. Acids and Acidulous Salts.

Action. Uses. Tonic.

D. gr. v.—ʒß. In Neuralgia, ʒß.—ʒij. or even ʒiv. two or three times a day.

For external application, there are two plasters, which, from the support such applications are calculated to afford, and from the moderately stimulant nature of the ingredients, are in general esteemed as strengthening plasters.

EMPLASTRUM FERRI, E. EMPLASTRUM ROBORANS. Iron or Strengthening Plaster.

Prep. Triturate *Red Oxide of Iron* (E.) ℥j. with *Olive Oil* f℥iij ß; add the mixture to *Litharge Plaster* ℥iij, *Resin* ℥vi. *Bees' Wax* ℥iij. previously melted with a gentle heat. Mix the whole thoroughly.

EMPLASTRUM THURIS, D. Frankincense Plaster.

Prep. Melt together *Litharge Plaster* ℔ij. and *Thus* (i.e. *Resin of Pinus Abies*) ℔ß. sprinkle in *Red Oxide of Iron* ℥iij. stirring. Make a plaster.

FERRUGO, E. Hydrated Sesquioxide of Iron. RUBIGO FERRI, D.

Besides the above Oxides, the E. P. contains a modification of the last in the form of Sesquioxide combined with 2 Eq. of water (Fe $1\frac{1}{2}$ O +2 Aq.=58). The Rust of Iron (Rubigo Ferri, D.) resembles it in also containing water in combination with the Sesquioxide. It has been adopted by the E. P., and has been introduced into practice, in consequence of its having been found to be an effective antidote against poisoning by Arsenious acid.

Prop. It is of a yellowish-brown colour, and though it can be dried without decomposition, it requires to be kept in a moist state, as in this state it combines so readily with Arsenious acid, that when prepared according to the following formula, and added in the proportion of 12 parts to 1 of Arsenious acid, and well shaken, the filtered liquor which previously contained the Arsenic, afterwards displays no traces of its presence, an insoluble Arsenite of the Protoxide of Iron having been formed.

Prep. Dissolve *Sulph. Iron* ℥iv. in *Aq.* Oij.; add commercial *Sul'* f℥iijß. and boil; then add *Nitric'* (D. 1380) f℥ix. in small portions, boiling for a few minutes after each addition, until the liquid acquires a yellowish brown colour, and gives with Ammonia a precipitate of the same colour. Filter, allow the liquor to cool, and add *stronger Aqua Ammoniæ* f℥iij ß. proceeding as directed in the process for Ferri Oxydum Nigrum, E. When kept as an antidote against poisoning by Arsenic it is preferable merely to squeeze it and preserve it in the moist state, in stoppered bottles.

In this formula the Protoxide is converted into the Sesquioxide by its taking some of the Oxygen of the Nit', *v.* Ferri Oxidum Nigrum. The Sul' is added to preserve the Sesquioxide formed in a state of solution, as Sulph. of the Sesquioxide. When Ammonia is added to this in excess, it combines with the Sul', and Sulph. Ammonia formed remains in solution, while the Sesquioxide is precipitated in combination with water and a little Ammonia, forming a Hydrated Ammoniaco Sesquioxide of Iron. When prepared with Potash or Soda, it is less efficacious in precipitating Arsenious Acid.

The yellow-coloured rust which forms when Iron is fully exposed to water and to the air, is essentially the same preparation as the above, though it is not to be substituted for it as an antidote.

Rubigo Ferri, D. Moisten *Iron Wire* in fragments, q. s. and expose to the air till it is converted into rust, then rub it in an iron mortar, and by pouring water on it separate the finer powder, and dry.

Tests. Entirely and easily soluble in Mur without effervescence: if previously dried at 180°, a stronger heat drives off about 18 per cent. of water. The magnet does not attract it. E.

Action. Uses. A good substitute for the Sesquioxide, and the best antidote to Arsenious acid.

D. gr. x.—3ß. as a Tonic; in much larger doses as an antidote.

Ferri Oxydum Nigrum, E. D.

Magnetic Oxide. (Fig. 23.) *Æthiops Martialis. Ferroso-ferric Oxide* of Berzelius. A compound of Protoxide and Sesquioxide of Iron. *F.* Oxide de Fer noir. *G.* Schwarzes Eisen Oxydul. Eisen Mohr.

Black Oxide (Fe_3O_4=116) or Martial Ethiops, long one of the esteemed preparations of Iron, was formerly made by moistening Iron-filings with water, and also by levigating the foregoing scales of the Oxide. It is probably composed of 1 Eq. Protoxide+2 Eq. Sesquioxide of Iron=116.

Prep. D. Wash with *Water* the *Scales of Oxide of Iron* which are found at the smith's anvil; when dry, separate them from impurities by a magnet; triturate and separate the finer powder, in the way directed to make prepared Chalk.

E. Dissolve *Sulph. Iron* ℥iij. in boiling *Aq.* Oiß. and add commercial *Sul'* f℥ij. and f℈ij.; boil, and gradually add pure *Nit′* f℥iv. boiling briskly for a few minutes after each addition. Again dissolve *Sul. Iron* ℥iij. in boiling *Aq.* Oiß. Mix the two solutions and immediately add of *stronger Aqua Ammoniæ* f℥ivß. in a full stream, stirring the mixture briskly. Collect the black powder in a calico filter and wash it with water, till Nitr. Baryta causes scarcely any precipitate; dry it at a temperature not exceeding 180°.

Fig. 23.

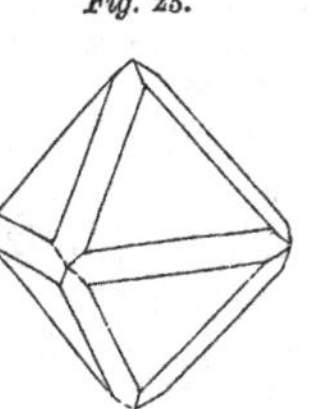

In the above formula, proposed by Wöhler, the first portion of the Sulph. of the Oxide is converted into the Sulph. of the Sesquioxide of Iron, by the addition of the Nit′, which yields a portion of its Oxygen, and Nitric Oxide gas escapes. As the object, according to Dr. Christison, is to obtain a compound, one half of the Iron of which is in the state of Protoxide, and the other in that of Sesquioxide, this is effected by mixing together the Sulphates of the Protoxide and Sesquioxide in the requisite proportions, and then precipitating them both by the addition of Ammonia in excess. The Oxides unite at once in the act of separation, and fall down in the form of a dark greyish-black powder, which, under exposure to the air, either with or without moisture, shows no tendency to undergo further oxidation.

Tests. Dark greyish-black; strongly attracted by the magnet; heat expels water from it; Mur′ dissolves it entirely; and Ammonia precipitates a black powder from this solution. E. P. The Black Oxide of the D. P. is blacker, and does not give off any water when heated. Wöhler considers the above Oxide to be composed of 2 Eq. of Protoxide and 1 of Sesquioxide of Iron, with two of water.

Action. Uses. Tonic; has the advantage of being a compound of the Protoxide, which is usually considered most efficacious.

D. gr. v.—℈j. two or three times a day.

Ferri Iodidum, L. E.

Iodide of Iron. *Ioduret and Hydriodate of Iron. F.* Iodure de Fer. *G.* Eisen Iodür.

Introduced into practice by Dr. A. T. Thomson, and described in his Obs. on the Prep. and Med. Employment of the Iodurets and Hydriodate of Iron.

Prop. Iodide of Iron (Fe I+5 Aq.=199) is of a dark-grey colour, with somewhat of a metallic appearance; its taste is acrid and styptic. It is often prepared in thin cakes, of a crystalline radiated structure, and light grey colour when fractured. If its solution be evaporated with as little contact of air as possible, "green tabular crystals are formed." It is very deliquescent, and readily dissolved in water, as also in Alcohol. The solution, when pure, is colourless, and when diluted, is not disagreeable. Heated, it volatilizes and readily fuses; but is then easily decomposed, Iodine escaping in vapour, and Iron being left in the form of Sesquioxide. From the absorption of Oxygen, the same change takes place on exposure to the air: water is absorbed to the extent of forming a dark-coloured solution, in which some Sesqui-iodide of Iron with a little free Iodine is held in solution, and a Sesquioxide of Iron precipitated. It is difficult to preserve it even in solution, unless a coil of Iron wire, as suggested by Mr. Squire, be introduced into it. Sugar also has been ascertained to have this preservative effect. Comp. Fe 14+I 63·3+Aq. 22·7=100.

Prep. L. E. Take *Iodine* ℥vj. (dry gr. 200, E.) *Iron Filings* ℥ij. (fine iron wire recently cleaned gr. 100, E.) *Aq. dest.* Oivß. (f℥vj. E.) Mix the Iodine with Oiv. of the water, add the Iron, heat them in a sand-bath (boil in a glass mattrass, first gently, to avoid the expulsion of Iodine vapour, afterwards briskly till concentrated to one-sixth its volume, E.) till the solution becomes pale green, pour off the liquid, wash the residue with the remaining boiling *Aq.* Oß. and add this to the other liquid, L. Filter. Evaporate to dryness in an iron vessel, at a temperature not exceeding 212°. (Put the filtered liquid quickly into an evaporator, with 12 times its weight of Quicklime round the basin, in an apparatus where it may be shut up in a small space, not in contact with the general atmosphere. Heat the whole apparatus, till all the water be evaporated, E.) Preserve the product in (small, E.) well-closed vessels (excluded from the light, L.)

The Messrs. T. and H. Smith now make a solution of Iodide of Iron in a Florence flask with ʒvj. of pure iron filings, ℥ij. ʒij. of Iodine and f℥ivß. of cold distilled water. Boil till the liquid loses its colour, and filter rapidly into another clean flask, and evaporate at a boiling heat. They obtain the compound either as a crystallized hydrate, or in an amorphous anhydrous form, according to the extent of the evaporation, and enclose without the smallest delay in small well-corked bottles. Mr. Kop recommends triturating 4 parts of Iodine with 2 parts of water, in a large dish; then to add at once, 1 part of iron filings in a state of fine division, and to continue the trituration.

Of these preparations that of the E. P. is preferable. The solution, like that of all the protosalts of Iron, is of a green colour. If this be quickly filtered and evaporated, and with as little access of air as possible, the salt may be obtained without much decomposition; but, as the Iron is apt to pass rapidly to the state of Sesquioxide, it is best prepared according to the Messrs. Smith's improved formula. As all the solid preparations are liable to change, they further recommend powdering their anhydrous Iodide as soon as it is taken from the flask, and then instantly to incorporate it with twice its weight of pure refined Sugar in powder, and to make it into a mass with honey. 4 grains will contain 1 grain of the Iodide. Keep in shallow corked bottles in a layer of some powder. (P. J. iii. 490.) Mr. Kop's preparation, it is said, may easily be administered in the form of pills. (P. J. v. 133.)

Tests. Entirely soluble in water, or nearly so, forming a pale-green solution, E. It gives off violet vapours when heated, leaving Sesquioxide of Iron; entirely soluble when recently made; but this solution in an ill-closed vessel quickly deposits Sesquioxide of Iron, and can be kept clear only in a vessel well closed and containing a coil of iron wire. L.

Inc. Acids, Alkalis, and their Carbonates, Lime-water and all such substances as are incompatible with Sulphate of Iron, such as vegetable astringents.

Action. Uses. Tonic, Deobstruent, Emmenagogue.

D. gr. j.—gr. v. or gr. x. in solution in water (a solution of gr. iij. in f℈j. is a convenient strength), in syrup, or in the saccharine pills.

FERRI IODIDI SYRUPUS, E. (*Liquor Ferri Iodidi, E.*) Syrup of Iodide of Iron.

Prep. Take *Iodine* (dry) gr. 200, *fine Iron Wire* recently cleaned gr. 100, and boil them in *Aq. dest.* f℥vj. at first gently (not to expel Iodine vapours), afterwards briskly, till about f℥ij. remain. Filter quickly, while hot, into a mattrass containing *powdered White Sugar* ℥ivß. dissolve the sugar with a gentle heat, and add *Aq. dest.* to make up f℥vj.

Sugar prevents the Protoxide of Iron passing to the state of Peroxide, as in the Ferri Carbonas Saccharatum, so it has been found to preserve the Iodide of Iron. A syrup was first suggested in Buchner's Repertor. fur die Pharmacie for 1839. The E. C. have introduced a formula of the Messrs. Smith modified from one proposed by Dr. A. T. Thomson, in the Trans. of the Pharm. Society, i. 47. This solution undergoes little change, even when preserved for some time. It ought to be "nearly colourless, or pale yellowish-green, and without sediment." Dr. C. cautions against employing common British Iodine in the same proportions, unless allowance is made for the moisture of the Iodine. The Syrup ought not to be diluted long before it is to be taken, and therefore the patient should himself make the mixture.

D. ♏xv.—f℈j. ♏xii. or xv. drops contain Iodide of Iron gr. j. It may be prescribed with vegetable astringents, and dilute mineral acids. (*t.*)

BROMIDE OF IRON is prepared in precisely the same way as the Iodide, substituting Bromine for Iodine, as in the first part of the process for making Bromide of Potassium, p. 74. Mr. Squire informs me that a coil of Iron wire traversing the whole column of the solution, is necessary to preserve it in a neutral and uniform state, and is perfect in its action. Others prefer the Syrup to the aqueous solution. The Bromide of Iron acts as an energetic Tonic.

TINCTURA FERRI SESQUICHLORIDI, L. FERRI MURIATIS (LIQUOR D.) TINCTURA, E.

Tincture of the Sesquichloride or Muriate of Iron. Steel Drops. *Tinctura Ferri Muriati. Tinctura Martis cum Spiritu Salis.* *F.* Teinture de Perchlorure de Fer. *G.* Salzsaure Eisentinctur.

Prop. The Sesquichloride of Iron (Fe $1\frac{1}{2}$ Cl=82) is one of the more powerful of the ferruginous preparations. Iron combines with Chlorine both as a Protochloride and a Sesquichloride (*Proto-chlorure* and *Per-chlorure de Fer* of the French). The first is white, in small scales, very soluble in water and alcohol, forming a green-coloured solution, and liable to change from the avidity with which the Iron absorbs Oxygen, and is precipitated as Sesquioxide; the Chlorine being left in the proportion to form the Sesquichloride. A Tincture of the Chloride is much used on the Continent. Both Chlorides used to be contained in the Tincture of the British Pharm. when it was made with the Black Oxide of Iron. A little of the Chloride exists in the present preparation. The Perchloride or Sesquichloride of the P. is volatile at a red heat. When the solution is concentrated, it yields either orange-yellow crystalline needles radiating from a centre, or large dark yellowish-red crystals. When sublimed, it is in brilliant scales, of a lively hue, very soluble in water, Alcohol, and Ether. Comp. Fe 34·15+Cl 65·85=100.

Prep. L. E. D. Pour upon *Sesquioxide* (*Red Oxide,* E.) *of Iron* ℥vi. (*Rust of Iron,* 1 part, D.) *Hydrochloric'* (*Mur'* E., 6 parts, D.) Oj. in a glass vessel, digest for 3 days, L. (till most of the oxide be dissolved, E.) frequently shaking. (Let the impurities subside, pour off the solution, evaporate slowly to one-third, D.) and add rectified spirit Oiij. (6 parts, D.) Filter, L. E.

In this formula, the Oxygen of the Oxide combines with the Hydrogen of the acid, and water is formed, while the freed Chlorine unites with the Iron in the same proportion, and thus Sesquichloride of Iron is formed, also a little Protochloride, in consequence of the Sesquioxide almost always containing a small portion of the Carbonate of Iron, of which the Carb' is expelled in effervescence. As the acid is in excess to keep the Chloride dissolved, the solution produces an acid reaction on test paper, and the taste is acid and astringent. It has a smell of Hydrochloric Ether, in consequence of the acid acting on the Alcohol, and is of a reddish or olive-brown colour. Sp. Gr. 0·992. The Tincture of commerce Mr. Phillips found to vary in strength from $9\frac{8}{10}$ to 20 grains of the Peroxide, but as above prepared a f℥ contains nearly 30 grs. of Sesquioxide of Iron. The presence of the Iron is readily indicated by its tests, p. 127. It is necessary to remember that a black inky mixture will be formed with any astringent vegetable preparation. The Chlorine forms a precipitate of Chloride Silver, when the Nitrate of Silver is added. It is also decomposed by the mucilage of Gum Arabic. The Perchloride is sometimes dissolved in Hoffmann's Liquor, and is then called Teinture de Bestucheff, the old Tinctura Nervina Bestucheff.

Tests. The strength and purity of this Tincture must be ascertained by its corresponding with the above characters, and by the quantity of the Sesquioxide precipitated by Liquor Potassæ.

Inc. Alkalis, Earths, and their Carbonates, Astringent Vegetables.

Action. Uses. Irritant, Caustic, Astringent Tonic.

D. ♏x.—♏xxx., even ʒj.—ʒij., in some suitable diluent.

Ferri Ammonio-Chloridum, L.

Ammonio-Chloride of Iron. *Ferrum Ammoniatum. Flores Martiales. F.* Chlorure Ferroso-Ammoniacal. Muriate de Fer et d'Ammoniaque. *G.* Eisenhaltiges Salzsaures Ammoniak. Eisen salmiak blumen.

This preparation was discovered by Basil Valentine in the 14th century, and has been employed in medicine under various names.

Prop. It is an orange-coloured powder, formed of small crystallized grains, which have a saline and astringent taste, with but little odour. It is deliquescent, very soluble in both water and Alcohol, and is considered to be only a mechanical mixture of Sesquichloride of Iron 15 parts, and Hydrochlorate of Ammonia 85 parts. Its nature is recognised by the effects of Potash or caustic Lime in evolving Ammonia, and the Iron and Chlorine will be revealed by their appropriate tests. Mr. Phillips has ascertained that it yields about 7 per cent. of Sesquioxide of Iron.

Prep. L. Digest *Sesquioxide of Iron* ℥iij. in *Hydrochloric Acid* Oß., in a proper vessel, in a sand-bath for 2 hours; add *Hydrochlorate of Ammonia* ℔ ij ß. first dissolved in *Aq. dest.* Oiij. Strain and evaporate to dryness. Rub the residue to powder.

In the first part of this process a Sesquichloride of Iron is formed, as in the foregoing Tincture. To this the Hydrochlorate of Ammonia being added, no change is observed to take place, and therefore the preparation is considered a mechanical mixture of Sesquichloride of Iron and Hydrochlorate of Ammonia. It is sometimes made by merely mixing these two together in the requisite proportions, dissolving in water, and then evaporating to dryness. It was formerly prepared by sublimation.

Tests. Totally soluble in proof Spirit and in water. Potash throws down Sesquioxide of Iron from the solution (in consequence of the Potassium combining with the Chlorine, and the Oxygen with the Iron), and if added in excess, disengages Ammonia, by decomposing the Hydrochlorate of Ammonia.

Inc. Alkalis and their Carbs., Lime-water, Astringent Vegetables.

Action. Uses. Tonic, &c.

D. gr. iij.—gr. x. in syrup, or bitter but not astringent extract.

Tinctura Ferri Ammonio-Chloridi, L. Tincture of Ammonio-chloride of Iron.

Prep. Dissolve *Ammonio Chloride of Iron* ℥iv. in *Proof Spirit* Oj. and filter.

This preparation is convenient for internal exhibition. f℥j. contains 5·8 grains of Sesquioxide of Iron.

D. ♏x.—f℈ß.

Ferri Sulphuretum, E. D.

Sulphuret of Iron. Iron Pyrites. *F.* Sulfure de Fer. *G.* Schwefeleisen.

Iron combines with Sulphur in several proportions. It is common in the form of the Bisulphuret, of a yellowish colour; hence has been thought to contain Copper. It was no doubt known to the ancients.

Fig. 24.

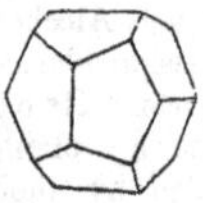

Prop. The natural Bisulphuret ($Fe\,S_2 = 60$) or Iron Pyrites, called also Mundic, is of a colour like Brass, in hard cubical crystals, which are not acted upon by any of the acids except the Nitric. Sp. Gr. 4·98. Often heated for the separation of the Sulphur by sublimation. If moistened and exposed to the air, Oxygen is absorbed, and the Sulphuret converted into Protosulphate of Iron.

Of the following processes, the first, though inferior, yields a product good enough for pharmaceutic purposes. It dissolves readily in Sul′ and H Cl′; H S′ gas is evolved, and Protosulphate of Iron remains in solution. The composition varies: the Protosulphuret consists of $Fe\,S = 44$, and the Sesquisulphuret of $Fe_2\,S_3 = 104$, the Bisulphuret, of $Fe\,S_2 = 60$.

Prep. E. Heat *Sublimed Sulphur* part j. and *Iron-filings* part iij. in a crucible on a common fire till the mass begins to glow. Then remove the crucible and cover it till the action, which increases considerably, ceases.

E. D. Take a *rod of Iron*, heat it white-hot in a forge, rub it with a *roll of Sulphur* (apply it to a mass, D.) (over a deep vessel, E.) filled with water to receive the (fused globules of, E.) Sulphuret. (When separated from the Sulphur and dried, preserve in closed vessels, D.)

By the first process a Protosulphuret is formed, if the mixture is in the proportion of 7 parts of Iron-filings to 4 of Sulphur. These rapidly combine, and heat is produced even to redness, and Sulphurous vapours escape: the redness will be preserved even when the crucible is removed from the fire. It usually contains an excess of Iron. The D. and second Ed. process yields a pure product, for which it is essential that the Iron be heated to a full white heat, when, the Iron and Sulphur combining together, bright sparks are emitted, and the melted Sulphuret falls into the water in light brown-coloured globules. Comp. Fe 63·4 + S 36·6 = 100.

Tests. Soluble nearly in diluted Sul′ with disengagement of Hydrosul′ gas. E.

Action. Uses. Antidote against Corrosive Sublimate. Employed as a ready means of obtaining Hydrosul′ gas by the addition of Sul′ or Cl H′.

Ferri Sulphas, L. E. D.

Sulphate of Iron. Sulphate of the Protoxide of Iron. *Ferrum Vitriolatum. Sal Martis. F.* Sulfate de Fer. *G.* Schwefelsaures Eisenoxydul. Eisenvitriol.

Vitriolated Iron, or Green Vitriol, was known to the ancients. It is mentioned in the *Amera Cosha* of the Hindoos (*Hind. Med.* p. 44), and it is used by them, as by the Romans in the time of Pliny, in making Ink. It is found in nature: the Sulphuret, absorbing Oxygen from the atmosphere, is converted into the Sulphate of the Protoxide of Iron; this is apt to be changed into the red-coloured Sulphate of the Sesquioxide. The Sulphate, being soluble, is found in some mineral waters. It is also made artificially on a large scale for use in the arts by exposing moistened Pyrites to the air, and is called Copperas or Green Vitriol.

Fig. 25.

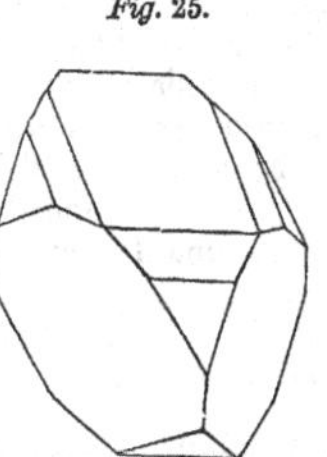

Sulphate of Iron ($FeO\,SO_3+7\,Aq.=139$) is a transparent crystallized substance of a bluish-green colour, and a styptic (which is also called an inky) taste. The crystals are modifications of the oblique rhombic prism. Sp. Gr. 1·82. They are soluble in a little more than their own weight of cold and in $\frac{3}{4}$ of their weight of boiling water. In the air they effloresce, and the salt, absorbing Oxygen, is converted into the Sulphate of the reddish-coloured Sesquioxide of Iron. Heated, it is first melted in its water of crystallization; this is afterwards expelled, and the salt reduced to the state of a dry white powder. (*v.* Ferri Sulphas exsiccatus, E.) At a still greater heat, the acid is expelled, and may be obtained in the form of anhydrous or glacial Sulph′, the latter portion being decomposed. The Iron is left in the state of the reddish-coloured Sesquioxide, the *colcothar* of old authors and of the D. P. Sulph. Iron is insoluble in Alcohol; its solution in water reddens Litmus; its Iron is precipitated on the addition of alkalis, alkaline earths, and their Carbonates, by the former as a Hydrated Protoxide, and by the latter as a Carbonate, which is soon changed into the red Sesquioxide. q. v. With Ferrocyanide of Potassium, a white precipitate is formed with the pure Sulphate of the Protoxide, but a blue one if the Sesquioxide be present: the same change of colour ensues when the former precipitate is exposed to the air. A black precipitate (Gallate of Iron) is formed when the Sulphate containing any of the Sesquioxide is added to an infusion or tincture of Galls, or of any other astringent vegetable. Comp. $Fe\,O\,25·9+S'\,28·8+Aq.\,45·3=100$.

Prep. Mix *Sulphuric′* ℥xiv. (7 parts, D.) with *Aq.* Oiv. (60 parts, D.) add *Iron-filings* ℥viij. (*Wire* 4 parts, D.) apply heat, (and when the effervescence is over, L.) filter (through paper, D.) Set the liquor aside to crystallize, (after due concentration, D.) and then concentrate the supernatant liquor to obtain more crystals. Dry them all. (If the Sulphate of Iron of commerce be not in transparent green crystals, without efflorescence, dissolve it in its own weight of boiling water acidulated with a little Sul′; filter, and set the solution aside to crystallize. Preserve the crystals in well-closed bottles, E.)

This process is introduced, as the Green Vitriol of commerce is usually impure. Concentrated Sul′ does not act on pure Iron, but the water of the dil. acid becoming decomposed, yields its O. to the Iron, while H. escapes in the form of gas. The Oxide of Iron formed unites with the Sul′, and the Sulphate of Iron is obtained.

Tests. Pale bluish-green crystals, with little or no efflorescence; entirely soluble in water; this solution does not deposit Copper upon Iron being immersed in it; its solution, first boiled with Nit′ and then precipitated by excess of Ammonia, yields on filtration a fluid which is colourless or very pale blue. L. and E. If it be of a deep blue, then Copper is present. The boiling in Nit′ is not always necessary, for Green Vitriol is usually a mixture of Sulphate of Protoxide and of Sesquioxide of Iron. Zinc may be similarly detected by adding Ammonia in excess to the Sesquioxidated solution; after filtering, expel the excess of Ammonia by heat, and any Zinc which is present will be deposited in flakes of the white Oxide.

Inc. Alkalis and their Carbonates, salts of Calcium and of Barium, Acetate and Diacetate of Lead, Nitrate of Silver, Vegetable Astringents.

Action. Uses. Astringent, Tonic, Emmenagogue.

D. gr. j.—gr. v. in pills with Bitter Extracts or Aromatic Confection.

Mr. Phillips warns from giving it in solution without first boiling the water, and expelling its atmospheric air, of which the Oxygen would peroxidize the Oxide.

FERRI SULPHAS EXSICCATUS, E. Dry or Anhydrous Sulphate of Iron.

Prep. Heat moderately in an earthen vessel, not glazed with lead, *Sulphate of Iron* q. s. till converted into a dry grayish white mass, powder this.

Convenient for exhibition in the form of Pills. By heat the crystals readily lose five-sixths of the water of crystallization, and the dried, therefore, is in this proportion stronger than the common Sulphate of Iron.

D. gr. ß.—gr. iv.

The dried Sulphate is employed in the E. P. in making the

PILULÆ FERRI SULPHATIS, E.

Prep. Take dried *Sulphate of Iron* 2, *Extract of Taraxacum* 5, *Liquorice-root Powder* 3, and *Conserve of Roses* 5 parts. Beat them into a proper mass and divide into 5 grain pills.

D. gr. x.—gr. xx.

Pharm. Prep. *Pilulæ Aloes et Ferri*, E. Containing 1 grain of the Sulphate in about 5 grains, with Aloes and Aromatic powder.

Pilulæ Rhei et Ferri, E. Containing 1 grain in 6 grains, with Extract of Rhubarb.

LIQUOR OXYSULPHATIS FERRI is a preparation, according to Mr. Tyson, P.J. i. 598, in constant use among the practitioners of Derbyshire, and considered one of the most powerful preparations of Iron. It is prepared by rubbing up *Ferri Sulph.* ʒij. or ʒiij. with *Acidi Nitrici* fʒiij., then gradually adding *Aq. Dest.* f℥j.ß. Filter. (A Sulphate of the Peroxide is formed.)

D. ♏v.—xij. 2 or 3 times a day.

CARBURET OF IRON.

Plumbago or Graphite.

This, commonly called Black Lead, is usually considered a Carburet of Iron; but, as the purest specimens are composed almost entirely of Charcoal, it is treated of with that substance. The specimens containing any notable proportion of Iron, must have some of the properties of Ferruginous preparations. Steel, being composed of pure Iron combined with a very small quantity of Carbon, is a true Carburet of Iron.

CARBONATE OF IRON.

F. Carbonate de Fer. *G.* Kohlensaures Eisenoxyd.

The name of Carbonate of Iron long held a place in the Pharmacopœias, and the preparation which was so called still does so in the D. P., but in the L. P. under the name of FERRI SESQUIOXYDUM, q. v. which contains about 4 per cent. of the Carbonate. The Carbonate of Iron is, however, officinal in the form of the *Mistura Ferri Composita* and in the *Pilulæ Ferri Comp.*, as likewise in the *Ferri Carbonas Saccharatum* and the *Pilulæ Ferri Carbonatis.* E. P.

Carbonate of Iron (Fe O, C O_2 = 66) is obtained by precipitating with an alkaline Carbonate a solution of any proto-salt of Iron, *e.g.* the Sulphate. A soluble Sulphate of Potash or of Soda is obtained in solution, and an insoluble Carbonate of Iron is precipitated. This is at first of a greenish-white colour, but becomes of a brownish-red colour in the air, and is converted into the Sesquioxide. (*v.* supra.) As the protosalts are more efficient as medicines than the persalts, it has been the practice with many to direct the *Mistura Ferri Composita* to be taken as much as possible when fresh made, or when of a greenish-colour. Sugar has long been prescribed with Ferruginous preparations, as in the old *Mars Saccharatus*, and with the Carbonate of Iron, the Mist. Ferri Comp. and Pil. Ferri Comp. Soubeiran states that the idea of preventing the oxidation of the Carbonate of Iron, and preserving a uniform medicine, first occurred to Dr. Becker, and that it was put into practice by M. Klauer, a Pharmacien of Mulhausen, who employed Sugar for this purpose, having discovered that it had the power of preventing the Protoxide of Iron from passing into the state of Peroxide by absorbing more Oxygen. Experiments have also been made by Drs. Clark and Christison, and Sugar is employed in preserving the Iodide, Bromide, and Chloride of Iron, also Iron-filings.

FERRI CARBONAS SACCHARATUM, E. Saccharine Carbonate of Iron. Klauer's FERRUM CARBONICUM SACCHARATUM (of the Continent).

Prep. Dissolve *Sulph. Iron* ℥iv. in *Aq.* Oij. and *Carb. Soda* ℥v. in *Aq.* Oij.; mix the two (cold) solutions, collect the precipitate on a cloth filter; wash it immediately with cold water, and without delay triturate the remaining pulp with *Pure Sugar* finely powdered ℥ij. Dry the mixture at a temperature not much above 120°.

Tests. Colour greyish-green, taste sweet and styptic, easily soluble in Muriatic acid with brisk effervescence. E. P. If the Carbonate of Iron is mixed with the Sugar as soon as possible after it is deprived of its water by filtration, it is found that it may then be dried even with the heat indicated, without undergoing conversion into the Sesquioxide of Iron. The action of the Sugar in this and other similar cases is, that it prevents oxidation. Klauer supposed that a definite compound is formed consisting of Protoxide and Sesquioxide of Iron, Carbonic acid, Sugar, and water. He found 80 per cent. of Protoxide and 20 of Sesquioxide. Fifty grains ought to yield 7·5 cubic inches of gas when decomposed by an acid.

Action. Uses. Tonic; and an excellent form for a Ferruginous preparation.

D. gr. x.—gr. xxx. It may be prescribed in the same doses in the form of

PILULÆ FERRI CARBONATIS, E. Pills of Carbonate of Iron.

Prep. Take *Saccharine Carbonate Iron* 4 parts, *Red Rose Conserve* 1 part, beat them into a proper mass and divide into 5 grain pills.

PILULÆ FERRI COMPOSITÆ, L. D. *Pilulæ Ferri cum Myrrhâ.* Compound Pills of Iron.

Prep. Beat powdered *Myrrh* ʒij. and *Carb. Soda* ʒj. together, add *Sulph. Iron* ʒj. (and brown Sugar, ʒj. D.) rub again; lastly, add *Treacle* ʒj. (Molasses, q. s. D.) Beat the whole in a vessel previously warmed until incorporated.

These pills are the usual form for prescribing the Carbonate of Iron, but the saccharine matter is in too small a proportion to prevent the oxidation. Therefore, they should be made only when required, or the *Pil. Ferri Carbonatis* be substituted. gr. xx. contain 1 gr. $\frac{7}{10}$ of the Protocarbonate, in which there is one grain of the Protoxide.

D. gr. v. or gr. x.—℈j. 2 or 3 times a day.

The *Pilules de Vallel* are similar. *Pilules de Blaud* are made with Carb. Potash and Gum Arabic, and are a close imitation of the following mixture.

MISTURA FERRI COMPOSITA, L. E. D. *Compound Mixture of Mars. Griffith's Antihectic or Tonic Mixture.*

Prep. Beat powdered *Myrrh* ʒij. [(bruised, E.) (finely powdered, ʒj. D.)] with *Spirit Nutmeg* f℥j. (f℥ß. D.) and *Carb. Potash* ʒj. (gr. xxv. D.) then add *Rose Water* f℥xviij. (℥vijß. D.) and then *Sulph. Iron* in (coarse, E.) powder ℈ijß. (℈j. D.) Put the mixture immediately into a well-stopped bottle.

Here the same changes take place; but Carbonate of Potash being employed instead of the Carbonate of Soda, we have a Sulphate of Potash left in solution. A hydrated Protocarbonate of Iron is suspended with Myrrh and the Spirit of Nutmeg as aromatics, while the Sugar no doubt assists in preventing the formation of Sesquioxide of Iron. Mr. Brande states that the best mode of making it is to triturate a fine piece of Myrrh into an emulsion with rose-water, add the Carbonate of Potash, Nutmeg, and Sugar, and lastly dissolve in it the Sulphate of Iron. If the bottle is kept full, the preparation will long remain unchanged: it should therefore be prescribed in draughts. Ince P. J. i. 252.

Inc. Acids and Acidulous Salts, Vegetable Astringents.

D. f℥j.—f℥ij. 2 or 3 times a day.

Soubeiran (ii. p. 434) describes some other preparations of the Carbonate of Iron, as the *Poudre Ferrugineuse*, in which Bicarb. Soda and Sulphate of Iron are mixed with Sugar, also *Poudre Ferée Gazeuze.*

Prep. Mix *Ferri Sulph.* gr. iij. carefully with pounded Sugar ʒijß. add *Bicarb. Soda* in powder grs. liv. and coarsely pounded *Tartaric'* ʒj.; add to a bottle of water, cork it immediately and agitate.

A sweetish acidulous preparation of Iron, made agreeable by the effervescence of the Carbonic acid. The Iron may be increased, as the 3 grs. of the Sulphate will produce 1¼ grs. of the Carbonate of Iron.

The *Aqua Chalybeata* of Bewley and Evans is also an elegant and agreeable preparation of this metal. *v.* Citrate of Iron.

FERRI ACETAS, D.

Acetate of Iron. *Extractum Martis Aceticum. Aceticum Martiale. F.* Acetate de Fer. *G.* Essigsaures Eisen.

This preparation is contained only in the Dublin Pharmacopœia, and unnecessarily in three different forms. But it was formerly more used in medicine; in the present day, it is chiefly known in the arts as the *Iron liquor.*

Prop. The Acetate of Iron of the D. P. is a liquid of a deep red colour, and is an Acetate of the Sesquioxide. This is evident from the ingredients employed, for the Carbonate so named as we have seen is only the Sesquioxide with a little Carbonate mixed with it; and therefore, after digesting in Acetic′, there will be a small portion of the Acetate of the Protoxide, and the great bulk consist of the Acetate of the Sesquioxide. It has an acid reaction and Chalybeate taste. The tests for Iron act as with the other persalts.

Prep. Digest *Carb. Iron* 1 part in *Acet′* 6 parts for 3 days. Filter.

Action. Uses. Tonic, Astringent.
D. ♏v.—♏xx. in water.

FERRI ACETATIS TINCTURA, D. Tincture of Acetate of Iron.

Prep. In an earthenware mortar rub *Acet. Potash* 2 parts, *Sulph. Iron* 1 part into a uniform mass; dry with a gentle heat. Triturate with *Rectified Spirit* 26 parts. Digest in a well-closed bottle for 7 days, agitating occasionally. Let it rest, pour off the clear fluid and keep it in well-closed bottles.

In this preparation double decomposition will take place, and Sulph. Potash and Acet. of the Protoxide of Iron be produced, but much of which will necessarily be converted into the Sesquioxide in the process of drying. The Acetate of Iron, as well as the excess of Acetate of Potash, is dissolved in Spirit of the Sp. Gr. ·840, the Sesquioxide and Sulphate being left. The Tincture is of a light red or claret colour, and forms an agreeable chalybeate. When evaporated to dryness, it yields saline matter which is whitish from the presence of Acetate of Potash.
D. f℈ß.—f℈i.

It is extremely liable to spontaneous decomposition, and is decomposed by the Acids, the Alkalis, and their Carbonates, also by vegetable Astringents.

FERRI ACETATIS TINCTURA CUM ALCOHOL, D. Alcoholic Tincture of the Acetate of Iron.

Prep. Rub *Sulph. Iron* and *Acet. Potash* āā ℥j. till they unite, dry them and when cold triturate with *Alcohol, by measure,* ℔ij. Digest in a well-stoppered bottle, agitating frequently, for 24 hours. Pour off the clear liquor and keep in well-closed vessels.

This preparation is very similar to the last, but instead of an excess of Acet. of Potash, there is an excess of Sulph. Iron. As above, the Acetate is a mixture of the Acetates of the Protoxide and Peroxide: the latter only is stated by Wood and Bache to be soluble in the Alcohol of the D. P. having the Sp. Gr. ·810. It is therefore a solution of the Acetate of the Peroxide of Iron, and forms a stronger and more uniform preparation. f℥j. evaporated, yields gr. x. of a crimson-coloured extract. Dr. Percival, who thought highly of the Chalybeate powers of these Tinctures, prescribed from f℈ß—f℈j. in asses' milk.

Ferrocyanide of Potassium. Potassii Ferrocyanidum, L.

Ferrocyanate of Potash. Ferroprussiate of Potash. Prussiate and Yellow Prussiate of Potash. F. Cyanure de Fer et de Potassium. *G.* Cyaneisen Kalium. Blausaures Eisenkali.

This salt is officinal for the purpose of making Diluted Hydrocyanic acid. There is no formula for preparing it, because it is found in a very pure state in commerce.

Prop. This salt is of a lemon-yellow colour, and transparent, without smell, but having a cooling saline taste. The crystals are large quadrangular, with more or less truncated edges and angles, derived from a primary octohedron; tough and flexible; Sp. Gr. 1·83. Soluble in 4 parts of cold and 2 of boiling water. (Gr.) Dr. Ure states that water at 60° takes up about ⅓, and at 212° its own weight of this salt; insoluble in Alcohol. Heated to 212°, they lose about 13 per cent., or their 3 Eq. of water of crystallization, and are reduced to a white powder. At a red heat they are decomposed, Nitrogen escapes, Carburet of Iron and Cyanide of Potassium are left; but if exposed to the air, the latter absorbs Oxygen, and becomes Cyanate of Potash. This salt is not precipitated by the alkalis, nor Sulphuretted Hydrogen, nor by Tincture of Galls, proving that the Iron is in a peculiar state of combination. With salts of the Peroxide of Iron, it produces precipitates of Prussian Blue, but white precipitates with salts of the Protoxide, which become blue after exposure to the air. With Lead, Zinc, Copper, &c., it forms different coloured precipitates, and therefore serves to ascertain the presence of these metals. Heated with diluted Sul', Hydrocy' is produced. Boiled with Binoxide of Mercury, Bicyanide of that metal is obtained in solution.

Ferrocyanide of Potassium is composed, according to Liebig, of (an imaginary radical) *Ferrocyanogen*, composed of 1 Eq. of Iron and 3 Eq. of Cyanogen, or $Fe\,Cy_3$, in combination with 2 Eq. of Potassium, $K_2\,Fe\,Cy_3+3\,H\,O$; or, according to other views, of 1 Eq. Cyanide Iron 54 + 2 Eq. Cyan. Potassium 132 + 3 Aq. 27 = 213.

This salt is prepared on a large scale by calcining animal matter, such as hoofs, chippings of horn, dried blood, pieces of hides, cellular membrane, the refuse of tallow melters, with Carb. Potash, in an iron pot with iron stirrers; throw into water, dissolve, filter, evaporate, and crystallize. Repeat the latter part of the process for purification.

Tests. Totally dissolved by water. A gentle heat evaporates 12·6 parts from 100 parts. It slightly (if at all) alters the colour of Turmeric. What it throws down from the preparations of Sesquioxide of Iron is blue, and that from the preparations of Zinc is white. When burnt, the residue dissolved by H Cl' is again thrown down by Ammonia: 18·7 parts of Sesquioxide of Iron are yielded by 100 parts. L.

Action. Uses. Sedative; but possessed of little activity; is rapidly absorbed and diffused through the secretions.

D. gr. x. to gr. xv.; but may be given in much larger doses.

Ferri (Cyanuretum, D.) Percyanidum, L.

Ferri Ferro-Sesquicyanidum, *Per.* Sesquiferrocyanide of Iron. (*Graham.*) *Ferri Ferrocyanas,* U. S. Prussian Blue. Ferro-prussiate of Iron. *F.* Bleu de Prusse. *G.* Cyaneisen. Berlinerblau.

This substance, so well known by the name of Prussian Blue, was discovered in 1710 by Diesbach, a preparer of colours at Berlin.

Prop. Prussian Blue is a light porous body, of a rich velvety blue colour, if dried when prepared at the temperature of the air; but, dried at a higher temperature, it is more compact (*G.*), and is usually seen in solid masses of its peculiar rich blue colour; devoid of taste and smell, and insoluble. It is decomposed by the alkalis, an alkaline Ferrocyanide produced, and Peroxide Iron precipitated. The strong acids also decompose it. Boiled with Red Oxide of Mercury, soluble Bi-Cyanide of Mercury is produced, with an insoluble mixture of Oxide and Cyanide of Iron. Great differences of opinion are entertained respecting the composition of this substance; a variety of names have, therefore, been applied to it by different chemists. The name Percyanide adopted by the London College, is objectionable, inasmuch as Prof. Graham states that Percyanide or Sesquicyanide of Iron, $Fe_2\ Cy_3$, is unknown in a pure state. He represents it as "consisting of 4 Eq. of Iron and 3 Eq. of the bibasic salt-radical, Ferrocyanogen, and therefore named a Sesquiferrocyanide. It contains Oxygen and Hydrogen besides, which cannot be separated without the decomposition of the compound." In conformity to the opinion of Berzelius, it is now generally considered a compound of 3 Eq. Cyanide of Iron and 2 Eq. of Sesquicyan. Iron ($3\ Fe\ Cy + 2\ Fe_2\ Cy_3$).

Prussian Blue may be obtained perfectly pure, on a small scale, by double decomposition between Ferrocyanide of Potassium and an acid solution of Persulphate or Perchloride of Iron. In the arts it is made by calcining Potash and animal matter, adding this to 12 or 15 times its weight of water, and precipitating the clear solution obtained with 2 parts of Alum and 1 of the Protosulphate of Iron.

Tests. It is pure if, after being boiled with dilute H Cl′, Ammonia throws down nothing from the filtered solution. L.

Action. Uses. Officinal, for making Bicyanide of Mercury. It has been accounted sedative, tonic, &c., but has little effect.

D. gr. v. 3 or 4 times a day.

FERRI MISTURA AROMATICA, D. Aromatic Mixture of Iron.

Prep. Take *Iron-filings* ℥ß. *Bark* (*Cinchona lancifolia*), in coarse powder ℥j. *Calumba* sliced ʒiij. *Cloves* bruised ʒij. Digest for 3 days in a close vessel, agitating occasionally with *Peppermint Water* q. s. to yield f℥xij. of the filtered liquor. Add *Compound Tincture of Cardamoms* f℥iij. and *Tincture of Orange-peel* fʒiij.

Dr. Neligan says of this combination of aromatic tonics holding in solution some Tannate of Iron, that, though unchemical, it is a most excellent tonic, much used in Dublin, in the various states of debility attended with anœmia, in doses of f℥j.—f℥ij. two or three times a day. It is often called *Heberden's Ink*, from its black colour.

FERRI ET POTASSÆ TARTRAS.

FERRI POTASSIO TARTRAS, L. FERRUM TARTARIZATUM, E. FERRI TARTARUM, D. Tartrate of Potash and Iron. Tartarized Iron and Chalybeated Tartar. Potassæ Ferro-Tartras. *F.* Tartrate de Potasse et de Fer. *G.* Eisenweinstein.

This preparation, in different forms, has long been employed in medicine ; one of these is familiarly known by the name of Steel Wine, which, though liable to variation, is a popular, and, when carefully prepared, an effective form for exhibition.

Prop. The Tartrate of Potash and Iron is usually in powder of an olive-brown colour, without odour, and of a mild chalybeate taste. It is now often prepared in shining brittle scales of a dark-brown colour. Exposed to moist air, it deliquesces, and is soluble in 4 times its weight of water; slightly in Alcohol, more so in Wine. The solution is of a light-brownish colour, and remains unchanged for a considerable time. It is remarkable for not being decomposed by Ammonia or its Carbonate at any temperature, nor by the other alkalis nor their Carbonates without the aid of heat. It is also unaffected by the Ferrocyanide of Potassium until a few drops of the stronger acids are added, when a deep blue colour is produced, showing that decomposition must have taken place. Vegetable astringents cause the usual black precipitate. It is considered by Mr. Phillips to be composed of 1 Eq. Tartrate of Sesquioxide of Iron 106+1 Eq. Tartrate of Potash 114=220, or 48·18 of the former, and 51·82 per cent. of the latter, and he found it to yield about 18 per cent. (Soubeiran states that it contains 13 per cent) of Sesquioxide of Iron.

Prep. The Colleges give different formulæ for preparing a Sesquioxide of Iron, which is then to combine with 1 Eq. of Cream of Tartar. The D. C. retain the old process for oxidizing the Iron. Mix *Iron-wire* 1 part with very finely powdered *Bitartrate of Potash* 4 parts and *Aq. dest.* 8 parts, or q. s. Expose to the air for 15 days in a wide vessel, stirring occasionally and adding water to keep the mixture moist, but not to cover the iron completely. Then boil with *Aq. dest.* q. s. Evaporate the filtered liquor to dryness in a sand-bath and keep the Tartar of Iron in well-closed bottles.

L. E. Mix *Sesquioxide of Iron* ℥iij. with *Hydrochlor'* Oß. and digest in a sand-bath for 2 hours. Add *Aq. dest.* cong. ij. set aside for an hour, then pour off the supernatant liquor. Add *Sol. of Potash* Oivß. or q. s., wash well the precipitate with water. (Take *Sulphate of Iron* ℥v. and make it into rust of Iron, as directed under Ferrugo, but without drying, E.) While moist boil it with *Bitartrate of Potash* ℥xiß. (℥ix. ʒj. E.) mixed in *Aq. dest.* Cong. j. (Oiv. E.) Let the solution cool, pour off the clear liquor. Should it be acid to Litmus tests, pour in solution of *Sesquicarbonate of Ammonia* Oj. or q. s. (*Carb. of Am.* in fine powder q. s. E.) till effervescence ceases. Over a sand-bath concentrate to the consistence of thin extract, or till the residue on cooling becomes a firm solid. (Preserve in well-closed vessels, E.)

The L. and E. C. have adopted the improved process of Soubeiran with slight modifications. In the L. process, Sesquichloride of Iron (*v.* p. 134) is produced. On the addition of the Potash, the Iron is precipitated in the form of Hydrated Sesquioxide of Iron. The E. C. directs this to be prepared as under Ferrugo, and to be kept moist, as in this state, when boiled with the Bitartrate of Potash 1 Eq. 40 of the Sesquioxide combines with 1 Eq. 66 of the Tartaric acid, and 1 Eq. Tartrate of Potash 114 remains as it was, or rather unites with the new-formed Tartrate of Iron, and thus perfectly prevents the action of some of the ordinary reagents on the Iron.

Tests. Totally soluble in water, and not attracted by the magnet: (if improperly prepared, a large portion is insoluble, and the magnet attracts the residual oxide). The solution does not change either Litmus or Turmeric; it is not rendered blue by Ferrocyanide Potassium, nor anything precipitated from it by any acid or alkali. P. L.

Inc. Strong acids, Lime-water, Acetate of Lead, Hydrosulphuric', Vegetable Astringents.

Action. Uses. Chalybeate Tonic, mild in taste, soluble and efficient.

D. gr. x.—ʒß. in pill, or solution with some aromatic.

Vinum Ferri, or Steel Wine, is omitted in the Pharmacopœias; but an efficient and uniform substitute can at any time be formed by dissolving a portion of the above preparation in Sherry Wine. It was formerly made by macerating for a month Iron-filings in Sherry Wine, and contained only 16 grs. of the Peroxide in a pint of Wine. Dose, ʒij.—ʒiv. thrice a day. Mr. Donovan macerates Rust of Iron of the shops, well levigated, ℥ij. with one pint of the best Hock, in a mattrass placed in a water-bath of 100° F. for one hour, constantly agitating, and the next day filters. Or, take *Tartrate of Protoxide of Iron* 1 part, and rub it up with *Tartaric'* 1 part in a porcelain or glass mortar, then add of *White Wine* 1000 parts, and filter the solution if necessary. The *Tartrate of the Protoxide of Iron* may itself be made by decomposing an Eq. of Protosulphate of Iron with an Eq. of neutral Tartrate of Potash, immediately washing the precipitate with water, collecting it on a strainer, pressing it strongly, and drying over a water-bath. (Soubeiran.)

FERRI AMMONIO-TARTRAS. Ammonio-Tartrate of Iron.

In the same way that Sulphate of Alumina, by combining with Potash, Soda, or Ammonia, will form either Potash, Soda, or Ammonia Alum, so various double salts may be formed by presenting different bases with Iron to the Citric and Tartaric acids. The Ammonio-Tartrate of Iron is one of these, which has for some time been prepared as a substitute for the Tartrate of Potash and Iron. It is of a dark-brown colour, in brilliant scales, or in angular grains, resembling Kino; its powder is of a brown colour, like Iron-rust. It has a mild ferruginous taste; soluble in rather more than its weight of water at 60°; not decomposed by boiling water; insoluble in both Alcohol and Ether. It is composed of 1 Eq. of Tartrate of Sesquioxide of Iron, 1 Eq. of Tartrate of Ammonia, and 4 of water. A formula for preparing it is given by Mr. Procter, junr. in Amer. J. of Pharm., by decomposing Bitart. Ammonia with fresh made Hydrated Sesquioxide of Iron. *v.* P. J. i. 291. It is an eligible preparation of Iron, and may be given in doses of gr. iij to gr. viij. in aqueous solution, or in pill, or with some confection.

FERRI CITRAS. Citrate of Iron.

M. Beral, in 1831, introduced to the notice of the profession several preparations in which Iron is combined with Citric acid. Of these the Citrate or Sesquicitrate of Iron is a combination of boiling *Cit′* ℥iv. and *Aq. Dest.* f℥iv. with moist *Hydrated Sesquioxide of Iron* (*v.* p. 130) about ℥viij. When cold, add water so as to make f℥xvj. It is obtained (by drying its solution spread out on glass) in thin and transparent laminæ of a beautiful garnet hue; has an acid and not disagreeable styptic taste. It dissolves slowly in cold water, but more readily in boiling water. M. Beral has also proposed a Protocitrate of Iron, which is white and pulverulent, with a strong chalybeate taste, and also a Citrate of the Magnetic Oxide (p. 130) of Iron. But the name *Citrate of Iron* is applied to a preparation in which the excess of acid has been counteracted by Ammonia, and which thus becomes converted into the following:

FERRI AMMONIO-CITRAS. Citrate of Ammonia and Iron. Ammonio-Citrate of Iron. *Citrate of Iron.*

This is prepared by adding Ammonia to the Citrate of Iron, so as to neutralize the excess of acid, and thus produce a double salt, then evaporate to dryness with a gentle heat. It differs sufficiently from the former to require to be designated by a different name, in order to prevent mistakes. It is, like the Citrate, in shining scales, of the same beautiful garnet-colour, but with a milder taste. Readily soluble in cold water, and may be given in doses of v.—viij. grs. in solution.

All these and some other preparations of Cit′ and of Iron, as Citrate of Iron with Zinc, with Soda, with Potash, and with Magnesia, are prepared of great purity and elegance by Mr. Bullock; the Citrate of Iron especially, in the form of Syrup, Wine, and Lozenges. For Citrate of Quinine and Iron, *v.* Cinchona.

AQUA CHALYBEATA. One of the most elegant and agreeable forms which has yet been proposed for the exhibition of Iron is an effervescing solution of Citrate of Iron flavoured with Orange-peel, called *Aqua Chalybeata* by Messrs. Bewley and Evans, and for which the name of Chalybeate Champagne has been proposed. It has been examined and reported upon by Dr. Ure and by Mr. Brande. It consists of a solution of Citrate of Iron highly charged with Carbonic acid gas, and flavoured with Orange-peel. It is sent out in six-ounce bottles, which afford on analysis gr. vij. $\frac{9}{10}$ of Peroxide of Iron, corresponding to gr. xiij.ß of the Citrate of Iron. Two ounces, or a wine-glass full two or three times a day, forms a dose which is grateful to the taste, and suitable to the most delicate in constitution, and to children, at the same time that it is effective as a tonic.

LACTATE OF IRON. FERRI LACTAS. Lactate of the Protoxide of Iron is a mild chalybeate, which is obtained in a greenish-white powder, or in greenish acicular crystals, by the direct action of a dilute solution of Lactic acid on Iron-filings. When it is dissolved, the Iron passes to a higher state of oxidation, and the solution becomes yellow. It may be prescribed in the form of Lozenges or of Syrup.

MALATE OF PROTOXIDE OF IRON is a form of preparation in use in the Prussian Pharmacopœia. It is procured by digesting one part of Iron nails or wire along with four parts of Apple-juice for some days, then evaporating the liquid to one-half, filtering and concentrating to the consistence of an extract.

ZINCUM, L. E. D.

Zinc. *F.* Zinc. *G.* Zink.

Zinc is considered to have been known as a distinct metal only since the time of Paracelsus. It has long been imported from China into India. The name *Tutenague*, by which Chinese Zinc was known in commerce, is evidently derived from the Tamul, *tutanagum* (Essay Hind. Med. p. 100), and it was at one time called Indian Tin (*Stannum Indicum*). The ores of Zinc were no doubt employed by the ancients in making Brass.

Zinc (Zn=32) is found in the state of an Oxide, but principally as a Sulphuret (*Blende*), and an impure Carbonate (*Calamine*). From both ores it is first converted into an Oxide by the process of roasting, and then reduced to the metallic form by the aid of Carbonaceous matter, when it may either be fused or sublimed. Until purified by a second distillation, it contains as impurities small portions of other metals, as Iron, Copper, Arsenic, &c.

Prop. Zinc is white, with a shade of blue, and of considerable brilliancy. Sp. Gr. from 6·8 to 7·2. At ordinary temperatures it has little ductility or malleability, but is hard and tough. When heated from 210° to 300°, it becomes ductile, and may be drawn into wire or beaten or rolled into thin sheets, and also at ordinary temperatures, if very pure. At 400° it becomes brittle, and may be powdered. It melts at 773°, and, on slowly cooling, crystallizes in four-sided prisms, with its fracture displaying a lamellated structure. At a white heat it may be volatilized in close vessels; but in contact with the air, it burns vividly, diffusing white fumes of Oxide. Exposed to the air, or kept under water, Zinc becomes covered with a thin film of Sub-oxide, which protects it.

Tests. Almost entirely dissolved by diluted Sulphuric acid (leaving only a scanty greyish-black residuum, E.). This solution is free from colour, and has the other characters of the Sulphate of Zinc. q. v.

Action. Uses. Zinc is not used medicinally. Zinc vessels for keeping articles of diet are not without danger.

Zinci Oxydum, L. E. D.

Oxide of Zinc. *Lana philosophica. Flores Zinci. F.* Oxide de Zinc. *G.* Zinkoxyd.

This Oxide has been long known, in an impure state, by the name of *tutty*, which appears to be of Oriental origin, as the Sulphate is still called *suffed tutia*, or white tutia; the Sulphates of Iron and Copper being called *green* and *blue tutia*. (Hindoo Med. p. 100.) So great is the affinity of Zinc for Oxygen, that it precipitates the other metallic oxides, when in solution, in the form of their respective metals.

Prop. Oxide of Zinc ($Zn\,O=40$), when pure, forms a light flocculent powder of a white colour, devoid of taste and smell, insoluble in water or Alcohol, but soluble in acids and in the caustic fixed alkalis. When heated to a low red heat, it acquires a yellow colour, which disappears again when it cools, unless Iron be present. At a low white heat it may be melted, and when this is increased it is volatilized. It is thrown down in a hydrated form, with some basic Zinc, when an alkali is added to a solution of a salt of Zinc. When pure, its neutral solutions in acids throw down a white precipitate on addition of Sulphuretted Hydrogen, or Ferrocyanide of Potassium. Comp. $Zn\ 80+O\ 20=100$.

Oxide of Zinc used to be prepared (as it still is by the D. P. process, and by some manufacturers, v. P. J. ii. 503) by the action of heat and air on the metal in a crucible, when it becomes oxidized by abstracting Oxygen from the atmosphere. So prepared, it is usually gritty, from containing small particles of metal, some of which may be got rid of by washing in water. The L. and E. P. therefore prepare it by precipitation.

Prep. L. E. Dissolve *Sulphate of Zinc* ℔j. (℥xij. E.) in *Aq. dest.* Oxij. (Oij.

E.) and *Sesqui-carb. Ammonia* ℥ viß. (℥vj. E.) in *Aq. dest.* Oxij. (Oij. E.) (filter, L.) Mix (collect the precipitate on a cloth, E.) Wash it thoroughly with water (squeeze and dry it, E.). Burn it for 2 hours in a strong fire (red heat, E.).

The precipitation was formerly effected with Ammonia; but this, like the other caustic alkalis, when in excess, redissolves the precipitate. Both the salts employed becoming decomposed, a Sulphate of Ammonia remains in solution, and the Carb′ combining with the Oxide of Zinc which is set free, an insoluble Carb. Zinc is precipitated. This being washed and heated, the Carb′ is expelled with any adhering Ammonia, and a pure Oxide of Zinc is left; but it is often sold without being calcined.

Tests. Yellowish-white; tasteless; entirely soluble in dil. Nit without effervescence: this solution is not affected by Nitrate of Baryta, but gives with Ammonia a white precipitate entirely soluble in an excess of the test. E. P. Carb. Zinc is often sold in the shops for the Oxide of Zinc (*p.*); and Mr. Redwood (P. J. ii. 506) states that another kind of Oxide is a mixture of basic Sulph. Zinc and of Hydrated Oxide, and that the Oxides of Zinc of commerce are either basic Carbonates or basic Sulphates, containing only from 64 to 67 per cent. of Oxide. Any Sulphate of Zinc will be detected by the Barytic salt; any basic Chloride, by Nitrate of Silver; Iron, by the yellow colour of the Oxide; and if Carbonates of Lime or Lead are used for adulterating, they will effervesce, and also remain undissolved as well as the Iron in Ammonia.

Inc. Acids and acidulous salts, also caustic alkalis.

Action. Uses. Ext. Dessicative. *Int.* Tonic and Antispasmodic.

D. gr. j.—gr. v. or more, twice a day, in pill.

UNGUENTUM ZINCI, L. E. UNG. ZINCI OXYDI, D.

Prep. L. E. D. Mix well together *Oxide of Zinc* ℥j. (prepared in the same way as Chalk, ℥ij. D.) *Lard* ℥vj. (Simple Liniment ℥vj. E. White Wax Ointment ℥vj. D.)

Mild siccative ointment.

ZINCI CHLORIDUM. Chloride of Zinc. *Butter of Zinc.* *F.* Chlorure de Zinc. *G.* Chlor-Zink.

This ($Zn\,Cl = 68$) may be obtained in solid pieces of a white colour, but it deliquesces rapidly. It is soluble both in water and Alcohol.

Action. Uses. Caustic; considered, on the Continent, superior to others, as inducing a more healthy action in the surrounding parts. Has been used as an Antispasmodic in doses of grs. j.—ij.

ZINCI CARBONAS IMPURUM, D. CALAMINA, L. E.

Calamine. Native Impure Carbonate of Zinc. *F.* Calamine. *G.* Kohlensaures Zinkoxyd.

Calamine, or Carbonate of Zinc ($Zn\,O, C\,O_2 = 62$) is found in various parts of the world. But the same name is applied to two very distinct ores, one being a Silicate of Zinc, and the other, which is alone officinal, a Carbonate of Zinc. It is found in compact or earthy masses, readily scratched with a knife, and breaking with an earthy fracture. It is also found crystallized. Sp. Gr. 3·4 to 4·4. The colour is various; but, as usually seen, it is grey, pinkish, or reddish-yellow. It dissolves in Nitric and other acids with effervescence, and is not rendered electric by heat, by which characters it is easily distinguished from the Silicate, which is also called Electric Calamine. The Carbonic acid and Zinc will be indicated by their respective tests. The crystallized variety is anhydrous.

CALAMINA PREPARATA, L. E. ZINCI CARBONAS IMPURUM PRÆPARATUM, D. Prepared Calamine.

Prep. L. D. Burn *Calamine* with a red heat, triturate and reduce to a very fine powder, as directed for Chalk.

By the process of roasting, some of the Carb′ and water are expelled, and a portion of Oxide of Zinc is formed. The mass being subjected to elutriation, it is often, like Chalk, prepared in little conical masses. It usually contains many impurities, as Oxide of Iron and other metals, and is adulterated with Sulph. Baryta, Carb. Lime, &c. Much of the Calamine of commerce has been shown by Dr. R. D. Thomson and Mr. Murdoch to contain no Zinc, but to be Sulph. Baryta coloured with Armenian Bole.

Tests. Almost entirely soluble in Sul′ with some effervescence, unless it has been too much burnt. The solution gives with Potash or Ammonia a precipitate soluble in an excess of the alkali. L. The impurities will remain undissolved—of a reddish-yellow, if Iron, and the solution will become blue with Ammonia, if Copper is present.

Action. Uses. A dessicant powder for excoriations in children.

CERATUM CALAMINÆ, L. E. UNG. CALAMINÆ, D. Cerate of Calamine. *Turner's Cerate.*

Prep. Take *Calamine* ℔ß. (prepared in the same way as Chalk, 1 part, E. impure *Carbonate of Zinc* prepared and dried ℔j. D.) Add it (well triturated, D.) to *Olive Oil* f℥xvj. mixed with *melted Wax* ℔ß. as soon as they begin to concrete. (*Simple Cerate* 5 parts, E. *Melted Yellow Wax Ointment,* ℔v. D.) Stir briskly till cold, L. (Mix thoroughly, E. D.)

ZINCI SULPHAS, L. E. D.

Sulphate of Zinc. *Sal Vitrioli. White Vitriol. F.* Sulfate de Zinc. *G.* Schwefelsaures Zinkoxyd.

This salt is found native in some places. It is known in India by the name of *suffed tootia*, or white vitriol. (*v.* p. 149.)

Prop. Sulphate of Zinc ($Zn\,O, S\,O_3 + 7\,Aq. = 143$) is a colourless salt, without odour, but having a disagreeable astringent and metallic taste. The crystals are transparent, and large or small, right quadrangular prisms terminated by four-sided prisms (sometimes six-sided, from

Fig. 26.

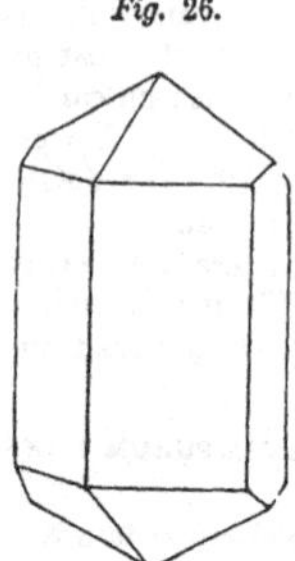

two opposite edges being truncated), often resembling those of Nitre, or, when small and acicular, like those of Epsom Salts. Exposed to dry air, they effloresce, and are soluble in $2\frac{1}{2}$ times of cold, and in less than their own weight of boiling water; insoluble in Alcohol. Heated, they melt in their water of crystallization, which becomes dissipated, except one Eq. which requires from 266° to 284° of temp. to expel it. At a still higher degree, the acid is expelled, and only Oxide of Zinc left. This Oxide is also precipitated when the caustic alkalis are added to a solution of Sulphate of Zinc, and is redissolved in an excess of the Alkali. The Carbonated alkalis throw down Carbonate of Zinc, and Ferrocyanide of Potassium a white gelatinous-looking precipitate, Sulphuretted Hydrogen a white Sulphuret of Zinc. Chloride of Barium and Acetate of Lead will throw down white precipitates of Sulphates of Baryta and of Lead. There are several varieties of this Sulphate of Zinc, forming Subsulphates; but the officinal is composed of $Zn\,O\ 28 + S'\ 28 + Aq.\ 44 = 100$.

Sulph. Zinc may be prepared either by roasting the Sulphuret of Zinc (*i. e.* Blende), exposing it moistened to the air, and purifying this as in the E. P.

Prep. L. D. Pour gradually *Dil. Sul'* Oij. (*Sul'* 20 parts and *Aq.* 120 parts, D.) upon *Zinc* in small pieces ℥v. (13 parts in a glass vessel, D.) The effervescence being finished (digest for some time, D.) filter. Evaporate till a pellicle begins to form, and set aside to crystallize.

E. Proceed as above or repeatedly dissolve and crystallize the impure *Sulphate of Zinc* of commerce until the product, when dissolved in water, yields no black precipitate with *Tincture of Galls*, and corresponds with the characters laid down for *Sulphate of Zinc* in the list of Materia Medica.

This process may answer when the White Vitriol of commerce is sufficiently pure, and it may be made so by precipitating the other metals, by immersing metallic Zinc in the solution. In the former process, water is decomposed, its Hydrogen escapes, and the Oxygen unites with the Zinc, which, in the form of the Oxide, is dissolved by the Sul', and the crystallized salt is obtained by evaporation.

Tests. Sulphate of Zinc, in its crude state, contains several metallic impurities, such as Copper, Lead, and Iron,—the last almost always, even in its purified state. "Totally dissolved by water: what is thrown down by Ammonia is white, and when the Ammonia is added in excess, it is again dissolved." L. P. In the E. P. a solution in six waters is directed to be boiled with a little Nitric acid, and treated as above with Ammonia: "No yellow precipitate (*Peroxide of Iron*) remains, or a trace only, and the solution is colourless." E. P. Magnesia as well as Iron will be detected by its insolubility, Copper by a blue coloured solution. "Arsenic or Cadmium may be detected by adding excess of Sul′ to the solution of the Sulphate, and then passing a stream of Hydrosulphuric′ through it: the Arsenicum and Cadmium are thrown down in the form of Sulphurets." (*p.*)

Inc. Alkalis, and their Carbonates, Lime-water, Salts of Barium and of Lead, Astringent Vegetables.

Action. Uses. Ext. Astringent. *Int.* Astringent; Tonic; in large doses, Emetic.

D. gr. j.—gr. ij. two or three times a day. gr. x.—℈j. in solution as an emetic.

Off. Prep. Liquor Aluminis Compositus.

Zinci Acetatis Tinctura, D.

Prep. Triturate together *Sulph. Zinc* and *Acetate Potash* āā 1 part, add *Rectified Spirit* 16 parts. Macerate for a week, agitating occasionally. Filter through paper.

Double decomposition takes place; Acetate of Zinc remains in solution, the Sulphate of Potash, which is at the same time formed, being insoluble in Spirit, is precipitated. The Acetate of Potash is in excess in the formula. The Acetate of Zinc may be obtained separate from the above, or from a solution crystallized in rhomboidal plates. The taste is astringent and metallic. The presence of the Acetic acid and of the Zinc may be detected by their respective tests. The Tincture or solution has necessarily the same properties, and in the latter form it is frequently prescribed extemporaneously.

Action. Uses. Astringent; used chiefly in collyria and injections.

The Citrate of Iron and Zinc is a new preparation, which is supposed to possess the beneficial effects of both metals.

Cuprum.

Copper. Venus of the Ancients. *F.* Cuivre. *G.* Kupfer.

Copper was, with Gold and Silver, one of the most anciently known of the metals. It is abundantly diffused in nature, being found native as an Oxide, a Sulphuret, and as a Sulphate, Carbonate, Arseniate, and Phosphate.

Copper (Cu=32) is a reddish-coloured metal, of a disagreeable smell when rubbed, and of an unpleasant taste. Sp. Gr. 8·85 to 8·95; very ductile and malleable, and possessed of considerable tenacity. It melts at 1996° (Daniell). Its crystals, obtained on the cooling of the fused metal, are rhomboidal, but those of native Copper are cubes or octohedrons. It oxidizes slowly in the atmosphere, becoming covered with a green crust of Subcarbonate of Copper. It decomposes water only at a bright red heat. Its oxygenation is much promoted by many acids, as the Nitric, and even by the weaker acids when exposed to the air, also by alkaline and fatty substances. There are two Oxides of Copper. One, a *Suboxide* or *Dioxide* (2 Cu O=72), also called Red

Oxide of Copper, which is found native in octohedral crystals; when hydrated, it is of a lively yellow colour. The *Protoxide* (Cu O=40), or Black Oxide of Copper, is, as its name indicates, of a black colour, but which is blue when hydrated. It combines readily with acids, and thus forms the base of the ordinary salts of Copper, which are of a blue or green colour when hydrated, but white when anhydrous.

Copper is easily recognized by its colour, taste, and smell when rubbed. It may be dissolved in Nitric acid, when the same effects will be produced by reagents as in solution of Sulphate of Copper.

Metallic Copper is not officinal, nor possessed of any action on the system, if pure; but oxidized, or combined with acids, is a powerful poison.

Antidotes. Evacuate Stomach; administer white of eggs, and milk, tepid diluents, antiphlogistics.

Cupri Sulphas, L. E. D.

Sulphate of Copper. *Cuprum Vitriolatum.* *Vitriolum Cæruleum.* Blue Vitriol. *Blue Stone.* *F.* Sulfate de Cuivre. *G.* Kupfer-Vitriol.

This salt is produced naturally in the water of many mines. It was no doubt employed by the ancients, as it was by the Arabs and Hindoos: by the latter it is called *neela tootia*, or blue vitriol.

Fig. 27.

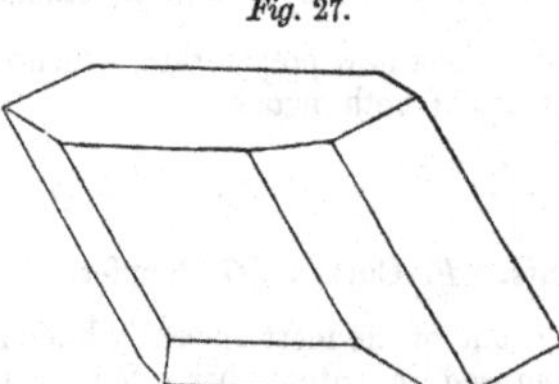

Prop. Sulphate of Copper ($Cu\,O, S\,O_3+5$ Aq.$=$ 125) is of a fine blue colour, without odour, but having a powerful styptic disagreeable taste. Usually seen in fragments of large rhomboidal crystals. Sp. Gr. 22. Exposed to the air, these effloresce, from losing a portion of their water of crystallization. They may be dissolved in four times their weight of cold and in double their weight of boiling water. When heated, the crystals first melt in, then lose their water of crystallization, and fall into a greenish-white crumbly powder; at a still higher heat, the acid is decomposed, sulphurous fumes escape, and the Oxide is left, of a brown colour. The solution is of a light or deep blue colour, according to its strength, from which the alkalis throw down precipitates of a more or less blueish-green colour; if a small quantity be added, then a Subsulphate is precipitated; but if a larger quantity, then a hydrated Oxide. An azure-blue precipitate is produced by Ammonia, which is redissolved by an excess of the precipitant. Sulphuretted Hydrogen throws down a brownish-black precipitate of Sulphuret of Copper.

Ferrocyanide of Potassium a reddish-brown one of Ferrocyanide of Copper. Arsenious acid, if a little alkali (as Ammonia) be added at the same time, throws down a grass-green precipitate of Arsenite of Copper. A polished Iron or Zinc plate, introduced into the solution, becomes covered with a metallic coating of Copper. By these tests the presence of Copper is indicated, that of the Sul′ will be revealed by its tests. (p. 32.) Comp. Cu O 34·48 S′ 34·48 Aq. 41·04=100.

Prep. Sulphate of Copper may be prepared by boiling Copper in diluted Sul′. This acid being decomposed Sulphurous Acid escapes. 1 Eq. Ox. combines with the metal, which dissolves in the remaining Sul′. More commonly by exposing the Sulphuret to air and heat, when both the Sulphur and metal becoming oxidated, combine to form the Sulphate. As the native Sulphuret contains both Iron and Copper, so does the salt produced contain Sulphates of Iron and of Copper. (Some lately sold not less than 52 per cent. Sulph. Iron, P. J. iv. 223.) A great portion of the Sulph. Iron is got rid of by fresh exposure to heat and air, when it is decomposed and the Iron peroxidized, and the Sulphate of Copper separated by solution and crystallization.

Tests. Totally soluble in water. In the air it becomes slightly pulverulent, and of a greenish colour. Whatever Ammonia throws down from its solution, an excess of Ammonia dissolves. L. P. If any Iron be present, it will become green in the air, and not be dissolved by the Ammonia.

Inc. Alkalis and their Carbonates, and many salts, as Borax, Chloride of Calcium, Tartrate of Potash, salts of Lead, Nitrate of Silver, Astringent Vegetable Infusions.

Action. Uses. Irritant, Escharotic, Astringent, Tonic. In large doses, Emetic.

D. gr. ß—gr. ij. Tonic. gr. iv.—gr. xij. Emetic.

Cupri Ammonio Sulphas, L. Cuprum Ammoniatum, E. D.

Ammoniæ Cupro-Sulphas. Ammonio-Sulphate of Copper. *F.* Cuivre Ammoniacal. *G.* Kupfer-Salmiak.

This preparation forms a beautiful azure-coloured powder, which has a strong odour of Ammonia, and a disagreeable Coppery taste. It is prepared by rubbing together Sulph. Copper and Sesquicarb. Ammonia; and as a portion of the latter remains in excess, the Ammoniacal odour and an alkaline reaction on test-paper are observed. From the volatile nature of the Ammonia, it escapes when the preparation is exposed to the air, or is too much dried, and is therefore apt to vary in its properties. When well made, it ought to dissolve in water; but if there be a deficiency of Sesquicarb. Ammonia, it is not only insoluble, but is further decomposed with precipitation of Dioxide (Disulphate?) of Copper. Many of the characteristics of the Sulphate of Copper are displayed by this preparation; but the solution of Arsenious acid renders it of a green colour, from combining with the Oxide of Copper, and forming the green insoluble Arsenite of Copper. As the composition of this substance has not been accurately determined by chemists, its mode of preparation may be first considered, and that is essentially the same in the three Pharmacopœias, with differences only in the quantities.

Prep. L. E. D. Rub together (in an earthenware mortar, D.) *Sulph. Copper* ℥j. (℥ij. E. 2 parts, D.) and *Sesquicarb.* (*Carb.* E. D.) *Ammonia* ℥iß. (℥iij. E. 3 parts, D.) till Carb′ is no longer evolved. Wrap the mass in bibulous paper and dry it in the air. (Preserve in closely stoppered bottles, E. D.)

On rubbing together the Sulph. Copper and Sesquicarb. Ammonia, effervescence ensues, from the escape of a great portion of the Carb′, and the mixture becomes of a deep blue colour, and moist from the water of crystallization of the salts; a portion of the Sul′ combines with the Am. set free; some Sulph. Ammonia, therefore, is formed. A portion of Carb′ combines with the Oxide of Copper set free, and forms Carbonate of Copper, which forms a mixture with the excess of Sesquicarbonate of Ammonia. Others consider it a double salt of Sulph. Ammonia and Subsulph. Copper combined together and mixed with the excess of Carb. Ammonia. Dr. Kane conceives it to be a compound of Sulph. Ammonia with the Ammoniacal Oxide of Copper, or Cuprate of Ammonia. The double salt of Sulph. Copper and Ammonia prepared by dissolving Sulph. Copper in Ammonia to saturation, and which may be obtained in a crystallized state, consists of 1 Eq. Sul′+1 Eq. Ox. Copper+2 Eq. Ammonia+1 Eq. Aq., and it is probable that the officinal salt resembles it nearly in constitution.

Tests. By heat it is converted into Oxide of Copper, evolving Ammonia. Dissolved in water, it changes the colour of Turmeric, and solution of Arsenious acid renders it of a green colour. L. P.

Inc. Acids, Potash and Soda, Lime-water.

Action. Uses. Irritant, Astringent, Tonic, Antispasmodic, Emetic.

D. gr. $\frac{1}{4}$—gr. v. gradually increased.

Liquor Cupri Ammonio-Sulphatis, L. Cupri Ammoniati (Solutio, E.) Aqua, D.

Prep. L. E. D. Dissolve *Ammonio Sulphate of Copper* (*Ammoniated Copper*, E. D.) ʒj. (1 part, D.) in *Aq. dest.* (*Water*, E.) Oj. (100 parts, D.) Filter (through paper, D.).

This solution is of a fine blue colour; but Mr. Phillips remarks, that unless the Ammonio-Sulphate of Copper retain some excess of Sesquicarbonate of Ammonia, the salt is decomposed, and one-half of the Oxide of Copper is precipitated.

Pilulæ Cupri Ammoniati, E.

Prep. Take *Ammoniated Copper* finely powdered 1 part, *Bread-crumb* 6 parts, *Solution of Carbonate of Ammonia* q. s. Beat into a proper mass and divide into pills, each to contain $\frac{1}{2}$ a grain of Ammoniated Copper, of which one may be taken two or three times daily, and the dose gradually increased.

Cupri Acetas. *Crystalli*, D.

Acetate of Copper, crystallized. *F.* Acetate neutre de Cuivre. Verdel cristallise. *G.* Essigsaures Kupferoxyd.

This salt is officinal only in the D. P. It is made principally at Montpelier in France by dissolving Verdigris in Vinegar with the assistance of heat, and afterwards crystallizing. The crystals are oblique rhomboidal or rhombic octohedral prisms, of a deep bluish-green colour and disagreeable metallic taste; efflorescent; soluble in water, and also a little in Alcohol. This salt has been employed for obtaining Acetic acid. It possesses similar medical properties with Verdigris, but is more virulent.

Ærugo. Diacetas Cupri impura, L. E. Cupri Subacetas, D.

Verdigris. Commercial Diacetate of Copper. *F.* Vert de gris. Acetate basique de Cuivre. *G.* Grünspan.

Ærugo or Verdigris must have been early known, from the employment of Copper vessels, as well as of Vinegar and sour Wines. It was in fact employed by the Greeks as a medicine, also by the Arabs, and it probably was so by the Egyptians. There is little doubt, however, that the term Ærugo, as well as the χαλκοῦ ἰὸς, included the Carbonate as well as this Acetate of Copper.

Prop. Verdigris is sold either in powder or in amorphous masses, of a pale blueish-green or of a bright blue colour, with an odour of Vinegar, and a disagreeable Coppery taste. It remains unchanged in the air; but when heated, water is first expelled, and then Acetic′, Oxide of Copper being left, with some metallic Copper. Verdigris is insoluble in Alcohol, and decomposed by water, being resolved into an Acetate which dissolves in it, and a dark-green powder, which afterwards becomes black, is left, and is a Subacetate or a Tribasic Acetate of Copper. Sul′ decomposes it with effervescence. There are several compounds of Acetic′ and Copper, but the bluish-coloured Verdigris (Acetate de Cuivre bi-basique) is that alone employed in medicine. The composition of this has been shown by Mr. R. Phillips to be 2 Eq. of Copper combined with 1 of Ac′ and 6 of water; that is, a Hydrated Diacetate ($2\,Cu\,O+A'+6\,Aq.=185$), or $A'\ 27{\cdot}57+Cu\ 43{\cdot}24+Aq.\ 29{\cdot}19=100$, the Acetate bibasique or bicuivrique of the French. But Verdigris, when of a green colour, also contains portions of the Subsesquiacetate and the Trisacetate, or Acetate sesquibasique and tricuivrique; sometimes also some of the Carbonate.

Prep. Verdigris may be made by acting on plates of Copper with Vinegar or with Ac′. Under the influence of air the Copper becomes oxidized, combines with the acid, and is scraped off as formed. In the South of France sheets of Copper are stratified with the refuse of the grape in making wine, and allowed to remain for a month or six weeks. Acetous fermentation takes place, and the Copper becomes coated with Verdigris, which is scraped off, and the operation repeated. The different scrapings form a paste, which is well beaten with wooden mallets, then packed in leathern bags.

Tests. Almost entirely soluble in Ammonia and in dil. Sul′ with the aid of heat; partially soluble in water. L. It is dissolved in a great measure by Mur′, not above five per cent. of impurity being left. The Sul′ forms a blue and the Mur′ a green solution.

Inc. Strong Acids, Alkalis, and their Carbonates.

Action. Uses. Detergent, Escharotic, Tonic, Emetic.

D. Usually applied externally only. Dose internally, gr.ß.

CUPRI SUB ACETAS PRÆPARATUM, D. Prepared Verdigris.

Prep. Triturate the *Sub-Acet. Copper* into powder and separate the finest parts as directed in Prepared Chalk.

The water converts the salt into a soluble Acetate (*v.* supra) and an insoluble Subacetate. Used only as an Escharotic and Stimulant.

LINIMENTUM ÆRUGINIS, L. OXYMEL CUPRI SUBACETATIS, D.
Mel Ægyptiacum. Unguentum Ægyptiacum.

Prep. Dissolve powdered *Verdigris* ℥j. in *Vinegar* (distilled, D.) f℥vij. Strain through linen. Add *Clarified Honey* ℥xiv. Boil to a proper consistence.

Action. Uses. Stimulant and slightly Escharotic. Applied with a camel's-hair brush to ulcers; or, diluted, employed as a gargle.

UNGUENTUM ÆRUGINIS, E. UNGUENTUM CUPRI SUBACETATIS, D.

Prep. Melt (White, D.) *Resinous Ointment* ℥xv. (℥xij. D.), sprinkle into it finely powdered *Verdigris* ℥j. (℥ß. rubbed up with *Olive Oil* ℥j Mix, D.) Stir the mixture briskly as it cools and concretes, E.

Similar in its nature and uses to the preceding, especially as an application to foul ulcers.

PLUMBUM, E.

Lead. Saturn of the Alchemists. *F.* Plomb. *G.* Blei.

Lead (Pb=104) is one of the most anciently known of the metals. It is found chiefly as Sulphuret or Galena, but a little in a metallic state also, as an oxide, and combined with several acids.

Prop. Lead is of a bluish-grey colour; fresh cut, is of great brilliancy, but soon tarnishes; has a slight taste, and a peculiar smell when rubbed; is soft, marks paper; has little tenacity, but may be beaten into thin sheets. Sp. Gr. 11·435. It melts at 612°, and boils at a red heat: on solidifying, it contracts, and may be crystallized in octohedrons. Exposed to the air, the surface becomes covered with a greyish pellicle. It may be preserved unchanged under perfectly pure water; but if any air be present, the Lead becomes oxidated, and some of it combining with Carb′, white Carb. Lead with Hydrated Oxide (p. 161) is formed. This change also readily takes place when rain or any other equally pure water is exposed to the action of air in Leaden cisterns or pipes. The vessels become coated with a white incrustation of a pearly lustre, and some of the fine crystals may be seen floating on the water, and being also (dissolved or) suspended, it thus becomes poisonous. Dr. Christison ascertained that the opinion of G. Morveau was correct, that the presence of any neutral salts, especially of Sulph. and Carb. of Lime, which are usually present in spring or river water, or a minute trace of Sul′ or Phosp′ prevents the continued corrosion of the Lead. The Oxide and Carb. being allowed to be deposited and to adhere with firmness to the Lead, and the other insoluble salts, such as Sulph. Phosph. and Carb. Lead, becoming added, protect the Lead with an impenetrable coating, instead of its being carried away by the water. The late Prof. Daniell ascertained that when water contains free Carb′, Lead is readily dissolved, and water cannot therefore be safely kept in or transmitted through that metal. (Morson, P. J. ii. 335.)* Though at ordinary temperatures Lead is little liable to oxidate, in a state of very fine subdivision, it will take fire when exposed to the air. It is also readily oxidated and dissolved by Nit′. (*v.* Acet. and Nitr. Lead for the effect of reagents.) It unites with Oxygen in several proportions, also with Sulphur, Phosphorus, Iodine, and Chlorine; likewise with numerous metals, and, in the state of Protoxide, forms with the acids and fatty substances various compounds. Zinc separates it from any of its solutions. Lead is obtained very pure, for chemical purposes, from the Nitrate; but, on the great scale almost entirely from the Sulphuret or Galena, by the process of roasting, when Sulphate and Oxide of Lead are formed. To these Lime and Carbonaceous matter are added, the former for the purpose of decomposing the Sulph. Lead, and the other will unite with the Oxygen of the Oxide of Lead, and thus both the Oxide and Sulphate are reduced to a metallic state.

Tests. Lead is usually sufficiently pure for Pharmaceutical purposes, but often contains Iron or Copper. These may be detected by dissolving it in dilute N′, and precipitating with a little excess of Sul′. On the addition of Ammonia to the filtered solution, it will become of a violet if Copper, and of a yellow colour if Iron, be present. If a plate of Zinc be introduced into solutions of Lead, this metal will be deposited in an arborescent form. Alkalis, combining with the acid of its salts, throw down the Hydrated Oxide; Sul′ and its salts, a white insol. Sulphate; Chromate of Potash, a yellow Chromate; Hydrosul′ and its salts, a black Sulphuret; Iodide of Potassium a yellow Iodide, and the Ferrocyanide of Potassium a white Ferrocyanide of Lead.

Oxides of Lead.

There are several Oxides of this metal. 1. The Suboxide or Dioxide of a dark grey colour, supposed by Berzelius to form the pellicle

* Water which contains less than 8000th of salts (such as Carbs. and Sulphs.) in solution (a 4000th or even a larger proportion of Muriates is insufficient) cannot be safely conducted in Leaden pipes without certain precautions. A remedy may be found either in leaving the pipes full of water and at rest for 3 or 4 months, or by substituting for the water a weak sol. of Phosph. Soda in the proportion of about a 25,000th part. Christison, Edin. Phil. Trans. xv. Part ii.; Pharm Journ. ii. 335; R. Phillips, Junr. P. J. iv. 304.

which covers Lead that has been exposed. 2. PROTOXIDE, of a yellow colour, commonly called *Massicot*, and of which Litharge and the Plumbi Oxydum Hydratum are two officinal forms. 3. BINOXIDE, sometimes called Peroxide, Red Lead, and Minium, Plumbi Oxydum Rubrum, E., supposed by some to be a compound of the Protoxide; and the 4th, a BROWN, BIN-, or PEROXIDE.

PLUMBI OXYDUM (SEMIVITREUM) L. D. LITHARGYRUM, E.

Oxide of Lead, fused. Litharge. *F.* Protoxide de Plomb. *G.* Bleioxyd.

This Oxide of Lead ($PbO=112$) was known to the ancients, being easily produced when melted Lead continues to be exposed to a current of heated air. The surface of the metal becomes rapidly covered with a scaly powder of a Sulphur-yellow colour, which is the Protoxide of Lead, and which, being skimmed off, is known in commerce by the name of *Massicot.* When the heat is continued to a bright red, some metallic Lead is separated, the Oxide is fused, though imperfectly, and on cooling becomes an aggregated mass, which readily separates into crystalline scales, of a greyish-red colour. These form the *Litharge* of commerce, which varies in colour, and is called *Gold Litharge* when of a red colour, owing to the presence of a little *Red Lead*, but *Silver Litharge* when lighter coloured. These are frequently obtained in the process of refining Gold and Silver by means of Lead, and in separating the Silver from Argentiferous Lead.

Prop. Litharge is nearly insoluble in water; tasteless; Sp. Gr. 9·42. It may be melted into a glass, but is readily reduced to a metallic state if Carbon be present. It is remarkable for its power of depriving many vegetable substances of colour. It is readily dissolved in diluted Nit′ or Acet′, as also in some other acids, and absorbs Carb′ from the atmosphere. Comp. $Pb\ 92{\cdot}85+O\ 7{\cdot}14=99{\cdot}99$.

Tests. The Litharge of commerce is liable to contain a little Iron, also Copper, Carb. Lead, Silica, and other earths. "Almost entirely soluble in dil. Nit′." L. P. 50 grs. dissolve entirely, without effervescence, in Pyroligneous acid f℥j.ß; and the solution, precipitated by 53 grs. of Phosph. Soda, remains precipitable by more of the test. E. P. If the whole is soluble in Nit′, it proves that neither Sulph. Baryta nor Sulph. Lead is present; and the want of effervescence, that there is no Carb. Lead. The Phos. Soda will detect all ordinary impurities, as in the case of Acet. and Carb. Lead. (*c.*) Copper and Iron will be detected by their respective tests.

Action. Uses. Litharge, like the other preparations of Lead, will affect those exposed to its influence; but it is only employed Pharmaceutically, as for making Diacetate of Lead, and by combining with oil to form the Lead Plaster, which is the basis of several others.

EMPLASTRUM PLUMBI, L. EMPLASTRUM LITHARGYRI, E. D. Litharge or Lead Plaster. Diachylon Plaster.

Prep. L. E. D. Mix *Oxide of Lead* rubbed to very fine powder ℔vj. [(*Litharge* ℥v. E.) (℔v. D.)] *Olive Oil* C j. (f℥xij. E.) *Aq.* Oij. (f℥iij. E.) boil over a slow fire (heat between 200° and 212°, D.) stirring constantly till the oil and oxide unite into the consistence of a plaster. Add a little boiling water, if that used at the beginning evaporate too much.

Action. Uses. Forms excellent strapping from its mildness, and is useful in surgical cases in keeping together the lips of wounds.

Off. Prep. Emp. Galbani, L. Emp. Hydrargyri, L. Emp. Opii. Emp. Saponis.

EMPLASTRUM RESINÆ, L. EMP. RESINOSUM, E. EMP. LITHARGYRI CUM RESINA, D. Emp. Adhæsivum. Resin Plaster, Adhesive or Sticking Plaster.

Prep. L. E. D. Take *Lead Plaster* ℔iij. [(*Litharge Plaster* ℥v. E.) (℔iij ß. D.)] *Resin* ℔ß. (℥j. E.) Melt the Plaster of Lead with a gentle heat, add the Resin in very fine powder and make a plaster. (Stir well till the mixture concretes in cooling, E.)

Action. Uses. This plaster, serving the same purposes as the Lead plaster, is more frequently employed on account of being more adhesive; but is objectionable in some cases, in consequence of being more irritant to the skin.

UNGUENTUM PLUMBI COMPOSITUM, L. Comp. Ointment of Lead.

Prep. Melt over a slow fire *Plaster of Lead* ℔iij. in *Olive Oil* Oj.; then mix *Prepared Chalk* ℥viij. and *Distilled Vinegar* f℥vj. When the effervescence subsides add this to the plaster. Stirring constantly till cold.

Action. Uses. The Acet. Lime formed is mixed with Lead Plaster and a mild Ointment produced, which is used as a dressing to ulcers.

PLUMBI OXYDUM HYDRATUM, L.

Hydrated Oxide of Lead.

Prop. This Hydrated Oxide ($2\,PbO+HO=233$) forms a white and heavy tasteless powder, which is officinal on account of possessing eminent decolorizing properties. It is sparingly soluble in pure water (1 part in 12,000 Aq.), which thus acquires an alkaline reaction; but it is not dissolved when neutral salts (p. 159) are present in the water. It is soluble in solution of Potash, and therefore an excess of this should not be used in preparing it. When evaporated, a crystallized alkaline compound is left. It is also soluble in Nitric and other acids, *v.* supra. This Oxide is also formed, as ascertained by Capt. Yorke, by the joint action of air and distilled water on Lead. Dr. C. has ascertained that a portion of this combines with the Carb′ of the atmosphere, and forms the Carb. Lead which he considers is in the proportion of $3\,Pb\,O$, $2\,C\,O_2+1$ Aq. united with the Hydrated Oxide; or rather, a compound of 2 Eqs. Carbonate of Lead in union with 1 Eq. of Hydrated Oxide of Lead.

Prep. Mix solution of *Diacet. Lead* Ovj. *Aq. dest.* Cong. iij. and solution of *Potash* Ovj. or q. s. to precipitate the oxide. Wash the precipitate with water till nothing alkaline remains.

Here the Potash combining with the Acet′, an Acet. Potash remains in solution. The Oxide in separating combines with some water, and is precipitated as Hydrated Oxide of Lead.

Tests. This preparation should possess all the characters of the Oxide of Lead, and dissolve entirely in Nit′ without effervescence.

Action. Uses. Employed only in the preparation of Disulphate of Quinine.

Plumbi Oxydum Rubrum, E.

Red Oxide of Lead. Red Lead. Minium. *F.* Oxide rouge de Plomb. *G.* Mennig Rothes Bleioxyd.

Red Lead or Minium, as this Oxide is called, was known to the Arabs and is the Suṛanj of Avicenna, commonly translated Cinnabar. It is prepared by the Hindoos, and is their *Sundoor.* Dioscorides (lib. v. c. 109) knew that Minium was distinct from Cinnabar. It is sometimes called Binoxide, also Deutoxide of Lead.

Prop. Red Oxide of Lead ($Pb_3\ O_4=344$), or Minium, forms a powder in scales of a bright red colour; tasteless, heavy (Sp. Gr. about 9), and insoluble in water. When heated, it melts, then gives out Oxygen, and becomes converted into the Protoxide. If placed on Charcoal, it is reduced, with the flame of the blowpipe, to a globule of Lead. Completely dissolved "in highly fuming Nitrous acid." E. Protonitrate of Lead is formed, as the acid by uniting with the excess of Oxygen, is converted into Nit′. It is only "partially soluble in dilute Nit′," E., being converted into two Oxides, a brown-coloured Peroxide, which is left, and a Protoxide, which is dissolved. The composition of the Red Oxide has not been definitively settled; indeed, the Red Lead of commerce is considered to be variable in composition. But 2 Eq. of Protoxide with 1 of the Binoxide, may be considered one of the most common proportions.

Prep. Obtained by exposing Massicot or the Protoxide to air and heat below what is required for fusion, when the quantity of Oxygen required is absorbed, and the yellow Protoxide converted into bright red coloured Minium.

Tests. Red Lead is not very liable to adulteration, but brick-dust or red-bole (insoluble in Nitrous acid) is sometimes mixed with it; also Red Oxide of Iron, which may be detected by testing the solution in Nitric acid with Tincture of Galls.

Action. Uses. Red Lead might be employed for the same purposes as the Protoxide. It is officinal in the E. P. for purifying concentrated Acetic acid, and for making Aqua Chlorinii.

PLUMBI CHLORIDUM, L.

Chloride of Lead. *Magisterium Saturni s. Plumbi.* *F.* Chlorure de Plomb. *G.* Chlor-Blei.

Chloride of Lead (Pb Cl=140) is found in nature, and known to mineralogists by the name of *Horn Lead.* It is met with in opaque horny masses, or in colourless transparent crystals; but the officinal Chloride of Lead is a white heavy powder, devoid of taste. It fuses at a temperature below redness, and becomes on cooling a semi-transparent mass. Sp. Gr. 5·133. It is soluble in 22 parts of boiling water, from which, on cooling, a part separates in small brilliant crystals: it is much less soluble in cold water. The presence of Chlorine is indicated by the white precipitate formed by Nitrate of Silver, and that of Lead by the foregoing tests. Comp. Pb 74·3+Cl 25·7=100.

Prep. Dissolve *Acet. Lead* ℥xix. in boiling *Aq. dest.* Oiij. and *Chlor. Sodium* ℥vj. in *Aq. dest.* Oj. Mix the solutions. When cool wash the precipitate with *Aq. dest.* and dry it.

The Acetic′ combines with the Soda formed by the Oxygen of the Oxide of Lead combining with the Sodium, and Acetate of Soda remains in solution, while the Chlorine of the Chloride being transferred to the Lead, forms an insoluble Chloride of Lead. Besides this, Mr. Phillips states that a double salt is formed, which is to a considerable extent soluble in water.

Tests. Totally dissolved by boiling water, the Chloride concreting almost entirely into crystals as it cools. On the addition of Hydrosul′ it becomes black, and by heat yellow. In the first case, a Sulphuret of Lead is formed, which is converted into the yellow Protoxide by heat.

Uses. Used only for making Morphiæ Hydrochloras.

PLUMBI IODIDUM, L.

Iodide of Lead. *F.* Iodure de Plomb. *G.* Iod-Blei.

Iodide of Lead (Pb I=230) has only recently been discovered and introduced into medicine. It may be formed by the direct action of Iodine on Lead, or as below. It is usually seen in the form of a fine yellow powder, which is without taste or smell, insoluble in cold but readily dissolved in boiling water, forming a colourless solution. But as this cools, it becomes deposited in brilliant scale-like crystals of a golden-yellow colour, in which form it is also sold, and is then very pure. It is first melted and then decomposed by heat, the Iodine being dissipated in violet-coloured vapours. It is soluble in Acet′ and Alcohol, also in solution of Potash. Comp. Pb 45·22+I 54·78=100.

Prep. L. E. Dissolve *Acet. Lead* ℥ix. (*Nitr. Lead* ℥ß. E.) in *Aq. dest.* Ovj. (f℥xv. E.) and *Iodide Potassium* ℥vij. (℥ß. E.) in *Aq. dest.* Oij. (f℥xv. E.) Mix the solutions. (Filter through linen or calico, E.) Wash the precipitate. Dry it. (Boil the powder in *Aq.* Cong. iij. acidulated with *Pyroligneous Acid* f℥iij. Keep the liquid near boiling, and let all insoluble matter subside. (Pour off the clear liquor, from which on cooling the Iodide will crystallize, E.)

By mutual decomposition, Iodide of Lead is formed and precipitated, while the Oxygen of the Oxide of Lead of the Acet. being transferred to the Potassium, Potash is formed, which, uniting with the acid, forms a soluble Acetate of Potash. Nitr. Lead is preferred in the E. P., because an excess of the Oxide of Lead is apt to exist in the Acetate, which is injurious because disposed to form compounds with the Iodide. The process of boiling in Pyroligneous acid and subsequent crystallization gets rid of any Oxide or Carb. of Lead. The same effect may be obtained by adding a little Acet′ to the solution of the Acet. Lead previous to pouring it into the solution of the Iodide of Potassium.

Tests. Totally dissolved by boiling water, and as this cools separating in shining yellow scales. It melts by heat, and the greater part is dissipated first in yellow (Iodide of Lead) and afterwards in violet vapours. L. P. Bright yellow: 5 grs. are entirely soluble, with the aid of ebullition, in f℈j. of Pyroligneous acid diluted with a f℥j.ß of distilled water, and golden crystals are abundantly deposited on cooling. E. P.

Action. Uses. Deobstruent, generally used externally, but sometimes internally.

D. gr. ¼ to gr. ij.

UNGUENTUM PLUMBI IODIDI, L.

Prep. L. Mix well together *Iodide of Lead* ℥j. and *Lard* ℥viij

Discutient in chronic enlargements of the joints and in scrofulous tumours.

PLUMBI CARBONAS, L. E. D.

Carbonate of Lead. *Subcarbonate of Lead. Cerussa. White Lead.* *F.* Carbonate de Plomb. *G.* Kohlensaures Bleioxyd. Bleiweiss.

Carbonate of Lead ($PbO, CO_2=134$) is one of the most anciently known of the metallic salts. It is found as a mineral in many Lead districts, and known as Cerusse and White Lead ore.

Prop. Carb. Lead is white, tasteless, and heavy; either in powder or in white amorphous masses. When found in nature, its crystals vary much in form from the right rhombic prism from which they proceed. Sp. Gr. about 6·25. It is insoluble in water. Heated, the Carb′ escapes, and the yellow Protoxide of Lead is left; on Charcoal and before the blowpipe, it is reduced to a globule of metallic Lead. It is soluble with effervescence both in the Acetic and the Nitric acids. Acetate and Nitrate of Lead being the result, will be affected by reagents like other preparations of Lead. Comp. Pb O 83·5 + C′ 16·5 = 100.

Carbonate of Lead may be prepared by the action of air and pure water on metallic Lead, or by passing a current of Carb′ gas through a solution of Diacet. Lead. The most ancient method, and that which makes the best White Lead for the use of painters, is by exposing sheets of Lead to the fumes of Vinegar or of strong Acet′. The latter is placed in an earthen vessel, and the former a little above it, either rolled spirally or cast into bars. The pots are arranged side by side, and imbedded in a mixture of new and spent tan (ground oak bark). The tan gradually heats or ferments, and begins to exhale vapour, the temperature of the inner parts of the stack rising to 140° or 150°, or even higher. The Acet′ is slowly volatilized, and its vapour passing readily through the gratings or folds of Lead, gradually corrodes the surface of the metal, upon which a crust of Subacet. is successively formed and converted into Carb., there being an abundant supply of Carb′ furnished by the slow fermentative decomposition of the tanner's bark. The White Lead, crushed and broken up, is transferred to mills, where it is ground up into a thin paste with water, washed, and dried.

Tests. "Dissolved with effervescence in dil. Nit′. What is precipitated from the solution by Potash is white, and is redissolved by excess of it: it becomes black on the addition of Hydrosul. It becomes yellow by heat, and with the addition of Charcoal, it is reduced to metallic Lead." L. P. If Sulph. Baryta be present, it will remain undissolved by the Nit′; if Chalk, it will be converted into Nitr. Lime. When the Lead has been precipitated by Potash, Lime will remain undissolved when an excess of the latter is added: or we may remove the Lead by means of H S′, and test for Lime by means of Oxalate of Ammonia. "It does not lose weight at a temperature of 212°: 68 grs. are entirely dissolved in 150 minims of Acet′ dil. with Aq. dest. f℥j.; and the solution is not entirely precipitated by a solution of 60 grs. of Phosph. Soda." Dr. C. states that 60 grs. of this Phosphate exactly decompose 67·05 grs. of pure Carb. Lead. Hence with 68 grs. of the Carb., as in the formula, some Lead must remain in solution if the Carb. be pure, and will be shown on adding more Phosphate to the filtered liquor. But if any of the ordinary adulterations be present, there will be an excess, not of Lead, but of Phosphate of Soda in solution.

Action. Uses. Dessicative and Astringent. Applied only externally, as by dusting on excoriations. Dr. A. T. Thomson considers it the only poisonous salt of Lead, the others becoming so by being first converted into Carbonate of Lead.

UNGUENTUM PLUMBI CARBONATIS, E. D.

Prep. E. D. Mix thoroughly *Simple Ointment* ℥v. E. (*White Wax Ointment* ℔j. D.) and *Carb. Lead* ℨj. (in very fine powder, ℨij. D.)

Action. Uses. Cooling and drying application to excoriations and burns, also to ulcers and eruptions attended with irritation.

PLUMBI NITRAS, E.

Nitrate of Lead. *F.* Nitrate de Plomb. *G.* Salpetersaures Bleioxyd.

Prop. Nit. Lead (Pb O, N O_5=166) crystallizes in octohedral and tetrahedral crystals, is soluble in about 4 parts of water. It increases the combustion of burning fuel, and is reduced to the state of metallic Lead. Heated, it evolves the fumes of Nitrous acid gas; Oxide of Lead being left.

Prep. With the aid of gentle heat saturate *dilute Nitric'* Oj. with *Litharge* ℥ivß. Filter and set aside to crystallize. Concentrate the remaining liquor to obtain more crystals.

Action. Uses. It will act as the other preparations of Lead, but is officinal in the E. P. for making Iodide of Lead.

PLUMBI ACETAS, L. E. D.

Acetate of Lead. *Plumbi Superacetas. Saccharum Saturni. Sugar of Lead. F.* Acetate acide de Plomb. *G.* Bleizucker.

Acet. Lead (Pb O, Ac′ + 3 Aq. = 190) seems to have been known at least since the cultivation of chemistry in the middle ages.

Fig. 28.

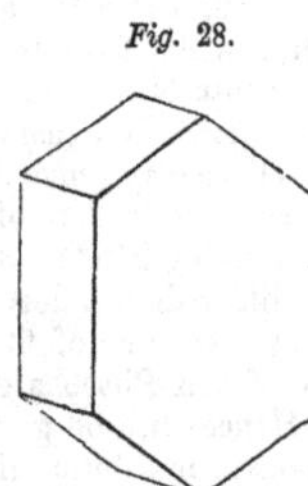

Prop. It is of a white colour, and of a sweet astringent taste, with a smell of Vinegar. The crystals form brilliant needles, or small tetrahedral prisms terminated by dihedral summits, and usually occur in an agglomerated mass; unalterable in the air, except when it is very dry, when they slightly effloresce; soluble in 4 parts of water and in Alcohol. The watery solution is capable of dissolving a portion of Oxide of Lead, which converts it into a Diacetate of that metal. Heated, the crystals fuse in their water of crystallization, which at a higher temperature is dissipated; the white mass which remains again melts, Acetic and Pyroacetic acids are given off, and a globule of Lead may finally be obtained. The Acet. may be partially decomposed by the Carb′ of the atmosphere, also by that contained in water; hence the hazy appearance when it is dissolved in it, and which may be removed by the addition of a little Acet′. It is decomposed by a number of acids, the Sulph′ causing the exhalation of the odour of Vinegar, also by the alkalis, numerous salts, infusion of Galls, most Vegetable principles, also Milk and Albumen. Comp. Pb O 58·9+Ac′ 26·8+Aq. 14·3=100.

Prep. L. E. D. Mix *Acet′* Oiv. (Oij. E.) with *Aq. dest.* Oiv. (Oj. E.) add to it *Oxide of Lead* (powdered) ℔iv. ℥ij. (Litharge ℥xiv. E. Carb. Lead (Cerusse) q. s. and Distilled Vinegar 10 times the weight of the Carb. D.) Dissolve with the aid of a gentle heat. (Digest in a glass vessel till the Vinegar becomes sweet· Pour off the liquor, add more Vinegar and proceed as before, D.) Strain, evaporate for crystallization (on cooling, E. Again evaporate and crystallize. Dry the crystals in the shade, D.)

In these processes the Acet′ combines either directly with the Oxide of Lead or by first expelling the Carb′. It is frequently prepared, as above, by dissolving Litharge in Pyroligneous acid; or by exposing Lead, half immersed in Acet′, to heat and air. The Lead being oxidized or converted into the Carbonate, will, on falling into the Acet′, be formed into the Acetate.

Tests. Dissolved by distilled water. By Carb. Soda a white precipitate (Carb. Lead) is thrown down from the solution, and by Iodide of Potassium a yellow one (Iodide of Lead); by Hydrosul′ it is blackened (Sulphuret Lead being formed). Sulph′ evolves Acetic vapours. By heat it first fuses, and is afterwards reduced to metallic Lead. L. P. "Entirely soluble in distilled water acidulated with Acetic′: 48 grs. thus dissolved, are not entirely precipitated by a solution of 30 grs. of Phosph. Soda." E. P. As 30 grs. of the latter salt will just decompose 47·66 grs. of Acet. Lead, if 48 grs. or a 144th part more of the latter be used, the solution will be affected, after filtration, by a further addition of Phosphate, provided the Acetate be tolerably pure.

Inc. Sulph′, H Cl′, Carb′, Cit′, and Tart′; Lime-water, Potash, and Soda (but in excess it redissolves); Hard Water from the salts and Liq. Ammoniæ Acet. from the Carb′, they contain.

Action. Uses. Irritant Poison; Astringent and Sedative in profuse discharges and Hæmorrhages. Lotion ℈j.—ʒv. or ℥viij. of fluid. Externally as an Astringent Collyrium and Wash.

D. gr. j. and ij.—gr. x. 2 or 3 times a day, especially with dil Acet′ or distilled Vinegar.

Antidotes. Evacuate the stomach with Sulph. Zinc and warm diluents; give Sulph. Soda and of Magnesia, and Phosphate of Soda.

CERATUM (UNGUENTUM, E. D.) PLUMBI ACETATIS, L. *Ung. Cerussæ Acetatæ. Unguentum Saturninum.* Cerate of Sugar of Lead.

Prep. L. E. D. Mix *Acet. Lead* powdered ʒij. (ʒj. E. D.) with *White Wax* ℥ij. and *Olive Oil* f℥viij. (Simple Ointment ℥xx. E. Ointment of White Wax ℔iß. D.) Dissolve the Wax in f℥vij. of the Oil, gradually add the Acetate triturated in the rest of the Oil. Stir the mixture as it concretes.

Action. Uses. Useful as a cooling application to burns, blistered surfaces, and irritable sores.

PILULÆ PLUMBI OPIATÆ, E. Acetate of Lead and Opium Pills.

Prep. Take *Acet. Lead* 6 parts, *Opium* 1 part, *Conserve of Red Roses* 1 part, beat into a proper mass and divide into 4 grain pills. They may also be made with twice as much Opium. Each pill contains 3 gr. Acet. Lead and ½ gr. of Opium.

In this pill chemical changes take place, Acetate of Morphia and Meconate of Lead being formed; but the pill has been so long employed, and its therapeutical value so fully confirmed, that the change would seem to be rather an advantage than a detriment. The E. C. do not indicate how the two kinds of Pill are to be distinguished, or how the change, if expressed, will differ from an ordinary prescription.

PLUMBI DIACETATIS, (SOLUTIO, E.) LIQUOR, L. D.

Plumbi Subacetatis Liquor D. *Aqua Lithargyri Acetati. F.* Sous-acetate de Plomb. *G.* Halb-Essigsaures Bleioxyd.

Diacet. Lead (2 Pb O, Ac′+10 Aq.=365, when solid and Hydrated), in the form of solution commonly called *Goulard's Extract*, was by himself called *Extract of Saturn*, that is, of Lead. It seems to have been known since the time of B. Valentine, and owes its formation to the solution of Acetate of Lead dissolving a further proportion of the Oxide, and thus becoming converted into the Diacetate of Lead.

Prep. L. E. D. Boil *Acet. Lead* ℔ij. and ℥iij. (℥vj and ʒvj. E.) and (finely, E.) powdered *Oxide of Lead* ℔j. and ℥iv. (*Litharge* ℥iv. E.) in *Aq. dest.* Ovj. (Oiß. E.) frequently stirring for half an hour. When cold add (distilled) water to make up Ovj. (Oiß. E.) Filter. (Preserve in well-closed bottles, E.)

In the D. P. the old and objectionable method, on account of its uncertainty, is still adopted.

Boil semi-vitrified *Oxide of Lead* 1 part, in *Distilled Vinegar* 12 parts, in a close vessel until only 11 parts remain. Leave the solution at rest till the impurities subside and filter it.

Prop. Colourless, with a sweetish and sub-astringent taste, having an alkaline reaction on test-paper. By careful evaporation out of the access of the air, crystals may be obtained of a tabular form, or the salt in an uncrystallizable mass (dry Extract of Saturn, of Goulard). When exposed to the atmosphere, some Carb′ is absorbed, and white Carb. of Lead is deposited; also on passing Carb′ gas into its solution, and this is one of the methods adopted in making Carb. Lead. q. v. It is decomposed by common water. It precipitates mucilage and most vegetable colours.

Tests. Sp. Gr. 1·260. L. A copious precipitate is gradually formed when the breath is propelled through it by means of a tube. E. The other properties are those of the Acet. Lead. When prepared with inferior Vinegar, it is of a brown colour.

Inc. Acids, Alkalis, Earths, Alum, Borax, Tartarized Iron and Antimony, Soap, Hard and Spring-water, Sulphuretted Hydrogen, Mucilaginous solutions and drinks.

Action. Uses. Astringent and Sedative. *Externally*, diluted with water; or it may be used in a milder form, in the following preparation:

LIQUOR PLUMBI DIACETATIS DILUTUS, L. Plumbi Subacetatis Liquor Comp. Diluted Solution of Diacetate of Lead. Goulard Water.

Prep. L. D. Mix *Solution of Diacetate of Lead* f℥iß. (*Subacetate of Lead* ℨj. D.) *Aq. dest.* Oj. *Proof Spirit* f℥ij. (℥j. D.)

Useful as a soothing, astringent, and sedative Collyrium or lotion.

CERATUM PLUMBI COMPOSITUM, L. Compound Cerate of Lead. *Goulard's Cerate. Ceratum Lithargyri Acetati.*

Prep. Melt *Wax* ℥iv. and mix with it *Olive Oil* f℥viij.; remove from the fire, and when first the mixture begins to thicken, add gradually *Sol. of Diacet. Lead* f℥iij. stirring constantly with a spatula till cool. Then mix with these *Camphor* ℨß. dissolved in *Olive Oil* f℥ij.

Action. Uses. This is commonly called Goulard's Cerate, and a soothing, astringent, and sedative application for irritable surfaces and in chronic Ophthalmia.

Pharm. Prep. Ceratum Saponis. Emp. Saponis.

BISMUTHUM, L. E. D.

Bismuth.. *F.* Bismuth. *G.* Wismuth.

Bismuth (Bi=72) is first mentioned by Agricola in 1520, having previously been confounded with Lead. It is usually met with in its metallic state, but also as an Oxide and a Sulphuret.

Prop. Bismuth is of a reddish-white colour, devoid of taste or smell. It is brittle, of a lamellar structure, and readily crystallizes in cubes or octohedra. Sp. Gr. 9·53—9·88. It melts at 497° F. or 507° (Gr.), and volatilizes at a full red heat. When exposed to the air, it tarnishes, but does not oxidate: at a high temperature, it burns with a pale blue flame, when Oxide of Bismuth (Bi O=80) is formed, and escapes in white fumes. Bismuth is with difficulty acted on by H Cl′ or by dil. Sul′, but readily by Nit′. *v.* infra.

Tests. Sp. Gr. 9·8. It is dissolved by dil. Nit′; the solution is colourless, when Subnitrate of Bismuth is precipitated from this solution by Ammonia, the liquor is free from colour. It also deposits a white powder when much diluted with water. L. and E. Bismuth is employed only for making the following preparation:

BISMUTHI TRISNITRAS, L. BISMUTHI SUBNITRAS, D. BISMUTHUM ALBUM, E. *Magistery of Bismuth.* Trisnitrate of Bismuth. *F.* Sous-nitrate de Bismuth. *G.* Wissmuth weiss.

The Trisnitrate of Bismuth (3 Bi O, N O_5=294) is a tasteless powder of a brilliant white colour, and consists of microscopic needle-shaped crystals. It is slightly soluble in water, and if the solution be heated, the salt is precipitated in the form of minute brilliant crystals.

Prep. L. E. D. Take *Bismuth* (finely powdered, E. D.) ℥j. (7 parts, D.) *Nitric′* (Dens. 1380, E.) f℥iß.; (dilute, 20 parts, D.) *Aq. dest.* (Water, E.) Oiij. 100 parts, D.)

Mix Nitric′ f℥iß. with Aq. dest. f℥j. (Take Nitric′, dens. 1380, f℥iß. E.;

dilute 20 parts, D.) Dissolve in it (gradually, E. D.) *Bismuth* ℥j. (with the aid of gentle heat, E. D., adding a little Aq. dest. as soon as crystals, or a white powder begins to form, E.) Pour off the solution (when complete, E.) add it to the rest of the water. Set aside for the powder to settle, pour off the liquid. (Filter through calico, E.) Wash the Trisnitrate of Bismuth (precipitate, E. D. quickly, E.) with distilled (cold, E.) water and dry it with a gentle heat (in a dark place, E.).

The Bismuth is oxidated at the expense of a portion of the Nit′, Nitric Oxide gas escapes, and the Oxide of Bismuth is dissolved by the remainder of the Nit′, and a Nit. Bismuth formed. When the solution of this is added to the water as above directed, decomposition ensues; most of the acid with a little Oxide of Bismuth (forming a Ternitrate) remains in solution, while the white-coloured precipitate is the Trisnitrate of Bismuth. Comp. Bi O, 81·64+N′ 18·36=100.

Tests. It is soluble in Nit′ without effervescence; solution colourless. Dil. Sul being added to the solution, nothing is thrown down. L. E. The first indicates that no Carbonate is present, and the second the absence of Lead.

Action. Uses. Irritant, Tonic, Antispasmodic.

D. gr. v.—℈j.

Stannum, L. E. D.

Tin. *F.* Etain. *G.* Zinn.

Tin is one of the most ancient known of the metals, being mentioned by Moses under the name Bedel. It was used by the Egyptians, probably obtained from the East by the trade with India. The Greeks and Romans obtained it through the Phœnicians from England. It abounds in Cornwall, and in the East from Mergui to the island of Banca. It occurs both as an Oxide and a Sulphuret, but chiefly the former, which is easily reduced to a metallic state by being heated with Charcoal, as this abstracts its Oxygen. It is brought into commerce in the form of Grain Tin and Block Tin. Malacca Tin and Banca Tin are the Eastern varieties; to these Mergui Tin has lately been added.

Prop. In mass, bluish-white, tarnishing but slightly, of a peculiar odour when rubbed; so malleable, as to be beaten into *sheet tin* and *tin-foil;* soft, fusible; Sp. Gr. 7·29; that of commercial specimens is often higher, from the impurities which they contain. It fuses at 442°, and becomes covered with a grey crust of the Oxide; burns at a red, and is volatilized at a white heat. There are several oxides and numerous salts of Tin, but none are officinal.

Tests. Tin, boiled with H Cl′, is almost entirely dissolved, a Protochloride being produced; solution colourless, but becomes purple on the addition of Chloride of Gold: the precipitate by Potash is white; when this is added in excess, the precipitate is redissolved. (L.) Hydrosulphuric acid gives a brown precipitate. Commercial Nit′ f℥iij. converts 100 grs. entirely into white powder; pure Oxide of Tin is formed, with much disengagement of Nitrogen; distilled water, boiled with this powder and filtered, yields no precipitate with solution of Sulph. Magnesia. E. P.

The tests of the L. C. being those which characterize Tin, will show its true nature. Those of the E. P. are intended to detect the presence of Lead; either no Oxide of Tin, or very little, will be formed, or the Sulph. Magnesia will cause a precipitate of Sulph. Lead if this metal be present.

PULVIS STANNI, L. E. D. Powder of Tin.

Tin may be employed either in filings, *Limatura stanni*, or preferably as granulated or powdered Tin, the officinal form, for preparing which the E. and D. Pharmacopœias give formulæ. When melted, it is agitated with an iron pestle, a birch broom, or by shaking it in a wooden box having its inside covered with Chalk, and sifting. This is nearly the D. formula.

Prep. Melt Tin in an iron vessel; pour it into a mortar previously heated rather above the fusing point of the metal; triturate briskly as it cools, ceasing as soon as a considerable proportion is finely pulverized; sift the product, and repeat the process with what is left in the sieve, E.

Action. Uses. Mechanical Anthelmintic.

D. ʒiv.—℥j. in treacle or confection.

ANTIMONIUM.

Stibium. Antimony. *F.* Antimoine. *G.* Antimon.

Metallic Antimony (Sb=65) is not officinal; but as so many of its preparations are employed, it is desirable to be acquainted with its characteristics. It was probably known to the Alchemists; but Basil Valentine (*Currus Triumphalis Antimonii*) made known the method of obtaining it. The Sesquisulphuret q. v. has been known from the most ancient times. Native Antimony occurs in France and Germany, also as, Oxide or White Antimony; the Sulphuret, or Grey Antimony; and as Sulphuretted Oxide, or Red Antimony. The Sulphuret is the most abundant ore, and that from which the metal is chiefly obtained. This used to be called Regulus of Antimony, and the Sulphuret, Crude Antimony. Antimony is obtained by heating the above Sulphuret with half its weight of Iron-filings or small Iron nails, when the Sulphur unites with the Iron, and the Antimony is set free. The melted Antimony collects in the bottom of the crucible, and may be run into moulds.

Prop. A bluish-white metal, usually lamellar in structure, and brittle in nature. Sp. Gr. about 6·7. It fuses at a temperature of about 800°, above which it may be volatilized, and in cooling may be made to crystallize in rhombohedra. It undergoes little change in the air, the surface only becoming tarnished and partially oxidized. Heated to a white heat, and suddenly exposed to the air, it burns with a white light; the vapour which escapes condenses in white needle-like crystals of Sesquioxide of Antimony, which were formerly called *Argentine Flowers of Antimony.* Antimony is dissolved by H Cl′ with the aid of heat, Hydrogen being disengaged. The Sesquichloride of Antimony, on being added to water, deposits a white precipitate, formerly called *Powder of Algaroth;* and an orange-red one of Sesquisulphuret of Antimony on the addition of Hydrosul′ or an alkaline Hydrosulph. Nit′ converts the metal into Antimonic acid, which is insoluble in the acid. There are three compounds of Antimony and Oxygen: Oxide of Antimony, Antimonious and Antimonic acids.

ANTIMONII OXIDUM, E. ANTIMONII OXYDUM NITROMURIATICUM, D.

Antimonii Sesquioxydum. Sesquioxide and Protoxide of Antimony. *Flowers of Antimony.* *F.* Oxyde d'Antimoine. *G.* Antimonoxyd.

Sesquioxide of Antimony (Sb 1½O=77) occurs native in Bohemia and Hungary, and is called White Antimony. It is produced when Antimony is burnt in the air.

Prop. As prepared below, the Oxide is white and tasteless; unalterable in the air, but becoming yellow by heat, and regaining its colour when cool. By a full red heat it may, like the native Oxide, be fused into a yellow liquid and afterwards sublimed, as before the blowpipe, when needle-like crystals will be deposited. If fused and exposed to the air, more Oxygen is absorbed, and Antimonious acid ($Sb\,O_2=81$) is formed, which is not volatile, less easily fused, and more inert as a medicine. The Sesquioxide is insoluble in water, but dissolved by H Cl′, Tart′, and Acet′, also in Bitart. Potash, when Tartar Emetic is formed. Comp. O 15·58+Sb 84·42=100.

Prep. E. D. Take *Mur′* (commercial) Oj. (100 parts, mix it with Nit′ 1 part in a glass vessel, avoiding the fumes, D.) Dissolve in the acid, *Sulphuret of Antimony* in fine powder ℥iv. (prepared Sulphuret 20 parts, D.), apply a gentle heat (gradually increase the heat, digest till the mixture ceases to effervesce, D.) Boil for half an hour. (1 hour, D.) Pour the fluid (cooled, D.) into Aq. Ov. (Cong. j. D.) Filter through calico. (Let the Oxide subside, D.) Wash the precipitate well with cold water, then with weak solution of Carb. Soda, and again with cold water (with cold water only, D.) till the water ceases to affect reddened Litmus. Dry the powder over the vapour-bath. (On bibulous paper, D.)

The two processes are essentially the same, consisting in the decomposition of the Sesquisulphuret by the H Cl′, and the formation of a Sesquichlor. Antimony, which is held in solution. Some Hydrosul′ (from the H. of the acid) is formed, which the Nit′ of the D. P. is intended to decompose. (Others consider the Chloride to be dissolved in the form of Hydrochlorate of the Oxide, water being decomposed, supplying Oxygen to the Antimony and Hydrogen to the Sulphur, c.) The quantity of Mur′, Dr. C. says, is large, being 3 times as great as is required to furnish the due proportion of Chlorine for forming the Sesquichloride, but it has been found to be necessary. Water being added to the solution, and having a greater affinity for the acid, a precipitate takes place of *Powder of Algaroth*, which is a Sesquioxide combined with some Sesquichloride of Antimony. The proportion of Chlorine is diminished by repeating the washings with water. The whole of the acid is removed by washing with sol. Carb. of Soda; while the Carb′ escapes. Mr. Tyson (P. J. i. 450) adds sol. of Carb. Ammonia as long as effervescence is perceived, and he obtains a straw-coloured Protoxide. If too much Nit′ be employed, according to the Dublin process, the oxidizement will be carried too far, and Antimonious acid formed, which is inert and insoluble.

Tests. Entirely soluble in Mur′, and also in a boiling mixture of water and Bitart. Potash; snow-white; fusible at a full red heat. E. P.

Action. Uses. Emetic, Diaphoretic, Expectorant. A good substitute for Antimonial or James's Powder.

D. gr. iij.—gr. x. in powder or pill. Mr. Tyson gives gr. $\frac{1}{10}$—gr. j.

Pharm. Prep. This Oxide forms the active ingredient in the officinal preparations, as Pulvis Antimonii Comp., the Oxysulphuret of Antimony, Tartar Emetic, Glass of Antimony, Kermes, and Golden Sulphuret.

PULVIS ANTIMONII COMPOSITUS, L. PULVIS ANTIMONIALIS, E. D.
"A mixture chiefly of Antimonious acid and Phosphate of Lime, with some Sesquioxide of Antimony and a little Antimonite of Lime." E. P.

Few empirical medicines have attained more permanent celebrity than the FEVER POWDER of Dr. James, commonly called JAMES'S POWDER, sometimes distinguished in prescriptions as the PULVIS JACOBI VERUS. As it was found impossible to make the Powder by following the Patentee's directions, and chemical analyses having ascertained that it consisted of Phosph. Lime and oxydized Antimony, the College adopted a formula suggested by Dr. Pearson for the preparation of a substitute.

Prep. L. E. D. Mix *Sesquisulphuret of Antimony* powdered ℔j. (Sulphuret of Antimony in coarse powder, E. Prep. Sulphuret of Antimony 1 part, D.) and *Horn shavings* ℔ij. (āā equal weights, E. 2 parts, D.) Put them into a red hot crucible [(iron pot, E. D.)] on the fire, stirring constantly till (they acquire an ash-grey colour, D. and E.) vapours (sulphurous, D.) no longer arise. (When cold, D.) Pulverise the product, and put it into a proper crucible. (With a perforated cover, E. Covered with another crucible having an orifice in the bottom, D.) Heat it with a gradually increasing heat till red hot (white hot, E.) for two hours. (When cold, E. D.) Rub the residue to a very fine powder.

The Sulphuret of Antimony, consisting of Sulphur and Antimony, and the horn shavings of Phosp. Lime cemented by gelatinous matter, become changed by the action of heat. The Sulphur, obtaining Oxygen from the air, escapes as Sulphurous acid; the Antimony being also supplied with Oxygen from the air, is converted into Antimonious acid and into a small portion, usually about 4 per cent., of Sesquioxide of Antimony. The animal matter of the horn is burnt off, and the earthy ingredients, or Phosph. Lime, with a small portion of the Carbonate, remain intermixed with the oxidized Antimony. During the second heating, some of the Sesquioxide formed becomes converted into more of the Antimonious acid, a little of which, combining with the Lime of the Carbonate, forms some Antimonite of Lime.

Prop. The powder produced is white, gritty, devoid of both taste and smell. The greatest portion is insoluble in water; but distilled water, as stated in the E. P., boiled on it and filtered, gives with Hydrosul′ an orange precipitate, in consequence of the water dissolving the Antimonite of Lime; but this effect will not be perceived if only Antimonious acid is present. The Lime will be revealed by its appropriate tests. H Cl′ digested on the residue, does not become turbid by dilution, but gives an abundant orange precipitate with Hydrosul′. E. P. But Dr. Pereira observes that the solution does become turbid sometimes on dilution, and deposits a white powder (Oxychloride of Antimony). The H Cl′ dissolves the Sesquiox. Antimony and some Phosph. Lime. After the precipitate of Sesquisul. Antimony, "if this be separated by filtering, and the solution boiled to expel any traces of Hydrosul′, a white precipitate (*Subphosph. Lime*) is thrown down on the addition of caustic Ammonia." (*p.*)

Comp. Antimonial Powder is variable in the proportion of its ingredients. Mr. Phillips found one specimen to consist of

Antimonious acid 35 + Phosph. Lime 65 = 100.
Another: Antimonious acid 38 + Phosph. Lime 62 = 100.

While James's Powder consisted of:

Antimonious acid 56 + Phosph. Lime 44 = 100.

Dr. Maclagan found James's Powder to consist of Antimonious acid 43·47 + Subphosph. Lime 50·24, Antimonite of Lime with some Superphosph. 3·40, and Sesquiox. Antimony 2·89 = 100; and Antimonial Powder, of Antimonious acid 50·09 + Subphosph. Lime 45·13 + Antimonite of Lime with some Superphosph. 0·8, Sesquiox. Antimony 3·98. Mr. Brande has found as much as 5 per cent. of Sesquioxide of Antimony.

Action. Uses. Diaphoretic, Emetic; but uncertain, as is James's Powder, which is given in doses of grs. v. to grs. xx.

D. gr. iij.—gr. vj. or gr. x. 100 grains have been given without any effect, but Antimonious acid is inert.

The foregoing Sesquioxide of Antimony, or Tartar Emetic in small doses, is a good substitute for both these powders. Mr. Tyson employs Protoxide of Antimony grs. ij. with grs. xviij. of Phosph. of Lime, but prefers a mixture of Phosph. of Lime and Sulphate of Potash āā grs. ix.

Chloride of Antimony (Sb $1\frac{1}{2}$ Cl = 119) is a soft solid, but becomes liquid by a gentle heat. It was called Muriate of Antimony and Butter of Antimony by old writers. It used to be employed as a caustic, and likewise for obtaining Powder of Algaroth, or Oxychloride of Antimony. (*v.* p. 173.)

ANTIMONII (SULPHURETUM, E. D.) SESQUISULPHURETUM, L.

Native Sesquisulphuret of Antimony. *Antimonium Crudum.* Grey Antimony. *Antimony.* *F.* Sulfure d'Antimoine. *G.* Dreifach Schwefel-Antimon.

Sulphuret of Antimony (Sb $1\frac{1}{2}$ S=89) is extensively diffused, being found in Hungary, at Borneo, Moulmein, and Pegu, in Persia and in Caubul. It has been employed from time immemorial in Asiatic countries for painting the eyebrows and eyelids. It is the στιμμι and Stibium of the ancients. By fusion in a covered crucible, it is separated from impurities, and then called Crude Antimony.

Prop. *Crude Antimony* is in roundish dark-coloured loaves, which when fractured, exhibit its peculiar and brilliant striated texture, and dark grey colour, often presenting some prismatic crystals. Sp. Gr. 4·6. It is readily reduced to powder, which is blackish, without taste and smell, but with a reddish hue when the Sesquisulphuret is very pure; insoluble in water; permanent in the air; fused at a moderate temperature, and volatilized in close vessels. Heated in the air, some Sulphurous acid is formed, as well as mixed Oxide of Antimony. It is soluble in solution of Potash (*v.* Antimonii Oxysulphuretum), also in H Cl′ (*v.* Antimonii Potassio-Tartras, and Tests). It may also be dissolved in Nit′, which, parting with its Oxygen to both the metal and the Sulphur, converts the one into an Oxide and the other into Sul′; and these, combining together, form a Sulphate of Antimony, which is dissolved in the remaining Nit′. Comp. Sb 72·8 S 27·2=100.

Tests. Seldom quite pure, being apt to contain some of the Sulphurets of Iron, of Arsenic, Lead, and Copper. "Striated; soluble entirely, with the aid of heat, in H Cl′; and deposits from this solution a white substance (Oxychloride of Antimony) on the addition of distilled water, leaving a liquid, which, when filtered, yields a reddish precipitate (Sesquisulphuret of Antimony) with Hydrosul′." L. P. The E. C. give the solubility in Mur′ as a sufficient test of purity. Any Iron or Arsenic will disappear in the P. processes in which this ore is used. The Iron will give a yellow tinge to the solution in H Cl′, and both Lead and Copper may be detected by their tests after the precipitation of the Antimony. The Arsenic may also be tested for by reducing it with a mixture of Charcoal and Carb. Soda, or by heating together equal parts of the Sesquisulphuret and Cream of Tartar for three hours, when an alloy is formed of Potassium and Antimony. This, when added to water, decomposes it: Hydrogen combined with Arsenic, or Arseniuretted Hydrogen escapes, which, being burned, the Oxide or a stain of metallic Arsenic is obtained.

By levigation and elutriation, as in the case of Prepared Chalk, the Sesquisulphuret is reduced to a fine powder, as when prepared in—

Antimonii Sulphuretum præparatum, D.

Prep. Reduce to powder *Sulphuret of Antimony* q. s. as directed in the preparation of Chalk. Preserve the most subtile particles for use.

Action. Uses. Diaphoretic, Alterative, Emetic; but uncertain, as it may be more or less oxidized by acid in the Stomach.

D. gr. x. or gr. xx.—ʒj.

Antimonii Oxysulphuretum, L. Antimonii Sulphuretum Aureum, E. Sulphur Antimoniatum fuscum, D.

Oxysulphuret and Golden Sulphuret of Antimony. *Sulphur Antimonii præcipitatum*. *F.* Soufre doré d'Antimoine. *G.* Goldschwefel.

Several Sulphurets of Antimony have long been employed in medicine. The present was known to Basil Valentine. Kermes Mineral is considered to have been discovered by Glauber, and made known through one of his pupils to La Ligerie, from whom, in 1720, the French government bought the secret of its preparation.

A preparation similar to this may be formed in several ways:—1. By boiling Sulphuret of Antimony with Carb. Potash or Soda. Or 2. With a caustic alkali. 3. By melting at a red heat a mixture of Sulphuret of Antimony and an alkaline Carbonate, and then treating the melted mass with boiling water. The Sulphuret, when boiled in a solution of Potash, becomes dissolved; but, on cooling, a reddish-brown powder is deposited, which is usually considered to be similar to the Kermes Mineral of old authors. But if we add H Cl′ before the deposit takes place on cooling, an orange-red precipitate is produced, which is the officinal preparation, and supposed to be analogous to the Golden Sulphuret of the older Materia Medica.

Prop. The Oxysulphuret of Antimony, called Golden Sulphuret in the E. P., is in powder and of an orange-red colour, devoid of smell, and with little taste. It is insoluble in water, but soluble, with the aid of heat, in alkalis. Acted on by H Cl′ or Nitro-H Cl′, it becomes dissolved, with the exception of a little Sulphur. Heated in a tube, Sulphur sublimes; in the air, it burns with a blue flame, with evolution of Sulphurous acid gas, leaving as a residue the oxidized metal. Boiled by Mr. Phillips in a solution of Tartaric acid, 12 per cent. become dissolved, which he considers to be the proportion of Sesquioxide of Antimony, and the Preparation to consist of 1 Eq. Sesquioxide Antimon. 77 + 5 Eq. Sesquisulph. Antimon 445 + 8 Eq. water 72 = 594; or, Sb 1½ O 13 + Sb 1½ S 75 + Aq. 12 = 100.

Prep. L. E. D. Boil together *Sesquisulphuret of Antimony* finely powdered ℥vij. L. (℥j. E.) (1 part, D.) in *Liq. Potassæ* Oiv. [(f℥xj, E.) (18 parts, D.)] *Aq. dest.* Cong. ij. [(Oij. E.) (none, D.)] with a slow fire for 2 (1, E. D.) hours, frequently stirring and replacing the water that evaporates. Filter [(immediately while hot, E.) (through cloth, D.)] pour in gradually dilute *Sul′* q. s. to precipitate the Hydrosulphuret of Antimony [(in excess, E.) (11 parts, or q. s. D.)

(collect it on a calico filter, E.)] Wash the precipitate with water, so that the Sulph. Potash may be removed. Dry the remainder with a gentle heat.

Of this process different explanations are given; but some of the Antimony of the Sesquisulphuret becomes oxidized at the expense of the Potash, the Potassium of which combining with the Sulphur set free, a double Sulphuret of Potassium and of Antimony is formed, which is dissolved in the undecomposed Liq. Potassæ with some of the remaining Sesquisulphuret and the Sesquioxide of Antimony which has been formed. On the addition of Dil. Sul′, some Sulph. Potash is at once produced. Water at the same time becomes decomposed, giving its Oxygen to the Potassium to be reconverted into Potash, and its Hydrogen to form some Hydrosul′, which escapes. The Potash which is the common solvent, and that which is formed, having combined with the Sul′, both the Sulphuret of Antimony in the form of a Hydrate and the Sesquioxide of Antimony are precipitated, with a little free Sulphur according to the E. P., and all are perhaps only mechanically mixed.

Tests. The L. C. states that it is "totally soluble in Nitro-Hydro Cl′, emitting Hydrosul′. But the E. C. more correctly states that it is tasteless: 12 times its weight of Mur′, aided by heat, will dissolve most of it, forming a colourless solution, and leaving a little Sulphur.

Action. Uses. Alterative, but uncertain; in large doses, Emetic.

D. gr. j.—gr. v.

Pharm. Prep. Pilulæ Hydrargyri Chloridi Comp., or Plummer's Pills.

Glass of Antimony (*Antimonii Vitrum*, L.) is prepared by partially roasting and fusing the Sesquisulph., by which a portion of it is converted into Protoxide of Antimony. It was formerly employed for making Tartar Emetic, but was an uncertain preparation, and apt to be mixed with Glass of Lead, and is therefore not now used.

Antimonii et Potassæ Tartras, D. Antimonii Potassio Tartras, L. Antimonium (Emeticum, D.) Tartarizatum, E.

Tartrate of Antimony and Potash. Potassio-Tartrate of Antimony. Tartarized Antimony. Tartar Emetic. *F.* Tartre émétique. *G.* Brechweinstein.

The discovery of this salt is attributed to Mynsicht (Thesaurus, &c. Hamburgh, 1631). It is a double salt composed of Tartrate of Potash and Ditartrate of Antimony, and was at first made with Cream of Tartar and Liver of Antimony.

Prop. Tartrate of Antimony and Potash (2 Sb 1½ O, Tar′+K O Tar′+3 Aq.=361), or Tartar Emetic, is usually seen as a white powder, but it crystallizes readily from a saturated solution in tetrahedra or in octohedra with rhombic bases. They are colourless and transparent, without smell, but have a nauseous, styptic, and slightly

Fig. 29.

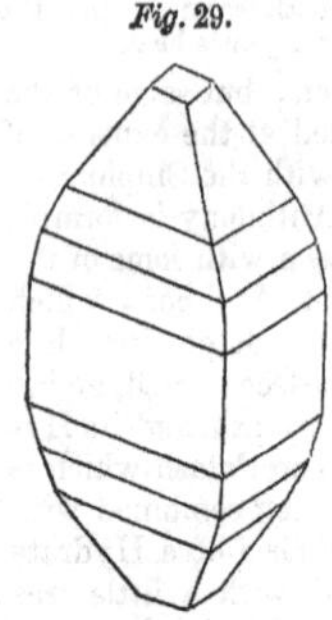

acid taste. Exposed to the air, they become opaque, and covered with a white powder, losing 4 or 5 parts per cent. of weight. The crystals are insoluble in Alcohol, but soluble in proof Spirit and in Wine, also in about 14 times their weight of temperate and about twice their weight of boiling water. The solution reddens Litmus, and, when diluted, soon undergoes decomposition. Heated, the crystals decrepitate, become charred, and leave a pyrophoric alloy of Antimony and Potassium. The solution is decomposed by the alkalis and alkaline earths, as well as their Carbs. (hence common water, containing Carb. Lime or of Magnesia, precipitates the Oxide Antim.); likewise by strong acids. Hydrosul′ throws down the Antimony in the form of an orange-red precipitate of Hydrated Sesquisulph., as do also the Hydrosulphates of Ammonia and of Potash. The juices of many plants also, and astringent decoctions throw down the Oxide of Antimony and an insoluble compound. So much is this the case, that powder of Cinchona and of Galls, as well as a decoction of the latter, prevent the emetic effects of Tartarized Antimony. The decoctions of Cinchona, Kino, and Ratanhy only partially neutralize its effects.

The Tartrate of Potash and of Antimony is composed of Tar′ 36·6 +K O, 13·3+Sb $1\frac{1}{2}$ O 42·6+Aq. 7·5=100. Or, in Equivalents:

1 Eq. of Tartrate of Potash	Tar′ 66 +	K O, 48	=114
1 Eq. of Ditartrate of Antimony	Tar′ 66 + 2 Sb	$1\frac{1}{2}$O, 154	=220
3 Eq. of Water . . .	H O 9 × 3		= 27
			361

Prep. Various methods have been adopted for preparing Tartar Emetic, all of them have for their object to prepare from the sesquisulphuret, a sesquioxide of Antimony, so that it shall be in a fit state to combine with the second Eq. of Tart′ in Bitart. Potash. The former, L. process obtained the sesquioxide from the *Glass* of Antimony. The present from a *Crocus* of Antimony. The D. P. obtain it from a chloride, by a process pronounced by M. Fleury as the best. This has been adopted by the E. P.

L. Take *Sesquisulphuret of Antimony* in powder ℔ ij. and mix it thoroughly with bruised *Nitr. Potash* ℔ ij. then add *HCl′* f℥iv. by degrees. Spread the powder on an Iron plate and set it on fire. When cold pulverise what remains very finely and wash it frequently with *boiling water* until it is free from taste. Mix the powder thus prepared with bruised *Bitart. Potash* ℥xiv. and boil for half an hour in *Aq. dest.* cong. j. Strain the liquor while yet hot, and set it aside to crystallize. Remove the first crystals and dry them, and let the liquor again evaporate that more crystals may form.

During the combustion in the first part of this process, the Oxygen of the Nitrate converts the Sesquisulph. into Sesquioxide of Antimony and Sulphuric acid. The latter combines with the Potash set free, and Sulph. Potash is formed, at the same time that the Sesquioxide formed combines with a part of the Sesquisulphuret, forming an Oxysulphuret or Crocus of Antimony. The H Cl′ combines with any free Potash, and prevents the formation of, or decomposes any, Sulphuret of Potassium which may be formed. The washing removes all these salts, and leaves the Sesquioxide in a state in which it readily combines with the Tar′ of the Bitart. Potash.

The E. and D. Colleges direct a Sesquioxide to be prepared from a Sesquichloride formed by the action of the H Cl′ on the Sesquisulphuret, (*v.* p. 172) Take of the precipitated Sesquioxide ℥iij. (Nitro-muriatic oxide of Antimony 4 parts, D.) and mix it with Bitart. Potash ℥iv. ʒij. (5 parts finely powdered, D.) Add to it Aq. f℥xxvij., boil for an hour, filter and crystallize by cooling. Evaporate the mother liquor for more crystals, which may require to be dissolved again and recrystallized.

The Sesquioxide of Antimony employed has adhering to it a little of the Chloride. On being boiled in water, it is resolved into Sesquioxide and H Cl′ from the decomposition of the water. The Sesquioxide unites with the Bitart. Potash, and as the Tartar Emetic crystallizes, the H Cl′ remains in solution and retains any Iron or other metallic impurity which may be present. (*c.*)

Tests. As Tartar Emetic in powder is apt to be adulterated, it is preferable to buy it in well-formed crystals. In this country it is most apt to be adulterated with Bitart. Potash and Oxide of Iron, and with the former most frequently; the latter gives it a yellow tinge. "Totally soluble in water, no Bitart. Potash remaining in the vessel; with Hydrosul′ a reddish-coloured precipitate (Hydrated Sesquisulphuret of Antimony) is obtained. Nit′ throws down a precipitate (Sesquioxide of Antimony), which is dissolved by an excess of it. Neither Chlor. Barium (there being no Sulphates), nor Nit. Silver (no Chlorides) being added to a solution, precipitates any thing." L. P. "It produces a white precipitate (unless the solutions be very dilute) with Nit. Silver, soluble in excess of water." (*p.*) Entirely soluble in 20 parts of water; solution colourless, and not affected by solution of Ferrocy. Potassium (if Iron be present, the solution will be yellowish, and become blue on this addition). A solution in 40 parts of water is not affected by its own volume of a solution of 8 parts of Acet. Lead in 32 parts of water and 15 parts of Acet′. E. P. This acid solution of Acet. Lead, suggested by M. Henry and adopted by the E. P., is so delicate as to detect less than one per cent. of Bitart. Potash; but Dr. C. states that he has experienced some difficulties in using this test.

Inc. Acids, alkalis, and their Carbs.; some of the earths, and metals, and their oxides; Lime-water, Chlor. Calcium, and Acetates of Lead; vegetable infusions, and decoctions, as of Cinchona, Catechu, &c.

Action. Uses. Irritant poison, Alterative, Diaphoretic and Expectorant, Nauseating Sudorific, Emetic, Contra-Stimulant. *Ext.* Counter-Irritant, Rubefacient.

D. Alterative, gr. $\frac{1}{16}$ to $\frac{1}{8}$. Diaphoretic and Expectorant, gr. $\frac{1}{8}$ to $\frac{1}{6}$: Nauseating Sudorific, gr. $\frac{1}{4}$ to $\frac{1}{2}$. Emetic, gr. j.—gr. ij. diluted. Contra-Stimulant, gr. j.—gr. iij. every 2 or 3 hours.

Antidotes. Excite vomiting by mechanical irritation in the fauces; or with draughts of warm water; or use the stomach pump. The best antidotes are astringent vegetable decoctions or infusions, as of Gall-nuts or of Cinchona, as these form insoluble Tannates with the Sesquioxide of Antimony.

VINUM ANTIMONII POTASSIO-TARTRATIS, L. VINUM ANTIMONIALE, E. LIQUOR TARTARI EMETICI, D. Antimonial Wine.

Prep. L. E. D. Take *Potassio-Tartrate of Antimony* (Tartar Emetic, E.) ℈ij. (℈j. D.) dissolve it in *Sherry Wine* Oj. (Dissolve the Tartrate in Aq. dest. boiling by measure ℥viij., filter and add to the liquid, rectified Spirit of Wine by measure ℥ij. D.)

Action. Uses. Alterative and Diaphoretic and Emetic.

D. ♏x.—fʒij. every 3 hours. Each f℥ contains 2 grs. of Tartar Emetic. f℥ß or f℥j. may be given in fʒj. doses, or a teaspoonful every 5 or 10 minutes, to act as an Emetic.

UNGUENTUM ANTIMONII POTASSIO-TARTRATIS, L. U. ANTIMONIALE, E. U. TARTARI EMETICI, D. Tartar Emetic Ointment.

Prep. L. E. D. Take *Potassio-Tartrate of Antimony* rubbed ℥j. [(ʒj. D.) (Tartar Emetic in very fine powder ℥j. E.)] *Lard* (Axunge, E.) ℥iv. (prepared Hog's Lard ℥j. D.) Mix. (Rub the salt into a very fine powder (Rub them carefully into a smooth and uniform mass, E.) then mix with the Lard.)

Action. Uses. Counter-Irritant. ʒß. applied twice a day by friction on the skin, produces a pustular eruption.

HYDRARGYRUM, L. E. D.

Argentum vivum et liquidum. Mercury. Quicksilver. *F.* Mercure. Vif-argent. *G.* Quecksilber.

Mercury (Hg=202), or Quicksilver, was known to the ancients. The Romans seem to have employed it as a medicine externally (p.133), as did the Arabs; but the Hindoos were probably the first to prescribe it internally. (*v.* p. 191.) It is found in China, at Almaden in Spain, and Idria in Carniola, and likewise in South America. It occurs occasionally in metallic globules; usually as the native Bisulphuret or *Cinnabar;* combined with Silver, forming a *Native Amalgam;* or with Chlorine, as in *Horn Mercury.* It is chiefly obtained from the Sulphuret by distillation with Lime or with Iron, which combining with the Sulphur, the metal distils over and is condensed.

Prop. Mercury is remarkable among metals for existing as a liquid at ordinary temperatures, and for its silver-like colour and lustre. It is without taste or smell. Sp. Gr. 13·568. Freezes at —40°; crystallizes in octohedra, becomes malleable, and has a Sp. Gr. of 14 from the contraction. It boils at 660°, and is converted into colourless vapour of great density. According to Faraday, it is converted into vapour at ordinary temperatures. It is unalterable in the air, except with the aid of heat, when it slowly combines with Oxygen, and forms the Red or Binoxide, and a greyish powder (Black Oxide or Æthiops per se), which is considered by some to be Protoxide, and by others a Suboxide. By increase of heat, the Oxygen is expelled. Mercury combines with both Chlorine and Bromine in two, and with Iodine in three proportions ; with Sulphur it forms a black and a red Sulphuret. It unites with several metals, especially Gold, Silver, Lead, Tin, Bismuth, Zinc, which it dissolves, and with which it forms amalgams. Cyanogen unites with it into a Bicyanide (or Cyanuret) of Mercury. All the acids combine with its Oxides and form salts, the Nitric acid most easily, even diluted and at ordinary temperatures. Solutions of these may be decomposed by introducing into them a piece of clean Copper, on which a thin layer of Mercury will become deposited. Sulphuretted Hydrogen will throw down a black Sulphuret of Mercury. Caustic Potash or Soda will give a grey precipitate with the salts of the Protoxide, and a reddish-yellow one with those of the Peroxide. Ammonia, on the contrary, deposits double salts, as in Hydrarg. Ammonio Chloridum. Most of these compounds will be treated of in the following pages.

Tests. Mercury may be adulterated with Lead, Tin, or Bismuth, &c. when it loses its lustre, especially if shaken. "A globule moved along a sheet of paper, leaves no trail," E., indicating that these are not present; but pure Mercury, if moist, will form this trail. "Entirely vaporizable ; soluble in dil. Nit′, but not in boiling H Cl′ ; the latter, after being boiled with it and cooled, is neither coloured nor precipitated by Hydrosul′. Sp. Gr. 13·5." L. "Pure Sul′ agitated with it, evaporates when heated without leaving any residuum." The several metals may be distinguished by their respective tests.

Hydrargyrum Purificatum, D.

The Mercury of commerce is now usually pure enough for Pharmaceutic purposes ; but the D. P. has a formula for its purification.

Prep. Mercury 6 parts ; distil 4 parts with a gentle heat. Better to distil with Iron-filings and receive in water.

Preparations of Mercury.

Though Mercury, in its metallic state, is considered to be inert, there are several valuable preparations in which the principal portion of it exists chiefly in that state. In these, the Mercury, by long trituration with dry powders, viscid confections, or greasy substances, gradually loses its fluidity and metallic lustre, and becomes what is called *extinct* or *killed*. It is, in fact, reduced to a dark grey mass, in which, when moist and well-prepared, globules cannot be distinguished even with a magnifier of moderate powers. But although a portion of the Mercury is oxidized, the greater portion, though finely divided, is in a metallic state; for if rubbed on Silver or Gold, the white mercurial stain will readily be displayed.

Hydrargyrum cum Creta, L. E. D.

Mercury with Chalk. Mercurius Alkalizata.

Prop. A heavy powder of a greyish colour; without smell; has a slight metallic and chalky taste; insoluble in water, but its Carbonate of Lime is readily acted on by acids.

Prep. L. E. Rub *Mercury* ℥iij. (2 parts, E.) and *Prepared Chalk* ℥v. (1 part, E.) until globules are no longer visible, L. Triturate the *Mercury* with *Manna* 2 parts, adding a few drops of water to impart the consistence of syrup. When the globules disappear add an eighth of the Chalk, continuing the trituration. When the mixture is complete add 16 parts of hot water; agitate, and when the sediment has fallen, pour off the liquid; repeat the washing once and again, to remove all the Manna. While the sediment is moist mix with it the rest of the Chalk, and dry the powder on blotting paper.

D. Prepare like Hydrargyrum cum Magnesia, substituting *precipitated Carbonate of Lime* for Carbonate of Magnesia.

The extinction of the Mercury is facilitated by the addition of the Manna and water. The great part of the metal is only minutely subdivided. Some think that a portion is converted into a Suboxide, and Mr. Phillips states (Transl. of Pharm.) that he has found a small portion of Binoxide of Mercury. That it is not in the state of the Protoxide is inferred from its not being acted on by Acetic′, which dissolves away the whole of the Chalk. Dr. Nevins has proved satisfactorily that a little (about ½ a grain in 100 grains of the Hydrarg. c. Creta) is in the state of Protoxide. Dr. N. dissolved away the Chalk with H Cl′, and thus converted the Oxide, if any were present, into Chloride of Mercury. After washing, the residue was digested in dil. Nit′ to remove metallic Mercury and the white powder or Chloride left. This became black when touched with Liq. Potassæ, &c. (P. J. iv. 412.)

Tests. Part is evaporated by heat; what remains is colourless and totally soluble in Ac′ with effervescence: this solution is not coloured by Hydrosul′. The ingredients can scarcely be so diligently triturated as that no globules shall be visible. L.

Inc. Acids and acid salts, Sulphates, Acetate of Lead.

Action. Uses. Mild Alterative and Cathartic, also Antacid.

D. gr. v.—℈ß. for adults; gr. ij.—gr. v. for children, in powder or some viscid substance.

Hydrargyrum cum Magnesia, D.

Prep. Rub together *Purified Mercury* and *Manna* āā 2 parts, *Carb. Magnesia* 1 part. Follow the direction of the E. P. for making Hydrarg. c. Creta, as the directions of the D. P. for making the present preparation have been adopted with the substitution of Carb. of Lime for Carb. Magnesia.

Action. Uses. Similar to the former. The Magnesia will make it more laxative in cases of acidity; but it is seldom employed.

Pilulæ Hydrargyri, L. E. D.

Pills of Mercury. *Pilula Cærulea.* Blue Pill.

These form a mass of a bluish colour and soft texture, in which most of the Mercury is minutely subdivided and a small portion oxidized.

Prep. L. E. D. Rub together *Mercury* ʒij. (2 parts, E.) and *Confection of Red Rose* ʒiij. (3 parts, E. D.) till globules can no longer be seen. Then add *Liquorice* bruised ʒj. [(in powder, E.) (extract of, in powder 1 part, D.) Beat the whole till incorporated. (Divide the mass into 5 grain pills, E.)

Steam power is now usually employed, which is an advantage, as the efficacy of the pill depends upon the extent to which the extinction is carried. It has been proposed to effect this by means of stearine, and then to add the Rose Confection. It ought to display no globules when rubbed on paper. Dr. Nevins has shown, as in the case of the Hydrarg. c. Creta, that a small portion, about $\frac{3}{4}$ gr. in 100 grs. of the Pill, is in the state of an Oxide. (P. J. iv. 412.) If washed with boiling water, this ought to give no indications of Sul' with Chloride of Barium, as acid is sometimes added to heighten the colour of the Confection.

Action. Uses. Alterative, Cathartic. Employed to affect the system with Mercury.

D. gr. iij.—v., or even gr. xv., to act as a purgative. gr. v., morning and evening, are prescribed to induce salivation, sometimes conjoined with a little Opium, to prevent the Mercurial acting on the bowels. Three grs. of the Pill contain 1 gr. of Mercury.

Unguentum Hydrargyri (E. D.) fortius, L.

Mercurial or Blue Ointment. *F.* Onguent Mercuriel double.

Mercury seems to have been employed medicinally by the Romans. Pliny says that Mercury is poisonous, "unless, indeed, it is to be administered in the form of an unction on the belly, when it will stay bloody fluxes." (Holland's Transl. lib. 33. c. 8.) It was subsequently employed by the Arabs, and thence re-introduced into European practice. It is called Ung. Hydrarg. fortius, L. P., to distinguish it from the next preparation.

Prep. L. E. D. Take *Mercury* ℔ij. *Lard* ℥xxiij. *Suet* ℥j. (Purified Mercury and prepared Hog's Lard equal parts, D.) Rub the Mercury with the Suet and a little of the Lard (in a marble or iron mortar, D.) until globules can no longer be seen. Add the rest of the Lard and mix. (This ointment is not well prepared if metallic globules can be seen in it with a magnifier of four powers. It may be diluted at pleasure with 2 or 3 times its weight of Axunge, E.)

Trituration produces extinction of the Mercurial globules, and with this some degree of oxidation. This has been denied by some skilful experimentalists; but, as it depends upon the mode and extent to which the trituration has been carried, the oxidation may not be observed when the trituration has been less effective. It has been ascertained that simple trituration is not sufficient, for globules continue to be observed with a magnifier of four or five powers. The complete extinction is best effected by the assistance of steam power, and by allowing the mixture to remain exposed to the air, and triturating occasionally; so that the operation is not completed for some weeks: also by triturating the mixture with some old Mercurial ointment, or, as Soubeiran likewise recommends, with Lard that has been exposed in thin layers in cellars from fifteen days to some months.

M. Guibourt and Messrs. Vogel and Boullay assert that, according to their experiments, the Mercury in the above Ointment remains in a metallic state, as does Mr. Watt (in *The Chemist*, No. 13). M. Guibourt found only a 500th part of the Ointment to consist of an Oxide of Mercury combined with a fatty acid. Dr. Christison, however, states that for the last eight years he has never failed to detect a sensible proportion of Oxide. On melting the ointment in a long tube, there is obtained a short column of Mercury at the bottom, and a long superstratum of yellowish, almost perfectly transparent oil. This, even when filtered, becomes intensely black with Sulphuretted Hydrogen; and if agitated with successive portions of dil. Ac′ at 150°, an acid liquor is obtained, which gives a copious black precipitate of Sulphuret of Mercury with the same reagent. Hence, Mercury must be present in the form of an Oxide combined with a fatty acid, and Dr. C. calculates in the proportion of about one per cent. of the ointment, or a fiftieth of the Mercury used. Mr. Donovan many years since proved that the superior stratum of melted ointment which contained only one-fifth of the original Mercury, was as energetic as ever in producing the effects of Mercury. He therefore conceived it as consisting partly of uncombined metal and lard, and partly of a chemical compound of the Protoxide and Lard: to the latter portion alone he attributed any medicinal efficacy. Dr. Paris, long since, in his Pharmacologia, recommended the adoption of Mr. D.'s ointment made with the Oxide of Mercury. Dr. Christison also concludes that the small proportion of Oxide either present at first, or formed during the process of rubbing the Ointment into the skin, is the only active part of the Mercury.

Tests. Mercurial Ointment is apt to be carelessly made, or with too little Mercury. Its colour should be compared with that of some genuine Ointment. Its Sp. Gr. 1·78, as recommended by Dr. Pereira, should be ascertained. When rubbed on paper, no globules should be visible with a magnifier of four powers; though innumerable ones may be seen with a powerful microscope, as represented in P. J. iii. 399. The fatty matter may be separated by means of boiling water or Ether, &c., and the residual mercury weighed.

Action. Uses. Rubbed on the skin, or taken internally, affects the constitution with Mercury, as indicated by salivation. Useful dressing to Syphilitic and other sores.

D. Contains equal parts of Mercury and Lard with Suet. ʒß—ʒj. rubbed morning and evening or more frequently, on the inside of the thighs or arms, or elsewhere, will speedily salivate. The patient should be kept warm and in the same clothing, the hand of the operator being protected with bladder, &c. On the Continent, grs. ij.—v. made into pills with Liquorice, salivate speedily. The practice is therefore worthy of adoption in extreme cases, especially in hot climates.

UNGUENTUM HYDRARGYRI MITIUS, L. D. Milder Mercurial Ointment.

Prep. L. Mix together strong *Mercurial Ointment* ℔j. and *Lard* ℔ij. D. To be made with twice as much Lard as the last. E. P. *v.* supra.

Action. Uses. Generally employed as a dressing to ulcers. ʒj. Mercury in ʒvj. of Ointment.

CERATUM HYDRARGYRI COMPOSITUM, L. Compound Cerate of Mercury.

Prep. Rub together stronger *Ointment of Mercury, Soap Cerate* āā ℥iv. *Camphor* ℥j. till incorporated.

Action. Uses. Applied to chronic enlargements of the joints, and to disperse indolent tumours.

LINIMENTUM HYDRARGYRI COMPOSITUM, L. Compound Liniment of Mercury.

Prep. Rub *Camphor* ℥j. with *Rectified Spirit* f℥j. then add *Lard* and the stronger *Ointment of Mercury* āā ℥iv. still rubbing. Gradually pour in *Solution of Ammonia* f℥iv. Mix the whole.

Action. Uses. A liquid form of Mercurial Ointment combined with stimulants, and, like the last preparation, employed to promote absorption, and hence to discuss indolent tumours, &c. ; sometimes to excite salivation more readily, by its stimulant action on the lymphatics.

EMPLASTRUM HYDRARGYRI, L. E. Mercurial Plaster.

Prep. Add *Sulphur* gr. viij. to *Olive Oil* f℥j. stirring till they unite. (Resin ℥j. Olive Oil f℥ix. E. melt together and let cool, E.) Triturate with these *Mercury* ℥iij. till the globules disappear. Add gradually *Plaster of Lead* ℥xij. (Litharge Plaster ℥vj. E.) melted with a gentle heat. Mix.

In the L. process Sulphuretted Oil is first produced, and a little Sulphuret of Mercury is afterwards formed, with most of the metal mechanically subdivided. The E.P. substitutes Resin for the Sulphur.

Action. Uses. Applied as a plaster, it stimulates the lymphatics of the part, whether this be a chronically enlarged joint, or node, glandular enlargement, or chronically diseased liver or spleen.

EMPL. AMMONIACI (ET HYDRARGYRI, E.) CUM HYDRARGYRO, L. D.

Prep. L. E. D. Add *Sulphur* gr. viij. gradually to *Olive Oil* f℥j. previously heated, stirring till they unite (the D. C. orders common Turpentine ʒij. instead of the Sulphur and Oil). Triturate *Mercury* ℥iij. with these till the globules disappear. Add gradually pure *Gum Ammoniac* ℔j. melted. Mix carefully. (Rub with a gentle heat till the ingredients unite, D.)

Action. Uses. Similar in its effects to the last, but usually considered more effective. Applied to discuss enlargements of glands and joints, or to indolent tumours.

HYDRARGYRI OXYDUM, (NIGRUM, D.) L.

Oxide or Protoxide of Mercury. Suboxide of some Chemists. *Hydrargyri Oxydum cinereum.* *F.* Protoxide de Mercure. *G.* Quecksilberoxydul.

The Oxide or Protoxide (Hg O=210) of Mercury is a dark grey powder, devoid of taste and smell; heavy; Sp. Gr. 10·69; insoluble in water; easily decomposed by light, and by a heat even of 212°, being resolved into metallic Mercury and some Binoxide, when it becomes of a yellowish or olive hue. Dissipated at 600°. Readily dissolved by Acetic′ or by dil. Nit, from which it will be again precipitated by the alkalis. When pure, insol. in H. Cl′; but this acid or the soluble Chlorides will give a white precipitate (Calomel) in solutions of its salts. A small portion of this Oxide has been detected in Hydrarg. c. Creta, Pil. Hydrarg. and Ung. Hydrargyri. Comp. Hg 96·2, O 3·8=100.

Prep. L. D. Mix *Chlor. Mercury* ℥j. (Sublimed Calomel 1 part, D.) with *Lime Water*, C. j. (Water of Caustic Potash 4 parts made warm, D.) Set by, and when the Oxide has subsided pour off the liquor. (Triturate together till a black Oxide is obtained, D.) Wash frequently with *Aq. dest.* till nothing alkaline can be perceived. Dry the Oxide wrapped in bibulous paper in the air. (With a medium heat, D.)

In the L. process, the Chlorine of the Calomel combines with the Calcium, and in that of the D. P. with Potassium: in one case Chlor. Calcium is formed, and in the other, Chlor. Potassium, which remain in solution; while in both cases Oxide of Mercury is precipitated; but a little Calomel often remains undecomposed. Mr. Donovan recommends that the Liq. Potassæ should be cold and in excess, the decomposition rapid, and that the drying take place in a dark place.

Tests. L. " By heat it is entirely dissipated, and is totally soluble in Acet′." Anything insoluble will be impurity. " Digested for a short time with Dil. H Cl′ and strained, neither solution of Potash nor of Oxal. Ammonia throws down anything." The Potash would throw down any Binoxide that had been dissolved as a yellow precipitate, and the Oxalate any Lime that the acid had taken up.

Action. Uses. Mild Mercurial; but uncertain in composition, and therefore seldom if ever used internally. Mr. Donovan recommended it to be employed for making Ung. Hydrargyri.

D. gr. j.—v. Employed sometimes in fumigations. Externally as an Ointment, 1 part to 3 or 5 of Lard; or as a Lotion in the BLACK WASH (*Lotio Nigra*), prepared by mixing *Calomel* ʒj. with *Lime-water* Oj., and shaken up when used.

HYDRARGYRI BINOXYDUM, L. HYDRARGYRI OXYDUM RUBRUM, D.

Binoxide of Mercury. Red and Peroxide of Mercury. Oxide of Mercury of some chemists. *Calcined Mercury.* *F.* Deutoxide de Mercure. *G.* Rothes Quecksilberoxyd.

The Red Oxide of Mercury ($Hg\ O_2=218$) has long been employed in medicine, being one of the preparations which was known to Geber.

Prop. Binoxide Mercury, prepared by calcination, is in red scales, but if by the following formula, is an orange-red powder, without smell, but having a disagreeable metallic taste. Sp. Gr. about 11·0. Nearly insoluble in water. "When very carefully prepared from the Nitrate, and boiled in five successive portions of distilled water, the water constantly contains about the same quantity of Mercury, and quite enough of it to give a black precipitate with Hydrosul′ a grey precipitate with Protochlor. Tin, and a yellow one with Bichromate Potash. Its solubility in boiling water is one grain in f℥xvj. or about a 7000th." (*c.*) Dr. Barker also found 0·62 parts soluble in 1000 of water. It is decomposed by light and heat, changing colour, and at a heat below redness giving out Oxygen, the Mercury becoming sublimed. It is readily dissolved by Nit′ and H Cl′, also by Ac′ and Hydrocyanic′. Comp. Hg 92·7+O 7·3=100.

It may be prepared by various methods, and no less than three of them are adopted in the Pharmacopœias. The oldest is very tedious, and probably now never employed, though still retained in the D. P.

Prep. D. Put into a glass vessel with a broad bottom and narrow mouth *Purified Mercury* q. s. Expose to a heat of about 600° F. till all is converted into red scales.

Here the Mercury is sublimed and oxidized by the Oxygen of the air, to which it is exposed, and then condensed in the long and narrow-necked bottle, reddish scales being slowly formed.

Prep. L. Dissolve *Bichlor. Mercury* ℥iv. in *Aq. dest.* Ovj., filter and add *Liq. Potassæ* f℥xxviij. Pour off the liquor, wash the powder thrown down in *Aq. dest.* till nothing alkaline can be perceived. Dry with a gentle heat.

The 2 Eq. of Cl of the Bichlor. take 2 of Potassium from the Potash, when 2 Eq. of Bichlor. Potassium are formed, and remain dissolved.

The 2 Eq. of O separated from the Potassium combine with the single Eq. of Mercury of the decomposed Bichlor., and the whole becomes precipitated as Binoxide of Mercury. Or we may suppose the water to be decomposed, and its elements combining with those of the Bichloride, a Hydrochlorate of the Binoxide of Mercury to be formed. On the addition of the Potash, which should be in excess, this will combine with the H Cl′, and the Binoxide combined with water is separated as a yellow-coloured Hydrate. This, when dried with a moderate degree of heat, becomes orange coloured and anhydrous.

Tests. L. When heated, it gives off Oxygen, and Mercury is left in globules, or is entirely dispersed. (Brick-dust, or the Oxides of Iron or of Lead will be left.) Does not emit Nitrous vapours when heated; is entirely soluble in H Cl′. Water in which it is boiled or washed yields no precip. to Lime-water (showing that no Corrosive Sublimate is present), or to Hydrosul′ &c. But this is not quite correct, according to the experiments of Dr. Christison. (*v.* supra.)

HYDRARGYRI NITRICO-OXYDUM, L. OXYDUM RUBRUM, E. OXYDUM NITRICUM, D. Nitric Oxide of Mercury. Red precipitated Mercury, or Red Precipitate.

Prep. L. E. D. Take *Mercury* ℔iij. [(℥viij. E.) (℥ij. D.)] *Nitric′* ℔iß. and *Aq. dest.* Oij. [(Dil. Nit′ dens. 1280 f℥ v. E.) (Dil. Nit′ by measure ℥iij. D.)] Dissolve the Mercury in the acid in a proper (glass, D.) vessel with a gentle heat (gradually increasing, D.) Evaporate the liquid and pulverize the residuum. (Evaporate till a dry white salt is left, E.) (Triturate this with the rest of the Mercury till a fine uniform powder be obtained, E.) Put this into a very shallow (porcelain, E.) vessel and apply a gentle heat, (constantly stirring, E.) gradually increasing, till red vapours (acid fumes, E.) cease to be evolved. (The residuum be converted into red scales, D.)

From the proportions of Mercury and acid employed, a Nitrate of Perox. Mercury is first formed. (*v.* p. 202.) This, for economy both of time and acid, is in the E. P. converted into a Nitrate of the Protoxide, by trituration with the rest of the Mercury. When heated, the Nit′ becoming decomposed, a part of it escapes in Nitrous fumes, and a part gives its Oxygen to convert the whole of the Mercury into the Binoxide of an orange-red colour; and, according to variations in the process, it is obtained in orange-red powder or in bright red scales. But these are valuable only as indications of the mode in which it has been made. The Oxide prepared in this manner contains a little Nit′, and is usually considered more acrid than the other forms. But when carefully prepared, the quantity of Nit′ is very small, and the preparation is essentially only Binoxide of Mercury.

Tests. On the application of heat, no Nitrous vapour is emitted, L. E. (Any Nitrate will be decomposed, and evolve Nitrous fumes.) Decomposed and entirely sublimed. (Any Red Lead will be left.) Completely soluble in Mur′, E. Neither Lime-water nor Hydrosul′ throws down anything from the water in which it has been boiled. L. The Lime-water will detect any Calomel, but as a small portion of this Oxide is soluble in boiling Aq. (*v.* supra), the Hydrosul' will always give a black precipitate.

Action. Uses. Irritant, Stimulant. As a powder sprinkled over indolent ulcers, or as a Caustic to repress exuberant granulations. A Lotion commonly called *Yellow Wash*, formed in the proportion of gr. 1—2 of *Calomel* to f℥j. of *Lime-water*, which contains Chlor. Calcium in solution and a precipitate of yellow Hydrated Binoxide, is prescribed in similar cases, and should be used only when shaken up. The Binoxide has been given internally in doses of $\frac{1}{8}$ to 1 grain in pills, but is objectionable.

UNGUENTUM HYDRARGYRI NITRICO-OXYDI, L. UNG. OXIDI HYDRARGYRI, E. U. HYDRARG. OXYDI NITRICI, D. Red Precipitate Ointment.

Prep. L. E. D. Take *Nitrico Oxide of Mercury* (Red Oxide, E.) ℥j. *Lard* (prepared Hog's, D.) ℥vj. (Axunge ℥viij. E.) *White Wax* ℥ij. L. D. Melt the Lard in the Wax, add the Oxide in very fine powder. Mix the ingredients thoroughly. (Triturate the Oxide and Axunge into a uniform mass, E.)

This Ointment, when fresh made, is of a bright scarlet colour; but the Oxide by degrees undergoes decomposition, as is evident from the colour changing first to a greyish-red and then to a bluish-grey.

Action. Uses. Stimulant, applied to indolent sores, and to chronic inflammation of the eye-lids.

HYDRARGYRI IODIDUM, L.

Iodide or Protiodide of Mercury. Subiodide of some Chemists. *F.* Proto-Iodure de Mercure. *G.* Quecksilberiodür.

Iodide of Mercury (Hg I = 328) has been only recently introduced into medicine, its employment having been first indicated by M. Coindet. It occurs as a heavy greenish-yellow powder, sublimed by heat, insoluble in water and in Alcohol, also in a watery solution of Chlor. Sodium, by which it is easily separated from the Biniodide; soluble in Ether and acids. It is decomposed by light, reddened by heat, but becoming yellow on cooling. By a higher degree of heat it is resolved into metallic Mercury and the red Iodide. Comp. Hg. 55·5 + I 44·5 = 100.

Prep. Rub together *Mercury* ℥j. and *Iodine* ʒv. adding gradually *Alcohol* q. s. till the globules disappear. With a gentle heat dry the powder immediately out of access of light. Keep in well-closed bottles.

The Mercury and Iodine being mixed in equivalent proportions, and triturated, considerable heat is produced, and sometimes even explosion; the Alcohol is added to dissolve the Iodine, and this being presented in a finely divided state to the Mercury, the latter becomes rapidly extinguished. It acts also on some Biniodide of Mercury which is formed at first, and facilitates its combination with metallic Mercury. The Iodide of Mercury may also be formed by precipitating or mixing together a solution of Iodide of Potassium with one of Protonitrate of Mercury having the slightest excess of Nitric′. (Soub. ii. 515.)

Tests. Yellowish when recently prepared, and when heat is cautiously applied it sublimes in red crystals, which afterwards become yellow, and blacken by access of light. It is not soluble in Chlor. Sodium, L. Any impurity will appear if the Iodide does not answer to all these characteristics.

Action. Uses. Irritant Poison. Alterative Stimulant in Syphilis occurring in scrofulous patients. Both the Iodides, if continued, will produce the effect of Mercurials.

D. gr. j,—gr. iij.

PILULÆ HYDRARGYRI IODIDI, L. Pills of the Iodide of Mercury.

Prep. Beat into a uniform mass *Iod. Mercury* ʒj. *Confec, Dog-rose* ʒiij. *Powdered Ginger* ʒj.

D. gr. v.—gr. xv., as 5 grains contain 1 grain of the Iodide.

UNGUENTUM HYDRARGYRI IODIDI, L. Ointment of Iodide of Mercury.

Prep. Melt *White Wax* ℥ij. and *Lard* ℥vj. add finely powdered *Iodide of Mercury* ʒj. Mix.

This Ointment may be rubbed in, or applied as a dressing to scrofulous sores.

HYDRARGYRI BINIODIDUM, L. E.

Biniodide of Mercury. Deuto- or Periodide and Red Iodide of Mercury. Iodide of some Chemists. *F.* Deuto-iodure de Mercure. *G.* Doppelt Iodquecksilber.

Biniodide of Mercury ($Hg\,I_2=454$) is a powder of a beautiful scarlet colour, insoluble in water. "By heat cautiously applied, it is sublimed in scales, which soon become yellow, and, when cold, red. It is partially soluble in boiling rectified Spirit, which affords crystals as it cools. It is alternately dissolved and precipitated by Iod. Potassium and Bichlor. Mercury. It is totally soluble in Chlor. Sodium," (L. P.), or, more precisely, in "40 parts of a concentrated sol. of Mur. Soda at 212°, from which it is again precipitated in fine red crystals on cooling." E. P. This serves to distinguish it from the Iodide of Mercury, which is insoluble in brine. It is remarkable for crystallizing in different forms according to the heat at which it has been sublimed, and also for the change of colour from yellow to red taking place upon merely touching the crystals with a hard body.

Prep. Rub together *Mercury* ℥j. (℥ij. E.) and *Iodine* ʒx. (℥iiß. E.) gradually adding *Alcohol* q s. (a little Rectified Spirit, E.) till globules are no longer visible (a uniform red powder be obtained, E.) Dry the powder with a gentle heat; keep in well closed bottles. (Reduce the product to fine powder, dissolve it in solution of *Mur. Soda* Cong. j with the aid of brisk ebullition. If necessary filter through calico, keeping the funnel hot. Wash and dry the crystals which form on cooling, E.)

Mercury and Iodine on being triturated together, the globules of the former quickly disappear, and heat is produced, and to an extent, if the quantities are considerable and the ingredients dry, so as to produce an explosion: hence Alcohol is added to keep the mixture moist. The Iodine, if moist, should be used in a proportionately larger quantity, or the Mercury will be in less proportion than is necessary to produce the Biniodide, and some of the yellow Iodide will be produced. The E. C. therefore directs it to be purified by boiling in brine, and crystallizing. The Biniodide may also be produced by acting on solutions either of Pernit. Mercury or of Bichlor. Mercury with sol. of Iod. Potassium, adding the latter by degrees, but slightly in excess; as, double decomposition taking place, the Biniodide precipitated is soluble in an excess of either of the salts employed in its production.

Tests. The tests given in the P's being characteristic of this salt, have already been given. (*v.* supra.) Dr. Pereira mentions in addition, that the presence of Bisulphuret of Mercury will be indicated by fusing it with caustic Potash, and then adding a mineral acid, when Hydrosul' will be evolved.

Action. Uses. Irritant Poison; Stimulant in scrofulous habits, but seldom employed. *Ext.* Caustic.

D. gr. $\frac{1}{16}$ to $\frac{1}{8}$ in pill or in Alcoholic solution.

UNGUENTUM HYDRARGYRI BINIODIDI, L. Ointment of Biniodide of Mercury.

Prep. Melt together *White Wax* ℥ij. and *Lard* ℥vj. add *Biniodide of Mercury* ℥j. finely powdered. Mix.

Action. Uses. Stimulant application to ulcerations of different kinds, but requires to be diluted.

HYDRARGYRI CHLORIDUM, L. CALOMELAS (E.) SUBLIMATUM, D.

Chloride of Mercury. Calomel. Protochloride and *Submuriate* and *Mild Muriate* of Mercury. *F.* Protochlorure de Mercure. Mercure doux. *G.* Einfach Chlorquecksilber.

The Chloride of Mercury ($Hg\,Cl=238$) occurs native in Carniola and in Spain, and is called Horn Mercury and Native Calomel; but it seems also to have been prepared artificially by the Hindoos at very early periods (Fleming and Ainslie), and prescribed internally. It has been known in Europe since 1608.

Prop. It is found crystallized in four-sided prisms terminated by four-sided pyramids. When prepared artificially, it may be obtained in similar quadrangular prisms covering a crystalline mass which is fibrous

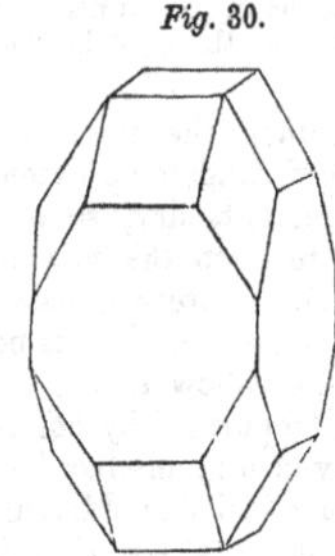
Fig. 30.

in texture, sparkling, and semitransparent, somewhat horny and elastic in nature. (Brande.) Sp. Gr. 7·2. When scratched, a yellow characteristic streak is observed. As usually seen, it forms a heavy tasteless powder, of different degrees of fineness as well as of whiteness; of a light yellowish or buff-colour if obtained by levigation, but when condensed in air, a pure white and impalpable powder is formed. It becomes of a darker hue when exposed to light, and when heated, yellowish. At a higher degree of heat it sublimes. It is insoluble in pure water, Alcohol, and Ether. Boiled in water for some time, under the influence of alkaline Chlorides, a portion is considered to be converted into Bichloride; but others conceive that a portion of it may become dissolved. The alkalis and Lime-water instantly render it black, from precipitating the grey Oxide and combining with its Chlorine; Ammonia produces a greyish powder. Chlorine converts it into Bichloride, as does boiling H Cl′, also setting free some metallic Mercury; Nit′ and Sul′ into Bichloride, and Nitrate and Sulphate of the Binoxide, Nitrous and Sulphurous fumes escaping in the respective cases. Many salts decompose it, and hence also water holding them in solution. Comp. Hg 85+Cl 15=100.

Prep. L. E. D. Calomel may be prepared by several processes, as by adding as much Mercury to Corrosive Sublimate as it already contains, by double decomposition of different salts, and subsequent sublimation or by precipitation. Boil *Mercury* ℔ij. (℥iv. E.) with *Sul′* ℔iij. Commercial f℥ij. and ʒiij. mixed with pure Nitric′ f℥ß. till dissolved, raising the heat till a dry (salt, E) *Bispersulph. Mercury* remains. Triturate this Persulphate of Mercury (25 parts, D.) in an earthenware mortar with *Mercury* ℔ij. (17 parts, D) till intimately mixed, then add *Chlor. Sodium* ℔iß. [(dried Muriate of Soda 10 parts, D.) (Mercury ℥iv. and Muriate of Soda ℥iij. E.)] Triturate till the globules (entirely, E.) disappear. Sublime (in a proper apparatus) [(in a sand-bath, E.) (with a gradually increasing heat, D.)] Reduce the Sublimate to very fine powder, wash carefully with boiling *Aq. dest.* (till solution of Iodide of Potassium (Caustic Potash, D.) does not affect the water, E. D.) Dry the Calomel.

In this process, a Bipersulph. Mercury is first produced, either with or without the aid of Nit′. (*v.* p. 201.) The salt produced is then rubbed up with a further quantity of Mercury (L. and D. P.), and converted into a Protosulph., or the Bipersulph. Mercury and Chlor. Sodium are all three triturated together until globules no longer appear; in either case the Bipersulphate is converted into the Protosulphate of Mercury, by the additional quantity of Mercury combining with the second Eq. of the Oxygen and Sul′ of the Bipersulph. When heat is applied, both the Chlor. Sodium and the Protosulph. of Oxide of Mercury are mutually decomposed: the Chlorine combining with the Mercury, the required Chlor. Mercury is formed and sublimed, while the Oxygen of the Protoxide of Mercury, combining with Sodium, forms Soda, which uniting with the Sul', a dry Sulph. Soda remains. A little Mercury is apt to rise, and some Bichlor. Mercury or Corrosive Sublimate to be formed; or some Calomel in subliming may become separated into these two. Manufacturers therefore sometimes add a little more of one or other of the ingredients, and sometimes the Bichloride to supply Chlorine. The next part of the directions, that of washing, is intended to get rid of it. This is best done by using cold distilled water, and gradually increasing its heat. The water of many springs, from containing various salts in solution, readily decomposes the Calomel. The Corrosive Sublimate being readily soluble in water, its presence will be detected either by Liq. Potassæ, which will cause a yellowish-red precipitate, or by Ammonia, which causes a white precipitate. (p. 198.)

Some variations are made in the latter part of the process, in order to obtain the Calomel of a white colour, and in the state of an impalpable powder. The first improvement was made by Mr. Jewell, by keeping the vessel into which the Calomel was sublimed full of water or of steam, by which Jewell's patent Calomel and Howard's Hydro-sublimate or Howard's Hydro-Calomel was obtained. The apparatus was improved by Mr. Ossian Henry of Paris, and has been generally adopted, being admitted into the French Codex to produce the "Mercure doux à la vapeur." Many English chemists have, however, been in the habit of subliming the Calomel into a large chamber full of air. (P. J. ii. pp. 586 and 657.) M. Soubeiran has now adopted this method, having first proposed a current of cold air, as had been done by Mr. Dann of Stuttgardt.

Calomel may also be obtained in a state of fine division by precipitation, as in the D. P., though the sublimed kind is preferred. It is called Scheele's Calomel, and by the French, Précipité blanc.

Calomelus Precipitatum, D.

Prep. Pour *dilute Nitric'* 15 parts on purified *Mercury* 17 parts in a glass vessel, when effervescence ceases, heat gently for 6 hours, agitating occasionally. Raise the heat till the mixture boils a little, pour off the liquid from the remaining Mercury. Mix it immediately with a previously prepared solution of *Mur. Soda* 7 parts in *boiling Aq.* 400 parts. Wash the precipitate with warm *Aq. dest.* till solution of Caustic Potash has no effect. Dry the powder.

The object here is first to prepare a Nitrate of the Protoxide of Mercury, which is effected at the expense of the Oxygen of the Nit'; but the boiling is injurious, as Mr. Phillips says it will produce a large proportion of the Peroxide, and, indeed, it is difficult to produce it free from this. On adding the solution of the Protonitrate to the hot solution of Chlor. Sodium (Muriate of Soda), Chlor. Mercury or Calomel is precipitated, and Nitrate of Soda remains in solution. A little water is always retained, and there is apt also to be some Corrosive Sublimate, as in the preceding process, and also some basic Nitrate. "The fixed alkalis and Lime-water render it dark grey, not black, as they do sublimed Calomel." (*c.* ex Gottling.)

Tests. Heat sublimes it without any residuum. L. and E. A whitish powder, which, on the addition of Potash, becomes black, and then, when heated, runs into globules of Mercury. The distilled water with which it has been washed gives no precipitate with Nitr. Silver, Lime-water, nor H Sul . L. If any Bichloride be dissolved, Nit. Silver will throw down Chlor. Silver, Lime-water yellowish Binoxide of Mercury, and Hydrosul′ a black Sulphuret of Mercury. Sulphuric Ether agitated with it, filtered, and then evaporated to dryness, leaves no crystalline residuum, and what residuum may be left is not turned yellow with Aqua Potassæ. E. Any Corrosive Sublimate will be dissolved by the Ether, and give a precipitate of the Binoxide of Mercury. The whitest Calomel is not necessarily fine in proportion, as some crystalline specimens are white, and the microscope reveals crystalline grains in other kinds of Calomel. White Precipitate (p. 198) mixed with Sulphate of Baryta has been sold on the Continent as Calomel, and also Sulphate of Baryta mixed with Calomel. (P. J. ii. 728.) That condensed by steam, or in a mass of cold air, is probably the best, and produces the effects of Calomel with most certainty.

Inc. Alkalis and their Carbonates, Lime-water, Alkaline Chlorides, Sal Ammoniac, Nit′, Metals and their Sulphurets.

Action. Uses. Alterative Stimulant, Sialogogue, Cathartic, &c., Antiphlogistic, Sedative. The sublimed Calomel is usually preferred, though the precipitated, from being finely subdivided, is an effective medicine.

D. gr. j. Alterative ; gr. iij.—gr. v. Cathartic ; gr. iij. with a little Opium, 2 or 3 times a day, will rapidly produce ptyalism ; gr. x.—gr. xx. acts as a Sedative in many cases.

PILULÆ HYDRARGYRI CHLORIDI COMPOSITÆ, L. PILULÆ CALOMELANOS COMPOSITÆ, E. D. *Pilulæ Plummeri.* Plummer's Pill.

Prep. L. E. D. Mix in fine powder *Chlor. Mercury* (Calomel, E. D.) and *Oxysulphuret of Antimony* (Golden Sulphuret, E.; Brown Antimoniated Sulphur, D.) āā ʒij. (1 part, E., ʒj. D., Guiacum 2 parts, E , ʒij. D.) Triturate them together, then with bruised *Resin of Guiacum* ℥ß. and lastly with *Treacle* ʒij. (2 parts, E., q. s. D.) till a uniform mass be obtained. [(Beat into a proper pill mass, E. D.) (Divide into 6 gr. pills, E.)] Balsam of Copaiba and Oil have been recommended for mixing up the ingredients.

Action. Uses. Alterative and Diaphoretic in doses of grs. v.; Cathartic in grs. xx. Dr. Plummer said of his Pill, that it is in vain to look for its beneficial effects unless the materials are well levigated together, and for a considerable time.

PILULÆ CALOMELANOS ET OPII, E.

Prep. Beat into a proper mass *Calomel* 3 parts, *Opium* 1 part, *Conserve of Red Roses* q. s. Divide into pills, each to contain 2 grs. Calomel.

Action. Uses. Diaphoretic and Antiphlogistic. A pill taken every 3 or 4 hours quickly produces ptyalism.

UNGUENTUM HYDRARGYRI CHLORIDI. Calomel Ointment.

Calomel is often prescribed in the form of Ointment, and with great benefit in various forms of Cutaneous eruptions. "Pommade de Mercure doux" is made with 1 or 2 parts of Calomel to 8 of Lard, and some oil may be added. Dr. Pereira recommends ʒj. to ℥j. of Lard, and Dr. A. T. Thomson the addition of ʒiv. of Tar Ointment, as one of the best applications in Leprous and other dry and scaly skin diseases.

HYDRARGYRI BICHLORIDUM, L. HYDRARGYRI MURIAS CORROSIVUM, D. CORROSIVUS SUBLIMATUS, E.

Bichloride of Mercury. Corrosive Sublimate. *Oxymuriate and Corrosive Muriate of Mercury. F.* Deuto- and Bi-chlorure de Mercure. Sublimé Corrosif. *G.* Doppelt Chlorquecksilber.

Corrosive Sublimate has been long known to, and prepared by, the Hindoos, being their *ruskapoor* (Hind. Med. p. 45). It seems also to have been known to the Chinese, and it was prepared by Geber in the 8th century. It is largely manufactured for use both in medicine and the arts.

Fig. 31.

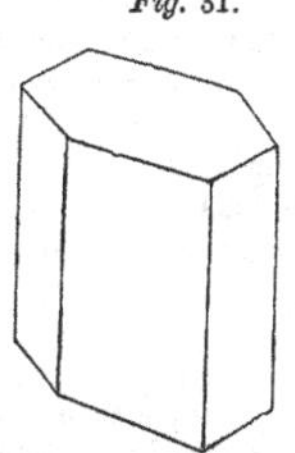

Prop. Bichloride of Mercury ($Hg\,Cl_2=274$) is white, with an acrid metallic and persistent taste, without smell. It is met with in small crystals, or in a semitransparent crystalline mass. Sp. Gr. about 5·2. It crystallizes in right rhombic prisms sometimes terminated by converging planes. These are readily powdered, and effloresce at the angles when some time exposed to the air. It is fused by heat, and then volatilized; is soluble in water, Alcohol, and Ether, requiring about three times its weight of boiling and about 16 times its weight of cold water. Its solubility is much increased by the presence of Chlor. Sodium and Hydrochlor. Ammonia. (*v.* p. 197.) Alcohol dissolves about one-third of its weight, and Ether still more, so as to be employed sometimes in separating it from its aqueous solution and from organic bodies. When exposed to light in contact with these, it is decomposed into Calomel and metallic Mercury. Nit′ and H Cl′ dissolve it without change. Potash, Soda, and Lime throw down the yellow Peroxide of Mercury, which afterwards becomes brick-red; and Ammonia a white precipitate (*v.* Hydr. Ammon. Chlorid. p. 198). The alkaline Carbonates precipitate a brick-red Carb. Mercury, Hydrosul′ throws down at first a greyish and then a black precipitate of Bisulphuret of Mercury; Ferrocy. Potassium a white Ferrocyanide of Mercury. Iod. Potassium causes a yellow precipitate, which by degrees becomes a bright red Biniodide of Mercury. Protochloride Tin, abstracting 1 Eq. of the Chlorine, becomes Perchlor. Tin in solution, and Calomel is precipitated; an excess abstracts more Chlorine, and metallic Mercury in a state of fine division is produced. Several of the metals, as Copper and Silver, decompose it, combining with the Chlorine and setting free the Mercury. Silver has lately been employed by Dr. Frampton. By triturating a grain of the Bichloride with several grs. of metallic Silver, a black powder was produced, and on heating this in the bulb of a small tube, a ring of metallic globules was obtained; so also in boiling metallic Silver in powder in a solution of Corrosive Sublimate; likewise when mixed with tea or a gelatinous solution. Mercury will combine with the second Eq. of Chlorine of the Bichloride, and convert it into the Chloride: this was one of the old methods for making Calomel. Gold, aided by Galvanic action, readily reduces it to a metallic state, and at the same time forms an amalgam. Thus by dropping the suspected liquor on a piece of polished Gold, or a sovereign, and touching the moistened surface with the point of a penknife, or as Dr. Pereira suggests, applying a key, so that it may touch simultaneously the Gold and the solution, the Bichlor. becomes decomposed, and a Silver stain is left on the Gold. Hence Dr. Buckler has suggested the reduction of the Mercury to the metallic state within the stomach by means of Iron-filings and Gold-dust, both being in a state of very fine division.

The action of vegetable and animal substances on Corrosive Sublimate is of considerable importance, from the combinations which take place in cases of poisoning. Most of the vegetable infusions and decoctions in use as medicines, as well as ordinary articles of diet, decompose it, especially when exposed to the action of light; the Bichloride also is decomposed when triturated with many fatty or volatile Oils, or boiled with Sugar; or it may combine with some of the vegetable principles, as in Kyanizing wood. The Gluten of wheat acts with apparently more energy, and more like animal principles. The greatest number of experiments have, however, been made with Albumen, into a solution of which, if a sol. of Corrosive Sublimate be dropped, a white flaky precipitate is thrown down, which, when dried, is hard, horny, and brittle. Ammonia rubbed up with this precipitate does not display any blackening, nor does Ac' leave a white insoluble residuum, both tests showing that no Calomel had been formed. Lassaigne has some time since shown that the precipitate consists of 6·45 Bichloride of Mercury with 93·55 of Albumen, and that it is soluble in an excess of Albumen as well as of sol. of Bichloride of Mercury. It is generally stated to be an inert and insoluble compound.

Bichlor. Mercury may be prepared by bringing together its constituents, and, as has been done by Dr. A. T. Thomson on a large scale, by passing Chlorine through Mercury heated to between 300° and 400°, and by other processes; but it is now chiefly prepared by acting on the Binoxide of Pernitrate of Mercury with H Cl′.

Prep. L. E. D. Dissolve *Mercury* ℔ij. (℥iv. E.) by boiling in a proper vessel with *Sul′* ℔iij. (commercial f℥ij. and f℥iij.) previously mixed with pure *Nit′* f℥ß. (with the aid of moderate heat) till a dry (Salt, E.) Bipersulph. Mercury remains. When cold triturate this (Persulph. Mercury 5 parts, D.) in an earthenware mortar (to very fine powder, D.) with *Chlor. Sodium* ℔iß. (Mur. Soda ℥iij. E. dried, 2 parts, D.) Sublime with a gradually raised heat [(in a proper apparatus, E.) (from a proper vessel into a receiver, D)].

By the first part of the process, a Bipersulph. Mercury is obtained (as above, and at p. 201); but no metallic Mercury is added, because what has been acted upon by the acids requires to be in the state of a Binoxide, in order that the Chlorine may combine with it in the same proportion. Thus as each Eq. of the Bipersulph. contains 2 Eqs. of Sul and 2 Eqs. of Oxygen combined with 1 Eq. of Mercury, it will require 2 Eqs. of the Chlor. Sodium to be decomposed in order that 2 Eqs. of Chlorine may be obtained to combine in the same proportion with the Eq. of Mercury set free, and form the Bichlor., which sublimes. The 2 Eqs. of Oxygen, of Sodium, and Sul′ set free, combining together, form 2 Eqs. of dry Sulph. Soda which remains behind.

Tests. The characteristics of Corrosive Sublimate having been fully given above, its purity may be ascertained by the following tests. It liquefies and sublimes entirely by heat. L. and E. It is totally soluble in water, L., and easily soluble in Sulphuric Ether. L. and E. Fixed impurities will remain after sublimation. Calomel will not be dissolved by water. Five or six parts of Ether will remove the whole of the Bichloride. The other tests of the L. P. will show that it has been properly made. "Whatever is thrown down from water, either by solution of Potash or Lime-water, is of a reddish colour, or if a sufficient quantity be added, it is yellow. Heated, this yellow substance emits Oxygen, and runs into globules of Mercury. L.

Inc. Alkalis and their Carbonates, Lime-water, Soap, Tartar Emetic, Nitr. Silver, the Acetates of Lead, Iodide of Potassium, Sulphuret of Potassium, many metals, infusions of bitter and astringent vegetables, as well as solutions of other vegetable and animal principles.

Action. Uses. Corrosive Irritant Poison, in doses of a few grains, producing depression of the nervous system and excessive Mercurialism. In smaller doses, an excellent Alterative in Syphilis and secondary Syphilis, and chronic cutaneous diseases.

D. gr. $\frac{1}{16}$ to $\frac{1}{8}$, in a pill, or in the following solution. *Ext.* As a lotion, gr. $\frac{1}{2}$—gr. ij. in Aq. Dest. f℥j.

Liquor Hydrargyri Bichloridi, L. Solution of Bichloride of Mercury.

The solvent power of water being increased both by Common Salt and Sal Ammoniac, the former used to be, as the latter is now employed in making a solution of Corrosive Sublimate.

Prep. Dissolve *Bichlor. Mercury* and *Hydrochlor. Ammonia* āā gr. x. in *Aq. dest.* Oj.

D. f℥ß—f℥ij. in some bland fluid. One fluid ounce contains ½ gr. of Bichlor. Mercury.

Antidotes. Albumen, as in white of Eggs, followed immediately by infusion of Galls or of Catechu; Milk; Gluten of wheat and wheaten flour; Protosulphuret of Iron, if administered immediately, or within 15 minutes after the poison has been swallowed; Iron-filings (with Gold-dust?); Antiphlogistic treatment, as with other irritant poisons.

HYDRARGYRI AMMONIO-CHLORIDUM, L. HYDRARG. PRECIPITATUM ALBUM, E. HYD. SUBMURIAS AMMONIATUM, D.

Ammonio-Chloride of Mercury. White Precipitate. *F.* Chlorure Ammoniaco-Mercuriel insoluble. *G.* Weisser Quecksilber präcipitat.

This salt was discovered by Raymond Lully in the 13th century, and is formed by precipitating a solution of Bichloride of Mercury with Ammonia. It is met with in masses, or as a heavy white powder, without smell, but having a metallic taste; insoluble in Water and Alcohol; decomposed by heat, and resolved into Calomel, Ammonia, and Nitrogen; so boiling water resolves it into Hydrochlor. Ammonia and yellow Binoxide of Mercury. Sul′, Nit′, and H Cl′ dissolve and at the same time decompose it. Sol. Caustic Potash, heated with it, expels Ammonia, forms and dissolves Chlor. Potassium, and leaves impure Binoxide of Mercury. Mr. Hennel, on analysing, found it to consist of 1 Eq. Binoxide of Mercury with 1 Eq. of Hydrochlorate of Ammonia. By Mr. Phillips it is considered to be a compound of 1 Eq. Binoxide of Mercury, 218, and 1 of Bichlor. Mercury, 274, with 2 Eq. of Ammonia, 34=526. Dr. Kane, however, states that on adding Ammonia a little in excess to the solution of Bichloride of Mercury, he found that one-half of its Chlorine is set free, and that the precipitate contains only Mercury, Chlorine, and Amidogen, the radical of Ammonia, $N\,H_2$ (*v.* p. 57) in the proportion of 2 Eqs. of each, so as to form a compound of 1 Eq. of Bichloride of Mercury, 274, with 1 Eq. of Binoxide of Mercury, 234=508.

Prep. L. E. D. Dissolve *Bichlor. Mercury* (Corrosive Sublimate, E.) ℥vj. with the aid of heat in *Aq. dest.* Ovj.; (take the liquor remaining after making Precipitated Calomel, D.) When cold add *Solution of Ammonia* f℥viij. (q. s. to throw down entirely the metallic salt, D.) stirring frequently. (Collect the powder on a calico filter, E.) Wash thoroughly (with cold (distilled, D.) water, E.) till free from taste. Dry it (on bibulous paper, D).

The description of the changes which take place on adding Ammonia to the Bichlor. Mercury, must depend upon the view taken respecting the composition of the precipitate.

Tests. Apt to be mixed with other white powders, as Carbs. Lime and Lead, Calomel, Starch, Sulph. Lime and Sulph. Baryta. "Totally evaporated by heat. Dissolved by H Cl′ without effervescence (but not if any Carb. is present). When digested with Acetic′, Iod. Potassium throws down nothing either yellow (Iod. Lead) or blue (Iod. Starch). The powder rubbed with Lime-water, does not become black (showing that no Protoxide of Mercury is present). When heated with solution of Potash, it becomes yellow, and emits Ammonia." L. No other white substance is known to do so. (Phillips.)

Inc. Acids, Alkalis, acid and metallic Salts, &c.

Action. Uses. Supposed to be that of other Mercurials, as the Bichloride, but is only used externally.

UNGUENTUM HYDRARGYRI AMMONIO-CHLORIDI, L. UNG. PRECIPITATI ALBI, E. UNG. HYD. SUBMURIATIS AMMONIATI, D.

Prep. L. E. D. Melt slowly *Lard* ℥iß. (℥iij. E.) add (when concreting, D.) *White Precipitate* ʒj. (ʒij. E.) Mix. (Stir briskly as the Ointment concretes in cooling, E.)

Action. Uses. Alterative Stimulant in Cutaneous diseases and indolent ulcers.

HYDRARGYRI BISULPHURETUM, L. HYDRARGYRI SULPHURETUM RUBRUM, D. CINNABARIS, E.

Bisulphuret and Red Sulphuret of Mercury. Cinnabar. *F.* Sulfure rouge de Mercure. *G.* Rothes Schwefelquecksilber.

Cinnabar was known to the Greeks. It has been discovered to be one of the pigments employed by the Egyptians. The Chinese as well as the Hindoos have from early times employed it in medicine, and the former have long been celebrated for their *Vermilion.* It was formerly called *Kinnabari* and also *Minium,* being often confounded with the Red Oxide of Lead. It occurs native, both massive and crystallized, and is the principal ore from which the metal is extracted at Idria, Almaden, and in China. It is prepared artificially for use both in medicine and the arts.

Prop. The Bisulphuret of Mercury ($Hg\ S_2 = 234$), the Sulphuret (Hg S) of some chemists, when in substance, is of a dark red colour, heavy, striated, gives a bright red streak when scratched; but when powdered, is of a brilliant red, commonly called *Vermilion.* Sp. Gr. 8·1. It is devoid of both taste and odour, is insoluble in both water and Alcohol, and in most of the acids, and is unalterable in the air. Heated, it becomes of a brownish-red; in the air, burns with a blue flame, yielding Sulphurous acid gas and metallic Mercury; but sublimes unchanged out of access of the air. Heated with Potash, globules of Mercury are given out, and the addition of H Cl′ evolves Hydrosul′. It may be made by the following process:

Prep. L. E. D. Melt *Sulphur* ℥v. (Sublimed 3 parts, D.) add (gradually) *Mercury* ℔ij. (19 parts, D.) Continue the heat till the mixture begins to swell up. Remove the vessel, cover it closely to prevent the mass taking fire, reduce to powder (when cold, E). Sublime it.

By the aid of heat, the ingredients combine, and would explode, if they were not covered and removed from the fire. A black Sulphuret is formed: by sublimation, it crystallizes and becomes of a dark red colour, without any change of composition.

Tests. Sublimed entirely by heat, L., and without any metallic globules being formed. Heated with Potash, it yields globules of Mercury; is not dissolved either by Nit′ or H Cl′, but is so by Nitro-Hydro-Cl′. Rectified Spirit with which it has been boiled is not reddened (showing absence of Dragon's-blood, &c.). Acetic′ digested upon it, yields no yellow precipitate with Iod. Potassium (showing that there is no Red Lead mixed with it). Red Sulphuret of Arsenic will be detected by the tests for Arsenic. (v. pp. 209 and 212.)

Action. Uses. Alterative, but seldom given internally. Still used by the Hindoos in fumigation; but the grey Oxide is preferable for such a purpose.

D. gr. x.—℈ß. For fumigation, ʒß.; but the Sulphurous vapours are irritating when inhaled.

HYDRARGYRI SULPHURETUM CUM SULPHURE, L. HYDRARGYRI SULPHURETUM NIGRUM, D. Sulphuret of Mercury with Sulphur. Black Sulphuret of Mercury. *Ethiops Mineral. F.* Sulfure noir de Mercure. *G.* Schwarzes Schwefelquecksilber.

The Black Sulphuret of Mercury has long been employed in medicine. It forms a black heavy powder, without taste or smell. Heated, it is entirely dissipated. Boiled in caustic Potash, a Sulphuret of Potassium is formed, and a black powder left, which, when sublimed, becomes red, and has all the characters of the foregoing preparation, whence Mr. Brande considers it, (and Mr. Phillips coincides with him,) as a compound of 58 parts of Bisulphuret of Mercury with 42 of Sulphur, mechanically mixed. The black substance which is thrown down by Hydrosul′ from solutions of the salts of Protoxide of Mercury, considered by some chemists to be a Protosulphuret, is by others accounted a Subsulphuret (Hg_2 S), resolvable into metallic Mercury and the Bisulphuret.

Prep. L. D. Rub together *Mercury* and *Sulphur* āā ℔j. (1 part in an earthenware mortar, D.) till globules are no longer visible, even with a magnifier.

It is preferable to moisten with a little water or with Hydrosulphate of Ammonia. (Geiger.)

Tests. Entirely evaporated by heat, no Charcoal nor Phosph. Lime being left. L. Both vegetable and animal Charcoal will thus be detected, and free Mercury by a silver stain when rubbed on Gold. Sulphuret of Antimony by boiling in H Cl′, and using the tests at p. 175.)

Action. Uses. Alterative, Diaphoretic in Cutaneous and in Glandular diseases.

D. gr. v.—ʒß., but is rather inert and little used.

Hydrargyri Persulphas, D.

Persulphate or Bipersulphate of Mercury. Sulphate of the Protoxide of some chemists. *F.* Deuto-Sulphate de Mercure. *G.* Schwefelsaures quecksilberoxyd.

The Bipersulphate of Mercury ($Hg\ O_2$, $2\ S\ O_3 = 298$) is officinal in the D. P. for pharmaceutic purposes; but though not mentioned in the L. and E. P., it is prepared in the first part of the processes for making both the Chlor. and Bichlor. Mercury, *v.* 192 and 197. It is a white crystalline salt, which is decomposed on being added to water, as in the next preparation, the yellow Subsulphate of Mercury.

Prep. Heat together in a glass vessel *purified Mercury* and *Sulphuric'* āā 6 parts, *Nitric'* 1 part. Increase the heat till the mass is white and dry.

Mercury and Sul' do not act upon each other when cold, but on being heated, the acid becomes decomposed; Sulphurous acid gas is given off, and the metal becomes oxidated at the expense of a part of the Sul' employed. It may thus be prepared without the aid of Nit'. The Nit' being added to facilitate the oxidation, and to diminish the quantity of Sul' and Sulphurous acids which must afterwards be driven off, is chiefly decomposed, and the Sul' combines with the Oxide when formed. The heat being continued, the Mercury is peroxidated and a Bipersulphate of Mercury obtained.

Action. Uses. Would no doubt act as other Mercurials, but is officinal for making Hydrargyri Chloridum and Bichloridum, and Oxydum Sulphuricum Hydrargyri.

Subsulphas Hydrargyri flavus. Hydrargyri Sulphuricum Oxydum, D.

Turbith or Turpeth Mineral. *F.* Sous-deuto-sulfate de Mercure. *G.* Mineralischer turpith.

Prop. Though without smell, it irritates the nostrils when snuffed up, has an acrid taste, but requires 2000 parts of temperate and 600 parts of boiling water for its solution. When heated, first Sulphurous acid and then Oxygen are given off, and lastly, Mercury is sublimed. Caustic Potash, when made to act on it, will afterwards deposit Sulph. Baryta on addition of the Nitrate. Some chemists consider it to be composed of 1 Eq. of Binoxide of Mercury with 1 of Sul': Mr. Phillips, of 3 Eq. of Sul' with 4 of Binoxide of Mercury. According to Soubeiran, Oxide of Mercury 80·09 Sul' 19·91 = 100.

Prep. Triturate together in an earthenware mortar *Persulph. Mercury* 1 part, warm *Aq.* 20 parts. Pour off the liquor. Wash the yellow powder with warm

Aq. dest. so long as drops of Liq. Potassæ cause any deposit. Then dry the Sulphuric Oxide of Mercury.

Water, from its great affinity for Sul′, decomposes the salt employed, abstracts the chief portion of its acid, separating it into an acid Sulphate, which remains in solution, and precipitates the Oxide of a lemon-yellow colour, still retaining a portion of the Sul′; which is thus considered a Subsulphate.

Action. Uses. Irritant Poison, Emetic, Errhine.

D. Too violent for internal exhibition, and used only as an errhine, gr. j. with grs. v. of some bland powder. Occasionally as an Ointment, 1 part to 8 of Lard, in some herpetic eruptions.

HYDRARGYRI NITRATIS UNGUENTUM, L. UNG. CITRINUM, E. UNG. HYDR. NITRATIS VEL UNG. CITRINUM, D. Citrine Ointment.

Citrine Ointment, or that of the Nitrate of Mercury, is a much-used and highly valued preparation, which was introduced into the Pharmacopœias as a substitute for one known as Golden Eye Ointment. When properly prepared, it is soft, of a bright yellow or lemon-colour, and of a strong Nitrous odour. It is apt, however, to change, especially if the directions for its preparation are not strictly followed. It is then hard, and becomes brittle and almost pulverulent, and its colour changed to bluish-grey or greenish, or of a mottled appearance, the metal becoming by degrees reduced.

Prep. L. E. D. Dissolve *Mercury* ℥j. (by weight ℥j. D.; ℥iv. E.) in *Nitric′* f℥xi. (℥xiß. D.; Dens. 1380 to 1390 f℥ixß. with the aid of a gentle heat, E.) Melt together (in a vessel capable of holding 6 times the quantity, E.) *Lard* ℥vj. (prepared Hog's ℥iv. D.; Axunge ℥xv. E.) and *Olive Oil* f℥iv. (f℥xxxviijß. E.; ℔j. by measure, D.) While hot mix the solutions thoroughly. (If the mixture do not froth up, increase the heat slightly till it does. Keep this ointment in earthenware or glass vessels, excluding the light, E. Make an ointment in the same way as the ointment of Nitric′, D.)

Difficulty having been experienced in making this Ointment, various suggestions have been made for its improvement, as diminishing the Lard to $\frac{1}{6}$, using Olive-oil alone, or substituting Almond, Rape, and Neat's-foot Oil, also Butter; though many have succeeded in making good Ointment, as we ourselves have done when abroad, by following the directions of the L. P. Mr. Alsop (in Pharm. Journ. i. p. 100) clearly pointed out that a due regulation of the heat necessarily generated when large quantities are made, is essential, and that a temperature of about 190° is the best for mixing the sol. of the Nitrate with the melted fatty matter, when strong effervescence takes place. It is equally necessary to attend to the Sp. Gr. of the acid, in order that the exact proportions of the College may be employed, and that the quantity must be increased when the Sp. Gr. is less than 1·5. The stirring usually employed he does not find essential. Mr. A. directs attention to the directions given by Dr. Duncan in his Dispensatory of 1794, as they embody these principles, and seem to have been originally proposed by Mr. Duncan of Edinburgh. The proportions have been adopted by the E. P., but the quantities are given above as corrected by Dr. Christison. (Disp. p. 530.) It yields Ointment of a fine golden-colour, and of the requisite softness, if kept from the light. Experiments have also been made in the Laboratory of the Pharmaceutical Society (*v.* P. J. iv. 450) shewing that perfectly good Ointment is made by attending to the L. P. directions, and that Ointment which was old, become hard, discoloured, and pulverulent, was restored to its original appearance by heating it with a little Nitric acid. When well made, Citrine Ointment, according to the explanation given by M. Soubeiran from the experiments of M. Boudet, contains Nitrate of the Binoxide and some Subnitrate of the Protoxide of Mercury, or "turbith nitreux"—less of the latter, and more of the former, as the heat is greater,—with some free Nitric acid (hyponitrique, *s.*). On the addition of the Mercurial solution, decomposition takes place, Binoxide of Nitrogen and Carbonic acid gas escaping. Some Elaidic acid formed combines with Oxide of Mercury, and forms some Mercurial Soap, or Elaidate of Mercury, there is also some Elaïdine and a small portion of a yellow matter soluble in Alcohol.

Action. Uses. Stimulant and Alterative application to the eyelids in chronic Ophthalmia, also in several cutaneous eruptions, and to foul and indolent ulcers. If long applied, it will produce the effects of a Mercurial on the system. It is usually necessary to dilute it with Oil when first applied to the eyelids; but it is more apt to spoil when diluted. When decomposition has taken place, it ought not to be applied, as it then becomes irritant.

LINIMENTUM HYDRARGYRI NITRATIS made with *Ung. Hydr. Nitr.* ℥ijß. *Cerati Simplicis* ℥vijß. *Olive Oil* f℥v., is officinal in the Manchester Infirmary, and is no doubt a good substitute for the weak Citrine Ointment, for which a formula existed in former Pharmacopœias.

HYDRARGYRI ACETAS, D.

Acetate (of Protoxide) of Mercury. *F.* Protoacetate de Mercure. *G.* Essigsaures Quecksilberoxydul.

The Acetate of Mercury (HgO, Ac′=261) has been long known to chemists, but was introduced into practice in consequence of the French Government having, in the middle of the last century, purchased the secret of *Keyser's pills*, which were vaunted as an antisyphilitic remedy. Some, however, suppose that he employed a mixture of the Acetates of the Protoxide and of the Peroxide of Mercury, and others that he employed the latter only.

Prop. It occurs in thin scale-like crystals, flexible, white in colour, without odour, but having an acrid metallic taste. Sparingly soluble in cold, and partially decomposed by boiling water, as it is also by boiling Alcohol. Light also decomposes and blackens it. Heat resolves it into Acet′ and Carb′ and Mercury. Sul′ disengages the odour of Ac′, and the alkalis precipitate the Black Oxide of Mercury from its solutions, while from that of the Acetate of the Peroxide a yellow precipitate takes place. Comp. Hg O 80·66+Ac′ 19·34=100.

Prep. Add *Nitric′* dil. 11 parts, to *purified Mercury* 9 parts. When effervescence ceases digest till the metal is dissolved. Dissolve *Acet. Potash* 9 parts in boiling *Aq. dest.* 100 parts, and acidulate the solution with *Distilled Vinegar* q. s. Add to this while boiling the *solution of the Mercury in the Nitric′* and filter as quickly as possible through a double linen cloth. Let it cool that crystals may form; wash these with cold *Aq. dest.*, dry them in paper with a very gentle heat. All through this process use glass vessels, D.

In the first part of this process, a Protonit. Mercury is intended to be obtained; but it is difficult to prevent some of the Pernitrate being formed, as is always the case when the acid is strong, or heat is employed. The former may be obtained by using a diluted acid without heat, and allowing the action to take place slowly, separating occasionally the crystals as they are formed. (*c.*) On mixing the acid solution of this Protonit, with a hot sol. of the Acet. Potash, double decomposition ensues, but no deposit takes place until the mixture cools, when the pearly crystals of the Protoacetate are deposited, and Nitrate of Potash remains in solution.

Action. Uses. Considered a mild Mercurial, but has occasionally acted with violence, in consequence probably either of being badly prepared, or having afterwards altered in composition.

D. gr. j.—grs. v.

HYDRARGYRI (CYANURETUM, D.) BICYANIDUM, L.

Bicyanide of Mercury. *F.* Cyanure de Mercure. *G.* Doppelt-Cyanquecksilber.

Bicyanide of Mercury (Hg 2Cy=254) was discovered by Scheele. It was introduced into the D. P. for making Hydrocyanic acid. Mr. Phillips, in his Translation of the L. P., mentions Acidum Hydrocyanicum as its officinal preparation; but, on turning to the formulæ for making this acid in the L. P., we find no mention made of Bicyanide of Mercury. It is of a dull white colour, without smell, but of a disagreeable metallic taste. Crystallized in anhydrous obliquely truncated four-sided prisms: permanent in the air, partially dissolved by Alcohol, requiring 8 times their weight of temperate but much less of boiling water. It is dissolved by Nit′, but decomposed by Sul′ and by Hydrosul′, the latter precipitating a black Sulphuret of Mercury from its solution. It is not affected by the alkalis, but its characteristics may be seen by the tests of the L. P. "Transparent and entirely soluble in water. H Cl′ disengages from its solution Hydrocyanic′ which is known by its peculiar smell; a glass rod moistened with the sol. of Nitr. Silver, and held over it, gives a deposit (Cyanide of Silver) soluble in boiling Nit′. Heated, it emits Cyanogen, and globules of Mercury are obtained. Comp. Hg. 79·6+Cy 20·4=100.

Prep. L. D. Mix *Percyanide of Iron* ℥viij. (Cyanuret of Iron 6 parts, D.) *Binoxide of Mercury* ℥x. (Nitric Oxide of Mercury 5 parts, D.) *Aq. dest.* Oiv. (add them to Aq. dest. warmed 40 parts, D.) Boil them together for half an hour, (continually stirring, D.) Filter (through bibulous paper, D.) Evaporate the liquor to obtain crystals. Wash the residue repeatedly with boiling *Aq. dest.* (Filter, D.) Again evaporate to obtain crystals.

It may also be prepared by accurately saturating with *Binoxide of Mercury* q. s. *Hydrocyanic′* distilled from *Ferrocyanide of Potassium* by acting on it with dilute *Sul′*, L.

In the first formula, on the ingredients being heated, the Cyanogen quits the Iron and combines with the Mercury, forming the Bicyanide of Mercury, which becomes dissolved, and is afterwards obtained in crystals by evaporation. The Oxygen of the Binoxide of Mercury, at the same time separating from this metal, combines with the Iron which has just been freed from the Cyanogen, and an insoluble Sesquioxide of Iron is formed.

In the second formula, which is in many respects preferable, on adding the Binoxide of Mercury to the Hydrocyanic′, the 2 Eq. of Oxygen of the one combine with the 2 Eq. of Hydrogen of the other, and so much water is formed, while the 2 Eq. of Cyanogen set free combine with the single Eq. of Mercury, and the required Bicyanide of Mercury is obtained.

Action. Uses. Irritant Poison. Sometimes used as a substitute for Corrosive Sublimate.

D. gr. $\frac{1}{16}$ gradually increased to gr. $\frac{1}{2}$, in pills or in solution.

ARSENICUM.

Arsenic. *F.* Arsenic. *G.* Arsenik.

The name Arsenic is ambiguous even in modern times, being applied sometimes to the metal, and sometimes to one of the compounds this forms with Oxygen (white Arsenic, or Arsenious acid). The same ambiguity occurs in old works; for the name Arsenikon (αρσενικον) is applied by Dioscorides to the yellow Sulphuret, while the red Sulphuret is distinguished by the name Sandarach (σανδαραχα). The Arabs call the former *zurneekh zurd* (yellow), and the second *zurneekh soorkh* (red). The name *zurneekh* is supposed by Sprengel to be a corruption of Arsenicon, but of this there is no proof. The Arabs were also acquainted with the white Oxide, which they call *sum-al-far*, *mouse-poison*, or Ratsbane, and also *shook*, *turab-al-hulk*, and *turab-al-kai*, windpipe-earth, and emetic-earth. But the Hindoos are also well acquainted with all three substances; Orpiment being their *hurtal*, Realgar their *mansil*, while white Arsenic they call *sanchya*. They were probably the first to prescribe it internally, as in Leprosy (Prof. H. H. Wilson), as they still do in that complaint and in intermittent fevers. Metallic Arsenic was first distinctly made known in Europe by Brandt in 1733. Geber seems to have been acquainted with it.

Arsenic (As=38) is sometimes found native in a metallic state, but it is most extensively diffused in combination with other metals, as Iron, Nickel, Copper, Cobalt, &c. It is separated from these by roasting in a reverberatory furnace, and collecting what is sublimed in a long horizontal chimney, or into one divided into numerous compartments. The exposure of the Arsenic to the heated air, oxidizes and converts it into white Arsenic, or Arsenious acid. q. v. This, being procured in an impure state, is first purified by sublimation, and then heated with Charcoal, which abstracts the Oxygen, and reduces the Arsenic to its metallic state, and enables it to be separated by sublimation.

Prop. Metallic Arsenic is of a steel-grey colour, has a metallic lustre, is crystalline in texture, and very brittle. Sp. Gr. 5·8. Heated in close vessels, it readily sublimes at a temperature of 360°, but Dr. Mitchell says, at a low red heat, and is again deposited in a bright metallic crust, shining like polished steel. Its vapour is remarkable for having a strong smell of garlic. Exposed to the air, it tarnishes, and becomes encrusted with a grey powder, which is considered to be an imperfect Oxide, or a mixture of Arsenious acid and metallic Arsenic. It is well known on the Continent as *fly-powder*. It is readily oxidated also in water, and even in Alcohol. Heated in the air, Arsenic easily burns, producing white fumes, which are sometimes called Flowers of Arsenic, but are those of the white Oxide, that is, of Arsenious acid. q. v. As it forms no salifiable base with Oxygen, some chemists have proposed placing it near the simple acidifiable substances rather than among the metals. It forms two distinct compounds with Oxygen: 1. Arsenious acid (As $1\frac{1}{2}$ O=50), which is officinal; and 2. Arsenic acid (As $2\frac{1}{2}$ O=58).

Acidum Arseniosum, L. Arsenici Oxydum album, D. Arsenicum album, E.

Arsenious Acid. White Oxide of Arsenic. White Arsenic. *F.* Arsenic blanc. *G.* Weisser Arsenic. Arsenichtesaure.

The substance commonly called white Arsenic has been long known. (*v.* supra.) It is found native, but is almost entirely obtained for use from the refuse ores of different metals, in which Arsenic is also contained, chiefly in Bohemia and Saxony, but also in Cornwall. It is usually purified by a second sublimation, and is sufficiently pure not to render necessary a repetition of the process, as in the Arsenici Oxydum album sublimatum of the D. P.

Arsenious acid (As $1\frac{1}{2}$O=50), called also Oxide, and sometimes

Sesquioxide of Arsenic, is colourless, with scarcely any taste (after a short time, a very faint sweetish taste (*c.*),* and devoid of smell both in its solid and vaporous state. It is found in commerce in masses, which, when recently prepared, are transparent and glassy, but in time become opalline and even opaque; often on breaking a piece which has become so, the interior will be observed still to have a vitreous and transparent appearance, but the fresh surface soon becomes like the exposed parts, and all are brittle and pulverulent. It may be crystallized in regular octohedrons, either on cooling a saturated solution obtained by boiling in water and evaporating, or by careful sublimation. The change in appearance from transparency to enamel-like opacity, is by some ascribed to mere difference of molecular arrangement. The opacity Mr. Phillips believes to be owing to the absorption of water from the atmosphere. These varieties differ from each other in density, the opaque having a less degree of specific gravity than the transparent variety, as ascertained by Messrs. Guibourt, Phillips, and Taylor, as evident in the following tabular view:

	Transparent.	Opaque.
M. Guibourt	3·7391	3·695.
Mr. Phillips	3·715	3·620.†
Mr. A. Taylor	3·798	3·529.

The solubility in water also of these varieties was said to differ by M. Guibourt, who found the transparent dissolved in 103 parts of water at (15° Cent.) 59° F., and in 9·33 parts of boiling water; and that the opaque variety dissolved in 80 parts of water at (15° Cent.), and in 7·72 parts of boiling water. Mr. Taylor (Guy's Hosp. Reports, vol. iv. p. 83), observing the great discrepancies in the statements of chemists respecting the solubility of Arsenic in water, submitted it to careful experiment; and he states that there is no observable difference in the solubility of the transparent and opaque varieties of Arsenious acid; that water at ordinary temperatures dissolves about $\frac{1}{1000}$ or $\frac{1}{500}$ of its weight, according to circumstances; that hot water at 212°, allowed to cool on it, dissolves less than $\frac{1}{400}$ of its weight, or about $1\frac{1}{4}$ grs. to each f℥; that water boiled for an hour on this substance, dissolves $\frac{1}{24}$ of its weight, or rather more than 20 grs. to each f℥.; that this water, on perfect cooling, does not retain more than $\frac{1}{40}$ of its weight, or 12 grs. to the f℥. It is observed as remarkable that the quantity retained in a cold saturated solution prepared by boiling water

* It is often differently described; as "acrid, nauseux," even by Soubeiran. Orfila likewise, as quoted by Dr. C., describes it as a rough, not corrosive, slightly styptic taste—persistent and attended with salivation. Dr. Christison and his friends in making experiments on it, "all agreed that it had scarcely any taste at all,—perhaps towards the close a very faint sweetish taste." So Dr. A. T. Thomson, Mr. A. Taylor, Mr. R. Phillips. There is little doubt but that in some cases the subsequent effects produced by the irritation of the poison have been confounded with the primary taste.

† Printed 3·260 in Transl. Pharmp.

should be so much greater (that is, 10 to 20 times more) than what cold water can dissolve, or even hot water without the continued boiling. He further confirmed what had been ascertained by Dr. Christison, that the presence of organic matter in a liquid is an obstacle to the solution of the poison, but viscid liquids, as gruel, may suspend a larger quantity than they can dissolve. The solution faintly reddens Litmus. Hence in searching for Arsenic in organic liquids, or in the stomach or intestines in cases of poisoning, it is proper first to dilute the liquid considerably with water, and, secondly, to boil the liquid thus diluted for at least 2 or 3 hours. Arsenious acid is soluble in Oil, also in Alcohol, which dissolves about 2 grs., but cold brandy not above 1 gr. in the ounce. When subjected to heat, Arsenious acid is volatilized at a temperature of about 380° (or 425°, Mitchell), but without the characteristic smell of metallic Arsenic, and is again deposited in sparkling octohedral crystals. If heated under the pressure of its own atmosphere, it melts and is transformed into a glassy-looking substance; but if heated with any Carbonaceous matter (or any of the easily oxidable metals heated to redness, *Paris*,) it becomes decomposed, from the Carbon abstracting its Oxygen and setting free the Arsenic, which being itself volatilized, will exhale a garlicky odour. It may be dissolved in some of the other acids, but when heated with Nitric', the latter is decomposed, and the former, by taking some of its Oxygen, is converted into Arsenic acid. It readily combines with Potash and Soda, forming soluble salts, and also with Lime and some metallic Oxides, forming insoluble and characteristic compounds, which will be noticed among the Tests. Comp. As 75·72+O 24·21=100.

Tests. It is entirely sublimed when heated. L. and E. Mixed with Charcoal and exposed to heat, it emits an alliaceous smell. It is dissolved by boiling water, and Hydrosulphuric acid, when added, throws down a yellow precipitate (Orpiment), and Lime-water a white one (Arsenite of Lime). L. Chalk, Sulphate of Lime, and Sulph. Baryta have been intermixed, and will be left when the Arsenious acid is sublimed. That of commerce usually contains only a little Oxide of Iron.

As Arsenic and most of its compounds are poisonous, and frequently employed both by suicides and murderers, it is necessary to be able to detect their presence. In suspected cases, any powder adhering to the coats of the stomach, &c., or left in the vessels employed, is to be searched for and kept apart; or the stomach and its contents may be boiled in distilled water, &c.; or the poison may be searched for in the blood and liquids and solids of the body, as it is not found in these or the bones naturally, as was at one time thought by Orfila. We may therefore have to treat it as a solid substance, or in the state of solution, either pure or intermixed with organic matters. We may often get a ready indication of the presence of Arsenic by the process of Reinsch, that is, boiling a small quantity of the suspected matter with Copper and Muriatic acid, when metallic Arsenic, if present, will be deposited on the Copper, which will become covered, as it were, with a thin coating of steel.

If Arsenious acid be obtained in a solid state, its characteristics may be shown—*a.* By its volatility when exposed to heat; and if this is effected in a tube, it will again be deposited in a cooler part in octohedral crystals. *b.* By a garlicky odour, which is that of metallic Arsenic, being exhaled when it is thrown on red hot coal or charcoal, in consequence of becoming deoxidized.

1. The most satisfactory and convenient test, and one delicate enough for medico-legal purposes, is the reduction of the Arsenious acid to the metallic state, followed by its subsequent oxidation. It may be performed, when the quantity is small, with Charcoal in a glass tube, which need not be above the $\frac{1}{8}$ of an inch in diameter. But when the quantity of poison is larger, it is preferable to use a Soda flux.* Heat the Arsenious acid and flux in the flame of a spirit-lamp, applied first to the upper part and then to the bottom of the tube. A little water escapes, and should be removed with a roll of filtering-paper, and then, holding the tube steadily in the flame, the heat should be raised so as to sublime the metal; it will then be obtained in a brilliant crust and distinct, even when weighing only the 300th of a grain. (*c.*)

2. A further proof that the metal is Arsenic, is afforded by the greyish-white combined with a crystalline appearance, observable in the cooler parts of the tube, and which may be further produced, as originally suggested by Dr. Turner, by converting the crust, or a portion of it, into Arsenious acid, by chasing it up and down the tube with a small spirit-lamp flame, till it is all converted into a white powder, among which the sparkling triangular facets of the octohedral crystals of Arsenious acid will be seen with the naked eye, and always with a glass of four powers. This crystalline powder may be dissolved in a few drops of distilled water, or, filing off the part containing the sublimate, boil the tube and its contents in another tube, and then apply the following tests.

When a clear solution in distilled water can be obtained, Arsenious acid may be detected by what are called the liquid tests:

3. Lime-water, when added to such a solution, gives a white precipitate of Arsenite of Lime; but there must not be any excess of acid, nor any free alkali, which should be neutralized with Ac′ or H Cl′. This test is, however, so indecisive, that it has been abandoned by Toxicologists.

4. Nitrate of Silver, dissolved in 10 parts of water, does not by itself occasion any precipitate; but if a little alkali, such as Ammonia, be

* Dr. C. recommends grinding crystals of Carbonate of Soda with $\frac{1}{8}$ of their weight of charcoal, and then heating the mixture gradually to redness. Mr. Taylor recommends neutralizing a solution of Tar′ with a solution of Carb. Soda, evaporating to dryness and incinerating in a closed platinum crucible.

added, forming an Ammoniaco-Nitrate of Silver, a lemon-yellow precipitate (becoming brown in the light) immediately takes place of Arsenite of Silver,—Nitrate of Ammonia remaining in solution. As the precipitate is soluble in Ammonia, and also in Nitrate of Ammonia, it is necessary to be careful in adding this preparation, the *Solutio Argenti Ammoniati*, E. (*v.* p. 217.) As Nitrate of Silver precipitates Chlorides, &c., and Chloride of Sodium is often present in organic liquids, it is desirable to get rid of it first, by adding plain Nitrate of Silver in excess, and then adding the Ammonia.

5. Sulphate of Copper, like the above, will not act on Arsenious acid until an alkali, such as Ammonia, has been added in just sufficient quantity to redissolve the metallic oxide which is at first thrown down: it then becomes Ammoniacal Sulphate of Copper. When this is added to an Arsenious solution, a grass-green precipitate of Arsenite of Copper takes place. Mr. Taylor (Guy's Hospital Rep. No. xiii.) has recommended this precipitate to be washed, collected, and dried, and then a small quantity of it, finely powdered, to be introduced into a minute tube, and very gently and carefully heated over the flame of a spirit-lamp, when a ring of small octohedral crystals of Arsenious acid will appear.

6. When *Sulphuretted Hydrogen* or *Hydrosulphuric acid* is passed through a solution of Arsenious acid, which has been previously acidified with a few drops of Ac′ or of H Cl′, a bright yellow precipitate takes place of Sesquisulphuret of Arsenic (Orpiment, as in the L. P. tests), which is soluble in Liq. Ammoniæ. The Oxygen of the Arsenious acid and the Hydrogen of the H S′ unite to form water. Excess of Hydrosulphuric′ (which should have been passed through a double-necked bottle holding water) must be got rid of by heat, and excess of any other acid neutralized with an alkali, and then the H Cl′ is to be added. Hydrosulphate of Ammonia is sometimes employed, with the addition afterwards of a few drops of acid to neutralize the Ammonia, but is objectionable except as a trial test.

This test is usually preferred to all the others, and is so delicate as to indicate Arsenious′ in 100,000 parts of water; indeed, Dr. Fresenius (Lancet, June & July, 1844) would almost rely on it exclusively; and it has the advantage of always acting. The precipitate may be finally heated with black flux or dry Carb. of Soda and Charcoal in a small tube, when the Arsenic will be deposited as a metallic crust, and may be reconverted into octohedral crystals of Arsenious acid, and thus enable the above to be distinguished from other yellow precipitates.

The three last tests, when they are characteristically developed, and concur, are considered by Dr. Christison to afford unimpeachable evidence of the presence of Arsenic; the more so, as the precipitates may be submitted to the demonstrative proof of reduction.

7. A new process for detecting the presence of Arsenious acid has been discovered by H. Reinsch, and fully reported on by Mr. A Taylor in Brit. and For. Med. Rev. No. xxxi. In this, if Copper foil cut

into pieces about an inch long and $\frac{1}{8}$ of an inch in width, or some fine Copper gauze, be heated near to boiling, then with a little (about $\frac{1}{10}$) of Muriatic′ in a solution containing Arsenious acid, this becomes decomposed, and a thin steel-like coating of metallic Arsenic is deposited on the Copper, and may be separated from it again by dissolving it off with Nitric acid, or by heating in a tube, when it will sublime in the form either of a ring of the metal, or as sparkling crystals of Arsenious acid. These may be dissolved in a little distilled water, and the liquid tests applied, if thought necessary. This process has the advantage of being readily applied, and is so delicate as to detect $\frac{1}{200000}$ ($\frac{1}{250000}$, *c.*) part of Arsenic, and so effectual, that Marsh's process fails to show the smallest trace of Arsenic in the residuary liquid.

8. A very delicate process, suggested by Mr. Marsh, of Woolwich, has been very generally employed since its invention. This depends upon the power of nascent hydrogen to deoxidize Arsenious acid. Of this the metal combining with the Hydrogen, passes off in the form of Arseniuretted Hydrogen gas, which may be burned so as to obtain the Arsenic in a metallic state, or as Arsenious acid, or it may be fixed by being passed into solutions of some of the liquid tests.

The suspected liquor is introduced into a suitable apparatus with pieces of Zinc and some Dil. Sulphuric acid. Water being decomposed, Hydrogen escapes along with some Arseniuretted Hydrogen gas if any Arsenic is present. If these two gases are burned at the end of a fine pointed tube, and a piece of glass or of porcelain be introduced into the flame, metallic Arsenic of a blackish colour will be deposited (and may be dissolved off with Nitrohydrochloric′) upon it; but if the porcelain, &c., be held above the flame, then Arsenious acid in a white crust will be deposited; or both deposits may be obtained by holding above the flame the open end of a tube $\frac{1}{4}$—$\frac{1}{2}$ inch in diameter and 10 inches in length. (Per.) We may obtain a solution of the acid by holding mica, moistened with a few drops of water, over the flame (Herapath, Med. Gaz. xviii, p. 889); or it may be moistened with Ammoniacal Nitrate of Silver, when the yellow precipitate would immediately take place; or the gas may be passed into a solution of Nitrate of Silver, as proposed by Dr. Clark.

When Arsenious acid is mixed with organic substances, as is usually the case in cases of poisoning, some difficulties are necessarily experienced. These, and the fallacies attending the use of the respective tests, are fully explained and provided for in works expressly devoted to the subject. The processes of Reinsch and of Marsh, or that by Hydrosulphuric acid, are employed with the necessary precautions, and among them it must never be forgotten that some of the substances employed as tests (such as Sulphuric′, Zinc, &c.) are apt to be themselves adulterated with Arsenic. (*v.* Christison on Poisons, 3rd and 4th Ed.; and Mr. Taylor's Papers and Reports in the Brit. and For. Med. Review; Pereira's Materia Medica.)

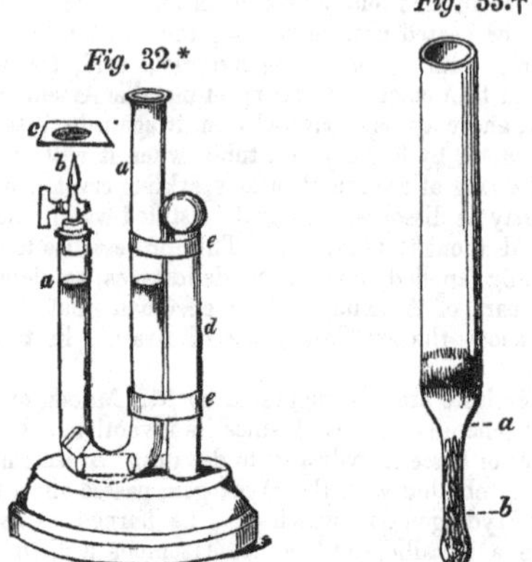

Fig. 32.* *Fig.* 33.†

Action. Uses. Irritant Poison, Antiperiodic, Alterative. *Ext.* Occasionally employed as a Caustic in Cancers and Cancer-like affections.

D. gr. $\frac{1}{16}$ or $\frac{1}{12}$ to $\frac{1}{8}$. Rub up gr. j. with Sugar grs. x., and make into pills with crumb of bread, and divide into 16 pills. The Hindoos usually prescribe it in a solid state with pepper, &c. But it is generally prescribed in the form of Liq. Potassæ Arsenitis.

Antidotes. Evacuate Stomach. Encourage vomiting by mechanical irritation, or prescribe an emetic of Sulph. of Zinc, or use the stomach-pump. Give frequent draughts of milk both before and after vomiting has begun, though not in large quantities, or demulcents or farinaceous decoctions. Large quantities of Magnesia and of Charcoal have been useful in some cases; but the most effectual antidote is the Hydrated Sesquioxide of Iron, or Ferrugo of the E. P. (*v.* p. 129); but it must be given in large quantities, and, as ascertained by Dr. Maclagan, in the proportion of 12 parts of the Oxide, in a moist state, to 1 of Arsenic; and it may therefore be given a spoonful every 5 or 10 minutes, and Reinsch's test employed upon the vomited matters, to ascertain the progress of the case. When the poison has been removed from the stomach, arrest inflammation by venesection, promote diuresis, and support the strength with Opium, administer occasional doses of Castor Oil.

* Marsh's apparatus. *a. a.* Bent tube containing suspected fluid, Sul′, and Zinc.
b. Stop-cock and jet.
c Plate of glass to receive the Arsenic.
d. e. Supports.

† Reducing tube. *b.* Charcoal and Arsenic. *a.* Metallic stain.

LIQUOR POTASSÆ ARSENTIS, L. LIQUOR ARSENICALIS, E. D. *Fowler's Solution.* Tasteless Ague Drop.

Prep. L. E. D. Boil *Arsenious Acid* broken in small pieces (in powder, E. D.) and *Carbonate of Potash* (from Tartar, D.) āā gr. lxxx. (gr. lx. D.) in a glass vessel in *Aq. dest.* Oj. (Oß. D. wine measure, D.) till they are dissolved. When cold add *Compound Tincture* (Spirit, ʒiv. D.) *of Lavender* fʒv. and then add *Aq. dest.* q. s. to fill accurately a pint measure. (℔j. by measure, D.)

When the Arsenious acid and Carbonate of Potash are boiled together, the Carbonic′ being expelled, the Arsenious acid combines with the Potash, and an Arsenite of Potash is formed and remains in solution. The Tincture of Lavender is intended only to give a little colour. f℥j. contains grs. iv. of Arsenious acid, and ♏lx. contain gr. ½ of the same acid. De Vallenger's colourless Solutio Mineralis Solvent. is thought by some to be a good form for exhibiting Arsenic, but its composition is unknown.

D. ♏iij.—v. increased to ♏xx. two or three times a day.

RED SULPHURET OF ARSENIC was in ancient times employed in medicine, as it still is in India, together with the Yellow Sulphuret. The Red is commonly known by the name of Realgar, and is a natural production, but is also prepared artificially. It is a Protosulphuret (As S=92), and usually met with in red vitreous masses, or as a red powder, being employed as a pigment. It acts as a poison, a part being converted into Arsenious acid in the stomach, "though a portion of the Oxide is subject to be converted into the Sulphuret, by H S′ gas evolved in the stomach after death." (*c.*)

YELLOW SULPHURET OR ORPIMENT (Auripigmentum), Sesquisulphuret of Arsenic (As 1½S=100) is a natural production, and also produced artificially, as in the above processes, by passing H S′ through solutions containing Arsenious acid. The Orpiment of the shops is a mixture of Sulphuret and of Arsenious acid, and is hence more rapidly poisonous than natural Orpiment. King's Yellow is another impure Sulphuret, of which the finest kinds are said to be imported from the East. Dr. Christison states that, according to his experiments, it contains a large proportion of Sulphuret of Arsenic, some Lime, and about 16 per cent. of Sulphur.

IODIDE OF ARSENIC. Arsenici Iodidum (As I_3) is an orange-red powder, without taste and smell, easily volatilized. It has been administered with benefit in doses of gr. ⅛ gradually increased to gr. ¼ in some chronic cutaneous diseases, as Lepra and Psoriasis.

SOLUTION OF HYDRIODATE OF ARSENIC AND MERCURY. Arsenici et Hydrargyri Hydriodatis Liquor of Mr. Donovan, of Dublin.* Iodo-Arseniate of Mercury of Soubeiran. Arsenic, Mercury, and Iodine being in some respects similar to each other in some of their effects, and occasionally prescribed in a solid form, Mr. Donovan was

* Donovan, in Dubl. Journ. of Medical Science. Nov. 1839, and Nov. 1842.

induced to propose the more perfect form of a chemical solution. This is of a yellow colour with a tinge of green, styptic in taste. Each f℥j. of solution (water) contains Protoxide of Arsenic, gr. $\frac{1}{8}$, Protoxide of Mercury gr. $\frac{1}{4}$, Iodine (converted into Hydriodic acid), gr. $\frac{4}{8}$, chemically combined together. Mr. D. gives the following directions for preparing it:

Triturate 6·08 grs. of finely levigated Metallic Arsenic, 15·38 grs. of Mercury, and 49·62 grs. of Iodine with f℥j. of Alcohol, until the mass has become dry, and from being deep brown has become pale red. Pour on Aq. dest. f℥viij. and after trituration for a few moments transfer the whole to a flask; add ʒß. of Hydriodic Acid prepared by the acidification of gr. ij. of Iodine and boil for a few moments. When the solution is cold, if there be any deficiency of the original f℥viij. make it up exactly to that measure with distilled water.

Action. Uses. Alterative, Stimulant. Effective in various obstinate Skin diseases, as Lepra, Psoriasis, &c.

D. ♏x.—f℥ß. three times a day in distilled water. *Ext.* f℥j. to Aq. Dest. f℥j. as a lotion.

Inc. Acids, many Salts, Opium, Morphia, and its salts.

ARGENTUM, L. E. D.

Silver. *F.* Argent. *G.* Silber.

Silver, one of the most anciently known of the metals, is found native and also combined with Sulphur in considerable quantities, also as a Chloride, and alloyed with other metals, especially Lead, Gold, Antimony, Arsenic, Copper. It is separated from its ores by the process of amalgamation. The Arabs are thought to have been the first to employ it in medicine. In its metallic state it is inert, but being little liable to alteration, or to be affected by reagents, it is much employed for surgical instruments, and for vessels for chemical purposes.

Prop. Silver (Ag=108) is remarkable for its whiteness and brilliancy, as well as for its malleability. Sp. Gr. 10·47. Unalterable in the air, with the exception of a little tarnishing from the formation of some Sulphuret of Silver. It melts at a bright red heat (1873° Daniell, 1830° Prinsep), but does not oxidize at any temperature, unless heated with some fusible siliceous substance, or acted on by Nit′. Boiling Sul′ converts it into a Sulphate, while H Cl′ has little action, though it combines with Chlorine, also with Cyanogen and Sulphur. The standard Silver of this country contains 18 parts of Copper to 222 of Silver.

Tests. Silver is sometimes mixed with Gold, usually with Copper, often with Lead. "It is entirely dissolved by Dil. Nit′ (any Gold will remain undissolved as a dark-coloured powder). This solution, on the addition of an excess of Chloride of Sodium, gives a white precipitate (Chloride of Silver), which an excess of Ammonia dissolves, and it should be free from colour (any Lead will be dissolved by the Nit′, be precipitated by the Chloride, but remain undissolved by the Ammonia). The Chloride of Silver being removed, and Hydrosulphuric acid added to the solution, it is not coloured by it, and nothing is thrown down (showing that both Lead and Copper are absent)." L. E. Chloride of Lead is more soluble in boiling water than in cold, and is partly deposited in acicular crystals as it cools. Iron, Copper, and Mercury reduce the solutions of Silver to a metallic state.

Pharm. Uses. Employed for making Nitrate of Silver.

Argenti Oxydum.

Oxide of Silver (Ag O=116) may be obtained by adding caustic Potash to a solution of Nitrate of Silver. ℨij. of the former to ℨiv. of the latter substance will yield about ℨiij. of the Oxide of Silver. The Oxide is thrown down of a brown colour, is soluble in Ammonia, and to a small extent in water, which then displays alkaline reaction. M. Sementini (Journ. de Pharm. viii. 93) inferred that it was to this Oxide that the antispasmodic properties of Nitrate of Silver were due. Mr. Lane (Med. Chir. Rev. 1840) has also argued that the Nitrate becoming Chloride of Silver in the stomach, and that being carried by the circulation to the cutaneous surface, is there converted into Oxide by the action of light and the strong affinity of Albumen; but that if the Oxide be prescribed, as it cannot penetrate the capillaries, its passage to the skin would not take place, and therefore the disfigurement or blue colour of the skin would be avoided, and we obtain the sedative effects of the Nitrate of Silver without its causticity. He has prescribed it for two months, Dr. G. Bird for four months. It has been prescribed in doses of gr. ß. to gr. j. 2 or 3 times a day in a pill with crumb of bread, gum, or with sugar. Dr. Stenhouse has shown that some of these reduce the Silver if aided by heat.

Action. Uses. An effective substitute for the Nitrate of Silver.

Chloride of Silver (Ag Cl=144) is always produced when Nitrate of Silver is added to any solution of a Chloride, and for which the former is always used as a test. It forms a curdy precipitate, at first white, afterwards becoming of a blackish-colour under the influence of light and moisture. It is insol. in water, and also in Nit′, but is soluble in Ammonia. Dr. Perry of Philadelphia, considering that Nitrate must in the stomach be converted into Chloride of Silver, inferred that this might be prescribed as an efficacious medicine to produce the alterative and tonic effects of Silver. He prescribed it in doses of gr. ½ to gr. iij. and gr. xij. 3 times a day, and states that, in less doses than 30 grs. no irritating effects result, but if that quantity be given at once it will produce emesis. (Brit. and For. Med. Rev. xii. 567.)

ARGENTI NITRAS, L. E. D.

Argenti Nitratis Crystalli et Argenti Nitras Fusum, D. *Argentum Nitratum.* Nitrate of Silver. Lunar Caustic. *Lapis Infernalis.* *F.* Nitrate d'Argent. *G.* Silbersalpeter.

Nitrate of Silver ($Ag\,O,\,N\,O_5=170$) was known to Geber, and has long been employed in medicine. Its two forms of *crystallized* and *fused*, still kept distinct in the D. P., were formerly supposed to possess different properties; but they differ only in molecular arrangement. In a crystallized state, Nitrate of Silver is white and transparent, in the form of hexangular tables or right rhombic prisms, of a powerfully metallic taste, and so bitter as formerly to have been called *Fel metallorum*, as also *Centaurea Mineralis.* Heavy, without water of crystallization, permanent in the air, soluble in its own weight of water at 60° F., and in half its weight of boiling water; readily so in hot Alcohol, but the greater portion is again deposited on cooling. Subjected to heat, it melts at 426° (at a higher heat it is decomposed), and is then run into moulds. Usually seen in sticks of a dark-grey colour; when fresh made, greyish-coloured, striated, and radiated in structure. The change in colour is probably owing to its becoming reduced at the surface, dependent on organic matter in the air; strong light is thought to reduce it, but Mr. Scanlan proved that if confined in a clean glass tube, hermetically sealed, and exposed to the light of the sun, it undergoes no change. It stains the skin of a blackish colour, as it does all organic matter, whether in solution or substance, and acts as a caustic on the latter. Its presence is readily distinguished by the white curdy precipitate which takes place on the addition of a Chloride or H Cl′ to its solution The Chloride of Silver deposited becomes black on exposure to light, is dissolved by Ammonia, but not by Nit′, and it is frequently employed in the P. as a test for Chlorides, and when ammoniated is one of the tests for Arsenious acid. Ferrocyanide of Potassium gives a white, and Hydrosulphuric′ a black, precipitate. Comp. Ag O 68·24 N′ 31·76=100.

Fig. 34.

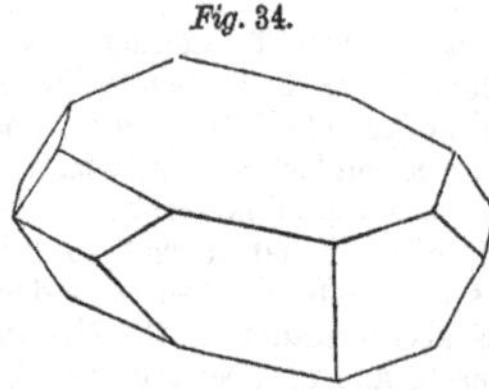

Prep. L. E. D. Dissolve (pure, E.) *Silver* (in sheets cut into small pieces, D.) ℥iß. (37 parts in a glass vessel, D.) in *Nitric′* f℥j. diluted with *Aq. dest.* f℥ij. (dilute Nitric′ 60 parts, D.) in a sand-bath (with gentle heat, E.) gradually increase the heat and evaporate to dryness. Fuse the salt with a slow fire in a crucible (earthenware or porcelain, E.) Expel the water, and when ebullition has ceased, L. pour into proper (cylindrical) (iron, E.) moulds [(previously heated and greased slightly with tallow, E.) (Preserve the product in glass vessels, E. D.)]

ARGENTI NITRATIS CRYSTALLI, D. Dissolve the metal as above, then evaporate and cool for crystals to form. Dry them without heat, and preserve in glass vessels in the dark.

The Nitric′ heated in contact with the Silver, becomes decomposed, and a portion of its Oxygen combines with the Silver; the Oxide of Silver which is thus formed unites with the Nit′ remaining undecomposed, and thus Nitrate of Silver is obtained. The Nitrogen of the decomposed Nit′ escapes in union with 2 Eqs. of Oxygen in the form of Nitric Oxide, which, uniting with a portion of the Oxygen of the atmosphere, fumes of Nitrous acid gas are observed. Crystals of Nitrate of Silver may be obtained, as in the D. P., by gradually evaporating the solution. By driving off the whole of the water, and continuing the heat, it is fused, and then run into moulds.

Tests. Apt to contain some reduced Silver, Nitrates of Copper, of Lead, of Zinc, and of Potash. "Originally white, blackened by exposure to light" (probably also from organic matter). "Entirely soluble in water," L. (except a very little black powder, E.). Copper put into the solution, precipitates Silver. Other characters as detailed under SILVER. If Copper be present, the Nit′ would produce a greenish or blackish Nitrate of Silver, and Ammonia will change this solution to a bluish colour. Chloride of Sodium will precipitate the whole of the Silver in the state of Chloride, which will be dissolved by Ammonia.

The Chloride of Silver being removed, and Hydrosulphuric′ added to the solution, it is not coloured by it. If Zinc should be present, a white Sulphuret of Zinc will be precipitated; and if Copper, then a black Sulphuret of this metal. The solution being evaporated, any saline impurity will be left, and may be tested.

Dr. Christison states that the Edinburgh College have adopted a plan which provides against all sorts of adulterations collectively, without indicating the nature of the impurity. "Grs. xxix. dissolved in Aq. dest. f℥j. acidulated with Nit′, and precipitated with a sol. of grs. ix. of Muriate of Ammonia, briskly agitated for a few seconds, and then allowed to rest a little, will yield a clear supernatant liquid, which still precipitates with more of the test." 9·12 grs. of the Muriate (*i. e.* Hydrochlorate) of Ammonia will precipitate 29 grs. of Nitrate of Silver. If 9 grs. be added to that quantity of pure Nitrate of Silver, a further addition of the test will cause further precipitation. The data put down in the formula allow about one per cent. of impurity." (*c.*)

Inc. Sul′, Phosph′, H Cl′ and Tar′ H Sul′, and the salts which contain them, Alkalis and their Carbonates, Lime-water. Ammonia in excess redissolves the precipitate first formed. Spring and River-water which contain any of the above. Astringent Infusions and other organic substances, as Albumen, Milk.

Action. Uses. Ext. Stimulant, Escharotic, may be used as a Vesicant. *Int.* Tonic Antispasmodic, Sedative. Very large doses act as a Corrosive Poison.

D. gr. ¼—gr. ij. or even more, made into pills. Readily decomposed in the stomach by H Cl′, Chlorides, &c. *Ext.* As a Lotion of various strengths.

Antidotes. Chlorides, Milk, Albumen. Evacuate Stomach. Antiphlogistic treatment.

LIQUOR ARGENTI NITRATIS, L. SOLUTIO ARGENTI NITRATIS, E.

Prep. Dissolve *Nitrate of Silver* ʒj. (gr. xl. E.) in *Aq. dest.* f℥j. (gr. 1600, E.) Filter. Preserve in well closed bottles (in the dark, L.).

SOLUTIO ARGENTI AMMONIATI, E. Solution of Ammoniaco-Nitrate of Silver.

Prep. Dissolve *Nitrate of Silver* gr. xliv. in *Aq. dest.* f℥j. Add gradually, an then cautiously, *Aquæ Ammoniæ* q. s. to nearly but not quite redissolve the precipitate at first thrown down.

This is a delicate test, commonly called Hume's test for Arsenious acid. (*v.* p. 209.)

ARGENTI CYANIDUM, L. Cyanide of Silver. *Cyanuret of Silver.*

The Cyanide of Silver (Ag Cy=134) is obtained as a white powder, heavy, without taste or smell, becoming of a violet hue by exposure to light and air. It is insoluble in water and caustic Potash, but soluble in caustic Ammonia. "Heated, it yields Cyanogen, and is reduced to Silver." L. H Cl′ and Hydrosul′ readily decompose it, Hydrocyanic acid (q. v.) being evolved. Comp. 80·6+Cy 19·4=100.

Prep. Dissolve *Nitrate of Silver* ℥ij. and ʒij. in *Aq. dest.* Oj. Add diluted *Hydrocyanic′* Oj. Mix. Wash the precipitate with Aq. dest. and dry it.

Cyanide of Silver is precipitated, as the Cyanogen of the Hydrocyanic combines with the Silver of the Nitrate. The Hydrogen of the acid combining with the Oxygen of the Oxide of Silver, some water is formed, which remains in solution.

Use. Employed to obtain Hydrocyanic acid (q. v.) extemporaneously.

AURUM.

Gold. *Sol. Rex Metallorum. F.* Or. *G.* Gold.

Gold (Au=200?) being always found native, was one of the earliest known metals, and highly esteemed for its many valuable properties.

None of its preparations are officinal, but some have been much used in modern times. The Greeks and sometimes the Arabs are supposed to have been the first to employ it medicinally. The Alchymists diligently investigated its properties for the purpose of finding the elixir of life and the universal remedy; but the Hindoos seem to have preceded them in this course.

Prop. The properties of Gold are well known, as that it is little acted upon by external agents, but soluble in *Aqua Regia*, or a mixture of Nit′ with 4 parts of Hydrochloric′, on account of the Chlorine it contains, with which Gold readily combines. Oxygen unites with it in several proportions. It is characterized by a purple precipitate being produced when Protochloride of Tin is added to a solution of Chloride of Gold, and by Protosulphate of Iron causing a brown precipitate which, with the aid of the blowpipe, may be fused into a globule of Gold.

PULVIS AURI. Gold-leaf, rubbed up with Honey, or, as in the Fr. Codex, with Sulphate of Potash, and then washed with water, is left in the state of a fine powder of a brown colour. It was submitted to experiment by M. Chretien and by M. Lallemand, and found to be mild in action but certain, of considerable benefit as an antisyphilitic, and in different affections of the Lymphatics, which it stimulates, in doses of gr. ¼ to gr. j. two or three times a day, or applied in friction on the tongue.

OXIDES OF GOLD. Oxygen combines with Gold in several proportions, but the nature of these has not been settled by chemists. A preparation, the *Purple Powder of Cassius*, which has been long employed, is supposed to owe its efficacy to the presence of Deutoxide of Gold. It is obtained by precipitating Chloride of Gold with Protohydrochlorate of Tin, when what is supposed to be a Deuto-Stannate of Gold is obtained. This Oxide is supposed to have been the active ingredient of some old preparations, as the Crocus Solis.

PERCHLORIDE OF GOLD, formed when Gold is dissolved in Nitro-Muriatic acid, is very liable to decomposition. In action and virulence it is analogous to Corrosive Sublimate. It readily combines with other metallic Chlorides, whence is obtained the

CHLORIDE OF GOLD AND SODIUM, which is now usually employed instead of the foregoing, as being more permanent in character and less costly. It is in elongated crystals, of a deep yellow colour, not alterable in the air, soluble in water. Composed of Chloride of Gold 69·3+Chloride of Sodium 14·1+Aq. 16·6=100. It is the most to be depended on of these preparations. It may be prescribed in pill with Liquorice powder or Starch, or in solution. One mode of administering it is to divide the first grain into 15 parts, the second into 14, then into 12, 10, and so on, giving one of the fractional parts every morning. It has been given in $\frac{1}{4}$ and also $\frac{1}{8}$ gr. doses; or it may be applied by friction to the tongue mixed with 3 times its weight of Iris root powder. (See Dict. Univ. de Matière Medicale of Merat and De Lens for a full article on this subject.)

VEGETABLE MATERIA MEDICA.

Many of the most valuable medicinal articles in use in the present day, as in ancient times, are yielded by the Vegetable Kingdom in all parts of the world. Some, therefore, are indigenous products, others obtained by foreign commerce. All have particular soils and climates where they can grow in full health and secrete the principles which make them useful as medicines in the fullest perfection. Some contain these diffused through their whole substance, when the whole herb or plant may be employed; others store them up only in particular parts or organs, and which therefore are alone employed. Or we may use in preference some proximate principle, separated either by nature or art from the rest of the vegetable matter. It is necessary, therefore, to be acquainted with the parts and products of plants, as well as with the best methods of preserving or preparing them for medicinal use.

The parts of Plants which are used officinally, and which it is desirable to know, as well for the above purpose as for understanding the Classification employed in Botanical arrangements, are in

Flowering Plants.

1. *The Organs of Vegetation.*

Radix. The root: usually sunk in the earth, serving to fix the plant and to absorb nourishment for its use. This it does through the naked extremities (*spongioles*) of its fibrils or radicles. Some roots are reservoirs of nutritious matter for the plants of the succeeding year; but most of these are rather Rootstocks.

Caulis. The stem is the part of the plant situated between the root and leaves, and which usually supports the parts rising above the ground. The differences are apparently so great among stems, that different names are applied to different varieties; they are, moreover, distinguished according to duration, into *annual*, *biennial*, and *perennial*. They differ remarkably in their mode of growth; some, called Exogenous, as those of all European trees, grow by the deposition of a layer of wood on the outside of that of the previous year, so that the oldest and most matured parts are in the centre, and the softer on the outside. Others, again, as Palms, grow only by additions to their centre, such additions being successively pushed outwards; so that the circumference becomes the hardest part of these stems. These are called ENDOGENS. The term Acrogen is applied by Dr. Lindley to those which are formed by the union of the bases of leaves and the original axis of the bud from which they spring, as in Ferns; among these are also included those which grow by simple elongation or dilatation, where no leaves or buds are produced.

The Stem of Exogens is distinguished into several parts, as—1. The *Pith* in the centre, which is seldom used officinally, except in the case of that of the Sassafras. 2. The *Medullary sheath*, which surrounds the Pith. 3. The *Wood*, formed of concentric layers; and 4. The *Bark*, which lies on the outside, and is connected with the pith by means of the Medullary rays. It is divisible into four layers,—the *Epidermis* on the outside, then the *Mesophlæum* and *Endophlæum*, counting from without inwards, with the innermost of all, commonly called *Liber*.

RHIZOMA, or Rootstock, differs so much in position from the Stem, that it used to be considered a kind of Root; but it is in fact a prostate thickened stem, which produces leaves from its upper and true roots from its under side. Many of the Radices or Roots of former Pharmacopœias are called Rhizomes in that of 1836, as that of Acorus Calamus, and as ought to have been that of Aspidium Filix Mas. A creeping stem, or root as it used to be called, is a kind of Rhizoma.

CORMUS. This term, or Corm, is applied to what is essentially a kind of stem, though it remains under ground, becomes of a roundish or ovoid figure, something resembling a Bulb in form, as in the Cormus of Meadow Saffron or Colchicum Autumnale.

BULBUS. A Bulb is roundish or ovoid, and consists of a flat fleshy disk, from the under surface of which true roots proceed, and on the upper surface arise fleshy coats, which are pressed close to each other either in an imbricate or tunicate manner, and enclose a true bud in their centre. Several are officinal, as those of the Onion, Garlic, Squill.

GEMMÆ. Leaf-Buds are the rudiments of new shoots, and are either naked or protected by particular coverings called Scales. None are mentioned in the Pharmacopœias, excepting those of Dyer's Oak, in an abnormal condition, Gemmæ Morbidæ.

FOLIUM. The Leaf is usually a broad and thin expansion of vegetable tissue, of a green colour. It varies, however, very much in form and figure, so as sometimes to be thick and fleshy. It consists 1. of the expanded part called *lamina* or blade, one or both surfaces of which may be covered by *stomata*, or breathing-pores; 2. the *petiole* by which it is attached to the stem, and where it is often supported by a pair of small leaves or *stipules*. The leaf may be simple or compound, that is, composed of several pieces united by a common petiole. It may have parallel or reticulate venation, and vary in a variety of ways. Many contain the most active principles of the plant, as the Senna, Cajaputi, and others.

It is not necessary for the object in view to enter into the intimate texture of plants, or to describe the characters of Membrane and of Fibre, of Cellular or of Woody Tissue, or the peculiarities of the Vascular or of the Laticiferous Tissue.

2. *Organs of Reproduction.*

FLOS. The Flower, when in the state of Bud (Alabastrus) is like the Leaf-Bud, surrounded by scales. It consists essentially of the *Stamens* and *Pistils*, or the parts concerned in fertilization; but the term is commonly applied to the Floral Envelope consisting of the *Calyx* and *Corol*, usually supported by a *Bract* or Floral leaf. The apex of the pedicle, or the part of the plant to which the Flower is attached, is called the *Receptacle*, sometimes called *Thalamus* in compound words, as Thalamifloræ. Besides these, it is necessary to notice the disposition of Flowers on a plant, which is called their *Inflorescence*. All the parts of the flower, either separately or in their aggregate state, are employed medicinally, also the bracts, with the ultimate ramifications of the plant, when the whole plant, as in the Mints and other Labiatæ, is officinal; also in the Cacumina or Tops of plants, as in the Rosemary, *Semina* Santonicæ, &c.

CALYX. The Calyx is the outer of a double whorl of floral envelopes, usually of a green colour; but when there is only a single whorl, then this is called the *Calyx*. Sometimes the term *Perianth* is applied when it is difficult to distinguish whether it be single or formed by the union of both calyx and corol. The Calyx is formed of one or more pieces, or *sepals*. Being exterior, it is necessarily always *inferior*, but being either unattached to or adherent to the interior parts, it is in the former case said to be *superior*, from being adherent to the ovary in its lower, and only free or visible as a distinct organ, in its upper part. All the outside of the flower is calyx, and is alone seen in the Clove.

COROLLA. The Corol is the inner of a double floral envelope, usually delicate in structure and brilliant in colour, consisting either of two or more pieces or *petals*, when it is said to be *polypetalous;* or if these be united into one piece, it is called *gamopetalous* by De Candolle, but *monopetalous* by most botanists. The Petals of Roses and of Corn Poppy are officinal.

STAMEN is the male organ, consisting of the *anther*, which is a case divided into cells, containing the *pollen*, a granular powder, or fecundating dust. The anther may be sessile or supported by a *filament*, or androphore, which term is usually applied where several areunited together. When the stamen is adherent to the sides of the calyx, it is said to be *perigynous*, or surrounding the ovary; but if united both with the calyx and the ovary, then it is said to be *epigynous*, or *upon* the ovary. But when it is quite free, the term *hypogynous* is used, indicating both its real and apparent position *below* the ovary.

DISK is a part not observed in all plants, but includes everything produced between the stamens and ovary: it used to be commonly called Nectary. It is annular, foliaceous, scale-like, or usually like small glands. It may be considered to represent a whorl of undeveloped leaves.

PISTILLUM. The Pistil is the female organ, surrounded by the stamen and floral envelopes, consisting of the *ovary* divided into one or more cells, and containing one or more *ovules* or the rudiments of future seeds. It is terminated by the *stigma*, which is properly a secreting surface and humid to receive the pollen. The stigma is either sessile on the ovary, or separated from it by the *style*. In the same way that adherence of the calyx to the ovary makes it appear superior, so the ovary is in that case said to be *inferior;* but when the calyx is free and inferior, the ovary is *superior*.

The only part of an immature Pistil which is officinal, is the Stigmata of the Saffron Crocus.

The suppression of any of these whorls of the floral series produces a difference of character, which is expressed by a name, as when the corol is absent, the flower is said to be *apetalous;* if the stamens are not developed, then the flower is said to be a *fertile* or *female* flower; but if the pistil is suppressed, then it is called a *male* flower.

FRUCTUS. The Fruit is the ovary arrived at maturity; with this some of the floral envelopes are occasionally united, and grow with it. The fruit consists of a *Pericarp* and of the *Seed* or *Seeds* enclosed within it. Some few, from being imperfectly covered, are called *naked seeds*. A fruit may be *simple* when produced by a single flower, or *compound* when formed out of several flowers. As the Pistil is considered by Botanists to be formed out of one or more modified leaves, which are then called *carpels*, so the fruit must be similarly constituted, and the number of cells and the partitions by which these are divided, must depend upon the number of carpels of which the fruit is composed; but it is observed that in consequence of some of the ovules becoming abortive, and the others growing inordinately, some of the cells become abolished. It is therefore necessary to examine the ovary to ascertain the normal number of cells. As the fruit is the pistil come to maturity, it bears upon it some traces of the style, and necessarily consists both of the seed and of its covering or pericarp. It ought never, however small, to be called simply a seed. Therefore, in the present Pharmacopœia, many of the fruits are correctly so called, instead of being incorrectly denominated seeds, Semina. Many fruits are officinal, also the rind (Cortex) of some, and the pulp (Pulpa) of others; and in the case of the *Mucuna* (Dolichos) *pruriens*, even the hairs with which it is externally covered.

SEMEN. The Seed is the ovule (Vegetable Egg) arrived at maturity, and contains the rudiments of a plant similar to that by which it has been produced. It is attached by a funiculus to the inside of the ovary. It consists—1. Of the Integuments, *Tunicæ Seminales*, or Matured Sacs of the ovules. 2. Of Albumen. 3. Of the Amygdala or Kernel. Some seeds are naked, as in the true Gymnosperms. A few are imperfectly covered. All are marked with the *hilum* or *umbilicus*,—the point where the seed was attached to the parent plant; often also with the *micropyle*, or minute hole which was the *foramen* of the ovule, and to which the radicle is always opposite. On the outside of the proper seed-coats we sometimes observe an *aril* (arillus), which is an expansion of the funiculus or of the placenta. That of the Nutmeg, known as Mace, is officinal. Sometimes the seed is covered with a hair-like substance, such as Cotton, officinal in the E. P.

AMYGDALA. The Almond or Kernel consists of the Embryo with or without Albumen. The *Albumen* or *Perisperm*, situated between the Seed-coats and Embryo, is a mass of cellular tissue filled with inorganic matter, which is, during germination, converted into nutriment for the young plant.

EMBRYO is composed of one, two, or more Cotyledons or seminal leaves of the young plant, which consists also of the Plumule and Radicle. The Embryo may be erect or pendulous, &c.

FLOWERLESS OR CRYPTOGAMIC PLANTS.

These plants, distributed into the natural groups of Ferns, Mosses, Lichens, Fungi, and Algæ, differ so much from others both in the parts of Vegetation and of Fructification, as generally to be treated of separately. Those therefore which afford any officinal plants will be mentioned in their proper places. It may suffice to state, that their substance is composed chiefly of Cellular tissue, and that, being destitute of organs of fructification, they are propagated by *spores*, which differ from seeds in being free in their cavities,—forming simple sacs, which separate into four distinct masses, and these germinate from any part of their surface.

CLASSIFICATION OF PLANTS.

BESIDES the parts of plants which are employed officinally, it is necessary to notice the mode in which plants are classified, as well for the purpose of understanding as for appreciating the advantages of the Natural Method of classification. Until of late years, the artificial system of Linnæus was adopted in this country; but botanists have for many years been studying the natural affinities of plants. The publication in 1789 of the "Genera Plantarum Secundum Ordines Naturales Disposita" of A. L. de Jussieu, proved upon how satisfactory and comprehensive a basis such a classification might be formed; and the series in which Jussieu first arranged them is probably as natural as any that has since been proposed. The corrections and additions made by Mr. Brown in 1810, and subsequently, to the characters of many of these orders, gave them a stability which they have continued to retain. The system of the celebrated De Candolle of Geneva is now usually followed, not because it is considered perfect, but because under it, in his "Prodromus Systematis Naturalis," the greatest number of plants have been arranged, and the authors of several Floras have adopted it. It is therefore the most convenient for study, and, being stable for a time, serves as a good basis for studying the affinities of plants, for future systems.

The parts of plants, or the compound organs of which they are formed, have been enumerated, but the internal structure requires also to be noticed. Dissection and the microscope show that plants are composed of Membrane and Fibre, formed into Cells and Ducts of different kinds, and into Woody, Vascular, and Laticiferous Tissue.

De Candolle* has laid it down as a fundamental proof of a classification being natural, that it arrives at the same results by considering either the organs of reproduction or those of nutrition.

He first divides all plants, as Linnæus had already done, into Phænogamous and Cryptogamous, or into Flowering and Flowerless Plants. The former have their organs of Fructification with their Envelopes disposed according to more or less of a symmetrical plan, while the Cryptogamic plants have their reproductive organs, if any, disposed without any order, and their integuments obscured and irregular. If we look to the parts of vegetation, the Phænogamous plants, moreover, are furnished with vascular tissue and with stomata, while the Cryptogamous have only cellular tissue, either during their whole life, or in their first foliaceous organs.

Phænogamous plants have been divided into *Dicotyledones*, or those which have two opposite, or several whorled Cotyledons; and into *Monocotyledones*, which have only one Cotyledon, or if more than one, these are arranged alternately. The Dicotyledons, moreover, grow by additions of new layers on the outside of their woody texture, and are hence called Exogenous in growth, while the Monocotyledons grow by additions to their centre, and are hence named Endogenous.

Cryptogamous Plants may also be divided into two classes, the Heterogamous and the Amphigamous, the first signifying that the fructification is unusual, having sexual organs visible under the microscope, but constructed upon a totally different plan. The *Amphigamæ* include those of which the fructification is doubtful, and such

* Sur la Division du Règne Végétal. 1833.

as display no sexual organs even under the microscope, though it is possible that the spores may have become fecundated even within the cells where they are produced. These Cryptogamous plants are also either *semivascular* or entirely *cellular*. The former, which are the same as the Ætherogamæ, have their first leaves formed of cellular tissue only and are without stomata; but at a later period of growth, have vessels and stomata. The true Cellulares are entirely without vessels and stomata, and consist of a homogeneous mass, in which the distinction of stem, leaves, and roots is obtained only by comparison. But both these groups are usually included under the general term *Acotyledones.*

Hence we observe that the divisions formed from the Organs of Reproduction correspond with those taken from the Organs of Nutrition.

The results may more clearly be represented in a tabular form, and for the sake of seeing the relative numbers of plants in each of these great subdivisions, we add the numbers of plants in each, which were known in the year 1830, as given by De Candolle.

From the Organs of Fructification.		From the Organs of Nutrition.		
I. Phanerogamæ	or	Vasculares	39,684	
Class 1. Dicotyledoneæ	,,	Exogenæ		32,264
2. Monocotyledoneæ	,,	Endogenæ		7,260
II. Cryptogamæ	,,	Cellulares	10,950	
3. Ætherogamæ	,,	Semivasculares		3,242
4. Amphigamæ	,,	Cellulosæ		7,723
			50,534	

These classes are subdivided into smaller groups, under which the several Natural Families are arranged; those which yield officinal plants will be mentioned in their appropriate places; v. p. 235, &c.

VEGETABLE PHYSIOLOGY.

The subjects which are attended to in Vegetable Physiology are, the mode in which the functions of Plants are performed, how they are enabled to grow, and how in perfecting their secretions they are influenced by the several agencies of Light, Heat, Air, and Moisture, as well as the nature of the nutriment afforded by the soil. This is interesting in a scientific point of view, and necessary to be studied, if it is wished to make any practical application of our knowledge of Plants, to the arts of Agriculture or of Horticulture, or to the cultivation of Medicinal plants. Also if we are desirous of knowing the season of the year best suited to the collection of the different parts of plants, or wish to judge correctly of the situations in which they are most likely to secrete the principles, which make them useful as Medicinal agents, in their most efficient state.

Plants not being endowed with voluntary motion, and unprovided with any *internal cavity* in which they may store up the fluids from which they are to derive nourishment, depend for these entirely on the soil in which they are fixed, and on the atmosphere by which they are surrounded. The nutriment absorbed by the aid of endosmose by the extremities of the root passes from cell to cell, or along the vascular tissue of the middle of the roots, and then, aided by capillary attraction and the void produced by evaporation from the surfaces of leaves, &c., ascends the stem chiefly in the course of the young wood; in early spring filling every part, that is, the cells, the fibres, and the vessels; but later in the season proceeding chiefly along the cells. In its course the watery fluid dissolves some of the organic matter stored up in the vegetable tissue, and is then denominated the *sap* of the plant. Arrived at the green shoots and surface of the leaves covered with stomata, it is exposed to the influence of light, heat, and air. About two-thirds of the moisture taken up, is now evaporated and exhaled; the remainder of course becomes inspissated. Some Carbonic acid is absorbed, and, as well as that obtained by the roots, becomes decomposed, the Carbon becoming fixed and the Oxygen set free.* Some water is also supposed to be decomposed, and its Hydrogen fixed, as also the Ammoniacal salts obtained from the soil, so as to give the supply of Nitrogen to the plant. Other decompositions also, and fresh combinations, probably take place among the elements of air and water, when the elaborated sap, consisting of fine granules floating in a limpid fluid, begins to descend by the under surface of the leaf and along the bark, composed as this is of cellular tissue, elongated fibres, and laticiferous vessels. In these the processes commenced in the leaf are probably completed, as the sap takes either a direct or a circuitous course downwards, and allows these proper juices to become deposited in the bark, or distributed horizontally by the medullary rays to the centre of the stem (thus forming the difference between young (Alburnum) and heart-wood); or the greater portion may be conveyed downwards, even as far as the root.

The proximate principles secreted by plants, though very various in nature, are found to be composed of only a few elementary principles, that is, of Carbon, Oxygen, Hydrogen, and Nitrogen; though their proportions are complex, and the equivalent numbers of the compounds high. To give a general view of the subject, these may be arranged in a tabular manner.

1. Compounds which contain Oxygen and Hydrogen in the same proportion as in water; sometimes called neutral Compounds or Hydrates of Carbon.	Starch, Dextrine, Cane Sugar, Grape Sugar, Gum, Cellulose.

* Plants in the dark and at night exhale Carbonic acid and absorb Oxygen, as do seeds during germination, and flowers during expansion, as in the Arum.

2. Neutral Azotized substances generally diffused through plants.	*Fibrine* insoluble and *Caseine* soluble in cold water. *Albumen* coagulated by heat. *Gluten*, a tenacious and elastic compound of Fibrine and of an Azotized principle.
3. Inflammable compounds, or Hydrurets, or those in which Hydrogen is in excess.	Ligneous tissue, Fixed Oils, Stearine, Margarine, Elaine, Volatile Oils (some of these contain no Oxygen, others contain Sulphur), Camphor, Balsams, Oleo-Resins, Resins, Wax,—Gum-Resins.
4. Vegetable Alkalis, composed of Carbon, Oxygen, Hydrogen, Nitrogen.	Morphia, Narcotina, Codeia, Quina, Cinchonia, Strychnia, Aconitina, Veratria, &c.
5. Vegetable Acids. Oxygen in excess, or in greater proportion than in water.	Citric, Tartaric, Pectic, Malic, Acetic, Tannic, Gallic, Oxalic, Meconic, &c. (Hydrocyanic acid is a compound of Hydrogen and the radical Cyanogen.)

Many acids, like the above alkalis, are peculiar to particular kinds of plants ; other principles are more general, but not universal, as Colouring matter. Some products, moreover, are obtained from plants by the processes of Fermentation, by the action of Heat, or the agency of Chemical reagents.

Some of these principles are supposed to contribute more directly than others to the nourishment and growth of the plant itself. Thus Cellulose, Starch, and Dextrine, are nearly identical in composition, and one may easily be changed into the other. For the Starch insoluble in water may be converted into the soluble Dextrine, and even into Sugar, by *Diastase*, a singular azotized substance, found in germinating seeds, also near buds and the eyes of potatoes when beginning to sprout. The Starch, that is stored up, may thus be converted into a soluble syrup, or spread out into a membrane, and form the walls of cells and vessels. The azotized principles, being generally present, are also supposed to contribute to the nutrition of the plant. They are, at all events, remarkable for corresponding in composition and properties with the principles of the same name, obtained from the blood and milk of animals. By a diminution or increase taking place in the number of equivalents of any of the elements composing these principles, there may ensue an excess of Oxygen, of Hydrogen, of Carbon, or of Nitrogen. Thus Ligneous tissue is supposed to increase its proportion of Carbon, as well as of Hydrogen, and hence to be more combustible than Cellulose. As the effects of the respiration of plants when exposed to light, is to fix the Carbon and set free the Oxygen, this would of itself leave an excess of Hydrogen; and to this is ascribed the properties of some substances found in the descending sap of the bark, as Chlorophylle, Latex, Resins, Essential Oils, Wax. The Fatty Oils are usually secreted only by the fruit, as in the sarcocarp of the Olive, Melia, and some Palms, &c.; but they are generally confined to the kernel of the seed.

When the proportion of Oxygen is increased, acidification takes place, and the several vegetable acids are formed; some found in a great variety of plants, others peculiar to individual species. But, as in the case of the mineral acids, Hydrogen is sometimes the acidifying principle, as in the case of Hydrocyanic acid. The formation of the oxyacids, it is supposed, must be favoured by nocturnal respiration, when some Oxygen is absorbed. It takes place in parts exposed to light, but which are not of a green colour, as in a variety of fruits, and in some roots. It has further been observed that the nascent parts of plants abound in azotized principles ; and though seeds contain a supply of Nitrogen, this is soon exhausted by the growing plant, and a fresh supply must be obtained from the soil, which it does chiefly from the Ammoniacal salts held in solution by the water absorbed by the roots. In connexion with these may therefore be mentioned the influence of the mineral contents of the soil. It is evident that as clay retains moisture, and sand allows water to percolate through it, such a mixture of the two as will hold the moisture without becoming wet, and yet allow the air to penetrate into the soil, must have a beneficial influence on vegetation. Other mineral substances, such as Gypsum, Oxides of Iron, and Alumina, are useful in fixing Carb′ and Ammonia. Other salts, being taken up by the water, must combine with the ordinary vegetable acids (No. 5), and are necessarily the source of the saline ashes (pp. 75 and 91) which remain when plants are burnt. Some are essential to the constitution of the plant, such as Phosphates to the grain of the Cereals, and Silex to the straw of Grasses.

Seeing, therefore, that the rate of absorption by the roots depends as well upon the moisture of the soil, as does the rate of evaporation from the leaves upon the dryness of the atmosphere, we require no further proof to perceive how much the functions of plants must be controlled by such physical agents as Light, Heat, Air, Water, and the Nutriment these last afford, nor how interesting and important must be the study of the functions of plants: for upon the due performance of these, not only depends the proper elaboration of medicines, but of all the principles which are to afford nourishment to all the vegetable feeders of the animal kingdom.

GEOGRAPHY OF PLANTS.

Finding that the growth of plants and the nature of their secretions are so much affected by the different physical agents, we may conclude that there are particular sets of plants fitted by nature for the particular circumstances in which they are placed. The Tropical Zone is characterized by brightness of light, great heat, and moisture. These are all favourable to the development of plants, which are accordingly characterized by vastness, the foliage by richness, and the inflorescence by brilliancy of colouring. From these regions, moreover, the rest of the world is supplied with aromatics and spices. Tropical climate is not terminated by an abrupt line; but, according to the influence of local causes, is extended into higher latitudes, carrying with it the peculiarities of tropical vegetation. So also in ascending mountains, the diminution of temperature being gradual, so is the disappearance of the vegetable forms growing at their base; and we find plants diminishing in number and in size as we ascend lofty mountains. Luxuriant vegetation, however, is not confined to tropical countries; for temperate climates can equally boast of beauty and variety of scenery, where the Pine tribe are conspicuous, Oaks with other catkin-bearing trees form valuable timber-trees, and the small Labiatæ the aromatics of northern regions. Between these extremes, there are many gradations of temperature, of moisture, and of dryness, all of which influence the nature of the vegetation and the secretions of plants; as, for instance, the tract of country which is beyond the reach of tropical influence, and yet not so cool or so moist as European regions, but where the atmosphere is clear and dry, the temperature hot, and the soil apparently barren. All this being favourable to the due secretion of vegetable products, we obtain from Persia, Arabia, and parts of Africa, many most important drugs. Therefore, in visiting or sojourning in different countries, when acquainted with the principles of geographical distribution, we know what groups of plants to expect, and what we may hope successfully to cultivate; so also in cultivating or collecting medicinal plants in our own country, we shall be better able to weigh the influences of soil and of aspect.

MEDICAL PROPERTIES OF NATURAL FAMILIES OF PLANTS.

The connexion between the medical properties of plants and their structure was a subject noticed by Cæsalpinus, Camerarius, Petiver, and Linnæus, but has been paid much more attention to since the publication by De Candolle of his "Essai sur les Propriétés Médicales des Plantes." As the author has elsewhere said, "In this work he has shown, that as the effects on the system of the different substances used as medicines must be owing either to their physical characters or their chemical composition, so must these depend on the peculiar organization of the vegetable, especially in the organs of nutrition, by which they are secreted. But as plants are classified from their organs of reproduction, and not from those of nutrition, it does not appear how we are led to the nature of the secretions formed by these, form a consideration of groupings founded on the examination of a different set of organs. To this it has been well replied, that though an artificial arrangement may draw its characters of classes from as small a number of organs as possible, the natural method is, on the contrary, the more perfect, in proportion, as the characters of its classes express a greater number of ideas. Hence those families which present the greatest number of points of analogy in the organs of reproduction, will also display them in the organs of nutrition, in which the secretions are chiefly performed. Thus the division of vegetables from the seeds into Acotyledons, Monocotyledons, and Dicotyledons, agrees with that taken from the existence and disposition of a vascular system. Hence the structure of the organs of reproduction may be a sufficiently certain index of the structure of those of nutrition; but as these determine the nature of the secretions or products of plants, so it follows that the properties of plants may be in accordance with their classification into natural families."

As examples, we may adduce the Gramineæ as yielding all our Cereal grains. The Palms afford Starch, Sugar, and Oil. The Coniferæ, Turpentine, Resin, and Tar, in whatever part of the world they are found. The Labiatæ yield Volatile Oil; the Solaneæ secrete narcotic, the Convolvuli, cathartic, and the Gentianeæ, bitter principles, both in hot and cold parts of the world. Numerous other instances will be adduced in the course of the work. Exceptions, no doubt, also occur; but agreements in the accordance of properties with structure are so numerous, that in no other way can we get so much information, or so readily find a substitute for a medicine or an equivalent for an article of trade, as by seeking for it in the families of plants which are already known to produce substances of similar properties in other parts of the world. This is no trivial advantage, if we consider only the immense extent and varied climates of the British dominions visited by the medical officers of the Royal and East India Company's army, as well as by the Royal and Mercantile Navy.

THE COLLECTING AND DRYING OF VEGETABLES.

As the medical and other properties of organized bodies depend not only on the peculiar secretions of each particular species (differing often in different parts or organs of each) but also on these secretions having been duly elaborated under the suitable influence of physical agents; so in the collecting of plants, we must pay attention not only to the genuineness of species and to their products, but also to the influence of Age, of Habit, of Season, of Situation, and Aspect, as well as to their being wild or cultivated. Care must, moreover, be taken that their secreted principles do not become decomposed by exposure to humidity or to too much light, heat, or air. The Colleges, therefore, give some general directions for the collecting of vegetables.

"Vegetables should be collected in dry weather, and when moist neither with rain nor dew. They should also be collected annually, and those which have been kept longer than the year should be thrown away." This has reference to herbaceous plants, which should be collected only when in full perfection, that is, when they contain the principles which make them useful as medicines in the fullest perfection. The period may therefore differ according as we seek only for mucilaginous principles, or those which present the concentrated essence of a plant in the form of an alkali.

Shortly after gathering (those which are to be used in a fresh state excepted) plants should be lightly strewed (or put in paper bags), and dried as quickly as possible with a gentle heat, in a dark airy place (taking care that the green colour is not injured by too much heat); pulverize immediately if required in powder, and preserve in proper vessels excluded from light and moisture. (Dry herbs and flowers for the preparation of oils and distilled waters should be used as soon as collected.) L. and D.

"Most Roots are to be dug up before the stalks or leaves appear." Dr. Houlton states that all roots should be taken up at the time that their leaves die, as they then abound with the proper secretions of the plant. Biennial roots at the end of their first year. All intended to be preserved should be dried as soon as possible after they have been dug up. The large true roots, especially the more juicy, dry better in their entire state than when sliced, and their juices are not then exposed to the influence of the atmosphere. The L. C. directs the roots which are to be preserved fresh, to be kept in dry sand.

"Barks should be collected at that season in which they can most easily be separated from the wood." This in general is the case in spring. The Oak is known to yield a larger quantity of Tannin when barked at that, than at any other season.

Leaves are to be gathered after the flowers have expanded, and before the seeds ripen. Mr. Battley's directions are that they should be freed from their stalks before being powdered or used medicinally. But it is sometimes preferable to allow them to dry while attached to the stalks. They may then be laid in thin layers in baskets of willow stripped of its bark, in a drying-room kept quite dark. They should then be exposed to a temperature of 130° to 140° F. for 6 or 8 hours. The leaves then having shrivelled, should be turned, and the same temperature continued until they crumble readily in the hand. The leaves so dried retain their green colour, and in a high degree their medical properties. The leaves so dried should be preserved in dry and clean jars closely covered, and powdered as required. Dr. Houlton believes that the juices of leaves are less liable to deterioration by being inspissated in their own cells than when they are formed into extracts, however carefully the process may be conducted.

"FLOWERS should be collected when just blown." But the petals of the Red Rose are directed to be gathered just before they blow.

SEEDS are to be collected when ripe, and are to be kept in their own seed-vessels.

PULPY FRUITS, if they are unripe, or if ripe and dry, are to be put in a moist place, that they may become soft; then press the pulp through a hair sieve; afterwards boil them over a slow fire, frequently stirring; lastly, evaporate the water in a water-bath, until the pulps become of a proper consistence. Press through a sieve the pulp or juice of ripe and fresh fruits, without any boiling water being used, as is necessary in the case of the bruised pods of Cassia.

SUCCI SPISSATI. Inspissated Juices. The freshly gathered herb is to be strongly pressed through a canvass bag, in order to obtain the juice; which is to be put into a wide, shallow vessel, and evaporated in vacuo or spontaneously, especially if a current of dry air be passed over, or by the aid of steam, or a water-bath placed under it. The mass when cold is to be put into proper glazed vessels, and moistened with strong Alcohol. Sometimes the supernatant liquid, being decanted off, is alone evaporated to the consistence of an extract. These inspissated juices are not distinguished by name in the L. P. from the extracts obtained by evaporating the watery or spirituous solutions of different vegetable substances, and which form true Watery or Spirituous Extracts.

The success with which the sensible and medical properties of plants may be retained, by attention being paid to the rules for their preservation, is well exemplified in the preparations of Mr. Battley and of Mr. Squire, the extracts of Mr. Hooper (v. P. J. ii. 638 and 723), the vegetable juices of Mr. Bentley, and the dried herbs and the officinal parts of indigenous plants as preserved by Mr. J. H. Kent, surgeon, of Stanton.

I. DICOTYLEDONES *vel* EXOGENÆ.

Exogenous Plants are such as grow by additions to their interior, and have two or more seed-leaves or cotyledons; hence they are also called *Dicotyledones.*

Exogenous plants are divided by De Candolle into four great subdivisions or subclasses, which are distinguished by the number of the parts of the flower, their union or separation, and by the insertion of the stamens. Thus, the groups are—

a. Thalamifloræ. A Calyx and Corolla. Petals distinct from one another. Stamens hypogynous or inserted below the Pistil, that is, into the Receptacle.

b. Calycifloræ. A Calyx and Corolla. Petals usually distinct. Stamens perigynous.

c. Corollifloræ. A Calyx and Corolla. Petals united into one, within which the Stamens are borne.

d. Monochlamydeæ, or Apetalæ. A Calyx only, or none.

De Candolle includes among Calycifloræ, all the orders with an inferior ovary, even though their petals are not distinct: Dr. Lindley arranges all these Monopetalæ under the head of Corollifloræ.

a. Thalamifloræ.

RANUNCULACEÆ. *Jussieu.* Crowfoots.

Herbaceous plants, or shrubs, often with climbing stems; juice watery. Leaves without stipules, usually alternate, with their bases or petioles expanded and half embracing the stem, the limb variously cut, sometimes abortive, and its place then supplied by the dilated petiole. Flowers usually complete, regular or irregular, solitary, racemose or paniculate. Calyx of 3, 5, or 6 sepals sometimes petaloid. Petals sometimes wanting, or equal to, or twice, or several times as many as the sepals, either flat or variously formed. Stamens numerous, anthers adnate, opening by a double cleft; ovaries single or many, free, one-celled, either one or many-seeded. Fruit consisting either of dry, single-seeded akenia, or of one or two-valved, usually few-seeded follicles. Seeds erect, horizontal or inverted Embryo minute at the base of a horny albumen.

Some of the Ranunculaceæ are found in most parts of the world where the soil and climate are not very hot and dry. They therefore indicate moisture of soil with moderate temperature of climate.

Properties.—A few of the Ranunculaceæ secrete a bitter principle, but most abound in an acrid principle, volatile in nature and destructible by heat, which makes the leaves and roots of several, useful as Rubefacients and Caustics when applied externally, or causes them to act as Irritants when taken internally. In others, peculiar Alkalis have been detected which make them useful as Sedatives and in larger doses to act as poisons.

RANUNCULUS ACRIS, *Linn.* Folia, D. *Polyand. Polygyn.* Linn.

Upright Meadow Crowfoot is an indigenous plant, very common in meadows and also on mountains throughout Europe.

Bot. Ch. Root fibrous; radical leaves palmato-partite, divisions subrhomboid, cut and acutely toothed, upper stem-leaves tripartite, segments linear; stem many-flowered; peduncles round, flowers yellow, a fleshy scale covering the nectarial pit; carpels smooth, lenticularly compressed, and margined, beak curved, much shorter than the carpel; receptacle smooth.—E. B. 652.

Action. Uses. As in R. Flammula.

RANUNCULUS FLAMMULA, *Linn.* Herba recens, D.

Lesser Spearwort is indigenous in wet places, flowering from June to the autumn, and is found also in other parts of the world.

Bot. Ch. The root is fibrous; the stem, at first decumbent and rooting at the base, is afterwards ascending; the leaves are entire, elliptical, lanceolate, or linear, occasionally serrated; the flowers numerous, yellow, much larger than the spreading calyx, with the nectarial pit at the base of the petals covered by a fleshy scale; carpels obovate, smooth, obsoletely margined, and terminated by a short point.—E. B. 387.

Prop. The recent herbaceous parts of this, as well as of other species of Crowfoot, are extremely acrid. The principle upon which this depends has not been accurately determined, but it has been ascertained to be of a volatile nature, so as to be destroyed by simply drying the plants, infusing or boiling them in water, or exposing them to heat.

Action. Uses. Acrid and Irritant. When applied to the skin, rubefaction followed by vesication is produced; and when brought into contact with the mucous membrane of the stomach, irritation and vomiting may be produced. Used as a Rubefacient and Epipastic, or as an Irritant in obstinate Scabies.

HELLEBORUS, L. E. D. Radix. The Root. HELLEBORUS NIGER, *Linn.* E. D. (H. officinalis Sibth. L.) *Polyandria Trigynia*, Linn.

Black Hellebore, so called from the colour of its roots, and Christmas Rose, from flowering in winter, is a native of the shady woods of the lower mountains of many parts of Europe.

Bot. Ch. The plant is herbaceous, with a perennial blackish-coloured rhizoma, tuberculated and scaly, from which descend numerous thickish radicles. The leaves, which sometimes make their appearance after the scape, are radical, with long, cylindrical, and spotted foot-stalks, pedately divided, with the lobes from 7 to 9, oblong lanceolate, sometimes cuneate-obovate, largely serrated towards their apices, and arranged apparently along the forked terminations of the petiole; they are stiff, almost leathery, of a dirty green colour, smooth above, paler and reticulate beneath. The scape is shorter than the petiole, furnished with two or three oval bracts, often simple and single-flowered, sometimes forked and two-flowered. The flower is large, terminal, white, with a tinge of pink, the most conspicuous part being the petaloid calyx: of this the sepals are 5, ovate, and permanent. The petals, 8 to 10, are small, greenish-coloured, tubular, tapering towards the base, with the limb tubular, bilabiate, and their outer margins terminated in a tongue-shaped lip. Stamens numerous, longer than the petals. Ovaries 6 to 8. Stigmas terminal, orbiculate. Capsules follicular, leathery. Seeds many, elliptical, umbilicated, arranged in two rows.—Jacq. Fl. Aust. t. 201. B. M. t. 8.

Hellebore root is usually imported in bags and barrels from Hamburgh, sometimes from Marseilles. (*p.*) French authors state that they are supplied from Auvergne and from Switzerland. The so-called roots, consist of the root-stock and of the radicles; the latter are chiefly recommended; the former some inches long, and half an inch thick, straight or contorted, is marked with transverse ridges, being the remains of the leaf-stalks, and on the under surface with long fibres, all more or less of a dark brown colour, internally with a white point in the centre. The odour of the dried root is feeble, but has been compared by Geiger with that of Seneka root. The taste at first sweetish, soon becomes bitter and nauseously acrid. Dr. Christison says he did not observe the roots to be acrid in February, and that the dried roots are not acrid. (Goebel and Kunze, 11. Tab. xxxi. fig. 1. a.)

Prop. Hellebore root has not yet been satisfactorily analysed. Feneulle and Capron found in it both a Volatile and Fatty Oil, a Volatile Acid, Resinous matter, Wax, a Bitter principle, Mucus, Ulmine, Gallate of Potash, Supergallate of Lime, and an Ammoniacal salt. They ascribe the activity of Hellebore to the union of the concrete oil with the volatile acid. As the root loses some of its efficacy by drying and also by long keeping, it requires to be frequently renewed. Water extracts some of its virtues, but Alcohol is the best menstruum.

Fig. 35.

HELLEBORUS OFFICINALIS of Dr. Sibthorp, (*Fig.* 35) found by him on hilly ground in Greece and the Levant, has been figured in Fl. Græca, t. 583. It was considered by Dr. S. to be the Black Hellebore of Dioscorides, being still used and called Zoptima by the Turks, and Σκαρφη by the Greeks. It had been discovered previously by Tournefort, and was called *H. orientalis* by Lamarck. Though this probably afforded the roots employed by the ancients, yet as it seems never to be brought to this country, it should not have been adopted as the officinal plant by the London College. It is intermediate in character between *H.*

niger and *H. viridis*, differing from the former in its rather leafy-branched, many-flowered stem, and from the latter in its coloured calyx, and from both in its leaves being pubescent on their under surface. Fig. 35. 1. A sepal with petals attached. 2. Sepals &c. removed to show the pistils with a stamen and petal.

HELLEBORUS VIRIDIS; Green Hellebore roots are often mixed with those of the Black Hellebore on the Continent, and are said to be efficient substitutes. H. FŒTIDUS, or Bearsfoot Hellebore, has its leaves still officinal in the United States. They are acrid, emetic, and cathartic, and were formerly employed as Anthelmintics. The roots of Actæa spicata are sometimes intermixed with Hellebore, and are figured with the above by Goebel and Kunze.

Action. Uses. The fresh root of Hellebore applied to the skin, induces inflammation, and vesication. Given internally, it acts as an irritant to the intestinal canal, producing vomiting and purging and in some cases inflammation of the rectum. Purgative Emmenagogue.

D. Hellebore is sometimes prescribed in fresh made powder, in doses of from grs. x. to ℈j., as a drastic purgative, but in gr. iij. to viij. for milder effects. Of the Infusion (ʒij. to Aq. ferv. Oj.) f℥j. every four hours. An Alcoholic extract is an efficacious preparation.

TINCTURA HELLEBORI (NIGRI, D.) L.

Prep. Macerate *Bruised Hellebore* ℥v. (℥iv. D.) in *Proof Spirit* Oij. for 14 (7, D.) days and strain.

D. fʒß. to fʒj. as an adjunct to draughts.

STAPHISAGRIA, L. E. D. Semina. The Seeds. DELPHINIUM STAPHISAGRIA, *Linn.* Stavesacre. *Polyandria Trigynia*, Linn.

Stavesacre is a plant of the south of Europe and of the Mediterranean islands, identified by Dr. Sibthorp as that employed in medicine by the Greeks. It is often confounded with *D. pictum* and with *D. Requienii.*

Bot. Ch. The plant is biennial, and hispid with a tall herbaceous stem. Leaves broad palmately 5-9 cleft, segments entire or trifid, pedicels tribracteate at their base. Flowers in a lax raceme, with 5 petaloid sepals. the upper one shortly spurred, petals four, united at the base and beardless, the two upper extended into appendages inclosed within the spur, the two lower spathulate, capsules 3 ovate ventricose. Seeds numerous.—Fl. Græca, t. 508.

Prop. Stavesacre seeds are imported from the south of Europe. Those from Germany are said to be very good. All are of an irregular triangular shape, rough dotted surface, and brownish colour, of little odour, but having a bitter and acridly burning taste. The properties of these seeds depend on an alkaloid called *Delphinia*, and on a volatile acid, which has been thought analogous to the acrid principle of other Ranunculaceæ, but is perhaps only the Malic with a Volatile Oil. The other constituents are a Fatty Oil, Gum, Starch, Azotized matter, Albumen, with several salts.

Delphinia was obtained by Couerbe by acting on the alcoholic extract of the seeds with boiling water acidulated with Sulphuric acid, and then precipitating the Delphinia from the solution by means of Ammonia. Or, boil the watery extract with Magnesia, refilter, boil with Alcohol, and evaporate the solution. This Delphinia is pure enough for medical purposes, but is still mixed with another substance called Staphisin. *Delphinia* is white, powdery but crystalline, very bitter and acrid, fusible like wax, nearly insoluble in water, soluble in Alcohol and Ether, forming salts with acids, which are also very bitter.

Action. Uses. Stavesacre seeds and Delphinia are acrid poisons; the former have been employed to kill pediculi. An Alcoholic solution rubbed on the skin, produces burning and tingling; and is hence used as a counter-irritant. Internally, the seeds have been given as emetics and cathartics, but are too violent; also sometimes in infusion as an Anthelmintic. In large doses, Narcotic.

Tinctura or Solutio Delphiniæ. Dissolve *Delphinia* ℈ij. in *Rectified Spirit* f℥ij. Useful as an embrocation in Neuralgic cases, and in chronic rheumatism. Or employ the following:

Unguentum Delphiniæ. Rub up *Delphinia* ʒß. with *Olive Oil* fʒj. and *Lard* ℥j. But the Aconite preparations are preferable.

Aconitum, L. E. D. Aconiti Folia, L. D. Leaves, E. Aconiti Radix, L. A. Napellus, *Linn.* E. (A. paniculatum, *Dec.* L. D.) Aconite. Monkshood. *Polyandria Trigynia*, Linn.

The name Aconitum, of Monkshood, is derived from the Greék *akonīton*, stated by Theophrastus to be a virulent poison. Dioscorides describes three or four kinds: of these, one agrees a little with the description of an Aconite, and may be *Aconitum Napellus*, Linn. found in the mountains of Italy and of Greece, where it is still known by the name of ακονιτον. It is also found in the mountainous pastures and cold hills of many parts of Europe, and is a doubtful native of this country, though sometimes found on the banks of rivers and brooks. As it is the most common species, and that alone procurable to any extent by druggists, it should have been retained as the officinal one, instead of *A. paniculatum* being selected by the L. and D. P. This notwithstanding that *A. Stoerkianum*, Reich (*A. intermedium*, Dec. and *A. neomontanum*, Willd), is supposed to have been the plant first submitted to experiment by Störk in 1762, and which is thought by some to be a variety of *A. paniculatum*. But this species has been proved by Dr. Fleming to be inert.

Aconitum Napellus is subject to a good deal of variation, whence numerous varieties and even species have been unnecessarily made It has a tapering rootstock, with one or more lateral pyriform tubers, and erect simple stem. The leaves are divided to the petiole into five wedge-shaped lobes, which are divided into pointed linear segments. The inflorescence is a spike-like raceme with deep blue flowers. The calyx consists of 5 petaloid sepals, of which the upper one is helmet-shaped, helmet semicircular, gradually tapering to a point. Wings hairy on the inside. Petals 5, the two upper ones converted into short sacks, which are horizontal, supported on long stalks and concealed within the helmet; the others small, linear, sometimes wanting. Filaments of stamens hairy with cuspidate wings. Ovaries three, when young diverging. Follicles with numerous three-sided seeds which are plicato-rugose at the back.

A. paniculatum may be distinguished by its more diffuse and spreading inflorescence, and by the more elongated helmets of its flowers.

Prop. The roots of Aconite, as seen in commerce, are of a tapering form, something like those of the carrot, of a dark brown colour externally, and white internally. The name has been derived from the resemblance of its roots to the napus or *navet*, the French turnip. Sometimes the lateral tubers remain joined to the root. They are of a lighter colour, smooth, and of a fleshy texture. The taste is bitter, then biting, followed by tingling and numbness in the lips and tongue, all which is shortly extended to the throat. These tubers form the roots or rootstocks of the following year, and are in perfection when the plant has done flowering. The leaves, flowers, seeds, &c., all have similar properties, producing, when applied to the surface or tongue, warmth, tingling, and numbness. This property is possessed by the leaves from their first appearance till the seeds begin to form: when these are ripe, it is entirely lost. (Geiger and C.) The seeds are then intensely acrid. The leaves and tubers ought, therefore, to be gathered when the flowers begin to fade, and the rootstocks early in autumn. These may be cut into thin slips and dried with care at a low temperature. Both will retain their properties unimpaired for a considerable time. (v. Fleming on Aconitum Napellus.)

Chem. Though Aconite has been so long employed as a medicine, its chemical composition is yet imperfectly known. An acrid volatile principle has been indicated by several chemists, though its real nature has not been ascertained. The discovery of the alkaloid, Aconitina, was made by Brandes, and confirmed by Geiger and Hesse. Dr. Pereira ascribes the acrid principle to the decomposition of Aconitina, and this seems probable from an inert extract being produced when too much heat is employed. A peculiar acid, the Aconitic, said to be identical with Equisetic acid, has been indicated by Peschier. Besides these, there is some extractive matter, albumen, a greenish wax, wax, some vegetable acids and salts,—among the latter the Aconitate of Lime is most abundant.

Action. Uses. Aconite is a powerful poison, a direct Sedative of the nerves of sensation, and useful in Antiphlogistic treatment. When a small piece is chewed, the flow of saliva is increased, and heat and tingling, followed by numbness, are experienced. Useful in Neuralgia, Rheumatism, and Diseases of the Heart.

D. Aconite leaves lose $\frac{5}{8}$ of their weight in drying, and may be prescribed in powder, in doses of gr. j.—gr. ij. gradually increased; but are uncertain.

ACONITI (SUCCUS SPISSATUS, D.) EXTRACTUM, L. and E.

Prep. L. D. Take fresh leaves of *Aconite* ℔j. sprinkle on them a little water, and bruise in a stone mortar; then express the juice and without straining (defecation, D.) evaporate to a proper consistence (in a water-bath, D.) with constant stirring.

This should be nearly an expressed juice of the plant, evaporated. It is, however, an uncertain preparation, and only slightly acrid, but when well made, causes numbness and tingling.

D. gr. j.—ij. gradually increased.

E. Take fresh leaves of *Monkshood* any convenient quantity; beat into pulp; express the juice; subject the residue to percolation with *rectified spirit*, so long as the spirit passes materially coloured; unite the expressed juice and the spirituous infusion; filter; distil off the spirit; evaporate the residue in the vapour-bath, taking care to remove the vessel from the heat so soon as the due degree of consistence shall be attained.

As the active properties of the plant are easily removed from the pulp by rectified Spirit in the process of percolation, this spirituous extract is strongly acrid, and energetic. Its properties were ascertained by M. Lombard of Geneva, who recommended the Spirituous Tincture and this Extract of the plant; but both must be carefully distinguished from the following.

EXTRACTUM ALCOHOLICUM ACONITI. This may be prepared by evaporating the following tincture to the consistence of an extract.

D. Internally, gr. $\frac{1}{8}$ in pills with *Liquorice* or any other mild powder; or, externally, in form of an ointment, made with 1 part of Extract and 2 parts of Lard, as recommended by Dr. Turnbull.

TINCTURA ACONITI. Macerate Recently dried and coarsely powdered *Aconite root* lb. j. in *Rectified Spirit* Ojß. for 14 days, and strain. This was the form recommended by Dr. Pereira, but, like the following, it must be employed with great caution.

Dr. Fleming has since made numerous experiments on the effects of Aconite, and prefers the following tincture for internal administration. Macerate *Root of A. Napellus*, carefully dried and finely powdered, ℥x. Troy, in *Rectified Spirit* f℥xvj. for four days; then pack into a percolator; add *Rectified Spirit* until f℥xxiv. of Tincture are obtained. It is beautifully transparent, of the colour of sherry wine; the taste is slightly bitter, followed by a sensation of tingling and numbness.

D. ♏iij. or ♏v. three times a day, and only to be very gradually, if at all increased. A very valuable preparation, if applied externally by brushing it on the surface, or rubbing it with the finger or with a sponge tooth brush.

Aconitina, L. The powerful sedative properties of Aconite being dependent on the presence of its alkali *Aconitina*, a formula has been given for its preparation; but considerable difficulties have been experienced in obtaining it in a pure state. It is probable that the *boiling*, which does not seem essential, may decompose the Aconitina. At all events, what is generally sold is impure, and comparatively inert. The L. C. describe it as "an alkali prepared from the leaves and root of Aconite. It is very soluble in Sulphuric Ether, less in Alcohol, and very slightly in water. It is entirely destructible by heat, without leaving as residue any salt of Lime. This substance being endowed with virulent properties, is not to be rashly employed." It is prepared, of the purest quality and possessed of its full properties, by Mr. Morson of Southampton Row, who has informed the Author that the alkali is contained in considerable quantities in the roots of the Himalayan Aconitum ferox. This is a white powder, without odour, easily fused, not volatile, soluble as above, and forms salts with acids. The Muriate has been submitted to experiment.

Prep., L. Take dried and bruised *Aconite Root* ℔ij., boil it with *Rectified Spirit* Cj. in a retort with a receiver adapted to it, for one hour. Pour off the liquor and again boil the residue with Cj. of *Rectified Spirit*, and *with that recently distilled*, pour off that also. Let the same be done a third time. Press the Aconite, mix all the liquors, strain and distil. Evaporate the residue to the proper consistence of an extract. Dissolve this in *Aq.*, strain. Evaporate the liquor with a gentle heat, that it may thicken like syrup. To this add *dil. Sul'* mixed with *Aq. dest.* q. s. to dissolve the Aconitina. Drop in *Sol. Ammonia* q. s. and dissolve the Aconitina precipitated in *dil. Sul'* and *Aq.* mixed as before. Then mix in *Animal Charcoal* q. s. frequently shaking for ¼ hour. Strain again, drop in *Sol. Ammonia* q. s. to precipitate the Aconitina. Wash and dry it.

The Aconitate of Aconitina being dissolved out by the Rectified Spirit, on the addition of Sul' a Sulphate of Aconitina is formed, which is decomposed by the Ammonia, and the Aconitina precipitated. This has all the properties of the Alcoholic Tincture, but in a more concentrated degree, proving that it is the active principle of Aconite: $\frac{1}{50}$th of a grain will kill a small animal, and nearly proved fatal to an elderly lady. (*p.*) If the solution or ointment be applied to or rubbed on the skin, great heat, tingling, and numbness, followed by soothing effects, are experienced. Dr. Fleming states that when the conjunctiva is slightly painted with the Ointment of Aconitina, contraction of the pupil speedily takes place, as has before been observed; but if it, or the Tincture of the root, is applied to the temple or forehead, the pupil occasionally becomes dilated. It should never be used except as an external application.

Tinctura vel Solutio Aconitinæ. Dissolve gr. viij. in Alcohol f℥viij. May be used as an embrocation, the proportion of the alkali being diminished or increased according to the effect produced.

Ung. Aconitinæ. Rub *Aconitina* gr. xvj. carefully with *Spir. Rect.* ♏xvj., then add *Axunge* or Hog's-lard ʒj. that it may form an Ointment.

Both preparations may be rubbed with the finger or a friction sponge (*p.*) over the pained part for some minutes, taking care to observe that the skin is not abraded.

Antidotes. If vomiting has not taken place, prescribe an emetic ; if time enough has elapsed for the poison to reach the intestinal canal, give a cathartic, and follow up, if necessary, with purgative injections. Tannic acid or Tannin will probably be useful, from forming an insoluble compound with the vegetable alkali; also an infusion of the stomach of the rabbit, and probably of some other herbivorous animals, as their gastric juice seems to neutralize the poison. Stimulants, as brandy and water, with Ammonia ; also strong Coffee may be prescribed to control the recent effects of the poison. Frictions with warm cloths and spirituous liniments should be applied generally. Sinapisms and bottles of hot water to the pericordia and extremities. If convulsions come on, abstract a little blood from the jugular vein. (Fleming.)

MAGNOLIACEÆ, *Dec.* Magnoliads.

Flowers hermaphrodite. Calyx of 3, seldom of 2, 4, or 6 sepals. Petals 6 or more, free. Ovaries several, one celled with two ovules, placed side by side, or one above the other, or with several pendulous ovules. Capsules dehiscent or indehiscent; united, one or few seeded. Seeds with a fleshy, often coloured aril, and a long funiculus. Embryo very small, at the base of the albumen.

Of these the tribe *Magnolieæ* has the ovaries arising from the torus in a spike-like manner, while the Illicieæ, have the ovaries disposed in a single whorl, the leaves full of pellucid dots, the stipules caducous or wanting. The species occur in a scattered manner in America, China, Japan, New Holland, and New Zealand. Many of these are remarkable for their aromatic properties, in consequence of the secretion of a volatile oil. The *Star-Anise*, or fruit so called from being arranged in a stellate manner and having the taste and odour of Anise, is well known in the East by the name of *Badian*. This name, having been introduced into Europe, has given origin to the term Badianifera. Star-anise is the fruit of *Illicium anisatum*, a native of China, or, according to Siebold and Zucc. of *I. religiosum*, which may be only a variety of the former.

DRIMYS WINTERI, *Dec.* (D. Aromatica, D.) CORTEX. D. Winter's Bark. *Polyandria Tetragynia*, Linn.

This tree derives its name from the bark having been used as a spice, and as a remedy for scurvy by Capt. Winter, who accompanied Sir F. Drake in his voyage round the world, and who brought some of it to Europe from the Straits of Magellan in 1579. Fig. 36.

The tree, of which there is a specimen in Kew Garden, varies in height from 6 to 40 feet, with knotty branches from the scars of old footstalks. The leaves are oblong, obtuse, with the under surface glaucous ; the peduncles almost simple, aggregated, often divided into elongated pedicles. The sepals (1) 2 to 3 or 2—3 fid. The petals, six in number, white, with the smell of Jasmine. Stamens numerous, with the filaments thickest towards the apex, and the anther cells separate (2). The carpels (3) are from 3 to 6 in number, of a light green colour, with a few black spots, containing several, usually 4, black annular seeds.

Winter's Bark is in flattish quills, some inches in length, from an inch to two inches in diameter, about one-sixth of an inch in thickness. It is smooth externally, of a pale or reddish-yellow colour, with

Fig. 36.

red oval spots; of an aromatic odour, and a warm, pleasant, spicy taste. Its properties depend partly on Tannin, but chiefly on a pale-coloured, warm, or rather pungent, Volatile oil. It is hence used as a substitute for Cinnamon or Canella-bark. The latter may be distinguished from it by being paler internally, and its infusion not being precipitated by Sulphate of Iron. G. and K. I. tab. fig. 5-7.

Action. Uses. Stimulant, Aromatic, and Tonic.

D. gr. x.—℈j. or ʒj.

Menispermaceæ, *Dec.* Moonworts.

Shrubs usually with sarmentaceous flexible stems and wood without zones. Leaves alternate, often peltate, without stipules. Flowers small, usually unisexual, often diæcious; racemose, or paniculate, floral envelopes in one or several rows, each consisting of three, seldom of 2, 4, or 5 pieces. Calyx 3, 6, or 12 leafed, seldom 4—10 leafed, free or united at the base. Petals often wanting. Stamens definite in number, free or united together; anthers opening externally. Ovaries several, free, one-celled, with a single ovule, or with several, or united at the base, or entirely into a many celled fruit. Fruit berried or drupaceous, straight or lunulate, single seeded. Seed inverse, straight or curved. Embryo large, usually curved, lying in thin albumen, radicle remote from the hilum. Figs. 37, 38, 39, and 40.

The Menispermaceæ are allied to Anonaceæ, through Bocagea, also to Berberideæ, and to Lardizabaleæ, with which they were at one time united. Being anomalous in some characters, they are considered by Dr. Lindley as more closely allied to Smilaceæ among Endogens. The plants of this family are confined chiefly within the tropics, both of Asia and of America, a few straggle beyond these limits, and some are found on the coasts of Africa.

Prop. The Menispermaceæ secrete a bitter principle along with a large proportion of starch both in their roots and stems. Many of them are also internally of a yellow colour; an acrid principle is added to these, especially in the fruits of some species. Hence some are useful as Tonics and Demulcent Diuretics. Cocculus indicus is poisonous.

PAREIRA, L. E. Radix, L. Root, E. CISSAMPELOS PAREIRA, *Linn.* Velvet Leaf. *Diœcia Monadelphia.* Linn.

PAREIRA is the root of a climbing plant, indigenous in Brazil, called *Pareira brava*, or Wild Vine, and Velvet Leaf in some of the West India islands. It was first made known by Marcgraf and Piso in their works, Hist. Nat. and Hist. Rer. Nat. 1648, by the name of Caapeba, the Portuguese call it Erva de nossa Senhora, and Ray mentions it in 1688, as "contra calculum excellentissima est." The root

Fig. 37.

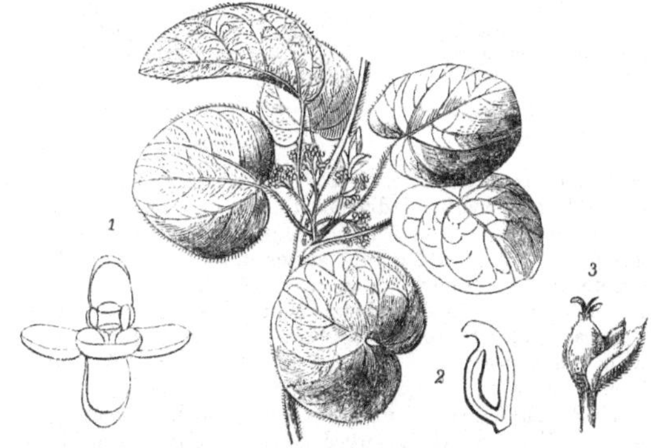

and also the stem, not only of this, but of other species, are employed. Aublet states, that the roots of *Abuta rufescens* pass for and are employed as White Pareira in Cayenne, and that Red Pareira is yielded by a variety of the same. Auguste St. Hiliare gives *Cissampelos glaberrima* as yielding the original Pareira of Brazil, where Martius states it is called Capeba and Sipo de Cobras.

Cissampelos Pareira (figs. 37 and 38), like others of the genus is diæcious, with round and smooth, or downy, twining stem. Leaves roundish, peltate, subcordate, aristate, smooth above, the under surface covered with silky pubescence. Flowers small, racemose, (fig. 37.) Racemes branched, with small bracts. Peduncles solitary, or in pairs, flowers hispid. (fig. 37, 1, fig. 38, 2.) Sepals 4.

Petals 4, united into a cup-shaped corolla. Stamens monadelphous with the 2 two-celled anthers opening horizontally at the top, (fig. 38, 1.) Racemes simple, with broad foliaceous bracts, Calyx, of one lateral sepal, with one petal in front of it, (fig. 37, 3.) Ovary solitary. Stigmas 3. Drupe hispid, scarlet, obliquely reniform, not compressed, wrinkled round its margin. Seed solitary uncinate. Embryo (fig. 37, 2 and 38, 3) long, roundish, enclosed in a fleshy albumen.

Fig. 38.

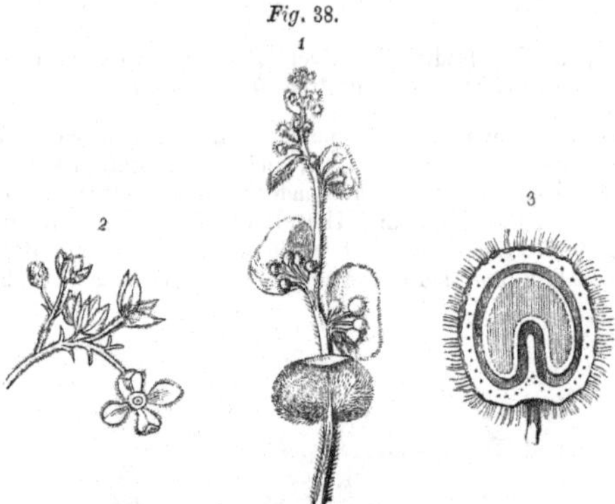

Pareira-root is found in commerce in pieces varying from a few inches to a foot in length, of different thicknesses, tortuous, more or less cylindrical, of a dark brown colour, furrowed longitudinally, exhibiting on the transverse section a number of concentric rings (which are sometimes very eccentric), and rays radiating from the organic centre. The root is without odour, and the taste is sweetish, with some aroma, and afterwards bitter. Some of the kinds found in shops are without any sweetness.

Chem. A soft resin, a yellow bitter principle, a brown-coloured matter, vegeto-animal matter, starch, nitrate of potash, with some other salts. (*Feneulle.*) Wiggers has announced a peculiar vegetable alkali, of a sweetish-bitter taste, which he has called *Cissampeline.* The active and useful properties seem to depend on the Bitter principle, Starch, and Nitrate of Potash.

D. and Adm. In powder, ℈j.—ʒj. The best form is the infusion, to which some of the extract may be added. Dr. Christison recommends "a solution obtained with cold water by percolation, as in the Edinburgh formula for obtaining the extract; because the product, as it does not contain the starch of the root, is less apt to decay."

INFUSUM PAREIRÆ, L. E. Infusion of Pareira brava.

Prep. Macerate *Pareira* ʒvj. in *Boiling Aq.* Oj. for 2 hours in a lightly covered vessel and strain (through calico, E.).

D. f℥iß.—f℥iij. Sir B. Brodie employs a decoction (Pareira ʒiv. Aq. Oiij. boiled to Oj.) f℥viii. to f℥xij. being given in the day.

Action. Uses. Mild Tonic and Demulcent Diuretic.

EXTRACTUM PAREIRÆ, L. E. Extract of Pareira brava.

Prep. Macerate bruised *Pareira* ℔ijß. in boiling *Aq. dest.* Cij. for 24 hours, boil down to Cj., strain while hot and evaporate to a proper consistence. (To be prepared like Extract of Liquorice, E.)

CALUMBA, L. E. D. Radix, L. D. The Root, E. COCCULUS PALMATUS, *Dec.* The Calumba Plant. *Diœcia Hexandria*, Linn.

Calumba-root was first made known as a medicine by F. Redi about 1677. Semedus mentioned it before 1722 among medicines from India. In works on Materia Medica in use in India, it occurs by the name of *Kalumb*. Dr. Berry first ascertained that it was the root of a plant (of which he figured the male) inhabiting the forests on the coast of Mozambique and Oibo in Eastern Africa, but where it is never cultivated. Sir W. Hooker in 1830 described and figured both male and female plants, from plants introduced by Capt. Owen into the Isle of France. Figs. 39 and 40.

The Calumba plant has a perennial root with several spindle-shaped fleshy tubers, (7) filled with longitudinal fibres or vessels, which are externally brown, with transverse warts, and internally of a deep yellow colour, devoid of smell, but very bitter. The stems are annual, herbaceous and twining, covered with glandular hair, hairy below. Leaves alternate, nearly orbicular, cordate at the base 5—7 lobed, lobes entire, wavy on the surface and margin, acuminate, hairy with long petioles. Racemes axillary. Flowers small, diœcious, green. Calyx of 6 sepals in two series with bracteoles. Petals 6, (1, 2) obovate half enclosing the 6 opposite stamens Anthers terminal, 2 celled, dehiscing vertically. Ovaries 3 (3) united at the base. Drupes (4) or berries about the size of a hazel-nut, densely clothed with long spreading hairs, tipped with a black oblong gland. Seeds (5, 6).—Bot. Mag. t. 2970-71.

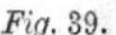

Fig. 39.

Calumba, in its officinal form, consists of transverse sections (8) of the root and its lateral tubers, which are flat, circular, about $\frac{1}{4}$ to $\frac{1}{2}$ an inch in thickness, and from $\frac{1}{2}$ to 2 or 3 inches in diameter. The cortical portion is 2 or 3 lines in thickness, covered externally with a brownish-coloured cuticle; the faces are of a greyish-yellow colour; the interior portion in concentric rings, easily distinguishable from the cortical, is soft, almost spongy, thinner towards the centre from shrinking there. The root is brittle, and therefore easily pulverized; the powder of a greenish-yellow tinge; its taste is bitter and mucilaginous, with a slightly aromatic odour. G. and K. ii. tab. v. fig. 5.

Fig. 40.

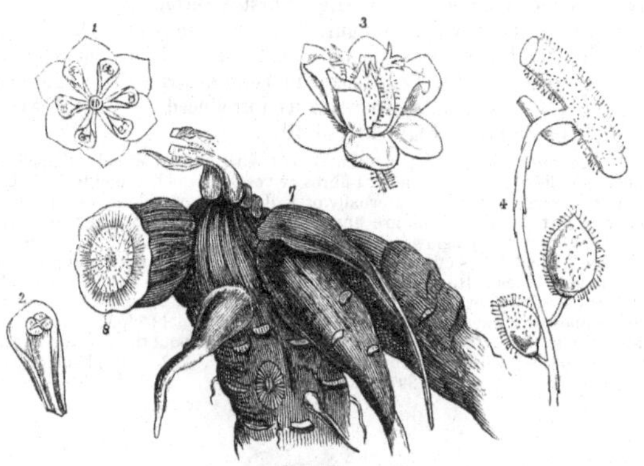

Chem. Calumba-root consists of one-third of starch, a yellow-coloured bitter substance (*Calumbine*), animal matter (Planche), resinous extractive (Buchner), a trace of volatile oil, &c. Calumbine was first obtained pure by Wittstock. When pure, it is colourless, but intensely bitter; it crystallizes in rhomboidal prisms, melts like wax; it is little soluble in water, but is dissolved by Alcohol or Ether, and by acids as well as alkalis, Acetic acid being the best solvent. Its composition is C 65·45, H 6·18, O 28·37, or $C_{12}\ H_7\ O_4$.

Calumba, when good, breaks easily, and, from the abundance of starch, gives a blue colour with Tincture of Iodine. Neither Sesquichloride nor Sulph. of Iron produce any change, as it contains no Tannin, as is also the case with Emetic Tartar and Gelatine, showing absence of Gallic acid. Infusion of Galls gives a greyish precipitate. (*p.*)

Calumba-root is sometimes adulterated with American or False Calumba, of which the infusion becomes dark green with the Sesquichloride of Iron; also with Bryony-root, which is distinguished by a permanent bitterness of taste with acridity.

Action. Uses. Stomachic and mild Tonic powder. gr. x.—ʒß. twice or thrice a day. As Calumba imparts its bitterness to water and Alcohol, both the Infusion and Tincture are officinal preparations.

INFUSUM CALUMBÆ, L. E. D. Infusion of Calumba.

Prep. Macerate *Calumba* sliced ʒv. (in coarse powder ℥ß. E.; ʒij. D.) in boiling *Aq. dest.* Oj. (℔ß. by measure, D.; cold water, E.) for two hours and strain, L. Moisten and percolate till f℥xvj. of infusion are obtained, E. Cold water and percolation remove the bitter principle with less of the starch.

D. f℥iß. twice or thrice a day. It soon undergoes decomposition.

TINCTURA CALUMBÆ, L. E. D. Tincture of Calumba.

Prep. Macerate *Calumba* sliced ℥iij. (℥ijß. D.) in *Proof Spirit* Oij. (℔ij. D.) for 14 days and filter. (Digest for 7 days, or prepare by percolation in moderately fine powder, which is first to be soaked for 6 hours with a little of the spirit, E.)

D. fʒj.—fʒij. as an adjunct to bitter draughts and mixtures.

COCCULUS INDICUS, E. D. FRUCTUS, D. The Fruit, E. ANAMIRTA COCCULUS, *Wight and Arnott.* (Cocculus suberosus, *Dec.*) The Cocculus Indicus plant. *Diœcia Monadelph.* Linn.

Cocculus Indicus, E. D. is the fruit of a climbing plant common in the mountainous parts of the Malabar coast, whence our supplies, and they are large, are now all derived through Bombay, Madras, and Ceylon. But formerly these berries reached Europe by the Red Sea and the Mediterranean, whence they were called Grana Orientis by Ruellius, 1536, Coque du Levant, by Pomet, &c. There is no proof, though it is probable, that they were known to the Arabs, the Mahizuhra (Fish-poison) of Rhases, Serapion, and Avicenna, and referred by Sprengel to these berries seems to have been a plant and its bark, "qua juvat in doloribus juncturatarum et contortione digitorum." (Serapion.) Plempius coined the name *Icthyoctonum*, to indicate "fish-poison." The plant yielding these berries was ascertained by Dr. Roxburgh. It was named *Anamirta paniculata* by Colebrook, and subsequently *A. Cocculus* by Wight and Arnott; the latter name has been adopted by the Edinburgh Pharmacopœia. It was the *Menispermum Cocculus* of Linnæus. The name Cocculus is probably derived from the Tamul *Kakacollie*, which signifies *crow-killing*, as does the Sanscrit *kakmare*.

The Cocculus plant is a powerful climber, with ash-coloured, deeply cracked, corky bark, whence the plant was called Cocculus suberosus. The leaves are stalked, large, broad-ovate, or rather roundish, truncated or somewhat cordate at the base, acute at the apex, firm in texture, soft and downy when young, with 5 digitate ribs,—petioles a little shorter than the leaves, tumid at both ends. Flowers diœcious, in lateral compound racemes. Calyx of 6 sepals in a double series with 2 close pressed bracteoles. Corolla none. ♂ Stamens united into a central column dilated at the apex: anthers numerous, covering the whole globose apex of the column. ♀ Flowers unknown. Drupes 1—3, 1-celled, 1-seeded. Seed globose, deeply excavated at the hilum. Albumen fleshy: cotyledons very thin, diverging, and each occupying a side of the hollow cavity that contains the embryo.

Cocculus berries were described by Dale as being kidney-shaped and something like bay-berries, but somewhat smaller. They are blackish-brown, and wrinkled externally, with the outer coat thin and dry, and within it is a white, woody, bivalvular shell, enclosing the whitish, semilunar, oily, and very bitter-tasted seed; which never fills the whole of the cavity, but, this in old seeds, is sometimes entirely empty. "The kernels should fill at least two-thirds of the fruit." E. P.

Chem. The kernels of Cocculus indicus were analysed by Boullay, most recently by Couerbe and Pelletier, who obtained *Picrotoxine*, Resin, Gum, Fatty acid, and a Waxy matter, Malic acid, Mucus, Starch, and Salts. In the shell they obtained two alkaloids, but in small quantity, which they called Menispermia and Paramenispermia. But the nucleus being the part used, it is necessary to notice only its active principle. *Picrotoxine* is colourless, crystallizes in needles, sometimes in silky filaments, in plates, and in rhombic prisms. (*c*) Its taste is intensely bitter. It is soluble in 150 parts of water at 57° F., and in 25 of boiling water, in 2 of Ether, and in about 3 parts of Alcohol. It is insoluble in both the fixed and the volatile oils. It does not combine with acids, but is soluble in Acetic acid, and as it forms combinations with alkalis, it is therefore considered to be of the nature of an acid. It consists of C_{12}, H_7, O_5. To obtain it, Dr. Christison recommends, to separate the oil first from the kernels by expression, then to exhaust the residuum by percolation with rectified Spirit, which is then to be distilled off. The residue is agitated with boiling water and a little H Cl′. The dissolved Picrotoxine may be obtained, on the water cooling after moderate concentration.

Action. Uses. Poisonous; used for taking fish and game, and employed by unprincipled brewers for adulterating porter, being unblushingly recommended by Childe and by Maurice in their books "on Brewing." It is used chiefly in the form of a bitter extract, known by the name of B. E., *black extract*, which is ostensibly prepared for tanners. (*v.* Cycl. of Pract. Receipts.) It produces giddiness, tetanic convulsions, and coma: applied externally, in powder, it destroys vermin, and is useful in scabies, ringworm, and porrigo.

UNGUENTUM COCCULI, E. Ointment of Cocculus Indicus.

Prep. Take any convenient quantity of *Cocculus Indicus*, separate and preserve the *kernels*, beat them well in a mortar, first alone, and then with a little axunge till it amounts altogether to five times the weight of the kernels.

An ointment of Picrotoxine may be formed in the proportion of *Picrotoxine* gr. x. to *Lard* ℥j.

PAPAVERACEÆ, *Jussieu.* Poppy-worts.

Annual or perennial herbs, seldom shrubs, usually with milky juice. Stem round, or forming a rootstock. Leaves alternate, simple or compound, without stipules. Flowers complete, regular, solitary, in racemes, or in scapes. Sepals two, very seldom three, deciduous. Petals 4, 6 or some multiple of the leaflets of the calyx. Stamens free, usually numerous. Ovary sessile, single-celled with intervalvular placentæ. Fruit a capsule or berry. Seeds numerous, furnished with albumen.

Papaveraceæ are allied, on one hand, to Berberideæ and to Ranunculaceæ, and, on the other, to Fumariaceæ and to Cruciferæ. They inhabit the temperate parts of the northern hemisphere; a few are found in tropical Asia, Australasia, Cape of Good Hope, and equinoctial America. The milky juice of most of the species is acrid and narcotic.

PAPAVER, *Linn.* *Polyand. Monogynia*, Linn.

Herbaceous plants with a white juice. Peduncles 1 flowered, naked, drooping before the expansion of the flower. Sepals two, convex, deciduous. Petals 4. (Fig. 39.) Stamens numerous. (39, 1.) Style wanting. Stigmas 4 to 20 radiating, sessile upon the disk which crowns the ovary. Capsule obovate, one celled, composed of 4 to 20 carpels united together, and opening by small valves beneath the crown formed by the stigmas. Placentæ opposite the stigmas, produced internally into spurious, incomplete dissepiments. Seeds numerous, reniform. (Fig. 39, 3, 4.)

RHŒAS, L. E. D. Petala. The Petals. PAPAVER RHŒAS, *Linn.* Corn Poppy.

The common Red, or Corn-Poppy, is found in corn-fields and on road-sides throughout Europe, and has probably been introduced with wheat. This species, or *P. dubium*, with its oblong capsules, is probably the *ῥοίας* of the Greeks. The root is fibrous, the stem many flowered and, like the peduncles, hispid with spreading hairs. Leaves pinnate or bipinnate, with oblong, lanceolate, jagged, toothed lobes. Petals a bright scarlet, often nearly black at the base. The filaments subulate. Capsule obovate, rounded at the base, smooth, with the margin of the 8 to 10 stigmas incumbent. The flowers expand in June and July.—E. B. 645.

The scarlet petals of this Poppy, which are officinal on account of their colour, become of a dull red colour on drying, and lose the somewhat heavy opium-like odour of the fresh flowers. "They should be dried quickly with the aid of a gentle heat and a current of air." They impart their colour to water, which is preserved in the form of the Syrup. This is supposed to have some slight narcotic properties, but is probably useful only as a colouring ingredient. This colour is blackened by alkalis, and rendered of a dark violet or brown tinge by Sesquichloride of Iron. The Petals consist of yellow Fatty matter 12, red Colouring matter 40, Gum 20, Lignin 28 in 100 parts. (Riffard, as quoted by Dr. Pereira.)

SYRUPUS RHŒADOS, L. E. SYR. PAPAVERIS RHŒADIS, D. Syrup of Red Poppy.

Prep. Red Corn-poppy petals (fresh, D.) ℔j. boiling *Aq.* Oj. (℥xx. by measure, D.) *Sugar* ℔ijß. L. E. q. s. D. Add the petals gradually to the water, heated in a water-bath, frequently stirring them; then, the vessel being removed, macerate for 12 hours, afterwards press out the liquor, strain, then add the sugar, and dissolve it.

PAPAVER, L. E. D. PAPAVER SOMNIFERUM, *Linn.* The Garden or White Poppy. Capsulæ maturæ, L. D. Capsules not quite ripe, E. Poppy-heads.

PAPAVER is applied in the Pharmacopœias to the capsules of the above plant, which are officinal as well as the inspissated juice or Opium, obtained from them. It appears to have been one of the early cultivated plants, as Homer is thought to allude to it as growing in gardens. Hippocrates mentions two kinds, the *black* and *white poppy*, so the Arabs and Persians distinguish the *khuskhash abiuz* or white, from the *khuskhash aswad*, or black poppy. The white Poppy is now cultivated in the plains of India, and the black, or rather deep-red, variety in the Himalayan mountains.* It was early cultivated, as it still is, in Egypt, also in India, Persia, Asia Minor, as well as in some parts of Europe.

Fig. 39.

* Mr. Hamilton says, "The opium is chiefly obtained from the single white poppy; I have also seen the red and purple colours, though only one is usually seen in a field. I hardly remember to have noticed any mixture of colour in one piece of ground. The kind here cultivated generally grows to a height of three feet."—Hamilton, *Travels in Asia Minor*, ii. p. 115.

The Garden Poppy (Fig. 39) is probably a native of Persia. It has, however, been so long grown in gardens in various parts, that it is sometimes found apparently wild, especially in the southern parts of Europe.

The roots are from 2—4 feet high, the stems are round and straight, glaucous, smooth, with a few hairs towards the white and tapering extremity and on the peduncles. The leaves are large, sessile, amplexicaul, smooth, of a glaucous green, margins wavy, cut and toothed. The flowers are large and terminal, drooping before flowering, with smooth concave sepals; 4 large petals, roundish in form, white or of a purplish colour with a darker coloured spot near the claws. The capsule is oval, or nearly globose, large, smooth, with parietal placentæ equal in number to the stigmas, which are covered with numerous white or brownish coloured, kidney-shaped seeds. Flowers in June and July, and the capsules ripen about two months later.—E. B. 2145.

Some consider that there are two distinct species instead of varieties of this plant. *P. officinale* (Gmelin) var. *album* is larger and less glaucous, with white petals and seeds, capsules ovate-globose and remaining closed under the crown of stigmas (fig. 39, 1), while *P. somniferum* (Gm.) var. *nigrum* has the flowers violet or red, seeds black, capsules globose opening by foramina under the stigmas (fig. 39, 2).

CAPSULÆ PAPAVERIS. Poppy-heads are officinal in the Ps., but the L. and D. P. direct them to be collected when ripe, the E. P. before they are quite ripe, as in this state they contain more of the narcotic principle. The French find those from the Levant or the southern provinces of France to be more powerful than those grown in the north. The seeds (*maw* seeds) ripen notwithstanding the separation of the capsules from the plant; and as they contain much oil, its presence adds to the demulcent properties of the decoction. Some of the properties of the following preparations depend on the presence of Morphia, especially if the poppy-heads be gathered unripe.

DECOCTUM PAPAVERIS, L. E. D. Decoction of Poppies.

Prep. Boil *Poppy Heads* sliced ℥iv. in *Aq.* Oiv. (Oiij. E.; ℔ij. by measure, D.) for ¼ hour. Strain.

Action. Uses. A Demulcent Anodyne fomentation, applied to swollen, painful, and inflamed parts, as the eye, abdomen, joints, &c.

SYRUPUS PAPAVERIS, L. E. SYRUP. PAPAV. SOMNIFERI, D. Syrup of Poppy.

Prep. Take *Poppy Heads* ℔iij. (sliced and without the seeds ℔iß. E.; ℥xvij. D.) boil them in boiling *Aq.* Cv. to Cij. and press strongly. (Infuse in boiling Aq. Oxv. (Cij. D.) for 12 (24, D) hours, boil down to Ov. and express strongly through calico, E.) Boil again the strained liquor to Oiv. (Oijß. E.) Strain while hot, set by for 12 hours, and boil the clear liquor to Oij. Add Sugar ℔v. L. (pure ℔iij. E.; ℥xxix. D.) dissolve it (with heat, E.) and make a syrup. See P. J. ii. p. 647 for making this Syrup with a cold infusion of Poppy-heads. The starch is not dissolved, and the albumen got rid of afterwards by boiling.

This is an excellent anodyne and narcotic syrup, when carefully prepared; but it is apt to undergo decomposition, and is liable also to be very carelessly made, as with extract of poppies and syrup, or with laudanum and treacle, and is hence very irregular in strength, and as it is often prescribed to children, becomes dangerous in consequence of their being apt to suffer from an overdose of an opiate.

D. f℈ij.—f℥ß. or more for adults. ♏x. to ♏xv. for infants.

Extractum Papaveris, L. E. Extract of Poppy.

Prep. Macerate (bruised, L.) *Poppy Heads* without the seeds ℥xv. in boiling *Aq.* Cj. for 24 hours. Boil down to Oiv.; strain while hot and evaporate to a proper consistence (over a vapour-bath, E.).

This extract has long been known, being the Meconion of the ancient Greeks. It is a good substitute for opium in many cases, being thought to allay pain and induce sleep, without producing nausea, or the irritability caused by opium.

D. gr. ij.—gr. x.

Opium, L. E. D. Capsulæ immaturæ Succus concretus, L. Capsularum succus proprius concretus, D. Concrete juice from the unripe capsules of Papaver somniferum, E. Opium.

Opium, obtained by making incisions into the unripe capsules of the Poppy, and inspissating the juice, seems to have been known from early times. Hippocrates is supposed to have employed it, and Diagoras condemned its use in affections of the eyes and in ear-ache. Dioscorides describes it: but opium does not appear to have been much employed until the time of the Arabs, except in the form of the confections called Mithridatica, Theriaca, and Philonium. The Arabic name *afioon*, the Hindu *aphim*, and the name *afooyung*, by which it is known in China, must all have proceeded from the original Greek name, which is itself derived from *οπος*, juice. The Sanscrit *apaynum* seems to have a similar origin.

Opium is obtained by a very simple process, consisting merely in making incisions in the evening into the capsules of the Poppy, shortly after the petals fall off, taking care not to penetrate into the interior, when a milky juice exudes, and either concretes upon the capsule, whence it may be taken off in little tear-like masses, or earlier in the morning in a softer state. Upon this will depend whether the grains run together, or remain separate even when pressed. When thus collected, they require nothing more than being dried in a warm and airy room, when the opium becomes of a brown colour, with a shining fracture, and has a strong and peculiar odour. Some opium which the author prepared in this manner in the Saharunpoor Botanic Garden in 1828-29, was pronounced by the Medical Board of Bengal to be like Turkey opium. Most of the opium made in the Himalayan mountains is similarly prepared, and is of very fine quality. Belon and Olivier describe the Opium of Asia Minor as formed by the assemblage of the small tears collected off the capsules. Dioscorides, however, describes the process as consisting in making incisions into the capsules when the dew has evaporated, collecting the juice in a shell, mixing the several portions, and rubbing them up in a mortar. Kæmpfer gives this as the Persian process, and M. Texier describes it as being adopted in Asia Minor; and it is certainly practised with the immense quantities collected in India in the provinces of Behar and Benares, and of which an excellent description has been given by Dr. Butter in the Journal of the Asiatic Society, v. p. 136. When this method is adopted, the mass will appear homogeneous; when it is omitted, it will appear to be composed of agglutinated tears. Both appearances may be observed in the opium of commerce. Dr. Butter describes the quantity of opium from each capsule as varying according to soil, irrigation, and to the quantity of dew which falls, but averaging about 1 gr. from each quadruple incision. The tears are of a reddish colour externally, but semi-fluid in the interior, and of a reddish-white colour. The juice is apt to be mixed with dew, and fraudulently with a little water, and will separate into a fluid portion (*passewa*) and into one which is more consistent, the former containing much the largest portion of the Bimeconate of Morphia. The whole of the day's collection is rubbed together in a mortar, so as to break down the grains, and reduce the whole to a homogeneous semi-fluid mass, which should be dried as quickly as possible in the shade, when it is called *pucka* or matured, being called *raw* in its former state. All samples of Opium brought for sale are submitted to a steam drying process, by which the quantity of fluid in each is easily ascertained. The Opium for the China investment contains about 30 per cent. of moisture: that for medical use in India is made quite dry.

The Opiums known in European commerce have been described under the following heads by Prof. Guibourt. That collected in Asia Minor, chiefly in Anatolia, is generally all included under the head of Turkey Opium, and most of it is exported from Smyrna; some of it, however, is taken to Constantinople, whence it is re-exported to other parts of Europe. Some Egyptian is imported into this country. The Persian is only known. The Indian kinds are exported to China.

Smyrna Opium, called also Levant Opium, is generally in flattened masses, and, in consequence of its original softness, without any definite regular form; weighing from a half to two pounds, and covered with the capsules of a species of Rumex. It is at first soft, of a distinct brown colour, becoming blackish and hard when dried, losing weight from evaporation of water, having the strong and peculiar odour of Opium. "When examined with a magnifier, it is seen to be composed of yellowish agglutinated tears." This is the purest kind of Opium, yielding about 8 per cent. of Morphia and 4 per cent. of Narcotine, and, on an average, about 12 per cent. of Hydrochlorate of Morphia.

An inferior kind is, however, also imported from Smyrna, which is more apt to be adulterated, is harder, of a darker colour, appears homogeneous, and may be seen covered either with Rumex capsules, or with the leaves of the Poppy.

The Smyrna Opium is produced at several places, at from 10 to 30 days' distance in the interior; but that grown at Caisar, about 600 miles from Smyrna, is the most esteemed for its cleanness and good quality. Mr. Hamilton states that much is produced at Bogaditza: it is made into lumps about four or five inches in diameter, round which leaves are wrapped.

Constantinople Opium, M. Guibourt conceives, may be collected in the northern parts of Anatolia. One kind is in small lenticular pieces about two inches in diameter, weighing from four to eight ounces, and always covered with a Poppy-leaf, of the midrib of which the mark may be seen on the middle of the pieces of Opium. Another variety is in large irregular cakes. Both are more mucilaginous than the Smyrna kind, and, though of good quality, the Constantinople is less uniform in the quantity of Morphia it contains, some yielding less, and others as much as, the best kinds of Opium.

Egyptian Opium is in flattened roundish cakes about three inches in diameter and covered with the remains of some leaf which M. Guibourt was unable to distinguish. It looks well externally, is homogeneous, has something of a reddish hue, not blackening by keeping, but softening on exposure to the air, and has somewhat of a musty smell. It is generally inferior, and M. Guibourt obtained only $\frac{5}{7}$ of the Morphia yielded by Smyrna Opium.

Persian Opium, which Dr. Pereira calls Trebizond Opium, from his specimens having been obtained from thence. The specimens in the King's College Museum were sent by Mr. Morson, to whom M. Guibourt was also indebted. This kind is of a black colour, apparently homogeneous in texture, and in sticks some inches in length, each wrapped up in a separate piece of paper, and tied with a piece of cotton.

Some Opium has been collected in Algiers. A new variety imported from Turkey has been described by Mr. Morson (P. J. iv. 503). It resembled the Constantinople, but was soft and light-coloured; contained much wax, caoutchouc, and about $6\frac{1}{2}$ per cent. of Morphia.

Besides these, some Opium is occasionally met with of European manufacture; and it might easily be produced in England if the summer was more regular. In the south of Europe the summer is probably too hot and dry. In India it can only be cultivated in the cold weather. Some good English Opium has been produced, but it is irregular in strength. The quantity of Morphia said to have been obtained from some specimens of French and of German Opium is enormous,—being from 16 to 20 per cent.

Indian Opium, not being known in European commerce, requires a very short notice. The *Saharunpoor Garden Opium*, sent home in 1844, is of a brown colour, shining fracture, with the strong and peculiar smell of opium, and yielded the late Professor Daniell, in one of the last analyses he made previous to his sudden and lamented death, 8 per cent. of Morphia.* The *Himalayan Opium* possesses similar sensible properties, and though liable to be adulterated, is, when pure, of very fine quality. The *Malwa Opium* is in flat circular cakes, average weight 1½ lb., of a rusty-brown colour, strong odour, and bitter permanent taste, varying much in quality. Some Malwa Opium lately analysed yielded only 2 per cent. of Morphia, was oily and mucilaginous, and appeared to have been obtained by expression of the capsules. Dr. Smyttan, late Opium Inspector at Bombay, obtained from 3 to 5 per cent. of Morphia from some varieties, and from 7¼ to 8 per cent. from finer kinds. Some *Kandeish Opium* yielded to Mr. E. Solly 72 per cent. of soluble matter, and about 7 per cent. of Morphia. The E. I. Company's Opium, which is that known under the name of Bengal Opium, and which is chiefly produced in the provinces of Behar and Benares, with some in that of Cawnpore, is also of different qualities: that intended for medicinal use in the hospitals in India is of very fine quality, of a brown colour, and fine smell, packed with great care in 4 lb. and 2 lb. squares covered with layers of talc, and further defended by a case of brown wax half an inch in thickness. This Dr. Jackson, lately Opium Inspector at Calcutta, informs me is the *Patna Garden Opium*, cultivated, prepared, and selected exclusively for the Dispensary, and that it yields about 7 to 8 per cent. and sometimes more (10½), of Morphia. It is of this kind that Dr. Christison says, "I have examined specimens little inferior to average Turkey Opium in the quantity of Morphia they contained."

The *Chinese Investment Opium*, which is highly esteemed by the Chinese, is made into cakes or balls, each containing about 4 lbs. and covered with a thick layer of poppy petals, made to adhere to the opium and to each other by means of a mixture of inferior kinds of Opium and water. It is of a dark brown colour, of the consistence of an extract when first cut into, containing 70 per cent. of solid matter, and about 2½ per cent. of Morphia.

Prop. Good Opium, when it has been some time made, is of a dark brown or blackish colour externally, and of a reddish-brown internally, either homogeneous in texture, or formed of agglutinated tears. Sp. Gr. about 3·36. The taste is strongly and permanently bitter, with some degree of acridity, and a little aroma. The odour is both powerful and peculiar. It is hard, and even becomes brittle, and breaks with a compact shining fracture, and produces a yellowish-brown powder. Some kinds, however, are soft internally, and others never become entirely dry. The London College, to preserve uniformity of strength, directs dried Opium only to be used for the several pharmaceutical preparations. Opium is softened by the application of heat,

* Some specimens of this opium, prepared when the author was Superintendent of the Saharunpore Botanic Garden, sent to the Medical Board of Bengal in 1829, were pronounced to be "very fine specimens, and to resemble Captain Jeremie's in almost every particular," and Captain Jeremie's they "considered equal, if not superior, to the finest Turkey opium that comes into the market at home."

and burns at a higher temperature. The effects of the ordinary reagents are, that water, either temperate or warmed, dissolves about two-thirds of good opium, and forms a solution of most of its active principles, and is of a bitter taste and of a reddish-brown colour. Rectified Spirit takes up four-fifths of the whole mass, including all the active properties of Opium. Ether dissolves much of what is left undissolved by water. Diluted acids take up all its active principles. The alkalis precipitate them from their solutions, but redissolve them when added in excess. They are also precipitated by baryta, lime, and magnesia, and their salts, also by the soluble salts of lead and of other metals, as well as by solutions of tannin and astringent vegetable substances. As some of these are apt to be prescribed with opium, it is essential to attend to the form of exhibition, for the precipitate may contain all the active principles, and the solution be inert; or it may be made active again by using an excess of ammonia or potash; or the active principle may be taken up by the acid of some of the salts used. But these various effects of reagents can be duly appreciated only when the composition of Opium is understood.

Several analyses were made of Opium before any just ideas were obtained respecting its constitution. Derosne in 1803 first obtained a saline body. Sertürner and Seguin, the first a Hanoverian, and the second a French apothecary, both discovered in 1804 another crystallizable substance, upon which subsequent experience has proved the narcotic power of opium to depend. In a second memoir of Sertürner published in 1817, he announced his discovery of the existence of *morphia* combined with *meconic acid.* This was confirmed by Robiquet. Since then, Geiger, Beltz, Pelletier, Couerbe, Schmidtz, and Mulder, have successively analysed and shown Opium to consist of a variety of principles. Of these, three are alkaline *Morphia*, *Codeia*, and *Paramorphia.* A fourth, *Narcotin*, though neutral to colours, forms salts with acids: of this a great portion is in a free state, and may at once be separated from Opium by ether; the remainder, as well as the whole of the *Morphia* and *Codeia*, are in combination with the *Meconic'* and some *Sulphuric acid* found in Opium. Other bodies are also acid, as the *brown acid extractive*, the *resin*, and *oily matter.* Some are neutral, as *Narcein* and *Meconin.* Narcotin is sometimes enumerated among the neutral principles, and Thebaina, called also Paramorphia, though alkaline, does not combine with acids. Besides these, there is a trace of Volative Oil (the odorous principle?), Gum, Bassorine, Albumen, Caoutchouc, Lignin, and Salts of inorganic bases.

MORPHIA, L. *Symb.* Mor. Fr. *Morphine.*

Morphia ($C_{35}H_{20}O_6N=292+2$ Aq.$=310$ when crystallized) is found in Opium in the proportion of 2 to 8 or 10 per cent., and is the principle upon which its medicinal properties chiefly depend. It crystallizes in shining flat six-sided prisms, usually in the state of a very white powder; is without smell, but has a very bitter taste.

L. P. "Very little soluble in cold water, little in boiling water, but very readily in alcohol (in 40 of cold anhydrous alcohol, and in 30 parts of ordinary alcohol at 212°; nearly insoluble in ether and the fixed and volatile oils). This solution exhibits alkaline properties, when tried with turmeric; and when the spirit is distilled from it, it yields crystals, which are totally destroyed by heat (about 6·33 per cent. being expelled; with further heat, it melts into a yellowish liquid, and in the air burns with a bright flame). On the addition of Nit′, Morphia becomes first red, and afterwards yellow. Tinct. of Sesquichloride of Iron gives it a blue colour. Chlorine and afterwards Ammonia being added to its salts, they are rendered of a brown colour, which is destroyed when more Chlorine is added. Morphia is also precipitated from its salts by solution of Potash (also by Ammonia and Lime-water), which, added in excess, redissolve it." It is precipitated from these by Tannic′, as by infusion of gall-nuts, Tannate of Morphia being formed. Morphia forms salts with S′, H Cl′, and Ac′: these are crystallizable, when pure colourless, and of a bitter taste.

Prep. Morphia being combined with Meconic acid, may be precipitated from a watery solution of Opium either by Ammonia or by Magnesia, which enter into combination with the Meconic acid. It may then be separated from the other insoluble matters by the agency of alcohol. The L. C. obtain it from Hydrochlorate of Morphia.

Prep. Dissolve *Hydrochlorate of Morphia* ℥j. in *Aq. dest.* Oj. than add *Sol. of Ammonia* f℥v. diluted with *Aq.* f℥j and shake well together. The precipitate is to be washed with distilled water, and then dried with a gentle heat.

The Ammonia, combining with the H Cl′, remains in solution, while the Morphia is precipitated, washed, and dried. The Codeia, which is usually present in Hydrochlorate of Morphia, is not thrown down by the Ammonia.

Tests. The characteristics of Morphia have been noticed above, and these can only be seen in pure Morphia. When quite white, there can be no colouring matter, and Narcotin, which is sometimes present, is insoluble in the solution of potash.

Action. Uses. Morphia possesses nearly all the action of Opium, but is less stimulating. Being, however, nearly insoluble in cold water, it is usually prescribed in the form of some of its salts, which are more certain in their operation. Both Morphia and its salts have frequently been employed endermically on the Continent.

D. gr. ¼ to gr. j. gradually increased. gr. j. in fine powder may be applied to the denuded skin.

MORPHIÆ (MURIAS, E.) HYDROCHLORAS, L. Hydrochlorate or Muriate of Morphia.

Hydrochlorate of Morphia (Mor. + H Cl + 6 Aq = 383) came into notice in 1831, with Dr. W. Gregory's method of obtaining Morphia. It must be distinguished, however, from what is commonly called *Gregory's Salt*, which is a compound of Muriate of Morphia and of Codeia. It is without colour or smell, is extremely bitter, in fine powder or in feathery acicular crystals, is soluble in about 16 parts of cold, and in its own weight of boiling water. This on cooling, congeals into a crystalline mass. It is also soluble in rectified Spirit. Dil. Sul′ decomposes it, as do the alkalis. Nit′ forms with it a reddish-yellow, and Sesquichloride of Iron a bluish coloured fluid. It is composed of Mor. 76·24 H Cl 9·66 Aq. 14·10=100.

Prep. It may be prepared by acting on Morphia with H Cl′, or more commonly by decomposing the Meconate of Morphia in Opium with some other salt, which shall produce an insoluble Meconate and a soluble Muriate of Morphia. The L. C. order Chloride of Lead. Dr. A. T. Thompson uses Chloride of Barium. The E C. prefer Chloride of Calcium, according to Dr. Gregory's original process. L. P. Macerate *Opium sliced* ℔j. in *Aq. dest.* Oiv. for 30 hours, and bruise it; afterwards digest for 20 hours more, and press it. Macerate what remains again, and a third time, in water, that it may become free from taste, and as often bruise and press it. (The Meconate of Morphia is dissolved in the successive portions of water.) Evaporate the mixed liquors, with a heat of 140°, to the consistence of a syrup. Then add *Aq. dest.* Oiij, and when all the impurities have subsided, pour off the supernatant liquor. To this add gradually *Chloride of Lead* ℥ij. or so much as may be sufficient, first dissolved in *boiling Aq. dest.* Oiv. till nothing further is precipitated. (The Chloride of Lead and water are respectively decomposed owing to the presence of the Meconate of Morphia. The Hydrogen and Chlorine form H Cl′ which uniting with the Morphia and Codeia form comparatively soluble Hydrochlorate of Morphia and of Codeia. The Oxygen of the water uniting with the lead forms an Oxide of Lead, which combining with the Meconic′ form an insoluble Meconate of Lead, which is precipitated together with a little Sulphate of Lead.) Pour off the liquor which holds the Hydrochlorate of Morphia in solution, and wash what remains frequently with *Aq. dest.* Then evaporate the mixed liquors as before, with a gentle heat, that crystals may be formed. Press these in a cloth, then dissolve them in *Aq. dest.* Oj. and digest with *Animal Charcoal* ℥iss. in a heat of 120° and strain. (This is useful in depriving the crystals of colour.) Finally, the charcoal being washed, evaporate the liquors cautiously, that pure crystals may be produced.

As some Hydrochlorate of Morphia remains in solution after the first crop of crystals have been obtained, it is directed to add to this liquor *Aq. dest.* Oj. and gradually drop in *Liq. Ammoniæ* q. s. to precipitate all the Morphia. (Hydrochlorate of Ammonia with Codeia is left in solution.) To this, washed with *Aq. dest.* add *Hydrochloric acid,* that it may be saturated, (thus Hydrochlorate of Morphia is again formed,) digest it with *Animal Charcoal* ℥ij. and strain. Lastly, the animal charcoal being thoroughly washed, evaporate the liquors cautiously, that pure crystals may be produced.

The E. C. uses Chloride of Calcium, instead of the Chloride of Lead, and therefore Meconate and Sulphate of Lime are formed instead of the analogous salts of Lead, and in place of Charcoal alternately acidulate with Muriatic acid and neutralize with finely powdered Marble, in order to remove the colouring matter. Dr. Christison states, that it is important not to employ too much water, and that about 4 times the weight of the Opium employed is sufficient to exhaust it, and that the Chloride of Lead or of Calcium should be added before instead of after concentrating the infusions, and that the evaporations should be conducted as quickly as possible, at a heat below 212°. By following this process the Edinburgh manufacturers obtain about 13 per cent. of very pure and white Hydrochlorate of Morphia from the recent soft Smyrna Opium.

Tests. "Snow-white: entirely soluble: solution colourless: loss of weight at 212° not above 13 per cent. 100 measures of a solution of 10 grs. in Aq. f℥ß. heated near to 212°, and decomposed with agitation by a faint excess of Ammonia, yield a precipitate which in 24 hours occupies 12·5 measures of the liquid." E. The precipitate thrown down in its solution by Nitr. Silver, is not entirely soluble in H Cl′ or N′, or in Ammonia, unless added in excess. The E. tests will detect any undue moisture, and ascertain the quantity of Morphia precipitated, and the absence of colouring matter. Narcotin, if present, would be detected by not being entirely soluble in an excess of Potash by which it had been precipitated from a solution.

Action. Uses. May be united with some medicines, and advantageously substituted for Opium in most cases as a sedative anodyne, diaphoretic, &c.

D. gr. $\frac{1}{4}$—gr. $\frac{1}{2}$;—a Narcotic poison in doses of gr. v.—gr. x.

Morphiæ Muriatis Solutio, E. Sol. of Muriate of Morphia.

Prep. Mix *Rectified Spirit* f℥v. with *Aq. dest.* f℥xv. and dissolve in the mixture, *Muriate of Morphia* ʒiß. with the aid of a gentle heat.

D. ♏x.—♏xl. Intended to be about the strength of Laudanum: 100 minims contain gr. j. of the Muriate of Morphia.

Trochisci Morphiæ, E. Morphia Lozenges.

Prep. Dissolve *Muriate of Morphia* ℈j. in a little hot water. Mix it with *Tincture of Tolu* fʒiv. and *pure Sugar* ℥xxv. and with a sufficiency of mucilage form a proper mass for making lozenges; each of which should weigh about 15 grains.

Action. Uses. Sedative, &c. Much used in this or the following form in combination with Ipecacuanha for allaying cough.

D. x—xx lozenges daily. Each contains about gr. $\frac{1}{40}$ of Muriate of Morphia. Those sold in London are marked Morphia gr. $\frac{1}{24}$.

Trochisci Morphiæ et Ipecacuanhæ. Morphia and Ipecacuanha Lozenges.

Prep. Dissolve and mix *Muriate of Morphia* ℈j. *Ipecacuanha* in fine powder ʒj. *Tinct. of Tolu* fʒiv. *pure Sugar* ℥xxv. and proceed as with the preceding Morphia lozenges. Useful for the same purposes.

D. x—xx lozenges daily. Each contains gr. $\frac{1}{40}$ of Muriate of Morphia, and gr. $\frac{1}{18}$ of Ipecacuanha.

Morphiæ Sulphas. Sulphate of Morphia is occasionally employed in medicine. A small portion exists naturally in Opium, and it may readily be made by acting on Morphia with dil. Sul′. In the United States' Pharmacopœia there is a Solutio or Liquor Morphiæ Sulphatis. The Sulphate may be prescribed in doses of gr. $\frac{1}{8}$ to $\frac{1}{4}$, but it does not appear preferable to the Hydrochlorate, though frequently employed on the Continent, especially endermically.

Morphiæ Acetas, L. E. Acetate of Morphia.

Vinegar has long been thought a good menstruum for dissolving the active properties of Opium, but the true Acetate of Morphia was introduced into practice by Majendie. When pure, it is seen as a colourless, snow-white powder, of an intensely bitter taste, imperfectly crystallized. It is apt to be decomposed from some of its acid escaping, and Morphia being left, which is insoluble. Hence in prescribing it is necessary to add a few drops of Acetic acid to its aqueous solution. It is soluble in rectified spirit, readily decomposed by heat as well as by dil. Sul′, with the disengagement of Acetic acid. Its solution is rendered reddish-yellow by Nit′ and blue by Sesquichloride of Iron. It is probably composed of Mor.+Ac′+Aq.=352, or per cent. Mor. 82·95 Ac′ 14·5, Aq. 2·55=100.

Prep. L. E. Mix *Acetic acid* f℥iij. with *Aq. dest.* f℥iv. and pour them upon *Morphia* ʒvj. to saturation. Let the liquor evaporate with a gentle heat, that crystals may be formed. The E. C. first obtain Morphia from Muriate of Morphia, as in the London process, p. 259, and then dissolve it in a slight excess of Pyroligneous acid.

Tests. "Very readily dissolved in water. Its other properties are such as have been stated of Morphia." "100 measures of a solution of grs. x. in *Aq. Dest.* f℥ß and *Ac′* ♏v. heated to 212°, and decomposed by a faint excess of Ammonia, yield by agitation a precipitate which in 24 hours occupies 15·5 measures of the liquid."

Action. Uses. Though liable to decomposition, it is preferred by some to the other salts of Morphia, in doses of gr. $\frac{1}{8}$ to $\frac{1}{4}$. A Syrup is lauded by M. Forget in doses of gr. $\frac{1}{15}$ in chronic bronchitis.

Citrate of Morphia was recommended by Dr. Porter of Bristol under the form of Liquor Morphiæ Citratis, and made by macerating Opium ℥iv. with Citric′ ℥ij. and Aq. Dest. Oj. It does not appear necessary, as the officinal preparations of Morphia seem sufficient.

Bimeconate of Morphia. Mr. Squire, on reflecting that the natural salts separated from the other ingredients existing in Opium, might prove the best therapeutic agents, has prepared a solution of the Bimeconate of Morphia, which is nearly of the same strength as Laudanum. Several practitioners have borne testimony to its being less exciting than Opium, and equally if not more efficacious than the other preparations of Morphia.

Codeia (*Codeine*) ($C_{35}H_{20}O_5N$=284+2 Aq. when crystallized from Aq.) was discovered by Robiquet in 1832. Its Hydrochlorate crystallizes with the Hydrochlorate of Morphia in the process p. 260; but as the Codeia is not precipitated by the Ammonia, it may be obtained by subsequent evaporation. It crystallizes in needles or right rhombic prisms, is alkaline in nature, and forms salts with acids, is soluble in water and in Alcohol, readily in Ether; insoluble in solution of Potash; does not become blue on the addition of Sesquichloride of Iron. Opium contains about ½ or 1 per cent. It has little taste; some state it to be excitant, and others hypnotic; it resembles Morphia in its effects, only three or four times as much is required.

NARCOTINA (*Narcotine*, Anarcotina, Beng. Disp.) ($C_{48}\,H_{24}\,O_{15}\,N=446$) was discovered by Derosne in 1803, and its properties investigated by Robiquet in 1817. Much of it is in a free state, and may be dissolved out of Opium by Ether. It exists in the proportion of 1 to 8 per cent. It crystallizes from Alcohol in bevelled pearly tables, but from Ether in regular rhombic prisms; is white, without odour, and insipid; is insoluble in cold water and in solution of Potash, very soluble in Ether, in Alcohol, and in volatile oils; neutral to vegetable colours, combines with diluted acids, and forms salts, as the Hydrochlorate, Sulphate, &c., which are very soluble and bitter. When pure, it does not form a blue solution with Sesquichlor. Iron, nor produce a brown colour when treated with Chlorine and Ammonia. It is not reddened by Nit′, but is so by S′ containing a trace of N′. From the decomposition of Narcotine, *Opianic acid* is formed, remarkable for its affinity for Ammonia, also *Cotarnin*, &c. (Liebig.) Narcotine may be obtained by macerating the Opium which has been exhausted by cold water in the process for obtaining Hydrochlor. Morphia with weak Pyroligneous or Muriatic acid, and precipitating with Potash. It may be separated from Morphia by Ether or by Potash,—the first dissolving the Narcotine and leaving the Morphia, while the Potash dissolves the Morphia, but leaves the Narcotine.

Action. Uses. It appears to be devoid of all narcotic properties. Dr. Roots prescribed its Sulph. in doses up to ℈j. as a substitute for the Disulph. of Quina for the cure of intermittents. It has been largely employed in India for arresting the paroxysms of intermittent and remittent fevers by Dr. O'Shaughnessy and other practitioners.

NARCEIA ($C_{28}\,H_{20}\,O_{12}\,N=298$) (*Narceine*) was discovered by Pelletier, and is in fine silky needles, which are slightly bitter, soluble in water, fusible at about its boiling point, neutral to test-paper, and not neutralizing acids. The diluted mineral acids produce a light blue (N′ a yellow) colour when brought in contact with it, as does Iodine.

MECONIN ($C_{10}\,H_5\,O_4=97$) is also white, crystallizes in six-sided prisms, is acrid in taste, fuses at 194°, is soluble in water, neutral to acids. If Chlorine gas be brought in contact with it when in a fused state, a blood-red fluid is produced, which crystallizes on cooling. It is remarkable in not containing any Nitrogen.

THEBAINA, or Paramorphia ($C_{25}\,H_{14}\,O_3\,N=202$) is alkaline in its relations, and forms crystallizable salts with diluted acids; most nearly resembles Narcotin, but is distinguished from it by crystallizing in short needles; fuses at 302, is much more soluble in Alcohol, and is acrid and not bitter in taste, little soluble in water.

Meconic acid ($C_7 H_2 O_7 = 100$), discovered by Sertürner, was studied by Robiquet. It is seen in the form of white, transparent, micaceous scales. It is soluble in water: when this solution is boiled, it is decomposed into Carbonic acid and *Metameconic* acid, which forms hard crystalline grains. By destructive distillation of Meconic acid, another acid, the *Pyromeconic*, is produced. Mec′ readily forms salts, and is remarkable for producing a deep red colour with the persalts of Iron, and a green precipitate of Mecon. Copper with Ammon. Sulph. Copper. It may be obtained by decomposing the Mecon. Lead or that of Lime (v. p. 260) with dil. H Cl′. Besides these, other principles of less importance, as Pseudomorphia and Porphyroxin, so named from being coloured purple when boiled with dil. N′, H Cl′ or S′.

Brown Acid Extractive has been little examined, and is no doubt a mixture of several substances, perhaps the result of some of the changes which have taken place, and is supposed to possess some of the narcotic properties of Opium. *Resin of Opium* contains Nitrogen, is brown, insipid and without odour, softens by heat, is soluble in alcohol, and in alkaline solutions, remarkable for its electro-negative properties. *Oily* or *Fatty Matter* of Opium is probably colourless when pure, commonly yellow or brownish, acid, its alcoholic solution reddens litmus, it combines with alkalis, and forms soaps from which it may again be separated unchanged by the action of acids. The nature of the *Odorous Principle* of Opium is unknown, as it has never been isolated. It may be a volatile oil, as it rises with water when this is distilled off Opium, which has the peculiar odour of Opium.—(Soubeiran, Traité de Pharm. i. p. 364, and Turner's Chem. p. 1159.)

Adulterations. Opium is of different degrees of value, according to its sensible properties, and the quantity of Morphia it contains; but it is subject to adulterations. First it may be mixed up with too much water, either intentionally, or in consequence of the dew having been very heavy. The quantity may be ascertained by the loss on evaporation. The most injurious fraud is that of washing out the soluble and most valuable parts of Opium, and bringing the residual mass for sale. In this case Butter states the Opium loses its translucency and redness of colour, also its adhesiveness. Sand, clayey mud, sugar, molasses, cowdung, Datura-leaves, the glutinous juice of Ægle Marmelos, and even pounded poppy-seeds, are employed to adulterate Opium. Malwa Opium often contains oil and other matters obtained by the expression of the poppy-heads. Some Opiums from which Morphia has been extracted have been occasionally met with in European commerce. To be enabled to judge of good Opium, one must be well acquainted with the different varieties of Opium, their respective colours, tastes, and textures, as well as the natural degree of moisture, and see that no mechanical admixtures are apparent, nor left on a filter. Several methods have been proposed for ascertaining the quantity of Morphia in Opium, but none of them are very satisfactory. In the E. P. it is proposed to ascertain it by the weight of the precipitate caused in an infusion of Opium by Carbonate of Soda. "A solution of 100 grs. macerated 24 hours in Aq. f℥ij. filtered, and strongly squeezed in a cloth, if precipitated by a cold solution of Carbonate of Soda ℥ß. in two waters, and heated till the precipitate shrinks and fuses, will yield a solid mass on cooling, which weighs, when dry, at least 11 grains, and, if pulverized, dissolves entirely in solution of Oxalic acid." Dr. Pereira has not found this satisfactory, but considers the process of Thibaumary as the best, in which Ammonia is employed to precipitate the Morphia in infusion of Opium. (v. Mat. Med. p. 1742.) Dr. Christison considers Dr. Gregory's method of obtaining Muriate of Morphia as the only certain one; but it requires about a pound of Opium to be operated on, which, if good, should not yield less than ten per cent. of a snow-white salt.

Tests. In cases of poisoning, the sensible appearances must necessarily differ according as a solid or a liquid preparation of Opium or one of Morphia has been employed. If any of the former are found, then the brownish colour, bitter taste, and peculiar odour, will indicate the presence of Opium either in a solid or liquid form. But in many cases the poison has entirely disappeared from the stomach, and the odour is alone recognizable,—especially on the first opening of the stomach. This odour is more perceptible, in any fluid containing it, on increasing the temperature (short of the boiling point, when some decomposition takes place). The other tests are of a chemical nature, and have been already enumerated, in describing the crystallizable ingredients of Opium; such as Nitric acid and Tinct. of Sesquichloride of Iron, both of which produce a red colour in a solution of Opium, the first from acting on its Morphia, and the second on the Meconic acid. In the case of organic mixtures, it is necessary first to make an aqueous extract of the contents of the stomach, &c., and then from that an alcoholic one. Dr. Christison has said that the evidence of Opium being present is irrefragable if the alcoholic extract present the peculiar bitterness of Opium,—if its watery solution, when acted on by Ammonia cautiously added, so as to avoid excess, yields a precipitate (Morphia) which becomes yellow with Nitric acid,—and if after the separation of this precipitate, the remaining fluid (then containing Meconate of Ammonia) gives, with Acetate of Lead, a precipitate (Meconate of Lead), which, when decomposed in water by Sulphuretted Hydrogen (Sulphuret of Lead being formed, Meconic acid is dissolved) imparts to the water the property of becoming deep cherry-red with Sesquichloride of Iron. Mr. Taylor finds that N′ detects gr. $\frac{1}{15}$ of of Mur. Morphia diluted in 300 parts of water; Sesquichlor. Iron gr. $\frac{1}{11}$ in 231 parts of water; and Iodic′ gr. $\frac{1}{100}$ in 1300 parts of water; but this last is open to fallacy with organic fluids. The *Iron test* for *Meconic′* is more delicate than the tests for Morphia. It may be first employed, and then Nit′ be added to the same quantity of liquid.

Action. Uses. Opium, applied *externally*, is at first stimulant, producing pain, as on the eye, and then sedative. When taken *internally*, in small doses, excitement is first produced, as apparent in the increased frequency of the pulse, and heat of the skin. This is soon followed by diminished sensibility, calmness, and sleep, with abatement of pain, suspension of mucous secretions, with the exception of that of the skin. But if the tendency to sleep be resisted, Opium, in moderate doses, and in those habituated to its use, in excessive doses, will produce intellectual excitement accompanied by bodily activity, soon to be followed by general debility, as is exemplified in Opium-eaters. In large doses, it is a narcotic poison. It is frequently employed as an anodyne and hypnotic, as a sedative, and to restrain inordinate discharges, as in diarrhœa and cholera, also as a diaphoretic, often as an antispasmodic, and even as a febrifuge. In Delirium Tremens it is beneficially given in large doses, and, combined with Calomel and sometimes with the addition of Ipecacuanha, even in inflammatory affections; though in general it is contraindicated when there is inflammation or much fever. It is no doubt the most important of all therapeutical agents, and that which is perhaps the most frequently employed.

D. Opium may be administered internally either in a solid or liquid form, or its effects obtained by using one of its salts; or it may be applied externally or introduced endermically. The medium dose is one grain, but is subject to every variation, being often sufficient in much smaller doses, and at other times requiring to be increased to an extraordinary extent. It may also be advantageously introduced into the rectum, either as a suppository, or in the form of an Enema. Externally it may be applied endermically or by friction in a liniment, or added to lotions, collyria, cataplasms, or plasters, by means of some of the following preparations.

Preparations of Opium.

Extractum Opii purificatum, L. Extractum Opii, E. Extractum Opii Aquosum, D. Purified Extract of Opium.

Prep. L. E. D. Take *Aq. dest.* Cj. (Aq. Ov. E.; boiling ℔j. by weight, D.) add a little (Oj. E.; the whole, D.) to *Opium* sliced ℥xx. (℔j. E. ℥ij. D.). Macerate for 12 (24, E.) hours to soften it, [(break down the fragments with the hand, and express the liquid with pretty strong pressure, E.) (triturate for 10 minutes, and in a little time pour off the liquor, D.)] Add gradually the rest of the water (Oj. E.; as much boiling *Aq.* as before, D.) to the residuum of the Opium. Triturate and mix well. (Macerate for 24 hours and express the liquid, repeating this process till the water is all used, E.) Set aside for the impurities to subside, L. (Pour off the liquor, repeating the process a third time. Mix the liquors and expose in an open vessel to the air for 2 days, D.) Filter (through the same (linen, D.) filter the successive infusions, E.) Evaporate (with vapour-bath, E.; low heat, D.) to a due consistence.

This extract is of a brownish colour, of a bitter taste, and without odour. The parts soluble in water (*v.* p. 258) with a little of the resin being taken up, and the insoluble, with some active principles however, left behind; this extract is considered to be less exciting than pure Opium, and is therefore sometimes preferred in some of the cases for which opium or the salts of Morphia are indicated. But as it is uncertain, the Morphia salts seem preferable in cases where constitutional disturbance is to be avoided, and crude Opium or Laudanum can alone be relied upon in urgent cases.

D. gr. ½ to gr. iij. or gr. v.

PILULÆ OPII SIVE THEBAICÆ, E. Opium or Thebaic Pills.

Prep. Beat into a proper mass *Opium* 1 part, *Sulphate Potash* 3 parts, *Conserve of Red Roses* 1 part. Divide into 5 gr. pills.

Each contains of Opium gr. j., that is, twice as much Opium as the Opiate pill of the last Latin ed. of the E. P. 1 to 2 pills for a dose.

PILULÆ SAPONIS (CUM OPIO, D.) COMPOSITÆ, L. (*v.* Sapo.)
Opium gr. j. in 5 grains of the Pill. gr. v.—gr. x. for a dose.

PILULÆ STYRACIS (E.) COMPOSITÆ, L. (*v.* Styrax.)
Opium gr. j. in 5 grains of the Pill. gr. v.—gr. x. for a dose.

PILULÆ CALOMELANOS ET OPII, E. (v. p. 194.)
Each Pill contains Calomel gr. ij. and Opium gr. ½.

PILULÆ PLUMBI OPIATÆ, E. (v. p. 167.)
Each Pill contains of Meconate of Lead gr. iij. and of Acetate of Morphia gr. ½ nearly.

PILULÆ IPECACUANHÆ COMPOSITÆ, L. (*v.* Ipecacuanha.)
gr. $\frac{3}{10}$ of Opium in 5 grains of the Pill.

TROCHISCI OPII, E. Opium Lozenges.

Prep. Reduce *Opium* ℥ij, to a fluid extract, as directed in Extractum Opii, E. mix it intimately with *Extract of Liquorice*, ℥v. reduced to the consistence of treacle, add *Tincture of Tolu* ℥ß. sprinkle into the mixture *powdered Gum Arabic* ℥v. and *finely powdered pure Sugar* ℥vj. Beat into a proper mass. Divide into lozenges of 10 grs.

Each lozenge contains about $\frac{1}{10}$ of a grain of the Extract; those of the shops usually contain about ⅛ of a grain of Opium. Both are useful, like the Morphia lozenges, in allaying troublesome cough.

PULVIS CRETÆ COMPOSITUS CUM OPIO, L. D. PULVIS CRETÆ OPIATUS, E. Compound Powder of Chalk with Opium.

Prep. Triturate thoroughly together *Comp. Chalk powder* ℥viß. (℥vj. E.) and *hard Opium* (Opium, E.) powdered ℈iv.

Two scruples of the L. and D. and 37 grs. of the E. prep. contain of Opium gr. j with Chalk, Tormentil, Cinnamon, and Long-pepper.

Action. Uses. Antacid, Astringent, Stimulating, and Narcotic. Useful in some Diarrhœas in doses of gr. x.—gr. xxx.

PULVIS KINO COMPOSITUS, L. D. (*v.* Kino.)
Contains of Opium gr. j. in 20 grains of the powder.

PULVIS IPECACUANHÆ COMPOSITUS, L. E. D. (*v.* Ipecacuanha.)
Contains of Opium gr. j. in 10 grains of the powder.

CONFECTIO OPII, L. D. ELECTUARIUM OPII, E. Confection of Opium.

Prep. L. D. Take finely powdered *Long Pepper* ℥j. *Ginger* ℥ij. *Carraway* ʒiij. *Tragacanth* ʒij. (add them to, D.) *hard Opium* ʒvj. (triturate with *Syrup* (℔j. D.) Preserve them in a close vessel, and when required, add them to warmed *Syrup* f℥xvj. and mix.

E Mix and beat into an electuary *Aromatic Powder* ℥vj. finely powdered *Senega* ʒiij. *Opium* diffused in a little *Sherry* ℥ß. *Syrup of Ginger* ℔j.

These preparations, intended as substitutes for the old Theriaca, differ a little in strength: that of the L. P. contains about Opium gr. j. in 36 grs. (25 grs. D. 43 grs. E.) of the Confection. . All are stimulant and anodyne, hence well-suited as additions in the treatment of Chronic Diarrhœas, &c. in doses of gr. x.—ʒj.

ELECTUARIUM CATECHU, E. (COMPOSITUM, D.) (*v.* Catechu.)
Astringents and Aromatics with Opium gr. j. in about 200 grs. of the Electuary.

TINCTURA OPII, L. E. D. Tincture of Opium. Laudanum.

Prep. L. E. D. Take *hard Opium* powdered ℥iij. (Turkish ʒx. D.; Opium sliced ℥iij. E.) and *Proof Spirit* Oij. [(℔j. by measure, D.) (Rectified Spirit Oj. and f℥vij. and Aq. f℥xiijß. E.)] Macerate for 14 days and strain. (Digest the Opium in the water near 212° for 2 hours. Break it down, strain, and express the infusion: macerate the residue in the spirit for about 20 hours: then strain and express very strongly. Mix the watery and spirituous infusions. Filter. Or, if of fine quality, slice the Opium finely; mix the spirit and water, and in f℥xiv. of the mixture, macerate the Opium for 12 hours: then break it down thoroughly, pour the fluid and pulp into a percolator, and let the fluid pass through. Without packing the Opium in the cylinder, add the rest of the spirit, continuing the process till Oij. are obtained, E.)

Laudanum is of a deep brownish-red colour, with the peculiar taste and odour of Opium. Dr. Christison says that by the E. process, that is, macerating the Opium first in hot water, then in rectified spirit, it may be made in 36 hours. Sp. Gr. 0·952 ; minims 19 contain about gr. j. of Opium. (Phillips.) Good Tincture should leave, when thoroughly dried up in the vapour-bath, from 17 to 22 grs. of residuum from f℥j. (*c.*) ♏xiijß. or about 25 drops, contain the active part of 1 gr. of Opium; but the London Tincture may sometimes be 16 per cent. stronger than the others. (*c.*) Some Morphia is contained in the residuum, and has been separated by Dr. Pereira. Mr. Haden used to make a substitute for Liq. Opii Sedativus by macerating the lees with Tar'. M. Martin, by fermenting the lees with sugar, obtained an extract possessed of narcotic properties.

Action. Uses. Laudanum is a powerful anodyne and narcotic, and the form in which the effects of Opium may most effectively be obtained either externally or internally.

D. ♏x.—fʒß ; but much larger doses may be exhibited in particular cases. Great precaution is required in prescribing it to children.

Infants have been killed by 4 drops, and unpleasantly deep sleep has been produced even by 2 drops. (*c.*)

VINUM OPII, L. E. D. *Laudanum Liquidum Sydenhami*, Ph. L. 1720. *Tinctura Thebaica*, Ph. L. 1745. Wine of Opium.

Prep. Macerate (Digest, E.) *Purified Extract of Opium* ℥iiß. (Opium (Turkey, D.) ℥iij. E.; ℥j. D.) *Cinnamon* bruised (in moderately fine powder, E.) and *Cloves* bruised of each ʒijß. (ʒj. D.) in *Sherry Wine* Oij, (℔j. by measure, D.) for 14 (7, E.; 8, D.) days. Filter.

Wine of Opium, Sydenham's Liquid Laudanum, differs from the former not only in the menstruum, but also in being made from the purified Opium and in the presence of the aromatics. Hence it is more agreeable both in taste and smell, and may be used in many cases for the same purposes. Dr. Paris has proposed adding the Opium to wine during its state of fermentation.

D. ♏x.—fʒj. Often dropped into the eye in Ophthalmia.

TINCTURA OPII AMMONIATA, E. Ammoniated Tincture of Opium. *Scotch Paregoric.*

Prep. Digest for 7 days *Opium* sliced ʒß. *Benzoic acid* and *Saffron* chopped of each ʒvj. *Anise Oil* ʒj. *Spirit of Ammonia* Oij. Filter.

The Spirit of Ammonia, E. being made with caustic Ammonia, and being in excess, first precipitates and then dissolves the Morphia. This preparation is three times stronger than English paregoric, the activity of 1 grain of Opium being possessed by 80 minims.

TINCTURA CAMPHORÆ COMPOSITA, L. Compound Tincture of Camphor. TINCT. OPII CAMPHORATA, E. D. Camphorated Tincture of Opium. Paregoric Elixir. English Paregoric.

Prep. Macerate together *Camphor* ℈iiß. (℈ij. D.) *Opium* in powder (sliced, E.) gr. lxxij. (℈iv. E.; ʒj. D. *Benzoic* gr. lxxij. (℈iv. E.; ʒj. D.) *Oil of Anise* fʒj. *Proof Spirit* Oij. (old wine pints, D,) for 14 (7, E.) days, and then strain.

Though this preparation is named from Camphor, Opium is its most powerful ingredient; hence it is described in this place. The L. name is advantageous, as enabling Opium to be prescribed without the knowledge of the patient. The presence of the stimulants is supposed to counteract the debilitating effects of Opium on the stomach, while the Benzoic acid determining to the mucous surface of the aerial passages, diminishes profuse secretion, and the Opium allays troublesome cough by diminishing sensibility. Hence it is much employed to allay the tickling of coughs, and likewise in some diarrhœas. Each f℥ß. or 240 minims contains of Opium gr. j. nearly.

D. fʒß.—fʒiv. Frequently added to cough mixtures.

ACETUM OPII, E. D. Vinegar of Opium.

Prep. Take *Opium* ℥iv. *Distilled Vinegar* f℥xvj. (℔j. D.) Cut the Opium into small fragments. Triturate into a pulp with a little of the Vinegar, then add the rest, and macerate for 7 days in a close vessel, agitating occasionally. (Pour off the supernatant liquor and strain, D.) Strain. (Express strongly and filter, E.)

Vinegar is one of the best solvents of the active properties of Opium. By some it is supposed that an Acetate of Morphia is formed at the expense of the Meconate, and it is preferred as capable of producing the anodyne and soporific with less of the disagreeable effects of Opium: may be given in doses of ♏x.—♏xxx.

The Black Drop is a celebrated nostrum, in which Opium is boiled with aromatics in verjuice (from the wild crab), and sugar is added and fermented. One drop is considered equal to two or three drops of Laudanum. The above or the salts of Morphia are the best substitutes.

Liquor Opii Sedativus, of Mr. Battley, is another secret preparation, which has long been esteemed in the Profession for its efficacy and its little disagreeable effects as an opiate. It is supposed by some to be an aqueous and by others an acetous solution of Opium. Mr. Cooley states that it is an impure Meconate of Morphia combined with extractive and such other matter as is soluble in temperate distilled water, and that we may produce it by the following formula:

Take *dry Opium* (Smyrna) *in powder*, 1 part, *clean washed* (silica) *sand*, 2 parts. Mix, and moisten with water, introduce into a percolator, and pass *Aq. Dest. at* 65° *or* 70° F. through the ingredients, until it passes both tasteless and colourless. Evaporate the liquor (by steam or water-bath) to the consistence of a *hard pill extract*. Take of this *hard extract* ℥iij. and *Aq. Dest.* ℥xxx. Boil for two minutes; let it cool; filter; then add *Rectified Spirit* ℥vj. and *Aq. Dest.* q. s. to make up nearly f℥xl. or one quart.

D. ♏v.—♏xx. ♏xx. are equal to about ♏xxx. of Laudanum.

ENEMA OPII, L. E. D. vel ANODYNUM, E. Opium Clyster.

Prep. Take *Tincture of Opium* ♏xxx. (f℥ß. to ℨj. E.; ℨj. D.) and *Decoction of Starch* f℥iv. (Starch ℨß. and Aq. f℥ij. E.; Aq. tepid ℥vj. D.) Mix. (Boil the Starch in the Aq.: when cool enough for use add the Tincture of Opium, E.)

Action. Uses. Laudanum in this form often relieves many painful affections of the intestinal canal and urinary organs.

LINIMENTUM OPII, L. E. LINIMENTUM SAPONIS CUM OPIO vel ANODYNUM, D. Liniment of Opium.

Prep, L. D. Mix *Soap Liniment* f℥vj. (4 parts D.) and *Tinct. of Opium* f℥ij. (3 parts, D.) E. Macerate *Castile Soap* ℥vj. and *Opium* ℥iß. in *Rectified Spirit* Oij. for 3 days; filter, add *Oil of Rosemary* fℨvj. and *Camphor* ℥iij. Agitate.

Action. Uses. The external friction of Laudanum not only relieves local pain, but produces the general soporific effects of Opium.

EMPLASTRUM OPII, L. E. D. Plaster of Opium.

Prep. Melt *Litharge Plaster* ℔j. (℥xij. E.) and *Abietis Resina*, powdered, ℥iij. add to it in powder *Hard Opium* ℥ß. (gradually, and mix thoroughly, E. D) *Aq.* f℥viij, L. Boil over a slow fire to a proper consistence, L.

Action. Uses. Applied to relieve Rheumatic and other pains.

UNGUENTUM GALLÆ COMPOSITUM, L. is an astringent application which contains Opium ℨß. in about ℥ij. of the ointment.

Inc. Several salts decompose Opium, and cannot therefore be prescribed with it, as alkalis in small quantity ; but in excess they redissolve the Morphia they have precipitated. Alkaline Carbonates, Lime-water, Astringents containing Tannic acid, Sulphates of Zinc, Copper, Iron, and Lead, Nitrate of Silver, Bichloride of Mercury.

Antidotes. In cases of poisoning by Opium or Laudanum, evacuate the stomach either by means of the stomach-pump, by tickling the throat, or prescribing such emetics as are at hand, as salt or mustard. The Sulphate of Zinc in large doses is the best emetic. Sometimes Tartar Emetic with Ipecacuanha is resorted to; or, in extreme cases, a solution of 1 gr. of Tartar Emetic may be injected into the veins, taking care that no air enters at the same time. During the whole of this time, the patient should be roused by loud talking, shaking and making him walk about. Apply the vapours of Ammonia or of Acetic acid to the nostrils. Cold affusions to the head and chest are of great efficacy. Distending the stomach with astringent infusions, as of Cinchona or of Gallnuts, will assist in decomposing the Opium. When the stomach has been freed of the poison, vegetable acids and venesection are useful; while such stimulants as Carbonate of Ammonia and Brandy and Coffee will be useful in rousing and supporting the patient. Sinapisms and irritants to the feet, &c., ought to be applied, artificial respiration not neglected.

CRUCIFERÆ, *Juss.* Cressworts.

Cruciferous plants are almost all herbaceous, more rarely perennials, seldom shrubby, the root sometimes turnip-shaped. Leaves usually alternate, simple or variously cut. No stipules or bracts. Flowers white, yellow, or purple, in racemes opposite the leaves, or in terminal corymbs, which become elongated. Calyx 4-leafed. Petals 4, cruciate, as are the sepals. Stamens 6, of which four are long and two short. Ovary sessile, 2-celled, or rarely one-celled, with two intervalvular placentæ, which, growing inwards, meet in the middle and form the spurious dissepiment. Fruit a silique, silicule, nut-like, or lomentaceous. Seeds without albumen, generally suspended by a funicle. Embryo curved, or with the radicle variously folded upon the cotyledons.

Cruciferæ are allied to Papaveraceæ and to Capparideæ, and are found chiefly in the temperate parts of the Northern hemisphere, but a few species in most parts of the world. They abound in mucilaginous, and the roots of some in saccharine principles. A fatty oil is stored up in the seeds of many, Sulphur is contained in some, and nearly all abound in an acrid principle, which makes them useful as Condiments, Rubefacients, and Stimulants.

COCHLEARIA, *Linn.*

Calyx spreading, equal at the base. Petals white, obovate, entire. Filaments not toothed, straight above. Silicle globose or ovate, valves very convex. Dissepiment thin, but broad. Seeds numerous. Cotyledons accumbent, or radicle bent up against their edges.

Cochlearia officinalis, D. Herba. Common Scurvy-grass. *Tetradynamia Siliculosa*, Linn.

This plant seems to have been first clearly described and figured by Lobel, and has long been employed medicinally in Europe, on the shores of which it is indigenous as also in moist situations on its mountains.

It is a small annual or biennial succulent plant, sending up a tuft of smooth, bright green, shining leaves, which are petiolate, cordate at base, roundish or subreniform. Stem-leaves, lower petiolate, upper sessile; amplexicaul ovate, margin irregular (dentate angular, Dec). Stem erect, racemes terminal. Pedicels twice as long as the ovate, globose silicles, partition broad ovate. Flowers white. E. B. 551.

The whole herb is officinal. When bruised, it emits a pungent odour, and has an acrid bitter taste. Its properties depend on a heavy volatile oil. It becomes inert when dried.

Scurvy-grass was long highly esteemed as an antiscorbutic.

Action. Uses. Stimulant and Diuretic.

Armoracia, L. E. D. Radix (D.) recens, L. Fresh root, E. Cochlearia Armoracia, *Linn.* Horse-radish. Fr. Cran de Bretagne.

This plant is supposed by some to be the *wild radish* of Dioscorides and the *Armoracia* of Pliny. It seems first clearly recognized by Brunsfels in 1530. It is a native of most hilly situations in Europe, and is much cultivated in this country, flowering in May.

Bot. Ch. Root perennial, long white, and tapering, pungently acrid, throwing up large pelcolate leaves, which are lanceolate crenate, smooth, deep green, much veined, and somewhat resembling those of the Water Dock. From the midst of these rises erect stems, 2 or 3 feet high, furrowed and branched towards the top. Stem-leaves small, sessile, the lower ones with the margin much cut, the upper lanceolate, toothed. Inflorescence a raceme. The silicles differ from those of Scurvy-grass, in wanting the dorsal nerve, they are inflated, almost globose, but often abortive. By Wettereau and others it has been separated from *Cochlearia*, (on account of the valves of the silique being without a dorsal nerve,) and formed into a new genus Armoracia, with the specific name of *A. rusticana.*

The fresh root, which is alone officinal, is thick and long, fleshy and white, emitting when scraped a pungent diffusible odour, and imparting a hot and acrid taste, with some sweetness. Its virtues depend upon a volatile oil, which is dissipated by drying and also by heat. Hence the root is usually used as a condiment in its fresh state. A little of the activity is communicated to water, but most completely to alcohol. Dr. Duncan states that the oil is in the proportion of four parts in a thousand, Gutret as much less. The oil is of a light yellow colour, heavier than water, very volatile. It tastes at first sweetish, soon becomes acrid, burning, and inflaming the lips and tongue, and will produce vesication when applied to the skin. The watery solution precipitates Acetate of Lead brown, and Nitrate of Silver black,

Fig. 42.*

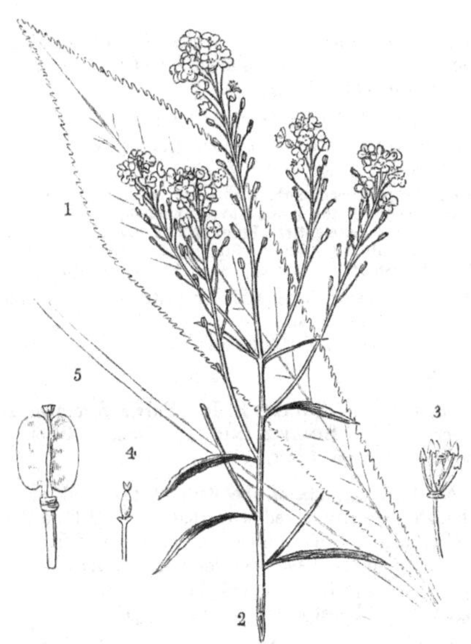

that is, Sulphurets of these metals, showing that the oil contains Sulphur. The other constituents of the root are Bitter Resin, Extractive, Sugar, Gum, Starch, Albumen, Lignin, and Salts.

Action. Uses. Externally. Rubefacient, Vesicant, Irritant. *Internally.* Stimulant, Masticatory, Diuretic.

INFUSUM ARMORACIÆ COMPOSITUM, L. D.

Prep. Macerate *Horse-radish* sliced, and *Mustard* bruised, of each ℥j. *Boiling Aq.* Oj. (℔j. D.) for 2 (6, D.) hours in a covered vessel; strain and add *Compound Spirit of Horse-radish* f℥j.

Action. Uses. Stimulant and Diuretic.

D. f℥j. to f℥ij.

SPIRITUS ARMORACIÆ COMPOSITUS, L. D.

Prep. Mix together (macerate for 24 hours, D.) *Horse-radish* sliced, *Dried Orange-peel*, āā ℥xx. (℔j. D.) *Nutmegs* bruised ʒv. (℥ß. D.) *Proof Spirit* C j. *Aqua* Oij. and let a gallon of fluid distil over.

* 1. Leaf. 2. Raceme. 3. Flower, with the Calyx and Corolla removed 4. Pistil. 5. Silicule.—E. B. t. 2223.

Action. Uses. Stimulant adjunct, especially to Diuretic infusions.
D. f3j.—f3iv.

CARDAMINE, L. D. Flores, L. D. The Flowers of CARDAMINE PRATENSIS, *Linn.* Cuckoo Flower. Common Bitter-Cress. *Tetradynamia Siliculosa*, Linn.

First figured by Brunfels in 1530 under the name of Gauchblüm; is indigenous in moist places throughout Europe, also in the north of Asia and in America.

Bot. Ch. Small bright green herb, with stem roundish, about a foot high. Leaves pinnate, leaflets of the lower leaves roundish, slightly angled, of the upper leaves linear-lanceolate entire. Flowers large, lilac-coloured. Petals three times as long as the calyx, spreading. Stamens half the length of the petals. Anthers yellow, style short Stigma capitate. Flowers in April.—E. B. t. 776.

The leaves and flowers have a bitter taste with some pungency; hence the name of Bitter-cress.
Action. Uses. Considered stimulant.
D. 3ij.—3iij.

SINAPIS, L. Semina, L. SINAPI, E. D. Pulvis Seminum, D. Flour of the Seeds, E. SINAPIS NIGRA, *Linn.* Black Mustard *Tetradynamia Siliquosa*, Linn.

Common Mustard consists of the flour of the seeds of the Black Mustard, though generally mixed with that procured from the seeds of the White Mustard, or Sinapis alba, and deprived of fixed oil by expression. E. Both species have been long used in medicine, being the ναπυ of Hippocrates, and the Sinapi of the Romans. The Black Mustard is indigenous in almost every part of Europe.

Bot. Ch. The root is thick and fleshy, the stem about 2 to 3 or 4 feet high, hispid below, with smooth round branches above. Lower leaves large, rough, lyrate, variously lobed and toothed; upper ones narrow, lanceolate, smooth, dependent. Calyx yellowish, equal at base, spreading. Petals obovate, yellow, spreading. Silique small, erect, or placed close to the stem, obtusely quadrangular, nearly even and smooth, tipped by a short quadrangular style, but without the proper, often seed-bearing beak of the genus; the valves convex, with one straight dorsal nerve, and a few lateral anastomosing veins. Seeds numerous, in a single row, small, round, blackish brown. Fig. 41.—E. B. t. 969.

SINAPIS ARVENSIS, Charlock, or Wild Mustard, E. B. t. 1748, has its seeds sometimes substituted for the Black Mustard. It may be distinguished by the long sword-like beak of the pods, and by the valves being three-nerved. The Black Mustard wanting these, has been removed to the genus Brassica by Koch and other botanists.

The Mustard plant is officinal on account of its seeds, or rather of the flour of these seeds, so well known as a condiment by the name of Mustard. But Dr. Pereira learnt that the best flour of Mustard is prepared by crushing the seeds of both Black and White Mustard between rollers, and then pounding them in mortars, when they are twice sifted

Fig. 44.

Fig 43.

to yield *pure flour of Mustard.* Dr. Christison's information confirms that of Dr. Pereira, that common flour of Mustard is adulterated, partly on account of the pungency of Black Mustard seed. "Two bushels of black, and three of white seed yield, when ground, 145 pounds of flour; which, to diminish the pungency and improve the colour, is mixed with 56 pounds of wheat flour and two pounds of turmeric; and the acrimony is restored without the pungency, by the addition of a pound of (capsicum) chilly pods, and half a pound of ginger." (*c.*) Dr. Thomson says he could detect no turmeric, nor cayenne pepper in the specimen which he examined. The *pure flour* of Mustard ought alone to be used officinally.

Though Black and White Mustard have both been minutely examined by several chemists, they still require further investigation. The former contains a Fixed Oil, Gummy matter, Sugar, a Colouring and a Peculiar green matter, a Fatty pearly matter; *Myronic acid* in combination with Potash, or Myronate of Potash (the Sulpho-Sinapisin of Henry and Garot) *Myrosyne*, *Sinapisin*, and some salts. The fixed oil of Mustard forms 28 per cent. It is mild in taste, with little odour, of a yellow colour, Sp. Gr. 0·917, thicker than olive oil, does not readily become rancid, makes an excellent soap, is sometimes used instead of rape oil, and has been employed as a purgative. (*Bussy.*) *Sinapisin* of Simon is in white, brilliant, micaceous, and volatile crystals, soluble in alcohol, ether, and oils, insoluble in acids and alkalis. *Myrosyne* is a substance analogous to vegetable albumen or the emulsin of bitter almonds. *Myronic acid*, composed of Carbon, Hydrogen, Oxygen, Nitrogen, and Sulphur, is bitter, without odour, uncrystallizable, and may be separated in an impure state by alcohol, when the fixed oil has already been removed by expression or by ether.

It is curious that we do not find among the above products, the acrid principle for which Mustard is so remarkable. In fact, it is not contained in the seeds; the acrimony is due to what is called *Volatile Oil of Mustard*, which is the result of the action of some of the constituents of the seed, that is, of the Myrosyne and Sinapisin, on one another, when water is added under 200° F. Bussy is of opinion that the Myronic acid, on the contact of Myrosyne and water, yields the volatile oil of Mustard. When flour of Mustard is exposed to dry heat, or acted upon by alcohol, no acridity is observed; but if water be added first, the pure volatile oil is obtained, which may be separated by distillation. The mineral acids check the formation of this volatile oil, so does the Carbonate of Potash, and also the vegetable acids when they are of the Sp. Gr. of at least 1022; though when once formed, the acids have no influence in preventing its effects.

The Volatile Oil of Mustard is white or of a lemon colour, extremely acrid and pungent, and exciting the secretion of tears. Sp. Gr. at 68° F. 1·015, boiling at 290° F. Soluble in alcohol and ether, and slightly in water, and separated from it with difficulty, in consequence of having nearly the same Sp. Gr. With Ammonia it forms a compound in which the oil of Mustard is destroyed. Sesquichloride of Iron produces an orange tint in infusion of Black Mustard seed. The oil consists of C 49·84, H 5·09, N 14·41, O 20·48, Sulphur 10·18. (Dumas and Pelouze.)

SINAPIS ALBA, *Linn.* E. White Mustard.

The white Mustard is mentioned in the E. P. as well as the black (v. supra). It is, like the former species, indigenous in most parts of Europe, and may like it have been employed by the Greeks, as it possesses nearly the same properties.

Bot. Ch. Root tapering, small. The stem 1—2 feet or more high, round, smooth, or slightly hairy. Leaves lyrate, irregularly lobed or pinnatifid, roughish. Sepals linear, green, horizontal. Flowers large, yellow. Silique hispid, spreading on nearly horizontal stalks, short, tumid or knotty from the prominent seeds, shorter than the sword-shaped beak; valves with 5 straight strong nerves. Seeds few on each side, large and roundish, pale yellow coloured, having in the interior a yellow mass, which is covered by a thin pellicle of what must be composed of condensed mucilage. Fig. 42.—E. B. t. 1677.

White Mustard seeds yield about 36 per cent. of fixed oil, and when macerated in water, a thick, mucilaginous, almost insipid liquor, while Black Mustard seeds give little mucilage, but a pungent taste to the water. According to the analysis of John, these seeds contain—1. An acrid volatile oil. 2. A yellow fixed oil. 3. Brown resin. 4. A very little extractive. 5. A little gum. 6. Lignin. 7. Albumen. 8. Phosphoric acid, and salts. Henry and Garot ascertained the presence of Sulphosinapisin,—a name which Berzelius has contracted into Sinapin. This is white and light, without odour, at first bitter in taste, but then like mustard; soluble in water, alcohol, and ether, and crystallizable. It consists of C 57·92, H 7·79 N 4·9 O 19·68, and of Sulphur 9·65. "Acted on by acids, oxides, and salts, readily yields Sulphocyanic acid" (*p.*), which strikes a red colour with the persalts, as for instance the Sesquichlor. Iron, and produces a white precipitate in a solution of Sulphate of Copper containing Iron. White Mustard does not furnish volatile oil; but, in certain circumstances similar to those with Black Mustard seed, a fixed acrid principle is produced, which, like the volatile oil of Black Mustard, did not previously exist.

The *fixed acrid principle* is an unctuous liquid of a reddish colour, without odour, but having a biting acrid taste, analogous to that of Horseradish root. It contains Sulphur. M. Faure states that this same principle is formed in small quantity when Black Mustard is treated with water. *Erucin*, which does not redden the salts of Iron, and contains no Sulphur, is another principle found by Simon.

Tests. As the common flour of Mustard is that usually employed, instead of the powder of Black, or the mixed powders of the Black and White Mustard, as in the E. P., so it often contains adulterations along with the true flour. But as Mustard flour is sometimes exhibited internally, it is desirable to have it in a pure form. The adulteration of wheat flour can be easily detected, by the test of the E. P. "A decoction allowed to cool is not turned blue by tincture of Iodine."

Action. Uses. Powerfully acrid and pungent. The seeds of the White Mustard, taken in an entire state, have their mucilaginous covering dissolved away by the juices of the stomach, and will then act as stimulants. Two or three tea-spoonfuls used to be given two or three times a day in dyspepsia. The uses of Mustard as a condiment, and of the young herb as a salad, are well-known. It is Stimulant, Diuretic in the form of Mustard Whey. Externally Rubefacient, &c. Much used in the form of the Mustard Poultice. (q. v.) Emetic in doses of a tea-spoonful to a table-spoonful in half a pint of water.

CATAPLASMA SINAPIS, L. D. Mustard Poultice or Sinapism.

Prep. Take of *Mustard Flour* and *Linseed Meal* āā ℔ß. warm *Vinegar*, as much as may be sufficient to make into a poultice. The D. P. orders 2 oz. of Horse-radish to be added, if it requires to be strengthened.

Here the Mustard flour is weakened by the addition of wheat flour. It may also be made with bread-crumb, or the Mustard flour may be spread on a poultice. The vinegar is worse than useless, as like other acids, it prevents the formation of the acrid principle. Messrs. Trousseau and Pidoux found boiling water unnecessary, as a cold poultice produces the same effect as a hot one, but takes a little longer time to produce its effect. The volatile oil is a powerful rubefacient and vesicatory, in the proportion of 1 part to 20 of proof spirit.

VIOLACEÆ, *Juss.*

Sepals 5, persistent, extended at the base. Petals 5, irregular; in some regular. Stamens equal in number to the petals, filaments dilated, sometimes united; connective elongated beyond the anthers. Ovary free, 1-celled, with three parietal placentæ. Style simple, thickened towards the apex. Stigma variously formed. Capsule three-valved, each valve bearing the seeds at its centre. Embryo straight, within a fleshy albumen.

The herbaceous species inhabit the temperate parts of the Northern, a few the Southern hemisphere, and within the tropics. The shrubby species in South America and India. The stems and leaves are mucilaginous, and contain Violine, which is similar in its nature and effects to Emetine. This has also been found in some of the shrubby species of Ionidium. Several of them (see Martius's Spec. Mat. Med. Braziliensis) are employed as substitutes for Ipecacuanha. The roots of Ionidium Ipecacuanha are the *false Ipecacuanha of Brazil,* and yielded Pelletier 5 per cent. of Emetine. Cuchunchully de Cuença, the roots of *Ionidium microphyllum,* are similar in properties. There are specimens in the Museum of K. C. from the Hon. Fox Strangways.

VIOLA ODORATA, *Linn.* E. D. Flores, D. Flowers, E. The March or Sweet Violet. *Pentand. Monog.* Linn.

This, the ἴον of the Greeks, is found wild on the borders of fields, in shady situations in many parts of Europe, but is cultivated on account of its flowers, which are so much esteemed for their agreeable odour and colour.

Bot. Ch. The plants are stemless, but give out runners. The leaves are broadly cordate, pubescent. Sepals obtuse, the lowest petal emarginate, the four upper ones roundish-obtuse, a little narrower. Stigma hooked, naked. The fruit bearing peduncles, prostrate, straight at the apex.

The flowers should be gathered soon after they have blown. The colour may be retained for some time if they are carefully dried, but for a still longer period if preserved in syrup. As the violet or purple colour is changed into red by acids, and green by alkalis, it is often employed as a test.

Action. Uses. The expressed juice and the syrup are slightly laxative ; and hence, besides being employed on account of its odour and colour, the Syrup is prescribed as a laxative for young, especially new-born children, with an equal quantity of almond oil, in doses of one or two tea-spoonfuls.

Violets, and other species, as V. tricolor, have also been employed as demulcent expectorants on the Continent. The seeds are stated to be purgative and emetic by Bichat, &c. as are also the roots.

SYRUPUS VIOLÆ, E. D. Syrup of Violets.

Prep. Fresh Violets ℔j. (petals ℔ij. D.) *Aq. ferv.* Oijß. (Ov. D. wine measure) *Pure Sugar* ℔vijß. (℔xij. D.) Infuse the flowers for 24 hours in a covered glass or earthenware vessel, strain without squeezing, and dissolve the sugar in the filtered liquor.

POLYGALEÆ, *Juss.*

Herbs or shrubs, sometimes climbing, some abounding in milky juice. Leaves alternate, entire, without stipules. Flowers axillary, solitary, spiked or racemed, pedicels often jointed at the base, tribracteate. Flowers irregular, æstivation imbricate. Sepals usually 5, irregular, often glumaceous, 2 interior sepals much larger than the others and petaloid. Petals usually three in number or 5, of which one (the keel) is anterior and larger than the rest, all united with the filaments. Stamens usually 8, seldom 4, united into a tube (cleft in front) by their filaments. Anthers innate, 1-celled, rarely 2-celled, opening at the apex by one or two pores. Ovary free, 2-celled, each cell with one pendulous ovule. Fruit compressed, capsular or drupaceous. Seeds pendulous, carunculated. Embryo straight within fleshy albumen.

The Polygaleæ are not very closely allied to any other order. In some respects they resemble both Violaceæ and Fumariaceæ, and in others even Leguminosæ. They are found in the temperate and warm regions of the whole world. Many of the species abound in bitter principle, as *P. vulgaris* and *P. amara*, others secrete a peculiar principle, which has been called Polygaline. Several species are officinal in Brazil and India.

SENEGA, L. E. Radix, L. The Root, E. POLYGALA SENEGA, *Linn.* D. Radix, D. Seneka Snake Root.

This root was introduced into practice in 1735, by Dr. Tennant, of Virginia, who learnt from the Senagaroo Indians that they employed it as an antidote against the bite of the rattle-snake. It is a native of the United States of America, chiefly in the southern and western sections, where the roots are collected in large quantities.

The plant (Fig. 43) is small, with a perennial branched root, from which arise several erect annual stems, smooth, simple, round and leafy, which are occasionally tinged with red in their lower portion. The leaves are sessile, alternate, oblong, lanceolate, of a bright green on the upper surface. The flowers are small, arranged in terminal spikes. Sepals 5, two of which are large, wing-like, and white. Petals 3, small, closed, with a beardless keel. Capsule elliptical, emarginate, covered by the persistent sepals.—Barton Am. Med. Bot. 11. t. 36. Fig. 43. 1. Sepals spread out with the Petals adpressed against each other. 2. Central Petal or Keel with the Stamens adhering. 3. A Seed.

Seneka roots are brought to market in bales of from 50 to 400 pounds: the pieces vary in thickness from a small quill to that of the little finger; head knotty, exhibiting marks of former stems, branched, twisted, with a projecting keel-like line along its whole length ; bark-like part is corrugated, cracked, of a yellowish-brown colour in the young roots, and brownish-grey in the old, resinous, and contains the

Fig. 45.

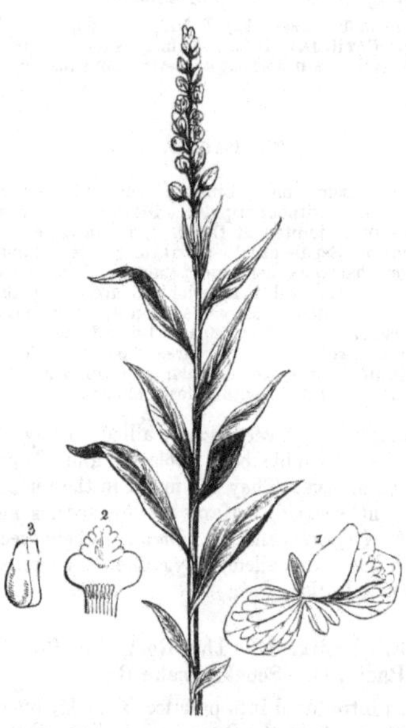

active principle; the central portion, or meditullium, is woody, white, and quite inert. (Goebel and K. ii. t. xx. f. 1.) The odour is peculiar, strong in the fresh root (Wood and B.); taste at first mild, becomes bitter and acrid, exciting irritation in the fauces and a secretion of saliva. Seneka has been analysed by various chemists: the latest, Quevénne, gives Polygalic, Virgineic, Tannic, and Pectic acids, Wax, Fixed oil, yellow Colouring matter, Gum, Albumen, Woody fibre, and various Salts. The *Polygalic acid*, Senegin of Gehlen, and Polygaline of others, is solid, brownish-coloured, when pure white, translucent, without odour, and at first insipid, but soon excites sneezing when powdered, and a disagreeable taste in the mouth with constriction in the fauces. It is insoluble in ether and oils, partially soluble in water, but readily so in alcohol. Given in doses of 6 or 8 grains to dogs, it caused vomiting, difficulty of breathing, and death in three hours.

Action. Uses. Seneka, as indicated by its acrid taste, is possessed of Stimulant properties, and increases many of the secretions, acting as a

Sialogogue, Expectorant, Diaphoretic, Diuretic, and Emmenagogue; and in large doses, as an Emetic and Cathartic.

The roots of Panax quinquefolium, or Ginseng, are frequently mixed with the Seneka.

D. Of the powder, gr. x.—gr. xx. But the decoction is the best form of exhibition.

DECOCTUM SENEGÆ, L. D. Decoction of Seneka Root.

Prep. Boil *Seneka-root* ʒx. in *Aq. dest.* Oij. (℔jß. D.) down to a pint and strain. The E. C. orders the *Infusum Senegæ* to be made by infusing for 4 hours *Seneka* ʒx. in boiling *Aq.* Oj. Strain.

D. f℥j. to f℥iij. 3 or 4 times a day.

The U. S. P. has a Syrup of Seneka root, and it forms an ingredient in their Mel Scillæ compositum.

KRAMERIACEÆ, *Lindl.*, now attached as an anomalous genus to Polygaleæ by Endlicher.

KRAMERIA, L. E. Radix, L. Root, E. Radix et Extractum, D.

KRAMERIA TRIANDRA, *Ruiz and Pavon.* The Rhatanhy Plant.

The Rhatanhy plant is a native of Peru, on the slopes of sandy mountains, especially near Huanuco, where it was discovered in 1779 by Ruiz, who found the root was employed by the ladies for rubbing the teeth and strengthening the gums.

The shrub is small but much branched, with the younger parts covered with silky hairs. The stems procumbent, and the roots horizontal or creeping, as said to be indicated in the name Ratanhia. These roots are long, much branched, with a dark reddish bark. The leaves are sessile, oblong-ovate, pointed, and silky. The flowers are solitary, in the axils of the upper leaves, with short stalks. The calyx consists of 4 spreading sepals, silky externally, but smooth, shining and lake-coloured in the inside, though this is not visible in dried specimens. Petals 5, unequal, the (two upper petals separate, spathulate; two lateral roundish, concave, *Lindl.*) three anterior clawed, with the claws united, limbs small, sometimes abortive, the two posterior sessile, thickish. Stamens 3, anthers opening by a double pore at the apex. The fruit is globular, leathery, indehiscent, about the size of a pea, covered with reddish-brown hooked prickles. One-celled, with one seed, the other being abortive. Seed inverse, suspended, without albumen.—Fl. Peruv. 1, t. 93.

Rhatanhy root is woody and branched; pieces vary in diameter from an inch to that of a quill; the cortical part is reddish-brown, fibrous, and easily separated from the central, reddish-yellow, woody part. The root is without smell, but has an extremely astringent taste without any bitterness. The cortical portion contains a much larger portion of the active principle than the interior; the smaller pieces, from the greater proportion of bark, are most efficacious. (G. and K. ii. tab. iv. fig. 2.) Besides the root, an extract is also officinal in the D. P. This is sometimes imported from S. America.

Rhatanhy Root consists of one-third of matters soluble in water. These consist of Tannin 42·6, Gallic acid 0·3, Gum, Extractive, and Colouring matter 56·6, and Krameric acid 0·5. The properties are no doubt, in a great measure, due to the Tannin, and, according to Peschier, to the Krameric acid, which he describes as being very styptic, not crystallizable, but forming salts with the alkalis, which do crystallize. M. Chevalier, on repeating the experiment, was unable to procure any of the acid. Water and Alcohol both take up the active properties, and become of a reddish colour.

Inc. The salts of Iron and other metals, Gelatine, mineral acids, Inf. Cinchonæ, Potassio-Tartrate of Antimony.

Action. Uses. Astringent, Tonic.

D. Powder, gr. x.—ʒß.

INFUSUM KRAMERIÆ, L.

Prep. Macerate *Krameria* ℥j. in boiling distilled *Aq.* Oj. for 4 hours in a lightly covered vessel and strain.

D. f℥iß.—f℥ij. twice or thrice a day. Decoction is also a good form for exhibition. Astringent taste, and of a reddish colour.

In the United States, a compound Tincture is prepared, with powdered root ℥iij. Orange-peel ℥ij. Serpentaria ℥ß. Saffron ʒj. in Proof Spirit Oj. It is a grateful astringent,

EXTRACTUM KRAMERIÆ, E. D.

Prepared as Extract of Liquorice, E.

D. gr. x.—℈j.

Has a reddish-brown colour, with a vitreous and shining fracture, and yields a blood-red powder, bearing a close resemblance to Kino. That imported from S. America used to be, and perhaps still is, employed for adulterating port wine.

SILENACEÆ, *Lindl.* CARYOPHYLLEÆ,* *Juss.* Tr. SILENEÆ, *Dec.*

Herbaceous plants, with opposite undivided exstipulate leaves and tumid nodes. Calyx free, with the sepals formed into a tube, 4 or 5 toothed. Petals 4 or 5, often slit. Stamens definite, hypogynous, inserted with the petals into the apex of a more or less distinct gynophore. Ovary one-celled, many seeded, with a free central placenta. Stigmas sessile 2—5. Capsule 2—5 valved. Seeds usually with the embryo curved round mealy albumen. Inhabitants of the temperate parts of the northern hemisphere, usually insipid, and possessed only of demulcent properties.

DIANTHUS CARYOPHYLLUS, *Linn.* Flores, D. Clove Pink or Carnation. *Decandria Digyn* Linn.

The Carnation has been cultivated in our gardens from the time of Gerard, is probably a native of the south of Europe, and found in many parts of the Continent on walls, &c. apparently in a wild state,

* The order *Caryophylleæ* (with which the Clove or *Caryophyllus* has nothing to do) as at present constituted contains so many others not included in the arrangement of *Dec.* that the author has thought it sufficient to notice one only of its tribes.

This plant is characterised by its solitary flowers, with depressed calycine scales, which are rhomboid, pointed, 4 times shorter than the tube. Leaves linear, acute, glaucous, scabrous at the base, with a smooth margin. Petals obovate, crenate, beardless, with elongated, much-branched runners. E. B. t. 214.

Action. Uses. The deep red coloured flowers are alone employed. They have a pleasant, aromatic, spice-like fragrance, with a bitterish taste, and were at one time considered stimulant. Their only use is to give a colour and flavour to some infusions and mixtures, or to a Syrup.

SYRYPUS DIANTHI CARYOPHYLLI. *Petals of Clove Pink* 1 pt. *Aq.* 4 pts. *Sugar* 7 prts. Valued for its rich colour and agreeable flavour.

LINEÆ, *Dec.* Flaxworts.

Annual herbs, or small shrubs. Leaves alternate or opposite, rarely whorled, entire, without stipules. Flowers regular, terminal, paniculate or corymbose. Sepals 5, 4 or 3, united at the base, imbricate in æstivation, persistent. Petals 5 or 4, twisted in æstivation. Stamens equal in number to petals, filaments united at the base into a ring, with intermediate teeth. Ovary with cells, equal in number to styles. Capsule globular, 3, 4, or 5 celled, each with two seeds; but these are separated from each other by secondary partitions, so that each seed is in a distinct cell, pendulous or inverted, without arillus (as in Oxalideæ) and usually without albumen. Embryo straight or curved, with flat cotyledons.

The Lineæ are closely allied to Geraniaceæ and to Oxalideæ, remotely to Sileneæ and to Elatineæ. They are found in temperate parts of the world, with a few in tropical regions. The Lineæ are remarkable for the tenacity of the fibre of their inner bark, also for the mucilaginous covering of the seed, and for the oil of the seeds. Some are bitter, and a few purgative.

LINUM USITATISSIMUM, *Linn.* L. E. D. Semina, L. D. Oleum e seminibus expressum, L. D. Seeds, E. Meal of the Seeds deprived of their fixed oil by expression, E. Linseed Meal. Expressed Oil of the Seeds. E. Flax. *Pentandria Pentagynia*, Linn.

Flax was cultivated in Egypt at very early periods. It is so at the present day from the north of Europe to the south of India; and it is not, therefore, easy to ascertain where it is indigenous.

The Flax plant is an annual, with a slender root, smooth, simple, erect stem, about a foot and a half in height and branched towards the top. The leaves are alternate, sessile, linear, lanceolate, smooth. The flowers, of a blue colour, are arranged in a corymbose panicle. The sepals are ovate, acuminate, slightly ciliated, but without glands, nearly equal to the capsule in length. The petals are obscurely crenate, of a purplish blue, large, deciduous. Capsule roundish, about the size of a pea, containing 10 seeds *(linseed)* small, oval, flattened, smooth and shining, of a brown colour but whitish in the inside; the seed-coat mucilaginous, the kernel oily and farinaceous.—E. B. 1357.

Flax, as it is well known, is prepared from the above plant, by steeping, stripping off the bark, and then beating, so as to separate the fibres. Linen and cambric are prepared from it, the latter differing from the former in its fineness, and in being obtained from plants which are more thickly sown. Linen as clothing is cool, from being a better conductor of heat than cotton ; but when the skin is covered with perspiration, or exposed to cold, it feels cold and chilly. The fibre of flax is a straight tube-like cylinder, and is therefore less irritating than the twisted fibre of cotton. Hence lint, which is prepared by scraping linen, is so much preferable to cotton for surgical dressings. Tow consists of the short fibres of the flax, which are removed in the process of hackling. It is used for a variety of purposes.

LINI SEMINA. Linseed, or the seeds of the flax plant, are small, compressed, oval-pointed, with sharp margin, brownish-coloured, smooth and shining on the outside, but white internally, without odour. The outside has a bland mucilaginous taste, as the skin of the seed is covered with condensed mucus ; the white part, or almond of the seed, has an oily taste, from containing fixed oil, which is separated by expression.

The seeds, analyzed by Meyer, consist in 100 parts, of 15·12 Mucilage (nitrogenous mucilage with acetic acid and salts, *p.*), chiefly in the seed-coat, 11·26 fatty Oil in the nucleus. In the *husk* Emulsin 44·38, besides in the husk principally Wax 0·14, acrid soft Resin 2·48, Starch with Salts 1·48. In the *nucleus*, besides the Oil, Gum 6·15, Albumen 2·78, Gluten 2·93, also Resinous colouring matter, 0·55, yellow Extractive with Tannin and Salts (nitre and the chlorides of potassium and calcium) 1·91, sweet Extractive with Malic acid and some Salts 10·88.

The condensed mucus which abounds in the testa of the seed is readily acted on by hot water, and a viscid mucilaginous fluid is formed, in which are two distinct substances, one completely dissolved, analogous to gum, and the other merely suspended, and considered by Berzelius as analogous to Bassorine. Alcohol produces a white flaky precipitate in mucilage of Linseed, and Acetate of Lead a dense precipitate.

Action. Uses. Emollient, Demulcent; may be employed in the form of

INFUSUM LINI (E.) COMPOSITUM, L. D.

Prep. Digest *Linseed bruised* ʒvj. (℥j. D.) *Liquorice-root sliced* ʒij. (ʒiv. D.) *boiling Aq. dest.* Oj. (℔ij. D.) in a lightly covered vessel for four hours, near the fire, and strain (through linen or calico, E.)

A simple infusion may be formed by merely steeping half an ounce of the seeds in a pint of boiling water, and rendering it more palatable by the addition of sugar and some aromatics, as mint, lemon-peel, &c. The decoction is more suitable for fomentation and enemata, as it separates more of the oil, but is on this very account less agreeable for internal use.

D. f℥jß. ad libitum.

Inc. Alcohol and metallic salts.

OLEUM LINI, L. E. D. Linseed Oil.

The oil contained in the kernel of the seeds, and obtained from them by expression, may be either cold-drawn, or, as usually seen, after the seeds have been subjected to a heat of 200°. The former, as in the case of cold-drawn castor oil, is paler, with less odour and taste, than Linseed oil prepared by heat. This is of a deep yellow or brownish colour, of a disagreeable smell and taste ; Sp. Gr. 0·932; soluble in alcohol and ether, differing from other oils, especially in drying into a hard transparent varnish,—a peculiarity which is increased by boiling the oil, either alone, or with some of the preparations of Lead.

Linseed Oil, according to Dr. Sace, is composed of *Margaric'* and *Oleic'* combined in equal equivalents with *Acroleine*. But the Oleic' of Linseed differs from that of other fatty bodies. The formula of the anhydrous acid is $C_{46} H_{38} O_5$. The Margaric' is as usual composed of $C_{34} H_{33} O_3$. The Glycerine obtainable from Linseed oil in large quantities is also similar to that procured from other fats.

Action. Uses. Emollient and Cathartic. Chiefly used externally.

D. f℥iv.—f℥j.

FARINA LINI, E. Linseed Meal.

Linseed, after having had the oil expressed from them, are in the form of a flat mass, commonly called *oil-cake*. This being reduced to powder, forms Linseed Meal, E., which is employed for making the Linseed Meal Poultice, and is an ingredient of the PULVIS PRO CATAPLASMATE, D. which consists of *Linseed Meal* 1 part and of *Oatmeal* 2 parts.

CATAPLASMA LINI, L. Linseed Meal Poultice.

Prep. Take of *boiling Water* Oj. and mix with it as much *Linseed powdered* as may be sufficient to make a poultice of the proper consistence.

Here the internal oleaginous and external mucilaginous parts being all ground up together, and their properties elicited by the hot water, an admirable mixture is produced for making an excellent and readily made emollient poultice.

The Linseed Meal sold in France has been found adulterated with some refuse oil seed powder, mixed frequently with a little bran, oatmeal, and almond powder, with the refuse of starch manufactories, and a little, often, rancid oil.

MALVACEÆ, *Brown.* Mallow Worts.

The Malvaceæ form herbs, shrubs, or trees, often with stellate pubescence. Leaves alternate, entire or lobed, often crenate or dentate, with stipulæ. Calyx free, usually of 5 sepals, united together at the base, valvate in æstivation, covered by an outer calyx or involucel. Petals 5, twisted, their claws often united together as well as to the stamen tube. Stamens numerous, monadelphous, or united into a tube. Anthers reniform, with a single polliniferous cell. Ovary formed by the union of several carpels, free or united with each other, or with the central axis. Styles equal in number to the carpels. Fruit either capsular or baccate, or nucamentaceous. Albumen usually small in quantity and varying in density. Embryo curved, with foliaceous, crumpled or twisted cotyledons.

From the name this might be supposed to be a European family, but the species abound in the tropics both as trees and herbs, and diminish in number and size, as they approach the poles. The species (about 600 in number) are almost all mucilaginous, and yield tenacious fibre. A few are employed as articles of diet.

MALVA, L. E. MALVA SYLVESTRIS, *Linn.* Herb of Malva sylvestris, E. Common Mallow. *Monadelphia Polyandria*, Linn.

The Mallow is found in most parts of Europe, by hedges, roads, and in waste places, flowering from June to August. It is the Μαλαχη κηπευτή of Dioscorides.

The root is perennial and branched. The stem erect or ascending, branched, the petioles and peduncles hirsute, leaves 5 to 7 lobed, plaited and with serrated margins, acute, peduncles axillary crowded and erect, even after flowering. Calyx usually surrounded by three narrow bracteoles, and much smaller than the petals, which are rose-coloured and purple-veined, the valves of the carpels margined, reticulated and rugose when ripe.—E. B. t. 671.

Prop. Common Mallow, like the round-leaved and other species, is without odour, but has a mild mucilaginous taste, imparting this property to water, as this dissolves the mucilage which forms its chief constituent, with a small portion of bitter extractive. Either this or *M. rotundifolia* was employed as an esculent vegetable by the Romans.

Action. Uses. Demulcent. Its infusion sweetened with Sugar may form a useful drink in some complaints. The Decoction may be similarly employed, either for fomentation or injection, or the herb may be formed into an emollient cataplasm.

DECOCTUM MALVÆ COMPOSITUM, L. Compound Decoction of Mallow.

Prep. Boil *Mallow dried* ℥j. *Chamomile dried* ℥ß. in *Aq.* Oj. for a quarter of an hour and strain. Use as a fomentation, &c. with appropriate additions.

ALTHÆA OFFICINALIS, *Linn.* Folia. Radix, L. D. Leaves, Root, E. Marsh Mallow. *Monadelphia Polyand.* Linn.

This plant (Fig. 44) is found in marshy situations both in this country and on the Continent, and is the Αλθαια of Dioscorides, the Guimauve of the French.

The root is perennial, tap-shaped, whitish, and the stems erect, soft and hairy. The leaves soft and woolly on both sides, unequally crenate, cordate or ovate in shape, the lower 5 and the upper 3 lobed; peduncles axillary, many flowered, much shorter than the leaf. Flowers of a pale bluish-colour. Calyx double, the exterior (v. 4) involucel, 6 to 9 cleft, the (3) interior 5 fid. Stamens (2) numerous, filaments united into a tube. Styles (1) numerous, united together near the base. Carpels arranged as in Malva.—E. B. t. 147.

Fig. 46.

Prop. Marsh-mallow roots, as usually seen, are whitish, being deprived of their epidermis; otherwise of a dirty yellow colour, but white in the inside, long, fusiform, fleshy, and, like the leaves, without odour, but having a bland, mucilaginous, even viscous taste.

Chem. Marsh-mallow roots analysed, yielded to Buchner, Mucilage and Starch in large proportions; hence Iodine strikes a blue colour; and the Sesquichloride of Iron forms in the decoction, a brown semi-transparent mass (*p*); *Altheine*, first discovered by M. Bacon, and since ascertained by M. Plisson to be identical with Asparagin, is crystalline, without odour, and nearly tasteless, soluble in water and in proof Spirit, but insoluble in Alcohol and Ether.

Action. Uses. Demulcent, Emollient. Used in the form of Decoction, Syrup, and Lozenge.

MISTURA (DECOCTUM, D.) ALTHÆÆ, E.

Prep. Boil *Althæa-root* (and Herb, D.) ℥iv. *Raisins stoned* (or opened) ℥ij. in *Water* Ov. (Ovij. D.) down to Oiij. (Ov. D.) Strain, and when the sediment has subsided, pour off the clear liquor for use.

A pleasant Diluent and Demulcent, of which a pint or two may be taken daily.

SYRUPUS ALTHÆÆ, L. E. D. Syrup of Marsh-mallow.

Prep. Boil *Althæa-root* fresh and sliced ℥viij. (℔ß. D.) in *Aq.* Oiv. (℔iv. D.) down to Oij. and express the liquor when cold. (Strain, E.) Set aside for 24 hours, that the dregs may subside. Then pour off the liquor, and add of *Sugar* ℔ijß. (℔ij. D.) and boil down to a proper consistence.

D. f℥j. to f℥iv. But chiefly added to mixtures to allay irritation of cough.

GOSSYPIUM, E. The Hairs attached to the Seeds of *Gossypium herbaceum* and other species of the genus. Raw Cotton.

Cotton has been characteristic of India from the earliest times. The first distinct notice of it is in the Book of Esther, i. v. 6, where its Sanscrit name *Karpas* is translated *green* in our Bible. Herodotus and Ctesias notice it, but it was not till the invasion of India by Alexander, that the Greeks were acquainted with the plant, as may be seen in Theophrastus and also in Pliny. Europe is now supplied chiefly from America, where two distinct species are indigenous: *G. Barbadense*, yielding the Cotton from the United States, and *G. peruvianum* or *acuminatum*, that which is produced in South America. India also has two distinct species, *G. herbaceum*, or the common Cotton of India, which has spread to the south of Europe, and *G. arboreum*, or Tree Cotton, which yields little if any of the cotton of commerce (very distinct from the species of Bombax often called Cotton-tree and Silk Cotton-tree).

The species consist of large or small shrubs, and one forms a tree. All have alternate leaves, which are more or less palmate or lobed, and usually covered, as well as the young branches, with little black dots, and the nerves below have one or more glands. The calyx is double, the exterior (involucel) is larger than the interior, divided into three large leaflets, cordate at the base, entire, toothed, or deeply cut along the margin. The interior or true calyx is one-leafed, cup-shaped, and with an obtusely quinquifid margin. The flowers are large and showy, more or less yellow or red, consisting of five petals united at their base, subcordate, flat and spreading. Stamens numerous, filaments united below and adhering to the petals, free above, with small kidney-shaped anthers. Ovary superior, oval, roundish or pointed, terminated by a style, which passes through the cylinder formed by the stamens, marked with three or five furrows towards its apex, and dividing into three, sometimes into five stigmas. The capsule is roundish, oval, or pointed, three to five celled, and three to five valved at the apex, with loculicidal dehiscence. Each cell contains from 3 to 7 ovoid seeds, from the seed-coats of which arises the filamentous substance, which by its twisting envelopes the seeds. Along with this Cotton there is often a shorter covering, called *fuzz* by planters.

Action. Uses. Cotton plants are mucilaginous, and have been used as Demulcents. The seeds yield Oil, which is sometimes expressed for burning in lamps. Cattle are, however, often fed on the seeds, which are also sometimes employed as manure for Cotton plants. Cotton wool is formed of tubular hairs, which in drying become flattened, and are transparent, without joints, and twisted like a corkscrew. Under water, they appear like distinct, flat, narrow ribands, with occasionally a transverse line, which indicates the end of cells.

This twisted nature of the Cotton fibre is probably the reason why Cotton cloth is not so well fitted as linen for surgical dressings. But being a worse conductor of heat than linen, it is well suited for inner clothing, where the object is to preserve uniformity of temperature, as it will retain heat, and prevent the body being so readily affected by external heat or cold. At the same time that it condenses less freely than linen the vapour of perspiration, but absorbs it readily when it has been condensed into the form of sweat. For these reasons probably, thick calico shirts, &c. have been introduced into the army for the use of soldiers. Cotton has been long a popular application to burns. Dr. Anderson (Ed. M. and S. Journ. 1828) directs it to be applied in thin layers, one over the other, and retained by the moderate pressure of a bandage. Pain is allayed, local irritation and blistering diminished or prevented, and constitutional disturbance proportionally obviated. M. Reynaud adopted its application in cases of Erysipelas, and M. Mayor employs it as a topical application with Calomel in cases of Ophthalmia. (B. and F. Med. Rev. xx. 463.)

BÜTTNERIACEÆ. *R. Brown.*

Theobroma Cacao, or Cacao-tree, though not officinal, is interesting in consequence of its seeds being largely employed in diet. The tree is a native of Mexico, but extensively cultivated in the West India Islands, and remarkable for its large and oval, yellow, cucumber-like capsules, hanging from the sides of the trunk and branches. These are divided into 5 cells, each filled with 8 to 10 ovoid seeds, piled one upon another, and covered by a membranous and succulent aril. There are several varieties of these seeds or *nibs*, which are more or less esteemed. The kernels of the seeds yield by pressure about one half their weight of a fatty oil, commonly called *Butter of Cacao*, at one time much lauded for its medical properties. The seeds, pounded, digested, and boiled with water, with the oil skimmed off, and sweetened with sugar and milk, afford a wholesome and agreeable beverage. The COCAO sold in the shops consists either of the roasted kernels and husks, or of the husks only, ground to powder ; it is sometimes made from the cake left after expressing the oil from the beans. "Much of the cheap stuff sold under this name, is very inferior, being made with damaged nuts that have been pressed for the oil, mixed with potato-flour, mutton-suet, &c." (*Cooley.*) *Flake Cocao* is Cocao ground, compressed, and flaked by machinery. CHOCOLATE (from the Indian name *chocolat*) is made by triturating in a heated mortar the roasted seeds without the husks, 10 lbs. with an equal quantity of Sugar, and about 1½ oz. of Vanilla, and 1 oz. of Cinnamon (*Cadet*) into a paste, which is put up in various forms. "The mass of the common Chocolate sold in England is prepared from the cake left after the expression of the oil, and this is frequently mixed with the roasted seeds of ground peas and maize, or potato-flour, to which a sufficient quantity of inferior brown sugar, or treacle and mutton suet is added, to make it adhere together." (*Cooley.*)

Action. Uses. Both Cocao and Chocolate form the basis of very nourishing and agreeable beverages (whence the name of Theobroma, or food for the gods) devoid of the stimulating properties of Tea and Coffee, but apt to disagree with some people and with many Dyspeptics, in consequence of the quantity of oily matter they contain.

CISTINEÆ, *Dec.*

The CISTINEÆ, or *Rock Rose tribe*, include plants, some of which used to be officinal in consequence of yielding LADANUM, a fragrant resin, formerly much celebrated, but now little employed. It is procured in the Levant from species of Cistus, such as *C. creticus*, *C. odoriferus*, &c., and can only be obtained pure in the situations where it is produced. It has a very agreeable smell, from the presence of a volatile oil. It was formerly employed as a stimulant, more recently as an expectorant, and continues to be esteemed by the Turks as a perfume, and used as a fumigation.

DIPTEROCARPEÆ.* *Blume.*

The Dipterocarpeæ, so named from some of the divisions of the calyx being extended into long wing-like bodies, require to be noticed, as one of the species, *Dryobalanops aromatica*, Gærtn., *D. Camphora* (*Colebr.*), has been selected, though incorrectly, in the D. P. as the plant yielding the Camphor of European commerce. This kind is produced by one of the tribe of Laurels, the *Camphora officinarum* of Nees v. Esenbeck (*v.* Laurineæ). But the kind called Sumatra or Borneo Camphor, as well as Liquid Camphor, is produced by the above tree, which is a native of Sumatra and Borneo. On the coast of the former island it is one of the largest of trees; and the same tree, it is said, which yields the oil would have produced the Camphor, if unmolested. This kind is not seen in European commerce, because the Chinese give eighty or a hundred times more money for it than that for which they sell their own Camphor. Specimens of both the Sumatra Camphor and of the Liquid Camphor are in the Museum of King's College, having been presented by Mrs Marsden.

Action. Uses. The Liquid Camphor or Oil might no doubt be beneficially employed for the same purposes as Cajaputi oil and Grass oil. The Sumatra Camphor does not appear to be preferable to that of China.

* Several other important products are yielded by the Dipterocarpeæ, as Wood-oil, which contains a principle analogous to Balsam of Copaiva. The Resin or Dammer of Shorea robusta. Indian Copal (sometimes mixed with Amber and sold as such), which is the Liquid Varnish inspissated of the Pineitree, or Vateria indica; of which the fruits yield to boiling water the esteemed and valuable vegetable Butter of Canara.

THEACEÆ, *Mirbel.* CAMELLIEÆ, *Dec.* : a tribe of TERNSTRŒMIACEÆ.

The genus Thea forms a small group of plants with Camellia, and is remarkable for containing the plant or plants which yield the different kinds of Tea imported from China. The question is still undecided whether all the kinds of Tea are made or can be made, from the same plant, by variations in the process; or whether it is preferable to have different varieties or kinds of plant for the distinct varieties of Tea, as, for instance, the Green and Black Teas; whether these were originally distinct species, or varieties owing to differences in soil, climate, or culture. The author has always been inclined to think the latter the more probable opinion, as fully detailed in his "Illustr. of Himalayan Botany," p. 107 to 128, and in his "Productive Resources of India," p. 257 to 311. Two plants are known in the gardens: one called *Thea viridis*, supposed to yield Green Tea, including—1. Imperial. 2. Gunpowder. 3. Hyson. 4. Young Hyson. 5. Twankay. This kind is capable of withstanding a greater degree of cold, and survives through the winter in the open air in this country, as may be seen in Kew Gardens. Green Teas we know are chiefly produced in the more northern districts of China. Some are factitiously coloured with Indigo and Sulphate of Lime, and Mr. Warrington has ascertained that of the Green Teas of commerce some are *unglazed*, others *glazed*. The former are of a yellow-brown tint, tending on the rubbed parts to a blackish hue without a shade of green or blue; while the *glazed* are faced or covered superficially with a powder consisting of Prussian Blue and Sulphate of Lime, or Caolin, with occasionally a yellow or orange-coloured vegetable substance. Indigo with Gypsum is sometimes used, as by the China tea-makers sent to Assam. Even the *unglazed* have a little Sulphate of Lime attached to their surface, either to act as an absorbent of moisture, or to give the bloom characteristic of the green teas of commerce. The *Thea Bohea* appears distinct as a species from the former, and has been supposed to yield the different kinds of Black Tea, that is, Pekoe, Lapsang, Souchong, Congou, Bohea, &c., the last being the inferior, and the Pekoes the best kinds of Black Tea. Plants collected in Chusan are somewhat intermediate in character. That growing wild in Assam is considered by some botanists to be another distinct species of Thea or of Camellia. But the information is too defective for any decisive opinion to be formed. If we compare the recent analyses* of Green and of Black Tea, it would appear that a less degree of heat and younger leaves being employed, would explain some of the differences between Green and Black Tea.

Some Tea has been manufactured in the Government Nurseries in Kemaon from plants grown from China seed, which has been pro-

* As that of Mulder, *v.* Pereira, Treatise on Food and Diet, p. 394.

nounced of the finest quality by the best judges, and compared with the Oolong Teas of the Ankoy district. Some prepared in August, 1845, in the Tea Nursery in the Deyra Doon, has also been pronounced of fine quality, and compared with Orange Pekoe.

The properties of Tea depend chiefly on the presence of Tannin, of a Volatile Oil, and of a principle called *Theine* ($C_8 H_5 N_2 O_2$), which has been found to be identical with *Caffeine*, and is a salifiable base. It may be obtained in white silky needles, has a mild bitter taste, is soluble in hot, but sparingly so in cold water and Alcohol. With S' and H Cl' it forms crystalline compounds, and is supposed to exist in Tea in combination with Tannic Acid. The quantity of Tannin is stated by Brande, and as appears by the taste, and in the analyses of Mulder, to be greater in Green than in Black Tea. Sir H. Davy and others have stated that Black Tea contains the largest proportion of Tannin. The volatile Oil is in larger quantity in the Green than in the Black Tea.

Tea is well known for its astringent and moderately excitant properties, chiefly affecting the nervous system, producing some degree of exhilaration, and of refreshment after fatigue. Its effects are well seen in the wakefulness produced, especially by Green Tea, in those unaccustomed to its use. But it is thought by some to act as a sedative on the heart and blood-vessels; or, as Dr. Billing explains it, Tea and Coffee are sedatives, and relieve the stupor produced by stimulants, or the drowsiness of fatigue, or other plethora, only by counteracting the plethoric state of the brain induced by the continued stimulation of action,—thus merely restoring the brain to its normal state. Liebig (*Anim. Chem.* p. 179) has suggested that *Theine*, as an ingredient of diet, may be useful in contributing to the formation of Taurine, a compound peculiar to Bile. Besides being useful as a diluent, it may often be prescribed as an agreeable and refreshing beverage; in some cases, especially when made strong, acting as an excitant, and at other times producing sedative and calming effects.

AURANTIACEÆ, *Correa*. AURANTIA VERA, *Jussieu*.

Trees or shrubs usually conspicuous for their beauty, and for having transparent receptacles of volatile oil immersed in their surface, commonly smooth, with the axillary branches often changed into straight and hooked spines. The leaves are alternate, articulated with the petiole, unequally or simply pinnate with one or many pairs, but sometimes the terminal leaflet is only produced. The petiole winged: sometimes the terminal leaflet being abortive, the dilated petiole supplies its place. Flowers regular, axillary or terminal, solitary, corymbose or in racemes, usually white or greenish yellow. The calyx is free, short, cup or bell shaped, 4 or 5-fid. Petals 3, 4 or 5, inserted into the base of a shortly stalked torus, which is more rarely formed into a hypogynous disk, free or connected at their base, subimbricate in æstivation. Stamens equal to, double, or some multiple of the number of petals; filaments free, or united into one or several bundles. Ovary 5 or many-celled. Style 1; crowned by a capitate stigma. Fruit dry or pulpy, with a thick valveless rind, two or many celled, often by abortion one celled, cells usually single seeded, seldom many seeded, filled with mucilage in vesicular cells. Seeds pendulous or nearly horizontal, marked with a longitudinal, branching raphe. Chalaza distinct. Embryo straight, without albumen. Cotyledons usually large and thick, a retracted superior radicle near the hilum, and the plumule conspicuous. Fig. 47.

The Aurantiaceæ are allied on one hand to Meliaceæ, and on the other to Xanthoxyleæ, once a tribe of Rutaecæ, and to Amyrideæ, a tribe of Terebinthaceæ. They are natives of tropical Asia, with a few species in Madagascar. Limonia Laureola is alone found in cold situations in the Himalayas; but many are cultivated in all parts of the world. A fragrant volatile oil abounds in many parts, with a bitter principle in the rind of the fruit, and an acid or saccharine juice in the fruit. Several species of the genus Citrus are officinal.

CITRUS, *Linn.* *Polyadelphia Polyandria,* Linn.

Flowers frequently with a quinary proportion of parts. Calyx urceolate, 3 to 5 cleft. Petals 5 to 8, or only 4. Stamens 20 to 60, their filaments compressed and more or less united at the base into several bundles, often 4 or 5 of them free. Anthers oblong. Style round, crowned by a hemispherical stigma. Fruit baccate, 7—12 celled. Seeds 4 to 8 in each cell, with numerous separate small bags of pulp. Seeds without albumen, seed-coat membranous, marked externally with the raphe and internally with the chalaza. Auricles of cotyledons very short. Trees or shrubs with axillary spines. Leaves compound, but often reduced to a single terminal leaflet which is jointed with the petiole, and often winged. v. Fig. 47.

AURANTII CORTEX, L. E. Fructûs Cortex (Tunica, D.) exterior, L. Rind of the Fruit of the Bitter Orange, E. AURANTII AQUA, E. Distilled Water of the Flowers. Orange Flower Water, E. AURANTII OLEUM, E. Volatile Oil of the Flowers. Neroli Oil, E. CITRUS VULGARIS, Risso. L. E. Seville or Bitter Orange.

This, which is called Citrus Bigaradia by Duhamel, and also by Risso in his work on Oranges, is supposed to have been introduced by the Arabs; because all the old established groves of Spain, as those at Seville, planted by the Moors, are of the Bitter Orange (*Macfadyen*).

The tree is erect in habit, smaller than that of the Sweet Orange, but the flowers more fragrant. The branches are spiny. Leaves elliptical, acuminate, slightly toothed. Petioles more or less winged. Flowers large white. Fruit uneven, more or less round, of a dark orange colour; rind with concave vesicles of oil; pulp acid and bitter.—Risso.

AURANTII CORTEX, L. E. D. Rind of the Bitter Orange.

The Rind of the Seville Orange is officinal, because it is more bitter than that of the Sweet Orange, with at the same time a considerable degree of aroma from the presence of volatile oil. But as the outer part alone possesses these properties, the white inner part should be removed when it is used officinally for the following preparations, either in its fresh state, or when intended to be dried. In the D. P., however, both the Sweet and Bitter kinds are included under the head of *Citrus*

Aurantium. In the L. P. the term *Aurantium* is applied first to *C. Aurantium*, but *Aurantii* Cortex is then referred to *C. vulgaris.*

CONFECTIO AURANTII, L. CONSERVA, E. Confection of Orange Peel.

Prep. Rub up in a stone mortar with a wooden pestle *fresh rasped Orange Peel* ℔j. (the outer rind of Bitter Oranges beat into a pulp, E.) add *Sugar* ℔iij. (White Sugar thrice their weight, E.) Pound till incorporated.

Action. Uses. Stomachic. An agreeable vehicle for prescribing tonic or purgative powders.

SYRUPUS AURANTII, L. E. D. Syrup of Orange Peel.

Prep. Macerate fresh *Orange Peel* (bitter, E.) ℥ijß. (℥viij. D.) in boiling *Aq.* Oj. (by measure ℔vj. D.) in a lightly covered vessel for 12 hours. Pour off the liquor and filter, then add *Sugar* (pure, E.) ℔iij. (xivß. D.) (dissolve with heat, E.) and make a syrup.

Action. Uses. An agreeable stomachic, useful as an addition either to disagreeable or to tasteless draughts.

INFUSUM AURANTII (E.) COMPOSITUM, L. D. Compound Infusion of Orange Peel.

Prep. Macerate for ¼ of an hour in a lightly covered vessel *dried Orange Peel* ℥ß. (℥ij. D.) *fresh Lemon Peel* ʒij. (ʒj. D.) *bruised Cloves* ʒj. (ʒß. D.) *boiling Aq. dest.* Oj. (℔ß. by measure, D.) Strain (through linen or calico, E.).

Action. Uses. Warm Tonic. Excellent vehicle for either acid, alkaline, or saline medicines, in doses of f℥iß. two or three times a day.

TINCTURA AURANTII, L. E. Tincture of Orange Peel.

Prep. Macerate for 14 (7, E.) days *dried* (bitter, E.) *Orange Peel* ℥iijß. in *Proof Spirit* Oij. Strain. (Express strongly and filter. Or this tincture may be prepared by percolation, E.)

Action. Uses. Tonic adjunct to draughts and mixtures, in doses of fʒj.—fʒiv.

AURANTII OLEUM, L. E. Oleum e Floribus destillatum. Volatile Oil of the Flowers, E. Oil of Orange Flowers. Oil of Neroli. In the list of Mat. Med. L. E. P.

A Volatile Oil being secreted in the flowers and other parts of both kinds of Oranges, is separated by distilling them with water. This is well known in France by the name of Neroli. It has a sweet aromatic odour different from that of the flower, and appears to Soubeiran to be a modification of the natural essential oil. Neroli contains a solid crystallizable oil, which has been called *Aurade* by Plisson, who discovered it. The Neroli obtained from the Bitter Orange is finer than that obtained from the Sweet Orange. But essential Oil, known as Oil of Orange, is also obtained by distillation from the leaves of the Orange, and also by expression of the grated rind.

AQUA FLORUM AURANTII, L. AURANTII AQUA, E. Orange Flower Water.

Prep. Take *Orange Flowers* ℔x. *Proof Spirit* f℥vij. *Aq.* Cij. Distil a gallon. (In the list of Mat. Med. of the E. P.)

Procured by the same process as that by which the Essential Oil is obtained, particularly in Italy and France. Besides Essential Oil, it also contains some Acetic′. It may be prepared extemporaneously by agitating some of the Volatile Oil with distilled water, and then filtering. But it is usually imported.

Mr. Squire (Br. An. of Med. i. p. 15) discovered that Orange Flower Water, which is imported from France in vessels of lead or Copper soldered with lead, contains some of the latter metal. This may be detected on the addition of a soluble Iodide, golden-coloured crystals of Iodide of Lead being deposited. He recommends the purification of the water by the immersion of a piece of Zinc wire, and then testing with Iodide of Potassium. The E. P. gives as its characteristics —"Nearly colourless ; unaffected by Sulphuretted Hydrogen." If either metal is present, a blackish-coloured precipitate of Lead or Copper will be produced.

Action. Uses. The Essential Oil is stimulant and antispasmodic. Orange-flower water is considered in France to be possessed of anodyne and antispasmodic properties, and is in constant use in doses of f℥j.—f℥ij. in nervous and hysterical cases.

AURANTIUM. FRUCTUS, L. AURANTII FLORES, L. AURANTII OLEUM ; Oleum è Floribus destillatum, L. Volatile Oil of the Flowers, Neroli Oil. E. AURANTII AQUA ; Orange Flower Water, E. Fructûs succus et tunica exterior. Flores. Folia, D. CITRUS AURANTIUM, *Risso*, The Common or Sweet Orange.

Like the Lemon, this is a native of India, being found in the forests on the borders of Silhet, and also on the Nielgherries, perhaps also in China. The Sanscrit *Nagrunga* and the Arabic *Narunj* are no doubt the European names of *Naranja* (Spanish), *Arancia* (Italian), whence we have *Aurantium* and *Orange.* The Orange is not mentioned either by the ancients or the Arabian medical authors. It is supposed to have been introduced into Europe after the middle ages.

The Orange-tree attains a height of 16 or 20 feet, and bears great abundance of fruit. It is remarkable, as well as others of the genus, for bearing the fruit at all ages at the same time with the flowers. Though a native of India, it does not ripen its fruit there until the winter, and hence has been able to travel so much further north than others of its compatriots. Leaves coriaceous, ovate-oblong, acute ; margins usually finely toothed ; petioles margined, sometimes winged. Petals 5, white. Stamens about 20, 5 of them often distinct and appressed against the stigma, the remainder in 5 bundles alternating with them. Fruit globose, rind thin, with convex oil vesicles, adhering loosely to the pulp, which is sweet.

Oranges are cultivated in the south of Europe and in the Azores, whence they are largely imported into this country. The parts which are officinal are the flowers, L. D., and their essence, L., also called Neroli Oil and distilled water, E. Juice of the Fruit and Leaves, D.

Aurantii Folia. Leaves of the Orange are officinal in the D. P. on account of their bitterness and the aromatic properties of the essential oil stored up in the vesicles with which their substance is studded. This essential oil may be obtained by distillation, as also a distilled water. Or an infusion of the leaves may be employed as a bitter and aromatic excitant and diaphoretic.

Aurantii Fructus. The ripe fruit of the Orange is well known for its extremely agreeable and refreshing juiciness, whence it is so much esteemed as a fruit even for the sick, and as a refrigerant. When of a small size, the fruit which falls off is dried, and forms the *Aurantii baccæ* or Curaçoa Oranges, so called from being employed in flavouring Curaçoa. The smaller ones are smoothed, and used for making issues. The rind or peel of the fruit is sometimes substituted for that of the Bitter Orange, as are also the flowers and their essential oil, also the oil expressed from the grated rind, likewise Orange flower water; all being used for the same purposes as those produced from the Bitter Orange.

CITRUS LIMETTA, *Risso.* var. BERGAMIUM. BERGAMII OLEUM, L. Oil or Essence of Bergamot. Oleum è fructûs cortice destillatum, E. BERGAMOTÆ OLEUM, E. Volatile Oil of the rind of the Fruit, E. Fr. *Limette et Bergamotte.*

This is the species which yields Oil of Bergamot. It appears to belong to the same species as the *Citrus acida* of Roxburgh, as this comprehends under it varieties of the sour Limes as well as the sweet Limes found in India. Latterly Risso and Poiteau have separated the C. Limetta from the C. Bergamia.

The leaves of the latter are oblong, more or less elongated, acute or obtuse, under side somewhat pale. Petiole more or less winged and margined. Flowers usually small, white. Fruit pale yellow, pyriform or depressed: rind with concave vesicles of oil: pulp more or less acid.—Wight and Arnott, Prod. p. 96.

Bergamot is the Volatile Oil of the rind of the fruit of the above variety, which is cultivated in the south of Europe, especially in the neighbourhood of Nice. Raybaud states that 100 fruits yield 2½ ounces of the oil by expression, which has a density of 0·88, is of a pale yellow colour, and very fragrant. It differs from the other volatile oils of this genus in containing Oxygen. It is believed to contain a mixture of oils, having the composition of Citrene (p. 298) with a Hydrate of such an oil, and an oxygenated oil formed by the action of the atmosphere. (*Liebig.*) From its agreeable fragrance, it has been much employed by perfumers, and has lately been made officinal in the L. and E. P. for the purpose of making an agreeable addition to mixtures and unguents. It may be substituted for the Oil of Lemons, or this may be used instead of that.

CITRUS MEDICA, *Risso.* The Citron. Fr. *Cedrat.* Fructûs succus, tunica exterior et ejus Oleum, D. Limones, E. The Lemon is probably intended by both Colleges.

Citrus Medica of botanists is the Citron, distinguished by its large ovoid fruit, with extremely thick rind and proportionately small quantity of acid juice. The name is erroneously adduced if intended to indicate the Lemon. Of the Citron Dr. Roxburgh states, that there are three varieties or species in the Calcutta Botanic Garden, reared from seeds obtained from the Garrow Hills, where they are found indigenous in the forests. The author has also found in the forests of the Deyra Valley, in 30° N. lat., apparently wild specimens of the Citron. It is the *Atruj* of the Arabs, who quote from Dioscorides the description of Μηδικα (μηλον μηδικον of Theoph.), Medica Mala or Median Apples. This seems to have been the only species of Citrus known to the ancients, and is that probably mentioned in the Bible by the name Tappuach, translated Apple in the English version. The rind is thick and spongy internally, tuberculated externally, and covered with numerous dots filled with essential oil (*huile de cedrat*). Its pulp is less acid and juicy than that of the Lemon. The rind of the fruit is preserved, and its essential oil separated; the juice may be employed for the same purposes as that of the Lemon.

LIMONES, L. E. D. Fructus, L. Fruit. Lemons (and Limes), E. *v.* Citrus Medica, D. LIMONUM CORTEX. Fructûs Cortex exterior, L. Lemon Peel. LIMONUM OLEUM. Oleum è Fructûs Cortice exteriori destillatum, L. Oil of Lemons. LIMONUM SUCCUS, L. Lemon Juice. CITRUS LIMONUM, *Risso.* Lemon. Fr. *Citron* ou *Limon.*

Lemons were unknown to the ancients and also to the Arabs, though noticed in Persian works on Materia Medica by the names *Leemoo* and *Neemboo*, and stated to be natives of India, where they are indigenous, and known by nearly the same names. The author has found the tree apparently wild in the forests at the foot of the Himalayan mountains. The annexed figure (47) is from one of these plants. Lemons are not the produce of the Citrus Medica, as seems to be implied in the D. P. Limes are produced by a distinct species, *Citrus acida.*

They form shrubs of from 10 to 15 feet in height, much branched, with stiff awl-shaped thorns. Leaves oval, oblong-oval; margin serrulate, or slightly toothed; petioles with a narrow leafy border, or simply margined. Flowers with 5, sometimes 4 petals. Stamens 20 to 30, in 4 or 5 bundles. Fruit of a light yellow colour when quite ripe, ovoid in shape, with a more or less nipple-like knob at the apex. Rind thin with numerous vesicles of oil, adhering closely to the pulp, which is very acid.

Though Lemons are originally natives of India, they are now imported into this country from the south of Europe and the Azores, each being separately rolled up in paper. The best plan " consists in

Fig. 47.

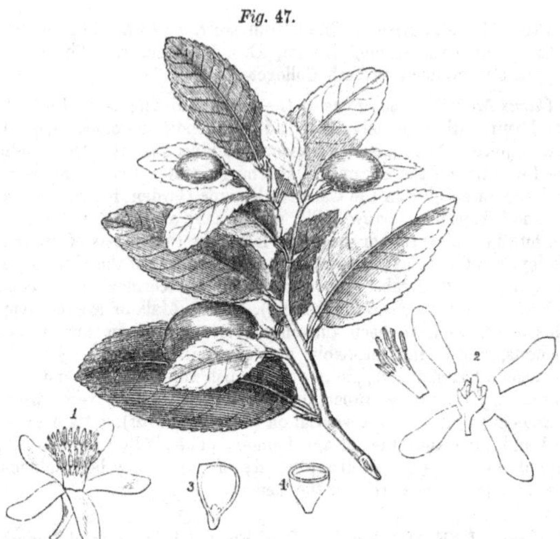

packing them with newly slaked lime in bottles or earthern-ware jars, the mouths of which are secured with corks and wax." (*c.*)

LIMONUM CORTEX. Lemon Peel is of a light yellow colour, but becomes of a brownish hue when dried. It is bitter and aromatic, from containing some Bitter Extractive which is insoluble in Ether, but soluble in Alcohol; and abundance of fragrant Volatile Oil stored up in the numerous vesicles with which the rind is studded. It forms an agreeable addition to different tinctures and infusions, and is an ingredient in the *Inf. Aurantii Compositum*, and *Inf. Gentianæ Compositum*, L.

LIMONUM OLEUM. Oil of Lemons. This, like the Oil of Orange, may be obtained either by distillation, or by simple expression of the finely grated rind. The latter is of the finest quality, of a light colour, and fine lemon odour, warm penetrating taste. Sp. Gr. 0·848 to 0·85 and higher, boils at from 330° to 353°, being a mixture of two oils, which may be separated by distillation to a certain extent. One of them, Citrene, has the Sp. Gr. 0·847, and boils at 330°; the other, Citrelene, has the Sp. Gr. 0·88, and boils at 345° to 353°. (Liebig, in *Turner's Chem.*) Both these and the Oil of Lemons have the composition of Oil of Turpentine, and are probably composed of $C_{10} H_5$.

They are therefore, when pure, true Hydrocarbons. Oil of Lemon absorbs Oxygen when exposed to the atmosphere. Acted on by Hydrochloric acid, it forms two compounds, one a liquid, the other an artificial Camphor composed of $C_{10} H_8 + H Cl$. The Oils of Oranges and of the Citron are identical in composition with Oil of Lemons.

LIMONUM SUCCUS. Lemon Juice is obtained by subjecting the pulp, freed of its rind and seeds, to pressure, either on a large scale, or for ordinary purposes. It is allowed to stand for a few days in a cool place, and then decanted and filtered. It however remains a little turbid, is sharply acid, with an agreeable flavour and a little of the odour of the Lemon. It consists of Citric acid (about 1·77 per cent.) dissolved in water with mucilage and extractive. It is apt to undergo decomposition, but with care may be preserved for a considerable time, as by corking up in full bottles the above juice, or pouring a layer of almond oil above it ; some subject it to a slight ebullition, or concentrate by freezing : " The British navy is supplied with it from Sicily, preserved by the addition of $\frac{1}{10}$ of strong brandy ; druggists in this country preserve it by adding about $\frac{1}{10}$ of spirit of wine, and filter off the mucilage which separates." (*c.*) A substitute may be formed for it by dissolving ℨxjß. of *Cit'* in *Aq. dest.* Oj., and flavouring with the smallest quantity of Oil of Lemons. (*Phillips.*)

Action. Uses. Refrigerant, Antalkaline, Antiscorbutic. Diluted with water, it forms a refreshing drink in hot climates, or in febrile and inflammatory complaints, made more agreeable with sugar, in the well-known form of Lemonade, or added to barley, rice water, &c. It is much employed in making effervescing draughts; in imitation of which *Effervescing Lemonade* is prepared by the soda-water manufacturers. The Citrates and Tartrates are converted into Carbonates in passing through the system, and will produce an alkaline reaction on the urine. Antiscorbutic: hence about ℥j. or ℥ij. are distributed to seamen in long voyages as a preventive, but f℥iv.—f℥vj. for the cure of Scurvy, or Citric acid is substituted for it.

Artificial Lemon Juice, may be made by dissolving Citric' or Tar' ℥iiß. Gum ℥ß. Fresh Lemon-juice ℨvj. fine Sugar ℥ij. Aq. ferv. Oij. Allow it to cool, and strain.

Lemonade. Macerate 2 *Lemons* sliced and *Sugar* ℥ij. in *Aq. ferv.* Oj. till cool, and strain.

Aerated or Effervescing Lemonade. Mix *Water* Oj. charged with five times its volume of *Carb' gas* with *Syrup of Lemons* f℥ij.

SYRUPUS (LIMONIS, D.) LIMONUM, L. E. Syrup of Lemons.

Prep. Take *fresh Lemon Juice* Oj. (℔ij. by measure, D.) having allowed the impurities to subside (subject it to the heat of boiling Aq. for ¼ of an hour and pass through a sieve, D.) dissolve in it with aid of gentle heat *Sugar* ℔ijß. (℥lviij. D.) Set it aside for 24 hours and pour off the clear liquor.

Action. Uses. An agreeable addition to diluent drinks or to draughts, in doses of fℨj. to fℨiv.

ACIDUM CITRICUM, L. E. D. Citric Acid. Concrete Acid of Lemons. *F.* Acide Citrique. *G.* Citronensaüre.

Citric Acid is that which gives the sour taste to the juice of the Lemon and Lime, and has been so named from Citrus. It is also contained in the juice of some other fruits, as in acid Grapes, in Tamarinds, in the Gooseberry, Red Currant, Cranberry, Bird-Cherry, usually mixed with some Malic acid, sometimes combined with Potash or with Lime. In the juice of Lemons and Limes it is in a free state, mixed only with mucilage and similar vegetable impurities, which prevent its crystallizing. It was first separated from these and obtained in a solid form by Scheele in 1781. To separate the Citric acid from the admixtures, Lime is presented to it, with which it combines, and is precipitated in the form of Citrate of Lime. This is separated and decomposed with diluted Sulphuric acid, when an insoluble Sulphate of Lime is formed, and the Citric acid becomes dissolved. Lemon Juice is sometimes imported instead of the Lemons, and the acid often in the form of Citrate of Lime.

Prep. E. D. Take of *Lemon Juice* Oiv. and gradually add to it made hot, *Prepared Chalk* ℥ivß. (or a sufficiency, E. D.) and mix. Set by, that the powder may subside; (Carb′ escapes and insoluble Citrate of Lime subsides, whilst most of the mucilage remains in solution;) pour off the supernatant liquor. Wash the Citrate of Lime frequently with warm water. (To get rid of the mucilage and other impurities.) Then pour upon it *Dil. Sul′* f℥xxvijß. (8 times the weight of the Chalk employed, D.) with *Aq. dest.* Oij. and boil for a quarter of an hour. (The S′ decomposes the Citrate, forms an insoluble Sulphate of Lime, and the Citric acid becomes dissolved.) Press the liquor strongly through linen, and strain it; evaporate the strained liquor with a gentle heat, and set it by, that crystals may be formed. (These crystals are of a dull brownish colour, and are to be rendered colourless only by a repetition of the last process, and therefore) Dissolve the crystals, that they may be pure, again and a third time in water, and strain the solution as often; boil down and set it aside.

The E. C. in their corrected edition order f℥xxxvj. of their diluted Sulphuric acid. Dr. C. says, in the proportion of nine parts of the concentrated acid for every ten parts of Chalk used. They also direct the juice to be boiled in the first instance, then set to rest and boiled again, before the Chalk is added. "It has been found of service for the subsequent purification of the acid to commence by clarifying the juice with Albumen." To ascertain that there is neither Sul′ nor Citrate of Lime in excess "Try whether a small portion of the liquid, when filtered, gives with solution of Nitrate of Baryta a precipitate almost entirely soluble in Nitric acid; and if the precipitate is not nearly all soluble, add a little Citrate of Lime to the whole liquor till it stands this test." Excess of Citrate is of course to be decomposed by an addition of Sulph′. The liquor and washings being concentrated to the density of 1130, the product is removed into shallow (leaden) vessels and evaporated until a pellicle begins to form. By a repetition of the process, the solution and crystals lose their colour. Berzelius recommends the addition of a little Nitric acid in the last step of the boiling.

Citric acid ($C_4 H_2 O_5$ or ($C_{12} H_5 O_{11}$ *Liebig*)$=$Cit′) is colourless and transparent, without odour, of a strong but agreeable acid taste, crystallizes in transparent short rhomboidal prisms terminated by four planes, apt to become moist in damp air, soluble in three-fourths of cold and half its weight of hot water. The solution spoils when it has been some time kept, becoming ropy from spontaneous decomposition. Cit' is also soluble in Alcohol. When heated with Sul', it is resolved into Carbonic oxide, Carb', Ac', and water. Nit' converts it into Oxalic', and when melted with caustic Potash, Ox', Ac', and water are produced. It is fused in its own water of crystallization, and at a higher temperature is decomposed. When obtained at ordinary temperatures, it crystallizes with 5 Eq. of water, two of which are water of crystallization; but when deposited from a solution cooled from 212°, the crystals contain only 4 Eq. Aq. three of which are basic, and 1 Eq. water of crystallization. The effects of heat on Citric' have been studied by several chemists. The decompositions according to Liebig (*Turn. Chem.* 1005) have been cleared up by Crasso. "Crystallized Citric' when exposed to heat, exhibits four stages of decomposition. During the first, the water of crystallization alone is given off, and the residue contains unaltered Cit'. The second stage is characterized by white vapours, and the production of Acetone, Carb. oxide, and Carb', while the residue consists of Hydrated Aconitic acid, which is therefore the true Pyrocitric'. In the third stage, the Aconitic', not being volatile, is itself decomposed, yielding Carb' and an oily liquid (Citricic', *Baup*), which Crasso proposes to call Itaconic acid. In the fourth period, empyreumatic oil is produced, and a voluminous coal remains behind. Citric acid forms numerous salts: those of the alkalis are soluble, and often prescribed in the form of effervescing draughts. The Citrates of Iron, also soluble, have already been mentioned at p. 147. The Citrates of Baryta, Strontian, Lime, Lead, and Silver are insoluble. If added in excess to Lime-water, no precipitate is observed until it is heated. ʒj. Cit' will saturate ʒij. of crystal. Carb. Soda.

Tests. Citric acid is apt to be adulterated with Tartaric acid; but the latter is easily detected by any of the soluble salts of Potash.

"A solution in four parts of water is not precipitated by Carbonate of Potash, E. No salt of Potash, except the Tartrate, is precipitated by solution of Citric'. What is precipitated from the solution by Acetate of Lead is dissolved by Nitric', L. It is totally dissipated in the fire, L., with the aid of the red Oxide of Mercury, E.

Inc. Alkalis and earths, Carbonates, most Acetates, Tartrate of Potash.

Action. Uses. Refrigerant, Antiscorbutic, Anti-alkaline. Substitute for Lemon Juice; employed for making effervescing draughts.

℈j. of the following Salts will saturate	Lemon Juice, or Sol. Cit'.	Citric Acid.	or Citric Acid ℈j. saturates.
Bicarb. Potash . .	fʒiijß.	gr. 14	29 grs.
Carbonate Potash .	fʒiv.	gr. 17	24 grs.
Sesqui-Carb. of Ammonia	fʒvj.	gr. 24	17 grs.
Carbonate of Soda .			41 grs.
Sesqui-Carbonate of Soda			24 grs.

GUTTIFERÆ, *Jussieu.* (Clusiaceæ, *Lindl.*)

Trees, rarely shrubs, sometimes parasitic climbers, abounding in yellow resinous juice, branches opposite, often four sided. Leaves decussately opposite, coriaceous, shining, without stipules. Flowers perfect or polygamous, usually terminal, sometimes axillary, articulated with their peduncles. Calyx 2, 4 or 6 leafed, persistent. Petals equal in number to leaflets of the calyx, alternating with them or opposite, occasionally numerous. Stamens usually numerous, free, united into a ring, or tube, or into separate bundles. Disk fleshy, angled or lobed. Ovary one to 5 or many-celled. Ovules solitary or twin, erect or ascending, in a single-celled ovary, about 4 or many attached to central placentæ. Stigma peltate or radiate. Fruit capsular, drupaceous, or baccate. Embryo without albumen, straight.

The Guttiferæ are allied to Ternstrœmiaceæ and to Hypericineæ, and are found in the tropical parts of Asia and of America. Many of the species yield a yellow resinous juice like Gamboge, useful both as a pigment and as a medicine. The fruit of some is edible, the seed oily, and the wood hard and useful as timber.

CAMBOGIA, L. Cambogia, D. Gummi-resina. (Stalagmites (Cambogia, D.) Cambogioides, L.) Gamboge. Cambogia (zeylanica) Ceylon Gamboge. Gummy resinous exudation of Hebradendron Cambogioides, *Gr.* E. C. (Siamensis). Siam Gamboge: probably from a species of Hebradendron inhabiting Siam, E.

Gamboge is stated by Murray (App. 4, p. 110) to have been first introduced to the notice of Europeans by Clusius, who received it from China in 1603. It is known in India by the name of *ossareh rewund*, or juice of Rhubarb. This substance is mentioned in Persian works on Materia Medica; but we are unable to ascertain when Gamboge came to be substituted for the real Extract of Rhubarb, which Dr. Falconer informs me he obtained in Tibet by the same name. Two kinds of Gamboge, the Siam and the Ceylon, are known in commerce. The former is commonly in cylinders, either solid or hollow. Specimens of both kinds were given to the author by G. Swinton, Esq. when Chief Secretary of the Indian Government, which had been sent to him officially from Bankok, as the produce of Siam. This form is no doubt given, by the Gamboge when in a fluid state being run into hollow bamboos, as described by Lt. White. I am indebted to Dr. Pereira for one of these imported a few years since. Kœnig learnt from a Catholic priest, who officiated as such to the Christians of Cochin-China, that the juice was obtained by breaking off the leaves and young shoots, and receiving the yellow juice as it issues in drops in suitable vessels, a cocoa-nut or a bamboo; also, that it formed a part of the tribute paid to the king of Siam. It is therefore most probably abundant, perhaps cultivated. The tree yielding this Gamboge is unknown to botanists. It may be the *Oxycarpus* (now Garcinia) *cochinchinensis* of Loureiro, who describes it as being both wild and cultivated in Cochin-China; and Rumph. (iii.

p. 58) describes it as exuding when wounded a yellow and viscid juice, which quickly dries up. Other species of Garcinia yield a yellow resinous juice. Dr. Malcolmson favoured the author with specimens of one of the Guttiferæ, which he collected near Rangoon, and of which the rind of the fruit yielded a yellow purgative juice.

STALAGMITES CAMBOGIOIDES, quoted in the L. P. &c., has been founded on a factitious specimen, which is still in the British Museum. This Mr. Brown ascertained to be formed of two plants joined together by sealing-wax, one being *Xanthochymus ovalifolius*, Roxb., and the other *Hebradendron cambogioides* of Graham.

The Ceylon Gamboge is found in the Bazaars of India, but is seldom met with in Europe. Mr. Charles Groves, now of Liverpool, informed me in 1832, that when engaged in the trade of Ceylon, he had sent a considerable quantity of the Gamboge of that island to London; but it was found to be unsaleable, from its inferior quality. Two trees yielding a Gamboge-like substance were first made known by Hermann in 1670: one, Goraka, *Garcinia Cambogia;* and the other, *Kana* (or eatable) *Goraka*, *Garcinia Morella* of later authors, *Stalagmitis* of *Moon's Cat.*, now *Hebradendron Cambogioides*. The latter (though it might have been referred to Garcinia with an amended character) was named and described by my friend, Professor Graham* of Edinburgh, from specimens and drawings sent him by Mrs. Col. Walker, who had seen the tree in different parts of the island of Ceylon. Col. Walker writes to Dr. Wight, that it is found in great abundance along the western and eastern coast in the neighbourhood of Battacola, but also inland, especially in low sandy ground, about Kanderaane, Negombo, and towards Chilau; also 100 miles inland, at so high an elevation as 2000 feet above the sea. Mrs. W. says, the Gamboge is collected by incisions into, or by cutting pieces off the bark about the size of the palm of the hand, early in the morning. The Gamboge oozes out in a semi-liquid state, but hardens on exposure to the air, and is scraped off by collectors next morning. She describes it as brilliant and excellent, and as good for water-colour drawing as any she ever used. Dr. Christison has shown that it has all but an identity of composition with that of Siam; and its medicinal effects were considered precisely the same by Dr. Pitcairn in Ceylon, and by Drs. Graham and Christison in Edinburgh. That procured in Indian bazaars, which is spongy in structure, was not found to be so good as a pigment by the E. I. Company's painters, when under the author's charge; nor did he find it so effective as a purge, in the hospitals at Saharunpore. Dr. Graham ascribes its inferiority, probably with truth, to the want of care in preparing the article for market.

* Since the above was written, we have to lament the death of Professor Graham, who was as much loved for his virtues as respected for his character.

The genus HEBRADENDRON has diœcious flowers. ♂ Calyx membranaceous (1) sepals 4 persistent. Petals 4. Stamens (2) monadelphous, column 4-sided, anthers terminal, (3) opening by the circumcision (4) of a flat and umbilicate terminal lid. The inflorescence of the female tree is similar to that of the male. Its flowers white and a little larger, with a germen in miniature of the fruit, and surrounded like it with several (ten?) abortive stamens; crowned by a lobed and muricated sessile stigma. The berry (5) is many or 4-celled, cells one-seeded. Cotyledons fleshy, united. Radicle central, filiform. Trees with entire leaves.

Fig. 48.

H. cambogioides (fig. 48) forms a moderate sized tree, with the leaves obovate, elliptical, abruptly subacuminate, the male flowers clustered in the axils of the petioles, on short single flowered peduncles. Sepals yellow on the inside, yellowish white externally. Petals yellowish white, red on the inside near the base. Berry about the size of a cherry (5), round, firm, with a reddish brown external coat, and sweet pulp. Ripe in July.

Besides the above species, there is probably another belonging to the same genus, or to the same group of a larger genus, which appears to yield a very good kind of Gamboge, and one which may prove a good substitute for either the Ceylon or Siam kind,—and that is *Garcinia pictoria*, Roxb. Fl. Ind. ii. p. 627. Dr. Roxburgh says, "I have frequently received samples of the Gamboge the produce of this tree, from my good correspondent, Mr. S. Dyer, the Surgeon at Tellicherry, and I have uniformly found it, even in its crude unrefined state, superior in colour, while recent, to every other kind I have yet tried; but not so permanent as that from China." This, Mr. Dyer, who is now in London, informs me he cannot understand; for he found it excellent as a pigment, and effective as a purgative, and, as far as he remembers, equal to the Gamboge then in common use. As Mr. Dyer has favoured me with a full-sized coloured drawing (*) of the foliage and fruit of this species, a woodcut is annexed, as it evidently belongs to the same genus as the above, to which indeed it has been referred with a query by Dr. Graham, and by Dr. Lindley in his Flora Medica, p. 114, where he has reprinted Dr. Roxburgh's description.

H. pictorium (fig. 49) is a tall tree with a pretty thick bark having considerable masses of gamboge on its inside. Leaves with short petioles, oblong ventricose, rather acute, from 3 to 4 inches long by 1½ or 2 broad. Flowers yellow, axillary solitary. Calyx (2) permanent of 2 pairs of concave obtuse sepals. Petals four. Stamens from 10 to 15, with their filaments united into four bodies, which are again united at the base into a narrow ring. "Anthers of the male flower 'peltate,' of the female 2 lobed and seemingly fertile." Germ superior, round, 4-celled, (3) one ovule in each attached to the axis a little above its middle. Stigma 4-lobed, permanent. Berry (1, 2, 3,) size of a large cherry, oval, smooth, very slightly marked with four lobes, crowned with the sessile, 4-lobed verrucose, permanent stigma. Rind leathery, of a reddish colour. Seeds 4, when all ripen (4, 5, 6,) oblong reniform. The filaments in the male flowers are described as being numerous and the anthers peltate. A native of the Malabar and Wynaad jungles.

Dr. Wight, who has paid considerable attention to the characters of the genera and species of the Guttiferæ, has in his Illustrations of Indian Botany,

Fig. 49.

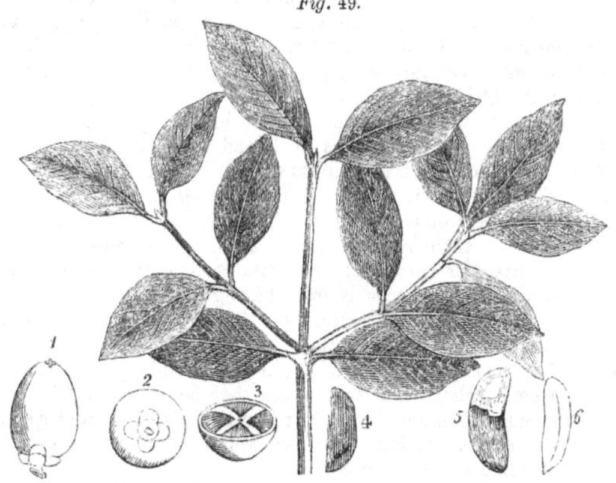

* While this sheet is passing through the press I have received a letter from Dr. Christison, informing me that he had at length got the Coorg or Wynaad Gamboge, from Mrs. Gen. Walker, but not the plant yet, though it has been seen in the jungles near Cannanore. He supposes it must be that of which the above is a wood-cut, and which Dr. C. saw when in London. This Gamboge has the composition of that of Siam, but with less gum, is "a capital purgative, and makes an excellent pigment, not fugacious as Roxburgh says." This is fully confirmed by a specimen of paper coloured with it which Dr. C. has had the goodness to send me. The new information will form a very desirable addition to Dr. C.'s admirable paper on the Gamboge.

p. 126, referred both of the above species to Garcinia, section Cambogia. Of the last species he says, "Though I consider this a distinct species, I am unable from an examination of Roxburgh's drawing and description to assign better characters. The difference of the anthers of the female flower affords the best mark, which in the former are, like the male, "peltate," in this 2 lobed and 2 celled, (the ordinary structure) and of course thus reduce the value of that character as a generic distinction.

It is evident that the foregoing facts respecting the Ceylon and Indian plants, afford no information regarding the plant which yields the Gamboge of Siam. Specimens of this might probably be obtained from Bankok by some of our countrymen at Sincapore. It is probably nearly allied to the above, as Dr. Christison has ascertained that the Gamboge of Siam is as nearly as possible identical in composition and properties with that of Ceylon. He indeed infers that the plant may possibly have been introduced from Siam with the religion of the Buddhists. It is well known, however, that the Buddhist religion travelled in an opposite direction, that is, from India and Ceylon to Siam, &c.

Prop. Siam Gamboge is usually seen (1) in cylinders, either solid or hollow in the centre, whence it is commonly called *pipe Gamboge*, varying in length, and in thickness from ½ to 2 inches, striated externally, evidently from the impressions of the bamboo mould into which it was run when soft. Sometimes these cylinders are doubled upon themselves, at others stuck together, all generally of fine quality. (2.) *Lump* or *Cake Gamboge* in round cakes or masses, several pounds in weight, most commonly inferior in quality to the former, and often mixed with impurities, as fecula and woody fibre. 3. *Coarse Gamboge*, formed of the fragments and inferior pieces of the other, which are, however, often mixed with impurities, and not entirely soluble in ether and water.

Ceylon Gamboge, though unknown in European commerce, is sometimes seen in irregular masses, often cavernous, or with many sinuous hollows, like the sponge, probably from having oozed out irregularly; the colour a uniform yellow, except on the parts exposed to light, where it is darker; brittle in texture. There seems no difficulty in obtaining it in a pure state, and if so, it might become an article of commerce from Ceylon. The pure pieces were found by Dr. C. to be identical in composition and purgative properties with the Gamboge of Siam. The specimen in King's College Museum was given to the Author by the late Dr. Malcolmson.

Gamboge is without odour, and has very little taste; but after a short time a little acridity and uneasiness are experienced in the fauces, and the fine dust, raised in pulverizing it, irritates the nostrils, so as to produce a flow of mucus. It is very brittle. "Fracture somewhat conchoidal, smooth, and glistening: a decoction of its powder, cooled, is not rendered green by Tincture of Iodine, but merely somewhat tawny," E. P., showing the absence of Starch. The colour becomes of a bright gamboge-yellow whenever it is rubbed, "and readily forms an emulsion or paste of the same hue when wetted and rubbed." A portion is dissolved by water, and the remainder forms a perfect emulsion, which is not easily deprived of its colour by filtration. Rectified Spirit dissolves a large portion, Ether about four-fifths, leaving only Gum, which has been called *Arabin*, from being the kind of which Gum Arabic is composed, and which has the composition of flour of Starch. The Resin dissolved by the Ether has been called Gambodic acid by Prof. Johnston. Its qualities are those of a fatty acid. (*Buchner.*) It may be obtained pure, and of a fine reddish-yellow colour, by distilling off the Ether. It will impart its colour to 10,000 times its weight of Spirit or water. Like other Resins, it is dissolved by solution of Potash (forming Gambodiate or Gambogiate of Potash, of a deep-red colour) as well as by the other caustic alkalis, from which it may again be separated by the addition of an acid. Comp. $C_{40} H_{23} O_8$ (*Johnston*). It also contains a little of a peculiar red-yellow colouring matter soluble in water and Alcohol.

Exposed to heat, it burns with a white flame, emitting much smoke, and leaving a spongy charcoal. In 100 parts of it, Braconnot found 19·5 parts of Gum, 0·5 of impurities, and 80 of a red, insipid, transparent, resinous substance, becoming yellow by pulverization.

The latest analysis is that by Dr. Christison:

Pipe Gamboge of Siam.		Cake Gamboge of Siam.		Ceylon Gamboge.	
Resin	72·2	Resin	64·8	Resin	75·5
Arabin	23·0	Arabin	20·2	Arabin	18·3
Moisture	4·8	Fecula	5·6	Cerasin	0·7
		Lignin	5·3	Moisture	4·8
		Moisture	4·1		

Tests. The characteristics, E. P. of good Gamboge have been given above. Iodine will detect Starch. Mechanical impurities can be seen. In external appearance it can only be confounded with the yellow resinous juices of some others of the Guttiferæ; of these that of Garcinia Cambogia, as described by Dr. Christison, is soft, of a pale lemon-yellow colour, and incapable of forming an emulsive paste with the wet finger. That of Xanthochymus pictorius has a pale yellowish-green colour and some translucency, and is not at all emulsive; also, as observed by Dr. Pereira, it may be confounded with *yellow gum* or rather resin of the Grass-tree, Xanthorrhæa hastile of New Holland. The presence of Gamboge may be detected by the effects produced by it on water, Alcohol, Ether, and caustic Potash. The *Gambogiate of Potash* gives, if the alkali be not in excess, with acids a yellow precipitate (*Gambogic acid*), with Acetate of Lead a yellow precipitate (*Gambogiate of Lead*), with Sulphate of Copper, a brown (*Gambogiate of Copper*), and with the salts of Iron, a dark brown precipitate (*Gambogiate of Iron*). (*p.*)

Action. Uses. Drastic Hydragogue, Purgative, Anthelmintic. Useful in obstinate costiveness, Amenorrhœa, Dropsy. Better given in combination than by itself, as in the following pill, originally introduced by Dr. G. Fordyce, and to which Morison's Pills are similar, with the objectionable addition of Cream of Tartar.*

D. gr. ij.—gr. v. in combination with Calomel, Scammony, &c.

PILULÆ CAMBOGIÆ COMPOSITÆ, L. PIL. CAMBOGIÆ, E. PIL. GAMBOGIÆ COMP. D.

Prep. Mix powdered Gamboge ʒj. (1 part E.) powdered *Aloes* (East India or Barbadoes, E.; Hepatic, D.) ʒiß. (1 part, E.) powdered *Ginger* ʒß. L. D. (Aromatic powder 1 part, E.) Add *Soap* (Castile, E.) ʒij. (2 parts, E.) and beat together (with Syrup q. s. E. treacle, D.) into a uniform mass.

D. gr. v.—℈j.

CANELLACEÆ, *Martius.*

Usually appended to Guttiferæ or to Meliaceæ.

CANELLA ALBA, *Murray.* Cortex, L. D. Bark, E. White Canella. *Dodecandria Monogynia*, Linn.

The name Canella, a diminutive of Canna, was at one time applied to the Cinnamon, whence its French name Canelle. When the present Canella was discovered in South America, it was supposed to be the true Cinnamon, and called by its then name. The earliest full, though not the first account was given by Monardes (Clus. Exot. p. 323), who states that in 1540 an expedition was sent by Pizarro to examine the province Cumaco, where this Cinnamon was said to be found. It was long confounded with Winter's Bark, and at one time called *Winterania Canella* and *Spurious Winter's Bark*, though both had been clearly distinguished by Sir Hans Sloane in Phil. Trans. 1692. v. Fig. 50.

Canella alba is a tree which is common in many parts of the West India Islands and in South America, frequently on the sea coasts, where it seldom exceeds twelve or fifteen feet, but in the inland forests it attains a more considerable height. It is propagated chiefly by wild pigeons feeding on its berries. The tree has a straight stem and branched top, and a good deal resembles the Pimento.

The bark is whitish, so that the tree is at once distinguished from others in the woods. The leaves are petiolate, alternate, but not regularly so, obovate, the younger ones pellucido-punctate, the older smooth, shining, of a thick consistence, without nerves, very entire and exstipulate. The flowers are arranged in terminal corymbs, small and of a violet colour, but seldom open. Sepals 3, imbricate, roundish. Petals 5, hypogynous, oblong, twisted in æstivation. Stamens united into a subcylindrical tube. (1) Anthers 21, linear, fixed longitudinally on the outside of the tube. Ovary free, but included within the stamen-tube, 3-celled. Style cylindrical. Stigma 2 lobed (2). Berry by abortion 1 or 2

---

* In the trial of Morison and others *v.* Harmer and Bell, the late Professor Daniell in analysing twelve of Morison's pills, No. 2, found of Resin of Aloes $5\frac{6}{10}$ grs. Resin of Gamboge $4\frac{1}{10}$ grs. pounded Colocynth 2 grs. Gum $4\frac{7}{10}$ grs. and Cream of Tartar $6\frac{7}{10}$ grs. Mr. Hume of Long-acre, found the same ingredients, with 8 grs. of Gamboge, in 50 grs. of the pills, that is about the same quantity as the pills varied in weight, ten of them weighing 20 grs. while ten others from the same box weighed 27 grs.

Fig. 50.

celled; cells 2—3 seeded; seeds one above the other (3), kidney-shaped, beaked, black and shining. Embryo within fleshy albumen in the beak of the seed, curved and roundish; cotyledons linear, radicle above, centripetal.—Sloane Jam. ii. t. 191, f. 2; Swartz, Lin. Trans. i. vol. viii. p. 102; fructif. Gærtner, i. 373, t. 77.

Prop. The Bark, being the only officinal part, is removed with an iron instrument, and then being deprived of its epidermis, is dried in the shade. It is in flat or quilled pieces, according to the part of the tree from which it has been removed, the thinner pieces drying into the quill form most readily. (Goebel and K. I. tab. iii. fig. 1—3.) The pieces are of a light buff-colour, paler internally; have an aromatic odour, a warm pungent taste, and are brittle, yielding a yellowish white powder. Boiling water takes up some of this bark, but Alcohol only dissolves its aromatic properties, becoming of a bright yellow colour. Distilled water affords a reddish-yellow, fragrant, and very acrid Essential oil, which is often mixed with and sometimes sold for Oil of Cloves. (*Browne.*) Petroz and Robinet also obtained Resin, which is aromatic, Bitter Extractive, a peculiar Saccharine substance, which will not undergo the vinous fermentation, and which has been called *Canellin*, Albumen, Gum, Starch, Lignin, and Salts. It may be distinguished from Winter's Bark by not being precipitated by Nitrate of Baryta, nor by infusion of Galls, nor by Sulphate of Iron, as it does not contain Tannin.

Action. Uses. Aromatic Stimulant. Adjunct to tonic and purgative compounds. Used as a spice in the West Indies.

D. gr. x.—ʒß. of the powder.

Off. Prep. Tinct. Gentianæ Comp. E. Vinum Gentianæ, E. Vinum Aloes, L. D. Pulvis Aloes cum Canella, D.

The HIPPOCASTANEÆ contain *Æsculus Hippocastanum*, or the Horse-chesnut, which being bitter and astringent, was at one time officinal, and employed as a tonic and febrifuge.

The MELIACEÆ, a tropical family, distinguished by the filaments of the stamens being united into a tube, contain many plants possessed of medicinal virtues. *Soymida febrifuga*, the *rohuna* of India, at one time officinal in the E. P., is a powerful East Indian febrifuge; so also species of *Khaya*, of *Cedrela*, of *Melia*, of *Heynea*, and of other genera, are employed for the same purposes in the countries where they are indigenous.

AMPELIDEÆ, *Kunth.* (VINIFERÆ, *Juss.*) Vineworts.

Shrubs usually twining and climbing, with water-like sap, stem and branches round or angled, with tumid joints. Lower leaves opposite, the upper alternate, stalked, simple, palmate or compound, usually with stipules. Petioles often converted into branched cirrhi. Flowers small, complete, sometimes by abortion unisexual, greenish coloured, mostly in umbels or racemose. Calyx small, entire, or 4 or 5 toothed, lined internally with the disk. Petals 4 or 5, valvate in æstivation. Stamens 4 or 5, opposite the petals, and inserted with them into the margin of the disk. Ovary 2, 3, or 6 celled, Ovules solitary, or two side by side, ascending or erect. Berry pulpy, 2 to 6 celled, or from the partitions not forming, one celled. Cells one or two seeded. Embryo small, erect, with hard albumen.

The Ampelideæ, so called from *αμπελος*, *ampelos*, a vine, are also sometimes called Vites and Vitaceæ, but these names are too similar to Vitex and to Vitices. The family is allied, in some respects, to Araliaceæ, and in others to Meliaceæ through Leea. They abound in the Tropics chiefly of Asia ; a few are found as far north as 30°, and still higher in North America. The species abound in acid with astringent or coloured juice, which is more or less grateful. The saccharine secretion of the Grape makes it highly esteemed as a fruit.

VITIS VINIFERA, *Linn.* L. E. D. Uva. Baccæ exsiccatæ demptis acinis, L. Fructus siccatus, D. Uvæ passæ. Dried Fruit, E. Grape Vine. Raisins. *Pentand. Monog.* Linn.

The Vine was early cultivated in Egypt, Palestine, and Greece. It is probably a native of Persia. It is found wild about Tinkaboon in Deilim about N. lat. 37°, on the southern shores of the Caspian (Royle, *Him.* p. 146). Humboldt also states that it grows wild on the coasts of the Caspian Sea, in Armenia, and in Caramania.

The Vine, like other cultivated plants, varies much in its growth and in the quality of its fruit. It sometimes attains a great size, climbing to the tops of the highest trees in Italy and in Cashmere, and lives to a great age, some vineyards being three or four hundred years old.

The Grape Vine is distinguished among the species of Vitis by having its leaves lobed and sinuato-dentate, naked or tomentose. The calyx is obscurely 5 toothed. The corol composed of 5 petals, cohering at the apex, and like a calyptra splitting at the base and falling off together. Stamens 5. Style wanting. Berry 2-celled, 4-seeded, cells and seeds often abortive. The great diversity in form has been summed up by De Candolle in the following words: The leaves are more or less lobed, smooth, pubescent or downy, flat or curled, pale or deep green. Branches prostrate, climbing or erect, tender or firm. Bunches loose or crowded, ovate or cylindrical; the berries red, greenish or white, watery or fleshy, globose, ovate or oblong, sweet, musky or austere. Seeds often varying in number, or fruit seedless.

Of the Grape-vine there are numerous varieties cultivated in different countries, as well as in the hot-houses of England. When unripe, the fruit is remarkable for the harsh acidity of its juice, which is then called *verjuice.* It owes this property to a little free Citric, Malic, and Tartaric acids, and to the Bitartrate of Potash. It also contains some Tannin and Extractive, as well as some Sulphates of Potash and of Lime, also Malate and Phosphate of Lime. This juice used to be employed in medicine, and still is so for making syrups and sherbets. Lieut. Burnes mentions that in Caubul they use grape-powder, obtained by drying and powdering the unripe fruit, as a pleasant acid.

Grapes as they ripen lose their acid taste, becoming sweet and delicious in flavour. They are wholesome as fruit, both to the sick and to those in health; allaying thirst in febrile affections, and being pleasant nutritious articles of diet. But they are a little acid, from containing Citric′ and Malic′, and some Bitartrates of Potash and of Lime. The sweetness is owing to the formation, at the expense of the acids, of some *Grape-Sugar* or Glucose, which differs from Cane-Sugar in being granular and not presenting crystalline faces, in being less sweet, and less soluble both in water and in Alcohol; differing also in its refractive powers. Composed of $C_6 H_7 O_7$, or $C_{24} H_{28} O_{28}$. When it undergoes fermentation, the whole is converted into Alcohol and Carbonic acid. Grape-juice also contains Gum, Extractive, Colouring matter, and a Glutinoid substance of the nature of ferment or yeast. This juice when expressed, is called MUST (Mustum).

RAISINS. *Uvæ passæ.* Grapes in their dried state are well known as Raisins, and are prepared by being dried in the sun or in ovens, or by steeping them in a weak alkaline ley formed from the ashes of the burnt tendrils. Some are prepared by partially cutting the stalk of the bunches before the grapes are quite ripe, and allowing them to dry upon the vine. They are chiefly prepared in Spain and in the Levant, hence called Valentias and Smyrnas; also in Affghanistan, whence they are taken to India. The best are the Muscatels, from the grape of that name. The Sultanas, like the *Bedanas* of the East, are without stones. The Malaga Raisins are large and fleshy, of a purplish-brown colour. Those of Calabria are similar. The Smyrna Raisins are of a yellowish-brown colour, slight musky odour, less sweet and agreeable than the former. The Corinthian Raisins, or, as they are commonly called, Currants, are produced by a small-sized grape which is abundant in the Ionian Islands.

Raisins differ from Grapes in containing less water and acid, and more Sugar. Besides their dietetical uses, they are demulcent, and are employed for improving the flavour of several officinal compounds as below, also for demulcent beverages. Though nutritious, they are apt to be indigestible.

Off. Prep. Decoctum v. Mistura Althææ (p. 287). Dec. Hordei Comp. Dec. Guaiaci Comp. Tinct. Cardamomi Comp. Tinct. Quassiæ Comp. Tinct. Sennæ Comp.

Besides the fruit in its *fresh* and *dried* state, the juice pressed out and fermented, yields Wine, Alcohol, and Vinegar. These may be treated of as the products of Fermentation at the end of the Vegetable Materia Medica. The Lees of the Wine, moreover, yield Tartar, that is, impure Cream of Tartar.

Tartar, or impure Bitartrate of Potash, enumerated above as one of the constituents of the juice of the Grape, has already been treated of at p. 86, where it is mentioned that as the Saccharine matter disappears and becomes converted into Alcohol, this salt, being insoluble in the Spirit formed, is deposited in the casks, and well known by the name of *argol* or *tartar*. This is chiefly composed of Bitartrate of Potash with a little Bitartrate of Lime. Besides its own particular uses (*v.* p. 86), it is important as the salt from which Tartaric acid is obtained.

Acidum Tartaricum, L. E. D. Tartaric Acid.

Tartaric acid, so named from Tartar, besides the juice of the Grape is contained also in Tamarinds and in some other acidulous fruits. Tartar, which in its purified state is so well known as Cream of Tartar, consists of two Equivalents of Tartaric acid in combination with one of Potash, thus forming a Bitartrate of Potash. One Equivalent of the acid is easily separated with the assistance of Chalk, and the Tartrate of Lime obtained is decomposed with Sul′, as in the case of Citric acid. The separation of the second Equivalent requires a little more complicated decomposition, to obtain a second portion of Tartrate of Lime.

Prep. Boil *Bitartrate of Potash* ℔iv. (in powder 10 parts, D.) in *Aq. dest.* Cong. ij. (100 parts heated, D.) Add gradually *Prepared Chalk* ℥xij. and ʒvij. (4 parts, D.) When effervescence has ceased add *Prepared Chalk* ℥xij. and ʒvij. previously dissolved in *Hydrochloric′* f℥xxvjß. or q. s. and *Aq. dest.* Oiv. (that is, a solution of Chloride of Calcium; the directions of the D. P. are substantially the same.) Filter and wash the Tartrate of Lime frequently with *Aq. dest.* till it is tasteless. Pour on it *dil. Sulphuric Acid* Ovij. and f℥xvij. [(Ox. and f℥vij. E.) (*Sul′* 7 parts diluted with *Aq.* 20 parts, D.)] boil for 15 minutes, (digest with a moderate heat for 3 days, frequently agitating, D.) Strain. (Wash the acid from the precipitate, D.) Evaporate (including the first acid liquor and the washings, D.) with a gentle heat to obtain crystals. Purify by repeating 2 or 3 times the solution, filtration, and crystallization. (Preserve in glass stoppered vessels, D.)—The second Eq. forming the excess of acid in the Bitartrate of Potash becoming saturated with the lime of the Chalk, the Carbonic acid escapes in effervescence: insoluble Tartrate of Lime is precipitated, and neutral Tartrate of Potash remains in solution. The other equivalent is procured by decomposing this Tartrate of Potash with the Chloride of Calcium (the old Muriate of Lime) in excess. By double decomposition Chloride of Potassium is formed, and remains in solution, while a fresh portion of Tartrate of Lime is precipitated. Both portions of this salt being mixed together, are decomposed with Sulphuric acid, which precipitates with the Lime as an insoluble Sulphate, while the Tartaric acid is liberated. It usually requires crystallization 2 or 3 times and to be purified with animal charcoal before the crystals can be obtained in a pure state, and free from colour.

Tartaric acid, discovered in 1770 by Scheele, is colourless, without smell, and pleasantly sour. Its crystals are large, clear, and more or less modified from their primary form, the oblique rhombic prism. They are permanent in the air, soluble in five or six times their weight of water at 60°, and in twice their weight at 212°; less so in Alcohol. The solution decomposes in keeping, a light and thin membranous-like matter being formed. The effects of heat, of acids, &c. are remarkable in producing a number of new compounds. When heated to about 400° it melts, loses one-fourth of its water, becomes deliquescent, and forms what has been called Tartralic′. A further degree of heat produces Tartrelic′ and anhydrous Tartaric′, which is insoluble and powdery. When subjected to destructive distillation, Carbonic acid and water are given off, and two pyrogenous acids are produced. One of these, the Pyrotartaric, is oily, and the other crystalline. A solution of Tar′ added to solutions of the earthy salts, will form white precipitates, as of Lime, Strontia, and Baryta, soluble in excess of acid. In the Acetate of Lead and Nitrate of Silver it also forms white Tartrates of these metals. It is most easily distinguished from other acids by a soluble salt of Potash, with which on addition it precipitates a Bitartrate of Potash, either as a powder or in crystals, according to the state of dilution. It is remarkable for forming double salts, of which the Tartrates of Potash and Soda, of Potash and Antimony, of Potash and Iron are officinal. Comp. $C_4\ O_5\ H_2$=Anhydrous Tar′+1 Aq. =Crystallized Tar′; or, $C_8\ H_4\ O_{10}$+2 Aq. (*Liebig.*)

Tests. Tar′ is apt to be adulterated with Bitartrate of Potash or with Lime. "Totally soluble in water. The solution throws down Bitartrate of Potash from any neutral salt of Potash. Whatever is precipitated from the solution by Acetate of Lead is dissolved in dil. Nit′." L. Any Sulphate remains insoluble. "When incinerated with the aid of red Oxide of Mercury, it leaves no residuum, or a mere trace only," E., showing the absence of Lime or of any fixed impurity.

Inc. Alkalis, Earths, and their Carbonates; salts of Potash, of Lime, and of Lead; Nitrate of Silver.

Action. Uses. Refrigerant. Being cheaper, it is often used as a substitute for Citric acid, especially in making effervescing draughts, its saturating power being nearly the same. The common Soda Powders are made with Tar' gr. xxv. and Bicarb. Soda ʒß, kept in separate powders, dissolved in water, and mixed at the time of being taken in a state of effervescence. So also the gentle aperient Seidlitz Powders are formed with Tartrate of Potash and Soda ʒij., Bicarb. of Soda ℈ij. dissolved in water and taken in effervescence with Tartaric acid ʒß.

OXALIDEÆ, *Dec.* Oxalids.

Herbs, often with bulbous or tuberous roots, seldom shrubs, very rarely trees. Leaves alternate, crowded, digitate, rarely pinnate, often sensitive. Stipules wanting. Calyx 5 partite or with 5 sepals, imbricate in æstivation. Petals 5, clawed, inserted into the receptacle, alternating with sepals, twisted in æstivation. Stamens 10, more or less monadelphous, alternate ones sometimes without anthers. Styles 5; stigmas capitate. Carpels 5, united together; or fruit capsular, 5 lobed, membranous with 5 cells, each cell with one or many ovules. Seeds inserted into central angle of cells, pendulous, often enclosed in a fleshy integument (arillus) which opens elastically at the apex. Embryo in the axis of fleshy albumen, radicle above.

The Oxalideæ are nearly allied to Geraniaceæ, and through the shrubby and arboreous genera to Zygophylleæ and Connaraceæ. They abound in tropical America and at the Cape of Good Hope. A few, and those most widely diffused, are found in temperate and warm parts of the world. Averrhoa in India and the Indian Islands. The herbaceous parts of many of the species, and the fruits of others, are acidulous from the presence of Binoxalate of Potash. The tubers of the stemless species abound in fecula and are esculent.

OXALIS ACETOSELLA, *Linn.* L. Wood-Sorrel. *Decand. Pentagyn.*

This elegant little plant is found throughout Europe in shady situations. It is distinctly noticed from the time of Charlemagne, and is supposed to have been known to the Alexandrian school. Mr. Bicheno considers it to be the true Shamrock.

Wood-sorrel is stemless, with a toothed creeping rootstalk, leaves ternate, leaflets obcordate, pubescent, drooping at night, the scape single flowered, longer than the leaves, bibracteate above the middle. Flowers white, or rose-coloured, with purplish veins and a yellow spot above the base. Calyx 5 sepalled, persistent, imbricate in æstivation. Petals 5, four times larger than the calyx, slightly adhering by their claws, oblong, obovate, slightly emarginate, spirally twisted in æstivation. Stamens 10, slightly monadelphous at the base, the 5 exterior shorter (the five interior ones twice as long as the calyx). Styles 5. Capsule oblong, 5 angled, 5 celled.

The Wood-Sorrel obtained its ancient name of Oxys (*Pliny*) from ὀξὺς, *acid taste.* It is, like the common sorrel, of an agreeably acid taste, but harsh to the teeth when chewed, owing to the presence of Binoxalate of Potash, which is secreted by this and several other plants, as by *Rumex acetosa*, the species of *Rheum*, and especially by *Cicer Arietinum.*

Action. Uses. Refrigerant, Antiscorbutic; allays thirst; occasionally used in salad, and its infusion as a substitute for Lemonade; highly extolled by Frank in Petechial Fevers; is injurious in cases of the Mulberry or Oxalate of Lime Calculus.

ACIDUM OXALICUM. Oxalic Acid. *Acid of Sugar.*

Oxalic acid ($C_2\,O_3$+3 Aq.=63 when crystallized) has obtained its name from the foregoing plant. It is said to be contained in a free state in Cicer Arietinum, but is probably in the state of Binoxalate. It is acid and powerful enough to blanch the boots in walking through a field of the plant. Some Lichens contain a very large proportion of Oxalate of Lime. It is now obtained in the largest quantities from the action of Nitric acid on several substances of the nature of Sugar and Starch, including these substances themselves. Hence it has been called Acid of Sugar. The Nitric′ becoming decomposed, these substances lose their Hydrogen, become oxidized, and converted into an acid, which is found to be composed of 2 Eq. of Carbon united with 3 of Oxygen. This is soluble in about its own weight of hot, and in about 8 times its weight of cold water, the solution being intensely acid. It readily crystallizes in quadrangular crystals, which are colourless and transparent, elongated, six-sided, and flattened, with two or four terminal planes, being derived from an oblique rhombic prism. The crystals effloresce in a dry atmosphere, and melt in their water of crystallization, are volatilized by heat, decomposed at a higher temperature, and with the aid of Sul′ into water, Carb′, and Carbonic oxide. Their acidity is powerful, acrid, and corrosive: hence it is a virulent poison. Numerous fatal cases have occurred from the resemblance of its crystals to those of Epsom Salts. But they may readily be distinguished by their crackling noise when dissolving in water; by the intensely acid taste and reaction of the solution; by its effervescing with the alkaline Carbonates, which give a white precipitate with Epsom Salts or Sulphate of Magnesia (p. 120). This is moreover distinguished by its nauseously bitter taste. The crystals of Ox′ also resemble those of Sulphate of Zinc (p. 152). Oxalic′ is distinguished by its powerful affinity for Lime, separating it even from Sulphuric′. The Oxalate of Lime formed is insoluble in an excess of acid. A soluble Oxalate will be detected by the solution of a neutral salt of Lime or of Oxide of Lead. The acid may be separated from the Lead by the action of Sulphuretted Hydrogen; and then being filtered and evaporated, it will crystallize. Insoluble Oxalates, the bases of which form insoluble compounds with Sul′, may be decomposed by the action of this acid, when the Ox′ will be separated.

Action. Uses. A virulent Poison, which is very speedy in its action. Acute pain is immediately experienced, followed by vomiting. Great depression of the circulation ensues, nervous symptoms, such as great debility, numbness, &c., sometimes followed by convulsions. "But death follows so speedily after the injection of large doses, few of those who have died survived above an hour, that the symptoms have not been fully made out." (*c.*) Irritation and corrosion of the stomach are observed.

Antidotes. Chalk, Whiting, or Magnesia mixed up with water should be administered as quickly as possible in large quantities. Evacuate the stomach. Large quantities of water may be also useful.

Binoxalate of Potash ($KO, C_2 O_3 + 2 Aq. = 138$), the salt which is contained in Wood-Sorrel and other plants, is often called *Salt of Sorrel* or *of Wood-Sorrel*, and very absurdly *Essential Salt of Lemons.* It may be obtained from the juice of the plant by evaporating and then redissolving, and subsequently crystallizing; or by neutralizing a portion of Oxalic acid with Carb. of Potash, and then adding an equal quantity of Ox′, when the salt is obtained in colourless rhombic prisms, having a sour taste, and requiring 40 parts of water for their solution. This salt may be used for the same purposes as the plant itself; but in large doses will act as a poison.

Quadroxalate of Potash ($KO, 4 C_2 O_3 + 7 Aq. = 255$) is usually sold in the shops for the above salt, being much used for removing iron-moulds and ink-stains. It may be made by a process similar to that just mentioned, but adding 3 more parts of Oxalic′ to the Oxalate.

AMMONIÆ OXALAS, E. Oxalate of Ammonia. Dissolve *Carb. Ammonia* ℥viij. in *Aq. dest.* Oiv., and add *Oxalic′* ℥iv. Boil and evaporate, that crystals may form. Used only as a test.

ZYGOPHYLLEÆ, *R. Brown.* Beancapers.

Herbs, shrubs, or trees. Branches often articulated at the joint. Leaves opposite, usually compound, impari, or abruptly pinnate, leaflets without dots. Stipules double, sometimes thorny. Flowers complete, on single flowered peduncles, often solitary, axillary or interpetiolar, without bracts. Calyx 4—5 partite, laciniæ imbricate in æstivation. Petals equal in number to divisions of calyx. Stamens double in number, filaments equal at the base, or inserted into the back of a scale. Ovary seated on a convex or depressed disk, more or less deeply furrowed, 4 to 5 rarely 10-celled. Cells 2 or many, rarely single seeded. Ovules pendulous or erect. Styles of the number of the cells, united into one. Stigma simple or 4 or 5 lobed. Fruit capsular, rarely fleshy, angled or winged, sometimes tuberculate or thorny, 4—5 or rarely 10 celled, capsules opening at the cells, or separating at the partitions into bivalved cocci, or indehiscent. Embryo with the radicle superior, enclosed in a horny albumen, or rarely without albumen.

The Zygophylleæ were formerly united with Diosmeæ and Rutaceæ under the general name of Rutaceæ. In some respects they approach Oxalideæ; in the scales of the stamens they resemble Simarubeæ. They abound in warm extratropical parts of the world. Some of the species of Guaiacum have their bark and wood abounding in resin, possessed of stimulant properties. Others of the species have a disagreeable odour.

GUAIACUM OFFICINALE, *Linn.* Lignum, L. D. Wood, E. Resina, L. D. Resin obtained by heat from the wood, E. Officinal Guaiacum tree. *Decandria Monog.* Linn.

Guaiacum was made known in Europe by the Spaniards, about the year 1508 (*Monardes*, c. xx.), having been previously employed in medicine by the natives of the West Indies and of South America, where the species are indigenous, and called *Guayacan.*

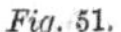

Fig. 51.

The officinal Guaiacum (fig. 51) is a large evergreen tree, from 40 to 60 feet in height, with deep penetrating roots and of a dark gloomy aspect. The wood is hard, heavy, of a greenish colour and remarkable for the direction of its fibres, being cross grained; the strata running obliquely into one another in the form of an X, (Browne, 1789,) or obliquely at an angle of 30° with the axis. The leaves are opposite, abruptly pinnate, with 2 sometimes 3 or 4 pair of leaflets, these are smooth, obovate, or oval obtuse, delicately veined. The flowers are borne on long single-flowered peduncles, 8 or 10 generally rising together from the axils of the upper pairs of leaves. The calyx is 5 partite, segments obtuse, a little velvetty. Petals 5, oblong, spreading, of a light blue colour. Stamens 10 (fig. 1,) with their filaments a little broader towards the base. Style and stigma simple. The fruit is a fleshy capsule of a reddish yellow colour, slightly pedicelled, almost truncate at the apex, 5 angled, 5 celled, (fig. 3,) or from abortion 2—3 celled. Seeds solitary in each cell, pendulous from the axis, (fig. 2,) radicle superior, cotyledons somewhat fleshy. Albumen cartilaginous. A native of the West India Islands, particularly Cuba, St. Domingo and the south side of Jamaica, flowering in April and ripening its seed in June.—*Sloane* Hist. t. 222, f. 3. Bot. Reg. New Ser. xii. t. 91.

G. sanctum, Linn., a native of Porto Rico, and *G. arboreum*, Humboldt and Bonpland, *Guayacan* of the natives of Cumana and Carthagena, are also said to yield some of the Guaiacum-wood, or Lignum Vitæ of commerce, which is also obtained from the Isthmus of Darien.

GUAIACI LIGNUM, L. D. Guaiacum Wood, E., known in commerce by the name of Lignum Vitæ, is imported in great logs generally without, but sometimes covered with, a smooth grey bark, from Jamaica, Cuba, St. Domingo, &c. It is remarkable for its weight (Sp. Gr. 1·33), hardness, and toughness, and is therefore much used in machinery, also for rollers, pestles and mortars, &c. It is distinguished by its cross-fibre (*v. supra*), and is surrounded with the alburnum or *sap-wood*, which is smooth, hard, and yellow, like box; while the *heart-wood* is of a dull brownish-green colour, from containing a large proportion of Guaiac. It is usually met with in shops in the form of shavings and turnings, which are, however, apt to be intermixed with those of other woods, as of box. The sawdust of Guaiacum as stated by Richard becomes green by exposure to the air. They become bluish-green by the action of Nitric′ or its fumes. But the cross-fibre should also be looked for.

The Bark, which is of a dark greenish colour with grayish spots, has sometimes been used officinally. It is acrid in taste, and has been thought by some to be as efficacious as the wood.

The wood is without smell, except when rubbed or heated; it has a slightly bitter and pungent taste, chiefly affecting the throat. It burns readily, even when the corner of a block is presented to a flame. It yields its virtues partially to water, a decoction becoming yellow in colour and acrid in taste. Geiger, from lbj. of the wood, obtained ℥ij. of the extract. Hagen obtained 3 per cent. of Guaiacum from the wood. It contains both an acrid principle and Guaiacum. "The former abounds most in the alburnum, the latter in the central wood: the more acrid alburnum ought perhaps to be preferred." (*c.*)

Browne states that all parts of the plant are possessed of active properties, the fresh bark being aperient and a purifier of the blood, the pulp of the berries emetic and cathartic, the leaves detergent, and employed in cleaning house-floors, and in washing linen.

Guaiaci Resina. Resin of Guaiacum or Guaiac (*Gum Guaiacum*) is the concrete juice, and is usually thought to be the only active part, even of the wood. According to Browne, it transudes frequently of its own accord, and may thus be seen concreted on the bark at all seasons of the year, but in greater abundance when the bark has been cut or wounded; also by heating in the fire billets of the wood which have been bored longitudinally, and receiving in a calabash the melted Guaiac at the other end. Likewise by boiling the chips in salt and water, and skimming off the Guaiac.

Guaiac may be seen in grains, sometimes agglutinated, but usually in homogeneous lumps. Sp. Gr. 1·2—1·23; but sometimes mixed with pieces of the wood and bark; of a brownish-green colour, sometimes with a tinge of red; fracture brilliantly shining, glass-like, and resinous; brittle, powder at first of a greyish colour, but becoming green like guaiac wood and resin generally, when exposed to light. It softens in the mouth: the taste, at first scarcely perceptible, is slightly bitter, but becoming acrid, produces burning in the fauces. The odour is slight, increased on pounding or on heating it, when it melts and evolves a balsamic odour. Water has but moderate action on it, dissolving about 9 per cent., chiefly Extractive. (*v.* Extract, *infra.*)

The fixed and volatile oils scarcely act upon it. Alcohol dissolves 91 per cent. of the peculiar substance called Guaiac, becoming of a deep brown colour. The Guaiac is precipitated on the addition of water, S′, and H Cl′. Ether also dissolves the resin, and separates *Guaiacic acid* from the Extract. Solutions of Potash and Soda dissolve it freely, as does Ammoniated Alcohol. Sul′ becomes of a rich claret-colour; Chlorine produces remarkable changes of colour in the Tincture, from green to blue, and from that to brown, so Nitric′; finally converting Guaiac into Oxalic acid. The changes of colour seem, as above, to be dependent on the absorption of Oxygen.

The Tincture imparts a blue colour to Gluten, and to substances containing it ; also to mucilage of Gum Arabic made with cold water, and to transverse sections of various roots: hence the E. C. employ slices of the Potato as a test of its purity.

Guaiac consists evidently of an extractive-like matter, which is taken up by water, and the Resin, which, having peculiar characters, has been called Guaiacin (Guaiacic acid, *p.*); Unverdorben considers this to be composed of 2 Resins, one soluble in Ammonia, and the other, which forms the largest portion of Guaiac, merely mixes with it. M. Thierry has by means of Ether separated from the extract of Guaiacum what he calls Balsamic Resin, and from it obtained an acid which he calls Guaiacic acid, and which resembles Benzoic and Cinnamic acids, but differs from them in being perfectly soluble in water. Besides the Balsamic Resin, the extract he states contains another resin, which is soluble in Ammonia. Dr. Ure, in an ultimate analysis of Guaiacum, found it composed of Carbon 67·88, Hydrogen 7·05, Oxygen 25·07=100. Prof. Johnston considers its composition to be $C_{40}\ H_{23}\ O_{10}$ and its Eq. 343.

Tests. "Fresh fracture red, slowly passing to green: the tincture slowly strikes a lively blue colour on the inner surface of a thin paring of a raw potato." E. P.

Action. Uses. Acrid Stimulant and Alterative, Diaphoretic. In large doses, irritant of the intestinal canal. Useful in chronic Rheumatism, Secondary Syphilis, Scrofula, and in chronic Skin Diseases.

D. gr. x.—ʒß. in powder or bolus, or in the following mixture:

MISTURA GUAIACI, L. E. Guaiacum Mixture.

Prep. Triturate *Resin of Guiacum* ʒiij. with *Sugar* ℥ß. then with *Mucilage of Gum Arabic* f℥ß., lastly add gradually *Cinnamon Water* f℥xix. (xixß. E.) constantly rubbing up.

An emulsion is formed with the aid of the Sugar and Gum, in which all the constituents of the Guaiac are suspended. It may be given in doses of f℥ß to f℥ij. two or three times a day.

DECOCTUM GUAIACI, E. DECOCT. GUAIACI COMPOSITUM, D.

Prep. Boil *Guaiac* (Wood, D.) *turnings* ℥iij. and *Raisins* ℥ij. E. gently in *Aq.* Oviij. (by measure ℔x. D.) till reduced to Ov. (half, D.) towards the end adding *Sassafras rasped* ℥j. (ʒx. D.) and *Liquorice Root bruised* ℥j. (℥ijß. D.; ℥ij. E.) Strain the liquor.

As water takes up only a small portion of Guaiacum, this would be a very inert preparation, were it not that the acrid extractive is one of the parts dissolved; and this therefore may prove a useful form in some cases. It is like the old *Decoction of the Woods*, and to which the *Dec. Sarzæ Comp.* L. also is very similar. Prescribed in doses of f℥ij.–f℥iv. it is useful in producing a diaphoretic effect in cases of chronic Rheumatism, &c., the patient being kept warm to favour the determination to the skin.

An *Extract of Guaiacum* is ordered in the French Codex, in which the *wood* is thrice boiled in *Aq. Dest.*; this allowed to stand for 12 hours to deposit, is decanted and evaporated to a soft consistence, the deposit being then mixed, and about a ⅛ part of Alcohol added towards the end of the process. The extract thus prepared, M. Thierry states, is very odorous, and yields more acid than when otherwise made, and he infers that it owes its properties to a peculiar balsamic resin and to this acid.

AQUA CALCIS COMPOSITA, D. Compound Lime Water.

Prep. Macerate *Guaiac-turnings* ℔ß. *Liquorice-root* cut and bruised ℥j. *Sassafras-bark* bruised ℥ß. *Coriander Seeds* ʒiij. and *Lime Water* Ovj. without heat for two days, in a closed vessel, shaking occasionally, then strain.

This, though called a Lime preparation, probably owes any efficacy it may possess to its similarity to the Decoction of Guaiacum. It is probable that some of the Lime combines with the acid, and some becomes deposited, as water when boiling dissolves less of it than when cold.

TINCTURA GUAIACI, L. E. D. Tincture of Guaiacum.

Prep. Macerate (Digest, E.) for 14 (7, E. D.) days *Resin of Guaiacum* bruised ℥vij. (iv. D.) in *Rectified Spirit* Oij. (by measure ℔ij. D.) Then filter.

Action. Uses. Rectified Spirit being a good solvent of Guaiacum, this is a good form for exhibition in chronic Rheumatism, &c., in doses of fʒj. to fʒiv. with milk or mucilage.

TINCTURA GUAIACI (AMMONIATA, E. D.) COMPOSITA, L. Compound or Ammoniated Tincture of Guaiacum.

Prep. Macerate (Digest in a well-closed vessel, E.) for 14 (7, E. D.) days *Resin of Guaiacum* bruised ℥vij. (iv. D.) in *Aromatic Spirit of Ammonia* (Spirit of Ammonia, E. Oij. (℔jß. D.) Then filter.

Action. Uses. Ammoniated Alcohol being an excellent solvent for Guaiacum, this Tincture has been much employed in chronic Rheumatism, &c., and is considered more efficacious than the other; but requires, like it, to be given in some viscid fluid.

Officinal Preparations containing Guaiacum. Pil. Hydrargyri Chloridi Comp. L. E. D. (p. 194.) Aqua Calcis Comp. (*v. supra.*) Decoctum Sarzæ Comp. L. Pulvis Aloes Compositus.

RUTEÆ, *Adr. Juss.* Rueworts.

Shrubs or herbs. Leaves alternate, often lobed, dotted with glands, without stipules. Calyx free. Petals free, twice as many as the divisions of the calyx, sometimes united together. Stamens twice or three times as many as the petals, (on the outside of a cup-like disk surrounding the ovary, *Lindl.*) filaments broader at the base. Ovary 3—5 lobed. Cells with 2, 4 or many pendulous ovules. Style single, often divided near the base. Capsule separating either at the partitions or in the middle of the cells into 2 valved carpels. Embryo straight or curved in the axis of albumen. Radicle superior.

The Ruteæ are closely allied to Zygophylleæ and are inhabitants chiefly of the temperate parts of the Northern Hemisphere. They secrete a volatile oil with bitter matter.

RUTA, L. E. Folia, L. D. Leaves and unripe Fruit, E. RUTA GRAVEOLENS, *Linn.* Common or Garden Rue. *Decand. Monog.* Linn.

The common Rue and *Ruta angustifolia*, natives of the South of Europe, were much employed and highly esteemed by the ancients, as they still are by Asiatic nations.

The Rue is a small branching under-shrub about 2—3 feet high. Stems straight, slightly striated, of a dull greenish colour. Leaves of a glaucous green, supra-decompound, leaflets thickish, dotted, oval oblong, tapering towards their bases, the terminal one obovate. Flowers in a terminal corymb. Calyx small, 4-fid, rarely 5-fid. Petals 4, in the upper flowers 5, yellow, oval, unguiculate, entire or denticulated, with their apices curved inwards. Stamens 8 or 10. Ovary marked with 2 crucial furrows. Capsule globular, warty in 4 or 5 obtuse lobes, each separable into two valves. Seeds dotted.

Every part of the Rue is distinguished by its strong and repulsive odour and its acrid and bitter disagreeable taste. The leaves have the strongest odour when the seed-vessels are well developed, but still green. A great portion of their peculiar characters is necessarily lost in drying. "The E. C., however, correctly adds the unripe fruit also, because the seed-vessel is covered with large oil-vesicles, which impart great activity to this organ." (*c.*)

Action, Uses. Rubefacient, Stimulant, Antispasmodic, Emmenagogue, Anthelmintic.

OLEUM RUTÆ, E. D. Oil of Rue.

Distilled with water from the herb and half-ripe ovaries of common Rue. It is of a light yellow colour, acrid in nature, and with a very disagreeable smell.

Action. Uses. Stimulant, &c. Used as an Antispasmodic and Emmenagogue.

D. ♏ij.—♏v. rubbed up with Sugar and water.

CONFECTIO (CONSERVA, D.) RUTÆ, L. Confection of Rue.

Prep. Rub together into a very fine powder *dried Rue* (leaves of Ruta graveolens, D.) *Caraway* (seeds, D.) *Bay Berries* āā ℥iß. *Sagapenum* ℥ß. *Black Pepper* (the fruit, D.) ʒij. (When the confection is to be used, L.) add *Honey* (despumated, D.) ℥xvj. and mix well.

The dried herb is less efficacious than when fresh, but being combined with substances having similar properties, this confection is sometimes useful in flatulent colic in doses of ℈j.—ʒj.

The *Syrup of Rue*, though not officinal, is kept by most druggists. It is made by dissolving *Oil of Rue* ♏xij. in *Rectified Spirit*, ʒiv. and then mixing with *Simple Syrup* Oj.

D. Often given by nurses, in doses of ½ to 2 tea-spoonfuls, in the flatulent colic of children.

EXTRACTUM RUTÆ, D. Extract of Rue.

Prepared as the simpler extracts; but the volatile oil being dissipated during the decoction, this extract is simply bitter and not preferable to others.

DIOSMEÆ, *Adr. Jussieu.* Diosmads.

Trees or shrubs, rarely herbs. Leaves without stipules, opposite or alternate, simple or compound, often dotted with glands. The calyx is free. Petals equal in number to segments of calyx, sometimes combined. Stamens equal to, or twice as many as the petals, the alternate ones opposite to them, then shorter or without anthers. Ovaries several, free or more or less united, 2-ovuled. Ovules affixed to the axial angle, collateral, or obliquely placed one over the other, very rarely with 4 ovules. Fruit separable into several carpels, which by abortion are often single-seeded; endocarp cartilaginous, free, two-lobed and elastic. Seeds inverse. Embryo included in albumen, or without albumen.

The *Diosmeæ* are closely allied to Zanthoxyleæ, and also to Ruteæ, with which they are indeed usually united, and are to be distinguished from them chiefly

Fig. 52.

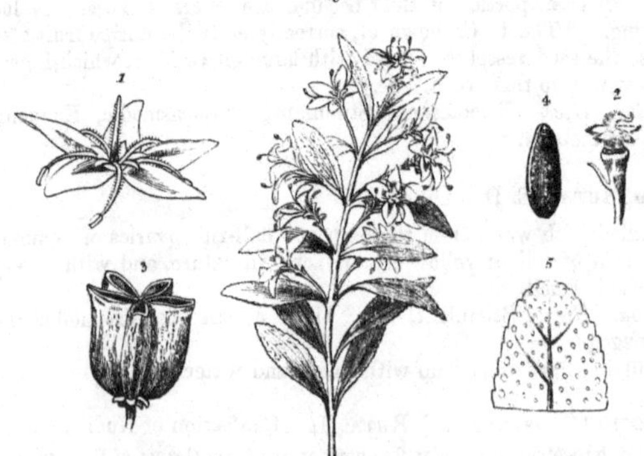

by the endocarp in the ripe capsule. They are found in South Africa and in New Holland, some in tropical America, and a few in equinoctial Asia, with only Dictamnus in the North of Asia and the South of Europe. They secrete volatile oil and Resin, as well as a bitter principle.

BUCKU, E. Diosma, L. Buchu Folia, D. Diosma crenata, *Dec.* L. D. Leaves of various species of Barosma, E. *Pentand. Monog.* Linn.

The leaves of one or more plants called *Bucku, Buchu*, or *Bookoo*, having been found by Mr. Burchell, the African traveller, to be employed by the Hottentots as a vulnerary and in the treatment of diseases of the urinary organs, became known in this country about the year 1823. Bucku was first introduced into the D. P., and then into the L. P. as the leaves of *Diosma crenata.* Sir W. Hooker, however, showed in the Bot. Mag. t. 3413, that the leaves of *D. crenulata* are those most common, and that those of *D. serratifolia* are also found in commerce. All these species have since been restored to the genus BAROSMA, to which they originally belonged. "Several species are collected by the Hottentots, according to Thunberg, especially *B. betulina* and *pulchella*, and even *Adenandra uniflora*, to which some *Agathosmas* and many others may no doubt be added." (*Lindley*, Fl. Med.) These are all included in the genus Diosma by De Candolle (Prod. i. p. 713.)

BAROSMA.

Calyx 5-fid or 5-partite. (1.) Petals 5, inserted into the base of the disk which lines the bottom of the calyx and has a short, scarcely prominent rim. Stamens 10, inserted with the petals and equal to them in length, 5 fertile, alternating with the petals, filaments filiform, subulate, with the anthers commonly terminated by a small gland, often becoming recurved, 5 opposite to and shorter than the petals, sterile, petaloid, indistinctly glandular at the apex. Ovaries 5, united into one, 5 lobed and auriculate at the apex, commonly with glandular tubercle. Style longer than the stamens. Stigma minutely 5 lobed. (2.) Fruit (3) composed of 5 compressed cocci, outwardly auriculate and covered with glandular dots. Seed (4) oblong.—Shrubs of the Cape of Good Hope. Leaves opposite or alternate, leathery, flat, dotted (6), especially near the margin, varying in shape. Flowers axillary, on single or 3 flowered peduncles, or fasciculate in single flowered peduncles.

B. crenata. Fig. 52. (*Diosma crenata,* Dec. L. D.) Leaves ovate and obovate, acute, serrated, dotted glandular at the margin. Flowers pink, solitary and terminal, on somewhat leafy pedicels.—Loddiges Bot. Cab. t. 404

B. serratifolia. (D. serratifolia, Dec. and Bot. Cab. t. 378.) Leaves linear-lanceolate, serrulate, smooth, dotted, glandular at the edges, three-nerved. Flowers white, on solitary lateral pedicels, bearing two leaflets above the middle.

B. crenulata, Willd. Hooker, B. M. t. 3413. (*D. crenulata,* Linn. *D. odorata,* Dec.) Leaves decussate, ovate, oblong, on very short petioles, very obtuse, minutely crenated, quite smooth and of a darkish green above, beneath paler, with a few obscure oblique nerves, dotted with oil vesicles, with at every crenature a conspicuous pellucid gland and a pellucid margin round the whole leaf. Peduncles axillary and terminal chiefly from the axils of the superior leaves, single flowered, often bearing a pair of small opposite leaves or bracts above the middle. Beneath the calyx are 2 or 3 pairs of small imbricated bracts.

Bucku leaves are smooth, leathery, and shining, serrate or crenate at their margins, studded with dots, *i. e.* vesicles filled with essential oil, of a light yellowish-green colour, of a strong, considered by some a disagreeable, odour; the taste warm and aromatic. They necessarily vary in form according to the species of plant from which they have been obtained. Those which are ovate or obovate by *B. crenata*, the linear-lanceolate ones by *B. serrulata*, and those which are ovate, oblong, and obtuse, by *B. crenulata*. All may be found intermixed among the Buchu leaves of commerce. They contain Volatile Oil, which is of a yellowish-brown colour and a penetrating odour, Bitter Extractive (*Diosmin*), Resin, Gum, Lignin, &c.

Action. Uses. Stimulant and Tonic in chronic affections of the Urinary organs attended with increased secretion of Mucus, in doses of the powder ℈j. or ʒß.

INFUSUM DIOSMÆ, L. INF. BUCKU, E. D. Infusion of Bucku.

Prep. Macerate (Infuse, E.; Digest, D,) for 4 (2 E.) hours in a tightly covered vessel *Diosma* or Bucku, ℥j. (the leaves of Diosma crenata, ℥ß. D.) in boiling *Aq. dest.* Oj. (by measure ℔ß. D.) Strain (through linen or calico, E. D.)

Action. Uses. Tonic and Diuretic in doses of f℥iß two or three times a day.

TINCTURA BUCKU, E. Tinct. BUCHU, D. Tincture of Bucku.

Prep. Digest (Macerate, D.) for 7 days *Bucku* ℥v. (leaves of Diosma crenata ℥ij. D.) in *Proof Spirit* Oij. (by measure ℔j. D.) Pour off the clear liquor, E. Filter. This tincture may also be made by percolation, E.

Proof Spirit is a good solvent for the active principles of Bucku.

D. fʒj.—fʒiv.

CUSPARIA, L. E. Cortex, L. Bark, E. Angustura. Cortex, D. GALIPEA OFFICINALIS, *Hancock*, E. G. CUSPARIA, *St. Hilaire*, L. (*Bonplandia trifoliata*, *Willd.* D.) Cusparia Bark.

Cusparia or Angustura Bark was introduced into England about the year 1788. It was subsequently ascertained that it was imported from Angustura on the Oronoco. Humboldt and Bonpland in their travels in South America, having ascertained that the bark was called *Cusparé* by the natives, called the tree which they supposed yielded it *Cusparia febrifuga;* but having sent home specimens to Willdenow, he named it *Bonplandia*, in compliment to one of the travellers, and *trifoliata* from the number of its leaflets. Aug. St. Hilaire having ascertained that, instead of being a new genus, it was only a new species of an old genus, GALIPEA, named the tree *G. Cusparia*, which it still retains. Dr. Hancock, however, who resided for some months in 1816 in the district where the Cusparia is produced, states that the above travellers did not themselves see the tree, but got branches of it without flowers from an Indian; that they afterwards thought they recognized the same plant, which they found growing in considerable forests. This Dr. H. considers to be a distinct species of the same genus as the plant which yields Cusparia, and which he calls *Galipea officinalis.*

GALIPEA, *Aublet.* Pentand. Monog. *Linn.*

Calyx cup-shaped, 5-toothed, often 5-angled. Petals 5, united below into a tube, which is often pentangular. Stamens 5, rarely 6, 7, 8 or 4, with filaments adhering to the tube of the corolla, rarely all fertile, usually 2—4 with abortive anthers. Ovaries 5, connected, 1-celled, supported by an urceolar disk. Styles 5, distinct or connected at the base, each terminated by an obtuse pentangular stigma. Capsules by abortion 1 or 2.—Shrubs or trees. Leaves simple and ternate. Natives of Tropical America.

G. OFFICINALIS, *Hancock*, E. A tree from 15 to 20 feet high, with smooth bark. Leaves alternate trifoliate, petiole about the length of the leaflets, which are oval, but tapering towards both the base and apex, from 6 to 10 inches long, smooth, shining, when bruised smelling like tobacco. Panicles cylindrical, contracted, stalked, longer than the leaves, with the branches about 3 flowered. Calyx hairy. Petals white, downy, 2 longer than the others. Stamens 7, of which only 2 are fertile. Carpels villous as they ripen, 2-seeded, one usually abortive. —Neighbourhood of the Oronoco, between 7° and 8° N. lat. *Orayuri* of the natives yields *Carony*, that is *Cusparia bark*, exported from Angustura.

G. CUSPARIA, (*S. Hilaire*), L. A forest tree, from 60 to 80 feet high, with fasciculate pubescence. Leaves alternate, trifoliate, long stalked, leaflets sessile, unequal, ovate, lanceolate, acute, gratefully fragrant, with scattered glandular and pellucid dots. Flowers in axillary racemes, which are almost terminal. Calyx and corolla white with fascicles of hairs, seated on glandular bodies on the outside. Stamens 6, only 2 fertile. Anthers with two short appendages. Seed solitary.—Forests of tropical America, between Cumana and New Barcelona; yields Cusparia or Angostura bark (Cusparé of the natives) according to Humboldt and Bonpland.

Cusparia or Angustura Bark is in pieces some inches in length, from half to two inches in breadth, and only one or two lines in thickness, more or less quilled, sometimes almost flat. It is covered with a thin yellowish-white, mealy epidermis, smooth or wrinkled; the inner surface is rather smooth, but separable into splinters of a dull brownish colour; the substance of the bark is compact, and of a dark cinnamon-colour. It is brittle, fracture short and resinous; powder of a greyish-yellow colour. The odour is strong and peculiar, the taste bitter, permanent, but slightly aromatic. It yields its properties to water and to Proof Spirit. Its properties depend on the presence of Gum, Resin, Volatile Oil, and a peculiar Bitter principle. The Resin is a little acrid, as is also the Volatile Oil, which has the peculiar odour of Cusparia Bark. The Bitter principle or Extractive has also been named *Angusturin* and *Cusparin*, being a neutral principle, crystallizable in tetrahedrons, easily fusible, soluble in rectified Spirit, in acids, and in alkaline solutions, and precipitated of a whitish colour by Tincture of Galls, sparingly soluble in water, insoluble in Ether and the volatile oils, bitter in taste and a little acrid.

Tests. Cusparia Bark may easily be distinguished from other officinal barks, and is not liable to be adulterated. The E. C. indicates the purity by stating that "its outer surface is not turned dark green, nor its transverse fracture red by Nit′." Some years since, several cases of poisoning occurred on the Continent from the substitution of what has been called *False Angustura Bark*, but which has been ascertained to be that of the *Strychnos Nux Vomica* (q. v.), and which Dr. Neligan, within the last two years, obtained as Angustura Bark at a druggist's in Dublin. The pieces of this are usually much thicker than Cusparia Bark, also harder, and more compact, covered with a ferruginous efflorescence, sometimes yellowish-grey, and marked with prominent white spots, without any aromatic odour, and having an intensely bitter taste. The transverse section, as indicated in the E. P., becomes bright red when touched with Nit′, in consequence of this acting on the *Brucia* in the bark, and also by the rusty spots on its epidermis becoming of a dark green when in contact with the same acid.

Action. Uses. Stimulant Tonic, Febrifuge, Antidysenteric.

D. Powder, gr. x.—℈ß. Extract, gr. v.—gr. xv.

INFUSUM (ANGUSTURÆ, D.) CUSPARIÆ, L. E. Infusion of Cusparia.

Prep. Macerate for 2 hours in a slightly covered vessel *bruised Cusparia* ʒv. (bruised Bark of Bonplandia trifoliata, ʒij. D.) in boiling *Aq. dest.* Oj. (℔ß. D.) Strain.

A stimulant and tonic in low states, in doses of f℥iß. It is of a dull orange-colour. Sesquichloride or Sulphate of Iron produces in it a dark greyish precipitate, Tincture of Galls a slate-coloured one; but no change is produced by Ferrocyanide of Potassium.

TINCTURA (ANGUSTURÆ, D.) CUSPARIÆ, E. Tincture of Cusparia.

Prep. Take *Cusparia* ℥ivß. in moderately fine powder and *Proof Spirit* Oij. Proceed as with Tinct. Cinchonæ, or more expeditiously by percolation.

Stimulant adjunct to bitter infusions in doses of fʒj.—fʒij.

ZANTHOXYLEÆ, made up chiefly of genera placed by De Candolle in *Rutaceæ* and in *Terebinthaceæ*, are distinguished by their flowers being usually unisexual, by the calyx being free, petals hypogynous, equal in number to divisions of the calyx, ovaries subdistinct or united, each with two ovules, fruit indehiscent, carpids opening by a vertical suture, embryo in the axis of albumen. Many of the Zanthoxyleæ are remarkable for secreting a bitter principle, *Xanthopicrine*, and also a volatile oil of aromatic pungency. Thus several species of Zanthoxyleæ are employed as stimulants: one formed the *Faghureh* of Avicenna. Species of Ptelea and Toddalia are bitter and febrifuge. Both the species of Brucea are likewise bitter and tonic. *B. antidysenterica*, the Woginos of Bruce, is most celebrated, because it was long supposed to yield the Bark which was known as false Angustura, and from which the alkali *Brucia* was obtained; but the former is now well ascertained to be the bark of Strychnos Nux Vomica.

SIMARUBEÆ. *Richard.* Quassiads.

Trees or shrubs. Leaves alternate, compound very rarely simple, without stipules and without dots. Flowers complete or unisexual from abortion. Calyx 4 or 5 partite, persistent, imbricate in æstivation. Petals equal in number to, alternate with, but longer than the divisions of the calyx; æstivation twisted, deciduous. Stamens twice as many as petals, with each filament arising from the base of a hypogynous scale. Ovaries seated on a short staminiferous stalk, 4 or 5 free, each with a single ovule suspended from the apex of the interior angle. Drupes distinct, one-seeded. Embryo without albumen.

The Simarubeæ are found in the tropical parts of America, with one species in the tropic-like forests at the base of the Himalayas, and a few simple-leaved species in tropical Asia and Madagascar. They are remarkable for their bitterness. Malombo Bark is thought to be a kind of Quassia.

QUASSIA, L. E. D. Lignum, L. D. Wood, chiefly of Picræna (Quassia, L. D.) excelsa, seldom of Quassia amara. E.

QUASSIA AMARA, *Linn. f.* Surinam Quassia. *Decand. Monog.* Linn.

Quassia Wood was first known in Europe about 1742, more fully in 1756, when Rolander returned from Surinam and gave some of the wood to Linnæus. Quassia amara, a native of Surinam, Guiana, and Panama, was first introduced into practice, and it was called Surinam Quassia; but the tree being of small bulk, and not very common, its place is supplied by the wood of Picræna excelsa. Mr. Lance informed Dr. Lindley that no Quassia had been exported from Surinam during the ten years he was at that place. The wood, as received by Dr. Pereira, is in cylindrical pieces, about two inches in diameter, very light, covered by a thin, greyish-white bark, all extremely bitter in taste. From its elegant pinnate leaves with winged footstalks, and its spike-like racemes of red flowers, it is often cultivated as an ornamental plant in the West India Islands.

PICRÆNA EXCELSA, *Lindley.* E. Quassia excelsa, *Swartz,* L. D. Simaruba excelsa, *Dec.* L. Jamaica Quassia. *Polygamia Monœcia,* Linn.

This tree attains a height of 50, 60, or even 100 feet in the woods of the lower mountains of Jamaica and other West India islands, where it is called Bitter Ash and Bitter Wood, and its wood has for some time been substituted for that of the Surinam Quassia, and is sometimes called Jamaica Quassia. Dr. Lindley has rightly formed it into a new genus, as it agrees with the characters neither of Quassia nor of Simaruba.

This tree, besides being lofty, is erect, often three feet in diameter, with a smooth dark gray bark. The wood is white, rather coarse-grained, bitter, but without smell; the bark is moderately thick, dark-coloured, and wrinkled.

Fig. 53.

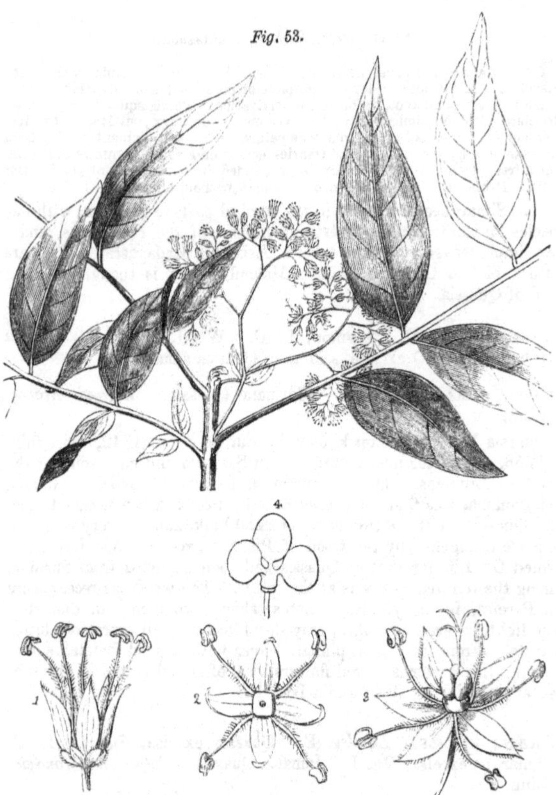

Leaves pinnate, with an odd one. Leaflets opposite, 4 to 8 pairs, stalked, oblong acuminate, unequal at the base. The flowers are small, of a pale yellowish green colour, polygamous, arranged in spreading pointed racemes, which are axillary towards the ends of the branches. The sepals are 5, minute. Petals 5, longer than the sepals. Stamens 5, about as long as the petals, rather shaggy. Anthers roundish. In the male, merely the rudiments of the pistil; in the female, ovaries 3, seated on a round tumid receptacle. Style 3-cornered, trifid. Stigmas simple, spreading. Drupes 3 (but only one coming to perfection), globose, 1-celled, 2-valved, distinct from each other, and placed on a broad hemispherical receptacle. When ripe, about the size of a pea, black and shining, nut solitary, globose, with the shell fragile. (Lindley.)

Quassia-wood is imported in logs covered with a dark grey bark, smooth in the younger and rough and irregular in the larger pieces, yellowish-white and fibrous in the interior. The wood is yellowish-white and glistening, without smell, but of a pure and intense bitterness, tough, and therefore pulverized with difficulty. It contains a bitter neutral principle, called *Quassine* ($C_{10} H_6 O_3$), which is intensely bitter, crystalline, sparingly soluble in water and in Ether, readily dissolved by Alcohol; also Gum, a little Volatile Oil, Lignin, and salts with a base of Lime, an Ammoniacal salt, and some Nitrate of Potash.

Action. Uses. A pure Bitter, useful as a Stomachic and Tonic.

INFUSUM QUASSIÆ, L. E. D. Infusion of Quassia.

Prep. Macerate (infuse, E.; digest, D.) for 2 hours in a lightly covered vessel cut *Quassia* ℈ij. (in chips ʒj. E; shavings of wood of Quassia excelsa ℈j. D.) in boiling *Aq. dest.* Oj. (℔ß. by measure, D.) Strain (through linen or calico, E.)

D. f℥iß. two or three times a day. A good vehicle for preparations of Iron.

TINCTURA QUASSIÆ, E. D. Tincture of Quassia.

Prep. Digest (Macerate, D.) for 7 days *Quassia* in chips ʒx. (dust of the wood of Quassia excelsa ℥j. D.) in *Proof Spirit* Oij. (by measure ℔ij. D.) Strain.

D. fʒß.—fʒij. as an adjunct to tonic draughts and mixtures.

TINCT. QUASSIÆ COMPOSITA, E. Comp. Tinct. of Quassia.

Prep. Digest for 7 days *Cardamom Seeds* bruised, *Cochineal* bruised āā ℥ß. *Cinnamon* in moderately fine powder, *Quassia* in chips āā ʒvj. *Raisins* ℥vij. in *Proof Spirit* Oij. Strain, express the residuum strongly and filter. Or obtain by percolation, as directed for Compound Tincture of Cardamom, rasping or powdering the Quassia.

D. fʒj.—fʒij. An aromatic tonic, and useful as an adjunct.

EXTRACTUM QUASSIÆ, E.

Prep. First make a watery infusion by percolation and without heat; and then evaporate, or prepare as Extract of Liquorice Root.

D. In form of pill, gr. v., or as a vehicle for metallic tonics, &c.

SIMARUBA, L. E. D. Radicis Cortex, L. D. Root-Bark, E. SIMARUBA AMARA, *Aublet.* S. OFFICINALIS, *Dec.* L. Quassia Simaruba, *Linn.* D. *Monœcia* or *Diœcia Decand.* Linn.

This tree, of which the bark of the root was first introduced into practice in 1713, is a native of Guiana and Cayenne. The same species is considered to be found in the mountains of Jamaica, where it is called Mountain Damson.

The tree attains a height of 50 or 60 feet, and considerable thickness, with long horizontally spreading roots. The bark in the young parts smooth and grey, and in the older blackish coloured, and somewhat furrowed. The Leaves are alternate pinnate; leaflets alternate, 2 to 7 on each side, nearly sessile, oval, lanceolate, acuminate, tapering towards base, very smooth and entire, firm,

Fig. 54.

coriaceous, of a deep green colour; petioles sometimes a foot and a half in length. The flowers are monœcious, disposed in a mixed scattered axillary panicle. The calyx is short, cup-shaped, 5-toothed or 5-partite. Petals 5, longer, twisted, and imbricate in æstivation. Male (fig. 3), stamens 10, alternate ones opposite the petals, and a little shorter; filaments each inserted (4) into a hairy scale, having a round or short gynophore (bearing rudiments of ovaries) sometimes wanting. Female flower (fig. 1) with ten scale-like rudiments of stamens; ovaries 5 on a short disk or gynophore, 1-celled, with a single ovule suspended to the inner angle. Styles 5, distinct at the base, united above, and separating again into 5 stigmata. Drupes (fig. 2) 5, or fewer by abortion, dark-coloured, spreading, one-celled, one-seeded. Seeds with a membraneous shell. Embryo straight; radicle above, retracted within cotyledons.—Aubl. Guian. 2, t. 311 and 312. Nees von Esenbeck, Ic. 382. v. fig. 54.

Dr. Lindley observes (Fl. Med. p. 208) that as the Jamaica tree is diæcious, it may be a distinct species. Dr. Macfadyan represents it, however, as agreeing with Aublet's figure of the Guiana plant. Its leaves, however, are described as being oblong obovate. Its small yellow flowers appear in April, those of the Guiana plant in November, which, moreover, is monœcious, the leaflets almost lanceolate, and the root-bark not warty. The bark and wood are said by Aublet to exude a bitter milky juice, and this is said by Dr. Wright not to be the case with the Jamaica plant. These may therefore prove to be distinct species, in which case the Jamaica plant may be called *S. officinalis*, and the Guiana *S. amara*; otherwise I agree with Dr. Lindley that it is improper to change Aublet's name. But they may prove identical, many trees flower twice in the year in warm countries, and the leaves vary in elongation according to age. Both ought to be carefully examined from genuine specimens.

The *bark of the root* is officinal, and sent to Europe from Jamaica. It is stripped off in pieces several feet in length, which are folded upon themselves, either flat or partially quilled, a few lines in thickness, light, tough, fibrous in structure, difficult to powder, of a pale colour, greyish throughout, epidermis a little warty, without odour, bitter in taste. Water and Alcohol both readily take up its virtues, which depend upon the presence of a principle nearly the same with *Quassine*, Volatile Oil, Resin, Ulmine, and several salts.

Action. Uses. Bitter Tonic. Useful in advanced stages of Dysentery and Diarrhœa.

INFUSUM SIMARUBÆ, L. E. D. Infusion of Simaruba Bark.

Prep. Macerate *Simaruba Bark* bruised ʒiij. (ʒß. D.) in *boiling Aq.* Oj. (Oß. D.) for two hours in a lightly covered vessel and strain.

D. f℥j.—f℥ij. as a Tonic. In larger doses it proves Emetic.

b. Calycifloræ.

RHAMNEÆ, *R. Brown.* Rhamnads.

Trees or shrubs, sometimes with the upper parts of branches climbing. Leaves simple, alternate, or sub-opposite, with stipules often converted into thorns. Flowers usually complete, small, greenish coloured, axillary, clustered, umbellate or cymose. Calyx 4—5 cleft, laciniæ valvate in æstivation; tube free, or with its base, sometimes the whole united with the ovary. Petals alternate with the divisions of the calyx, often scale-like, inserted into the throat of the calyx or the margin of the disk. Stamens 4 to 5, opposite to the petals. Ovary usually immersed in or surrounded by a glandular disk, 2, 3, or 4-celled; with one, rarely two seeds, placed side by side; ovules erect or ascending. Style single, with 2, 3, or 4 stigmas, sometimes divided to the base. Fruit free or covered by calyx, capsular, 2 to 3-celled, or by abortion 1-celled, drupaceous, and indehiscent, or forming a berry, or capsular, 2—3 coccous. Seeds solitary, erect, with sparing fleshy albumen. Embryo straight, with large flat cotyledons, and a short inferior radicle.

The Rhamneæ are found both in the temperate and tropical parts of the world. They vary in properties: some secrete a bitter principle, with which acridity is sometimes united; hence these act as stimulants to some of the functions. Some are useful for the colour (sap-green) yielded when acted on by Lime and alkalis. The fruits of a few are edible, as the Jujube, the Lotus, and the Ber of India.

RHAMNUS CATHARTICUS, *Linn.* L. E. D. BACCÆ, L. D. The Fruit, E. Common or Purging Buckthorn. *Pentand Monog.* Linn.

Buckthorn is indigenous in hedges and woods, and found in Europe generally, flowering in May and June, and ripening its fruit in the autumn. It has long been employed in medicine, and thought to be the ῥαμνος of Dioscorides, but without proof.

Bot. Ch. A spreading shrub 8 to 10 feet high. Old branches forming thorny terminal spines. Leaves sub-opposite, ovate, cordate at base, acute, toothed, with 4 to 6 marked veins parallel to and converging to midrib. Stipules linear. Flowers polygamous, often diœcious, in clusters between the leaves, small, of a yellowish green colour. Calyx 4-cleft, tubular at base, persistent and adherent to fruit. Petals 4, a little yellowish, male flowers with a stamen opposite to each, and a rudimentary pistil. Female flower, ovary globular, with 4 single seeded cells, and a 4-cleft stigma. Fruit, a small round berry, which becomes black when ripe, containiug usually 4 smooth hard seeds, which are ovate, triangular, and keeled.

Buckthorn Berries are small, round, of a black colour and shining when fully ripe; they contain a greenish pulp, which has a bitter and disagreeable taste and nauseous smell. This is composed of a green Colouring matter, of Acetic acid, Mucilage, Sugar, Azotized matter. (*Vogel.*) According to Hubert, the purgative properties are dependent on the presence of *Cathartine;* but his experiments are not considered conclusive by Soubeiran. A similar property is possessed by the inner bark.

Action. Uses. Hydrogogue Cathartic; but apt to create nausea and griping.

SYRUPUS RHAMNI, L. E. D. Syrup of Buckthorn.

Prep. Take fresh *Juice of Rhamnus* (Buckthorn, E.) berries Oiv. (by measure ℔ijß, D.) let it stand for 3 days that impurities may subside. Filter. To the filtered fluid Oj. (℥x. D.) add *sliced Ginger* and *bruised Pimento* āā ʒvj. (ʒiij. D.) Macerate with a gentle heat for 4 hours. Strain. Boil down the residue to Oiß. (℔. by measure; make a syrup, D.) Mix the liquors, add *Sugar* (pure, E.) ℔iv. and dissolve (with aid of heat.)

D. Generally employed instead of the berries or expressed juice, in doses of f℥ß.—f℥j.

The TEREBINTHACEÆ of Jussieu have been divided by modern botanists into several orders, such as *Anacardieæ*, *Burseraceæ*, *Amyrideæ*, and *Connaraceæ*. But these are all so closely allied to one another, and participate in so many of the same properties, and have so much the same geographical distribution, that it is sometimes convenient to speak of the whole as forming one family under the name of Terebinthaceæ. As it contains however only one of the Turpentine-yielding plants, the rest belonging to Coniferæ, the name is objectionable.

The products of the *Terebinthaceæ* have been stated by M. Fée to consist—1. Of fixed Oil in the almond of the seed. 2. Essential Oil, which is combined with Resin in the Turpentine of the Pistacia. 3. Resin, which flows naturally or from incisions made into the trunks, &c. of most of the species, usually combined with a little Volatile Oil. 4. Gum, seldom found pure, but frequently combined with the Resin, as in Myrrh, &c.

ANACARDIEÆ. *R. Brown.* Anacards.

Trees or shrubs, with a gummy, viscous or acrid resinous juice, which becomes black in drying. Leaves alternate, simple or compound, without pellucid dots. Stipules wanting. Flowers often unisexual, small, greenish-coloured, axillary or terminal, spiked or paniculate. Calyx free, rarely united with the ovary, 3—5 fid, sometimes deeply divided. Petals equal in number to divisions of the calyx, inserted into an annular or orbicular disk, imbricate in æstivation, rarely valvate, sometimes wanting. Stamens inserted with petals, equal to them in number or double, seldom more. Ovary single and one celled, or several, distinct, all but one abortive, often reduced to a style only. Ovule single, attached to a funicle ascending from the base of the cell. Fruit indehiscent, one-seeded, usually drupaceous. Seed erect or inverse. Embryo without albumen.

Anacardieæ have a single fertile ovary. They abound within the tropics, with a few species (as of Rhus in Europe and N. America) extending to higher latitudes. Anacardieæ abound in resinous juice, with volatile oil, or acrid principle, which is employed in varnishing and lacquering. As medicines these juices act as stimulants, or are sufficiently acrid to be poisonous. The seeds of many abound in oil. The bark of some is astringent. The fruit of a few is edible.

PISTACIA, *Linn.* *Diœcia Pentand.* Linn.

Flowers diœcious, without petals. *Males* in amentaceous racemes, each supported by a scale-like bract. Calyx small, 5 cleft. Stamens 5, opposite to the calycine divisions, nearly sessile, 4 cornered. *Female flowers* in more lax racemes. Calyx 3—4 cleft. Ovary 1, rarely 3 celled. Stigmas 3, thickish, spreading, recurved. Drupe dry, ovate, with a bony, commonly 1-seeded nut, sometimes showing laterally, 2 abortive cells. Seeds solitary, erect, without albumen. Cotyledons fleshy, containing oil, with a superior lateral radicle. Trees with pinnated leaves, extending from the Mediterranean region to Affghanistan. In properties resembling the Burseraceæ.

PISTACIA VERA. The Pistachio Nut tree extends from Syria to Bokhara and Caubul. It has long been introduced into the south of Europe, and is remarkable for its green-coloured kernels enclosed within a reddish-coloured testa, and where the funicle can be well seen. The kernels are oily, and an oil used to be obtained from them. They are pleasant tasted, often eaten at our desserts, either raw or after having been fried with pepper and salt. They have also been made into a demulcent emulsion.

PISTACIA TEREBINTHUS, *Linn.* L. E. D. TEREBINTHUS CHIA. Resina liquida, L. Liquid Resinous Exudation, E., of the Chian or Pistacia Turpentine Tree.

The Terebinth or species of Pistacia yielding the Turpentine of Chio, was well known to the ancients. It is the *Alah* of the Old Testament, translated *Oak*, *Terebinth*, &c., τερμινθος of the Greeks, and the *butm* of the Arabs. It is found in the south of Europe, Asia Minor, Syria, and the north of Africa.

A tree 20 to 40 feet in height. Leaves pinnate, with an odd one. Leaflets about 7 or 9, ovate, lanceolate, round at the base, acute, mucronate, reddish coloured when young, afterwards of a dark green colour. Inflorescence a large compound panicle. Scales of the male flower covered with brown hairs. Anthers yellowish. Stigmas of a crimson colour. Fruit purple, roundish, about the size of a large pea. Horn-shaped galls are produced on these trees.

The Turpentine of this tree is obtained in the island of Chio by making transverse incisions into the bark of the trees, of which each yields only a few ounces, and the whole island not more than 1000 pounds. The harvest is from July to October. The juice issuing from the wounds is allowed to fall upon smooth stones, from which it is scraped and purified by being melted in the sun and strained into bottles. It used to be taken chiefly to Venice, where it was in request for making the far-famed Theriaca. Chian Turpentine is a pellucid liquid of a yellowish colour, having the consistence of honey, tenacious, with an agreeable terebinthinate smell, and moderately warm taste. Exposed to the air it thickens, and becomes hard from the loss of its Volatile Oil. A Resin is said by Belon to be produced by the same tree, and the small kernels of its fruit are edible. From its scarcity and high price, it is usually adulterated with other Turpentines, which it resembles in properties.

Action. Uses. Excitant and Diuretic, like the other Turpentines. (*q. v.* Coniferæ.)

PISTACIA LENTISCUS, *Linn.* L. E. D. Mastiche Resina, L. D. Concrete Resinous Exudation, E., of the Mastic or Lentisk-Tree.

Mastic and the tree yielding it (Σχῖνος of the Greeks) were well known to the ancients, from the latter being a native of the Grecian Archipelago and of the Mediterranean region.

The Mastic shrub, about 10 or 12 feet high, is distinguished from the Terebinth tree by having its leaves pinnate without an odd one. Leaflets 8 to 10 in number, usually opposite, small, oval, lanceolate, petiole winged. Both male and female flowers small, in axillary racemes near the ends of the branches. Fruit small, roundish, of a brownish red colour when ripe.

This shrub is cultivated in the Isle of Chio, whence the Mastic is chiefly obtained by transverse incisions being made in the trunk and principal branches in the month of July. Some of it adheres to the tree in the form of tears, and some falls on the earth, and is collected in August. The best kind is in small roundish or oblong tears, of a pale yellow colour, transparent, dry, and brittle; hence usually covered with a light white powder from attrition ; becoming soft and ductile when chewed ; breaks with a vitreous fracture, and has a mild resinous taste and an agreeable odour, especially when rubbed or heated. An inferior kind, in masses of agglutinated tears, is darker coloured, and mixed with impurities.

Mastic melts when heated, and burns at a higher temperature, diffusing an agreeable odour. It is insoluble in water, but completely soluble in Ether. About nine-tenths (a resinous acid which has been called the Masticic') are soluble in cold Alcohol, and the remainder, which is soluble only in hot Alcohol, has been called *Masticine.* To this, Mastic owes its ductility when in a moist state. Besides these there is a trace of volatile oil.

Action. Uses. Little used, except as a masticatory in the East, sometimes in fumigation ; most frequently by dentists for stuffing decayed teeth. It forms one ingredient of *Tinct. Ammoniæ Comp.* (p. 59); a substitute for *Eau de Luce.* Commonly used as a varnish dissolved in Alcohol or Oil of Turpentine.

RHUS TOXICODENDRON, *Linn.* Folia, L. D. Leaves of the Poison Oak. *Pentand. Trigyn.* Linn.

The species of Rhus are known for their astringent and resinous, and some for their poisonous properties. The present species was brought into notice in this country by Dr. Alderson of Hull, in 1793, by following up the experiments of Du Fresnoi, made at Valenciennes in 1788.

Botanists differ whether this be distinct from another plant which has been called *Rhus radicans,* Poison Vine or Poison Ivy, or whether both are only varieties of one species. *Rhus Toxicodendron* forms a shrub of a few feet in height, and has its leaflets irregularly indented or deeply sinuate, hence it is called *Poison Oak.* When older or growing in favourable situations it shoots up and throws out lateral fibres, which take hold of the trees up which it climbs. The leaflets are entire or rarely toothed. Bigelow states that he has frequently observed shoots from the same stock, having the characters of both varieties. Elliott and Nuttal insist upon their being distinct species. All three are American botanists. More extended observation is required to establish the distinctness of species The leaves are long-stalked, trifoliate, with the lateral leaflets sessile, the terminal one stalked, they are broad, ovate or rhomboidal acute; leaflets smooth, sometimes a little pubescent, entire or irregularly toothed or lobed. The flowers are small, greenish white, diœcious, produced in lateral, usually axillary panicles. The *male* flowers have a small 5-partite calyx, with erect segments. Petals 5, oblong, recurved. Stamens 5 with the rudiment of a style. The *female* flowers are smaller, have 5 abortive stamens, and a globose one-celled ovary, with a short erect style terminating in 3 stigmas. The drupes are roundish, of a pale green colour, juiceless, 1-celled, containing from abortion a bony one-seeded nut.—Bigelow's Med. Bot. 111, t. 42.

When wounded, this plant, like other species of the family, exudes a milky juice, which becomes black on exposure to the air, and, as in the case of the Cashew and Marking nuts, forms an indelible stain on cotton or linen. The juice applied to the skin produces inflammation, as sometimes does linen marked with the juice of the above nuts. At night and in the shade especially, a gaseous body is exhaled, which is also of an irritant nature, producing itching, redness, and great swelling. Cases of a similar kind have occurred with gardeners touching the plant and then rubbing the eye; and Sir D. Brewster has described similar effects produced by one of the varnishes of this family of plants. These properties depend upon a volatile acrid principle, which seems to be a Hydrocarbon. The other principles have not been well ascertained; but there is Tannin, Gallic acid, a little Resin, Gummy substance, green Fecula, according to Von Mons. The leaves are alone officinal: they are without smell, but have an astringent and acrid taste when fresh.

Action. Uses. Acrid, Stimulant. Useful in some cases of Paralysis, twitches and pricking of the affected limb being experienced.

D. gr. j.—gr. v. till pricking commences. A Tincture or an Extract of fresh leaves preferable (*c.*); or frictions with oil in which the leaves have been digested.

BURSERACEÆ. *Kunth.* Balsamads.

Trees or shrubs with resinous juice. Leaves alternate, impari-pinnate or ternate, sometimes with pellucid, transparent dots, and usually with two deciduous stipules. Flowers small, regular and complete, or unisexual from abortion, in axillary or terminal racemes or panicles. Calyx free, 3—5 fid, persistent. Petals equal in number to divisions of the calyx, inserted below the orbicular or annular disk. Stamens inserted with the petals, and double their number. Ovary free, sessile, 2—5 celled. Ovules in pairs, side by side in each cell, suspended from the apex of the central angle. Style simple or wanting. Stigma undivided or 2—5 lobed. Fruit hard, bony, one to 5-celled, often single-seeded, epicarp dry, usually resinous, sometimes splitting into valves. Seeds without albumen, pendulous, cotyledons wrinkled or plaited; radicle small, superior, straight, turned towards the hilum.

The Burseraceæ are distinguished among Terebinthaceæ by their many-celled germen, and by their ovules being pendulous and in pairs; also by their plaited convolute cotyledons. They are all found in tropical parts of the world. The juice of these plants is famed for its balsamic odour and stimulant properties: *e. g.* Balsam of Gilead, Olibanum or Frankincense, Myrrh, Bdellium, Elemi, &c.

OLIBANUM, L. D. Gummi Resina. BOSWELLIA SERRATA, *Roxb.*

The name Olibanum seems to be derived from the Greek λίβανος, and this probably from the Hebrew *lebona.* This is very similar to the Arabic *luban*, which signifies *milk*, or the juice exuding from a tree, and is applied especially to what used in early times to be called Thus, and more recently, Olibanum. Two kinds of Olibanum are known in commerce, one Indian, the other African.

Fig. 55.

The Indian is imported in chests chiefly from Bombay, also from Calcutta, but the place producing it is not well ascertained. Mr. Turnbull, of the Medical Service, many years since sent some resin of the *Salai* tree collected in the hills near Mirzapore, which in the London market was recognized as Olibanum. Mr. Colebrooke determined that *luban* or Olibanum is produced by a tree called *salai.* The author has also collected a very fragrant resin from the *saleh* tree of North-west India, which bears a very close resemblance to common Olibanum. This tree is *Boswellia glabra*, Roxb., the former is *B. thurifera* of Colebrooke, called *B. serrata* in many works; but as Messrs. Wight and Arnott say, "we dare not quote here *B. serrata*, Stack. extr. Bruc. p. 19, t. 3, the leaves being usually described as ovate, oblong, and acuminate." Both species were collected by Col. Sykes in the Deccan as the Olibanum-tree. Dr. O'Shaughnessy states that he has received fine specimens of Olibanum from the Shahabad district, where it is called *salegond*, and at Chandalgur *gunda barosa.*

This tree (Boswellia thurifera, fig. 55) grows to a large size in hilly situations, from the Coromandel coast to the central parts of India. It is much branched but bare of leaves in its lower parts, but these are crowded and alternate towards the ends of the branches, unequally pinnate. Leaflets oblong, obtuse, serrated, pubescent. Stipules none. Inflorescence in single axillary racemes near the ends of the branches, shorter than the leaves. Flowers on short pedicels, of a pinkish white colour. Flowers bisexual, (v. 55, 1.) Calyx small, 5-toothed. Petals 5, obovate, tapering to the base, inserted under the margin of the disk; æstivation slightly imbricate. Disk surrounding the base of the ovary, cup-shaped, fleshy, crenulated. Stamens 10, inserted under the disk. Ovary sessile, 3-celled, with 2 ovules in each, attached to the axis. Style terminated by a capitate 3-lobed stigma. Fruit capsular, 3-angled, 3-celled, 3-valved, septicidal (splitting at the angles into valves). Seeds solitary in each cell, girded by a membranous wing. Cotyledons intricately folded, multifid.

Indian Olibanum, which is now the most esteemed, is in roundish or oblong tears, of a reddish or a light yellow colour, usually covered with whitish powder, from attrition of the pieces against each other, translucent within, of a warm bitterish taste, and having a balsamic odour, especially when warmed or burnt. Sp. Gr. 1·22. Analysed by Dr. O'Shaughnessy, a fine specimen gave of Resin 37 parts, Volatile Oil 28 parts, Gum 4, Gluten 11, in 100 parts. But the quantity of Volatile Oil is necessarily much less when it has been exposed and become dry, as seen in commerce. Braconnot obtained only 8 per cent. of Oil, of Resin 56, Gum 30, matter like Gum 5·2, loss 0·8 = 100.

AFRICAN OLIBANUM (F. *Encens d'Afrique*, G. *Africanischen Weihrauch*) imported into Venice and Marseilles from Suez, and obtained from Arabia and the east coast of Africa, is mentioned by Dr. Pereira as African or Arabian Olibanum, and as occurring in smaller tears than the Indian variety, yellowish or reddish, and intermixed with crystals of Carbonate of Lime. One kind of African Olibanum is no doubt produced on the hills of the Somauli coast westward from Cape Guardafui, and carried to the Arabian coast chiefly by native boats from Maculla. This tree, partially described by Capt. Kemthorne of the

Indian Navy, has been identified by Mr. Bennett of the British Museum with *Plöslea floribunda* of Endlicher, but which appears to the author to be nothing but a species of Boswellia, which he would therefore call *B. floribunda.* The specimens are covered with little resinous exudations, as are the leaves of a plant collected in the island of Socotra by Lt. Wellsted, which also appear to be those of a Boswellia.

Action. Uses. Stimulant. Sometimes used in chronic affections of mucous membranes, but chiefly in plasters, and as a fumigation.

MYRRHA, L. E. D. Gummi Resina, L. Gummy-Resinous Exudation, E. Balsamodendron (Protium) Myrrha, *Nees.* L. E. Myrrh. *Octand. Monog.* Linn.

Myrrh is first mentioned in Exod. xxx. 23, by the name of *Mor* or *Mur.* The Arabic name also is *Mur.* The Greeks called it μυῤῥα, and also Σμυρνα. Herodotus mentions it as produced in the South with Frankincense, &c. Dioscorides states the variety called *Troglodytica* to be the best. In the Periplus of the Red Sea, Arrian mentions Myrrh with Olibanum as exported from the coast of Barbaria, that is, the modern Berbera. Bruce learnt that it was produced, as well as Frankincense, in the country behind Azab, or in that of the Dankali. The embassy to Abyssinia under Major Harris met with it on the hills in the comparatively flat country which extends from Abyssinia to the Red Sea near the Straits of Bab-el-Mandeb, or from the Doomi Valley to the banks of the Hawash. Mr. Johnston (Trav. i. p. 249) met with it in nearly the same locality. Both authors describe the Myrrh as exuding from wounds made in the bark, and that it is collected in January and March (H.), but chiefly in July and August, and in small quantities at other times of the year (J.), and exchanged for Tobacco with the merchants who proceed to Berbera, &c., whence it is exported to the coast of Arabia.

Dr. Malcolmson writes to the author from Aden, that it is exported in native boats from different ports in the Red Sea, but chiefly from Berbera, Zela, and Massowah, and adds "there is no Myrrh produced in Arabia."

Myrrh, it is well known, now reaches Europe chiefly from Bombay. having been imported from the Arabian and Persian Gulfs. It used formerly to be obtained also from Turkey. Some Myrrh, however, appears to be produced in Arabia, as Ehrenberg and Hemprich found a small tree in Arabia near Gison, on the borders of Arabia Felix, from off which they collected some very fine Myrrh. There is still considerable uncertainty respecting the plant or plants which yield Myrrh, though it is probable they all belong to the genus Balsamodendron.

BALSAMODENDRON, *Kunth.* Amyris. Protium. *Wight and Arnott.*

Flowers often unisexual. Calyx 4-toothed, persistent. Petals 4, linear, oblong, induplicately valvate in æstivation. Stamens 8, inserted like the petals under the margin of the annular disk or torus, which is cup-shaped, fleshy, deeply crenated. Ovary 2-celled. Style short, obtuse, 4-lobed. Drupe globose or ovate, nut thick and very hard, bony, 2-celled, (one of the cells by abortion, often obliterated,) cells one-seeded.—Balsam producing trees. Leaves with 3 to 5 sessile leaflets, which are without dots.

B. Gileadense, K. Unarmed. Leaves palmately 3-foliolate, petiolate, smooth; leaflets obovate, oblong, very entire, glabrous; pedicels short, single-flowered, with the calyx broad, shallow, and campanulate.—This includes the *Amyris Opobalsamum* of Forskal; but fresh specimens and recent observations are required to distinguish whether this is different from the *Balessan* of Bruce. Other species which are found in the Peninsula of India and are *spinescent*, used to be included under it, but are considered distinct by Dr. Arnott, who does not now unite them with *Protium* (as in the Prod. Fl. Ind. Penins.) from which indeed they differ much in habit.

The *Balessan* of Bruce was found by him at Azab, and said to extend to the straits of Bab-el-Mandeb. Gerloch found it at Bederhunin, a village between Mecca and Medina. Forskal found his Opobalsamum at Haes in Arabia Felix. At Aden it is called Beshan. Dr. Roth, in the Appendix to Harris' Abyssinia (ii. p. 414) mentions B. Opobalsamum as occurring in the Adel country and the jungles of the Hawash along with the Myrrh tree. The bark when wounded exudes the fragrant and far-famed Balsam, which has been called Balm of Gilead, but which seems to have been only cultivated in Palestine near the town of Jericho. It is now never obtainable in Europe in a pure state, and therefore seldom if ever employed medicinally, though it is no doubt possessed of stimulant properties

Fig. 56.

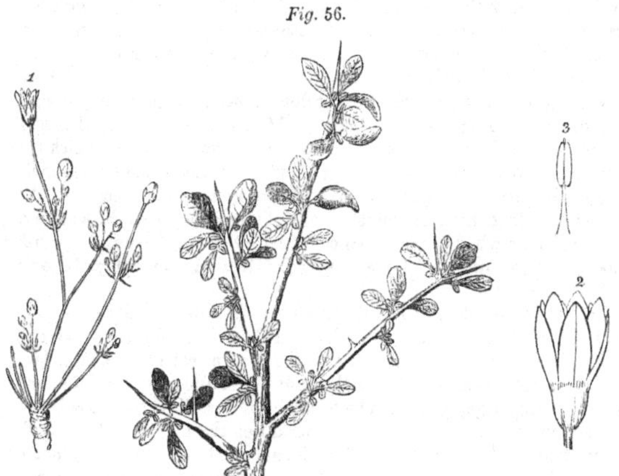

56. Balsamodendron Myrrha. 1, 2, 3. B. Kataf.

B. Myrrha (fig. 56). Stem shrubby, arborescent; branches squarrose, spinescent. Leaves ternate; leaflets obovate, obtuse, obtusely tooth-letted at the apex, the lateral smooth. Fruit acuminate (Nees).

Bark pale ash-gray, approaching white. Wood yellowish-white; both it and the bark have a peculiar odour. Leaves on short stalks. Flowers unknown. Fruit ovate, smooth, brown, somewhat larger than a pea; surmounted at the base by a four-toothed calyx, and supported on a very short stalk.

The author has adopted the above description, as translated by his friend, Dr. Pereira, who has closely followed the account given by Nees von Esenbeck in the folio work *Beschreib. Officin. Planzen*, where he says that his friend Dr. Ehrenberg collected from off this tree "*sehr schöne Myrrhe*," and that the description is taken from the specimens of the plant collected by Ehrenberg at Gison, on the borders of Arabia Felix.

Though it is not very probable, from modern information, that any large quantity of Myrrh is produced in Arabia, yet it is possible that some may be produced there, and from the same species of plant as that which yields the large quantities of Myrrh in Africa. Indeed, it is stated in the Appendix to Harris' Abyssinia (ii. 414), probably on the authority of Dr. Roth, that the "*Balsamodendron Myrrha* (*Karbeta* of the natives) grows on the borders of Efat and in the jungles of the Hawash, and in the Adel desert. The resinous gum called *Hofali* is collected for exportation. *B. Opobalsamum* (Besham) grows commonly with the former, and grows even at Cape Aden." Unfortunately there are no specimens of either plant in the Herbarium collected by the embassy and sent to the India House.

On examining the specimens in the British Museum of *B. Opobalsamum*, or *B. Gileadense*, now united into one species, it appears to the author that the specimens vary sufficiently to require careful examination and detailed description of good and complete specimens, before we can distinguish them as varieties, or determine them to belong to separate species; also whether the above *B. Myrrha* itself differs sufficiently to constitute a distinct species. The drawing of Nees, of which the annexed woodcut (fig. 56) is a copy, appears to the author to resemble some of the Arabian specimens of *B. Opobalsamum* very closely. All are very distinct, both in foliage and inflorescence, from *B. Kataf* (v. 56, 1, 2, 3), of which a specimen collected by Forskal is in the British Museum. The species referred to by Dr. Roth are probably the same as the following.

Mr. Johnston, also, in his Travels through Adel to Abyssinia (i. p. 249), in treating "of the tree that yields this useful drug, Myrrh," says "there are in the country of Adel two varieties, one a low, thorny, ragged-looking tree, with bright green leaves, trifoliate, and an undulating edge, is that which has been described by Ehrenberg." (v. the annexed figure, 56.) "This produces the finest kind of Myrrh in our shops." This may be either the above *B. Myrrha* or one of the forms of *B. Opobalsamum*. "The other is a more leafy tree, if I may use the expression, and its appearance reminded me exceedingly of the common hawthorn of home, having the same largely serrated, dark-green leaves, growing in bunches of four or five, springing by several leaf-stalks from a common centre. The flowers are small, of a light-green colour, hanging in pairs beneath the leaves, and in size and shape resemble very much the flowers of our gooseberry-tree. It belongs to the Octandria Monogynia, the eight stamens being alternately long and short, the former corresponding to the four partial clefts in the edge of the one-leafed calyx. The fruit is a kind of berry, that when ripe easily throws off the dry shell in two pieces, and the two seeds it contains escape. The outer bark is thin, transparent, and easily de-detached; the inner, thick, woody When wounded, a yellow turbid fluid (the gum-myrrh) immediately makes its appearance. Naturally, the gum exudes from cracks in the bark of the trunk near the root, and flows freely upon the stones immediately underneath. Artificially, it is obtained by bruises made with stones."

This plant, judging from the specimens deposited by Mr. Johnston in the British Museum, corresponds exactly with one, also in the same collection, obtained by Mr. Salt in Abyssinia, *Balsamodendron Kua* of Mr. Brown's MSS., and of which Mr. Salt says, he obtained from it a gum much resembling the Myrrh.

B. Africanum, *Arnott.* (*Heudelotia africana*, Guill. et Per.)

This species, first found on the west of Africa, occurs also in the Abyssinian collection, having been found in the flat country of the Adel.* It yields African Bdellium, or that imported into France from Guinea and the Senegal, according to M. Perrotet. M. Adanson, likewise, in his Travels in the Senegal, mentions it by the name of Niotout, as producing Bdellium. It may also yield the Bdellium which is exported from the west coast of Africa. Dr. Malcolmson writes to the Author that "Bdellium (of which he sends a specimen) is produced in Africa by a tree similar to the Myrtle. None is obtained in Arabia. It is very similar to Myrrh, and sometimes sold for it."

One kind of Bdellium is produced in India, which the author was informed was yielded by a tree called *googul* by the natives (v. Himal. Bot. 177), and which is the *Amyris Commiphora* of Roxb. referred by Messrs. Wight and Arnott to Balsamodendron. Dr. Walker, in his account of the drugs produced near Aurungabad, states "a gum-resin, called by the natives *googool*, is produced by a tree (Dr. W. calls it Amyris Bdellium? Roxb.) which grows in the neighbourhood of Umber, a town twenty miles to the westward of Aurungabad. Roxburgh imagined that Googool was identical with Myrrh."

But the whole of the species of this genus require to be carefully examined from good and authentic specimens accompanied by their respective products, before the several doubts can be resolved.

* Mr. Johnston immediately recognized it as one of the trees yielding gum-resin. The leaflets are like those of *B. Kua*, Br.

Prop. Myrrh is imported from Bombay. It is generally in pieces of irregular form and size, formed apparently by agglutinated tears, dry, and covered with a fine dust, commonly of a reddish-brown colour; brittle, fracture irregular, conchoidal, shining, with the surface apparently dotted with volatile oil, often varied with opaque, whitish, semicircular marks; the smaller pieces angular, shining, semi-transparent; taste bitter and aromatic, the smell peculiar and balsamic. Other kinds are also met with, probably derived from the same source, as it sometimes reaches this country with the different qualities intermixed (*Myrrh* in sorts) or when the finer pieces are picked out (Turkey or picked Myrrh). But when the process is adopted abroad, the inferior qualities may come separately or remain intermixed with other gums or resins, especially Bdellium, from the careless manner in which all are collected by the natives. The specimens brought by the Abyssinian embassy have *granular* fragments mixed with roundish tears; and some of a pale, even whitish, colour may be seen on the same piece of bark with ordinary coloured Myrrh. The specimens which have not been exposed to the air are darker coloured, moister, and of a more powerful and agreeable smell.

Indian Bdellium is sometimes sold for and considered as an inferior kind of Myrrh. It is in roundish pieces of a dull dark-red colour, more moist than Myrrh, and not brittle like it, softening even with the heat of the hand; bitter and a little acrid in taste, with a less agreeable odour. It often has portions of the birch-like bark adhering to it.

Myrrh, when heated, softens, then burns partially, leaving a black spongy ash. Triturated with water, it forms an emulsion: the Alcoholic tincture is rendered opaque on the addition of water. It is a Gum-resin with volatile Oil, and salts of several acids combined with Potash and Lime. The Gum, about 63 per cent., consists of two kinds, one-half being *Bassorin* or insoluble, and the remainder *Arabin*, or soluble Gum. This being dissolved in water, readily suspends the Resin and Oil. The Resin, about 28 per cent., is also of two kinds, one soft, odorous, and soluble in ether, while the other is hard, without odour, soluble in alkalis. The volatile Oil, about 2·5 per cent., passes over if distilled with water; it is at first without colour, but becomes yellowish, has the odour and taste of Myrrh, is soluble in Alcohol, Ether, and the fixed Oils. Upon it and the Resin the properties of Myrrh chiefly depend.

Action. Uses. Stomachic, Excitant, Stimulant Expectorant, Emmenagogue.

D. gr. x.—℈ß.; but usually united with tonics or with purgatives, as Aloes.

Tinctura Myrrhæ, L. E. D. Tincture of Myrrh.

Prep. Macerate (Digest, D.) for 14 (7, E D.) days *bruised Myrrh* ℥iij. (℥iijß. E.) in *rectified Spirit* Oij. (Oj. and f℥xiij. E. by measure ℔ß. and Proof Spirit by measure Oiß. D.) Strain (or much more conveniently, pack the

Myrrh in a percolator, add the Spirit and when f℥xxxiij. have passed through, agitate well to dissolve the oleo-resinous matter which lies at the bottom, E.)

Action. Uses. As rectified Spirit is the best solvent for the Resin and volatile Oil of Myrrh, this is an effective preparation in doses of f℥ß.—f℥j. But it is usually only employed as an adjunct, or as an external stimulant application to foul and indolent ulcers.

Off. Prep. Tinct. Ammoniæ Comp. L. Mist. Ferri Comp. L. D. Pil Ferri Comp. L. D. Pil. Assafœtidæ Comp. E. Pil. Galbani Comp. L. D. Pil. Aloes et Myrrhæ, L. E. D. Pil. Rhei Comp. E. Tinct. Aloes et Myrrhæ, E. Decoct. Aloes (Comp. D.) E.

Elemi, L. E D. Concrete Resinous Exudation from one or more unascertained Plants, E. Resin of Amyris elemifera, *Linn.* L. D. Elaphrium elemiferum, *Royle*, yields Mexican Elemi.

Elemi has long held a place in works on Materia Medica, but great uncertainty has always prevailed respecting both the country and the plant producing it. C. Bauhin mentions it as Elemi and Elemni officinarum. J. Bauhin (1650) describes it as being in cylindrical pieces, of the consistence and colour of yellow wax, with the odour nearly of Fennel, and wrapped up in a leaf which appears to be that of Canna indica. Ray (1688) states that many consider it to be the gum of the Ethiopic Olive, and the word Elemi to be derived from Elea. He refers to a passage "ex Andreæ Bacci autoris gravissimi lib. 5. de Vinis Apulis," who states that in Apulia the Olives grow to a great size, and on account, he thinks, of the great and continued heat, exude a very excellent gum, "quod Chirurgi gummi Elemi appellant." There appears no evidence to connect Elemi with the Ethiopian Olive of Dioscorides, nor does there seem any proof that it was ever obtained from thence. Indeed, J. Bauhin, as early as 1650, says of Elemi, "De loco ejus natali nihil adhuc certi accessimus."

From America, however, a resin was early received, which had the characteristics of and was compared with Elemi. Marcgraaf, indeed, describes a Brazilian tree, called *Icicariba*, which exudes a very fragrant resin smelling like freshly bruised Anethum. His editor De Laet says that large quantities of this resin are brought from New Spain, and that it is also collected in Brazil. The American Elemi seems for a long time to have been the only kind procurable, except a factitious preparation made with yellow resin, turpentine, and oil of spike, for making which directions are given by Savary and others. M. Guibourt states, as Rutty had done, that both kinds of Elemi come from America, and that of Brazil, which Geoffroy calls the *false*, is in the present day the most esteemed, on account of its strong but pleasant odour; owing to the presence of much volatile oil. If this kind does actually come from Brazil, it is most probably procured from the above Icicariba of Marcgraaf, *Icica Icicariba*, Dec., as Martins, in his *Syst. Materiæ Med. Brazil.* p. 118, mentions "Elemi occidentale, which in stimulant and healing powers excels the other balsamic resins of

Brazil," as obtained from this tree. He also enumerates other species of Icica indigenous in the same region, which pour out balsams which when dry are known as different kinds of Elemi and of Anime.

But Elemi has, for some years at least, been imported direct from Mexico, and the author has received specimens from Dr. Budd, to whom they were given by R. Cotesworth, Esq., who imports Elemi as an article of commerce from Mexico. These specimens were accompanied by specimens of the trunk, branches, leaves, and fruit; but unfortunately, mostly all detached from their points of insertion. The materials are, however, sufficient to determine that they belong to the genus *Elaphrium*, and that the species is a new one, which the author has named *E. elemiferum*, from its produce. To this probably early accounts refer, as the Elemi-like produce from New Spain.

ELAPHRIUM ELEMIFERUM, *Royle*.

Twelve feet high? stem three inches in diameter; wood white, spongy; bark about a line in thickness, rugose, of a reddish-brown colour, but covered with a grey epidermis and lichens; branches tortuous; twigs smooth, somewhat angular, striated, and flexuose; leaves exstipulate, unequally pinnate; rachis winged; leaflets 8 to 10 pairs, opposite, without dots, very variable in form, ovate, obtuse, even roundish, entire or ovate-acuminate, irregularly toothed, the terminal one usually elongated acute, the lateral ones, especially the lower pairs, are sometimes ternately or pinnately cut, with their petioles also winged; at other times all the leaflets are rhomboidal and deeply cut into acute segments, all are smooth and shining on the upper surface. Flowers not seen. Drupes ovoid and rather acute, composed of a thick and tough epicarp, which splits into two valves, and displays the blackish apex of the seed, of which the lower part is enveloped in a reddish-yellow aril-looking body (membrana tenui (pulpa molli testi, *Jacq.*) vestita, *Kunth.* in ch. Generis), which exhales a strong odour of Elemi when scraped; seed single, ovate, one being abortive; cotyledons contortuplicate; radicle above.—Native of Mexico, near Oaxaca.

Elemi has also been imported, for the last two or three years at least, from Manilla into the London market. This is in masses of a light yellowish colour, internally soft, and about the consistence and appearance of thick honey, smelling strongly of fennel, and in this respect resembling that imported from Mexico. Though it is possible that this might be conveyed as an article of commerce from Acapulco to Manilla, yet M. Perrotet obtained, in the Philippine Islands, a produce like Elemi from a Terebinthaceous tree.

Canarium commune (C. zephyrinum of Rumph, H. A. ii. t. 47), cultivated on account of its kernels in the Spice islands, and extending even to Ceylon, yields a resin, which Rumph describes as white and tenacious, of the consistence of suet, becoming by degrees yellow, and when fresh, exhaling a strong odour; that of the wild plant, he describes as "substantia, colore et odore adeo similis est Gummi Elemi, ut pro eo haberetur."

Dr. Pereira has received from Dr. Christison the resin of *Canarium balsamiferum* of Ceylon, which in odour and general appearance strongly resembles Elemi.

If the Molucca resin should be found to be exported, it will account for our having for many years received one kind of Elemi from Holland, as was long since stated to be the case by Pomet, and as has been traced out by Dr. Pereira, whence he rightly concluded that it was the produce of a Dutch settlement.

Hence it appears that the different kinds of Elemi are produced—

BRAZILIAN ELEMI, by *Icica Icicariba.* Marcg. Ic. p. 98. Piso, Ic. p. 59. Martius, Pl. Med. t. 22.

MEXICAN ELEMI, by *Elaphrium elemiferum.* Royle.

MANILLA ELEMI, probably by *Canarium commune*, Linn. Rumph. H. A. 2. t. 47. Kœnig. An. Bot. t. 7. f. 2.

The *Amyris elemifera*, Linn , which is adduced in the L. and D. P. is made up of two or three distinct plants, the above *Icica Icicariba*, and a plant figured by Catesby, Carol. 2. t. 33. f. 3. a native of Carolina, and also *Amyris Plumieri*, Dec. Plum. ed. Burm. t. 100, a native of the West India Islands.

Elemi necessarily varies according to the source whence it has been obtained; but it has generally something of a waxy appearance, is of a light, changing to a deeper, yellowish colour, with occasionally a tinge of green; soft or hard, dry or moist, according to the time it has been exposed, and the volatile oil has become evaporated. The smell of the different kind varies in fragrance, being more or less agreeable, but in some smelling strongly of fennel, especially when the resin is freshly imported and moist; in others, this odour is mixed with that of lemons. Factitious Elemi is of a dark yellow colour, is something like yellow resin, and with more of a terebinthinate odour. The Elemi analysed by Bonastre yielded 60 parts of Resin, 24 of a sub-resin, or *Elemine*, insoluble in cold Alcohol, 12·5 of volatile Oil, 2 of Bitter Extractive, and 1·5 of impurities. But the volatile Oil must be in much larger proportion in recent specimens.

Action. Uses. Stimulant. Formerly an ingredient of the ointment of Arcæus. The Turpentine must destroy the Elemi odour in

UNG. ELEMI (D.) COMPOSITUM, L. Comp. Elemi Ointment.

Prep. Melt *Elemi* ℔j. L. D. in *Suet* ℔ij. (prepared Lard ℔iv. and White Wax ℔ß. D.) Mix with them immediately *Common Turpentine* ℥x. *Olive Oil* f℥ij. Strain while hot through a sieve or linen.

LEGUMINOSÆ, *Juss.* Leguminous Plants.

The *Leguminosæ* are so named from the fruit of all consisting of a legume or pod, and form one of the largest of the natural families of plants. They may be divided into several very natural groups, but are usually treated of under the head of three sub-orders named Mimoseæ, Cæsalpineæ, and Papilionaceæ.

Herbs, shrubs or trees. Leaves alternate, compound. Stipules 2. Calyx free, often with unequal divisions, the odd segment being anterior. Petals inserted into the calyx, sometimes into the receptacle, with the odd petal always posterior, equal in number to the divisions of the calyx, or fewer from abortion. Stamens double the number of the petals, or numerous, perigynous, and in some hypogynous like the petals. Ovary simple, superior, one-celled, becoming a legume or lomentum. Embryo almost always without albumen.

Sub-order : MIMOSEÆ, *R. Brown.* Mimosads.

Shrubs or trees, rarely herbs, unarmed or furnished with prickles or thorns. Leaves often irritable; abruptly, usually bi- to tripinnate, sometimes imparipinnate; the leaflets in some being abortive and the petiole dilated vertically (forming phyllodia); these appear to have simple leaves. Flowers are complete or unisexual, regular, usually spiked or capitate. Calyx free, 4—5 parted or divided, laciniæ and petals *valvate*, rarely imbricate. Petals equal in number to and alternate with divisions of the calyx, inserted into its base, or into the receptacle, free or united more or less into a tube. Stamens *hypogynous*, very rarely subperigynous, usually numerous, free or monadelphous. Legume usually bivalved, one-celled, sometimes divided into several cells by transverse partitions, or into single seeded joints. Embryo straight, cotyledons large, radicle short, straight, plumule inconspicuous.

The Mimoseæ are found in tropical parts of the world, with comparatively few species in the north, but great numbers in the south temperate zone. They are characteristic of hot dry parts of the world. Many of the species exude gum and secrete astringent principle in their bark, wood, and fruit.

ACACIA, *Dec.* *Polygamia Monœcia*, Linn.

Flowers polygamous. Calyx 4 to 5 toothed. Petals 4—5, either free or united together, and forming a 4—5 cleft corolla. Stamens varying in number from 8—200, distinct or united into bundles. Legume continuous, dry, bivalved. Seeds without pulp. Shrubs or trees, unarmed or provided with stipular thorns or scattered prickles. Flowers yellow, white, or occasionally red, in globular heads or elongated spikes. Several of the species exude gum, and store up astringent matter in the wood, bark, and legumes.

ACACIA, L. GUMMI ACACIÆ, E. GUMMI ARABICUM, D. Gum of various species of Acacia, E., of A. vera, *Willd.* L., of A. vera and A. arabica, D. Gum Acacia or Gum Arabic.

Gum being an exudation from many trees, especially in warm and dry climates, must have been known from the most remote antiquity; and hence we find it mentioned by early Greek writers. It probably formed an article of commerce from Africa into Europe in ancient as it does in modern times. Gum being required for use in the arts as well as in medicine, large quantities are imported from the west and the east coast of Africa, from Egypt, Arabia, India, New Holland, and from the Cape of Good Hope.

Gum Arabic, so called from being supposed to be produced in Arabia, is imported in immense quantities from Africa into Aden; none is collected and very little produced in Arabia. (Malcolmson.) It used to be produced in Upper Egypt and Nubia. M. Pallme describes the gum as being collected in Kordofan, especially in the district of Bara, in November, December, and January, it "is of the finest quality, and is erroneously named Gum Arabic;" from 10 to 14,000 hundred weight being conveyed on camels from Bara to Dongola on the Nile, whence it is conveyed to Cairo, and thence distributed to Europe. As the whole of the arid desert country is covered with Acacia trees, much is also conveyed to the ports of the Red Sea, and from thence to the opposite coast of Arabia, whence it is re-exported to Bombay, and from thence to this country. This is probably yielded partly by the *Acacia vera* and *A. arabica* mentioned in the Pharmacopœias, but chiefly by *A. Seyal*, *A. Ehrenbergii*, and *A. tortilis.*

M. Pallme says that the Gum-tree of Kordofan differs materially in the shape of the tree, its leaves, and spines, from the Mimosa nilotica, that is, *Acacia vera.*

Mr. Johnston, when near the Hawash, had given to him a lump of soft Gum Arabic, nearly a pound in weight, and of most agreeable flavour, like a green ear of corn. He mentions at the same time, that the trees were without exception the long-thorned Mimosa, and tall enough to ride under. It was probably the A. tortilis or A. Seyal.

Gum Senegal is exported from Portendic, Sierra Leone, and the French settlements on the Senegal, being produced chiefly in the desert country to the north of the Senegal.

Acacia Verec is stated by the authors of the Fl. de Senegambie to yield the pale and fine varieties, *A. albida* (*A. Senegal*, Willd.), and *A. Adansonii* the inferior reddish varieties. *A. Seyal*, *A. vera*, and *A. arabica*, being found in Senegambia, probably also yield some of the gum exported from the western coast.

Barbary Gum is exported from Mogador on the west coast of Africa, and is produced in a similar kind of country. It is an inferior kind, and, moreover, a mixture of two or three kinds.

Acacia gummifera is thought to yield some of this gum. Jackson gives *attalet* as the name of the tree which produces gum.

East India Gum is exported to Europe chiefly from Bombay, having been previously conveyed there from the coast of Arabia; so that it is chiefly of African origin. But some of Indian origin is also exported from Calcutta by the name of Babool Gum, which is that of the *Acacia arabica*, and is of good quality, but Gum is yielded also by *Acacia Serissa* and *A.* now Vachelia *farnesiana*, also by species of other genera.

Gum is also imported from the Cape of Good Hope, yielded by *Acacia Karroo;* and *A. decurrens* yields gum in New Holland.

As space cannot be afforded for all, we will restrict ourselves to the description of the officinal species.

ACACIA VERA. Willd. (*Mimosa nilotica*, Linn.) A middling sized tree, with spines in pairs, straight, sharp, about a quarter to half an inch long. Leaves bipinnate, and as well as the branches smooth; these are covered with a reddish brown bark. Two pairs of pinnæ, leaflets small, 8 to 10 pairs, oblong linear, with a gland between the pinnæ. Flowers in yellow globose heads, from 2 to 5 in the axillæ of the leaves, and stalked. The legume is moniliform, short, straight and containing but few seeds. A native of Egypt and extending across Africa to the Senegal. An astringent extract, known to the ancients by the name of *ακακια*, was prepared from the legumes of this and probably of other species. The Author obtained it in the bazars of India by the old name of *akakia.*

ACACIA ARABICA. Willd. (*A. nilotica*, Delil.) Usually only a small, but growing to a tree 40 feet in height. Spines in pairs, usually short. Leaves bipinnate, with 4 to 6 pairs of pinnæ, with a gland between the first and between the last pairs. Leaflets from 10 to 20 pairs, minute, oblong, smooth. Flowers yellow, fragrant, in globose stalked heads, axillary and subternate. Legume stalked, moniliform, long and curved, compressed, contracted on both sutures between each seed. A native of Egypt, extending across Africa to the Senegal; found in Arabia and in every part of India.—Roxb. Corom. Pl. ii. t. 149.

Gum, dissolved in water or in the form of a mucilage, is a very generally diffused principle of vegetables. It flows from the several Acacias in a liquid state, but soon hardens, and may be seen from a perfectly colourless substance to different shades of yellow, even on the same tree. It is in dry, semitransparent, roundish masses, of the size of a small nut, or larger, often in fragments, rugose at the surface, brittle, friable; fracture vitreous; without odour; of a mild, slightly sweetish, viscous taste. Sp. Gr. about 1·31 to 1·52. It is soluble in water, having a slightly acid reaction on litmus. The finer pieces are often separated after being imported, and form the picked or Turkey Gum; the inferior kinds are in larger or irregular pieces, of a deeper colour, more mixed with impurities, and less soluble. It is insoluble in Alcohol, which, indeed, precipitates it from its watery solution. Sesquichloride of Iron forms with it a brown jelly. Diacetate of Lead and Silicate of Potash also cause a white precipitate in this solution. Boiled with Sul′, a variety of Sugar is produced; but if with Nit′, Mucic and Oxalic acids. Subjected to heat, it loses 17·6 per cent. of water. It yields 3 per cent. of ashes, composed chiefly of Carbonate of Potash and Carbonate of Lime, with a minute portion of Oxide of Iron, and is composed of 79·6 of soluble Gum or *Arabin*, which displays the characteristics of pure Gum, and is composed of $C_{12} H_{11} O_{11}$ or $C_{24} H_{22} O_{22}$ (*Liebig*), and is therefore identical in composition with Cane Sugar. But Gum also contains a little Nitrogen.

Adulterations. The inferior kinds of Gum are apt to be intermixed with the finer kinds of African Gum, especially when powdered; but in this state, starch also is apt to be added, but may be detected with Iodine.

Action. Uses. Demulcent. Used in its solid form, or in powder, or in the form of Mucilage.

MUCILAGO ACACIÆ, E. Mucilago Gummi Arabici, D. MISTURA ACACIÆ, L. Mucilage.

Prep. Take bruised *Acacia* ℥x. (℥ix. E.; ℥iv. D.) add it gradually and mix with boiling (cold, E.) *Aq.* Oj. (f℥iv. D.) Dissolve (without heat, but stirring occasionally; strain through linen or calico, E.)

Action. Uses. Demulcent. May be taken *ad libitum.* Employed pharmaceutically to suspend powders, or to make a mixture with oily and resinous substances. The E. process is the best, as heat is injurious. The Dublin preparation is unnecessarily thick.

MISTURA ACACIÆ, E. Acacia Mixture.

Prep. Steep in hot *Aq. Sweet Almonds* ℥j. and ʒij. peel and beat them to a smooth pulp in an earthenware or marble mortar, first with *Pure Sugar* ʒv. and then with *Mucilage* f℥iij. add gradually *Aq.* Oij. stirring constantly. Strain through linen or calico.

Action. Uses. Demulcent: very similar to the Almond Mixture of the L. P.

TROCHISCI ACACIÆ. Acacia Lozenges.

Prep. Mix *Gum Arabic* ℥iv. *Starch* ℥j. and *Pure Sugar* ℔j. with *Rose-water*, and make into a mass fit for forming into lozenges.

CATECHU. Ligni Extractum, L. D. Extract of the Wood, E. Acacia Catechu, *Willd.* CATECHU. Extract of the Wood of the Catechu Acacia.

The early history of Catechu is obscure: it must have been known in India from very early times, as it is one of the ingredients of the compounds which they chew with the leaf of the Betle Pepper. But the Persian works on Materia Medica in use in India do not quote any Arabic or Greek names for it. It was known to Garcias ab Horto, who supposed it to have been the Lycium of Dioscorides. But this the author has found to have been the extract of Barberry-root.*

A great variety of extracts are now known, which are prepared

Fig. 57.

* See a Paper on the Lycium of Dioscorides, by the Author. Linnean Trans. vol. xvii. p. 83.

from the wood, bark, and fruit of various plants. The L. and D. Colleges restrict themselves to that of the *Acacia Catechu.* The E. P. mentions also that prepared from the Betel-nut or seed of the *Areca* Catechu Palm, and also refer to that of *Uncaria Gambir.* The Acacias, however, are most noted for the secretion of astringent principle in the wood, bark, and legumes of various species, which hence form articles of commerce, and are employed in tanning. But none are more valuable than the officinal species yielding Catechu.

ACACIA CATECHU. *Willd.* A tree from 15 to 20, but sometimes 30 feet high, with hard and heavy wood of which the interior is of a dark red or brownish colour, and the sap-wood white. Branches with stipulary thorns. Leaves bipinnate. Pinnæ 10 to 15 pairs. Leaflets 30 to 50 pairs; linear oblong, unequal and auricled on the lower side at the base; petiole angular, often armed in arid situations with a row of prickles on the under side, with one large urceolate gland below the lowest pair of pinnæ and smaller ones between the 2 to 4 terminal ones. Inflorescence a spike, 1 to 3 together in the axillæ of the leaves. Flowers numerous white. Calyx downy, 5-fid. Petals united into a 5-fid corolla. Stamens numerous, distinct, double the length of the corolla. Ovary shortly stipitate. Style the length of the stamens. Legumes straight, thin and flat and smooth, with about 4—6 seeds. A native of the jungles and low hills of many parts of India.—Roxb. Corom. Plants, 11, t. 175.

Catechu (called *Kut* and *Kutch* by the natives of the East and Cutch, and Terra Japonica in commerce) is properly an extract prepared from the wood of the above tree; but the term is now applied also to other extracts similar in appearance and properties. It should be confined to these, and the term Kino applied to astringent *natural* exudations. The mode of preparing Catechu, by cutting into chips the inner brown coloured wood, and making a decoction which is afterwards evaporated to a proper consistence, was first accurately described by Mr. Ker, as practised in Behar; so it is on the confines of Nepal. The Author has seen the same process in north-west India. We have evidence that it is so prepared on the Malabar coast, and also in Ava, from the same tree.

There is no proof that any Catechu is obtained from *Butea frondosa*, q. v. Some is prepared from the kernels of *Areca Catechu*, q. v., and a kind called Gambir from the leaves of *Uncaria Gambir*, q. v.

Catechu is seen either in square or roundish pieces or balls, varying in colour, from a pale whitish or light reddish-brown to a dark brown colour; either earthy in texture, or lamellated, or presenting a smooth shining fracture. Some kinds are hence more friable than others; all are without smell; the taste is bitter, astringent, followed by a little sweetness. The pieces are generally of a darker colour externally than they are in the inside. Some of the kinds are covered with Rice husks, others are enveloped in leaves, which Dr. Pereira has ascertained to be those of *Nauclea Brunonis*, a native of the Malayan peninsula.

The *pale* variety is usually distinguished from the *dark*-coloured, and said to be imported from Calcutta; but we have obtained both kinds in the bazaars there, the pale being imported from the upper provinces, and the dark from Pegu and Singapore. The dark-brown Catechus are obtained from Bombay ; but both kinds may no doubt be prepared from the same tree, as a greater degree of, or longer continued heat, and greater exposure to light, is said to produce the dark colour. The dark are heavier, more dense in texture, and have a resinous fracture.

The largest portion of good Catechu is taken up by water, especially when boiling, the infusion being of a light or reddish-brown colour, according to its strength : it reddens litmus, and is strongly astringent in taste. It yields a precipitate with the salts of Alumina, also with Acetate of Lead, and one of a blackish-green colour with the salts of the Sesquioxide of Iron. From forming a curdy precipitate with a solution of Gelatine, Catechu is applicable to the tanning of leather, for which it is now much employed.

Sir H. Davy, in analysing the Dark and Pale Catechu, or the Bombay and Bengal, as they were called, obtained from

	of Tannin	Extractive	Mucilage	insol. residuum.		
Dark Catechu	109	68	13	10	=	200
Pale „	97	73	16	14	=	200

The Tannin of Catechu is very similar in properties to that obtained from Galls. The principle called Extractive by Sir H. Davy, has by others been called Resinoid matter, Resinous Tannin, and of late *Catechine* and *Catechuic acid.* This is most easily obtained by treating Gambir with cold water: the Tannin being dissolved, the insoluble residue is impure Catechine, which may be purified by solution in Alcohol and subsequent crystallization, when it appears as a white powder, but is in silky needles, and has something of a sweetish taste, producing a green colour with the salts of Iron. It is composed of $C_{15} H_6 O_6$.

Tests. Catechu being of such different qualities, and liable to be mixed with mechanical impurities, means are adopted for ascertaining the quantity of Tannin; as by ascertaining the weight of the precipitate made by Gelatine. The E. C. states that "the finest qualities yield to Sulphuric Ether 53, and the lowest qualities 28 per cent. of Tannin dried at 280°." Dr. Pereira having remarked that the Catechine would be dissolved as well as the Tannin by the Ether, Dr. Christison has stated that it is necessary to deduct from the dry residuum of the Ethereal solution what is left when it is acted on by cold water; as this will dissolve the Tannin and leave the Catechuic acid.

Action. Uses. Powerful Astringent. Applied externally, or taken internally.

D. gr. x.—gr. xxx. or more of the powder ; or allow a piece of pale Catechu, as pleasanter tasted, to dissolve in the mouth ; or soak a lump of Sugar in the Tinct. of Catechu, in relaxation of the throat, &c.

INFUSUM CATECHU (E.) COMPOSITUM, L. D. Infusion of Catechu.

Prep. Macerate for 1 (2, E.) hour in a vessel lightly covered in boiling *Aq. dest.* Oj. (f℥xvij. E.; ℔ß. D.) bruised *Catechu* ʒvj. (ʒijß. D.; powdered Catechu ʒvj. E.) bruised *Cinnamon* ʒj. (ʒß. D.) Strain (through linen or calico, E. D.; add Syrup f℥iij. E.)

Action. Uses. Powerful Astringent, in doses of f℥iß. three or four times a day.

TINCTURA CATECHU, L. E. D. Tincture of Catechu.

Prep. Macerate (Digest, E.) for 14 (7, E. D.) days *Catechu* (in fine powder, E,) ℥iijß. (iij. D.) bruised *Cinnamon* ℥ijß. (℥ij. D.) in *Proof Spirit* Oij. (by measure ℔ij. D.) Strain. (Express strongly the residuum. Filter or prepare by percolation, E.)

Action. Uses. As Proof Spirit dissolves both the astringent and resinoid principles, this Tincture is strongly astringent, and useful as an adjunct to Chalk Mixture, &c. in doses of fʒj.—fʒij.

ELECTUARIUM CATECHU (E.) COMPOSITUM, D. Compound Catechu Confection.

Prep. Take *Catechu* ℥iv. *Kino* ℥iv. (iij. D.) *Cinnamon* ℥j. (ij. D.) *Nutmeg* ℥j. E. Pulverize, then mix *Opium* diffused in a little *Sherry Wine* ʒiß. and *Syrup of Red Roses* (of Ginger, D.) reduced to the consistence of Honey Oiß. (by weight ℔ij¼. D.) Mix. (Beat into a uniform mass, E.)

Action. Uses. Aromatic Astringent, with gr. ij.ß. of Opium in ℥j. D.

D. ℈j.—ʒij.

CÆSALPINEÆ, or Cæsalpiniads, *Brown.*

The Cæsalpineæ abound in tropical and warm parts of the world; a few, as Cercis Siliquastrum, spread into more northern latitudes. Some are highly ornamental. The wood of many is red-coloured and astringent. Hymenæa Courbaril yields a resin, the *Gum Anime* of the shops. The leaves and fruit of some are purgative, as of the Cassia Sennas and of the Tamarind.

Calyx 5-toothed or bilabiate, deciduous or withering on the plant. Corol irregular, imbricated, subpapilionaceous, or nearly regular, spreading, of 5 petals, which are free, inserted into the bottom of the calyx. Stamens 10, or fewer from abortion, often unequal, perigynous, or inserted with the petals, usually free, sometimes united. Ovary free, placenta unilateral. Seeds without albumen. Embryo straight. Leaves alternate stipulate, impari- or abruptly pinnate, sometimes single.

HÆMATOXYLON, L. E. D. Lignum, L. D. Wood, E. HÆMATOXYLON CAMPECHIANUM, *Linn.* Logwood. *Decandria Monogynia*, Linn.

Logwood is noticed by Monardes for its medical uses; but it has also been long employed in the art of dyeing. It is a native of the coast of Campeachy, but is now common in the West Indies, as also in India.

Fig. 58.

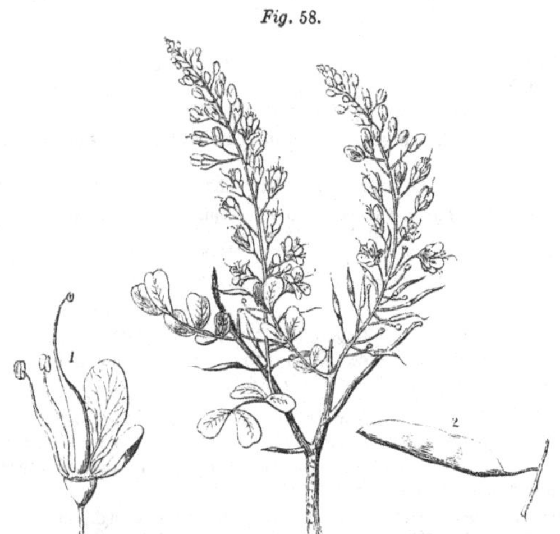

A tree (fig. 58) of moderate size, stem generally crooked, furnished with spines in arid, but unarmed in moist situations. Leaves 2—4 from the same point, pinnate; leaflets 2 to 4 paired, obovate or obcordate. Flowers in racemes, shortly stalked, yellow. Sepals 5, united at the base into a permanent cup, the lamina of which are purplish and deciduous. Petals 5, obovate, a little larger than the sepals. Stamens 10, hairy at the base. Legume (2) small, compressed, lanceolate, pointed at each end, 2-seeded, sutures indehiscent, valves bursting longitudinally in the middle.—*Sloane*, Hist. 2, t. x. f. 1 to 4.

The sap-wood of this tree, being light-coloured, is rejected, but the interior red-coloured wood is imported in logs, chiefly for the use of the dyer. These are externally of a dark colour, internally yellowish-red. The wood is hard, close-grained, and tough; usually in chips. Sp. Gr. 1·057. It has a slight but rather pleasant smell when in mass, which is compared to that of Iris-root; the taste is slightly bitter and astringent, with a little sweetness. Both water and Alcohol take up its active principles; acids render its decoction of a brighter red, and throw down a slight precipitate. Alkalis produce a purplish-colour. Alum, Acetate of Lead, and the salts of Iron throw down precipitates, the last of a bluish-black colour; and Gelatine, reddish-coloured flakes. Analysed by Chevreul, it yielded Volatile Oil, an Oleaginous or Resinous matter, a brown substance containing Tannin, Glutinous matter, several salts, and a peculiar azotized, crystalline colouring substance, *Hæmatine*, which is sometimes deposited in the form of crystals in the wood, and may sometimes be obtained by evaporating red ink. It has a subastringent and slightly bitter taste.

Action. Uses. Mild Astringent and Tonic.

DECOCTUM HÆMATOXYLI, E. D. Decoction of Logwood.

Prep. Boil *Logwood* in chips ℥j. (℥jß. D.) in *Aq.* Oj. (Oij. D.) down to f℥x. (Oj. D.) and add towards the end, *Cinnamon* powder ʒj. and strain.

Action. Uses. Astringent: given in Diarrhœas in doses of f℥j.–f℥ij.

EXTRACTUM HÆMATOXYLI, L. E. D. Extract of Logwood.

Prep. Macerate *Logwood* in coarse powder ℔jß. (℔j. E.) in *Aq. dest.* Cij. (Cj. E.) for 24 hours. Then boil down to a gallon (Oiv. E.); strain the liquor while hot and evaporate to a proper consistence. Some is imported.

Action. Uses. Astringent in doses of grs. x.—ʒß.

TAMARINDUS, L. E. D. Leguminis Pulpa, L. D. Pulp of the Pods, E. TAMARINDUS INDICA, *Linn.* Fruit of the Common Tamarind Tree. *Monadelphia Triandria*, Linn.

The Tamarind is a native of India, and has been long used there as an article of diet and in medicine. The Arabs, on becoming acquainted with it, called it *Tamr hindee*, that is, "the Indian Date," whence, no doubt, the Latin name is derived.

A lofty tree with crooked branches, remarkable for its light and elegant foliage. Leaves abruptly pinnate, with 10 to 15 pairs of leaflets, which are small, narrow, oblong, obtuse. Stipules small, deciduous. Flowers in lateral and terminal racemes, of a yellow colour variegated with red. Calyx turbinate, at the base, limb bilabiate, reflexed, upper lip tripartite, lower broad 2-toothed. Petals 3, unilateral, the middle cucullate. Stamens 2—3, united together, fully developed, 7 very short and without anthers. Ovary stalked, style subulate. Legume pendulous, broad and thickish, more or less curved, having externally a hard but brittle scabrous rind, which does not separate into valves, but under it run some woody fibres, and there lies some acidulous reddish-brown pulp. Seeds from 3 to 12, covered by a membranous coat, flattened, bluntly 4-angled, smooth, hard and brown coloured, inserted into the convex side of the legume. There is no solid foundation for the distinction into two species; *T. orientalis* being supposed to be 6—12 seeded, and *T. occidentalis* 3—4 seeded.

Tamarinds are imported either simply dried, as from India, where there are two varieties, one a dark and the other a light-coloured fruit. In the West Indies, the outer shell having been removed, they are preserved either between layers of moist sugar or in syrup. Preparations are also made from them with sugar in India, which are employed in making sherbets. They are also used in preserving fish, which is hence called Tamarind Fish. The proper officinal part is the pulp stored up between the seeds and husk.

Tamarinds have a powerful acid taste, but when preserved, they are sweet and acidulous, and then form a dark-coloured adhesive mass, containing pulp, stringy fibres, seeds, and sugar. Vauquelin, in one of the first analyses he published, found of Citric′ 9·4, Tartaric′ 1·55,

Malic' 0·45, Bitartrate of Potash 3·25, Sugar 12·5, Gum 4·7, Pectin 6·25, Parenchymatous fibre 34·35, with Water 27·55=100.

Action. Uses. Refrigerant, Laxative. A Syrup of Tamarinds diluted with water makes an excellent refrigerant drink. An Infusion may be similarly used, as also Tamarind Whey, made by boiling Tamarind pulp ℥ij. in Milk Oij.

Off. Prep. Inf. Sennæ c. Tamarindis, D. Inf. Sennæ Comp. E. Confect. Sennæ, L. Confect. Cassiæ, L.

CASSIA, *Linn.* *Decand. Monog.* Linn.

Calyx of 5 sepals, which are united at the base, and more or less unequal. Petals 5, also unequal. Stamens 10, free, the 3 upper short, rarely fertile; the 7 others bearing anthers, but often unequal. Anthers opening at the apex by 2 pores. Ovary stalked, usually arched. Legume usually compressed, many-seeded.—Trees, shrubs, or /herbs, of tropical countries. Leaves simply and abruptly pinnated, leaflets opposite. Petioles often glandular. Flowers yellow.

CASSIA, L. Cassiæ Pulpa, E. Leguminum Pulpa, L. D. Pulp of the Pods, E. Cassia Pulp. CASSIA FISTULA, *Linn.* Purging Cassia.

This has been unfortunately named, as it is constantly confounded with the Cassia yielded by the family of Laurels, with which the present product has nothing to do. The plant is a native of India, where it is constantly employed in medicine by the natives, and thus became known to the Arabs. The tree has been introduced into the West Indies, whence Cassia pods are now imported, as well as from India and the north of Africa.

One of the most showy of trees, having something of the foliage of the Ash, with the inflorescence of the Laburnum. Leaves from 12 to 18 inches long, with from 4 to 8 pair of opposite, ovate, rather pointed leaflets, smooth on both sides, of a light green colour, from 2 to 6 inches long and 1 to 3 broad. Stipules minute. Petioles round, without glands. Racemes 1 to 2 feet long, pendulous, without bracts. "The three lower filaments longer than the others, with oblong anthers opening by two lines in the face, the other 7 clavate, with pores at the small end." (Lindley.) Ovary slender, smooth, one-celled, with numerous seeds, and without any transverse separations. Legumes cylindrical, 1 to 2 feet long, smooth, somewhat obtuse, indehiscent, marked externally with 3 longitudinal bands, one being opposite to the two others, divided into a number of spurious cells by transverse partitions. Seed one in each cell, surrounded by a soft blackish-coloured pulp. On account of this peculiarity of the legume, this plant is sometimes but unnecessarily placed in a separate genus, *Cathartocarpus.*

The pods being officinal on account of the pulp, those are to be chosen which are heavy, and in which the seeds do not rattle. The pulp is of a blackish colour, viscid, with a rather mawkish sweet taste, and a slight sickly odour. The L. P. contains directions for separating the pulp, by washing it out from the bruised pods with boiling water, straining, and then evaporating to a proper consistence. Soubeiran states that four ounces of pod give one ounce of pulp. This, analysed by Vauquelin, yielded Sugar, Gum, Extractive, Vegetable Jelly, Gluten, Parenchyma, Water. M. Henry states that the Sugar possesses the nauseous taste peculiar to the pulp, and he has announced the presence of a principle having many of the properties of Tannin.

Action. Uses. Laxative ; in large doses, Purgative.

D. ʒj.—ʒij. as a laxative; but apt to create flatulence, &c. Usually given in combination, as in Conf. Sennæ, and in

CONFECTIO (ELECTUARIUM, D.) CASSIÆ, L. Confection of Cassia.

Prep. Dissolve *Manna* ℥ij. in *Syrup of Rose* f℥viij. (of *Orange-peel* ℔ß. D.), add *Cassia-pulp* ℔ß. *Tamarind-pulp* ℥j. evaporate to a proper consistence.

Action. Uses. Laxative in doses of ʒij.—℥j. for adults.

SENNA, L. E. D. Folia. Leaves of CASSIA LANCEOLATA, C. OBOVATA, and of other species. (*C. Senna*, Linn.), D. Senna-leaves.

Senna has been distinctly known only since the time of the Arabs; but they refer to the legumes only, though the leaves have long been employed in the East. There is, however, great uncertainty respecting the species of Cassia which yield the different commercial varieties. This is owing partly to all the Senna countries not having been thoroughly explored, and partly to species having been formed from imperfect specimens, and others from leaves collected out of different samples of the Sennas of commerce. The following species seem to be clearly distinct. The Author has changed the name of Forskal's *C. lanceolata*, in consequence of the great confusion which has arisen from this name having been applied by so many authors to the sharp-leaved Senna, which is imported in such large quantities both from Arabia and Egypt, and to which he restricts it.

1. C. FORSKALII (C. lanceolata, *Forsk.* and *Lindley*, Fl. Med. p. 259). Leaflets in 4 or 5 pairs, never more; oblong and either acute or obtuse, not at all ovate or lanceolate, and perfectly free from downiness even when young; the petioles have *constantly* a small round brown gland a little above the base. The pods are erect, oblong, tapering to the base, obtuse, turgid, mucronate, rather falcate, especially when young, at which time they are sparingly covered with coarse scattered hairs. (Lindl. *l. c.*) Collected by Dr. S. Fischer in Palm-grounds in the valley of Fatmé, flowering at the end of February. Forskal describes this as being distinguished "glandulâ supra basin petioli." It was found by him at Surdud and about Mor. It is called *Suna* by the Arabs, and probably yields some of the Arabian Senna of commerce.

2. CASSIA LANCEOLATA (Fig. 59, taken from the Author's "Illustrations of Himalayan Botany," t. 37.) This is a bushy annual, of about 2 to 3 feet in height, extremely leafy, and of most luxuriant inflorescence in a cultivated state. The stems are erect, round, smooth, a little flexuose towards the apex. The leaves alternate, abruptly pinnate. The leaflets 5 to 8 pairs, with short petioles, ovato-acute in the lower and lanceolate-acute in the upper parts of the plant, "slightly mucronulate, smooth above, rather downy beneath (especially in young leaves), with the veins turning inwards and forming a flexuose intramarginal line; petioles without glands; stipules softly spinescent, semihastate, spreading, minute." Racemes axillary and terminal, erect, rather longer than the leaves. Ovary linear, downy, falcate, with a smooth recurved style. Legumes (3) pendulous, membranous, flat, only slightly protuberant over the seeds, oblong, sometimes elliptical, nearly straight, with the upper margin a little curved, tapering abruptly towards the base, and rounded at the apex, of a brown colour, containing from 5 to 8 white rugose seeds (2). These are figured by Gærtner, ii. t. 146. It is probably the *Cassia medica* of Forsk. p. cxi., and agrees

Fig. 59.

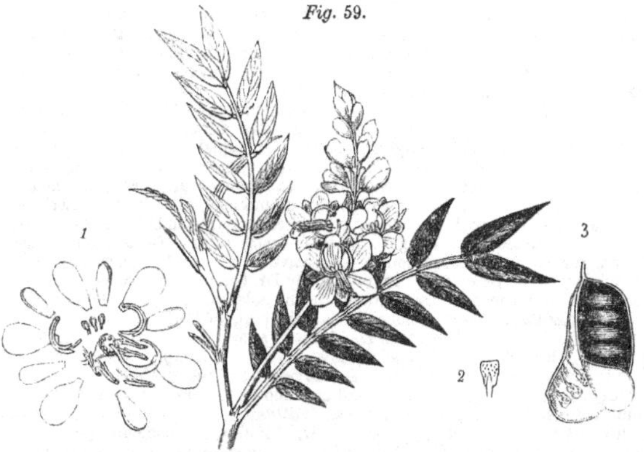

with his specimen of "Senna Meccæ Lohajæ inveniebatur foliis 5—7 jugis, lineari-lanceolatis," p. 85, of which Forskal states large quantities are yearly exported from the district of Abu-arisch to Jidda. This species includes:

a. Tinnivelly Senna, cultivated by Mr. Hughes in the south of India; also that cultivated by the Author at Saharunpore, *C. lanceolata*, Royle, Him. Bot. t. 37, and by Dr. Wight near Madras. v. fig. 60, B. and spec. in Brit. Mus. It is the *Cassia officinalis* of Gært. and Roxburgh, Fl. Ind. ii. p. 346, which name ought to have been retained, or the above *C. medica*, Forsk. instead of *C. elongata* being coined, especially as this was formed from the leaves of a cultivated Indian Senna found in commercial samples. It is cultivated by Dr. Gibson, near Poona.

Dr. Burns writes that he has found the lanceolate Senna wild near Kaira in Guzerat. His cultivated specimens, if picked, would form good Senna.

b. C. lanceolata of most authors. *C. acutifolia*, Hayne, ix. t. 41. Nees and Eberm. t. 345. St. and Church. Pl. 30, as C. Senna. These best represent the form of Alexandrian Senna (v. 60. A., a small leaf), and specimen in Brit. Mus. from Senaar. (*Kotschy.*) It is found in the valleys of the desert to the south and east of Syene or Assouan, and collected for the trade to Cairo, forming 3-5ths of Alexandrian Senna.

c. C. acutifolia, called of Delile, Esenbeck and Eberm. t. 346. (fig. 60, C.) The leaflets are narrower and more tapering towards the apex than the foregoing, as might be expected in a poorer soil and drier climate. Some of the Indian specimens in Dr. Rottler's Herbarium closely resemble this variety; also African specimens from Tajowra to the south of the Straits of Bab-el-Mandeb.

The Author is unable to distinguish these by any permanent characters, nor dried Senna-leaves cultivated at Saharunpore from good specimens of *Bombay Senna* (that is, ordinary Indian Senna) imported here from India; nor these from *Suna Mukki* sent him by Dr. Malcolmson from Aden, and which he states are "the produce of Africa, but in appearance exactly resemble the Arabian Suna. In the market both are sold as one kind, and bring the same price."

3. C. OVATA of Merat, Dict. de Mat. Med. ♄. 613. *C. æthiopica*, Guibourt. Is probably a distinct species, as it is said to have a gland at the base of the petiole and another between each pair of leaflets. The leaflets are in 3 to 5 pairs, exactly oval acute, slightly pubescent below; the follicles are thin, pale yellow-coloured, 1-3rd smaller than those of *C. obovata*. It is said to be found both in Nubia and Fezzan, and to furnish exclusively the Senna of Tripoli, Sené de Tripoli. It is extremely like a variety of *C. lanceolata*. The figure of C. Senna in Stevenson and Churchill, Med. Bot. t. 30, quoted by Dr. Pereira as representing this plant, is referred to by Dr. Lindley as a good representation of

C. acutifolia of Delile. But Merat and De Lens say of it: "Nous ne le connaissons que par les feuilles et les fruits qu'on en voit dans la commerce." M. Guibourt calls it *C. æthiopica;* but instead of referring to *C. lanceolata* of Colladon, Pl. xv. f. e. as representing this species, he says that it is exactly represented by the Sené de Nubie of Nectoux, pl. 2.

4. C. OBOVATA, *Colladon.* Hayne, ix. 42. Nees and Eberm. 347. Diffuse herbaceous plant. Leaves equally pinnate, glandless. Leaflets 4-6 pairs (somewhat villous, Roxb.) obovate, obtuse, but slightly mucronate, unequal at the base, the terminal pair more cuneate and larger. Stipules triangular, narrow, and tapering, rather stiff and spreading. Flowers yellow, in racemes. Bracts ovate, cordate, acuminate, concave, single-flowered. Legumes broad, membranous, smooth, lunate in shape, rounded at each end, with an elevated crest over each side on both valves, so as to form an interrupted ridge along the middle of each valve. Seeds 6 to 8, wedge-shaped, rugose as in C. lanceolata.—A native of Africa, from Senegal (*Fl. de Senegambie*) to the Nile; found in Fezzan by Dr. Oudney (*R. Brown*), in Egypt from Cairo to Assouan, Nubia; found in the Adel country near Sultalli (*Mission to Abyssinia*); Desert of Suez; Syria; dry parts of India, as Kaira (*Burns*); Guzerat, Dekkan (*Col. Sykes*), near Delhi, and Valley of Rungush, near Peshawar (*Falconer*); high dry uncultivated lands of Mysore (*Roxburgh. Wight*). It has been cultivated in Italy (*Sene d'Italie*), and forms 3-10ths of Alexandrian Senna.

This species is very distinct, in its obtuse obovate leaves and crested legumes, from the preceding acute-leaved species. *C. obtusa* Roxb. was probably described from young legumes, as the author, like Dr. Lindley, has compared good specimens from Mysore with others from Africa. The *obtusata* (fig. 60, H.) of Hayne does not seem to differ sufficiently from his *C. obovata* (fig 60, G.) It is possible, however, that there are two very similar species in Africa. Mr. H. Grant, late of the India House, has favoured the author with a specimen in flower, collected by him in February at Philæ, which has upon it *both* obovate (H.) leaflets and some which are ovate and acute (G.)! Lieut. Wellsted's collection contains a specimen from the coast of Arabia, of which the leaflets are obtuse, elliptic, and hairy.

The Sennas of commerce may be arranged as follows.

1. TINNIVELLY SENNA, first cultivated in the district of that name, in 12° of N. lat. by the late Mr. Hughes, from seed probably obtained from Arabia or picked out of *Suna Mukki*, as was done by the author when he cultivated Senna at Saharunpore. (*v.* Himal. Bot. p. 186, t. 37, and Trans. Med. Soc. of Calcutta, v. p. 433.) The author also grew Senna from Tinnivelly seed sent to him by Sir C. now Lord Metcalfe; but he did not find the smallest difference between the two when grown in the same situation. The Tinnivelly Senna is well-grown and carefully picked; the leaflets are of a fine rather lively green colour; thin, but large, being from one to two inches in length, truly lanceolate. This kind is "highly esteemed in this country, and is quickly displacing all the other sorts in this (that is, Edinburgh), and many other cities in Britain." (Christison.) Dr. A. T. Thomson says of it, it is mild in operation, certain as a purgative, and operates without griping. It is now cultivated by Mr. Hughes' successor.

Saharunpore Senna, the same kind of Senna, cultivated at Saharunpore, differed only in the leaflets being smaller, as might be expected from the more northern latitude (30°). These the author prescribed in the hospitals at Saharunpore, and found them effective as a purge, and operating without producing inconvenient nausea or griping. Mr. Twining, after trying them in forty-five cases in the General Hospital at Calcutta, says, in his report to the Medical Board:

"From these trials, I am disposed to consider the Senna now under trial equal to the best I have ever seen."

Madras Senna. Senna is now imported also from Madras, the produce of that Presidency. In 1843-44, I find 11,536 lbs. were exported to this country, having been previously imported into Madras from Tinnivelly, where it is cultivated by the natives, and is of the same nature as Mr. Hughes' Senna, though not so well grown nor so carefully picked. Dr. Christison says of it, the leaflets are longer than those of Bombay Senna, and not so taper-pointed, but otherwise differ only in being better preserved, and being more active, are more esteemed.

Dr. Searle, in a communication to the India House, says of this Senna, that "now furnished to the profession by the Madras Government is in my experience as good quite as the Alexandrian," "every leaf of the Indian being of the genuine spear-shaped species."

2. Bombay or Common Indian Senna, *Suna Mukki* of the natives, is first imported into Bombay from the Arabian Gulf:

316,728 lbs. in 1837-38. 570,426 lbs. in 1838-39.
Re-exported to Great Britain, 262,284 lbs. in 1838-39.

That this Senna forms a large, if not the largest proportion of what is consumed in this country, is not only evident from the above importation, but also from a comparison with the whole quantity of the other Sennas imported, as given by Dr. Pereira.

	1838.	1839.
From East Indies	72,576 lbs.	110,409 lbs.
From other places	69,538 „	63,766 „

Some of this Senna is no doubt produced in Africa, as stated above by Dr. Malcolmson; a good deal of it in Arabia, probably by *Cassia lanceolata*, and some perhaps by *C. Forskalii*. The leaflets are thin, lanceolate, usually entire, about an inch or an inch and a half in length, narrower than either the Tinnivelly or Saharunpore Senna, probably from growing in a poorer soil and drier climate. They are of a pale-green colour, often with dark brown-coloured leaflets intermixed, also some pods, and many leafstalks, with occasionally other impurities, The good specimens of this Senna are, however, of excellent quality, and its commercial and medical value would be much increased, if the finest leaflets were picked out. It is in constant use in hospital practice in India, and generally highly approved of. The author prefers them for all purposes to the following kind as found in commerce.

3. Alexandrian Senna is an excellent kind, when the genuine lance-shaped leaflets have been picked out; but that commonly employed in this country, is a mixed and very impure kind, being made up of the leaflets, much broken, of *C. lanceolata* and of *C. obovata*, with some pods and broken leaf-stalks, and also with leaves of other plants. It should be used only after having been carefully picked, as directed in the E. P. Picked Alexandrian Senna is of a pale green colour,

with a faint smell. The leaflets are broad-lanceolate, the two sides unequal; they are thicker and shorter than the Indian Sennas.

The lanceolate Senna of Upper Egypt, Nubia, and Senaar, yields two crops annually, the plants being cut down in spring and autumn, dried in the sun, when the leaves are stripped off, packed in bales, and sent to several *entrepôts*, and finally to Boulac in the vicinity of Cairo. Of the lanceolate Senna *five* parts are here mixed with *three* parts of the leaflets of *C. obovata*, brought from other parts of Egypt and even from Syria, and also with the leaves (*two* parts) of *Cynanchum Argel.* This mixed Senna is that exported from Alexandria. On the Continent a further addition is made of the leaves of *Colutea arborescens* and of *Coriaria myrtifolia.*

Dr. Pereira states that, "under the name of *heavy senna* he has met with *argel* leaves, which were sold at a higher price than ordinary senna," and Dr. Christison mentions what indeed may often be seen, that is, Argel leaves left intermixed even in what is called Picked Alexandrian Senna. This Senna is often called "Sené de la Palthe."

4. Tripoli Senna, is brought from Fezzan to Tripoli. This has the general appearance of Alexandrian Senna, but is less esteemed, though it is a more pure Senna, probably because the leaflets are more broken down, and all the leaf-stalks have not been removed. The leaflets are shorter and less pointed than in lanceolate Senna—indeed, more ovate; hence this Senna is said to be produced by *C. ovata* (C. æthiopica); but it also contains leaflets of *C. obovata*, which species was found in Fezzan by Dr. Oudney.

5. Aleppo Senna is now seldom imported into this country. It consists of the leaflets of *C. obovata*, as do some other kinds, such as Italian Senna. Dr. Ainslie says that the *obovate* is the only kind of Senna met with in India, meaning the Peninsula of India; for it is not met with in the Bengal Presidency; nor, according to Dr. Searle, is it used in that of Madras at the present day. It is less effective as a purgative, and apt to create nausea and griping.

Adulteration. Commercial Senna is prepared for use by picking out the leaflets, and rejecting the leaf-stalks, also extraneous matter, as dust, date-stones, &c., as well as the leaves of other plants. The legumes, however, possess the cathartic properties of the leaves to a considerable extent, and were alone used by the original Arabs; and there is no reason to believe that the stalks are inert. The most important adulterations are, however, the leaves of other plants. Those of the *Argel* (D.) may be distinguished by being lanceolate, equal on the two sides of the midrib, thick, leathery, and paler. They operate very dubiously as a cathartic, but occasion griping and protracted sickness. (*c.*) Those of *Tephrosia Apollinea* (F.) are obovate downy, and the veins proceed transversely from the midrib to each margin of the leaf without forming a marginal vein. The leaves of *Colutea arborescens*, or Bladder Senna, are ovate, but equal at the base. Those of *Coriaria myrtifolia* (E.) are astringent, usually broken down, and marked on each side of the midrib with a strong lateral nerve. As the systematic adulteration of Senna in Egypt with the leaves of other plants is objectionable, and has been so noticed by the Pharmaceutical Society, the most efficient method of stopping it would be to purchase only the pure African and Arabian Sennas which come to us by Bombay, instead of (unless it has been picked) that which is called Alexandrian from its place of export (*v.* P. J. ii. p. 63.) In India, a good substitute for Senna is afforded by *Rae Suna.* (*v.* COMPOSITÆ.)

Fig. 60.

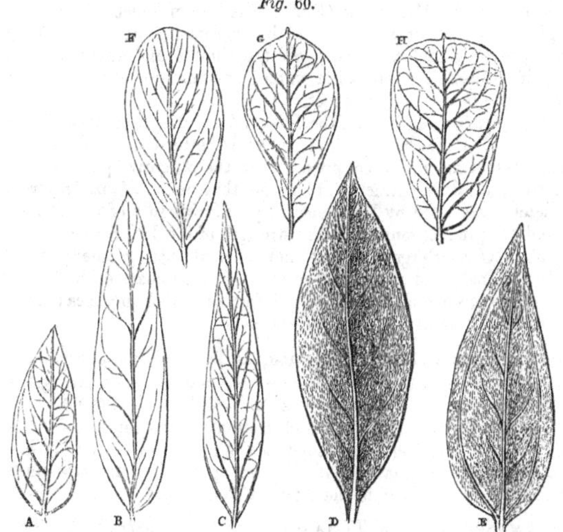

A. Cassia lanceolata. B. Tinnivelly Senna. C. C. acutifolia D. Cynanchum Argel. E. Coriaria myrtifolia. F. Tephrosia Apollinea. G. C. obovata. H. C. obtusata.

Prop. Senna has a faint sickly smell, the taste is slightly mucilaginous, bitter, and nauseous. Alexandrian Senna, analysed by MM. Lassaigne and Feneulle, yielded Mucilage, Albumen, Chlorophyll, Fixed Oil, a little Volatile Oil, yellow Colouring Matter, and some Salts. But its properties are supposed to depend upon *Cathartine*, which is described to be a deliquescent uncrystallizable matter. The pods are composed of the same principles, with the exception of the Chlorophyll. M. Heerlein has lately experimented upon this Cathartine, and describes it as a dark-brown clear extract, with an unpleasant odour, and an acidulous, bitter, also unpleasant taste ; perfectly soluble both in Alcohol and water. He considers it to be merely an extract, containing a free Vegetable acid, a Salt with an alkaline base, and a Brown Bitter Extractive. He further found that it does not contain the purgative principle of the Senna. gr. v. = ℨij. of Senna-leaves, and even gr. x., given to patients were inefficacious; also four doses of ℈j. each taken by himself at an interval of an hour and a half were also without effect.

The active principles of Senna are extracted both by rectified and by proof Spirit, and both by cold and by hot water; but long boiling injures its properties as a medicine. A very useful set of experiments has been made by Mr. Deane (P. J. iv. 61), from which he finds that though the best result was obtained by macerating Senna in a weak Spirit, yet that cold water extracts the soluble and active portions, nearly if not quite as well as hot water; and that picked Alexandrian Senna of the best quality is superior to all the others, from the quantity of extract it contains. Of the East Indian, that from Tinnivelly is best; and that the common East Indian is better than small Alexandrian. But it has not been proved that the purgative property is in proportion to this Extract. He found that Senna 7½ oz. troy were completely exhausted by Dil. Spirit (1 part Spirit to 5 of water) 20 fl. oz., and the product four times the strength of the Inf. Sennæ L. P.

Action. Uses. Purgative; safe and efficient, acting chiefly on the small intestines, and producing copious loose evacuations in doses of ℨß.—ℨij. as in some of the following forms, or with Bohea Tea (Paris), or with Coffee, as in the French *Café au Séné.*

INFUSUM SENNÆ (E.) COMPOSITUM, L. D. Infusion of Senna.

Prep. Macerate for an hour in a slightly covered vessel *Senna* ℥xv. (℥iß. E. ℥j. D.), cut *Ginger* ℈iv. (ℨj. D.) in boiling *Aq. dest.* Oj. (℔j. D.); strain.

This infusion has the odour and taste of Senna, and is of a clear brown colour. It is much employed as a purgative in doses of f℥iß.—f℥iij., often in combination with a saline purgative, and a warm or purgative Tincture, forming the common *Black Dose.*

INFUSUM SENNÆ (CUM TAMARINDIS, D.) COMPOSITUM, E. Compound Infusion of Senna.

Prep. Infuse for 4 hours (occasionally stirring, E.) in a covered vessel not glazed with lead, *Senna* ℨj. *Tamarinds* ℥j. bruised *Coriander Seeds* ℨj. *Muscovado* or *Brown Sugar* ℥ß. (℥j. D.) boiling *Aq.* f℥viij.; strain through linen or calico. The same may be made with 2 (or 3, E.) times the quantity of Senna.

This infusion is sometimes preferred on account of the combination of aromatics and Sugar, with the cooling effects of the Tamarinds, &c. Acts as an effective purgative in doses of f℥iß.—f℥iij.

ENEMA CATHARTICUM, E. D. Cathartic Enema.

Prep. Infuse for 1 hour in boiling *Aq.* f℥xvj. *Senna.* ℥ß. then dissolve *Sulph. Magnesia* ℥ß. *Sugar* ℥j.; add *Olive Oil* ℥j. and mix by agitation. E.

Dissolve *Manna* ℥j. *Sulph. Magnesia* ℥ß. in *Comp. Dec. of Chamomile* f℥xx. and add *Olive Oil* f℥j. D.

Employed as a laxative Enema.

TINCTURA SENNÆ COMPOSITA, L. E. D. Comp. Tinct. of Senna.

Prep. Macerate for 14 (7, E.) days *Senna* ℥iijß. (℥iv. E ℔j. D.) bruised *Caraways* ʒiijß. (ʒv. E. ℥iß. D.), bruised *Cardamoms* ʒj. (ʒv. E. ℥ß. D.), *Raisins* (stoned) ℥v. (℥iv. *Coriander* bruised ℥j. powdered *Jalap* ʒvj. *Sugar* ℥iiß. E.) in *Proof Spirit* Oij. (by measure Cj. D.); strain. (Express the residuum and filter. Or prepare by percolation, as directed for Comp. Tinct. of Cardamom. If Alexandrian Senna be used, free it of Cynanchum leaves by picking, E.)

A warm and stimulant purgative. That of the E. P. is made more effective by the Jalap, and by the corrective effects of the Sugar. Usually prescribed as an adjunct to the Infusion, in doses of fʒj., sometimes alone in doses of f℥ß.

SYRUPUS SENNÆ, L. E. Syrup of Senna.

Prep. Macerate in boiling *Aq.* Oj. (Oj. and f℥iv. E.) with heat for 1 hour (12, E.) *Senna* ℥ijß. (℥iv. E.), bruised *Fennel* ʒj. L. Filter. (Infuse *Senna* alone in the water; express strongly, so as to obtain at least Oj. and f℥ij. of liquid, E.) Add (while hot, E.) *Sugar* ℥xv. and *Manna* ℥iij. L. Boil down to a proper consistence; add to the infusion of *Senna, Treacle* concentrated in the vapour-bath as much as possible ℥xviij. E. Stir carefully, and when the mixture is complete, remove it from the vapour-bath. Carefully pick Alexandrian Senna, E.

The Syrup obtained by the E. process is said by Dr. Christison to be far superior to that obtained by the London formula, as the Infusion is added after the treacle has been concentrated, and not boiled down with all the ingredients. Both preparations have scarcely the taste of Senna, cause little sickness or griping, and are effective as purgatives. A "Concentrated Syrup of Senna" has for some time been prepared in the metropolis. The following, or "Fluid Extract of Senna," is strongly recommended by Dr. Christison (*v.* P. J. iii. 115 and 248), and is prepared by several druggists.

"Take of *Tinnivelly Senna* ℔xv. avoirdupois, and exhaust it with boiling water by displacement: (about 4 times its weight of water is sufficient). Concentrate the infusion in vacuo to ℔x.; dissolve in it *Treacle* ℔vj. previously concentrated over the vapour-bath till a little of it becomes nearly dry on cooling; add of *Rectified Spirit* (Sp. Gr. ·835) f℥xxiv; and, if necessary, add *water* to make *fifteen* (16 oz.) *pints.* Every f℥. will correspond to Senna ℥j. avoirdupois.

D. fʒij. for an adult. It tastes like treacle, the feeble mawkish one of Senna being covered; and it operates usually without producing either nausea or griping. Dr. C. informed the author, when in London, and again by letter in July, 1842, that a nobleman who had been in the habit of taking this Syrup made from Tinnivelly Senna, immediately discovered when the Alexandrian had been used, in consequence of the severe griping and its ineffectual teazing effect as a purgative, though taken in the same doses. There can be no doubt of the superiority of the Tinnivelly and other pure Sennas.

CONFECTIO (ELECTUARIUM, E. D.) SENNÆ, L. Confection of Senna.

Prep. L. E. Rub together *Senna* ℥viij. and *Coriander* ℥iv. Pass through a sieve ℥x. of the powder; boil the residue with Aq. Oiij. (Oiij. and ½ E.), *Figs* ℔j.

and *Liquorice* ℥iij. down to one-half. Express, strain, and evaporate in water-bath till f℥xiv. remain. In this dissolve *Sugar* ℔ijß. and make a syrup; rub in gradually *Pulp of Prunes* (*Cassia, Tamarinds*, L.) āā ℔ß.; then throw in the sifted powder, and mix (triturate to a uniform pulp, E.).

D. Boil *Pulp of French Plums* ℔j. and *Pulp of Tamarinds* ℥ij. in *Treacle* by measure ℔ß. to the thickness of Honey; add very finely powdered *Senna* ℥iv. and when cold *Essential Oil of Caraway* ʒij. Mix well.

Action. Uses. A mild but useful purgative in doses of ʒj.—ʒiv.

Andira inermis, *Kunth.* (*Geoffroya inermis*, Swartz), D. Cortex. The Bark of the Cabbage Tree. *Diadelphia Decandria*, Linn.

This tree is a native of the West Indies and of Guiana, where it is called Worm Tree from its uses, and *Wild* and *Bastard* Cabbage-tree, to distinguish it from *Areca oleracea*, the true Cabbage Palm.

This tree produces good timber. Leaves pinnate. Leaflets 10—17, oblong, lanceolate acute, about 4 inches long, of a dark green colour and smooth. Stipules lanceolate, persistent. Flowers of a reddish lilac colour, paniculate, with short pedicels. Calyx urceolate, 5-toothed, ferruginous, pubescent. Corolla papilionaceous. Stamens diadelphous. Ovary containing 3 ovules. Legume stalked, size of a large plum, hard, roundish, 1-celled, 1-seeded when ripe, divisible into 2 valves according to Swartz.—Wright, Phil. Trans. 1777, t. x.

A variety of *Andira retusa* in Surinam is said to have similar properties, and to be used on the Continent. This Legume is compared to a Drupe, or plum with 2 furrows, and the genus therefore in this respect resembles the Amygdalous Rosaceæ, and in its flowers the Papilionaceous tribe.

The Bark is officinal on account of its anthelmintic properties. It is in long half-quilled pieces, fibrous in structure, of a greyish colour externally, and brownish internally, having a disagreeable smell, a bitter, acrid, and mucilaginous taste. It contains Gum, Starch, Resin, Salts; but its properties seem to depend on an alkaline body, which is very bitter, and has been called *Jamaicine.*

Action. Uses. Cathartic; in large doses, Emetic and Narcotic. Formerly used as an Anthelmintic against lumbrici (Ascaris lumbricoides) in doses of ℈j.—ʒß.

Decoctum Geoffroyæ, D. Decoction of Cabbage-tree Bark.

Prep. Boil *Cabbage-tree Bark* bruised ℥j. in *Aq.* Oij. down to Oj.; strain, and add *Syrup of Orange-peel* f℥ij.

Action. Uses. Cathartic and Anthelmintic in doses of f℥j.—f℥ij. It is better to prescribe it with warm water and Castor Oil.

Copaiba. Resina Liquida, L. Fluid Resinous Exudation of several species of Copaifera, E., of C. Langsdorffii, *Dec.* L., of C. officinalis, *Linn.* D. Copaiva.

Copaiba was first described by Marcgraaf and Piso in 1648; but the species is uncertain, as the latter gives no figure, and the former only one of the fruit (supposed by some to be of *Copaifera bijuga*, Willd). Jacquin, in 1763, described a species of Copaifera from Martinique, which he named *C. officinalis*, and which probably yields the little

Copaiba obtained from the West Indies. It has, however, been ascertained that several species yield the Copaiba of commerce. The Wood-Oil of some species of *Dipterocarpus* yields a substance closely resembling Copaiba.

COPAIFERA, *Linn.* *Decand. Monog.* Linn.

Calyx without bracts, 4-parted, divisions small, spreading; corolla none; stamens 10, separate, nearly equal, declinate; ovary compressed with two ovules; fruit a legume, stalked, obliquely elliptical, coriaceous, somewhat compressed, 2-valved, 1-seeded; seed elliptical, enclosed in a 1-sided aril; embryo straight; radicle somewhat lateral.—Trees or shrubs of tropical America. Leaves alternate, equally or unequally pinnate; leaflets opposite or alternate, coriaceous, somewhat unequal, ovate, either dotted or not. (*Lindl.*) Stipules generally wanting; bracts caducous; flowers in compound axillary and terminal spikes.

Fig. 61.

C. Langsdorffii, Desf. (*v.* fig. 61). Leaflets 3—5 pairs, equal-sided, obtuse, with pellucid dots; the leaves ovate, the upper elliptical; petioles and peduncles slightly downy.—As space will not allow of the other species which yield Copaiva being described, a list with their habitats is subjoined.

Species	Habitat	Species	Habitat
1. C. Langsdorffii, *Desf.*	San Paulo and Minas.	6. C. nitida, *Mart.*	Minas Geraes, Cujaba, and Goyaz.
2. C coriacea, *Mart.*	San Paulo and Minas.		
3. C. guianensis, *Desf.*	Rio Negro, Para.	7. C. Beyrichii, *Hayne.*	Rio and mountains of Estrella.
4. C. multijuga, *Hayne.*	Para.	8. C. officinalis, *Linn.* C. Jacquini, *Desf.*	West Indies, and Venezuela.
5. C. Martii, *Hayne.*	Para, Maranhao.		

Copaiva, though usually called a Balsam, is not correctly so named, as it contains no Benzoic acid. It is an Oleo-Resin, which varies more or less in colour, odour, specific gravity, and medical virtues, according to the species from which it is obtained. (*Martius.*) The species 1 and 2 yield the best Copaiva in the district of San Paulo, 3 in Guiana, 4 and 5 in Para, 6 in Minas Geraes, 7 in mountains of Estrella and at Rio, 8 in Venezuela and the West Indies. Other species are capable of, and no doubt yield some of the Copaiva of commerce. Dr. Christison mentions Copaiva sent from British Guiana, obtained from plants growing further north than the above (except No. 8), and near the Orinoco, the species of which botanists have not yet determined. The species growing in the hot and moist parts of Brazil form large trees, and yield very fine Copaiva; those of the drier and interior districts, as Minas Geraes, &c., are shrubby in nature, and yield less, but also a more resinous balsam; that of the West Indies is darker-coloured, turbid, more acrid in taste, and smells more of turpentine. It is in all these places obtained by making deep incisions into the trunk of the trees, chiefly at the end of the rainy season, when it flows out so abundantly, that 12 lbs. are said by Piso to be obtained in a few hours.

Prop. Balsam of Copaiva is a liquid (Sp. Gr. ·095) of an oily consistence, transparent, of a pale straw-colour, of a strong odour, and disagreeable, nauseous, acrid taste. It becomes more dense and darker-coloured, if kept exposed to the air. It is soluble in Alcohol, Ether, and Oils; but, like other Oleo-Resins, is insoluble in water. With alkalis it forms soaps, which are precipitated when much diluted with water. Analysed by Stolze and Gerber, it yielded of Volatile Oil from 32 to 34 parts, Yellow Resin (*Copaivic acid*) 38 to 52, Viscid Resin 1·65 to 2·13, the rest being water and loss in 100 parts. The *Volatile Oil* may be separated by distillation with water. (*v.* Oleum Copaibæ, E. P.) The Resin which remains consists of two parts: one, *Copaivic acid*, hard, brittle, and crystallizable, having an acid reaction on Litmus, and forming compounds with bases: like Colophane or Pinic acid, it consists of $C_{40}\ H_{30}\ O_4$. The other, *soft, brown*, or *viscid Resin*, which is more abundant in old than in fresh Copaiva, has little affinity for bases, and may be separated from the other by being insoluble in Naphtha.

Tests. The E. C. characterize Copaiva as "Transparent: free of Turpentine odour when heated: soluble in two parts of Alcohol: it dissolves a fourth of its weight of Carbonate of Magnesia, with the aid of a gentle heat, and continues translucent." An inferior kind, or some adulterated with Turpentine or fixed Oils, is occasionally sold. A greasy spot will be left, if any fatty oil is present, when a little is dropped on bibulous paper. Many inferior kinds are sold.

Action. Uses. Stimulant of Mucous Membranes, especially of the Urinary passages. In large doses, Cathartic and Diuretic. Diminishes Mucous discharges, as of Gonorrhœa, &c.

D. ♏xv.—f℥ss or even f℥j. two or three times, swimming on some fluid, or made into an emulsion; or in capsules, where the Copaiva is enclosed in a thin layer of Gelatine; or in form of pill, with $\frac{1}{16}$ of Magnesia, as in the U. S. P.

OLEUM COPAIBÆ, E. Volatile Oil of Copaiva.

Prep. Distil *Copaiva* f℥j. with *water* Ojss. and separate the Oil.

Action. Uses. This oil is colourless, but acrid in taste, with a strong odour of Copaiva, soluble in Alcohol, Ether, &c. It consists, like Oil of Turpentine, of $C_{10}H_8$. It is preferred by many to every other form of Copaiva, in doses of ♏x.—♏xxx.

PAPILIONACEÆ, *Linn. Nat. Ord.* Papilionads.

Calyx 5-dentate, or bilabiate, deciduous or withering on the plant. Corol papilionaceous, or subpapilionaceous, with the 5 petals inserted into the bottom of the calyx, usually free, sometimes united with one another or with the stamens, imbricate. Stamens 10, inserted with the petals, united together, or 9 united into a bundle, and the tenth remaining free. Ovary free, placenta on one side. Seeds without albumen. Embryo curved, or bent back upon the cotyledons, rarely straight.

The Papilionaceæ may be distinguished by their irregular flowers, number and insertion of the stamens, and by the leguminous fruit, as well as by their habit; but the most doubtful may be distinguished from Rosaceous flowers by the odd segment of the calyx being anterior. Papilionaceæ are found in all parts of the world: their geographical distribution is best studied in their several tribes. Uses various.

BALSAMUM PERUVIANUM, L. E. D. Fluid Balsamic Exudation of Myrospermum Peruiferum, *Dec.* E. (*Myroxylon peruiferum*, Linn. fil.) L. D. Balsam of Peru.

BALSAMUM TOLUTANUM, L. E. D. Concrete Balsamic Exudation of Myrospermum Toluiferum, *Ach. Rich.* E. of Myroxylon Peruiferum, *Linn. fil.* L. Toluifera Balsamum *Mill.* D. Balsam of Tolu.

The Balsams of Peru and Tolu were first made known by Monardes in the year 1580. It is still uncertain whether they are the produce of the same or of different trees. Mutis sent a branch in flower of the first in 1781 to the younger Linnæus.

Balsam of Tolu has been long supposed, as stated by Ruiz, to be produced by the same tree. Miller having, however, grown it from seeds sent him from the province of Tolu behind Carthagena, considered it distinct, and continued to call it *Toluifera Balsamum.* Collected by Humboldt and Bonpland, it was described by Kunth as *Myroxylon Toluiferum.* Ach. Richard, having examined the characters of these species, finds them all to belong to the genus *Myrospermum* of Jacquin, in which they have been continued by other botanists.

MYROSPERMUM, *Jacquin.* *Decand. Monog.* Linn.

Calyx campanulate, slightly 5-toothed; petals 5, subpapilionaceous, the upper largest; stamens 10, free; ovary stipitate, oblong, membranous, with 2—6 ovules, terminated with a lateral filiform style; legume with a winged stalk, which is very broad at the apex, and which supports an oblique, indehiscent, 1-celled, 1—2-seeded samaroid fruit; seeds involved in balsamic juice; cotyledons fleshy; embryo curved.—Trees with abruptly (*Dec.*) impari-pinnated (*Kunth*) leaves; leaflets marked with round and some linear dots.

M. PERUIFERUM, *Dec.* (*Myroxylon peruiferum*, Linn. fil. &c.) A tall and much branched tree, with a smooth warty bark, which is thick and filled with resin, hence has a grateful smell and aromatic taste. Leaflets alternate, of 3 to 5 pairs, with an odd one, subequilateral, oblong, obtuse, emarginate, rounded at the base, sometimes subcordate, coriaceous, smooth; midrib prominent below, and with the flexuose rachis rather hairy; racemes axillary; of the 5 petals the upper or standard broad and roundish, the others linear-lanceolate; stamens spreading; styles deciduous; legumes straw-coloured, pendulous, stalked, linear, oblong, coriaceous, about 2 inches long, its wing very thick on one side, on the other not veined, 1-celled, 1-seeded; seed reniform, involved in liquid yellow balsamic juice, which hardens into resin.—(Lam. Illustr. t. 341. f. 1. a.—g.) The Quinquino or Balsam of Peru Tree, a native of tropical forests on the banks of the Maranao in Peru, near Bagota, and also, or a nearly allied species, in Mexico, according to Hernandez, who says it was employed as a substitute for Syrian Balsam. De Candolle inquires, as there is so much variation in the leaves, whether several species are not confounded together. Guibourt mentions having received some Balsam of Peru from near San Saladra in Guatemala.

Balsam of Peru is imported from several parts of the western coast of South America, and it is probably obtained by making incisions into the stem, as originaly described by Monardes; but one kind is also described as being procured from decoction of the branches; but no good and recent information has been obtained on this subject. Ruiz states that the balsamic juice, when received in bottles, may be preserved in a liquid state for some years, and is then called *White Liquid Balsam;* but that which remains in the tree and is obtained by boiling the branches, is a dark-coloured liquid, called *Black Peruvian Balsam.* But further information is required. This is the kind commonly met with, of a viscid syrupy consistence, Sp. Gr. 1·15, of a reddish-brown almost black colour, of a strong balsamic odour, and a bitterish rather acrid taste. According to M. Stolze, it is composed in 100 parts of 23 of Brown Resin, 69 of a peculiar Oil, Benzoic acid 6·5, Extractive Matter and loss 1·5. It is soluble in Alcohol, and in about 5 parts of rectified Spirit, burns with a good deal of smoke. Boiled with water, this becomes charged with its acid and a little of the Oil. Balsam of Peru has also been analysed by Fremy and Wernher. The former calls its oil, *Cinnameine*, which Richter says is composed of two oils. (*v.* Turner's *Chemistry.*) Fremy considers the acid to be Cinnamonic, that is, Cinnamic acid, and the Resin a Hydrate of Cinnameine.

MYROSPERMUM TOLUIFERUM, *Ach. Rich.* A tree very like the former. Branches warty, smooth; leaflets equilateral, from 7 to 8, thin, membranous, ovate, oblong, acuminated, rounded at the base, shining, and as well as the rachis, smooth. The leaves only are figured by Nees von Esenbeck in T. 322; but they are so similar to those of the plant he has figured as *M. peruiferum*, that they might belong to the same plant.—Mountains of Turlaco near Carthagena; banks of the Magdalena and the high savannahs of Tolu.

Balsam of Tolu is probably obtained by incisions made in the bark of the above tree. It is imported from Carthagena, &c. in vessels of different kinds, sometimes in small ovoid gourds. Ruiz states that it is only the Balsam of Peru in a dried state. According to Fremy, it is composed of the same constituents, and will necessarily act in a similar manner with reagents. It is usually in a solid state, dry and friable, of a yellowish-red or reddish-brown colour; but when fresh, it is soft and of the consistence of thick honey, from containing more oil, of a fragrant balsamic odour, and warm sweetish taste.

Action. Uses. Stimulant, Expectorant. The Balsam of Peru is a useful application to indolent and also to phagedenic ulcers. Stimulant Expectorant in Chronic Catarrhs, in doses of ♏xv.—♏xxx. made into emulsion with Mucilage or Yolk of Egg. The Balsam of Tolu is more frequently prescribed.

TINCTURA (TOLUTANA, E.) BALSAMI TOLUTANI, L. D. Tincture of Tolu.

Prep. Digest *Balsam of Tolu* ℥ij. (℥j. D. ℥iijß. E.) in *Rectified Spirit* Oij. (Oj. wine measure, D.) with a gentle heat, E. (in a close vessel, D.) until the balsam is dissolved, and filter, L. D.

D. f℥ß. to f℥j. made into an emulsion.

SYRUPUS (BALSAMI TOLUTANI, D.) TOLUTANUS, L. E. Syrup of Tolu.

Prep. Boil *Balsam of Tolu* ℥x. in *boiling Aq.* Oj. for half an hour in a lightly covered vessel, stirring; strain; then dissolve in it *Sugar* ℔ijß. The E. and D. Cs. prepare this syrup by adding *Tinct. of Tolu* f℥j. gradually to *Simple Syrup* ℔ij. E. ℔ijß. D.

Action. Uses. Stimulant, in doses of f℥j.—f℥ij. Sometimes added to flavour draughts.

Ph. Prep. Tinct. Benzoini Comp. Trochiscus Opii, and Troch. Morphiæ et Ipecacuanhæ.

PTEROCARPUS, *Linn.* *Diadelphia Decandria*, Linn.

Calyx 5-toothed, obscurely bilabiate; petals 5, papilionaceous, petals of the keel free; stamens 10, filaments variously united; ovary stipitate, with few (2 to 4) ovules; legume suborbicular, compressed, indehiscent, surrounded by a membranous wing, usually rugose in the middle; 1—3 celled, each cell 1—3 seeded; seeds kidney-shaped—Trees or shrubs. Leaves unequally pinnated; racemes axillary or terminal, and paniculate. Natives of the tropical parts of India, of the west coast of Africa, and of tropical America; they secrete and exude reddish-coloured juice, which is usually astringent, and hardens in the air.

PTEROCARPUS, L. E. PTEROCARPUS SANTALINUS, *Linn. fil.* D. Lignum, L. D. Wood, E. Red Sandal Wood.

Sandal-wood is mentioned by Serapion and other Arabs, and distinguished into white, yellow, and red. The last is known in commerce as Red Saunders Wood. This is called *rukta* (red) *chundun* (Sandal) in India, and the name is applied to the wood of *Pterocarpus santalinus*, and also to that of *Adenanthera Pavonina.* Lignum Pavona was an old dye wood. That of *Pterocarpus dalbergioides* is said to yield the Andaman Red Wood.

P. SANTALINUS, *Linn. fil.* (*Santalum rubrum*, Kœnig.) A lofty tree. Leaflets 3, rarely 4 or 5, alternate, roundish, retuse, smooth above; racemes axillary, simple, or branched; petals long-clawed, crenate, undulate; standard yellow, streaked with red; filaments 10, diadelphous (triadelphous, 5, 4, and 1, *W. and A.*); legume suborbicular, stalked, 1-seeded; the wing somewhat membranous, waved.—A native of the Pulicat (Paulghat) mountains; also of Ceylon.

This tree was pointed out by Kœnig as yielding Red Sandal-wood. Its wood is dark-red with darker-coloured veins, heavy and compact, capable of taking a fine polish; when moistened with water, it is said to produce a fine red colour; and a reddish-coloured juice exudes from its bark, which Kœnig considered a kind of dragon's-blood. The imported wood is similar in appearance, is without odour, has a feeble taste. and sinks in water. Alcohol and Ether readily extract its colour, as do alkaline solutions. Pelletier found it contained Woody Fibre, Extractive, Gallic acid, and about 17 per cent. of a peculiar colouring matter, which he called *Santaline*, and which is somewhat allied to the Resins in properties.

Action. Uses. Used only as a dye, and to give colour to the Tinct. Lavandulæ Comp.

KINO, L. E. D.

Kino is well known as an astringent substance, in small and shining, brittle, angular fragments of a deep-brown colour, which appears to be a natural exudation of some one plant, from the uniformity of its appearance. Several kinds of Kino are, however, met with in commerce, as well as described in books, as that of *Butea frondosa* from India, at one time acknowledged by the D. C., and which has no doubt been sometimes imported as Kino. Botany Bay Kino, produced by *Eucalyptus resinifera*, or Brown Gum Tree, at one time acknowledged by the E. C., a Jamaica, and a Columbian Kino, are mentioned; and an extract of Rhatany is sometimes enumerated with them. But genuine Kino has been supposed to come from the west coast of Africa. There is no doubt, however, and the fact may be easily ascertained by any one making inquiries in the proper channels, that the best is now imported into this country from Bombay.

Kino seems to have been first introduced into European practice by Dr. Fothergill, in 1757, who states, in a paper in Med. Obs. and Enq. i. 358, that he was indebted for information respecting it to Dr. Oldfield, and that the substance was obtained from the river Gambia, whence he called it *Gummi rubrum astringens Gambiense.* Previous to this, Moon, in his travels into Africa, mentions a red gum as issuing from incisions in trees, and which he mistook for Dragon's-blood.

Mungo Park discovered that the tree which yielded this substance was called Pao de Sangue (Blood-tree) by the Portuguese. His specimens were determined by Mr. R. Brown to belong to *Pterocarpus erinaceus*,—a tree which has since been well figured and fully described in the Flore de Senegambie.

P. ERINACEUS (*Poiret*, Illustr. t. 602, f. 4.) L. E. A tree 40 or 50 feet in height, with the bark exuding a peculiar blackish-coloured juice; leaflets 11—15, alternate, ovate, oblong, obtuse, or emarginate, above smooth, on the under surface covered with dense but short tomentum; flowers yellow; stamens 8—10, monadelphous or irregularly diadelphous; legumes orbicular, membranous, undulate at the margin, and terminated on one side by a sharp point (the base of the style), in the centre covered with stiff bristles, 2-celled or 1-celled; each cell 1-seeded.—A native of Senegambia. Its wood is reddish-coloured When the bark of its trunk or branches is injured, a reddish-coloured juice exudes, which quickly hardens in the air, becoming of a blackish colour. This brilliant, friable, and astringent substance, though like Kino, does not seem to be collected. "Nous ne l'avons pas vu extraire pour les usages pharmaceutiques sur les bords de la Gambie," (Fl. de Senegambie, i. p. 230, tab. 54,) and no Kino is known to be imported here from the coast of Africa.

The origin of the name Kino has not yet been satisfactorily ascertained. It was introduced into the E. P. 1774 as *Gummi Kino*, and into the L. P. 1787 as *Resina Kino*. I have long been of opinion that the name was derived from the Indian *kuenee*, or *kīni*, applied to a similar exudation from the bark of *Butea frondosa*, of which the Sanscrit name is *Kin-suka* (*Himal. Bot.* p. 195, and *Proc. Royal Asiatic Soc.* p. 50, May, 1838), because this *Butea gum* had been sent as Kino to the above Society from Bombay. An old specimen in the India Housè is marked *Gum Cheena.* Dr. Pereira, several years since, found this "in the warehouse of an old drug firm in London a substance marked *Gummi rubrum astringens*," which he was told had formerly fetched a very high price. It is, however, very distinct from the Kino of commerce, which, for many reasons, the Author was inclined to think was the produce of *Pterocarpus Marsupium.*

Dr. Pereira states that what he calls *East Indian Kino* is always regarded in commerce as *genuine Gum Kino*, and that an experienced East Indian broker assured him it was the produce of the Malabar coast. He also traced it to Bombay, and to Tellicherry, on that coast. In the official reports of the commerce of Bombay, the Kino exported to this country appears to have been previously imported from the Malabar coast. The author's attention was again especially turned to this subject on finding in the India House specimens of Kino marked from Anjarakandy, which he recognized as being identical with the present Kino of commerce; but was unable for some time to ascertain the locality of Anjarakandy, until informed by Mr. Dyer (*v.* p.304) that it was the name of a farm within a few miles of Tellicherry,—that is, near the very place to which Dr. Pereira had traced the East Indian Kino.*

* Since then I have discovered that this was formerly one of the East India Company's plantations, under the superintendence of Mr. Brown, and was visited by Dr. Buchanan, in January, 1801 (*Mysore* II. p. 544), when he states, numerous valuable experiments were carrying on in the plantation.

Having thus determined the place, the next point was to ascertain the plant which yielded this kind of Kino, as well as its mode of preparation. This was effected by writing to Dr. Wight, stationed at Coimbatore ; and though he did not at first succeed, Dr. Kennedy afterwards sent him specimens of the flower, leaves, and fruit, also a small portion of the wood and of the gum. On inspecting these, Dr. Wight states, "the specimens received along with the letter leave no doubt that the Malabar Kino is the production of *Pterocarpus Marsupium.*"

Dr. Kennedy writes that he is informed by his friend Mr. J. Brown of Anjarakandy, that "the juice is extracted when the tree is in blossom, by making longitudinal incisions in the bark round the trunk of the tree, so as to let the gum ooze down into a receiver formed of a broad leaf so placed and fixed in the bark as to prevent the gum from falling on the ground. From the leaf it is made to run into a receptacle placed under the leaf to receive the gum. When this receptacle is filled, it is removed, the gum is dried in the sun until it crumbles, and then filled into wooden boxes for exportation."

Dr. Gibson had already stated (see the above *Proc.* p. 59) that "Kino was the produce of *Pterocarpus Marsupium* (*beula* or *bia*), a tree very common below the Ghats," also that the Kino is exported in considerable quantities from the Malabar coast. Dr. Roxburgh, however, was the first to direct attention to this tree, which he states exudes a red juice, which hardens into a strong, simply astringent, brittle gum-resin, of a dark red colour, strongly resembling that of the *Butea frondosa;* so that the same analysis might serve for both. He further observes, that the specimen of the gum Kino tree (*P. erinaceus*) in the Banksian herbarium is exceedingly like this plant. The specimens of the Indian and of the African Kino were, as we have seen, the produce of two distinct species of *Pterocarpus.*

P. Marsupium, Roxb. (Fig. 62.) A lofty tree, with the outer coat of the bark brown, inner red, fibrous and astringent ; leaves subifarious, alternate, leaflets 5—7 alternate, elliptic, emarginate, above shining and of a deep green colour, from 3 to 5 inches long ; panicles terminal ; petals white with a tinge of yellow, long clawed, all waved or crested on the margins ; stamens ten, united into one body near the base, but soon splitting into two bodies of five each ; ovary generally two-celled, legume long-stalked, the under three-fourths orbicular, the upper side straight ; the whole surrounded with a waved veined membranous wing rugose and woody in the centre, generally one, sometimes two-celled ; seed solitary, kidney-shaped. Roxb. Corom. Pl. ii. t. 116 ; Fl. Ind. iii. p. 234. A native of the Circar mountains and forests of the Malabar coast, apparently also in those at the foot of the Himalayas, according to Buchanan Hamilton.

Kino is in small, irregular, somewhat angular, glistening fragments, of a dark-brown or reddish-brown colour, brittle, and affording a powder which is lighter-coloured than the masses. It is without odour, and has a bitterish, highly astringent, and ultimately sweetish taste. It is not softened by heat ; cold water dissolves it partially, boiling water more largely, and the saturated decoction becomes turbid on cooling, and deposits a reddish sediment. Alcohol dissolves the greater

Fig. 62.

portion. It consists chiefly of a peculiar modification of Tannin, with Extractive matter, and, in some of the varieties, of a minute proportion of Resin. According to Vauquelin, it contains no Gallic acid, but Tannin and peculiar extractive 75, red Gum 24, insoluble matter 124. Its aqueous sol. is precipitated by Gelatine (with which it produces a green colour, in consequence of the presence of a little *Catuchine*), by soluble salts of Iron, Silver, Lead, and Antimony, by the Permuriate of Mercury, and by the Sulphuric, Nitric, and Muriatic acids. The alkalis favour its solubility in water, but essentially change its nature, and destroy its astringent property.

Action. Uses. Powerful astringent. Useful in restraining mucous discharges, &c.

TINCTURA KINO, L. E. D. Tincture of Kino.

Prep. Digest *Kino* bruised ℥iijß. (℥iij. D.) in *Rectified Spirit* Oj. (Proof Spirit Oij. old wine measure, D.) for 14 (7, E. D.) days and filter. This Tincture cannot be conveniently prepared by percolation.

Uses and D. Astringent adjunct to Chalk Mixture, &c., in doses of f℥j.—f℥ij.

PULVIS KINO COMPOSITUS, L. D. Compound Powder of Kino.

Prep. Rub up *Kino* ʒxv. *Cinnamon* ʒiv. and *Hard Opium* ʒj. separately to very fine powder, then mix them.

Uses and D. Astringent and anodyne. Useful in chronic diarrhœa and dysentery in doses of gr. x.—℈j. Grs. xx. contain of Opium gr. j.

Butea frondosa. This plant, as mentioned above, yields by incisions made in its bark an astringent gum, which was at one time supposed to be the genuine Kino of commerce. It is no doubt possessed of similar properties, is frequently used as such in India, and useful like it in Diarrhœas and advanced stages of Dysentery. It is also used in the art of tanning. It has been occasionally sent to this country as Kino, and Dr. Pereira found it in an old drug firm marked *Gummi rubrum astringens.* Its Sanskrit name is *Kin-suka.* It is commonly known in India as *Kīni ke gond*, and also by the name *Kumrkus.* Its chemical characteristics are very like those of the Kino of Pterocarpus. It is very carelessly collected, and therefore often contains impurities. It is remarkable for containing a beautiful red colouring matter, difficult of separation. Analysed by Mr. E. Solly, a portion in the crude state yielded about 50 per cent. of Tannin; but when purified by simple solution in water, so as to separate the impurities, 100 parts contained 73·26 parts of Tannin, 5·05 of difficultly-soluble Extractive, and 21·67 of Gum, with Gallic acid and other soluble substances. The colour and properties of Tannin vary with the exposure and season of collection. (*v.* Roxburgh, Fl. Ind. iii. p. 245, and Proc. R. Asiatic Soc. May, 1838.)

Scoparium, L. E. Cacumina (D.) recentia, L. (Fresh) Tops, E., of Cytisus Scoparius, *Dec.* L. E. of *Spartium Scoparium*, Linn. D. Tops of the Common Broom. *Diadelph. Decand.* Linn.

The common Broom, by some, and Spanish Broom by others, is supposed to be the σπαρτιον of Dioscorides.

A shrub with angular, unarmed branches. Leaves trifoliate, the upper ones simple, stalked, leaflets oblong. Flowers yellow, axillary, solitary, stalked. Calyx bilabiate, the upper lip often entire, the inferior subtridentate. The standard large ovate. Keel very obtuse, enclosing the stamens and pistils. Stamens all united together. Legume of a dark brown colour, flat, compressed, hairy at the margins, containing about 15 seeds. Indigenous in sandy and uncultivated places throughout Europe.

Broom-tops, like the rest of the plant, have a bitter nauseous taste, and, when bruised, a peculiar odour. Their properties are supposed to depend on the presence of *Cytisine*, and the seeds are the most effective part. The ashes contain about 30 per cent. of Carbonate of Potash and other salts.

Action. Uses. Emetic and Cathartic in large doses; but used only as a Diuretic in small doses. May be given in dropsies, in powder or in Extract, in doses of gr. x.—ʒß.; or in

Infusum Scoparii, L. Infusion of Broom Tops.

Prep. Macerate *Broom Tops* ℥j. in boiling *Aq. dest.* Oj. for 4 hours in a lightly covered vessel and strain.

It may be prescribed in doses of f℥j. to f℥iij. three times a day.

DECOCTUM SCOPARII (E.) COMPOSITUM, L. Compound Decoction of Broom Tops.

Prep. Boil *Broom Tops, Juniper Fruit,* (Dandelion, L.) āā ℥ß. *Bitartrate of Potash* ʒijß. in *Aq.* Ojß. down to a pint and strain.

Efficient Diuretic in the same doses as the Infusion.

EXTRACTUM SPARTII SCOPARII, D. Extract of Broom Tops.

To be prepared from Broom Tops as directed for Extract of Gentian.

GLYCIRRHIZA, L. E. D. Radix (D.) recens, L. Root (E.) fresh L. of GLYCIRRHIZA GLABRA, *Linn.* Liquorice Plant. *Diadelph. Decand.* Linn.

Liquorice, the produce probably of more than one species, was known to the ancients by the name *γλυκύρριζα*. It was employed by the Arabs, and well-known in the East, a produce of Mooltan, &c.

GLYCYRRHIZA, *Linn.* *Diadelph. Decand.*

Calyx naked, tubular, 5-cleft, bilabiate; with the two upper lips united more than the others. Standard ovate-lanceolate, straight; keel 2-petalous, or 2-parted, straight, acute. Stamens diadelphous. Style filiform. Legume ovate or oblong, compressed, 1-celled, 1—4 seeded.—Perennial, herbaceous plants of the tribe *Loteæ*, with very sweet roots. Leaves unequally pinnated. Racemes axillary. Flowers blue, violet, or white. Natives of the South of Europe, and some of the northern parts of Asia.

G. GLABRA. *Lin.*—The roots running to a considerable distance. Leaflets about 13, oval, slightly emarginate, viscid underneath, stipules wanting. Racemes axillary erect, shorter than the leaves. Flowers distant, pale lilac. Legumes compressed, smooth, 3—4 seeded. Native of the South of Europe, Syria, foot of Mount Caucasus, cultivated at Mitcham in Surrey, &c.—St. and Ch. 111, 134. *Liquiritia officinalis,* Nees von E. 327.

G. ECHINATA. *Lin.*—Leaflets oval, lanceolate, mucronate, glabrous; stipules oblong, lanceolate; spikes of flowers capitate, on very short peduncles; legumes oval, mucronate, 2-seeded, echinated by bristles. This is sometimes called Russian Liquorice. It is found in Greece and Southern Russia, extending, it is said, into Tartary and Northern China.—Sim's Bot. Mag. 252; Nees, 328.

Species of Glycyrrhiza no doubt also extend into Affghanistan, whence Liquorice-root, *Jeteemudh,* is imported into India. These species may, or may not, be distinct from the preceding.

Liquorice-roots, or, rather, underground stems, when fresh, are roundish, plump, and smooth. They may be preserved thus for some time, if kept in sand; but when dry, they are wrinkled, of a brown colour externally, yellowish and fibrous internally, with considerable sweetness, still more conspicuous when powdered, but in either case also a little acrid. The roots consist of Lignin, Starch, Albumen, Wax, Asparagin, Resinous oil, Colouring matter, Phosphates and Malates of Lime and of Magnesia, and a peculiar principle which has been called *Glycion* and *Glycyrrhizin*, or Liquorice Sugar, and upon which depends the sweetness of Liquorice, while its acridity is connected with the Oleo-resin. *Glycion* is very sweet, of a yellow colour, and transparent, but uncrystallizable. It seems to partake partly of the nature of acids, and partly of those of alkalis. It is soluble both in water and in Alcohol. This principle has been found in a few other sweetish-tasted roots.

Action. Uses. Demulcent in Catarrhs, Urinary and Bowel complaints. Useful in sweetening and flavouring medicines. Powder and Extract employed in making pills, the former in covering them.

DECOCTUM GLYCYRRHIZÆ, D. Decoction of Liquorice.

Prep. Boil *bruised Liquorice-root* ℥jſs. in *Water* Oj. (wine measure) for ten minutes and strain.

Useful Demulcent *ad libitum.* In this and similar preparations, the Liquorice root should be decorticated, as in the outer part the acrid principle connected with the oleo-resin is chiefly contained.

EXTRACTUM GLYCYRRHIZÆ, L. E. D. Extract of Liquorice.

Prepared in the same way as Extract of Gentian. The E. C. directs the root to be first cut into chips, dried, and reduced to moderately fine powder. When well prepared, this extract is of a brown colour, very sweet, and not at all acrid. Dr. Christison remarks that boiling is unnecessary, indeed injurious, cold water and the process by percolation yielding often 40 to 58 per cent. of very fine extract.

TROCHISCI GLYCYRRHIZÆ, E. Liquorice Lozenges.

Prep. Dissolve *Extract of Liquorice, Gum Arabic* āā ℥vj. *Pure Sugar* ℔j. in *boiling Aq.* q. s. and then evaporate over a vapour-bath to the proper consistence for lozenges.

COMMERCIAL EXTRACT OF LIQUORICE. This is recognized in the E. P. by the same name, *Glycyrrhizæ Extractum*, as the Pharmaceutical preparation. It is not noticed by the other colleges. It is commonly known by the names of *Liquorice*, *Extract of Liquorice*, and *Liquorice Juice.* It is prepared in large quantities in the south of Spain, in Italy, and in Sicily, and brought to a proper consistence by evaporating the Decoction in copper vessels. It is then formed into roundish or flattened sticks, of a brownish-black colour, often covered with Bay-leaves. The finest is that marked Solazzi. What is called *Refined* Liquorice, in black, shining, pipe-like cylinders, is a mixture of Liquorice and of Gum or Gelatine. Liquorice is so well known as not to require detailed notice.

Off. Prep. Infus. Lini Comp. L. E. D. Decoctum Hordei Comp. L. E. D. Dec. Sarzæ Comp. L. D. Dec. Mezerii, E. D. Dec. Guaiaci Comp. E. D. Aqua Calcis Comp. D. Conf. Sennæ, L. E. D. Dec. Aloes, Comp. L. E. D. Tinctura Aloes, L. Tinct. Rhei Comp. L. D. Troch. Opii, E. Troch. Lactuarii, E.

TRAGACANTHA, L. E. D. Succus concretus, L. Gummi, D. Gummy Exudation of Astragalus gummifer, *Lab.* and other species, E., A. verus, *Olivier*, L. and E., A. creticus, *Lamarck*, D. Tragacanth.

The τραγάκανθα of Dioscorides was no doubt a plant of the same genus as that which now yields Tragacanth. Sibthorp considers *Astragalus aristatus*, L'Hert. to be the plant. Arab authors describe it by the name *kusera* or *kutira*, for which, in India, *kuteera* is substituted. This is produced both in the North-west and in the Peninsula of India by *Cochlospermum Gossypium*. Tournefort adduced *A. creticus* of Lamarck, a native of Mount Ida in Crete, to be the plant yielding Tragacanth. Labillardiere describes his *A. gummifer*, a native of Mount Libanus in Syria, as one of the plants ; while Olivier states his *A. verus*, inhabiting Asia Minor, Armenia, and northern Persia, as yielding the largest quantity of Tragacanth. Dr. E. Dickson, Physician to the Consulate at Tripoli, when travelling in Koordistan, collected specimens of the plants which he ascertained to yield Tragacanth. These he gave to Mr. Brant, British Consul at Erzeroum, by whom they were sent to Dr. Lindley, who determined that the white or best variety of Gum Tragacanth is yielded by *Astragalus gummifer*, and the red or inferior kind by his *A. strobiliferus*. Dr. Dickson, when in England, favoured the Author with the following observations : "Besides the two last-named species, I observed also a third variety that gave Gum Tragacanth, which, unfortunately, I lost when my things were robbed at Hassan-kalek. From the Koordish mountains being covered with many species of Astragalus, I should think it not unlikely that other varieties of this genus may hereafter be discovered yielding the Gum."

ASTRAGALUS, *Dec.* *Diadelph. Decand.* Linn.

Calyx 5-toothed. Corolla with an obtuse keel. Stamens diadelphous. Legume 2-celled, or half 2-celled in consequence of the dorsal or lower suture being turned inwards, *Dec.*

Sectio Tragacanthæ.—Petioles permanent, thornlike. Stipules adhering to the petioles.

A. VERUS. *Oliv.* L. E. Flowers yellow, axillary, in clusters of 2 to 5, sessile. Calyx tomentose, obtusely 5-toothed. Leaflets 8 to 9 pairs, linear hispid. A native of Anatolia, Armenia and Northern Persia, yields Tragacanth, which is collected from July to September. Used in Persia, exported to Europe and also to India.—Oliv. Voy. 3, t. 44 ; Nees von E. 329.

A. GUMMIFER. *Labill.* E.—Flowers 3 to 5 axillary, sessile. Calyx 5-cleft, together with the legumes woolly. Leaflets 4 to 6 pairs, oblong, linear, smooth. A native of Mount Lebanon, also of Koordistan, where it yields white Tragacanth, but which Labilladiere represents as vermicular in form.

A. CRETICUS. *Lam.*—Flowers axillary, sessile, aggregate. Calyx 5-partite, with feathery setaceous lobes rather larger than the corolla. Leaflets 5 to 8 pairs, oblong, acute, tomentose. Mount Ida in Crete, where it yields a little Tragacanth.

A. ARISTATUS. *L'her.*— Peduncles very short, usually 6-flowered ; calycine teeth long and setaceous ; leaves with 6—9 pairs of oblong, linear, mucronate, pilose leaflets; legumes scarcely half bilocular. A native of the Alps of Europe, also of Greece. Sibthorp stated that this species yielded a gum called τραγοχανδα in Greece, which was exported to Italy. Landerer has lately ascertained that Tragacanth was yielded by this species on the hills near Patrass, and exported to Venice and Trieste, or as Levant Tragacanth to Marseilles and Ancona: (c)

A. DICKSONII. *Royle.* (*A. strobiliferus, Lindley,* not of Royle, Him. Bot. p. 199.)—Flowers capitate in an ovate, sessile, axillary strobile. Bracts imbricate, pinnated, tomentose. Calyx feathery, 5-cleft. Segments of the corolla equal. Leaflets 3-paired, woolly, oval, awned at the apex, narrow at the base. *Lindley.* As Dr. L. has inadvertently named this species *A. strobiliferus,* there being already one of that name, the Author has named it after the discoverer of the plant, who also found that it yielded a reddish coloured Tragacanth.

Tragacanth exudes from the above plants either naturally or from wounds, and hardens in various forms. It is imported into this country from Smyrna, the Levant, and also from Greece. It is found in commerce either of a white or a reddish-yellow colour, in broad thin flakes, or in tortuous vermicular pieces; the former is the best, and most common here. It is white or greyish, semitransparent, tough, horn-like, and tasteless, and being a little elastic, is with difficulty reduced to powder, unless heated to 110°. In contact with cold water, it absorbs a certain portion, swells, becomes adhesive and diffused. It does not dissolve except in boiling water, when some change is supposed to take place, a great portion, however, separating again. It is insoluble in Alcohol. Tragacanth appears to consist of two distinct gummy principles. Bucholz and Guerin Varry found of common Gum or Arabin from 53 to 57 per cent., and of Bassorin 33 to 43 parts, with water and a little Starch, the presence of the latter producing a blue colour with Iodine. The *Arabin* rather resembles than is identical with Gum Arabic, for, as first pointed out by Dr. Duncan, its mucilage is not precipitated by Silicate of Potash. The *Bassorin* is like that found in Gum Bussorah, and other imperfectly soluble gums. It is sometimes called *Tragacanthine,* is solid, colourless, without odour or taste, insoluble in water, but absorbing it and swelling up. Nit′ converts it into Oxalic′ and into Mucic acid. Sul′ changes it to a saccharine substance, which is not susceptible of Alcoholic fermentation. Guibourt, however, considers Tragacanth to consist of a peculiar Mucilaginous principle, with a little Starch and Ligneous fibre.

Action. Uses. Demulcent. Useful from its viscidity.

MUCILAGO (GUMMI, D.) TRAGACANTHÆ, E. D. Mucilage of Tragacanth.

Prep. Macerate *Tragacanth* (powder of, D.) ℥ij. in *boiling Aq.* f℥ix. (f℥viij. D.) for 24 hours, triturate to dissolve the gum and strain. It requires a little skill to prepare it. Soubeiran states that it is more viscid if made with the entire than with the powdered gum.

Action. Uses. Demulcent. Used also in making pills and suspending heavy powders.

PULVIS TRAGACANTHÆ COMPOSITUS, L. E. Compound Powder of Tragacanth.

Prep. Rub together *Starch* ℥jß. and *pure Sugar* ℥iij. then add *Tragacanth* bruised and *Gum Arabic* bruised āā ℥jß. and mix all carefully together.

Action. Uses. Demulcent in doses of ℨß—ℨj. Used also as a vehicle for other medicines. Or a Syrup may be made with *Tragacanth* ℨj, to *Syrup* Oij., which is much commended.

MUCUNA, *Adans.* L. E. Leguminum Pubes, L. D. Hairs from the Pod, E. Mucuna pruriens, *Dec.* L. E. *Dolichos pruriens*, Linn. D. Cowhage or Cowitch. *Diadelph. Decand.* Linn.

The strigose hairs of the plant called *kiwach* in India, as well as those of *Rottlera tinctoria*, are used in India as an Anthelmintic, whence the practice was probably introduced, as well as its corrupted name. Sir W. Hooker has distinguished the East Indian plant, *M. prurita*, from *M. pruriens*, which is indigenous in the West Indies.

MUCUNA, *Adans.* (*Stizolobium*, Willd.) *Diadelph. Decand.* Linn.

Calyx with 2 long caducous bracteoles, campanulate, 2-lipped, upper lip entire, lower trifid. Vexillum shorter than the wings and keel. Keel terminated by a polished acute beak. Stamens diadelphous, alternately longer. Legume hispid, oblong, few-seeded, with partitions of cellular substance between the seeds. Seeds oval, roundish or reniform, with a narrow oblong line, the hilum.—Twining plants of the tribe *Phaseoleæ*. Leaves trifoliolate; leaflets hairy on the under surface. Racemes axillary with large purplish, white or yellow flowers.

M. PRURIENS, *Dec.*—Leaflets ovate acute, the middle one rather rhomboidal, the latter ones oblique at the base. Racemes lax, many flowered, interrupted, 1—1½ foot long. Flowers with a disagreeable alliaceous odour, standard flesh-coloured, wings purple or violet, keel greenish white. Calyx hairy, pink, with lanceolate segments. Legume about 3 inches long and roundish, as thick as the finger, with somewhat keeled valves, densely covered with strong and stiff, sharp-pointed brown hairs.—Native of the West Indies. Bot. Reg. 1838, t. 18; Steph. and Churchill, iii. t. 179.

M. PRURITA, *Hook.*—Leaflets smaller, more obtuse, the middle one truly rhomboidal, the lateral ones dilated on the upper edge. Raceme ovate, compact, more often 3-flowered. Flowers dark purple. Calyx with short triangular teeth. Legumes oblong, much broader, curved and compressed, without any raised keel on the back of the valves, densely covered with sharp stinging hairs; which, white and soft when young, become brown and stiff when ripe.—Native of the East Indies.—Hooker, Bot. Misc. ii. 348; Suppl. t. 13.

The pods of the Kiwach, when young and tender, form articles of diet in India. When ripe, they are of a brownish colour, and covered with innumerable sharp prickle-like hairs, which penetrate into and irritate the skin.

Action. Uses. Mechanical Anthelmintic. Useful in expelling lumbrici and ascarides, by sticking into their bodies, when pressed against the intestinal parietes, and thus irritating and dislodging them.

D. The pods being dipped into treacle or honey, have the hairs scraped off until they have the consistency of an electuary, when a table-spoonful may be given to adults, and a tea-spoonful to children, followed by a purgative of Castor Oil, &c.

Rosaceæ, *Endlicher*. Roseworts.

The Rosaceæ, *Dec.*, like the Leguminosæ, are divided into several groups, which by botanists are treated of as distinct orders. Among these, the Amygdaleæ, true Roseæ, and Pomaceæ, contain officinal species. They may all be distinguished from Leguminosæ by the odd division of the calyx being anterior.

Herbs, shrubs, or trees. Leaves alternate, pinnately, or digitately compound, sometimes simple. Stipules adherent to the petiole, commonly foliaceous. Flowers regular, usually perfect, cymose or corymbose. Calyx free, 4—5 fid, laciniæ often doubled, the external series alternating with the internal. Petals 4 to 5, inserted into the throat of the calyx, imbricate. Stamens inserted with the petals, usually numerous, free, incurved in bud. Ovaries several, free, inserted into the bottom or into the tube of the calyx, with a single ovule, seldom more. Ovules pendulous or ascending. Style 1 to each ovary, terminal or more or less lateral. Fruit consisting of several single seeded carpels, covered by fleshy tube of calyx (Roseæ and Potentilleæ), rarely a single carpel within the hardened tube of the calyx (Sanguisorbeæ), or several follicular capsules, which are one- to many-seeded and opening by a ventral suture (Spireæ). Seeds erect or pendulous. Embryo without albumen, straight, radicle, superior or inferior.

The Rosaceæ of Endlicher include several tribes, as Roseæ, Dryadeæ, Spiræaceæ, and Neuradeæ. They are found in the temperate and cold parts of the northern hemisphere; a few only occur in the plains of tropical countries. An astringent principle is found in most parts of many of the species, a highly fragrant volatile Oil is also secreted by the Roses, &c. In others, the carpels being berried, or the receptacles fleshy, a highly grateful fruit is afforded.

Geum urbanum, *Linn.* D. Radix. The Root of Common Avens. *Icosand. Polygyn.* Linn.

This plant, indigenous in hedgerows and woods throughout Europe, has been long employed in medicine.

Roots perennial. Stems herbaceous, about 2 feet high, erect. Radical leaves interruptedly pinnate and lyrate; stem-leaves sessile, ternate; stipules large, rounded, lobed and cut. Flowers small, erect. Calyx 10-cleft, in 2 rows, the outer smaller, in the fruit reflexed. Petals 5, yellow, obovate. Stamens numerous. Fruit composed of little nuts, each terminated by the persistent lower parts of the jointed styles, which become hooked. Those in this species are much longer than the glabrous upper joint.

The rootstock is fusiform, a few inches in length, brownish-coloured externally, and of a reddish hue in the inside, with numerous radicles depending from it. It is astringent in taste, with some degree of fragrant aroma, whence it used to be called *Radix caryophyllatæ*. Analyzed by Tromsdorff, it yielded Tannin of two varieties, about 41 per cent., Resin, and some Volatile Oil, Bassorin, Gum, and vegetable Fibre.

Action. Uses. Astringent and Tonic; sometimes employed as a febrifuge; occasionally in Diarrhœas, &c., in doses of ℈ß.—℈j. of the powder; or in decoction (℥j.—Aq. Oj.), f℥jß. every 3 or 4 hours.

TORMENTILLA, L. E. D. Radix, L. The Root, E. Potentilla Tormentilla, *Sibthorp*, L. E. (*Tormentilla officinalis*, Sm.) D. Root of Common Tormentil. *Icosand. Polygyn.* Linn.

Tormentil has been long employed. Some suppose it was known to the Greeks.

Root large, perennial, irregularly shaped. Stems slender, spreading, often procumbent or straggling. Leaves sessile or shortly stalked, ternate, the lower leaves quinate on long petioles; leaflets oblong, acute, deeply serrated, a little hairy; stipules smaller, deeply cut. Flowers yellow. Calyx concave, usually 8-parted, in two rows, the exterior smaller. Petals 4, sometimes 5. Stamens numerous. Style lateral. Fruit consisting of numerous small nuts collected upon the flattish dry receptacle; in this species these nuts are longitudinally wrinkled. Seeds suspended.—Common on heaths and meadows throughout Europe.—E. B. t. 863; St. and Ch. i. t. 26.

The root is tuberous and knotty, with numerous radicles, of a dark-brown colour on the outside, and reddish internally, with little smell, but having a strong astringent taste. It contains about 17 per cent. of Tannin, with Colouring matter, Gum, and a little Volatile Oil. It is employed in tanning in the north.

Action. Uses. Astringent. Useful in Diarrhœa or Chronic Dysentery, in doses of ʒß—ʒj.

DECOCTUM TORMENTILLÆ, L. Decoction of Tormentilla.

Prep. Boil *Tormentil* bruised ℥ij. in *Aq. dest.* Ojß. down to one pint and strain.

Action. Uses. Astringent in doses of f℥iß. two or three times a day; or used as an astringent lotion.

ROSA, *Tourn.* Rose. *Icosandria Polygynia*, Linn.

Calyx urceolate, contracted at the mouth, ultimately succulent; limb 5-cut. Segments imbricated, often pinnately divided. Petals 5, obcordate, deciduous. Stamens numerous, inserted with the petals into the rim of the calycine tube. Carpels numerous, inserted into and enclosed within the fleshy tube of the calyx, each thickly covered with hairs and having a lateral style on the inner side; styles all passing through the contracted mouth of the calyx. Fruit globular or ovate, formed of the above fleshy and coloured tube of the calyx enlarged, enclosing within it numerous hard and bristly little nuts with inverted seeds.—Shrubs, often scandent, leaves usually impari-pinnate, leaflets serrated; stipules attached to the sides of the petiole.

Some species of Rose, being indigenous in Greece, were no doubt known to the Greeks, and *R. canina* is supposed to be their κυνοροδον. But the Hundred-leaved and the Damask Rose, natives of, and cultivated in, the East, were also known and highly esteemed. But the term ροδον seems to have been also applied to the Oleander, or Rose-Bay, called at one time Rhododendron.

ROSA CANINA, *Linn.* Fructûs Pulpa, L. Rosæ fructus, E. D. Hip of Rosa canina and of several allied species deprived of the carpels, E. Pulp of the Fruit, L. Fruit of the Dog Rose.

Common in hedge-rows, &c., in Europe; is supposed to have been the κυνόῤῥοδον of the Greeks.

This is a variable species, and several of its varieties have obtained distinct names. Shoots assurgent with uniform hooked prickles, and chiefly without setæ. The leaves are without glands, naked or slightly hairy, the serratures simple or compound. Flowers of a rose red colour. Sepals pinnate, deciduous. Styles remaining distinct.—E. B. 992; St. and Ch. 11, 100.

The fruit, or rather the inferior part of the calyx, become succulent, is of an ovoid form, of a scarlet or crimson colour, containing within its hollow the true fruit or woolly carpels, which require to be carefully removed, as their setæ are very irritant. The pulpy part has a sweetish acidulous taste. When dried, it yielded to Bilz 25 per cent. of Gum, 30·6 of uncrystallizable Sugar, of Citric′ 2·95, of Malic′ 7·77, with several Salts, a little Tannin and Volatile Oil.

CONFECTIO ROSÆ CANINÆ, L. CONSERVA ROSÆ FRUCTÛS, E.

Prep. Take pulp of *Dry Roses* ℔j. expose it to heat in an earthen vessel; add gradually powdered *Sugar* ℥xx. and rub till thoroughly incorporated.

E. Beat up the pulp of *Hips* with three times their weight of *White Sugar* gradually added.

Action. Uses. Acidulous Refrigerant. Chiefly valuable as a vehicle for other medicines.

ROSA GALLICA, *Linn.* Petala, L. D. Petals, E., of the Red Rose.

The Red, or French, called also the German and Austrian Rose, is a native of the middle and south of Europe, and may have been known to the ancients. Dr. Christison states that the true Red Rose of pharmacy is a variety, considered by some a distinct species, and called *Rosa provincialis*, which was probably introduced into Europe by the Crusaders, from its native country, Barbary. It is cultivated at Mitcham.

A dwarfish, stiff, short-branched bush, with the shoots armed with nearly equal uniform prickles and glandular bristles intermixed. Leaflets stiff, elliptical, rugose. Flowers several together, large, erect, with leafy bracts. Sepals ovate, leafy, compound. Fruit oblong.—Nees von E. 303; St. and Ch. iii. 99.

The petals are alone officinal, and look velvety, are of a purplish-red colour, with whitish down; with little scent when fresh, but this becomes developed as they dry. The half-blown buds are preferred: from these the calyxes and claws being cut off, they are quickly dried and sifted to get rid of impurities. They should be kept in well-closed vessels, and in the dark. In this state, they have a rose-like odour, will long retain their colour, and have a slightly bitter, astringent taste. Analysed, they have been found to contain a little Tannin, Gallic acid, Colouring matter, a little Volatile Oil, with other vegetable matters and some Salts, together with a little Oxide of Iron. The Infusion strikes a black with ferruginous salts.

INFUSUM ROSÆ (E. ACIDUM, D.) COMPOSITUM, L. Infusion of Roses.

Prep. In a glass (or porcelain not glazed with lead, covered, E.) vessel infuse dried *Petals of Rosa Gallica* ʒiij. (deprived of their claws ℥ß. D.) in boiling *Aq. dest.* Oj. (by measure ℔iij. D.) then mix in *dilute Sul'* fʒjß. (by measure ʒiij. D.) and macerate for 6 hours (4 hours, E.; half an hour, D.) Strain (through linen or calico, E.; when cool, D.) and dissolve in it *Sugar* ʒvj.

Action. Uses. Slightly Astringent and Tonic. The colour imparted to water is heightened by the acid. A much approved vehicle for saline purgatives, Quinine, &c. The presence of Sulphuric' must never be forgotten in prescribing it.

D. f℥iß. every three hours. Makes a good gargle with acids or Alum and Honey.

CONFECTIO ROSÆ GALLICÆ, L. CONSERVA ROSÆ, E. D. Conserve of Red Roses.

Prep. In a stone mortar, L. D., beat *Petals of Rosa Gallica* (buds, rejecting the claws, D.) ℔j. Add gradually *Sugar* ℔iij. (twice their weight, E.) and thoroughly incorporate.

Action. Uses. Slightly astringent in doses of ℈j. or ℈ij. but chiefly useful in making pills.

MEL ROSÆ, L. E. D. Honey of Roses.

Prep. Macerate in boiling *Aq.* Oijß. (by measure ℔iij. D.) for 6 hours dried *Petals of Rosa Gallica* ℥iv. Strain (let the impurities subside, pour off the clear liquor, E.) Add *Honey* ℔v. and boil down to the consistence of syrup in a vapour-bath, (removing the scum, E. D.)

Action. Uses. Mild Astringent, and being pleasant-tasted, is applied to Aphthæ, and used as a vehicle in gargles.

SYRUPUS ROSÆ GALLICÆ, E. Syrup of the Red Rose.

Prep. Proceed as for Syrup of Damask Rose, employing *Dried Red Rose Petals* ℥ij. boiling *Aq.* Oj. pure *Sugar* ℥xx.

Action. Uses. Slightly astringent; but chiefly used for colouring and flavouring medicines.

ROSA CENTIFOLIA, *Linn.* Petala, L. D. Petals, E. Rosæ Oleum. Volatile Oil of the Petals, E. Attar of Roses. The Hundred-leaved or Cabbage Rose.

This Rose has long been cultivated in Europe, having been introduced from the East. It is said to be indigenous in the Eastern Caucasus. The Persians also have a *sud-burg* (or hundred-leaved Rose), and the ancients were acquainted with one having many petals. Of the above species there are many varieties.

A bush with erect shoots, these are rather thickly covered with nearly straight prickles, scarcely dilated at the base intermixed with glandular bristles, all of different forms and sizes, the large ones falcate. Leaflets 5 to 7, oblong or ovate, glandular at the margin. Flowers several together, drooping. Buds short, ovate, with leafy bracts. Sepals in flowering, spreading not deflexed, leafy, more or less pinnate and with the peduncles glandulously viscid. Fruit ovate.—Cultivated at Mitcham, &c.—Nees von E. 302; St. and Ch. iii. 99.

The petals of this species are well-known for their fragrance, on which account, as well as for the beauty of the flowers, they are extensively cultivated, and consequently numerous varieties have been produced; so that it is difficult to say which is a species and which only a variety. In many parts of India, *Rosa damascena*, or Damask Rose, is cultivated for the purpose of yielding the *Attar* of Roses, as well as Rose-water. As the species of Rose are but few in India, perhaps the same may be cultivated in the extensive Rose Gardens of Ghazipore, which is the great mart for Attar in India. The petals should be collected just when fully blown, and if quickly dried, will long retain much of their fragrance, especially if preserved with salt. Besides the Volatile Oil, these petals contain a slightly laxative principle, with some of the same constituents as the other Roses.

SYRUPUS ROSÆ (L. D.) CENTIFOLIÆ, E. Syrup of Roses.

Prep. Macerate *dried* (fresh, E.) *Petals of Rosa centifolia* ℥vij. (℔j. E.) in *boiling Aq.* Oiij. (by measure ℔iv. D.) for 12 hours. Strain. Evaporate the strained liquor in the water-bath till only Oij. (℔ijß. D.) remain. Then add *Sugar* ℔vj. (℔iij. E.; q. s. D.) and make a Syrup. The E. College direct the Syrup to be prepared without concentration, but the Sugar to be dissolved in the liquor with the aid of heat.

Action. Uses. Slightly laxative. Given to infants in doses of f℥j.—f℥iv.

AQUA ROSÆ, L. E. D. Rose Water.

Prep. Mix *Petals of Rosa centifolia* ℔x. (viij. D.) with *Proof Spirit* f℥vij. L. (Rectified Spirit f℥iij. E.; ℥ß. to each ℔ Aq. D.) and *Aq.* Cij. (q. s. to prevent empyreuma, D.) Distil off a gallon. (Fresh petals are to be preferred, but those preserved, by having been beaten up with twice their weight of Muriate of Soda, may also be employed, E.)

Action. Uses. An agreeable vehicle for lotions and for active medicines.

OLEUM ROSÆ, E. Volatile Oil or Attar of Roses.

This is officinal in the E. P., and imported from India and the Levant. It is too well-known for its delightful fragrance, to require a detailed description. 100,000 roses distilled with water yield only about 180 grains of Attar. It varies in colour, becomes solid below 80° F. Sp. Gr. 0·832 at 90° F. Soluble in Alcohol, and a little taken up by water, as in Aqua Rosæ. It consists of two principles, one being a *solid* the other a *liquid volatile oil.* The former is scarcely soluble in Alcohol. In distilling Rose-water in this country, some of the crystalline Volatile Oil is sometimes obtained. As it is added by the perfumers to many scents, so it may be employed in imparting an agreeable odour to ointments and lotions.

POMACEÆ, *Juss.* Apple Tribe.

The Pomaceæ form a tribe of Rosaceæ in the system of De Candolle, but by many botanists they are separated into a distinct family. They may be distinguished from other Rosaceæ by their leaves being usually simple, the tube of the calyx adherent to the ovary; and thus, including the carpels, a fleshy fruit is eventually formed, which is crowned by the limb of the calyx, and is well known in the Apple, Pear, Quince, &c. whence it is called a Pome. This may be from 2—5 celled, each cell formed of cartilaginous or bony membrane, and containing 2 or more erect ovules.—They mostly inhabit the North temperate Zone and the great mountainous range of India. They are chiefly remarkable for their edible fruit when cultivated, abounding in saccharine matter with a pleasant acidity. In a wild state, they are austere or astringent and acid. By distillation of the seeds of some of the Pomaceæ a very little Hydrocyanic acid is obtained.

CYDONIA, L. Semina. CYDONIA VULGARIS, *Pers.* Seeds of common Quince. (*Pyrus Cydonia,* Linn.) *Icosandria Pentagynia,* Linn.

The Quince (κυδωνια) was known to the ancients and Arabs; the seeds (*bihee dana*) are employed medicinally in India, being imported from Caubul and Cashmere, where the tree is cultivated.

A moderately sized, much branched but crooked tree. Leaves ovate, obtuse at the base, quite entire, with their lower surface, as well as the calyxes and pedicels tomentose. Flowers few, of a white or rose-colour in a kind of umbel. The pomes closed, globose or oblong, 5-celled; *cells many seeded,* cartilaginous. Seeds enveloped in condensed mucilage.—Nees von E. 305; St. and Ch. ii. 114.

The fruit of the Quince is of a yellow colour, downy, and remarkable for a fine odour. The ancients used it as a medicine, but it is now chiefly employed for flavouring other fruits, or as a preserve. It contains some Astringent matter, with Malic acid, Sugar, and Azotised matter. (*Soubeiran.*) The seeds are oblong, pointed, convex on the outside, and with one or two flat sides, according to the pressure of neighbouring seeds. Their testa, or thick seed-coat, is covered with condensed mucilage, which, according to Bischoff, as quoted by Dr. Pereira, is lodged in very fine cells, and becomes easily dissolved out when submitted to the action of boiling water.

DECOCTUM CYDONIÆ, L. Decoction of Quince Seeds.

Prep. Boil with a gentle heat for 10 minutes *Quince Seeds* ʒij. in *Aq. dest.* Oj. Strain.

Action. Uses. Demulcent. Chiefly applied externally. It is analogous to the Mucilage of Linseed, or Linseed Tea, being viscid and insipid. It has been proposed to evaporate it to dryness, and powder the residue, which will readily afford mucilage with water.

Dr. Pereira considers *Quince Mucilage* as a peculiar substance, and calls it *Cydonin.*

AMYGDALEÆ, *Juss.* Almond Tribe. *Icosandria Monogynia*, Linn.

The Amygdaleæ form shrubs or trees, which are unarmed or have thorny branches. The leaves are simple, often glandular towards the base. Stipules deciduous. The stamens are about 20 in number. The ovary is free and single, and thus forms the distinguishing characteristic. It is one-celled with two pendulous collateral ovules. Style terminal or sublateral. Drupe with a fleshy or fibrous sarcocarp, a hard bony nut, which is usually single-seeded.—They are found wild in the mountainous parts of the North temperate Zone, but are now cultivated in most parts of the world with moderate climates. The fruit of many is edible, and the kernels abound in oil, many exude gum, but they are remarkable for secreting Hydrocyanic acid.

PERSICA VULGARIS, *Miller* (*Amygdalus Persica*, Linn.), D. Folia. The Leaves of the common Peach.

The Peach was known to the ancients, and called Persian Apple. It has been introduced into India from the north, and is no doubt a native of the Hindoo Khoosh mountains, &c.

A small tree. Leaves lanceolate, acutely and often doubly serrate, with a short petiole, not equal to half the transverse diameter of the leaf. Fruit a fleshy indehiscent drupe, the stone irregularly marked with furrows and small holes. The outer covering may be velvety or quite smooth. The Cling-stone and the Free-stone peaches are known in Persia as well as in Europe.—*Loudon's Arboretum*, t. 86.

The fruit of the Peach is well known as a delicious and wholesome fruit. The flowers and leaves, as well as the kernels, exhale the odour of bitter almonds ; but the leaves only are officinal in the D. P. By distillation they yield a volatile Oil, which contains a little Hydrocyanic acid.

Action. Uses. Sedative, Anthelmintic ; but seldom if ever used now. They may be given in infusion (*Dried Leaves* ʒß. to *Aq.* Oj.) in doses of f℥ß. two or three times a day.

AMYGDALUS COMMUNIS, *Linn.* L. E. D. Nuclei. Kernels. *Var.* α. AMYGDALÆ DULCES. Sweet Almonds. *Var.* β. AMYGDALÆ AMARÆ. Bitter Almonds.

The Almond is mentioned in the Bible. Both varieties are found in the countries from Syria to Affghanistan. Both were known to the ancients and to the Arabs.

A small tree with lanceolate leaves, which are glandularly serrate, young leaves folded flat; petioles glandular, equal in length to, or larger than the transverse diameter of the leaf. Flowers nearly sessile, solitary, appearing earlier than the leaves. Tube of the calyx campanulate. Fruit a dry drupe, ovoid compressed, externally tomentose, when ripe bursting irregularly. Within this is contained a hard but brittle shell, within which is enclosed a kernel, well known as the Almond.—St. and Church. i. t. 43. Nees, 312, 313.

Though a few botanists have considered the Sweet and Bitter Almonds to be distinct species, the generality describe them as varieties of one species. Nees von Esenbeck indeed states that both are sometimes obtained from the same tree. De Candolle enumerates several varieties, as are to be found in all cultivated plants.

Var. *α. dulcis.* The Sweet Almond has ash-green leaves, with the glands on the base of the leaf and lower serratures. The style much longer than the stamens. Shell hard, but some sweet almonds have very fragile shells, and are called *kaghuzee*, that is, *papery*, in the East.

Var. *β. amara.* The Bitter Almond has the petioles of the leaves studded with glands, and the style equal in length to the stamens, shell hard or brittle.

Sweet Almonds are sometimes sold with the brittle shells on them, and are then called *Shell Almonds.* Almonds are imported into this country from the south of Spain and of Italy. They are known by the names of Jordan, Valentia, and Italian Almonds. They are introduced into India from Persia and Affghanistan. The Almond is ovoid, being rounded at one end and pointed at the other ; flattened ; of a cinnamon-colour, from the tough testa with which the kernel is enveloped. When blanched, they are found to be composed almost entirely of two large and conspicuous cotyledons, white in colour, without smell, and of a mild agreeable taste. When old or worm-eaten, they have an unpleasant or rancid taste. Analysed by Boullay, they were found to contain 54 per cent. of a bland Fixed Oil, 24 of *Emulsin*, 6 of liquid Sugar, 3 of Gum, Water 3·5, Lignin 4·0, and Acetic' 0·5, the seed-coats, 5 per cent., contain a little Tannin. The *Emulsin* has also been called *Synaptase*, also the Vegetable Albumen of Almonds. It is white, and owing to its presence the Oil becomes suspended in water in *Almond Emulsion.* When the Oil has been expressed, we have, as in the case of Linseed, a *cake* left, which, being dried and powdered, is known under the name of Almond powder.

Action, Uses. Dietetical, Demulcent, and Emollient.

Confectio (Conserva, E.) Amygdalæ, L. (Amygdalarum, E.D.) Almond Confection or Paste.

Prep. Take *Sweet Almonds* ℥viij. (℥j. D.) and having macerated in cold water and removed the skins, beat well with *powdered Gum Arabic* ℥j. (ʒj. D) and *Sugar* ℥iv. (℥ß. D.) into a uniform pulpy mass. (Or it may be longer preserved if the ingredients are kept in powder and mixed when required, L.)

A pleasant-tasted Confection, useful only for making

Mistura Amygdalæ, L. (Amygdalarum, E. D.) Almond Emulsion or Milk.

Prep. L. E. Add gradually *Aq. dest.* Oj. (Oij. E.) to *Confect Almond* ℥ijß. (℥ij. E.) triturate constantly and strain through linen (or calico, E.); or, as follows: E. D. Take *Sweet Almonds* ℥j. and ʒij. (℥jß. D.) steep in hot water and peel them (Bitter Almonds ℈ij. D.) beat them to a smooth pulp with *pure Sugar* ʒv. (℥ß. D.) *Mucilage* f℥ß. E. adding gradually *Aq.* Oij. (by measure ℔ijß. D.) Strain.

Dr. Pereira recommends *Sweet Almonds* ʒiv. *powdered Gum Arabic* ʒj. *White Sugar* ʒij. *Water* f℥vjß. Blanch the Almonds, beat them with the Sugar and Gum, the Water being gradually added.

Action. Uses. Demulcent and Emollient, or as a vehicle for other medicines.

OLEUM AMYGDALÆ, L. (AMYGDALARUM, D.) Expressed Oil of either the Sweet or Bitter Almond.

A bland oil, apt to become rancid, of a pale yellow colour, very liquid, Sp. Gr. 0·917—0·920 ; consisting of Margarine 24, Elaine 76 parts in 100.

Action. Uses. Laxative and Emollient, like Olive and other fixed Oils.

AMYGDALÆ AMARÆ. Bitter Almonds are usually found shelled, smaller, and commonly imported from Mogadore. These, like the Sweet Almond, are without smell, but have a strong and peculiar bitter taste. Like them also, they contain bland fixed Oil, with *Synaptase*, and readily form a white emulsion with water. They also contain a small portion of an albuminous but very peculiar principle, called *Amygdalin*, soluble in water and in boiling Alcohol, colourless, and crystallizable; this contains Nitrogen, and has a pure bitter taste, but no smell (C_{40} H_{27} N O_{22}). It should be remarked that neither volatile Oil nor Hydrocyanic acid is mentioned as a constituent of Bitter Almonds. Indeed, both have been proved by chemists not to exist in them, though they may easily be obtained from them. This is by the mutual action, with the assistance of water, of one principle, Synaptase, when in solution, upon another, the Amygdalin; the one being supposed to bear the same relation to the other that Diastase does to Starch, or acting as Yeast does upon Sugar. The result, made immediately evident by the smell, is the production of the Oil of Bitter Almonds, which is a true Essential Oil, containing Hydrocyanic acid.

Action. Uses. Sedative, Poisonous. Bitter Almonds, even in small doses, disagree with many, producing derangement of the digestive functions, and a kind of nettle rash. They have proved fatal to men, children, and small animals. They flavour the Almond Emulsion of the D. P., but are seldom employed medicinally.

Oil of Bitter Almonds, though not officinal, requires to be noticed, as it is sometimes employed therapeutically, and is, moreover, a fearfully powerful poison. This when pure is considered to be a Hydruret of Benzoyle, a limpid colourless liquid, with a powerful odour ; but it is not in this state so poisonous. Exposed to the atmosphere, it absorbs Oxygen, and is converted into Benzoic acid. Ordinarily, Oil of Bitter Almonds is of a yellow amber-colour, has an odour of Hydrocyanic acid in addition to its own, which is usually considered rather agreeable. It has a bitter and burning taste, from containing Hydrocyanic acid (8·5—14·33 per cent.) with some other substances. This acid may, however, be separated from the oil, which is powerfully poisonous, heavier than water, soluble to a small extent in water, but very readily in Alcohol and Ether.

Action. Uses. Poisonous, like Hydrocyanic acid, and sometimes used for the same purposes, in doses of ♏¼ to ♏j.; also for flavouring.*

PRUNA, L. E. D. Drupæ exsiccatæ. Dried fruit of PRUNUS DOMESTICA, *Linn.* Prunes.

The common Plum-tree is supposed to be the κοκκυμηλεα of Dioscorides; but this may have been *Prunus Cocumilia* of Tenore, a native of Calabria, a species which is supposed by some to be the original of the former. The astringent juice of *Prunus spinosa*, or the Sloe, inspissated, is substituted for the ancient *akakia.*

A small tree with smooth branches and elliptical leaves. Flower buds formed of one or two flowers. Petals white, oblong-ovate. Drupes fleshy, ovate-oblong. Nut smooth or furrowed, without small holes.—Many varieties of it are cultivated every where in Europe. It occurs apparently wild in some places, but it is thought to be originally a native of Asia.

The fruit in a dried state forms the Prunes or French Plums of the shops, which are prepared in France chiefly from the St. Catharine and the Green-gage varieties, and "in Portugal from a sort which derives its name from the village of Guimaraens, where they are principally dried." A variety (the *Quetsche*) is also dried in Germany. The black Plums "*dits à médecine*," are prepared from the small black Damascus Plums, and are more acid and laxative. (Merat and De Lens.) Prunes are composed, like the other fruits of this family, of a large proportion of water, with about 20 per cent. of solid matter, consisting of Sugar, Gum, Malic acid, some Azotised matter, Pectin, and Ligneous fibre.

Action. Uses. Demulcent, Dietetical, Laxative. Given entire or in decoction, or their prepared pulp. This forms an ingredient in the Electuarium Sennæ.

LAURO-CERASUS, E. D. Folia, D. Leaves, E. of *Prunus Lauro-cerasus*, Linn. E. D. *Cerasus Lauro-cerasus*, Loisl. and Dec. The Cherry Laurel.

This shrub, so common in every garden almost in England, is a native of Asia Minor, especially near Trebizonde, whence it was introduced into Europe by Clusius about 1576.

A small tree or smooth evergreen shrub. Leaves with short petioles, oblong, acuminate, remotely serrated, shining on the upper surface, with 2 or 4 glands beneath and coriaceous in texture. Racemes simple axillary, about the length of the leaves. Petals white, roundish, spreading. Stamens 20. Drupe destitute of bloom, round, black, about the size of a small cherry.—Nees von E. 317; St. and Ch. ii. t. 117.

---

* Dr. Pereira remarks, that though its strength is variable, it is in general four times the strength of officinal Hydrocyanic acid, and that f℥ij. of the Oil in *Rectified Spirit* f℥vj. form an useful essence for flavouring and scenting.

This plant, being commonly called Laurel, and found in every shrubbery, must not be confounded, as it usually is, with the true Laurel or Sweet Bay (*v. Laurus nobilis*), which does not possess any of its deleterious properties. It should also be distinguished from the true Portugal Laurel. The leaves are alone officinal. In their dried state they are bitter and astringent, without aroma, as in the fresh state, until they are bruised, when the ratifia odour peculiar to so many of the *Amygdaleæ* is exhaled, from the formation probably of an essential Oil and Hydrocyanic acid, in the same way as in the Bitter Almond from the reaction of different principles on each other. Dr. Christison has made the important observation, that the buds and unexpanded young leaves in May or June yield 6·33 grs. of Oil in 1000.

Fig. 63.

The proportion sinks to 3·1 grs. in July, and goes on gradually diminishing to only 0·6 in the subsequent May, when they are twelve months old, and when the new unexpanded leaves of the same plant give ten times as much. By distillation with water, the Essential Oil is obtained, which exactly resembles that of Bitter Almonds; but the distilled Oil is alone in general employed, and is officinal in the E. and D. P.

Action. Uses. Poisonous, Sedative. The powdered leaves have been given in doses of from gr. iv.—gr. viij., and in cataplasms with Flour or Linseed-meal are sometimes applied to sores.

AQUA LAURO-CERASI, E. D. Water of the Cherry Laurel. Laurel-water.

Prep. Take *fresh leaves* of *Prunus Lauro-cerasus* lbj. *Aq.* Oijß. (by measure lbiij. Distil lbj. and add *Compound Spirit of Lavender* ℥j. D.) Chop down the leaves, mix them with the water, distil off a part, agitate the distilled liquid well, filter if any milkiness remains after a few seconds of rest, and then add the Spirit of Lavender, E. These directions are intended to obtain uniformity of preparation. It is, however, always uncertain. It is stronger when fresh made, or from young leaves; but as some opacity is created when there is excess of oil, this may be got rid of by filtering.

Action. Uses. Poisonous, Sedative, in the same cases as Diluted Hydrocyanic acid, in doses of f℥ß.—f℥j.

ACIDUM HYDROCYANICUM (E.) DILUTUM, L. ACID. PRUSSICUM, D. Diluted or Medicinal Hydrocyanic Acid: that is, diluted with 50, L. (30, E.) parts of water. F. *Acide hydrocyanique.* G. *Blausäure.* Prussic Acid.

Hydrocyanic Acid is so named from being a compound of Hydrogen and Cyanogen, and is called Prussic acid from having been first obtained from Prussian Blue. (p. 144.) Scheele in 1782 first obtained it in a diluted, and Gay-Lussac in 1815 in a pure state. It was employed in medicine in 1809 by Brera, in 1817 by Majendie, and in 1819 by Dr. A. T. Thomson in this country. But its effects had long previously been obtained from the employment of the above Laurel-water.

Hydrocyanic′ is of vegetable origin, being contained, as above mentioned, in the Distilled Oil and Waters both of the Bitter Almond and of the Cherry Laurel. It may also be obtained from many others of the Amygdaleæ, as kernels of Peaches, of various Plums and Cherries; also from some of their flowers, and likewise from Apple-pips. It is, however, usually obtained by decomposing some of the compounds of Cyanogen.

CYANOGEN (C_2 N=Cy=26), so named from *κυανος*, *blue*, and *γεννάω*, *I generate*, because it is an essential constituent of Prussian Blue. Though a compound body, it acts the part of a simple body in entering into and separating from chemical combination, and is usually adduced as a type of organic radicals (p. 16). It is composed of equal volumes of Carbon and of Nitrogen, or of 2 Eq. C+1 Eq. N=26, and is therefore a Bicarburet of Nitrogen, and forms a colourless permanent gas, with a penetrating and peculiar odour. Its compounds are called Cyanides or Cyanurets; of these some are found in various animal secretions. Hence some of the salts which are employed in making Hydrocyanic acid are obtained from animal matter. Cyanogen is interesting to us as combining with Hydrogen to form this acid.

Hydrocyanic Acid (Cy H=27) called also Cyanide of Hydrogen, was obtained by Gay-Lussac in a pure anhydrous state. This solidifies at the zero of F., but readily becomes liquid with heat, when it is transparent and colourless; Sp. Gr. nearly 0·697 at 64° F.; tasting at first cool, then acrid; but it can be tasted only with the greatest caution. It has a strong and very peculiar odour, differing from that of the Oil of Bitter Almonds. If a few drops be placed on paper, a part volatilizes so rapidly as to freeze the rest. It boils at 79° or 80°, when the acid rises in the state of vapour, which is combustible, and will form explosive mixtures with Oxygen. It rapidly decomposes, becoming of a reddish-brown colour, and finally exhaling an ammoniacal odour. Dr. Christison states, however, that he has kept it unaltered at 32° for three weeks. It has a feeble reaction as an acid, and forms Hydrocyanates, which are liable to decomposition. It is very soluble in Alcohol and in water, the solution in the latter forming

Medicinal or Diluted Hydrocyanic Acid. Differs from the former chiefly in its strength, having the same characteristic taste and odour, though in a less degree. The odour is so peculiar as to be enumerated among its tests. "But care must be taken not to confound it with the odour of Bitter Almond Oil, as many do ; for that odour is decidedly different, and depends much more on a true Essential Oil than upon the concentrated Hydrocyanic acid." (*c.*) It differs in being more easily preserved, especially if made from the action of S′ on Ferrocyanide of Potassium, or has a small quantity of some other acid mixed with it, or is kept in a dark-coloured bottle, or in one covered with paper, and well stopped. The medicinal acid of the shops has been found to vary from 1·4 to 5·8. That of the L. P. is now directed to be of the strength of 2 per cent., that of the E. and D. P. 3·3 per cent., or of the strength of Vauquelin's acid. Fearful consequences have ensued from the want of uniformity in its preparations; hence the Colleges give directions for ascertaining its strength. (*v.* Tests.)

Strength of the Medicinal Acid. In the D. P. the Sp. Gr. only is indicated as a criterion of its strength, and that of course is always lower as the acid is stronger ; but this method is not sufficiently precise for ordinary use. In the L. P., H Cy′ is directed to be prepared of such strength that 100 grains of it will exactly precipitate 12·7 grains of Nitrate of Silver dissolved in water. This precipitate is readily soluble in boiling Nitric acid, and 5 parts of it correspond to 1 of real acid. Dr. Christison states that so rigorous a test would exclude nine-tenths of the acid even of respectable shops, and that "irregularity within certain limits may exist without the slightest danger or inconvenience in medical practice." A variation of one-eleventh is therefore allowed in the E. P. "Fifty minims of the acid diluted with *Aq. dest.* f℥j. agitated with 390 minims of a solution (containing $\frac{1}{40}$) of Nitrate of Silver, and allowed to settle, will again give a precipitate with 40 minims more of the test." Dr. Ure has suggested ascertaining the quantity of Red Oxide of Mercury which a given weight (say 100 grs.) of this acid will dissolve ; and as the Eq. of the Oxide, 216, is to 2 Eq. 54, of the acid in Bicyanide of Mercury, as 4 to 1, so we have only to divide by 4 the weight of Oxide dissolved, and the quotient will represent the quantity of anhydrous acid present.

Prep. A simple process is that originally recommended by Mr. Everitt, and adopted by the L. C. for extemporaneously obtaining Dil. H Cy′. *Cyanide of Silver* gr. xlviijß. are to be added to *Aq. Dest.* f℥j. with which *Hydrochloric acid*

gr. xxxixß. had been previously mixed, and well agitated together in a close phial. (The Hydrogen of the acid combining with the Cyanogen of the Cyanide, Hydrocyanic′ in the proportion of 2 per cent. is formed. The Chlorine set free combining with the Silver, forms a white precipitate of Chloride of Silver.) After a short interval, the clear liquor is to be poured off, and kept out of the light.

The D. C. direct *Bicyanide* (*Cyanuret*, D.) *of Mercury* ℥j. *Hydrochloric′* f℥vij. *Aq. Dest.* f℥viij. both by measure, to be mixed together, and f℥viij. to be distilled from a glass retort into a cooled receiver, and to be kept in a cool and dark place. Sp. Gr. ·998. The strength of this acid is 1·6 according to Barker, but according to Mr. Donovan it is 2·82 per cent. (*n.*) The Society of Apothecaries used to obtain it by a process similar to this: Sp. Gr. 0·995, indicating 2·9 per cent. of real acid. By a process similar to this, but without water, Gay-Lussac obtained the anhydrous acid, by passing it through pieces of Chalk and of Chloride of Calcium.

The method which is most generally adopted, on account of its cheapness, easy management, and the preservation of the acid, is also given in the L. and E. P. Dissolve (in a retort, L.; mattrass, E.) *Ferrocyanide of Potassium* ℥ij. (℥iij. E.) in *Aq. dest.* Oß. (f℥xj. E.) then add to *Sulphuric′* ℥jß. (f℥vj. E.) previously mixed with *Aq. dest.* f℥iv. (f℥v. E.) and allowed to cool. Pour *Aq. dest.* f℥viij. into a cooled receiver and fit on the retort, and distil over into this water with a gentle heat in a sand-bath f℥vj. of acid fluid. (Distil over f℥xiv. or till the residuum begins to froth up, E.) Lastly, add *Aq. dest.* f℥vj. or as much as may be sufficient for the product to measure (f℥xx. L.; f℥xvj. E.) that 12·7 grs. of Nitrate of Silver dissolved in Aq. dest. may be accurately saturated by 100 grains of this acid.

When the Sul′ is added to the solution of the Ferrocyanide of Potassium, water is decomposed: its Hydrogen unites with some Cyanogen of the Ferrocyanide, and forms Hydrocyanic′; while the Oxygen uniting with some of the Potassium, generates Potash, which with the Sul′ forms a Bisulphate of Potash; and at the same time a new insoluble salt is formed, composed of Cyanogen, Iron, and Potassium. The constituents of this new salt have been differently deduced by Gay-Lussac and Mr. Everitt; but the former having attended most to theoretical considerations, and the latter to the results derived from numerous experiments, his numbers are preferred, and are those adopted by Mr. Phillips. But Dr. Pereira, in repeating the experiments, agrees with Gay-Lussac in the colour of the salt, which the latter states to be white, while Mr. Everitt found it to be yellow. This he observed to be owing to the admission of air; for when the action was made to take place without the admission of the air, the salt precipitated was always white.

The proportion of the substances submitted to distillation, and the results where 6 equivalents of Sulphuric acid are heated with 2 equivalents of Ferrocyanide of Potassium, are:

Submitted to Distillation.		*Results of Distillation.*	
6 eq. Sulphuric acid	240	3 eq. Bisulphate of Potash	384
4 — Cyanide of Potassium	264	3 — Hydrocyanic acid	81
2 — Cyanide of Iron	108	1 — Cyanide Potassium 66 } yellow salt	174
12 — Water { 6 in the acid, 6 in the salt }	108	2 — Cyanide Iron 108 } yellow salt	
		9 — Water	81
	720		720

Tests. Besides ascertaining the strength of H Cy′, it is necessary to determine its purity. "Colourless; entirely vaporizable, with a peculiar odour; slightly and transiently reddens litmus; unaffected by Sulphuretted Hydrogen (showing absence of Bicyanide of Mercury, &c.) Solution of Nitrate of Baryta causes no precipitate, E. The presence of any other acid is indicated by the Iodo-cyanide of Mercury and Potassium being reddened," L. in consequence of the formation of Biniodide of Mercury. The slight effect on Litmus, as well as the Nitrate of Baryta prove that there is no Sul′ nor H Cl′.

To detect this acid in cases of poisoning, its odour is a delicate and usually very conspicuous test. The contents of the stomach may be washed and filtered; or, in cases of mixed fluids, these may be treated with animal charcoal without heat, neutralized with Sul′ if they are alkaline from Ammonia disengaged during putrefaction, and about 1-8th part distilled over in a vapour-bath, applying the following tests.

1. Supersaturate with solution of Potash, then add a solution of the salts of the mixed Peroxide and Protoxide of Iron (as common Green Vitriol, or the Tincture of the Sesquichloride), when a greyish-green precipitate is formed, which, on adding a little Sul′, becomes of a deep Prussian Blue colour.

2. Nitrate of Silver causes a precipitate of Cyanide of Silver, which is soluble in boiling Nitric acid.

3. Sulphate of Copper with Hydrocyanic′ supersaturated with Potash forms a greenish precipitate, which, on the addition of a little Hydrochloric acid, becomes nearly white.

Consult works on Forensic Medicine, as of Christison and of Taylor.

Action. Uses. Sedative and Anodyne. Powerful Narcotic Poison. Useful in chronic Coughs and in affections of the Heart, in painful Neuralgia and Stomach complaints, as Gastrodynia. Externally as a wash to allay pain and irritation in chronic Skin Diseases.

D. ♏ij.—♏v. L., or ♏j.—♏iv. E. and D. in an ounce or so of water or Emulsion. In Lotion, f℥ij. in Oß. of Rose- or common Distilled water.

Antidotes. In cases of poisoning, it acts so quickly and energetically, that few antidotes can be effectively employed. 1. Cold affusion is usually most quickly available, and applied chiefly as cold *douche* to the Head and Spine, is particularly useful. 2. *Ammonia* or its Carbonate, cautiously inhaled, or in solution taken internally, or rubbed externally, is useful from its stimulant properties. 3. *Chlorine*, in the form of gas much diluted, cautiously inhaled, or Chlorine-water, in doses of a tea-spoonful or two, or the Chloride of Lime or of Soda, may be prescribed. Bleeding from the jugular vein is strongly recommended when there is cerebral congestion; artificial respiration ought not to be neglected. Give as a chemical antidote solution of Carbonate of Potash followed immediately by a solution of the mixed Sulphates of Iron.*

* The Messrs. Smith of Edinburgh published in the *Lancet* of 5 Oct. 1844, the detailed instructions for preparing Prussic acid antidotes, which depends on

MYRTACEÆ, *R. Brown.* Myrtle Family.

Trees or shrubs, often with angled branches. Leaves opposite, rarely alternate or verticillate, entire, with transparent dots, transverse veins, uniting into one which runs parallel with their margin, usually without stipules. Flowers perfect, regular, with 2 bracts. Calyx adherent by its tube to the ovary, limb 4 or 5 cleft, valvate in æstivation, sometimes entire opening like a lid. Petals inserted into the throat of the calyx, equal in number to its divisions, imbricate or twisted in æstivation, sometimes united with the cap-like limb of the calyx. Stamens numerous, inserted with the petals, sometimes only equal to or double their number, filaments distinct or connected into several bundles, curved inwards in æstivation, anthers ovate, opening by a double fissure. Ovary inferior, or half inferior, covered with a fleshy disk, several, 2, 4, 5 or 6, sometimes single-celled, placenta central. Style and stigma single. Fruit either dry or fleshy, dehiscent or indehiscent, usually surmounted by the limb of the calyx. Seeds straight. Embryo without albumen, straight, curved, or twisted spirally. Cotyledons often united into a homogenous mass with the radicle.

The Myrtaceæ, elegant in appearance, are allied on one hand to Pomaceæ, and on the other to Melastomaceæ, and through these to Lythrariæ, &c. They abound in the tropical parts of New Holland and of America, fewer in Asia and Africa, rare in the south of Europe. They secrete much astringent matter, as well as grateful Volatile Oil. Some therefore are employed as astringents, others as spicy aromatics. The fruit of some being berried, with a grateful acid and sweetish secretion, form edible fruits.

Tribe 1. LEPTOSPERMEÆ, *Dec.* Sub-tribe, *Melaleuceæ.*

CAJUPUTI, L. CAJUPUTI OLEUM, E. CAJEPUT, D. Oleum e foliis destillatum, L. Volatile Oil of the Leaves of MELALEUCA CAJAPUTI, *Maton and Roxb.* (*M. minor*, Sm.), L. E. *Polyadelphia Icosandria*, Linn. Cayaputi Oil, or Oil of Cajeput.

Cajuputi Oil, pronounced *Kayapootee* (meaning *Arbor alba*) in the East, appears to have been known only since the time of Rumphius, who describes two trees. 1. *Arbor alba major*, H. A. ii. t. 16. 2. *Arbor alba minor*, H. A. ii. t. 17. f. 1. In 1798, Mr. Smith, of the Calcutta Botanic Garden, was sent to the Molucca Islands to obtain the true sort of Cayaputi plant. He obtained several, which were introduced into the above Garden, and have since been distributed all over India. It is curious that this species, though a native of Molucca, is able to stand the cold of N. W. India, probably owing to the thick-

the presentation to the acid of Iron in such a state of oxidation as to form with it the comparatively inert Prussian Blue. They have since given the following in the *Pharm. Journ.* v. 35, July, 1845.

Dissolve in a phial *Carbonate of Potash* (Salt of Tartar) grs. xx. in water f℥j. or f℥ij. Dissolve quickly in a mortar *Sulphate of Protoxide of Iron* in *water* f℥j. and add of *Tinct. of Sesquichloride* (Muriate) *of Iron* fʒj. and put in another phial. To prevent delay, let the dispenser go at once and himself give to the person who has taken Prussic acid, first the Potash solution and immediately afterwards that of the salts of Iron.

ness of its bark. Mr. S. having also sent specimens to this country, they were ascertained by Dr. Maton to be those of the second kind, and named *Melaleuca Cajaputi* in the London Pharm. for 1809, a name which Dr. J. E. Smith afterwards unnecessarily changed to *M. minor.* The other species, which the Malays also call Cayaputi, is the *Melaleuca Leucadendron*, of which the leaves are larger, more falcate, 5-nerved, and smooth, but possess little or no fragrance, and are not known to yield any of this celebrated Volatile Oil.

Fig. 64.

Melaleuca Cajuputi, Roxb. (fig. 64), forms a small tree with an erect but crooked stem covered with thick, rather soft, light-coloured bark; branches scattered, with slender twigs which droop like those of the Weeping Willow. Leaves alternate lanceolate, acute, slightly falcate, 3 to 5 nerved, while young silky, and diffusing a powerful odour when bruised. Spikes terminal and from the extreme axils, downy as well as the calyx and branchlets, while in flower there is only a scaly conic bud at the apex, which soon advances into a leafy branchlet. Bracts solitary, 3 flowered. Calyx urceolate, (fig. 64, 3,) limb 5 parted. Petals 5, white, scentless. Stamens from 30 to 40, in five bundles; filaments 3 or 4 times longer than petals. Anthers incumbent with a yellow gland at the apex. Style long. Stigma obscurely 3-lobed. Ovary ovate, and like the capsule 3-celled, (f. 64, 2 and 1,) many seeded, lower half united with, but the capsule is enclosed within the thickened tube of the calyx. Seeds angularly wedge shaped. A native of the Molucca Islands, especially of Boerou, Manipe and of the S. of Borneo. It is called *Daun. kitsjil*, but also Cajuputi.

The leaves are collected on a warm dry day in autumn, and placed in dry sacks, in which they nevertheless become heated and moist. They are then cut in pieces, macerated in water for a night, and then

* In Murray, App. Medicam. we have it named *Caieput* s. *Kaiuput Oleum.*

distilled. Two sackfuls of the leaves yield only about 3 drachms of the oil. This is clear and limpid, of a light green colour, very volatile, diffusing a powerful odour, having a warm aromatic taste, something resembling that of Camphor, followed by a sense of coolness. Sp. Gr. 0·914 to 0·927; soluble in Alcohol. It boils at 343°. When distilled with water, a light and colourless Oil first comes over, and then a green-coloured and denser Oil, which, with less odour, is more acrid. Comp. $C_{10} H_9 O = 77$. Some adulterations are occasionally practised with this Oil, especially at the time when it reached a high price, chiefly with the Oils of Rosemary and of Camphor; but it is now commonly met with in a pure state, and without any admixture of Copper.

Action. Uses. Diffusible Stimulant, Antispasmodic; externally in Rheumatism. Surprise was excited in India when Cayaputi Oil was stated to be a cure for Cholera, as Oil of Peppermint is as useful.

D. ♏iij.—♏v. given on a lump of Sugar.

Tribe *Myrteæ.*

CARYOPHYLLUS, L. E. D. Flores nondum explicati, exsiccati, L. Dried undeveloped Flower of CARYOPHYLLUS AROMATICUS, *Linn.* L. E. (*Eugenia caryophyllata*), D. Cloves. *Icosand. Monog.* Linn.

Though it is doubtful whether the ancients were acquainted with the Clove, it is curious that the Arabs give *kurphullon* as its Greek name. P. Ægineta and Myrepsius seem to have known it; yet the author is of opinion that the ancients were not well acquainted with any substances produced further east than the coasts of the Bay of Bengal.

The Clove-tree is an evergreen and, like others of the Myrtaceæ, elegant in appearance. It is like the Pimento. The wood is hard and covered with a smooth grey bark. The leaves opposite and decussate, ovate-lanceolate, tapering towards both ends, about 4 inches long, somewhat leathery, shining, and minutely dotted, diffusing a clove-like fragrance when bruised. Panicles short, trichotomously divided, jointed at every division. The calyx tube is cylindrical, of a dark purple colour, adhering to the ovary, divided into 4 ovate concave segments. Petals 4, overlapping each other and of a globular form when in bud, afterwards spreading, roundish, whitish and said to exhale a grateful odour. Within the calyx and at the top of the ovary is a quadrangular disk, surrounding but not embracing the base of the short obtuse style. Stamens in 4 bundles, filaments long, yellow. Ovary nearly cylindrical, 2-celled, with many small ovules in each cell attached to the sides of the dissepiment. Fruit a large elliptical berry, containing a single seed, by the growth of which the second cell and numerous ovules have been obliterated. Embryo large, elliptical, dotted. Cotyledons unequal, sinuose, the larger one partly enveloping the smaller, including the superior radicle.—See the Bot. Mag. t. ii. 749, for a full description; Dict. des Science, Nat. Bot. for detailed dissections. A native of the Moluccas, but confined by the Dutch to Amboyna and Ternate. It has, however, been introduced into the Isle of France, India, the West Indies and Guyana.

Cloves, the unexpanded flower-buds, are picked by hand or with long reeds, and then quickly dried in the shade. The best are obtained from the Moluccas. They have some resemblance to a nail, (whence the French name of *clou de girofle*), are usually of a dark-brown colour, with a pleasant odour, a warm and aromatic, even burning, taste. They have considerable weight. The best will exude a little Oil when pressed or scraped. Their active properties are extracted by water and by Alcohol. They contain of Volatile Oil 18, of a peculiar Tannin 13, Gum 13, Resin 6, Extractive 4, Lignin 28, water 18=100. The Oil is officinal. The Resin has been named *Caryophyllin*, and is obtained in brilliant satiny crystals, without taste and without smell, fusible, volatile, insoluble in water, soluble in Alcohol; isomeric with Camphor. The dried berries, called *Mother Cloves*, are still imported into China.

Action. Uses. Stimulant, Carminative. Used as a condiment and as a corrective, for flavouring medicines.

OLEUM CARYOPHYLLI, L. E. D. Oil of Cloves.

Oil of Cloves is usually imported from Amboyna. In America it is said to be distilled from the Cloves grown in Cayenne. The Cloves from which oil has been distilled are apt to be intermixed with others. The Oil, when recent, is clear and colourless, but by degrees becomes of a dark-brown colour, which is its ordinary appearance. Its odour is strong, and its taste warm, aromatic, and even acrid. Sp. Gr. 1·05 to 1·06, and being thus heavier, sinks in water. It is best distilled with salt and water, as in the case of other heavy oils, and requires repeated cohobation. M. Ettling, according to Soubeiran, finds this Oil composed of—1. A Hydrocarbon, like Essential Oil of Turpentine, which is lighter than water. 2. An oxygenated Oil, which is heavy, Sp. Gr. 1·079, and has some of the properties of an acid (*acide eugenique* of Dumas), and composed of $C_{24} H_{15} O_5$. 3. Stearoptene, which is also sometimes met with in distilled water of Cloves. This Oil is sometimes mixed with Oil of Pimento.

Action. Uses. Aromatic Stimulant, Carminative, and used as a corrective in doses of ♏ij.—♏v.

INFUSUM (CARYOPHYLLORUM, D.) CAROPYHYLLI, L. E. Infusion of Cloves.

Prep. Macerate in a slightly covered vessel for 2 hours *bruised Cloves* ʒiij. (ʒj. D.) in boiling *Aq. dest.* Oj. (by measure ℔ß. D.) Strain (through linen or calico, E.)

Action. Uses. A clear infusion, with the odour and taste of Cloves, incompatible with preparations of Iron. Useful as a warm carminative, or as a vehicle for other medicines, in doses of fʒiß.

Off. Prep. Ferri Mistura Arom. D. Inf. Aurantii, Comp. L. E. D. Sp. Ammoniæ Arom. L. D. Sp. Lavandulæ, Comp. D. Vinum Opii, L. E. D. Conf. Aromatica, L. D. Pil. Colocynth. Comp. D. Elect. Scammonii, L. D.

PIMENTA, L. E. D. Baccæ immaturæ exsiccatæ, L. (Fructus, D.) Unripe Berries of Eugenia (Myrtus, *Linn.* L. D.) Pimenta, *Dec.* E. Pimento. *Icosand. Monog.* Linn.

A native of South America and the West Indies. In the latter it is much cultivated in regular walks. Besides Pimento, it is also called Allspice and Bay-berry tree.

An elegant tree about 30 feet high, foliage dense and evergreen, branches round, twigs compressed, the younger as well as the pedicels pubescent. The leaves are petiolate, oblong or oval, marked with pellucid dots, smooth. The peduncles axillary and disposed in terminal trichotomous panicles. Calyx and petals 4-fid, the latter reflected greenish white. Stamens numerous. Ovary 2—3 celled; cells many ovuled. Berry spherical covered by the roundish persistent base of the calyx, which when ripe is smooth, shining and of a dark purple colour; one-, rarely 2-celled; two-seeded. Embryo roundish, cotyledons united into one mass, radicle scarcely distinct.—Nees von E. 298. *Myrtus* or *Myrcia pimentoides* is figured in t. 297, and yields ovate Pimento.

Browne (*Nat. Hist. of Jamaica*) describes the berries as being gathered before they are ripe, because they then lose their aromatic warmth, and acquire a taste like Juniper-berries, and are much eaten by birds. When gathered, they are dried with care in the sun. They are round, rugose, unequal in size, and of a brownish colour, and consist of the pericarp, in which the virtues chiefly reside, and of two dark-brown seeds. The odour is strongly fragrant, and the taste warm and aromatic. This depends on a Volatile Oil, which is separated by distillation. There is also some Fixed Oil, a pungent Resin, Extractive, Tannin, Gallic acid, &c.

Action. Uses. Stimulant Aromatic. Carminative in doses of gr. x.—ʒß.

OLEUM PIMENTÆ, L. E. D. Oil of Pimento.

Obtained in the proportion of 1 to 4 per cent. by distilling bruised Pimento with water. It resembles and is sometimes sold for Oil of Cloves, or employed to adulterate it. Dr. Pereira describes it as consisting of two Volatile Oils, one *light* (Hydrocarbon), the other *heavy* (Pimentic acid). It produces a red colour with Nitric' and a bluish-green with Tinct. of Sesquichloride of Iron; thus resembling Morphia in these particulars.

Action. Uses. Stimulant Carminative in doses of ♏iij. to ♏vj. Rubefacient externally.

SPIRITUS PIMENTÆ, L. E. D. Spirit of Pimento.

Prep. Prepare as Sp. Myristicæ, L.; as Sp. Caraway, using bruised Pimento, ℔ß. E. Macerate for 24 hours *bruised Pimento* ℥iij. in *Proof Spirit* Cj. and *Aq.* q. s. to prevent Empyreuma. Distil Cj. D.

Action. Uses. Carminative, in doses of fʒj.—fʒiv. Used chiefly as an adjunct.

AQUA PIMENTÆ, L. E. D. Distilled Water of Pimento.

Prep. Take *Oil of Pimento* ℥ij. L. or *bruised Pimento* ℔j. (℔ß. D.) *Proof Spirit* f℥vij. L. (Rectified Spirit f℥iij. E.) *Aq.* Cij. (q. s. to prevent Empyreuma, D.) Distil Cj.

Action. Uses. Carminative. Much used as a vehicle for other medicines in doses of f℥iß.

Off. Prep. Syrupus Rhamni, L. Emplastrum Aromaticum, D.

GRANATEÆ, *Don.* Pomegranates.

This order was instituted by the late Professor Don for the Pomegranate, which was usually included, as it still is by Dr. Lindley, among Myrtaceæ. It is chiefly distinguished by its leaves not being dotted, by the want of the marginal vein, by the peculiarities of its fruit, and by the seeds being involved in pulp, and by its cotyledons being convoluted, also by the absence of aromatic properties, and the geographical distribution being beyond the range of tropical Myrtaceæ. Dr. L. thinks that the several variations are not greater than occur in genera of other families, without their being raised to the rank of orders.

PUNICA GRANATUM, *Linn.* L. E. D. Granatum, L. Fructûs Cortex. Baccæ tunica exterior, D. Rind of the Fruit. Granati Radix, E. Radicis Cortex, D. Bark of the Root. Flores, D. Flowers of the Pomegranate. *Icosandria Monogynia*, Linn.

The Pomegranate, a native of the mountainous countries from Syria to the north of India, must always have been an object of attention. It is the *rimmon* of the Bible, and the *rooman* of the Arabs. It was well known to the Greeks and Romans.

Fig. 65.

Stem arborescent and irregular, in arid situations rather thorny; the leaves usually opposite, often fascicled, oblong, inclining to lanceolate, quite entire, not dotted, smooth, shining, and of a dark green; flowers commonly solitary, of a brilliant scarlet; calyx thick and fleshy, adhering to the ovary, turbinate, 5 to 7 cleft; petals 5 to 7, crumpled; stamens numerous, often double; style filiform; stigma capitate; fruit of the size of a large apple, with a thick leathery rind, and crowned by the tubular limb of the calyx; cells several, arranged in two strata, separated from each other by an irregular transverse diaphragm, lower division of 3 cells, the upper of from 5 to 9 cells; seeds numerous, involved in pellucid pulp, with foliaceous, spirally convolute cotyledons.—Nees von E. 301.

The parts of this plant which were employed by the ancients, still are so in the East, and are officinal in the D. P. Thus the Flores, D. are the *Balaustion* of the ancients. In India, *buloositoon* is given as the Greek name of the double flower. They are devoid of odour, but have a bitterish and astringent taste, tinge the saliva of a reddish-colour, contain Tannin, strike a black with ferruginous salts.

The Rind of the Fruit (*Granatum*, L., *Baccæ Tunica exterior*, D.), especially of the *wild* plant, is extensively employed as an astringent and as a dye in the East. It is of a reddish-brown colour and smooth externally, but yellow on the inside; usually in irregular fragments, dry, hard, and leathery, of a very astringent taste. It contains of Tannin 18·8 per cent., with 10·8 of Extractive, and 17·1 of Mucilage, and is used for tanning in some countries.

The Bark of the Root (*Radicis Cortex*) was employed as an anthelmintic by Dioscorides and by Celsus, and still is so in India. It was reintroduced into practice by Drs. Buchanan and Anderson. The root itself is heavy, knotted, and of a yellow-colour; its bark often sold in strips, sometimes with parts of the root still adhering to it. On the outside of a greyish-yellow colour; on the inside, yellow, something like that of the barberry. It has little smell; when chewed, colours the saliva yellow; has an astringent taste, without any disagreeable bitterness. It has been analyzed by Mitouart and Latour de Trie, and others; but the source of its peculiar anthelmintic powers has not been discovered, and the subject requires further investigation. It contains Tannin (about 20 per cent.), Gallic acid, Resin, Wax, Fatty matters, and Mannite. "An infusion yields a deep-blue precipitate with the salts of Iron, a yellowish-white one with solution of Isinglass, and a greyish-yellow one with Corrosive Sublimate, and Potash or Ammonia colours it yellow." (*c.*) It is apt to be adulterated with the barks both of box and of barberry. The former is white and bitter, but not astringent; the latter yellow, very bitter, and not thus affected by the above four reagents.

Action. Uses. All parts are astringent, the rind of the wild fruit especially so, and useful in Diarrhœa and advanced stages of Dysentery; the Flowers in infusion slightly astringent; the Bark of the Root astringent, but remarkably useful as an Anthelmintic against tænia.

D. It may be given in doses of ℈j. in powder; or a decoction may be formed by steeping for 12 hours *of fresh Root-bark of Pomegranate*

℥ij. in *Aq.* Oiſs., and boiling down to Oj. Of this f℥ij.—f℥iv. may be given in the morning fasting, and repeated every two hours, until three or four doses have been taken; pursuing the same course another day, if not efficient at first, with occasional doses of Castor Oil.*

LYTHRARIÆ, *Juss.* Loosestrifes.

Herbs, shrubs, or trees, often with 4-cornered branches; leaves opposite; calyx tubular or bell-shaped; petals inserted into the throat of the calyx, and alternating with its lobes; stamens equal in number to, or twice or thrice as many as the petals; ovary free, 2 or 4 celled, each with many ovules; seeds apterous or winged, without albumen.—They are found both in temperate and tropical parts of the world. Many of them secrete Tannin, some Colouring matter, and a few a little Volatile Oil.

LYTHRUM SALICARIA, *Linn.* Herba, D. Herb of Purple Loosestrife. *Dodecandria Monog.* Linn.

This plant is indigenous over all parts of Europe, &c., in wet places and banks of ditches.

Stem 2—4 feet high, quadrangular; leaves opposite or whorled, lanceolate from a cordate base, varying in length, nearly smooth. Flowers in whorls on a leafy spike. Bracts O. Calyx tubular, with 8 to 12 teeth, erect, twice as long as the other. Petals 6, alternating with these. Stamens usually twice as many as the petals, alternately shorter, inserted into the base of the calyx. Style filiform. Stigma capitate. Capsule oblong, covered by the calyx, 2-celled, many-seeded.—E. B. 1061.

This plant, so showy from its long spikes of purple flowers, has little activity, though it contains some Tannin. It is without odour, and has a mucilaginous and moderately astringent taste.

Action. Uses. Gentle Astringent. Occasionally prescribed in Dysentery in doses of ʒj.—ʒij. or in infusion.

CUCURBITACEÆ, *Juss.* Gourd Family.

Annual or perennial succulent herbs, climbing with tendrils. Leaves alternate, palmate, more or less rough. Flowers usually unisexual. Calyx with its tube united to the ovary, limb 5-parted. Corolla usually so closely united with the calyx as to appear a continuation of it, limb 5-parted with reticulated veins. (*Male*) Stamens 5, free or united in pairs, with the fifth remaining free. Anthers sinuose. (*Female*) Ovary 3—5 celled, or spuriously 1-celled, with numerous ovules, placentæ parietal. Style 1. Stigma thick and lobed. Fruit or pepo usually fleshy, crowned by the remains of the calyx, and from the partitions becoming pulpy, 1-celled. Seeds flat, with a membranous or horny integument, often thickened at the margin. Embryo without albumen. Cotyledons leaflike.

The Cucurbitaceæ abound chiefly in warm parts of the world, but a few are found in temperate climates. A bitter, often purgative, principle is secreted by many of them; several by cultivation yield edible fruit, but even in these the rind continues bitter. The seeds contain much bland fixed oil.

---

* Dr. Budd, Physician of King's College Hospital, informs the author that he has often prescribed the bark of the root of Pomegranate, and that he considers it as efficacious as Turpentine, and much safer, producing only a feeling of weight in the stomach or nausea, apparently only from the quantity taken. Dr. Budd insists, as some others have done, upon the *fresh* root only being employed.

COLOCYNTHIS, L. E. D. Peponum Pulpa exsiccata, L. D. Pulp of the Fruit, E. CUCUMIS, *Linn.* (now CITRULLUS) COLOCYNTHIS, Colocynth. *Monœcia Monadelphia*, Linn.

The Colocynth (κολυκυνθις of the Greeks and *Hunzal* of the Arabs) has been used in medicine from the earliest times, and is one of the plants supposed to be the *Pakyoth* or *wild gourd* of Scripture.

Fig. 66.

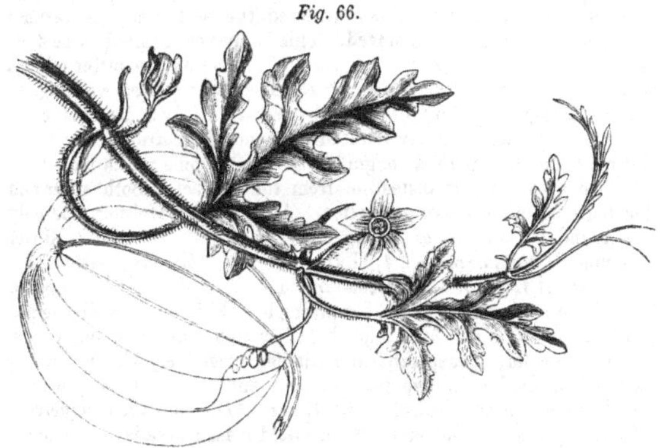

Citrullus (*Cucumis*, Linn.) Colocynthis.

Annual Herb. Roots thick, whitish. Stems procumbent, angular, hispid. Leaves cordate-ovate, divided into many lobes; lobes obtuse (but rather acute in the Linnæan specimen, and as represented by St. and Ch. iii. t. 138, from a plant grown in Chelsea Garden from seed sent from the Mediterranean, and from which fig. 66 is taken), of a bright green on the upper surface, whitish below, and muricated, from being covered with small white hairs and often hair-bearing tubercles. Petioles as long as the lamina. Tendrils short. Flowers axillary, solitary, stalked. Calyx with 5 subulate segments. Female flowers with the tube of the calyx globose, and somewhat hispid, the limb campanulate, with narrow segments. Petals small, yellow, with greenish veins, scarcely adherent to each other and to the calyx. Fruit globose, smooth, about the size of an orange, with a thin but dense rind, 6-celled, pulp very bitter. Seeds ovate, not marginate, whitish, sometimes brownish, bitter.—Extending from the south of Europe to Syria and the south of India,* north of Africa, Egypt, and Nubia. It

* Found in various parts of India, as on the sandy lands of Coromandel (*Roxburgh*), Peninsula (*Wight*), Deccan (*Col. Sykes*), sea-shores of Guzerat (*Gibson*), Kaira (*Burns*), near Delhi (*Mackintosh and Rankin*), in Bengal Dispensatory also (*Falconer*). The author had heard of it in this direction, but on sending for *Indrayun* and *Bisloombha* plants, which are Arabian and Indian names for the Colocynth, a species of a nearly allied species, with oval instead of globular fruit, was obtained. This he named *C. Pseudo-Colocynthis*, and figured (Himal. Bot. t. 47. fig. 2). A good supply of Colocynth may therefore be obtained for the public service or for commerce, from India, whence it has sometimes been imported.

is possible that in some of the localities usually cited, some nearly allied species may be found instead of the true Colocynth.

Colocynth is imported in two forms. 1. *Unpeeled*, from Mogadore, in its entire state, and covered by its hard yellow rind. 2. *Peeled*, from the Levant, North of Africa, and South of Spain; with the rind peeled or pared off, and the pulp dried, when the fruit is ripe. It then appears in the shape of white balls, which are light, porous, and spongy, but tough, usually with the seeds forming about ¾ of the whole weight. The smaller variety of fruit is considered the best, and is sometimes imported with the seeds removed. This is always required to be done before any preparations can be made. The seeds are bitter; but a good deal of the bitterness may be removed by repeated washings in water. The pulp is without odour, but is nauseously and permanently bitter. It is with difficulty reduced to powder (*Poudre de Coloquinte*), and may therefore with a magnifier be seen in pills which have been made up with the pulp instead of from the Extract. Both water and Alcohol extract its active properties. Analyzed by Meisner, the pulp was found to contain of Fixed Oil 4·2, Bitter Resin, 13·2, Bitter Principle (*Colocynthin*) 14·4, Extractive 10, Gummy matters 30, Phosphate of Lime and Magnesia, 5·7, Lignin 19·2. The *Colocynthin* is not, however, a pure vegetable principle. Examined by Herberger and Braconnot, it was of a reddish-yellow colour in mass, but yellow when in powder, transparent and friable, excessively bitter, burning like resins, soluble in five parts of cold and in less boiling water; equally soluble both in Alcohol and Ether. Acids and the deliquescent salts precipitate it as a coherent and viscid mass; alkalis do not precipitate it, neither does Gall-nut when it is quite pure. It contains Nitrogen, and, according to Braconnot, restores the colour of Litmus reddened by acid. (*Soubeiran.*)

Action. Uses. Colocynth is a powerful Hydrogogue Cathartic, but an irritant Poison in large doses.

EXTRACTUM COLOCYNTHIDIS (SIMPLEX, D.) L. E. Colocynth Extract.

Prep. Boil gently for 6 hours *dried pulp of Colocynth* ℔j. in *Aq. dest.* Cij. (Cj. till reduced to ℔iv. by measure, D.), occasionally replacing the evaporated water with *Aq. dest.*) While hot, strain and evaporate to a proper consistence.—From 45 to 65 per cent. of Extract are obtained. The D. C. order too little water. (*Phillips.*)

Action. Uses. Cathartic, but seldom prescribed alone, though it may be so, in doses of gr. v. to ℈j.

EXTRACTUM COLOCYNTHIDIS COMPOSITUM, L. D. Compound Colocynth Extract.

Prep. Macerate *Colocynth* pulp in small pieces ℥vj. in *Proof Spirit* Cj. with a gentle heat for 4 days. Filter, and add *purified Extract of Aloes* (hepatic, D.) ℥xij. *Scammony* in pieces ℥iv. *Soap* (hard, D.) ℥iij. Evaporate to a proper consistence, and towards the end add *Cardamoms* powdered ℥j.

Action. Uses. As Colocynth taken alone is much more griping and irritating than when prescribed with other Cathartics, the Compound Extract or Pill is a safe and energetic purgative, and probably more frequently prescribed than any other. The addition of a little Calomel makes it still more useful.

PILULÆ COLOCYNTHIDIS, E. D. Compound Colocynth Pill.

Prep. Pulverize together *Socotrine* (*Hepatic,* D.) *Aloes* (℥j. D.) 8 parts, *Scammony* (℥j. D.) 8 parts, *Sulphate of Potash* (ʒj. D.) 1 part. Mix them with finely-powdered *Colocynth* 4 parts (℥ß. D.), add *Oil of Cloves* (ʒj. D.) 1 part, and take *Rectified Spirit* (*Syrup,* D.) q. s. to beat the whole into a proper pill mass. Divide into 5 gr. pills.—Dr. Christison states that nothing keeps the pill so long soft as Spirit.

Action. Uses. Nearly the same as the foregoing, but more eligible for prescription from its constituents, in doses of gr. v.—gr. xv.

PILULÆ COLOCYNTHIDIS ET HYOSCYAMI, E. Colocynth and Henbane Pills.

Prep. Beat together *Colocynth Pill mass* 2 parts, *Extract of Henbane,* 1 part, adding, if necessary, a few drops of *Rectified Spirit.* Divide into 5 gr. pills, of which one to three form a dose.

Action. Uses. The addition of the Henbane deprives the pill of its tendency to gripe and irritate, and therefore makes this form applicable to all the same cases as the above.

ENEMA COLOCYNTHIDIS, L. Colocynth Enema.

Prep. Mix and rub together *Compound Extract of Colocynth* ℈ij. *Soft Soap,* ℥j. *Aq.* Oj.

Action. Uses. Cathartic Enema in obstinate constipation.

A *Tincture* and a *Wine of Colocynth* are employed sometimes on the Continent: A little of the former, or ℈j. of the powder, mixed with lard, and rubbed on the abdomen, will sometimes produce full cathartic effect.

ELATERIUM, L. E. D. Pepones (Fructus, D.) recentes, L. The Fresh Gourd. Fæcula, D. Feculence of the Juice of the Fruit, E. Folia, D. *Momordica Elaterium,* Linn. Squirting Cucumber.

This plant was known to the Greeks, and called Σίκυς ἄγριος, and sometimes also Ἐλατήριον, a name which was also applied to the feculence of the juice of its fruit. By Richard it has been formed into a genus ECBALIUM, and the species called E. Elaterium. (*E. agreste Rchb.*)

Annual, with hispid, scabrous trailing stems, which are glaucous and without tendrils. Leaves cordate, somewhat lobed, crenately-toothed, very rugose, on long bristly stalks. Flowers monœcious. ♂ Calyx 5-toothed. Corolla yellow, 5-parted. Stamens triadelphous, with connate anthers. ♀ Filaments 3, sterile. Style trifid. Stigmas bifid. Ovary 3-celled, with many ovules. The fruit ovate, 1½ inch long, muricated, when mature, being freed from its petiole, and contracting with elasticity, it forcibly projects the juice and seeds from a basilary orifice. Seeds of a brown colour, compressed, reticulate.—A native of the south of Europe, cultivated in England.—Esenb. and Eberm. 272. St. and Ch. i. 34.

Fig. 67.

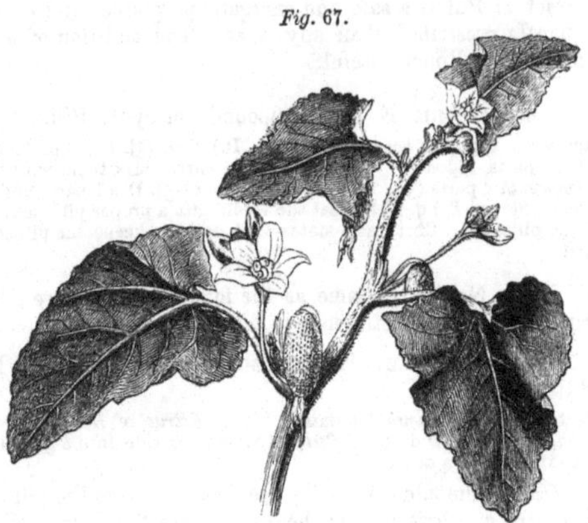

Ecbalium (*Momordica*, Linn.) Elaterium. *Rech.*

Elaterium is the feculence deposited from the juice of the fruit, when separated and allowed to stand. Dr. Clutterbuck proved that it is contained only in the juice around the seeds, which is of a gelatinous consistence. The rest of the fruit is comparatively inert. When the fruit is sliced and placed upon a sieve, a limpid and colourless juice flows out, which after a time becomes turbid, and then deposits a sediment. This, when dried, is light and pulverulent, of a light yellowish-white colour tinged with green, and is genuine Elaterium, of which Dr. C. obtained only 6 grains from 40 of these Cucumbers; and found $\frac{1}{8}$ of a grain to produce powerful cathartic effects. In conformity to these experiments is the method now adopted for obtaining Elaterium. The processes of the three Colleges are nearly the same, though that of the E. P. is the best; because if only the *quite ripe* fruit is collected, the greater part of the active principle would in most cases be expelled by the peculiar method in which this plant discharges its seeds.

ELATERIUM, E. EXTRACTUM ELATERII, L. D. Elaterium, improperly called an Extract.

Prep. Take of the *Fruit of Momordica Elaterium* before it is quite ripe (when ripe, L. D.) any convenient quantity. Cut the fruit and express the juice gently through a (very, L.) fine sieve. Allow the liquid to rest (for some hours, L. D.) till it becomes pretty clear. Pour off the supernatant liquid, which may be thrown away, and dry the feculence with a gentle heat. (See Pereira's Materia Med. for greater details.)

Tests. Colour pale grey ; when exhausted by rectified Spirit, the solution concentrated, and poured into hot diluted Liquor Potassæ, deposits on cooling minute, silky colourless crystals weighing from a seventh to a ninth of the Elaterium, E. The Spirit dissolves the active principle (*Elaterin*) with Chlorophyll : the latter is retained.

Elaterium is in thin cakes of a pale-grey or of a greenish-grey colour, often marked by the substance upon which it has been dried ; light and friable, with little odour, but with an acrid and bitter taste, which is possessed by other parts of the plant, as the *Leaves*, which, though officinal in the D. P., are seldom employed medicinally. An inferior kind is also met with, which is more compact, and of a darker colour, either brownish or of an olive-green. This is probably prepared by expressing the whole juice, and then evaporating to dryness. Dr. Pereira describes *Maltese Elaterium* as in larger flakes, and of a paler colour, often chalky, sometimes mixed with Starch, hence effervescing with acids, and becoming blue with Iodine. Elaterium, carefully prepared, consists only of the feculence deposited when the juice has been exposed, and some change is supposed to take place from the influence of the air. Alcohol is its best menstruum, dissolving from 50 to 60 per cent. of good Elaterium. It was first analysed by Dr. Paris, who discovered an active principle which he named *Elatin.* This was found by Mr. Morris of Edinburgh, and by the late Mr. Hennel, to be composed of a peculiar principle, *Elaterin*, and of a green Resin. Elaterium also contains Bitter matter, Starch, Woody fibre, and Saline matters; but the proportion of Elaterin is very variable, from 15 to 25, in different specimens, according probably to the method of preparation and the goodness of the fruit Elaterin may be obtained by the E. P. process for ascertaining the purity of Elaterium.

Dr. Christison, who states he witnessed the experiments of Mr. Morris Stirling, describes *Elaterin* as consisting of very delicate colourless, striated, satiny, prismatic crystals, with a rhombic base, permanent in the air, without odour, but of an intensely bitter and somewhat acrid taste. It fuses a little above 212°, and by a strong heat is decomposed with the evolution of ammoniacal smoke. It is soluble in rectified Spirit, Ether, fixed oils, and weak acids, but not in water or weak alkalis.

Action. Uses. Powerful Hydrogogue Cathartic ; apt to create nausea and vomiting, and in larger doses will act as an irritant poison, producing inflammation of the intestinal canal. Useful from procuring copious watery evacuations in Dropsy, and as a revulsive in Cerebral affections.

D. Good Elaterium will act effectively in doses of $\frac{1}{8}$ or even $\frac{1}{16}$ of a grain ; but being generally inferior, gr. ß.—gr. j. or more, is prescribed every other day with a bitter Extract. *Elaterin* dissolved in Rectified Spirit may be given in $\frac{1}{16}$ grain doses.

UMBELLIFERÆ, *Juss.* Umbellifers. *Pentand Digyn.* Linn.

Herbaceous annuals, or with perennial root-stocks, having round often furrowed fistular stems. Leaves simple, most frequently deeply cut, with petioles sheathing at the base. Flowers complete, sometimes unisexual, white, purplish, or yellow, the external one sometimes rayed in umbels (fig. 72), often supported by involucres and involucels. Calyx (68 E *o*) adherent to the ovary, limb superior, entire, 5-toothed, or 5-parted, sometimes wanting. Petals (E. *p.*) 5, entire, or 2-lobed, sometimes inflected at the point, inserted on the outside of the disk, which invests the upper part of the tube of the calyx and crowns the top of the ovary, subimbricate or valvate in æstivation. Stamens (E. *e.*), 5 inserted with the petals, and alternating with them, replicate in æstivation. Ovary inferior, 2-celled, each with a single pendulous ovule. Styles 2 (*s*), distinct, with the base thickened into a stylopodium which covers the top of the ovary (and thus forms an epigynous disk), diverging at top, one towards the centre, and the other towards the circumference of the umbel. Fruit (A. D.) usually crowned by the limb of the calyx and the two persistent styles, formed of two carpel-like bodies (which are called *mericarps*, or half-fruits, and each has one-half of the calyx attached to it, adhering to each other by one side (*commissure*), separable and pendulous from the filiform but double central column (*carpophore*, 70. 12), and externally marked each with 5 primary (68 A.), sometimes with 4 secondary ridges, which are separated by channels or *valleculæ*, below which are usually placed the *vittæ* (71. 6) or receptacles of volatile oil. Seed (D. *e. p.*) solitary, single in each, inverse closely united to the pericarp, seldom free. Embryo straight, short, at the base of horny albumen, which is either flat or curved inwards.

Fig. 68.

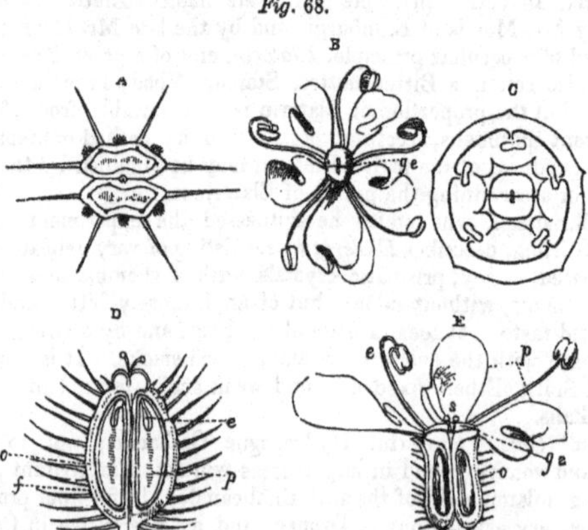

68. *Daucus Carota*, v. p. 425. C. Plan of the flower. B. Flower seen from above. *g. e.* Disk. E Vertical section of the flower. *p.* Petals. *e.* Stamens. *o.* Ovary adherent to the calyx. *s.* Styles and stigma. *g. e.* Disk. D. Vertical section of the fruit. *f.* pericarp. *o.* Seed. *p.* Albumen. *e.* Embryo. A. Horizontal section of the fruit: primary ridges projecting into prickles, and alternating with the bristly secondary ridges.

The Umbelliferæ are allied to Araliaceæ and to Saxifrageæ, also to

Ranunculaceæ, and likewise to Corneæ. They are natives chiefly of the northern parts of the northern hemisphere, many of them in the Persian region and in the Himalayan mountains. Volatile Oil is the chief secretion of this family, and abounding much in the fruits commonly called *seeds.* These are frequently employed as Carminatives. When it is diffused through the herbaceous parts, the plants are employed as culinary herbs. A Gum-resin exudes from some of them in the warm and dry Persian region, while others, growing chiefly in moist situations, are possessed of poisonous properties. These may be distinguished from the wholesome species by the absence of the aromatic odour.

Tribe *Ammineæ.* Fruit laterally compressed or didymous.

CARUM CARUI, *Linn.* D. Fructus, L. E. Semina, D. Common Caraway.

Caraway was known to the Greeks, being a native of most parts of Europe.

Biennial, about 2 feet high. Root fusiform. Leaves bipinnate. Leaflets cut into linear segments. Involucre wanting, or of one leaf. Involucel none. Cal. obsolete. Pet. obcordate, with a narrow acute inflexed point. Fruit aromatic, oblong, a little curved, brownish-coloured. Carpels with 5 filiform ridges. Interstices with single vittæ.—Stylopodium depressed.—Meadows and pastures: cultivated in Essex. E. B. t. 395.

The fruits of the Caraway, or Seeds, as they are commonly called, have a pleasant odour and a warm aromatic taste, owing to the presence of about 5 per cent of Volatile Oil, which may be dissolved by Alcohol or distilled off with water.

Action. Uses. Stimulant Carminative, much used in Confectionary. Its Oil and Spirit as Corrective Adjuncts, the water as a vehicle.

OLEUM CARUI, L. E. D. Oil of Caraway.

Obtained by distilling with *Aq.* the (bruised, E.) fruit of *Carum Carui.*

SPIRITUS CARUI, L. E. D. Spirit of Caraway.

Take *bruised Caraways* ℥xxij. (℔ß. E. ℔j. D.) *Proof Spirit* Cj. (Ovij. E.) *Aq.* Oij. L. (q. s. to prevent empyreuma, D.) Mix. (Macerate for two days (24 hours, D.) in a covered vessel, add *Aq.* Ojß. E.) with gentle heat, L. Distil Cj. (Ovij. E.)

AQUA CARUI, L. E. D. Caraway Water.

Prep. Take *bruised Caraway Seeds* ℔j. pour on *Aq.* q. s. to prevent empyreuma. Distil Cj. Same as Aq. Anethi, L.

PIMPINELLA ANISUM, L. E. D. Fruit of Pimpinella Anisum. Anise.

Anise (*ανισον*) being a native of the Grecian Archipelago, was well known to the ancients.

Stem about a foot high, smooth. Radical leaves heart-shaped, rather roundish, lobed, incised; stem leaves biternate. Segments linear, lanceolate, rather wedge-shaped, acuminate. Umbels on long stalks, many rayed without involucres. Flowers small, white. Calyx obsolete. Petals obcordate, with an inflexed point. Fruit ovate, 1½ line long, covered with a few scattered hairs. Carpels with 5 filiform equal ridges. Interstices with 3 or more vittæ. Stylopodium tumid. Styles of the fruit recurved.—Nees and Eberm. 275.)

The fruit, commonly called *Aniseed*, is ovoid, of a greenish-grey colour, and slightly downy; the taste is warm, sweetish, and aromatic; the odour penetrating but agreeable, in both resembling the Star-anise (p. 244). It is cultivated in Malta and the south of Spain, and also in Germany. The kernel contains 3·5 per cent. of fixed oil, and the inner firmly-adhering seed-coat about 3 per cent. of *Volatile Oil of Anise*, on which its properties chiefly depend.

Action. Uses. Agreeable Carminative, and much used for flavouring condiments.

OLEUM ANISI, L. E. D. Oil of Anise.

This, obtained by distillation with water from Aniseed, is of a bright yellow colour. It has the strong odour and taste of Anise. Much is imported from abroad; but that which is said to come from the East Indies is probably produced by some other plant. It solidifies very readily at 50°, from containing a large proportion of Stearoptene.

Action. Uses. Stimulant Aromatic, Stomachic. In flatulent Colic, in doses of ♏v.—♏xv.

SPIRITUS ANISI, L. (COMPOSITUS,) D. Spirit of Anise.

Prep. Mix *bruised Anise* ℥x. (and bruised seeds of *Angelica* āā ℔ß. D.) *Proof Spirit* Cj. and *Aq.* Oij. (q. s. to prevent empyreuma. Macerate for 24 hours, D.) with gentle heat, L. Distil Cj.

Tribe *Seselineæ.* Section of the fruit rounded or roundish.

FŒNICULUM VULGARE, *Gærtn. Dec.* L. F. OFFICINALE, *All.* E. (*Anethum Fœniculum*, Linn.), D. Common Fennel.

Fennel being found all over Europe, was known to the Greeks, and called *μαραθρον.*

Biennial. Stems 3—4 feet high, roundish at the base, filled with pith. Leaves decompound. Segments capillary and elongated. Involucre wanting. Umbels large, of 13 to 20 rays, many concave. Calyx obsolete. Petals yellow, roundish, entire, with a broad, obtuse, inflexed lobe. Fruit oblong, "scarcely 2 lines long, oval, of a dark or blackish aspect." (*p.*) Carpels with 5 perennial obtusely-keeled ridges. Interstices with single vittæ. Stylopodium conical.—Sandy and rocky ground, particularly near the sea.—E. B. 1208.

The fruit of wild Fennel, or *seed*, has a strong, rather disagreeable odour, and an aromatic but acrid taste. Its properties depend upon a Volatile Oil of a pale yellow colour. Dr. Pereira has remarked that this species is not employed in medicine, and that the Colleges err in quoting it; but as decoction of Fennel *seeds* is sometimes employed as an Enema in the flatulent colic of children, those of the wild plant are well suited for this purpose. But for internal exhibition, the Sweet Fennel is alone eligible. Some botanists consider this only as the cultivated variety of the Fœniculum vulgare; but others account it a distinct species.

FŒNICULUM DULCE. *C. Bauh. Dec.* Sweet Fennel.

Stem somewhat compressed at the base. Radical leaves somewhat distichous. Segments capillary, elongated. Umbels of 6 to 8 rays. *Dec.* This is, moreover, a smaller plant, and an annual; but its fruit is much larger, some nearly 5 lines in length, less compressed, somewhat curved and paler, with a greenish tinge.—A native of the south of Europe, cultivated in gardens as a pot-herb and for garnishing (*Finnochio dulce,* turionibus edulibus). Dr. Pereira long since favoured the author with the two kinds; the more agreeable taste and odour of this kind are, as he describes, very decided. Care therefore must be taken in determining which kind of fruit is used in the following preparations.

Action. Uses. Stimulant Carminative. Sometimes used in Flatulent Colic.

OLEUM FŒNICULI, (E.) DULCIS, D. Oil of Sweet Fennel.

Prep. Distil the (bruised, E.) fruit of *Fœniculum dulce* E. D. with *Aq.*

AQUA FŒNICULI, E. D. Fennel Water.

Prep. Take *bruised seeds of Fœniculum dulce* ℔j. *Aq.* q. s. to prevent empyreuma. Distil Cj. Take *Fennel* and prepare as Aq. Anethi, E.

Tribe *Angeliceæ.* Fruit much and dorsally compressed, with a double wing on each side.

ARCHANGELICA OFFICINALIS, *Hoffm.* (*Angelica Archangelica,* Linn.) E. D. Root, E. Semina, D. Garden Angelica.

This plant has long been employed in medicine; but is a doubtful native of this country.

A biennial plant. Root large, pungently aromatic. Stem 3 or 5 feet high, hollow, striated, rather glaucous. Foliage, stalks, and even flowers of a bright green. Leaves 2 or 3 feet wide, bipinnated or biternate. Leaflets ovate, lanceolate, sharply and closely serrated, all sessile, partly decurrent, terminal one trifid. Petioles much dilated at the base. Umbels terminal, globular, with dense secondary umbels. Involucre of 2 or 3 linear bracts, secondary one of about 8 linear-lanceolate bracts. Calyx minutely 5-toothed. Petals ovate, entire, acuminate, incurved. Fruit nucleated. Carpels or half-fruits with 3 dorsal thick-keeled ridges and 2 marginal ridges dilated into broad wings. Interstices without vittæ. Seed free, with numerous vittæ.—Native of watery places in the northern parts of Europe.—E. B. t. 2561. Nees and Eberm. 279—80.

The whole plant when bruised diffuses a strong and rather grateful odour. The root, when wounded in the spring, exudes an odorous yellow juice; when dried, the root is wrinkled, of a greyish-brown externally, and white in the inside, has a warm and bitterish taste. The stem and leaf-stalks, cut in May, when they are tender, are made into a preserve with Sugar. The fruits have the same odour and taste, depending on the presence of a Volatile Oil and of a Resin, as well as of a Bitter Extractive. The other ingredients of the root are Gum, Starch, water, and woody fibre. "The best way to preserve it, is to pulverize it, and to pack the powder firmly in bottles." (*c.*)

Action. Uses. Aromatic Stimulant, but little used; Stomachic. An infusion of the Root or Fruits (ʒij.—Aq. Oj.) may be given. The Fruits form an ingredient of the Spir. Anisi Comp. D.

Tribe *Peucedaneæ.* Fruit much and dorsally compressed, with a single wing on each side, which is flat or thickened towards the edge.

ANETHUM, L. Fructus, L. Fruit, E. of ANETHUM GRAVEOLENS. *Linn.* Common Dill.

Dill (ανηθον), a native of the south of Europe and of the Oriental region, was well known to the ancients.

Annual, 1 to 2 feet high, every part smooth and glaucous, stem finely striated. Leaves tripinnated, with fine capillary segments like those of the Fennel, petioles broad and sheathing at the base. Umbels long-stalked without general or partial involucres. Calyx-margin obsolete. Petals varnished, yellow, roundish, entire, involute. Fruit lenticular flat, of a bright brown colour on the rather convex back, surrounded by a pale membranous margin. Carpels or half fruits with equidistant filiform ridges, the 3 dorsal acutely keeled, the 3 lateral more obsolete, and passing into the margin. Vittæ broad, solitary, filling the whole channels, 2 on the circumference.—Much cultivated in the East, but also in this country.—St. and Ch. iii. t. 137.

The flattened elliptical fruits, commonly called seeds of the Dill, with their brown and slightly convex backs and pale membranous margin, are easily distinguished from the other officinal fruits. Both the plant and the fruit are much used in the East as condiments and articles of diet. The plant is hence mentioned in the New Testament among the things tithed; but it is translated Anise. The carpels have a bitter but aromatic taste, owing to the presence of volatile oil which is stored up in the vittæ, making them useful as Carminatives.

OLEUM ANETHI, E. Oil of Dill.

Prep. Distil with *Aq.* bruised fruit of *Anethum graveolens.*

Action. Uses. Odour and taste aromatic. Used for making Aqua Anethi. Carminative in doses of ♏v.

AQUA ANETHI, L. E. Dill Water.

Prep. Mix bruised *Dill Seeds* ℔jß. (℥xviij. E.) *Aq.* Cij. *Proof Spirit* f℥vij. (rectified, E. f℥iij. E.) Distil Cj.

Action. Uses. Aromatic, and much given to infants to relieve Flatulence, and used as a vehicle for active medicines.

The FÆTID GUM RESINS, as they are called, or *Opopanax*, *Assafœtida*, *Sagapenum*, *Galbanum*, and *Ammoniacum*, are all produced within the limits of the Persian region of botanists; and though the plants producing them are not accurately known, they are yet all supposed to belong to this tribe of Umbelliferæ; but Galbanum officinale belongs to Silerineæ.

OPOPANAX, L. D. Gummi-Resina, L. D. Gum-Resin, supposed to be produced by Opopanax Chironium, *Koch*, L. (*Pastinaca Opopanax*), Linn. D. Opopanax.

Opopanax is described by Dioscorides as the produce of πάνακες Ἡράκλειον, a plant of Bœotia and Arcadia, which has been identified with the above Opopanax Chironium, referred by Sprengel to the genus Ferula. It is found also in the open fields of the south of France, of Italy, Sicily, and Greece, and according to Merat and De Lens, also in Syria and the East. Dodoens first grew this plant from seeds found attached to pieces of Opopanax; and he states that when wounded in warm weather, especially near the root, a juice exudes, which concretes into a gum resembling Opopanax. But there is no proof that this plant yields the Opopanax of commerce which reaches India, either from the Persian Gulf or the coast of Arabia, and is called *juwa sheer*, or the milk of *juwa*, as it is by Serapion. Dr. Lindley describes (Fl. Med. p. 100) the fruit of a species of Ferula, which he names *F. Hooshee*, and of which the produce (which, however, is not collected) is said to resemble the Opopanax, according to a letter from Mrs. Macneil. It is imported into this country from Turkey; in the time of Mathioli it was obtained from Alexandria.

Opopanax occurs in irregular-shaped but usually angular pieces of a reddish-yellow colour, sometimes speckled with white from the interior having become recently exposed; of a strong, rather fœtid odour, and of a bitter acrid taste. Sp. Gr. 1·62. It is composed chiefly of Resin and Gum, with 5·9 per cent. of Volatile Oil, which may be separated by distillation. It will form an emulsion with water.

Action. Uses. Antispasmodic. It formerly enjoyed a high repute, as its name indicates, and was an ingredient of the Theriaca.

SAGAPENUM, L. D. Gummi-Resina, L. D. Gum-Resin from an unknown species of Ferula, L. Sagapenum.

Sagapenum, like Opopanax, has been known to us since the time of the Greeks. Dioscorides describes it as the produce of a Ferula growing in India, and we have no more recent information. It is the *sukbeenuj* of the Arabs, who give *sagafioon* as its Greek name. It reaches India from the Persian Gulf or the coasts of Arabia. It is difficult by price-currents and commercial reports to distinguish the routes which small articles of commerce pursue. It is imported into Europe from the Levant, and also from Alexandria. It is probably a product of Persia. Willdenow was of opinion that it was produced by *Ferula persica*, which Olivier thought produced Ammoniacum, Dr. Hope that it yielded Assafœtida, and which probably does yield some kind of fœtid Gum-Resin, as Michaux sent its seeds from Persia as those of Assafœtida. But it is preferable to leave this as a subject for investigation, than to stop inquiry by promulgating imperfect information.

Sagapenum is of a brownish-yellow or olive colour, chiefly in amygdaloidal masses, sometimes in tears: these are more or less transparent, soft, and of a waxy consistence. It has an alliaceous odour and acrid taste, similar to, but less powerful than, Assafœtida. Pelletier found it to be composed chiefly of Resin and Gum, with about 11·8 per cent. of Volatile Oil; but Brandes found only 3·7 of the last.

Action. Uses. Antispasmodic, but considered less powerful than Assafœtida, in doses of gr. v.—℈j.

PILULÆ SAGAPENI COMPOSITÆ, L. Compound Sagapenum Pills.

Prep. Rub into a mass *Sagapenum* ℥j. *Aloes* ʒß. *Syrup of Ginger* q. s.

Action. Uses. Stimulating Purgative and Emmenagogue, in doses of gr. v.—℈j.

Off. Prep. Pil. Galbani Comp. L.

ASSAFŒTIDA, L. E. D. Gummi Resina, L. D. Gummy-resinous Exudation (E.) of NARTHEX (*Ferula*, Linn.) ASSAFŒTIDA, *Falconer*. Assafœtida.

Assafœtida, a product of Persia and Affghanistan, is mentioned in the ancient Sanscrit Amera Cosha. The ancients highly esteemed a gum-resin which the Romans called *Laser*, and the Greeks ὀπος κυρηναϊκος, or the Cyrenaic Juice, from being produced in that region. The plant σιλφιον yielding it was an Umbellifer, and is represented on the coins of Cyrene. It has been discovered of late years, and named *Thapsia Silphium.* This Laser had become scarce even in the time of Pliny, who as well as Dioscorides describes another kind as obtained from Persia, India, and Armenia, which was probably the same that was known to the Hindoos. Avicenna describes *hulteet* as of two kinds: one, of good odour, from Chiruana (Cyrene?), and the other fœtid, the present *Assa-fœtida.* The term *assa* is no doubt of oriental origin, since it is applied to other gum-resins. Thus Benzoin is called *hussee-looban;* it used to be called *Assa dulcis* in old works. Dr. Lindley has received the seeds of a Ferula called *hooshee. Anjedan*, the fruits or seeds (φυλλον of the Greeks), is usually translated *Laserpitium.* The plant is called *Angoozeh* by the Arabs. The root of *Silphion* is described by Arrian as affording food to herds of cattle on Paropamisus.

Assafœtida is produced in the dry southern provinces of Persia, as in the mountains of Fars and of Beloochistan, but chiefly in Khorassan and Affghanistan; likewise to the north of the Hindoo Khoosh range of mountains, where it was found by Burnes and also by Wood's expedition to the Oxus. (*c.*) Dr. Falconer found it in Astore, introduced the plant into the Saharunpore Botanic Garden, as mentioned in the author's "Product. Resources of India," p. 223, and has obtained from it a small quantity of Assafœtida. He also sent home numerous seeds, which were distributed from the India House to several gardens; but the author has not heard whether any plants have been produced from them. But he has no doubt that some of those which the author is informed by his friend Dr. Christison are still in the Edinburgh Botanic Garden, were produced from these seeds, and not from those sent by Sir John M'Neill. The Assafœtida is conveyed on camels into India across both the Punjab and Bhawulpore, and is sold in large quantities at the Hurdwar Fair. It is also conveyed down the Indus and by the Persian Gulf to Bombay.

Two or three kinds of Fruit called Seeds are met with, which are said to be those of the Assafœtida plant; but there is no proof that more than one plant yields Assafœtida. Dr. Falconer, an excellent botanist, after examining the original specimens, considers the plant he saw in Astore to be the same as that figured by Kæmpfer; and Dr. G. Grant, who saw the plant at Syghan, says, as stated by Dr. Christison, that its roots, leaves, and flowering stem correspond on the whole with Kæmpfer's description, except that the root is deeply divided, like the outspread hand. The E. P. assign *Ferula persica* as probably yielding some Assafœtida. There is no doubt that its seed has been sent from the north-west of Persia as those of the Assafœtida plant; but there is no proof, nor indeed is it probable, that it yields any of the Assafœtida of commerce. The gum-resins of these Umbelliferæ are too similar to each other, for any but experienced pharmacologists to determine between *inferior* Assafœtida and varieties of Sagapenum or other Gum-resins.

As Dr. Falconer, the author's friend and successor as Superintendent of the East India Company's Botanic Garden at Saharunpore, has had excellent opportunities for examining the Assafœtida plant, both in its native sites and as cultivated by himself, he has favoured the author with the following full account of this important plant, which he conceives belongs to a genus allied to but distinct from Ferula.

NARTHEX (Falc. MSS.)

Calycis margo obsoletus. *Petala* ? *Stylopodium* plicato-urceolatum. *Styli* filiformes demum reflexi. *Fructus* a dorso plano-compressus margine dilatato cinctus. *Mericarpia* jugis primariis 5 : 3 intermediis filiformibus, 2 lateralibus obsoletioribus, margini contiguis immersis. *Vittæ* in valleculis dorsalibus plerumque solitariæ (valleculis lateralibus nunc sesqui- vel bi-vittatis), commissurales 4—6 variæ inæquales, exterioribus sæpe reticulatim interruptis. *Semen* complanatum. *Carpophorum* bipartitum. *Umbellæ* pedunculatæ compositæ. *Involucrum* utrumque nullum.—Genus inter Peucedaneas, calycis margine edentato, fructus vittis magnis, commissuralibusque inæqualibus et involucro utroque nullo distinctum; Narthex nuncupatum, a vocabulo νάρθηξ apud Dioscoridem Ferulæ attributo.

N. Assafœtida (Falc.) Caule tereti simplici, petiolis dilatatis aphyllis instructo, foliis radicalibus fasciculatis, petiolis trisectis, segmentis bipinnatisectis, laciniis lineari-lingulatis obtusis, inæqui-lateralibus integris vel variè sinuatis decurrentibus.—Asa fœtida Disgunensis, Kæmpf. *Amœnit. Exot.* p. 535. Ferula Asafœtida, Linn. *Mater. Med.* p. 79. De Cand. *Prod.* iv. 173. Lindl. *Flor. Med.* p. 45. Fig. 69, 70, et 71.

Habit. in apricis inter saxa in valle "Astore" vel "Hussorah" dictâ prope Indum, ultra Cashmeer: indigenis Daradris "Sip" vel "Sup." Legi fructigerum prope Boosthon 21° die Septembris, 1838.

Descrip. A tall perennial plant, 5 to 8 feet high. *Root* fusiform, simple, or divided, a foot or upwards in length, about 3 inches in diameter at the top, with a dark greyish transversely corrugated surface: the summit invested above the soil with dark hair-like fibrous tegmenta, the persistent exuviæ of former years: cortical layer thick and tough, white or ash-coloured in the section, readily separable from the central core, and, like the latter, abounding in a white, milky, opaque, excessively fœtid, alliaceous juice. *Leaves* collected into a fascicle above the root, numerous, large, and spreading, about 18 inches in length in the adult plant, of a light green colour above, paler underneath, and of a dry leathery texture: the petioles terete amplexicaul, and channeled at the base, trifurcated a little above it, the divisions united at an angle with each other, like the legs of a tripod, and bipinnately sected: the leaf-segments linear-ligulated, more or less obtuse, entire or sinuately lobed, variable in their offset, being either alternate or opposite, for the most part unequal-sided, and decurrent along the divisions of the petiole, forming a narrow winged channel upon the latter. *Midrib* prominent on the under side, veins slender and anastomosing by numerous reticulations. The leaves observed in a young growing plant (Fig. 69) were about 9 inches in length, the leaf segments being from 2 to 4 inches long by 4 to 6 lines in width. *Stem* erect, terete, simple, striated, about 2 inches in diameter at the base, solid throughout, the spongy medulla being traversed by scattered tough fibrous bundles of vessels, invested with alternate, vaginating, dilated, aphyllous petioles, and terminating in a luxuriant head of compound umbels. General as well as partial involucre entirely wanting. *Umbels* 10 to 20 rayed, emitted from the dilated spherical head of a common peduncle, the rays 2—4 inches in length. *Partial umbels* with very short rays aggregated into round capitula varying from

Fig. 69.

69. Narthex Assafœtida. *Falc.* Plant grown in H. E. India Company's Botanic Garden at Saharunpore.

10 to 20 rays in the fertile, and from 25 to 30 in the barren umbellulæ. *Flowers* small; barren generally mixed up with the fertile flowers (?). Border of the *Calyx* obsolete, being reduced to very minute denticular points. (Petals in the barren flowers small oblique, unequal-sided, acute, without an elongated acumen?). *Stylopodia* urceolate and plicated, with a sinuous margin. *Styles* filiform, reflected in the ripe fruit, rather short and slender, attached by a broad base. *Fruit* from 7 to 15, ripening on the partial umbels, supported on short stalks. *Mericarps* varying from broad elliptical to elliptical obovate, 5—6 lines long by 3 to 4 lines broad, flat, thin foliaceous, but somewhat convex in the middle, with a dilated border, generally unequal-sided, of a dark reddish-brown towards the centre, lighter towards the margin, perfectly smooth, with somewhat of a glossy surface. Dorsal primary ridges 5: the 3 middle ridges filiform, slightly crested towards their confluence at the apex: the lateral ridges more obsolete, situated close to the margin, immersed in the substance of the border, but distinctly seen on the surface of commissure, and confluent with the middle nerve of the latter. The dilated border as wide as the space occupied by the 3 middle ridges. *Vittæ* in the dorsal furrows large and broad, occupying the entire width of the valleculæ, stretching from base to apex, usually solitary, but sometimes double in one or other of the middle furrows, and generally double or dichotomous, with a small branch in the broadest side of the margin, turgid with a fœtid juice: vittæ of the commissure varying from 4 to 6, very unequal and variable: one very slender vitta, which is frequently dichotomous in two fine threads confluent at the apex, being placed close on either side of the middle nerve; another of the size of the dorsal vittæ, situated more outwards, and

Fig. 70.

70. 9. Ovary. Style and Stylopodium enlarged. 10. Partial Umbel with fertile flowers. 11. Umbel of barren flowers. 12. Partial Umbel in fruit with persistent Carpophores.

a third at the inner side of the dilated border, over the edge of the seed, more slender, but frequently subdivided and interrupted so as to cover the border with a beautiful network of anastomosing ramifications. *Seed* flattened, with plain albumen. *Carpophores* bipartite, persistent, twice the length of the pedicells. *Flowers* white?

"The plant above described I believe to be the true "Asafœtida disgunensis," or "Hingiseh" of Kæmpfer. It does not appear to have been met with by any other botanist since it was examined *in situ* by that excellent and careful observer a century and a half ago.

I have compared my materials with Kæmpfer's description and figures (Amœn. Exot. p. 537), and with his original specimens contained in the Banksian collection in the British Museum, and found them, so far as a comparison could be instituted, to agree in every essential respect. The leaves "instar Pœoniæ ramosa" as represented in his figures, have the segments more obtuse and sinuated, and more alternate in their offset than they are represented in my drawing; but he describes them as being very variable in form, and some of the numerous leaf-specimens in his herbarium correspond with the figure which I have given. Kæmpfer mentions the umbellulæ as having only 5 or 6 rays, whereas I found them as numerous as 25 or 30 in the sterile capitula, and from 10 to 20 in the fertile ones. But he states that he never saw the plant in flower, and his description was probably drawn from the ripe state, in which the partial umbels occasionally present no more than 7 fruit-bearing stalks. There are two mericarps in his herbarium, of one of which I have given a representation (fig. 71. 5) agreeing exactly in form and in the development of the dorsal juga with those met with by me in the Astore plant: but Kæmpfer's specimens are

Fig. 71.

71. 1—4. Mericarps of the natural size. (*a.*) Dorsum. (*b.*) Commissure. 5. Mericarps in Kæmpfer's Herbarium, dorsal aspect. 6. Transverse section of Mericarp enlarged. 7. Seed, natural size. 8. Two Petals of a barren flower enlarged.

glued down on paper, and they seem to have undergone some decay or alteration by which the vittæ have been emptied, so that their number and size cannot be distinctly made out. But they appear to be solitary in the dorsal valleculæ, and there is no indication of the numerous striæ represented in the figures of the fruit given in the Amœnitates, which may have confirmed authors in the belief that Kæmpfer's Assafœtida plant belonged to a species of Ferula. These mericarps are perfectly smooth, and exhibit nothing of the "quadatenus pilosum sive asperum," described in the Amœnitates, p. 538. Dr. Lindley, in his Flora Medica, p. 45, after an abridgment of Kæmpfer's description, states, (it is not mentioned upon what evidence,) the vittæ of the back to be "about 20 or 22, interrupted, anastomosing, and turgid with Assafœtida: of the commissure 10." This account will apply to the fruit of a species of Ferula, but is entirely at variance with the characters presented by the fruits of the plants observed by Kæmpfer in Persia, and by myself in Astore.

Kæmpfer in his description says: "Folia sero autumno ex vertice progerminant, sex septem, et pro radicis magnitudine plura vel pauciora: quæ per brumam luxuriose vigent adultoque vere exarescunt." From the information which I gathered on the spot, confirmed by subsequent observation upon the growing plants introduced into the Botanic Garden at Saharunpore, the leaves of the Astore Assafœtida plant make their appearance in spring, and not in autumn, surviving through the winter, as stated by Kæmpfer, respecting the Persian form. With these slight discrepancies, his description might serve for the Astore plant.

Narthex, both in the characters of the flowers and fruit, and in its "Pæony-leaved" habit, differs widely from any known species of Ferula, and appears to constitute a distinct and well-marked genus.

In the Dardoh or Dangree language (the Dardohs being the Daradi of Arrian) the plant is called "Sip" or "Sup." The young shoots of the stem in spring are prized as an excellent and delicate vegetable.

The species would appear to occur in the greatest abundance in the provinces of Khorassan and Laar in Persia, and thence to extend on the one hand into the plains of Toorkestan on the Oxus north of the Hindoo Khoosh mountains, where it seems to have been met with by Sir Alex. Burnes,* and on the other to stretch across from Beloochistan, through Candahar and other provinces of Affghanistan to the eastern side of the valley of the Indus, where it stops in Astore, and does not occur in great abundance. The whole of this region, which constitutes the head-quarters of the gum-bearing Umbelliferæ, possesses the common character of an excessively dry climate, indicated in Berghaus's hygrometric map in Johnson's Physical Atlas by a belt of white.

Besides the gum-resin, the fruit of *Narthex Assafœtida* is imported into India from Persia and Affghanistan, under the name of "Anjoodan," being extensively employed by the native physicians in India: "Anjoodan" being the epithet applied to the seed of the "Heengseh," or "Hulteet," by Avicenna, also quoted by Kæmpfer, and used by the Indo-Persian and Arabic writers generally in describing the Assafœtida plant. Another Umbelliferous fruit is also imported with it, and sold under the name of "Dooqoo" (a word evidently connected

* Burnes mentions the plant as an annual, probably in consequence of the annual decay of the stems. He states that sheep browse on the young shoots.

with the δαυκος of the Greeks), being recommended as an excellent substitute for "Anjoodan," which it closely resembles in its general appearance. This I found to be the fruit of a species of true Ferula; it is one of the two Assafœtida-like fruits mentioned by Dr. Royle as occurring in the bazaars of northern India. The species of Ferula yielding this fruit may furnish some one of the obscurely-known gum-resins resembling Assafœtida produced in Persia.

I have examined another kind of Umbelliferous fruit in the collection of Dr. Royle, labelled as "the seed of the wild Assafœtida plant collected and brought to England by Sir J. Macneill from Persia," which differs widely from the fruit both of Narthex and of Ferula, and belongs to another tribe of the order." H. F.

Assafœtida is obtained by making incisions into or taking successive slices off the top of the root, and then collecting the produce, which is then united in masses, and in this state is usually met with in commerce. It is at first rather soft, but becomes hard, of a yellowish or reddish-brown colour. When broken, an irregular, whitish, somewhat shining surface is displayed, which soon becomes red. The mass is composed of various-shaped pieces, some like tears pressed together, and in some parts agglutinated together by darker-coloured gum-resin. Some parts are cellular. By thus becoming red on exposure to the air, and its intolerable alliaceous odour, Assafœtida may be readily distinguished. The taste is garlicky, bitter, and acrid. It is best preserved covered by bladder. It is powdered with difficulty, even when become hard; softens by heat, and burns with a clear flame. Assafœtida is composed of Resin 65 parts, Volatile Oil 3·6, Gum 19·44, Bassorin 11·66, Salts 0·30. (*Pelletier.*) Brandes obtained less Resin, Volatile Oil 4·6, and 10·5 of various salts and impurities. The Oil is at first colourless, but becomes yellowish-brown, has an exceedingly offensive odour, a bitter and acrid taste, and contains some Sulphur. Water will dissolve the Gum, and form an emulsion with the other ingredients. Alcohol or Rectified Spirit is a good solvent, but an emulsion is formed when the solution is added to water. Ether dissolves the Oil and all the Resin, except about 2 per cent. of a peculiar kind. Ammonia also takes up the active ingredients.

Action. Uses. Stimulant, Antispasmodic; thought to be Emmenagogue and Anthelmintic. Much used as a condiment in the East. Useful in Spasmodic and Convulsive diseases, as Hysteria and Chorea, also in Hooping Cough, Flatulent Colic, and in Chronic Cough.

D. grs. v—℈j. in pill or in some of its compounds every 3 or 4 hours. The Emulsion acts quickly and may be formed readily by adding the Tincture to Aqua Pulegii.

MISTURA ASSAFŒTIDÆ, L. D. Assafœtida Mixture or Emulsion.

Prep. Rub up till well mixed *Assafœtida* ʒv. (ʒj. D.) *Aq.* Oj. L. (*Aq. Pulegii* gradually added, by measure ℥viij. Make an emulsion, D.)

Action. Uses. Antispasmodic. Sometimes prescribed in Hysteria in doses of ℨiv. to f℥iss. Useful in the treatment of feigned diseases.

ENEMA FŒTIDUM, E. D. Fœtid or Assafœtida Enema.

Prep. Take *Cathartic Enema* and add to it *Tinct. Assafœtida* f℥ij.

Action. Uses. Stimulant. Used in Flatulent Colic and Ascarides.

TINCTURA ASSAFŒTIDÆ, L. E. D. Tincture of Assafœtida.

Prep. Take *Assafœtida* (in small fragments, E.) ℥v. (℥iv. rub it up in Aq. ℥viij. by weight, D.) Macerate for 14 days (7, E.) in *Rectified Spirit* Oij. (℔ij. by weight, D.) Strain. (Not easily made by percolation, E.)

Action. Uses. Antispasmodic. Prescribed in Hysterical cases, &c. as Spir. Ammoniæ fœtidus (p. 61), in doses of fℨj.—fℨij.

EMPLASTRUM ASSAFŒTIDÆ, E. Assafœtida Plaster.

Prep. Melt together *Assafœtida* ℥ij. *Galbanum* ℥j. Strain them. Add melted *Litharge Plaster* ℥ij. and *Bees' Wax* ℥j. Mix thoroughly.

Action. Uses. Useful externally in the foregoing class of cases.

Off. Prep. Spir. Ammoniæ Fœtidus, L. E. D. Pilulæ Galbani Comp. L. (*Pil. Assafœtidæ*, E.) Pil. Aloes et Assafœtidæ.

AMMONIACUM, L. E. D. Gummi-Resina, L. D. Gummy-resinous Exudation of DOREMA AMMONIACUM, *Don.* L. E. of Heracleum Gummiferum, *Willd.* D. Ammoniacum. Gum Ammoniac.

Ammoniacum is described by Dioscorides, 3. c. 88 (or 98) as the produce of a plant called *Agasyllis*, *Metopium* of Pliny, which grows in Cyrenaic Africa near the temple of Jupiter Ammon, whence it derives its name. Mr. Don supposed this to be a corruption of Armoniacum: it is so written in some old books. Jackson in his account of Morocco states that the Ammoniacum plant, which he calls *Feshook*, grows in Morocco, near Al-Araish. The Hon. Fox Strangways favoured Dr. Lindley as well as the author with the fruit of a *Ferula* which was marked as that of *Fusogh*, or Gum Ammoniac, obtained by him from Tangier. Some of these were sent to Dr. Falconer, by whom they were grown in the Saharunpore Botanic Garden, and the plant found to be identical with *Ferula Tingitana.* Dr. Lindley had previously determined the fruit to be that of the same plant. In his Flora Medica he refers the *Feshook* of Jackson, t. 7. to *F. orientalis*, with a query. But the Ammoniacum of commerce of the present day is a product of Persia, and obtained from Bombay, having been previously imported there from the Persian Gulph, whence probably a portion is also carried up the Red Sea, and thus reaches Europe by the Levant. Capt. Hart (Trans. Med. Soc. of Calcutta, i. p. 369) found the plant in the plains between Yezed-khast and Kumisha, on the road from Shiraz to Ispahan, or on the border of the provinces of Fars and of Irak Ajemi. Lt.-Col. Johnston saw the plants growing at Mayer and Yezde-Khast, and collected specimens of the plant with its fruit and gum. Lt.-Col. Wright obtained specimens at the same place, which he gave to the Linnean Society, and which Mr. Don described. M. Fontanier (Merat and De Lens, i. p. 25) also obtained it at Yezd-Cast in Faristan, which appears to be the same place. Major Willock informs the author that the *ooshak* plant is only to be met with in the province of Irak, in dry gravelly plains, where it is exposed to an ardent sun. Sir John M'Neil found it on the low hills near Herat, and Dr. Grant at Syghan to the north of Bamean, where the same dry climate prevails. In the same kind of country, but more to the eastward, many other Umbelliferæ were found by Dr. Falconer, together with *Narthex Assafœtida* and *Prangos pabularia.* M. Fontanier says Gum Ammoniac exudes naturally at the axils of the umbel and upon the tumid apices of the peduncles. Willdenow concluded erroneously that Ammoniac was produced by *Heracleum gummiferum.*

Root large, perennial. Stems 7 to 9 feet high, about 4 inches in circumference at the base, clothed with glandular down (*Don*), smooth (*Fontanier*), glaucous, with the habit of Opopanax Chironium. Leaves large, petiolate, somewhat bipinnate, two feet long; pinnæ usually 3 pairs, each pair rather remote: lower leaflets distinct; superior ones confluent, deeply pinnatifid; segments oblong, mucronate, quite entire, or rarely a little lobed, coriaceous, veined beneath, 1—5 inches long, and ½ to 2 inches broad. Petiole ribbed, pubescent, much dilated, and sheathing at the base. Umbels proliferous, racemose, partial umbels globose, on short peduncles, usually disposed in a spicate manner. Neither general nor partial involucre. Peduncles terete, woolly. Flowers sessile, immersed in wool. (*Lindley.*) Margin of calyx 5-toothed, teeth acute membranous. Petals white, ovate, with an inflexed point. Disk large, fleshy, cup-shaped, with a plicate, rather lobulate margin. Stamens and styles yellow, the latter complanate, recurved at the apex. Stigmas truncate. Ovary densely woolly. Fruit elliptic, compressed from the back, surrounded by a broad flat edging. Mericarps with 3 distinct filiform ridges near the middle, and alternating with them 4 obtuse secondary ridges (two of the primary ridges confluent with the margins). Vittæ 1 to each secondary ridge, 1 to each primary marginal ridge, and 4 to the commissure, of which 2 (the exterior ones) are very small.—*Don and Lindley.**

As the plant abounds in juice, this readily exudes on the slightest puncture: M. Fontanier says, spontaneously. Capt. Hart states that when the plants have attained perfection, or about the middle of June, innumerable beetles pierce it in all directions. The juice soon becomes dry, and is picked off. The finest pieces being kept separate, form the *Ammoniacum in tears* of commerce, which vary in size, are yellowish externally, and of a white, opaline, or waxy appearance when fractured.

* The late energetic traveller, Aucher-Eloy visited the same localities, one of them Yezdikhast, and obtained fragments of a plant which has been named *Disernestum gummiferum* by Jaubert and Spach, in Illustr. Pl. Orient., who state that it is allied to Siler and to Agasyllis, and is hence placed in the tribe *Silerideæ.* v. Walper's Repert. ii. p. 939. It appears to the author to be only the above Dorema Ammoniacum described from imperfect specimens. Plant of considerable height, finely hairy when young, but becomes smooth with age. The inferior leaves are very large, doubly compound. Partial umbels sessile or pedunculated on a leafless panicle, with the gum-resin collected especially in the axils of the partial umbels. Petals white. Disk cup-shaped, crenate, plicate. Fruit oval, compressed dorsally with narrow wings. Mericarps 6 or 9 ribbed, ridges and vittæ delicate and fine.

These, when pressed together, form lump or Amygdaloidal Ammoniacum, in which the tears appear agglutinated together by a softer material, often mixed with some of a darker cclour. In the inferior kinds the tears are less abundant, and impurities, as sand, fruit of the plant, &c., are intermixed.

Ammoniac is rather hard, but readily softens by heat, has a powerful and peculiar smell, and the taste bitter and acrid. Sp. Gr. 1·207. It consists, according to Bucholz, of 22·4 parts of Gum, 72 of Resin, 1·6 of Bassorin (Gluten?), and 4 of Volatile Oil. But much less Oil has been obtained by other chemists, and by some none at all. Ammoniac forms an emulsion with water. Alcohol dissolves its Resin and Oil, but becomes milky on the addition of water.

Action. Uses. Stimulant Expectorant. Antispasmodic chiefly in chronic Catarrhs.

D. gr. v.—gr. xx. Usually taken with Squill or in Emulsion.

MISTURA AMMONIACI, L. D. Ammoniac Mixture or Emulsion.

Prep. Rub up *Ammoniacum* ʒv. (ʒj. D.) with *Aq.* Oj. (Aqua Pulegii by measure ℥viij. D.) add gradually till thoroughly mixed (milky. Strain through linen, D.)

Action. Uses. Water dissolves the Gum, and thus suspends the Resin of the Ammoniac; when an Emulsion is formed. The addition of a little Vinegar assists in making it smoother. Stimulant Expectorant in chronic Catarrhs, &c., in doses of fʒiv. to f℥iß.

EMPLASTRUM AMMONIACI, L. E. D. Ammoniac Plaster.

Prep. Mix *Ammoniacum* ℥v. with *distilled Vinegar* f℥viij. (ix. E. *Vinegar of Squills,* ℔ß. D.) Evaporate with a gentle heat (over the vapour-bath, E.), stirring assiduously.

Action. Uses. The Vinegar does not dissolve but softens the Gum-resin, which may then be applied as a poultice, and forms an adhesive stimulant and resolvent plaster; often of use when applied to indolent swellings.

Off. Prep. Pilulæ Scillæ Comp. L. E. D. Emplastrum Ammoniaci cum Hydrargyro, L. E. D. (*v.* p. 186.) Emp. Saponis.

GALBANUM, L. E. D. Gummi Resina, L. D. Concrete Gummy, resinous Exudation of an imperfectly ascertained plant, E. Galbanum.

Galbanum is supposed to be the same substance as the *Chelbenah* of Scripture (v. Cycl. of Biblical Lit.), as this word is very similar to the χαλβανη of the Greeks, which was known to Hippocrates, and is described by Dioscorides, who gives μετωπιον as an additional name. Theophrastus had long previously stated that it was the produce of a *Panax*, Dioscorides of a *Ferula* of Syria. But this word had a wide geographical signification in anicent times. Arabic and Persian authors seem to have been well acquainted with the plant, as they give *kinneh* and *nafeel* as its names, and *barzud* as that of the gum-resin. D'Herbelot states that this is the same as the *pirzed* of the Persians, who call the plant yielding it *giarkhust.* But whatever the plant may be, it is unknown to botanists. Lobel attempted to ascertain it by sowing some of the seeds which he found attached to Galbanum, and obtained *Ferula Ferulago*, a native of North Africa and of Asia Minor, but which is not known to yield Galbanum; while Bubon *galbaniferum*, which is sometimes adduced, is a native of the Cape of Good Hope. The late Professor Don having found some seeds sticking to Galbanum, named the plant yielding them, though yet unknown, *Galbanum officinale*, belonging to the tribe *Silerinœ*, which is admitted in the L. P. But these fruits may, or not, be those of the Galbanum plant. Dr. Lindley has in consequence suggested another plant, of which he received the fruits from Sir John Macneil. This grows at Durrood, near Nishapore, in Khorassan, and yields a gum-resin of which a specimen seen by Dr. Pereira did not correspond with any known gum-resin. The author has several such in his collection. Dr. Lindley has named the plant which is admitted into the E. P. *Opoidia galbanifera*, tribe Smyrneæ. But there is equal uncertainty about all.

Galbanum is imported from India and the Levant, having probably been brought down the Persian Gulf. It is usually met with in masses of a brownish-yellow colour, more or less translucent and shining, sometimes in small tears, which are of a paler or even yellowish colour. In the former kind, the tears are of a reddish-yellow, often agglutinated together by a darker coloured substance, often mixed with pieces of the stalk, fruits, sand, &c. It is soft, can only be powdered in cold weather, has a bitter and even acrid taste, with a peculiar but not alliaceous odour. It consists of Resin 65·8, Gum 22·6, Bassorin 1·8, Volatile Oil 3·3, with a trace of Malic acid, &c. Its properties depend chiefly upon the Oil, which may be separated, of a yellowish colour, by distilling with water. When Galbanum is distilled by itself at a temperature of 250° F. a bluish-coloured Oil is obtained. Galbanum forms an emulsion with water, and is dissolved by proof Spirit.

Action. Uses. Antispasmodic; less powerful than Assafœtida; Expectorant.

D. gr. x.—℈j. in substance, or made into an emulsion.

PILULÆ GALBANI COMPOSITÆ, L. D. (ASSAFŒTIDÆ, E.) Compound Galbanum Pills.

Prep. Mix well and beat into a pill mass *Galbanum* ℥j. (3 parts, E.) *Myrrh* (3 parts, E.) *Sagapenum* āā ℥iß. L. D. *Assafœtida* ℥ß. (3 parts, E.) *Syrup* q. s. L. D. (*Conserve of Red Roses* q. s. or 4 parts, E.)

Action. Uses. Antispasmodic. Emmenagogue in doses of gr.x.—℈j.

TINCTURA GALBANI, D. Tincture of Galbanum.

Prep. Digest for 7 days *Galbanum* cut very small ℥ij. in *Proof Spirit* by measure ℔ij. Strain.

Action. Uses. Antispasmodic adjunct to draughts in doses of f℥j.—f℥ij.

EMPLASTRUM GALBANI, L. D. EMP. GUMMOSUM, E. Galbanum Plaster.

Prep. Melt *Galbanum* ℥viij. (℥ß. E. ℔ß. D.) *Common Turpentine* ℥x. L. (*Ammoniacum* ℥ß. E. Strain). Melt also *Resin of Spruce Fir* powdered ℥iij. L. (Yellow *Wax* ℥ß. E. ℥iv. D.) and *Lead Plaster* ℔iij. (℥iv. E. ℔ij. D.) Add the mixtures and mix carefully together.

Action. Uses. Stimulant and Discutient application to indolent tumours, &c.

Pharm. Prep. Emp. Assafœtidæ.

Tribe *Daucineæ.* Fruit dorsally sub-compressed or round. Carpels with 5 primary ridges, the lateral ones on the inner face; and 4 secondary ridges forming rows of prickles.

DAUCUS CAROTA, *Linn.* Fructus, L. (Semina), D. var. sylvestris, D. Radix recens, L. Fresh Root, var. sativa, E. Fruit of the Wild and Root of the Cultivated Carrot.

The Carrot has long been cultivated in the East, where it is called *jugur* and *gajur.* The Arabs give *istufleen* as its Greek name,—a corruption, no doubt, of the σταφυλῖνος of Dioscorides.

Stems 2 to 3 feet high, hispid. Leaves tripinnate; leaflets of the upper leaves linear, lanceolate, acute, of the lower leaves broader; leaflets of the general involucre pinnatifid, with linear segments, or the partial one linear, entire, or trifid. Calyx (fig 68, *c.*) 5 toothed. Petals white, except in a central neutral flower, which is red, as mentioned by Dioscorides, obcordate (D.), with an inflexed lobe, exterior usually radiant and bifid. The umbels at first flat, become afterwards hollow, from the incurvature of the pedicels. Fruit dorsally (A.) compressed. Carpels with bristly primary (A. D.) ridges; secondary ridges winged, the wings divided often to the base into a row of simple prickles, which are equal in length to the diameter of the fruit.—Common on roadsides and in pastures throughout Europe and the Oriental Region.—Fig. 68, p. 408, a section of the flower and fruit from Adr. Juss. with the several references.—E. B. 1174.

The root of the cultivated Carrot is too well known from its fusiform shape, yellow colour, sweetish taste, and nutritious nature, to require description. It is officinal on account of its succulent nature being favourable for making poultices, which are moderately stimulant. The seeds are larger but milder in consequence of being cultivated.

The *wild variety* has fusiform roots, which are small, yellowish, and woody, with a bitter and acrid taste, but with the peculiar odour of the Carrot. The fruits, commonly called seeds, have this odour in a more marked degree, from their natural mode of growth being best suited for the development of the peculiar secretions of the plant. The Volatile Oil, which is secreted especially in the fruit, is diffused in less quantity over the whole plant, and even in the cultivated root. These fruits have an aromatic odour, warm and pungent taste, and yield by distillation a Volatile Oil.

Action. Uses. Fruits Carminative and Diuretic in doses of ℈j.—ʒj.

CATAPLASMA DAUCI, D. Carrot Poultice.

Prep. Boil the root of Garden *Carrot* q. s. in *Aq.* till soft, and then make into a poultice.—If the raw Carrot be scraped down, and applied as a poultice, it will produce an irritant action.

Tribe *Cumineœ.* Fruit compressed laterally, ridges all apterous.

CUMINUM. Fructus L. Fruit, E. of Cuminum Cyminum, *Linn.* Common Cumin.

Cumin (*kumoon* of the Arabs) is probably a native of Asia, and was made known to the Greeks from Egypt. It is extensively cultivated in the East, but has long been introduced into the south of Europe. England is chiefly supplied from Sicily and Malta.

Cumin is an annual, from 1 to 2 feet high, with much-divided leaves, having the segments long and setaceous. The umbels, both general and partial, from 3- to 5-rayed, with involucres of 2—4 simple or divided leaves. Involucels halved, of 2—4 leaves, finally reflexed, and exceeding in length the pubescent fruit. Flowers white or pink. Calyx with 5 lanceolate, setaceous, permanent teeth. Petals oblong, emarginate, with an inflexed point. Fruit contracted from the sides. Mericarps with wingless ribs, the 5 primary ones minutely muricated; 4 secondary ones prickly. Channels oblong, striated, minutely aculeate under the secondary ridges, with 1 vitta in each. Seed somewhat concave in front, and convex on the back.—Esenb. and Eberm. 288.

Cumin seeds, or rather fruits, are of a light-brown colour. The odour is aromatic, dependent on the volatile Oil which is stored up in the seed-coat. Taste warm, bitterish, and aromatic, but not so agreeable as anise. The albumen is insipid. 16 cwts. of the fruits yield about 44 lbs. of the Oil, which has a Sp. Gr. 0·945, pale yellow colour and is limpid, of a disagreeable smell and acrid taste.

Action. Uses. Stimulant Carminative. Condiment in India. Seldom used in medicine. Dose gr. xv.—ʒß. Formerly much employed as an external application in Emplastrum and Cataplasma Cumini, and by Jews in the process of circumcision.

Tribe *Smyrneœ.* Fruit turgid, compressed laterally. Carpels with primary ridges only.

CONIUM MACULATUM, *Linn.* Conii Folia, L. E. D. Conii Fructus, L. Leaves and Fruit of Hemlock.

There is little doubt of this being the κωνειον of the Greeks, and the *cicuta** of the Romans, as has been long supposed, and as has been well argued by Dr. Pereira. The objection that it is not so clearly described as to be readily distinguished from other Umbelliferæ, would apply to accounts of many officinal plants, even in comparatively modern works. It is the *shokran* of the Arabs, who give *chuniun* and

* This must not, from the similarity of name, be confounded with *Cicuta maculata. Cicuta virosa* occurs in Cashmere, where it is called *Zehr-googul*, or poison-turnip. Persian, *Salep-e-Shaitan*, or Devil's Salep.

kunion as its Greek name, and give *bunj-roomee*, or Turkish *bunj*, as another name. The name *bunj* is applied to Henbane, while Datura is *bunj-dushtee*. It was re-introduced into practice by Störck.

Root biennial, fusiform, whitish, a little fleshy. Stems 3 to 5 feet high, erect, round, smooth, spotted with dull-coloured purple spots. Leaves large, shining, of a deep green colour, tripinnate, on long furrowed petioles, sheathing at the base. Leaflets lanceolate, pinnatifid with the lower lobes incised, the others toothed. Umbels numerous, terminal, composed of many general as well as partial rays. General involucre of from 3 to 7 leaflets, ovate, cuspidate with membranous edges, partial involucre of 3 leaflets on one side ovate lanceolate, shorter than the umbels. Margin (*c*) of calyx obsolete. Petals 5, (*d*) white, obcordate with inflexed apices. Stamens (*d*) 5. Ovary ovate, 2-celled. Styles (*c*) 2, spreading. Fruit (*c*) ovate, compressed laterally. Carpels or half fruits with 5 prominent, (*c*, *b*) equal undulated, primary ridges, of which the lateral ones are marginal. No secondary ridges. The channels with many striæ but no vittæ.—Hedges and waste places throughout Europe; found in Greece by Sibthorp, in Cashmere by Falconer. Fl. in June and July. Fruit ripe in August and September. Fig. 72.—Esenb. and Eberm. 282; St. and Ch. 13.

Fig. 72.

72. *Conium maculatum*. *d*. Flower. *c*. Fruit. *b*. Transverse and *a* vertical section of fruit.

This plant in its first year has a long slender root and a few radical leaves. In the second year it throws up its characteristic and spotted stem, on all parts smooth, but possessed of a strong and fœtid odour, compared to that of mice. It may be known by these characters, and by its unilateral partial involucres and wavy crenated ridges of the fruit.* It is generaliy stated that the best time for collecting the leaves is when the plant is in full flower, or just before the forming of the fruit, as in other cases. This, Dr. Christison, who has paid great attention to the subject, doubts, as he has found that its poisonous properties are considerable in November and March of its first year. The fruit is more active than the leaves, and more so when green than when ripe and dry. The leaves especially should be carefully dried, in a dark airy room, at a temperature of about 120°, and preserved in well-closed, dark, and dry vessels. They should retain much of their natural deep-green colour, have a nauseous and somewhat acrid taste, with a peculiar mouse-like odour. This should be readily evolved when the plant is rubbed up with caustic Potash.

Several analyses have been made of Conium. Giseke in 1827 succeeded in concentrating its active principle with Sulphuric′. Geiger in 1831 detached it in the form of a volatile and oleaginous alkali, possessed of powerful poisonous properties. The experiments of Geiger have been confirmed by Dr. Christison. Both found that the distilled water, though having the peculiar odour of Hemlock owing to the presence of its Volatile Oil, was yet not poisonous; but if the full-grown green fruits or leaves be distilled with water and caustic Potash, a strongly *aklaline* and poisonous *liquid* passes over, which is *Conia.* This has been found in the fruit and leaves, and exists in them as a salt, though its acid (Coneic′) is unknown. It may be detached by presenting an acid, as the Sul′, which combines with the Conia; or this, by the action of Potash, may be detached from its acid, and then distilled over with water at 212°. *Conia,* called also *Conine* and *Conein,* is a colourless and transparent oily-looking body, lighter than water, having a powerful mouse-like odour, and a very acrid taste. It is strongly alkaline. Water takes up but little of it, but with a fourth of its weight it forms a Hydrate of Conia. Alcohol and Ether dissolve it readily, as do diluted acids, which indeed combine with it. Its vapour produces white fumes with the vapour of H Cl′, Hydrochlorate of Conia being formed. Exposed to the air, it becomes brownish; a resin and Ammonia are formed. It boils at 370°. Liebig determined it to be composed of $C_{12} H_{14} N O = 108$. ($C_{16} H_{16} N_2$ *Greg.*) Subjected to a strong decomposing heat, an empyreumatic oil

* This plant should be distinguished from Æthusa Cynapium, or Fool's Parsley, and also from *Anthriscus vulgaris* and *sylvaticus* as well as *Myrrhis odorata* and *temulenta.* The other poisonous umbellifers are the above *Æthusa Cynapium,* Fool's Parsley; *Cicuta virosa,* Water Hemlock; *Œnanthe crocata,* Hemlock Water Dropwort; *Œnanthe apiifolia,* which is probably only a variety of *Œ. crocata.*

is yielded, which is very poisonous. It is most easily obtained by cautiously distilling over a muriate of lime bath, a mixture of strong solution of Potash with the alcoholic extract of the unripe fruit. (See Dr. Christison's able paper, Trans. Roy. Soc. Edin. 1836.)

Tests. Hemlock may be recognized by the characters which have been given of the plant and fruit, and the efficiency of the officinal products by triturating them with *Liq. Potassæ*, as directed in the E. P. By this it has been ascertained that some preparations contain no Conia, either from defective preparation, or from subsequent change. This accounts for some of the discrepant statements respecting the efficiency of Conium.

Action. Uses. Narcotic Poison, supposed to excite convulsions and fatal coma, &c. From Dr. Christison's experiments, it exhausts the nervous energy of the spinal chord and voluntary muscles, occasioning merely convulsive tremors and slight twitches, and eventually general paralysis of the muscles, and consequent stoppage of the breathing. Hemlock has long been employed as a Deobstruent and Alterative in glandular and visceral enlargements. Scirrhous, Cancerous, and Scrofulous diseases have been greatly relieved by it. (*v.* Bayle.) Useful also as an Antispasmodic in Hooping and other Coughs. It has been tried in Tetanus. As an Anodyne and Hypnotic, allays pain and irritation, and promotes sleep. But Dr. Christison says the whole subject requires to be investigated anew.

D. Powder fresh and well-dried gr. iij.—gr. v. 2 or 3 times a day.

EXTRACTUM (SUCCUS INSPISSATUS) CONII, L. E. Extract of Hemlock.

Prep. Take fresh leaves of Conium and proceed as for Extract of Aconite, L. D. v. p. 241. The Edinburgh process is essentially the same; but the evaporation is to be made in vacuo, with the aid of heat, or spontaneously in shallow vessels exposed to a strong current of air; as a temperature of 212° destroys the Conia. When the extract is triturated with *Liq. Potassæ*, a strong odour of Conia should be disengaged.

This extract, when well prepared, is of a fine deep-green colour, and may be kept good for some time. Evaporating the E. Tincture also forms an excellent Extract. (*c.*)

D. gr. iij. 2 or 3 times a day, and gradually increased.

TINCTURA CONII, L. E. D. Tincture of Hemlock.

Prep. Macerate for 14 (7, D.) days dried leaves of *Conium* ℥v. (℥ij. D.) bruised *Cardamoms* ℥j. in *Proof Spirit* Oij. (by measure ℔j. D.) Strain. L. D. Fresh leaves of *Conium* ℥xij. *Tinct. of Cardamom* Oß. *Rectified Spirit* Oiß. Express the juice from the leaves, transmit Rectified Spirit through the residuum, mix the watery and spirituous fluids, and filter the product. E.

Action. Uses. The Tincture obtained by the process of percolation Dr. Christison considers the best of all preparations for medical use.

That made by Mr. Squire, by adding rectified Spirit to the expressed juice, without extracting the residuum, is also a good preparation.

D. f℥ß.—f℥j. of the L. P. ♏xv.—♏xxx. gradually increased of the E.P.

PILULÆ CONII COMPOSITÆ, L. Compound Pills of Hemlock.

Prep. Rub together into a mass *Extract of Hemlock* ʒv. powdered *Ipecacuanha* ʒj. *Mixture of Acacia,* q. s.

Action. Uses. Anodyne Expectorant in spasmodic Coughs in doses of gr. v.

UNGUENTUM CONII, D. Hemlock Ointment.

Prep. Take fresh leaves of *Conium* ℔ij. and boil them in *prepared Hogs' Lard* ℔ij. till crisp. Express through linen.

Action. Uses. As oil and fatty matters take up some of the active properties of Hemlock, this is an efficient application to foul or painful ulcers, &c.

CATAPLASMA CONII, L. D. Hemlock Poultice.

Prep. Take *Extract of Conium* ℥ij. mix it in *Aq.* Oj. and add *bruised Linseed* q. s. to thicken, L. Take dried leaves of *Conium maculatum* ℥j. *Aq.* by measure ℔ß. Boil down to ℔j. strain, and add powder of the same q. s. to make a poultice, D.

Action. Uses. Soothing application to Cancerous and other sores.

Tribe *Coriandreæ.* Fruit contracted from the side, didymous or globular. Ridges apterous.

CORIANDRUM, L. E. D. Fructus, L. (Semina) D. Fruit of CORIANDRUM SATIVUM, *Linn.* Coriander.

Coriander was the *gad* of the Hebrews, and was well known to the Greeks by the name κορίον. It has long been cultivated throughout the East, and is so now in Europe, as well as in this country.

Stems annual, from 1 to 2 feet high, round, striated, smooth. Leaves bipinnate, cut; leaflets of the lowermost wedge-shaped, of the others divided into linear segments. Calyx of 5 teeth. Petals white, often with a tinge of pink, obcordate, with an inflexed lobe, the exterior ones radiant and bifid. Fruit globose. Carpels with the primary ridges obsolete, the 4 secondary ones prominent, keeled. Interstices without vittæ. Commissure with 2 vittæ. Seed excavated in the front, covered with a loose membrane.—Occasionally found wild in this country, but must have escaped from cultivation.—St. & ch. 94.

Coriander is much esteemed on account of its fruit (seeds as they are commonly called) both in the East, where it is much employed as a condiment, being an ingredient, for instance, of Currie Powder, and also in Europe, where it is required by confectioners and distillers. Coriander is well known by its globular form, and by its two carpels adhering firmly together, and forming a greyish-coloured fruit about the size of white pepper. It has a peculiar odour, and a warm aromatic taste, dependent on the presence of a yellowish-coloured volatile

Oil, which necessarily possesses in a high degree the qualities of Coriander.

Action. Uses. Stimulant Carminative in doses of ℈ß.—ʒj. Chiefly employed as an adjunct.

Off. Prep. Confectio Sennæ, L. E. Inf. Sennæ cum Tamarindis, D. Tinct. Sennæ Comp. E.

The ARALIACEÆ are closely allied to *Umbelliferæ;* but as they do not contain any officinal plants, a detailed notice of them is not here necessary. *Panax quinquefolium*, of which the root, called *Ginseng*, is so highly esteemed by the Chinese as to be considered a *panacea*, used to be sold for its weight in gold. It has a feeble odour, and a sweet, slightly aromatic taste, abounds in fecula, and can only be useful as a Nutrient and Demulcent. It is found both in the northern parts of China, in Tauria, as well as in North America. A nearly allied species, *Panax Pseudo-Ginseng*, was found in the Himalayas by Dr. Wallich. *Aralia nudicaulis*, a native of North America, has roots which are slightly fragrant and of a sweetish aromatic taste, is sometimes called *False*, and used as a substitute for, *Sarsaparilla*. *A. spinosa*, called *Angelica* or *Toothache-tree* in North America, is a stimulant Diaphoretic.

c. *Corollifloræ.* Lindley. (*v.* p. 235.)

CAPRIFOLIACEÆ, *Juss.*

The Caprifoliaceæ are so closely allied to Rubiaceæ as to be included with them in one close alliance both by Lindley and Endlicher. They are distinguished from them by having leaves without stipules, but are divided into two groups, LONICEREÆ VERÆ and SAMBUCEÆ, the latter having corollas which are rotate, or with short tubes, and the raphe of the seeds turned inwards. They are found in temperate and cold parts of the world, and are not possessed of very active properties.

SAMBUCUS, L. E. D. Flores, L. D. Flowers, E. Baccæ et Cortex interior, D. Berries and inner Bark of SAMBUCUS NIGRA, *Linn. Pentand. Trigyn.* Linn. Common Elder.

The Elder, indigenous in Europe, was known to the Greeks, and called ακτη by Dioscorides.

Arborescent; much, but always oppositely branched; young branches filled with spongy pith. Leaves pinnate; leaflets usually 2 pairs with an odd one, ovate, serrate, cuspidate. Cymes large, terminal, with 5 principal branches, Calyx limb 5-cleft. Corolla cream-coloured, rotate, 5-lobed, finally reflexed, with a faint smell. Stamens 5. Stigmas 3, sessile. Berry globular, black, 3—4 seeded.—St. and Ch. ii. 79.

Though so many parts of the Elder are officinal, few are possessed of any active properties. A little of the Volatile Oil, upon which the odour of the flowers depends, may be separated by distillation; but

Dr. Pereira has shown that the *Oleum Sambuci*, L. D. which is sold as Oil of Elder, is a spurious preparation made by boiling Elder leaves in Rape Oil. The taste of the inner bark is slightly astringent, the odour is slight, but it possesses Cathartic and also Emetic properties. A decoction of it has sometimes been prescribed in Dropsy. It has been given in infusions by M. Delens, ʒij.—ʒiv. in four cups of water reduced to three, and taken during the day in cases of Scrofula and of Leucorrhœas, for the improvement of the constitution.

Aqua Sambuci, L. E. Elder Flower Water.

Prep. Mix *Elder Flowers* ℔x. (or *Oil of Elder* ʒij. L.) with *Aq.* Cij. *Proof Spirit* f℥vij. (*Rectified Spirit* f℥iij. E.), and distil Cj.

Action. Uses. Used as a vehicle, and for flavouring medicines.

Unguentum Sambuci, L. D. Elder Ointment.

Prep. Boil *Elder* (*Flowers*) ℔ij. in *Lard* ℔ij. till they become crisp. Press through linen. L. *Fresh Elder Leaves* ℔iij. in *prepared Hogs' Lard* ℔iv. *prepared Mutton Suet* ℔ij. Prepare as Savine Ointment, D.

Action. Uses. Mild cooling Ointment. The L. prep. is the White Elder Ointment, the D. prep. the Green Elder Ointment of the shops.

Succus Spissatus Sambuci, D.

Prepared from the fresh ripe Elder Berries, as Succus spissatus Aconiti, D.

Action. Uses. This purple-coloured juice is Refrigerant and slightly laxative in doses of fʒj.—fʒij.

Rubiaceæ, *Juss.*

Herbs, shrubs, or trees, with roundish or 4-cornered stems and branches, with nodose joints. Leaves opposite, occasionally verticillate, with interpetiolary stipules, which are either distinct or variously adherent, simple, bifid, or multifid. Flowers complete, occasionally unisexual from abortion, almost always regular, variously arranged. Calyx-tube united with the ovary, limb superior or half-superior, 2—6 toothed or cleft, sometimes entire, disappearing on the fruit. Corol monopetalous, inserted into the tube of the calyx, tubular, or campanulate, seldom rotate, 4—5 or 6-cleft, valvate or imbricate, twisted. Stamens inserted into the tube of the corol, and alternate with its lobes. Ovary inferior, surmounted by a disk, 2- to several-celled, 1- to many-ovuled. Style simple. Stigma 2- to many cleft. Fruit capsular, berried, or drupe-like, cells 1- to many seeded. Seeds various in situation. Embryo small, usually with horny albumen.

Several botanists, as Ray, Brown, &c., have distinguished the tribe Stellatæ of the *Rubiaceœ* of Jussieu and De Candolle as a distinct family, by the name Stellatæ, called *Galiaceœ* by Dr. Lindley. These are distinguished by their weak, often quadrangular stems, whorled leaves, absence of stipules, and didymous fruit, also by their being found in temperate climates. Several of them are a little bitter, some secrete red colouring matter.

Rubia Tinctorum, *Linn.* Radix, D. The Root of Dyer's Madder. *Tetrand. Monog.* Linn.

Madder was employed by the ancients, as species of Rubia still are in the East. *R. Mungista* or *cordifolia* yields the *munjeet* of India.

Root perennial, long, horizontal, of a reddish-brown colour. Stems herbaceous, four-sided, with the angles aculeate. Leaves 4 to 6, in a whorl, petiolate, lanceolate, smooth above, margin and keel aculeate, rough. Peduncles axillary, trichotomous. Tube of the calyx ovate-globose; limb scarcely any. Corol 5-partite, rotate, lobes gradually callous, acuminate. Stamens short. Styles 2, short. Fruit didymous, somewhat globose, succulent.—A native of the south of Europe and of the Grecian Archipelago and north of Africa.—Nees von Esenb. and Eberm. 255.

Madder is cultivated in France and Holland, and largely imported for the use of the dyer either whole or ground. The roots are branched, about the thickness of a quill, of a reddish hue externally, and consist of an easily separable bark, with an external woody part of a reddish colour. It has little smell, with a slightly bitter astringent taste. The colouring matters, as *Alizarin*, &c., which it contains, are a subject of great interest both to the chemist and to the dyer. It is of little importance in medicine.

Action. Uses. Slight Astringent and Tonic. When taken internally, colours the bones red. It was at one time considered Emmenagogue (*Home*) in doses of ʒß.—ʒij. every 3—4 hours.

CINCHONACEÆ.

2. The rest of the Rubiaceæ, named CINCHONACEÆ by Dr. Lindley, are distinguished by their opposite, rarely verticillate leaves, and intermediate or interpetiolar stipules. They are found in the tropical parts of the world, and on the Andes of Peru, and on the mountains of India as high as the belt of tropical vegetation. They are remarkable for secreting *bitter* and *astringent* principles: hence many are employed as febrifuges, in addition to those which are conspicuously so from secreting the alkalis *Quinia* and *Cinchonia*. Some, like the Stellatæ, secrete *red colouring matter;* while some, with herbaceous stems and perennial roots, are equally remarkable and useful from secreting *Emetine*. The officinal plants all belong to *Pentandria Monogynia* of Linnæus.

IPECACUANHA, L. E. D. Radix, L. D. Root, E. of CEPHAELIS IPECACUANHA, *Tussac.* Ipecacuan. Ipecacuanha.

Ipecacuanha, a name adopted from the language of the South Americans (by whom it is also called *Praya de Mato*), has been applied to a variety of Emetic roots, but is restricted in the Pharmacopœias to the roots of the above Cephaëlis. This was first distinctly noticed in the Nat. Hist. of Brazil of Piso, and Marcgraaf, p. 101, and p. 17, (1648,) as a *brown*-coloured Ipecacuan, and distinguished from another of a *white* colour, with the plant something like Pulegium. It was first brought into notice in Europe by Helvetius, about 1686. The plant yielding it was long unknown. Dr. Gomez was the first (Memoria sobre ipecacuanha fusca du Bresil, 1801) to describe and figure the genuine plant; but having left specimens with Brotero of Coimbra, and he with Tussac of Nantes, the former, without the permission of and acknowledgment to Gomez, described it in the Linnæan Trans. vi. p. 137. t. 11. 1802, as a species of Callicocca. But this genus is identical with the CEPHAELIS of Swartz, to which it was referred by Tussac and published in the Journ. de Bot. of M. Desvaux, iv. p. 204 (1813). The subject has since been investigated by Merat, Richard,

Martius, and A. St. Hilaire. Merat, Guibourt, and others distinguish it into three varieties, *Brown*, *Red*, and *Grey Ipecacuan*, depending on the colour of the epidermis.

Fig. 73.

Cephaëlis Ipecacuanha (fig. 73) has a perennial root, simple, flexuose, or with a few diverging branches a few inches in length, about the thickness of a quill, knotty, with transverse rings; when fresh, of a pale brown colour externally. Stem suffruticose, ascending, often rooting near the ground, at length erect, somewhat pubescent towards the apex. Leaves from 4 to 6 or 8, on a stem, opposite, oblong, obovate, acute, roughish above, finely pubescent beneath. Stipules erect, 4—6 cleft. Peduncles solitary, axillary, downy, erect when in flower, drooping when in fruit. Flowers collected into heads, and inclosed by a large 1-leafed involucre, which is deeply 4 to 6 cleft. Segments obovate. Bracts one to each flower, obovate, oblong. Calyx minute, with 5 blunt short teeth. Corolla white, funnel-shaped; tube downy on the outside and at the orifice; limb with 5 ovate reflexed segments. Stamens 5, with filiform filaments and linear anthers, which project a little beyond the corolla. Ovary surmounted by a fleshy disk. Stigma bifid. Berry about the size of a coffee-bean, of a dark violet-colour, crowned by the remains of the calyx, 2-celled, 2-seeded, with a longitudinal fleshy dissepiment. Nucleus plano-convex, furrowed on the flat side. Flowers from November to March, and ripens fruit in May.—A native of shady places in the forests of Brazil from the province of Rio Janeiro to that of Pernambuco.—Fig. *v.* Gomes l. c. l. 2. Linn. Trans. vi. t. 11. Martius, Spec. Mat. Med. Bras. 4. t. 1. St. Hilaire. Pl. Us. de Brazil. pl. 6. Nees and Eberm. 258. St. and Ch. 62.

This plant yields the Brown or Grey, *annulated*, or true, sometimes called Brazilian or Lisbon Ipecacuanha, which is in general alone met with in this country. It is collected at all seasons of the year, but chiefly from January to March, and is imported from Rio Janiero, Bahia, and Pernambuco.

The roots of *annulated* Ipecacuanha, as met with in commerce, are of a greyish or light-brown colour, have sometimes attached to their upper part a straight cylindrical part, by which it was connected with the stem, are 2 or 3 inches in length, simple or branched, variously contorted, about the thickness of a small quill, composed apparently of a series of transverse but unequal rings, separated by nearly parallel grooves, giving the whole a knotted appearance. The rings are composed of an external *cortical* portion, which is horny but brittle, and are apparently strung upon a slender, tough, and whitish ligneous portion, which is called *meditullium*, forming about $\frac{1}{5}$ of good Ipecacuan. This part has little odour or taste, and is comparatively inert. The cortical portion has a peculiar nauseous odour, and slightly bitter, somewhat acrid taste. This is said to be more evident in the fresh plant, and the odour is extremely disagreeable to many when the root is powdered. The active properties are taken up by water, Alcohol, Proof Spirit, or Wine. Pelletier analyzed Ipecacuanha root, and found in the cortical portion, of a peculiar principle which has been called *Emetine*, but is rather Emetic Extract, 16 parts, odoriferous Fatty matter 2 (this consists of an odorous volatile oil, and of a scentless fixed fatty matter), Wax 6, Gum 10, Starch 42, Lignin 20 parts, loss 4 parts=100. The Red variety contained only 14 per cent of Emetine. The Meditullium contains only about 1 per cent. of Emetic Extract, and about 67 per cent. of Ligneous fibre.

The Emetic Extract, or *Matiere Vomitive* as it was first called, was afterwards found to be impure, and good Ipecacuanha to contain only about 1 per cent. of pure *Emetine.* This is colourless, uncrystallizable, but alkaline in its properties, without odour, and nearly without taste; fusible about 120°, sparingly soluble in water, but much so in Alcohol or proof Spirit. It forms bitter salts with acids, and is precipitated when in solution by Tincture of Galls. Sesquichloride of Iron imparts a greenish colour to the decoction of Ipecacuanha. Emetine is composed of $C_{35}\,H_{25}\,O_9\,N$. Impure Emetine, which is that most frequently met with, is of a yellowish-white colour.

A kind of *Ipecacuanha*, known as *striated* and also as *Black* or *Peruvian Ipecacuan*, is yielded by a different plant of this family, with much larger joints, the *Psychotria emetica*, of the tribe Coffeæ, a native of New Granada, which indeed was at one time supposed to yield the true kind. Pelletier found it contained about 9 per cent. of impure Emetine.

A third kind of Ipecacuanha, and that referred to by Piso, is the white kind, distinguished by its *white* colour, *amylaceous* nature, and *undulated* appearance. This is yielded by *Richardsonia scabra* (*R. braziliensis* of other authors), a plant of the tribe *Spermacoceæ* of this family, a native of Brazil, New Granada, Vera Cruz, &c. *R. rosea* yields a similar product. This kind contains only about 6 per cent. of impure Emetine.

Several other Rubiaceous plants are emetic in nature, and some of other families, as some species of *Polygaleæ*, of *Asclepiadeæ*, and of *Euphorbiaceæ*, and of *Violaceæ*, *Ionidium Ipecacuanha*. *I. parviflorum* yields the Cuichuncully de Cuenҫa, for which I am indebted for specimens to the Hon. Fox Strangways.

Action. Uses. Irritant, Nauseant Emetic, Expectorant, Diaphoretic, Sedative. Useful as an Expectorant and Diaphoretic in Catarrh, or as a Diaphoretic in febrile affections of various kinds, or to cause a determination to the skin in Diarrhœa and Dysentery. Emetic to cut short the accession of an ague, &c., evacuate the stomach, or give a shock to the system. Nauseant Sedative in Hæmorrhage, &c.

D. gr. xv.—℈j. or even ʒß. of the powder as an Emetic; often conjoined with Tartar Emetic gr. j., assisting its action with warm water or Camomile Tea. gr. ij. as a Nauseant. gr. j.—gr. ij. as an Expectorant and Diaphoretic. gr. $\frac{1}{2}$ as an Alterative in Dyspepsia, and of Emetine gr. $\frac{1}{16}$.

VINUM IPECACUANHÆ, L. E. D. Ipecacuanha Wine.

Prep. Macerate *Ipecacuan* in moderately fine powder ℥ijß. (℥ij. D.) in *Sherry Wine* Oij. (old wine m. D.) for fourteen (7 E.) days, and filter.

Action. Uses. Expectorant and Diaphoretic in doses of ♏x. to ♏xxx. Emetic in doses of fʒij. to fʒiv. Often given to children in doses of ♏xx. to fʒj.

SYRUPUS IPECACUANHÆ, E. Syrup of Ipecacuanha.

Prep. Take of *Ipecacuan* in coarse powder ℥iv. *Rectified Spirit* Oj. *Proof Spirit* and *Aq. Dest.* āā f℥xiv. *Syrup* Ovij. Digest the *Ipecacuan* in *Rect Sp.* f℥xv. at a gentle heat for 24 hours; strain, squeeze the residue, and filter. Repeat this process with the residue and proof Spirit, and again with the water. Unite the fluids, and distil off the Spirit, till the residuum amount to f℥xij. Add to the residuum *Rect. Sp.* f℥v. and then the Syrup. Dr. Christison says this process is unnecessarily complex, and that a Syrup made from the Alcoholic Extract, as directed in the Parisian Codex, is probably as good.

Action. Uses. Expectorant fʒj.—fʒij. Emetic for infants fʒß—fʒj.

PULVIS IPECACUANHÆ COMPOSITUS. Compound Ipecacuan, or Dover's, Powder.

Prep. Triturate thoroughly together *Ipecacuan* in powder and *Opium* āā ʒj. (℥j. E.) *Sulphate of Potash* ʒviij. (℥viij. E.) (Triturate the Salt and Opium to powder, and mix the Ipecacuan, D.)

Action. Uses. Diaphoretic in doses of gr. v.—gr. x., sometimes repeated at short intervals. One of the most valuable Sudorifics, the Opium apparently causing a determination to the skin, and the Ipecacuan its relaxation. Dr. Dover directed his powder to be given in a glass of white wine posset, covering up warm, and drinking about a quart of the posset while sweating. It is often necessary to avoid drinking too soon, to prevent vomiting.

PILULÆ IPECACUANHÆ (ET OPII, E.) COMPOSITÆ, L. Compound Ipecacuan (and Opium) Pills.

Prep. Mix *Comp. Ipecacuan powder* ʒiij. *Squill* fresh dried, *Ammoniacum* āā ʒj. *Mucilage* q. s. to make into a mass of proper consistence. The E. C. orders *Comp. Ipecac. powder* ʒiij. *Conserve of Red Roses* ʒj. to be beaten into a proper mass, and divided into gr. iv. pills.

Action. Uses. Diaphoretic Expectorant, as the foregoing, in doses of gr. v.—gr. x.

Ipecacuanha and Emetine are both sometimes introduced into the system by friction in the form of liniment or ointment.

Pharm. Prep. Trochisci Morphiæ et Ipecacuanhæ (p. 261).

COFFEA ARABICA, *quhwa* of the Arabs (of the tribe and subtribe *Coffeæ*) is a native of Arabia Felix and of the borders of Abyssinia. From the former it has been introduced into, and is cultivated in various countries. It is too well known as an article of diet, to require detailed notice here, and is remarkable for *Caffeine* having been found to be identical with *Theine* (*v.* p. 292) and for its stimulant influence on the nervous system, especially in those unaccustomed to its use. Hence it is sometimes employed as a Stimulant and Antisoporific, and to counteract the effects of Opium and of other Narcotic poisons.

CINCHONA, L. E. D. Pale, Yellow, and Red Barks produced by different species of CINCHONA, several of which are yet unascertained.

Cinchona, so named by Linnæus in compliment to the countess of Chincon, lady of the then viceroy of Peru, who was cured by, and brought from thence to Europe in 1639, some of this not more celebrated than valuable bark. But the native name *quinquino*, p. 368, does not differ much from Cinchona. Its history is obscure: the natives of the country are supposed to have been unacquainted with its virtues, which are thought to have been discovered by the Jesuits, by whom it was chiefly made known in Europe. The plant or plants yielding this were long entirely and are still in a great measure unknown. The first notice is by Dr. Arrot, in Phil. Trans. for 1737. In the same year Condamine, and in 1740 Joseph de Jussieu obtained specimens from near Loxa. About 1772 Mutis sent a few specimens to Linnæus from about Santa Fé de Bagota, and with Zea published some inaccurate information respecting the officinal species. At first a species called *Cinchona officinalis* was supposed to yield all the barks of commerce. Ruiz and Pavon commenced in 1777, and carefully examined the vegetation of Chili and of Peru, published accurate and valuable information respecting several kinds of Bark and of species of Cinchona, from 1792 to 1801, in their *Flora Peruviana* and *Quinalogia.* Humboldt and Bonpland, in their *Plantæ Æquinoctiales*, have accurately described some of the species, and Pöppig has given most valuable information respecting the species and barks of the more southern Cinchona countries. The species made known by these travellers and botanists have been elucidated by Mr. Lambert in his works on the genus Cinchona, and by De Candolle in his Prodromus; but most fully and clearly by Professor Lindley in his Flora Medica, 1838, having as materials to work with, the foregoing works, and a most extensive series of dried specimens taken out of a Spanish prize, and collected by Mutis in 1805, near Loxa and Santa Zé, belonging to Dr. A. T. Thomson, as well as the collection of Mr. Lambert, which besides several unpublished species, contained nearly a complete set of those described in the Flora Peruviana of Ruiz and Pavon. From these he has described twenty-one species of Cinchona, and mentions five others. Three more have been described by Martius, from Brazil, and one by Mr. Bentham from British Guiana. (*v.* Walper's Repert.)

Many species at first referred to Cinchona, have been since found to belong to other nearly allied genera, as to *Exostemma*, *Remijia*, *Buena*, *Pincneya*, *Danais*, *Luculia*, *Hymenodictyon*. Some of these yield febrifuge barks, but which are now ranked with *false* Cinchonas.

The genus Cinchona (fig. 74) is characterized by having the calyx (*a*) turbinate, 5-toothed. Corolla (*b*) of one petal, hypocrateriform. Anther hairy or smooth inside the limb, divided into 5 oblong lobes, valvate in æstivation. Stamens 5; filaments short, and inserted (*b*) in the middle of the tube; anthers linear, almost entirely helmeted. Stigma bifid, somewhat club-shaped. Capsule (*c*) ovate or oblong, marked with a furrow on each side, 2-celled (*d*), crowned by the calyx, splitting (*e*) from below upwards into 2 cocci. Seeds numerous, erect, imbricated (*f*) upwards, compressed, girded by a membranous (*g*) lacerated wing.—The species are shrubby or arboreous, conspicuous for their ornamental appearance. The bark of all is bitter, astringent, and aromatic, remarkable for secreting the alkalis Quinia and Cinchonia. The leaves are short-petioled, with plane margins. Stipules foliaceous, free, deciduous. Flowers paniculate, corymbose, terminal, white or roseate, some purplish.

The true Cinchonas are supposed to be confined exclusively to the Andes, within the boundaries of Peru, Columbia, and Bolivia, extending from La Paz, about 20° of S. lat. to the mountainous regions of Santa Martha on the northern coast, in 11° of N. latitude. But the extent to which they spread to the East has not been well ascertained. Mr. Mornay some years since mentioned to the author that a species of Cinchona with dark red-coloured bark, and yielding *Quinine*, was found on the mountains of Paraguay. Mr. Bentham has described *Cinchona Ronaima*, from British Guiana. The elevations vary, and must do so according to latitude and the local influences, which produce great modifications of climate. The species occur from 1200 to 10,000 feet, but the principal from 6000 feet to the latter height. The temperature on an average of these may be 59° or 62°. The best Barks are found in dry rocky soil at the greatest elevations and the coldest climates.

The bark is peeled about May by the Indians, who are called *Cascarilleros*, either by cutting down the trees and taking the bark from the branches, or "the bark is cut from the trees as they stand." It is made into bundles and carried out of the forests to be carefully dried. The thinner curl inwards, and form *quilled* bark; the larger and thicker barks, probably to facilitate the drying, have the epidermis stripped off. When dry, the bark is conveyed to the coast, and packed in chests or in serons, each package usually containing only one kind of bark, though the species and varieties of those imported are very great. The several kinds are imported from various ports along the western coast of South America, originally from Loxa and Payta, which is near it, but latterly also from Valparaiso, having first been imported there from more northern ports. Also from Arica, the produce of the neighbourhood of La Pas, and of Apolobambo, and from Lima the produce of the forests of Huanuco. Some is carried across the country and exported from Buenos Ayres.

Some finds its way to the northern port of Carthagena, the produce of the forests near Santa Fé de Bagota.

Though many species of Cinchona are now known, and likewise a great variety of Barks, there is considerable difficulty in determining which of the latter are produced by the known species of the former. The difficulty is only to be determined by those who, like Pöppig, will collect the bark and a specimen of the plant at the same time, and submit both to careful examination by qualified observers. The greatest mistakes have occurred in consequence of Mutis and Zea having taken for granted, that the Barks of the north were identical with those of the south, and consequently that the trees which yielded the Yellow and Red Barks of Carthagena must be the same as those which produced the Yellow and Red Barks of Lima,—a mistake by which they have not only deceived themselves, but led into error Humboldt, Lambert, Don, and the authors of the London Pharmacopœia of 1836.

As the plants, so the barks of these Cinchonas have been carefully studied. They might be arranged according to the plants which yield them, if our knowledge was more perfect; or according to their chemical characters, as has been done by Goebel and Geiger; or according to their physical characters, as of colour, by which the L. and D. P. mention the *Pale*, *Yellow*, and *Red* Barks: the E. C. substituting *Crown* and *Silver* Bark for the first. This method has been chiefly adopted, as by Bergen* in a monograph pronounced by Dr. Duncan to be "the most perfect specimen of Pharmacography," under nine different heads: 1. China rubra, or Red Bark. 2. China Loxa, or Crown Bark. 3. China Huanuco, or Silver Bark. 4. China regia or Yellow Bark. 5. China flava dura, or hard Carthagena Bark. 6. China flava fibrosa, or fibrous Carthagena Bark. 7. China Huamalies, or Rusty Bark. 8. China Jaen, or Ash Bark. 9. China Pseudo-Loxa, or inferior pale Bark. Then by M. Guibourt,† who arranges them under the divisions of Grey, Yellow, Red, White, and False Cinchonas, and enumerates 37 different varieties. Dr. Pereira adopted these general heads, and has incorporated much of the description

* Bergen's work is entitled "Versuch einer Monographie der China," Hamburgh, 1826. An abstract of his account of the different kinds of Bark is given in Wood and Bache's Disp. United States.

† Histoire abrégée des Drogues simples. Tome ii. p. 44—107.

of Bergen, first in the Medical Gazette, No. 45 and 46 for 1837-38, and then more fully in his valuable Elements of Materia Medica. His synonymes are particularly valuable, in consequence of his having exchanged specimens both with Bergen and with Guibourt, and thus identified the German, French, and English names, by a comparison of the several kinds of Barks. Dr. Christison has been able to confirm many of these results, from having also received some of the same Barks. We shall therefore follow these well established classifications.

Dr. Pereira's arrangement, which we have reduced to a tabular form, is as follows:

1. Genuine Cinchonas with epidermis normally brown.

Barks.	*Plants.*
A. Cinchonæ pallidæ, or Pale Barks.	Quinquinas gris, *Guibourt.*
1. Crown or Loxa Bark. Cinchona Coronæ, E.	Cinchona Condaminea, *H.* and *B.*
2 Grey or Silver Cinchona. Cinchona cinerea, E.	„ micrantha, *Ruiz* and *Pavon.* (*C. scrobiculata*, H. and B.)
3. Jaen or Ash Cinchona. (*C. lancifolia*, L.	„ ovata. *Fl. Peruv.*
4. Huamalies or Rusty Bark.	„ pubescens, *Vahl.* (*purpurea. Fl. Peruv.*)
B. Cinchonæ flavæ. Yellow Barks.	
5. Royal Yellow or Calisaya Bark. (C. cordifolia, L. and D.)	Species unascertained, E.
C. Cinchona rubra.	
6. Red Bark. (C. oblongifolia, L. D.)	Species not ascertained, E.

2. Genuine Cinchonas, with whitish and micaceous Epidermis.

A. Pale Barks.	
7. White Loxa Bark.	
B. Yellow Barks.	
8. Hard Carthagena Bark.	Cinchona cordifolia, *Mutis.*
9. Fibrous Carthagena Bark.	C. cordifolia?
10. Cusco Bark.	
11. Orange Cinchona of Santa Fé de Bagota.	Cinchona lancifolia. *Mutis.*
C. Red Barks.	
12. Red Cinchona of Santa Fé.	
13. Red Cinchona with white epidermis.	Cinchona magnifolia. *Fl. Peruv.* (*C. oblongifolia. Mutis.*)

3. Barks falsely called Cinchonas, and not yielding the Cinchona alkalis.

St. Lucia Bark.	Exostemma floribundum.
Jamaica Bark.	„ caribæum.
False Peruvian Bark.	„ peruvianum.
Brazilian Bark. (*Quina de Pianhy.*)	„ Souzanum.
Pitaya Bark.	„ Malinea? racemosa? &c.

In the following account we have dwelt chiefly on the characters of the officinal barks, and given only short notices of the others; but instead of adhering to the above arrangement, have thought it desirable to bring together the pale, yellow, and red barks of the two divisions, in order that they may more readily be compared.

CINCHONÆ PALLIDÆ, the Pale Barks, the *quinquinas gris* of Guibourt, are almost always quilled, moderately fibrous, more astringent than bitter. Their powder is of a greyish fawn-colour, and they contain *Cinchonia* and little or no *Quinia*. An infusion of Pale Bark does not deposit any Sulphate of Lime on the addition of a solution of Sulphate of Soda.

1. Crown or Loxa Bark. Cinchona Coronæ, E. (*Cinchona lancifolia,*) L. (*C. officinalis,*) D. Cortex, L. D. Bark of Cinchona Condaminea, *H.* and *B.* E. Quinquina gris-brun de Loxa, *Guibourt.* China Loxa, *Bergen.*

This kind of Bark is considered to have been one of the first introduced into Europe. It is always in quills or cylindrical tubes strongly rolled, usually single, from 6 to 15 inches in length, and varying from 2 lines to an inch in diameter, and from half a line to 2 lines in thickness. The epidermis is entire, of a light or dark grey, sometimes even of a brownish colour, often covered with white crustaceous lichens. The outer surface is marked with numerous longitudinal wrinkles, of little depth, or crossed by transverse cracks, which often run entirely round the bark, dividing it into rings, the edges of which are somewhat elevated. The inner surface is smooth and uniform, and of a cinnamon-brown colour, as is the powder. The middle-sized quills are probably the best. The taste is astringent and bitterish, with a little aroma. The odour is compared by Bergen to that of Tan, but it has a slight degree of aroma. It is imported in chests and serons, and is collected in the woods round Loxa and on the neighbouring mountains of Peru. There seems no doubt of this bark being produced by Cinchona Condaminea of *H.* and *B.*

Some *Cascarilla fina* is said to be yielded by *C. nitida, Fl. Per.* which grows on the lofty mountains of the Andes, and some of the *Quina fina de Loxa* by *C. lucumæfolia, Pavon.*

C. Condaminea (fig. 74) is a tree about 18 feet in height, with opposite branches, which are horizontal in the lower parts, but form above an acute angle with the stem, smooth, as high as the inflorescence. Leaves quite smooth, usually ovate, lanceolate, sometimes only lanceolate, at other times ovate, generally furnished at the axils of the veins underneath with a pit, which is either naked or ciliated. Petioles smooth. Stipules oblong, obtuse, membranous. Peduncles panicled, corymbose in the axils of the upper leaves, forming a large loose thyrse, covered with a thick short down. Tube of the calyx downy, like the pedicels; limb very short, urceolate, 5-toothed, pubescent. Tube of the corolla slender, about four times as long as the tube of the calyx, tomentose; limb very shaggy internally. (*Lind.*) It grows in the mountains near Loxa, and also near Guancabamba in Peru. It always grows on micaceous schist, and at an elevation of from 5700 to 7500 feet, and thus occupies a belt of 1800 feet. This species yields the *Cascarilla fina de Uritusinga,* which forms the Crown Bark of commerce.

Fig. 74.

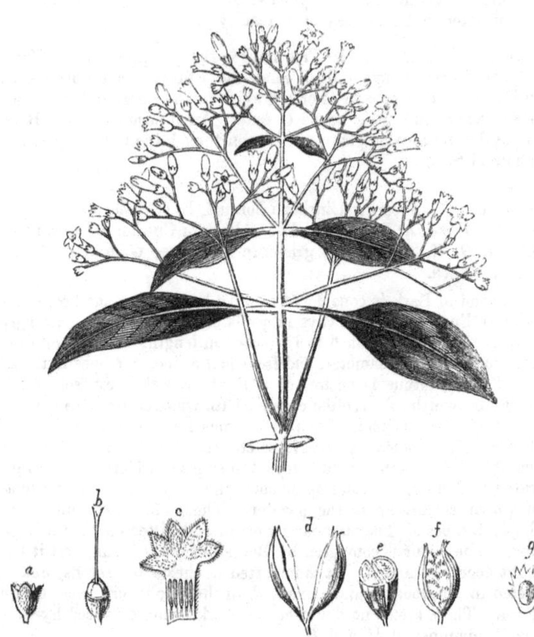

Cinchona Condaminea, v. p. 438 and 441.

2. Grey or Silver Bark. Cinchona cinerea, E. Bark of Cinchona micrantha, *R.* and *P.* China Huanuco, *Bergen.* Quinquina de Lima, *Guibourt.*

This kind of pale bark was first introduced into Europe about the beginning of the present century. It obtained its name of Lima from its place of export, and that of Huanuco from the city in central Peru, near which it is produced. It resembles the former kind in dimensions, but is rather longer and coarser, occurs in quills covered by a greyish epidermis. Many of the smaller quills have a more or less spiral form, a large oblique slit is observable at the edge of most of the complete quills. It is less wrinkled longitudinally than Crown Bark, and the transverse fissure less generally runs entirely round, and with the edges not elevated. On the inside it is rather more of a red colour, more or less uneven and fibrous, but its powder is nearly of the same Cinnamon-brown, and similar to it in odour and taste. Bergen says the odour of the bark is like that of clay, and in this respect different from that of all other varieties. According to the satisfactory information of Pöppig, this bark is produced by Cinchona micrantha, *R.* and *P.* (the C. scrobiculata of *H.* and *B.*) which grows on the high mountains of Peru.

3. Ash or Jaen Cinchona, apparently produced by *Cinchona ovata*, Fl. Per., is distinguished by its thin light coat, readily pulverized, cracks few, quills mostly crooked. Colour dark cinnamon-brown. (*Bergen.*) From one of its names, it would seem to come from near St. Jaën de Bracomoras.

4. Huamalies or Rusty Bark, comes from Lima, and is the produce of *C. pubescens, Vahl* (the *C. purpurea* of Fl. Peruv.) it is distinguished by its coat, thin and spongy longitudinal wrinkles and warts, which penetrate to the cortical layers; under surface even, colour red brown. (*Bergen*, as given by Pereira.) Yielded to Goebel and Kunze 38 per cent. of Cinchonia and 28 per cent. of Quinia.

Pöppig discovered that the bark called *Cascarilla nigrella* in Peru, and much esteemed there, was obtained from *Cinchona glandulifera* of *R.* and *P.* when growing on high mountains. It came formerly from Lima, and appears equal to the finest from Loxa. Another kind, inferior in quality, called *Cascarilla provinciana nigrella*, is produced by the same tree when growing in warm valleys.

7. White Loxa Bark is sometimes found intermixed with the Crown or Lima Cinchona, but is to be distinguished by its white epidermis.

B. Cinchonæ flavæ. Yellow Barks.

5. Cinchona flava, E. D. (*Cinchona cordifolia*, L.) Species of Cinchona not ascertained. Cinchona Calisaya (Colli-salla, *Pöppig*) regia. Quinquina jaune royal, *G.* König's China. *Bergen.* Yellow Bark. Royal Yellow Bark.

The Yellow Bark of commerce appears to have been first introduced into Europe about the year 1790. The tree producing it is still unknown. The reference to the *C. cordifolia* of Mutis is erroneous, in consequence of that author having mistaken the bark of his tree for the above Royal Yellow Bark of commerce. M. Guibourt recognized it, however, as the hard Carthagena bark of commerce, No. 8. Yellow bark is imported from southern ports, such as Coquimbo, as originally stated in the United States Dispensatory, but chiefly from Arica, whence it is often first conveyed to Lima. It is said by the Messrs. Gibbs to be produced in the province of La Paz in Bolivia, in a plain bounded by mountain ranges, and elevated 14,000 or 18,000 feet above the level of the sea (*Pereira*), also around Opolobambo. It is imported in chests and serons, and two kinds are known, *quilled* and *flat* Yellow Bark.

The *quilled*, called *Calisaya rolada*, or *rolled*, is in pieces from 3 and 4 to 18 inches in length, and from a $\frac{1}{4}$ to 2 or 3 inches in diameter, and varies in thickness from ⅛ to ⅓ of an inch, in general only singly quilled. The epidermis is brownish, often mottled by whitish or yellowish lichens, and is marked by longitudinal and transverse fissures, generally easily separated from the bark; sometimes in the larger pieces very rough from the furrows and cracks. Its inner surface is smooth but fibrous, and of a yellow cinnamon colour. The transverse fracture is short but splintery, and the powder contains spiculæ which are irritating to the skin. The *flat* Calisaya, or *Calisaya plancha*, appears to have been derived from the trunk and larger branches; it may be quite flat or slightly curved, and being destitute of epidermis, is of a yellowish-colour on both sides, but more fibrous in structure. The powder is of a yellow orange-colour; taste less astringent but more bitter than the pale bark: that of the flat Bark is less bitter than the quilled. Yellow Bark contains a large proportion of Quinia, and very little Cinchonia. Sulphate of Soda produces an abundant precipitate of Sulphate of Lime in its infusion.

8. *Hard Carthagena Bark*, China flava dura, *Bergen*, the *Quina jaune* of Humboldt, the *Quina amarilla* of Mutis, and the produce of his *Cinchona cordifolia*, which grows in the forests of New Granada, and is that which has been mistaken for the foregoing yellow Calisaya Bark. G. and K. i. taf ix. fig. 1 to 4. It is distinguished by its epidermis being velvety, greyish-white, thin and soft, or warty, longitudinal furrows irregular, few transverse fissures; under surface uneven or splintery; colour dull citron-yellow. (*Bergen.*) Goebel and Kirst obtained in ℔j. about 56 grs. of Quinia in this and the following, and of Cinchonia 43 grs, in the former, but none in the fibrous bark.

9. *Fibrous Carthagena Bark*, China flava fibrosa. G. and K. taf. ix. fig. 5—8. Occurs with the foregoing, whence Dr. Pereira suspects it may be produced by the same species, either at different seasons or in different localities. Coat thin, soft, of moderate thickness, or rubbed off; under surface even, but rough to the touch; colour pure ochre-yellow.

10. *Cusco Bark*, first described by Guibourt, is the *China rubiginosa* of Bergen, may by some be mistaken for Yellow Calisaya Bark. Its epidermis is shining, pale grey, without fissures, the naked surface is orange-red. But Dr. Pereira has pointed out that it may be distinguished by Sulphate of Soda not producing any precipitate in its solution. Guibourt obtained of Cinchonia about ℥j. from a pound of Bark.

11. *Orange Cinchona of Santa Fé.* Quinquina de Carthagene spongieux, G. l. c. p. 78, is the *Quina naranjada* of Mutis, and produced by his *Cinchona lancifolia*, which is so frequently stated to be the species yielding pale bark.

C. *Cinchonæ Rubræ*, or Red Cinchonas.

6. CINCHONA RUBRA, E. D. (*Cinchona oblongifolia*) Cortex, L. D. Red Cinchona Bark from an undetermined species, E. Rothe-China, *Bergen.* Quinquina rouge, or Cascarilla colorada of the Spaniards.

Red Cinchona Bark was early known, but not distinguished in England until 1779. It is imported from Lima in chests, but the species yielding it is unknown. C. magnifolia, Fl. Peruv. (the *C. oblongifolia* of Mutis) yields a different and inferior kind of Bark, the *Quina nova*, the Red Cinchona of Santa Fé. No. 12. It is received in quilled, but most frequently in flat or rather curved pieces, varying in length from a few inches to two feet, from one to five inches in breadth, from a quarter to three-quarters of an inch in thickness. The pieces are usually covered with epidermis, which is of a greyish or reddish-brown colour, sometimes mottled with purple or with white from adhering lichens, though these are less frequent than in others; but it is often rough, wrinkled, and warty, forming the variety *verruqueux* of G. On the inside its surface is coarsely fibrous, and of a deep cinnamon-brown, but it looks of a red colour when placed near the other Barks, and especially in the thicker pieces. The fracture is short, fibrous, and splintery. The powder is of a reddish-brown colour; the taste is powerfully bitter, with a slight degree of aroma. It is more scarce than the other kinds, and is remarkable for yielding both Quinia and Cinchonia.

12. *Red Cinchona of Santa Fé.* (*Per.*) Quinquina nova of authors. Quina roxa or Azahar of Santa Fé. An inferior kind of Carthagena Bark of a red colour, with white epidermis, containing little or no Quinia or Cinchonia. It is produced by *Cinchona magnifolia*, Fl. Per. (*C. oblongifolia*, Mutis), G. and K. taf. xi. f. 6 to 9.

13. *Red Cinchona*, with a white micacious epidermis, mentioned by Guibourt and Pereira.

The physical character of these several Barks is important only as indications of their richness in the principles which make them so valuable as medicinal agents. So little was known of the chemical nature of these Barks, even in 1802, that Seguin concluded that the active principle was of the nature of Gelatine. In 1803 Dr. Duncan indicated the presence of a peculiar principle, which he named *Cinchonia*, and which Gomez isolated in 1810. But Pelletier and Caventou in 1820 determined the alkaline nature of *Cinchonia*, discovered *Quinia*, and elucidated in general the chemical constitution of the Cinchona Barks, and that these alkalis were in combination with Kinic, called also Cinchonic acid (and perhaps also with a little Sul'), that they also contain Tannin, Colouring matters of a peculiar nature, one called Cinchonic red, the other Cinchonic yellow, also a little Volatile Oil, a Green Fatty matter, Calcareous salts, Starch, Gums, Ligneous Fibre. It has also been observed that the Gum abounds in the Pale, and is deficient in the Yellow and Red Barks, whence it has been inferred that these are produced by older parts of their respective trees; also that Cinchonia predominates in the Pale, Quinia in the Yellow, and that both Quinia and Cinchonia are found in nearly equal quantities in Red Bark. Pelletier with Coriol discovered a third alkali, which has been named *Aricina*, from the port Arica, whence Cusco Bark, from which it was extracted, is obtained.

Of these principles, the Volatile Oil has been obtained of a thick consistence, acrid taste, and with the odour of Cinchona Bark. The Fatty Matter is of the nature of a concrete oil, and is capable of forming soaps with alkalis; it differs in colour according to that of the Bark. The Tannin, or Tannic acid, called by some Soluble Red

Colouring Matter, has a brownish-red colour, and forms a green precipitate with ferruginous salts. It precipitates also with Tartar Emetic and with Gelatine. It absorbs Oxygen, when, according to Berzelius, it is converted into Cinchonic Red, or Insoluble Red Colouring Matter. (*P.* and *C.*) It is of a reddish-brown colour, insipid, is soluble in Alcohol, and also in solutions of alkalis and of their Carbonates; but not so in water or Ether. It is precipitated by Tartar-Emetic, and is to be noted for uniting with Quinia and Cinchonia, and forming compounds of a brownish-red colour, which are decomposed by the alkalis, but are little soluble in cold water, though dissolved by boiling water, rectified Spirit, and weak acids. Cinchonic Yellow is precipitated by Diacetate of Lead, but not by Tartar-Emetic, and is soluble in water, Alcohol, and Ether. Cinchonic or Kinic acid is separated in the form of a syrup-like liquid, has an acid taste, is soluble both in water and Alcohol, and in many points resembles Acetic acid: by the action of heat it is converted into Pyrokinic acid. In the Bark it is combined with Quinia and Cinchonia, and these Kinates, as also that of Lime, are soluble in water, but insoluble in Alcohol. Kinovic acid was discovered by Pelletier in Cinchona nova, or the Red Bark of Santa Fé. (*p.*) It is in some respects analogous to Stearic acid.

Cinchonia ($C_{20}\,H_{12}\,N\,O$) exists in both *pale*, *grey*, and *red* barks, in the form of Kinate of Cinchonia; when pure, is white and crystalline, nearly insoluble in cold, but soluble in 2500 parts of boiling water, very soluble in Alcohol, from which it may be obtained in brilliant, 4-sided, oblique crystals. It is slightly soluble in the fixed and volatile oils. Its bitter taste is very obvious when in solution. As an alkali it neutralizes the strongest acids. One of its salts, the Disulphate of Cinchonia, which is obtained in short prismatic crystals, soluble in Alcohol, less so in water, is sometimes employed for and is an efficient substitute for the similar preparations of Quinia.

QUINA, L. Alkali obtained from Yellow Bark (*Cinchona cordifolia*, L.) Quinia. Quina. *Quinine.*

Quinia is the alkali to which Yellow Bark owes its chief medical properties. It exists also in Red Bark, and in both as Kinate of Quinia, but may most readily be obtained by precipitating it with Ammonia from a solution of the Disulphate of Quinia. No formula is given for it in the L. P., nor is it used medicinally.

Quinia ($C_{20}\,H_{12}\,O_2\,N$), when pure, is seen as a white powder; but it may with care be crystallized from its Alcoholic solution in silky needles. It is without odour, but is intensely bitter; though, from its insolubility in water, the taste is not readily developed. It is very soluble in Alcohol, and also in Ether, but requires about 200 parts of boiling water. When precipitated from water, it falls as a Hydrate. It is fusible about 300° F., first parting with its water, and swelling if afterwards brought in contact with it. It readily combines with and neutralizes acids, forming salts, which are all bitter, moderately soluble, and crystallizable. Ammonia and Nutgalls both cause a precipitate in the solutions, the former separating it as Quinia, the latter as Tannate of Quinia.

Tests. Soluble very easily in Alcohol, but not in water, unless an acid be added. It alters the colour of Turmeric, tastes bitter, and is entirely destroyed by heat. L.

QUINÆ DISULPHAS, L. QUINÆ SULPHAS, E. D. Disulphate or Sulphate of Quinia, commonly called Sulphate of Quinine.

This salt has now almost entirely superseded all the other preparations of Cinchona, and is that commonly spoken of as Quinine. As Quinia exists in Bark combined with Cinchonic acid, it must be detached from it, and made to unite with Sulphuric′. This is effected by various processes. In these Yellow Bark is employed, as richest in Quinia, and as this is less mixed with the other alkalis. The Bark is first exhausted of its bitterness by repeated maceration in water acidulated with Sul′ or with H Cl′, then some more powerful base is presented to the acid, which precipitates the Quinia. This is finally purified and combined with Sul′.

The D. C. adopt the prize method of M. Henry, which, however, has the disadvantage of employing so expensive a menstruum as rectified Spirit. Digest *Yellow Cinchona-bark* in coarse powder ℔s. iv. in *Aq. dest.* Oviij. (old Wine) acidulated with *Dil. Sul′* f℥ij. with moderate heat and frequent stirring for four hours, and then strain. Repeat the process three times with fresh portions of water and mix them all together (a very soluble Sulphate of Quinia is formed from the excess of Sul′). Add to the united liquors, of *fresh burnt and slaked Lime* enough to saturate the acid. (Sulphate of Lime and Quinia are precipitated.) Dry the precipitate on blotting-paper, then digest it for 6 hours with frequent agitation in *Rectified Spirit* Oiij. by measure, and filter. Repeat the process three times with an equal quantity of Spirit (the Quinia is all dissolved by the Spirit and thus separated from the Sulphate of Lime.) Distil the spirituous liquors to dryness in a water-bath. Add gradually to the residuum *Dil. Sul′* till there is a slight excess of acid. (Sulphate of Quinia is formed, but usually requires the addition of animal charcoal and subsequent filtering to deprive it of colouring matter). Obtain crystals by cooling and evaporation.

L. P. Mix with *Aq. dest.* Cvj. *Sul′* ℥iv. and ʒij. then add of bruised *Yellow Bark* ℔vij. boil one hour, strain. Again boil the residue in *Acid* and *Water* in the same proportions, strain. Then boil the *Cinchona* in *Aq. dest.* Cviij. for 3 hours and strain. Wash well the residue with boiling *Aq. dest.* Mix the liquors and add *moist Hydrated Oxide of Lead* nearly to saturation. Pour off the liquor and wash the precipitate with *Aq. dest.* q. s. Boil down the liquors for ¼ hour, strain. Add gradually *Sol. Ammonia,* q. s. to precipitate the Quinia. Wash this till nothing alkaline is perceptible. Saturate what remains with *Sulphuric′* ℥iv. ʒij. diluted. Digest with *purified Animal Charcoal* ℥ij. strain. Wash the Charcoal thoroughly and evaporate the liquor cautiously to form crystals.—The Bark is exhausted by the acid and water, and the solution contains Sul′, Kinic′, and Quinia, with extractive and colouring matter. On adding the hydrated Oxide of Lead, the Sul′ combines with it and insoluble Sulphate of Lead is precipitated; the Kinic′ and Quinia remaining in the solution poured off. When Ammonia is added a Kinate of this alkali is formed, which remains in solution and the Quinia is precipitated. On the second addition of the Dil. Sul′. Disulphate of Quinia is formed and purified by means of the Animal Charcoal and recrystallization.

The E. C. directs *Yellow Bark* in coarse powder ℔j. to be boiled for one hour in *Water* Oiv. in which *Carbonate of Soda* ℥iv. has already been dissolved. Strain and express strongly through linen or calico; moisten the residuum with water and express again; and repeat this twice. (The colouring principles, extractive, and the Kinic acid are removed by the boiling alkaline solution.) Boil the residuum for half an hour in *Aq.* Oiv. acidulated with *Sul'* ʒij. strain, express strongly, moisten with Aq. and express again. Repeat the process twice with Aq. and Sul' ʒij. in divided portions. (The residual Quinia with some Cinchonia is all removed by combining with the Sulph'.) Concentrate the whole acid liquors to about a pint; when cool filter and add *Carbonate of Soda* ℥iv. (Sulphate of Soda is formed and remains in solution, impure Quinia being precipitated.) Collect the impure Quinia on a cloth, wash it slightly and squeeze out the liquor with the hand. Break down the moist precipitate in *Aq. dest.* Oj. add of Sul' nearly fʒj. heat to 212° and stir occasionally. (Sulphate of Quinia is formed.) If necessary add a little more Sul' Carb. of Soda and Aq. dest. Filter. Evaporate and crystallize, repeat the process with Animal Charcoal and pure crystals will be obtained if the details are minutely followed (*c*).

As the proportion of Quinia and of Cinchonia procurable from the different kinds of Bark is a subject of interest to both the practitioner and practical chemist, the author has borrowed the following tabular view from the Dispensatory of his friend, Dr. Christison; but, as he observes, the results are so discrepant, that it is scarcely possible that the same kinds of Bark have been examined by different experimentalists. The proportion of the two alkaloids in 1000 parts of the chief Barks of commerce are here given according to the analyses of four esteemed authorities.

	Crown-Bark.	*Cinch.*	*Quin.*
Von Santen	*Fine quills*	0·0	0·5
Soubeiran	*Fine quality*	12·3	trace
Michaëlis	*do.*	2·4	1·0
Goebel	*do.*	2·6	2·0
Von Santen	*Medium quills*	0·0	2·1
Soubeiran	*Low quality*	9·2	0·0

	Grey-Bark.	*Cinch.*	*Quin.*
Von Santen	*Fine quills*	24·33	0·0
Do.	*Medium quills*	27·3	0·0
Soubeiran	*Fine quality*	9·2	0·0
Michaëlis	*do.*	10·0	3·6
Goebel	*do.*	21·3	0·0
Michaëlis	*Inferior*	6·4	4·2

	Yellow-Bark.	*Cinch.*	*Quin.*
Von Santen	*Stripped quills*	0·0	15·0
Do.	*Stripped flat*	0·0	14·6
Soubeiran	*Quilled*	0·0	17·2
Michaëlis	*do.*	0·0	20·0
Goebel	*do.*	0·0	11·0
Soubeiran	*Flat.*	0·0	21·3
Michaëlis	*do.*	0·0	37·0
Goebel	*do.*	0·0	12·3

	Red-Bark.	*Cinch.*	*Quin.*
Von Santen	*Thick quills*	24·0	0·8
Soubeiran	*Fine quality*	6·1	11·5
Michaëlis	*do.*	4·2	8·3
Goebel	*do.*	8·4	5·2
Soubeiran	*Pale Red*	6 1	8·6
Von Santen	*Fine quills*	9·0	7·5
Do.	*Flat*	11·8	1·5

	Carthagena Bark.		
Von Santen	*Hard*	4·0	3·2
Goebel	*do.*	5·5	7·3
Von Santen	*Woody*	4·4	3·0
Goebel	*do.*	7·0	5·4

	Ash-Bark.		*Rusty-Bark.*	
	Cinch.	*Quin.*	*Cinch.*	*Quin.*
Von Santen	0·0	trace	12·4	0·0
Goebel	1·6	1·2	5·1	3·6
Michaëlis	1·6	10·4	6·3	3·6
Do. *thin quills*	—	—	0·0	1·0

Disulphate of Quinia, or Sulphate of Quinia, as it is considered by other chemists, is usually seen as a light flocculent mass of white, silky, slightly flexible, needle-shaped crystals, interlaced with each other, and grouped in small star-like tufts. They are without odour, but the taste is intensely bitter. Exposed to the air, the crystals effloresce, losing six parts of their water of crystallization. At 212°

they become luminous, especially if rubbed: at 240° they melt, losing two more equivalents of water, then become red, and at last ignite and burn away, leaving no residuum. According to Baup, they require 740 parts of cold and about 30 parts of boiling water for solution, giving a bluish tinge to the water. They require only 60 parts of Rectified Spirit at ordinary temperatures, and are very soluble in diluted acids, especially Dil. Sul′. This salt is composed of 2 Eq. Quinia+1 Eq. Sul′+8 Aq.=436.

Neutral Sulphate of Quinia, is readily obtained by adding to the Disulphate as much Sul′ as it already contains, by which its solubility is greatly increased, and in this form it is usually prescribed. By evaporation, rhombic crystals may be obtained, which are soluble in 10 parts of water at 60°, and will melt in their own water of crystallization at 212°. They consist of 1 Eq. Q.+1 Sul′+8 Aq.=274.

Tests. Disulphate of Quinia, from its high price, is apt to be adulterated with earthy and alkaline salts, as Sulphate of Lime, Gum, Sugar, Starch, Fatty matters of a crystalline nature, Sulphate of Cinchonia, Caffein, Salicin. "Totally dissolved in water, especially when acidulated. Quinia is thrown down by Ammonia. On evaporating, the liquid which remains ought not to taste of Sugar. 100 parts lose 8 or 10 parts of water when heated. It is entirely consumed by a red heat. It is turned of a green colour if Chlorine be first added to it, and then Ammonia." L. The Starch, Sulphate of Lime, and Fatty matters, will remain undissolved by the cold water. Sugar will be detected by its taste. Gum and alkaline earths will remain undissolved by Alcohol. Earthy impurities, moreover, will resist the action of heat. The E. test is intended to determine generally whether or not the salt is pure. "A solution of gr. x. in Aq. Dest. f℥j. and Sul′ ♏ij.—iij., if decomposed by a solution of Carb. Soda ʒiv. in two waters, and heated till the precipitate shrinks and fuses, yields on cooling a solid mass, which, when dry, weighs 7·4 gr. and in powder dissolves entirely in solution of Oxalic acid." Though sufficient for most impurities, this, as Dr. Christison states, will not detect Caffein or Sulphate of Cinchonia. But the first is too dear to be used. Salicin is turned red by Sul′, and cannot therefore be very easily employed; but the Sulphate of Cinchonia is probably frequently mixed. If the salt be dissolved in 40 parts of boiling water, most of the Sulphate of Quinia will be deposited on cooling, but the Sulph. Cinchonia will be retained, and may afterwards by evaporation be obtained in short rhombic prisms, or pearly scales: or precipitate the two with caustic Soda, and then redissolve the Sulphate of Quinia in Ether (*M. Calvert*); or "precipitate a solution of the suspected salt in water by Carb. of Potash, collect the precipitate, and boil it in Alcohol. The Cinchonia crystallizes as the liquor cools, while the Quinia remains in the mother liquor." (*p.*) The Spirit will also dissolve any Sugar that may have been added, and leave the Sulph. of Potash. (*Med. Gazette.*)

The E. C. give a test for ascertaining the quality of Yellow Bark which is similar in principle to that for ascertaining the purity of the Disulphate. "A filtered decoction of 100 grains in Aq. dest. f℥ij. gives with f℥j. of concentrated solution of Carbonate of Soda, a precipitate which, when heated in the fluid, becomes a fused mass, weighing when cold 2 grains or more, and easily soluble in solution of Oxalic acid."

Aricina or *Cusconin* is the alkali found in Cusco Bark, which is very similar in many of its properties to Cinchonia. M. Guibourt, indeed, says he could obtain only this alkali from the above Bark. Aricina is distinguished by the discoverers as containing 1 more Eq. of Oxygen than Quinia, as being soluble in Ether, and rendered of a green colour by Nit′. Dr. Pereira, from analogy in properties, inferred that the three alkalis are oxides of a compound base, which he calls Quinogen, and that Cinchonia is a Monoxide, Quinia a Deutoxide, and Aricina a Tritoxide of Quinogen.

Quinoidine, obtained by Sertuerner in the mother liquor of the first, has been proved by Liebig (*Lancet*, May, 1846) to be *Quinine* in an *amorphous* state, which if pure, may be used for the same purposes, or converted into a salt of any acid, as suggested by Mr. Bullock. Both Quinia and Cinchonia, when heated with Potash, yield *Quinoleine*, an oily volatile liquid, bitter and strongly alkaline, which is analogous to Conia.

The *Lactate* and *Valerianate of Quinia* have also been beneficially employed in suitable cases. The latter is lauded by M. F. Devay (Br. Retros. xi. 123) for its neurosthenic properties, and is said to be a more powerful antiperiodic than the Sulphate. The Citrate of Quinine and Iron, prepared in crystalline scales, is soluble in water without the addition of any acid, and is given in doses of gr. ij.—gr. v. in the same cases as Quinine.

Inc. Alkalis, Earths, and their Carbonates, &c.; Astringents.

Action. Uses. The Cinchona Barks are slightly Astringent, eminently Tonic and Antiperiodic; hence they are frequently prescribed to strengthen, in diseases of debility or in convalescence from acute diseases, but especially for arresting the accession of Intermittent and Remittent Fevers, and attacks of Periodic Neuralgia and of Rheumatism, either in the form of Powder, in doses of gr. x.–gr. xxx., in Infusion, Decoction, Extract, or Tincture: sometimes all three are united in one of the watery preparations. But as the properties depend chiefly on the alkalis, though partly also on the astringent principle, the former, especially in the form of the Sulphate of Quinia, have nearly superseded all the other preparations. A similar preparation of Cinchonia may no doubt be and is used for many of the same purposes. The Disulphate of Quinia, often converted to a Sulphate by a few drops of Sul′, is prescribed in doses of gr. j.—gr. v.; but scruple doses have been given in obstinate periodic attacks of Neuralgic pain or of Ague.

INFUSUM CINCHONÆ, L. E. D. Infusion of Pale (of any, E.) Cinchona.

Prep. Take powdered Pale or Grey *Cinchona Bark* (species according to prescription, E.) ℥j. boiling *Aq. dest.* Oj. (cold, by measure ℥xij. D. rub it up gradually with the water, D.) Macerate for 6 (4, E. 24, shaking it occasionally, D.) hours in a lightly-covered vessel. Strain (through linen or calico, E. Pour off the clear liquor, D.)

Action. Uses. Water dissolves the Kinates of the Cinchona alkalis and the Tannin; but as some of the alkalis remain behind united with the insoluble Colouring matters, this is necessarily a weak preparation. It is well suited as a Tonic from its lightness for delicate states of the stomach and constitution, in doses of f℥j.—f℥ij. every 3 or 4 hours.

DECOCTUM CINCHONÆ, E. Decoction of Bark.

Prep. Boil for 10 minutes *Crown, Grey,* or *Yellow Cinchona* ℥j. bruised, in *Aq.* f℥xxiv. Cool, filter, and evaporate to f℥xvj.

Action. Uses. By the continued action of boiling water, much of the active principle is extracted, but it is also deposited on cooling, in consequence of the Cinchonic Red uniting with the alkalis, and forming compounds insoluble in cold water. Both the Infusion and Decoction would be much improved by being made with Acidulated water, which would retain the alkalis in solution. This may be employed for the same purposes and in the same doses as the Infusion.

DECOCTUM CINCHONÆ CORDIFOLIÆ, L. Decoction of Yellow Bark.

Prep. Boil for 6 hours in a lightly-covered vessel powdered *Yellow Calisaya Bark* ʒx. in *Aq. dest* Oj. and while hot, strain.

The Dec. of Pale and of Red Bark are similarly prepared, and used as the Dec. Cinchonæ, E.

TINCTURA CINCHONÆ, L. E. D. Tincture of Bark.

Prep. Take bruised *Yellow Bark* (any species, E.) ℥viij. (℥iv. D.) *Proof Spirit* Oij. (wine measure ℔ij. D.) Macerate for 14 (7 D.) days, and strain. (Or much more expeditiously, and with less loss, prepare by percolation, the bark being in fine powder, E.)

Action. Uses. Prescribed as a Tonic with the Infusion or Decoction in doses of fʒj.—fʒiij. Proof Spirit being a good solvent of the active principles of the Cinchonas, especially if acting by percolation, an excellent Extract is yielded on distilling off the Spirit.

TINCTURA CINCHONÆ COMPOSITA, L. E. D. Compound Tincture of Bark.

Prep. Macerate for 14 (7, E.) days *bruised Yellow Bark* ℥iv. (℥ij. D.) *dried Orange-peel* (bitter, bruised, E.) ℥iij. (℥ß. D.), powdered root of *Aristolochia Serpentaria* ʒvj. (ʒiij. D.), chopped *Saffron* ʒij. (ʒj. D.), bruised *Cochineal* ʒj. (℈ij. D.) in *Proof Spirit* Oij. (by measure ℥xx. D.) Strain (express strongly and Filter. Or prepare by percolation, the bark being in fine powder, the same way as Comp. Tinct. Cardamom, E.)

Action. Uses. Stimulant Tonic; more agreeable from the presence of the stimulants; sometimes called Huxham's Tincture of Bark. Used as the above in doses of f℈j.—f℈iv.

Pharm. Prep. Vinum Gentianæ, E. Mistura Ferri composita, D.

Prof. Donovan has (Pharm. Journ. iv. 125) recommended a Syrup of Cinchona, of which the active ingredients are Dikinate of Quinia, with the natural Tannin of Bark. It promises to be an efficient and pleasant preparation.

EXTRACTUM CINCHONÆ, E. Spirituous Extract of Bark.

Prep. Take any of the varieties of *Cinchona*, especially the Yellow or Red, in fine powder ℥iv. *Proof Spirit* f℥xxiv. Prepare by percolation, and evaporate in an open vessel to the proper consistence over a vapour-bath.

Action. Uses. The active principles having been extracted by, and again deposited from the Spirit, this Extract forms an efficient preparation in the form of pills in doses of gr. v.—gr. xx.

EXTRACTUM CINCHONÆ CORDIFOLIÆ, L. Extract of Yellow Bark.

Prep. Take bruised Yellow *Cinchona Bark* ℥xv. and *Aq. dest.* Civ. boil down in Cj. of the Aq. to Ovj., while hot, strain. In the same way boil the bark four times, and strain. Then mixing all the liquors, evaporate to a proper consistence.

EXTRACTUM CINCHONÆ LANCIFOLIÆ, L. EXTRACTA CINCHONÆ, D.

Prepared as *Ext. Cinch. Cord.* L. The D. process is exactly the same, using of *Pale Bark* coarsely powdered ℔j. *Aq.* by measure ℔vj. (This extract should be kept soft for pills, and *hard* that it may be reduced to powder, D.)

EXTRACTUM CINCHONÆ OBLONGIFOLIÆ, L. Extract of Red Bark.

Prepared in the same way as *Ext. Cinch. Cord.*

Action. Uses. The watery extracts are convenient for exhibition in pills in doses of gr. v.—℈ß. They are best prepared in vacuo; but in efficiency are inferior to the Spirituous Extracts in the same doses.

UNCARIA (sometimes called Nauclea) GAMBIR, a plant of this family and of the tribe Cinchonaceæ, a native of the Malayan Peninsula and of the Indian Archipelago, is interesting as yielding large quantities of the kind of Catechu known by the names of Terra Japonica and of Square Catechu, and which in Indian commerce is called *Gambeer*. It is a powerful astringent, much used in tanning, and in medicine as a substitute for the Catechu of the Acacia.

VALERIANEÆ, *Dec.* Valerianads.

Herbaceous stems, often with perennial root-stocks. Stem-leaves opposite, radical ones clustered. Calyx tube adherent to the ovary; limb superior, 3—4 toothed, or pappose. Corolla inserted into the top of the ovary, tubular, 3—5 lobed, unequal, or irregular, often spurred or gibbous at the base. Stamens 1—5, free, inserted into the tube. Ovary 3-celled, 2 usually abortive. Ovule single, pendulous. Style simple. Stigmas 2 or 3. Fruit dry, indehiscent. Embryo straight, without albumen. Radicle superior.—Allied to Dipsaceæ and these to Rubiaceæ. The perennial species secrete a volatile oil possessed of strong odour and stimulant properties. They inhabit temperate parts of the world.

VALERIANA OFFICINALIS, Linn. L. E. D. Var. *Sylvestris*, L. D. Radix, L. D. The Root, D. of Wild Valerian. *Triandria Monogynia*, Linn.

Some of the Valerians have been used in medicine from the earliest times. Dioscorides describes three kinds of Nard or Valerian besides the φου. The Spikenard of the ancients, *Nardostachys Jatamansi* (Him. Bot. t. 54.) a produce of the Himalayas, is still highly esteemed in the East. *Valeriana celtica* and *Saliunca* are even imported by the Red Sea from Austria for perfuming their baths. (*v.* Illustr. Himal. Bot. p. 242.) *V. Dioscoridis* is supposed to be the φου of that author, and the officinal or wild Valerian was no doubt early introduced as a substitute for it.

Root perennial, tuberous. Stem 2 to 4 feet high, smooth, furrowed. Leaves all pinnate, or pinnately cut; leaflets lanceolate-dentate, in 7 to 10 pairs, terminal one very little, if at all, larger than the others. Inflorescence a corymb, becoming at length somewhat panicled. Bracts ovate-lanceolate. Calyx-limb involute during flowering, then unrolled into a deciduous pappus, consisting of many plumose setæ. Corolla roseate; tube funnel-shaped, gibbous at the base; limb 5-lobed. Stamens 3. Fruit smooth, compressed, 1-celled, 1-seeded, crowned by the limb of the calyx expanded into a feathery pappus.—Ditches and damp places throughout Europe.—E. B. 698. Esenb. and Eberm. t. 254.

The tuberous root-stock with its numerous radicles is the officinal part; that which grows in dry pastures is more fragrant, and that of the wild more so than that of the cultivated plant, whence this is directed to be used in the L. P. It has a bitter acrid taste, and a powerful penetrating odour, which is considered disagreeable by most people. It consists of Volatile Oil about 1 per cent, Resin 6, Resinous Extractive 12·5, Extractive 9·4, and of Woody Fibre 71 per cent. (*Trommsdorff.*) The Oil, upon which the properties of Valerian depend, is of a greenish colour, has a strong penetrating odour, and a camphoraceous, aromatic taste. When fresh, it contains an Oil, *Valerole*, which is crystallizable, and some passes into Valerianic acid by being oxygenated in the air. It also contains a Hydro-Carbon, *bornéene* ($C_{10}H_8$), identical with the Oil obtained from Borneo Camphor, and finally a Camphor which is identical with the Borneo Camphor. (*Gerhardt.*) When the root is distilled with water, there comes over with the Oil an acid fatty matter, *Valerianic acid.* This is an oily fluid, with a disagreeable smell; Sp. Gr. 0·944, boiling at 270°. It forms soluble salts of a sweet taste with bases. This acid may also be produced by the oxidation of the Hydrated Oxide of Amyle, or Oil of Potato Spirit. As there has been a great demand for this acid of late for making Valerianate of Zinc, the Messrs. Smith (P. J. v. 110) have given a process for obtaining it. This consists in boiling the root with a little Carb. Soda (℥j. for lbj. of root). To the strained liquid add Sul (fℨij. for each lb. of root), and distil. When about ¾ have passed over, neutralize the distillate with Carb. Soda (ℨij. for lbj. of root). Concentrate the Valerianate of Soda, decompose it with

Sul′, and obtain the Valerianic′ set free, by means of a separate distillation. The active properties of Valerian Root may be extracted by alkalized water, Spirit, or Ammoniated Spirit.

Action. Uses. Diffusible Stimulant and Antispasmodic. The Volatile Oil is recommended by many. The Valerianate of Zinc in doses of gr. j.—gr. ij. has been employed of late as an Antispasmodic Tonic.

D. Of the powdered root, gr. xx.—gr. xl. Of the Volatile Oil, ♏iij.—♏v.

INFUSUM VALERIANÆ, D. Infusion of Valerian.

Prep. Digest for 1 hour the root of *Valeriana sylvestris* in coarse powder ʒij. in boiling *Aq.* by measure ℥vij. When cold, strain.

Action. Uses. Moderate Stimulant in doses of f℥j.—f℥ij.

TINCTURA VALERIANÆ, L. E. D. Tincture of Valerian.

Prep. Take bruised (powdered, D.) *Valerian root* ℥v. (℥iv. D.), *Proof Spirit* Oij. (by measure ℔ij.) Macerate for 14 (7, D.) days. Strain. (Proceed by percolation or digestion, as for Tinct. Cinchonæ, E.)

Action. Uses. Stimulant adjunct to draughts in doses of fʒß to fʒiv.

TINCTURA VALERIANÆ COMPOSITA, L. (AMMONIATA), E. D. Ammoniated Tincture of Valerian.

Prep. Take *Valerian root* bruised ℥v. (℥ij. D.) *Aromatic Spirit of Ammonia* Oij. (Spirit of Ammonia Oij. E., by measure ℔j D.) Macerate for 14 (7, D.) days, and strain. (Proceed by percolation, as directed for Tinct. Cinch. E.

Action. Uses. Antispasmodic, and more Stimulant from presence of Ammonia, may be given in doses of fʒß.—fʒij.

COMPOSITÆ, *Adans.* Synanthereæ, *Auct.* Asteraceæ, *Lindley.*

Herbs usually perennial, sometimes under-shrubs, rarely trees; with either watery or milky juice. Leaves alternate or opposite, sometimes whorled, simple, entire, or variously cut, in some compound, without stipules; petioles sometimes furnished with stipulary auricles. Flowers complete, or by abortion unisexual, or truly so, in heads, supported by a many-leaved involucre, and seated on a flat or conical receptacle, with bracts or paleæ, or each included in an involucre, and collected into a common head. Calyx-tube adherent to the ovary; limb wanting, rim-like, toothed, scale-like, or divided into a pappus. Corol inserted into the upper part of the tube of the calyx, monopetalous, limb regular, 5-fid, valvate, or irregular or strap-shaped. Stamens 5, inserted into the tube of the corol, and alternate with its lobes; filaments jointed towards the apex; anthers united into a tube; appendix above, and 2 setæ below. Ovary single-celled, with a single erect ovule. Style 1. Stigmas 2. Fruit (achænium) indehiscent, dry. Embryo without albumen, erect; radicle below, turned towards the hilum.—The Compositæ are extremely numerous in species, some of which are found in almost all parts. They are generally devoid of active medical properties. A few are cultivated as articles of diet, and others yield fixed oil, which is stored up in their seeds. The medicinal plants are bitter and stimulant, the latter from the presence of volatile oil. A few, as some of the species of Lactuca, are narcotic, others are acrid and irritant.

A. CYNAROCEPHALÆ.—Florets hermaphrodite, all tubular, with 5, or rarely 4, equal teeth, with a convex or hemispherical top. Stigma jointed to the style.—Several of these are bitter, as *Centaurea Centaurium;* a few are a little odorous, as *Centaurea moschata.* The Safflower, *Carthamus tinctorius,* is valued on account of the colouring matter procurable from its florets.

CNICUS BENEDICTUS, *Linn.* Folia, D. The Leaves of Blessed Thistle.

The name of this plant indicates the esteem in which it was held.

An annual, with angled, branched, woody stem. Leaves amplexicaul, semi-decurrent, nearly entire, irregularly pinnatifid, or toothed; apices of lobes prickly. Heads solitary, terminal, enveloped in leaf-like bracts. Involucre ovate; scales extended into a spiny pinnated appendage. Florets yellow, about 20 to 25 in number, those of the ray sterile, slender. Fruit small, longitudinally striated, with a lateral scar. Pappus triple, the outer being the crenated margin of the fruit, the intermediate of ten long and the inner of ten short setæ, all alternating with each other.—South of Europe, the Levant, Persia.—St. and Ch. 128. Nees von E. 223.

This plant, devoid of odour, is possessed of considerable bitterness. The Leaves contain Gum-Resin, Bitter Principle, a little Volatile Oil, several Salts, &c.

Action. Uses. Tonic in the form of cold infusion (ʒiv.—℥j. in Aq. Oj.). Diaphoretic. In large doses its decoction causes vomiting; hence given to assist the action of emetics.

ARCTIUM MINUS (*Schkuhr*) (*A. Lappa*, Linn.), D. Radix et Semina, D. Root and Seeds (Fruits) of Lesser Burdock.

The Burdock has been long employed in medicine. The Lappa of Linnæus has been divided into two or three species, of which *A. minus* is most common, and the medicinal species or variety. It is called Lappa minor by De Candolle and others.

Stem erect, 3 feet high, branched, furrowed, leafy. Leaves stalked, large, cordate-ovate, undulated, the radical ones very large. Heads many-flowered, racemose; involucre globose, imbricated, with scales terminating in hooked points, connected by a cobweb-like down, inner ones coloured, subulate, rather abruptly shorter than the florets. Florets with their anthers and stigmas purple. Receptacle flat, with rigid subulate scales. Fruit compressed, oblong. Pappus short, pilose, distinct.—Waste places in Europe.—Nees and Eberm. 227.

The root is about the thickness of the thumb, tapering, fleshy, without odour, but having a sweetish rather astringent taste. It contains *Inuline*, Bitter Extractive, and Potash Salts. The fruits are bitter, a little acrid, and aromatic.

Action. Uses. Mild Diaphoretic, Alterative, and Diuretic; hence a Decoction (℥ij. of the root to Aq. Oij.) has been prescribed as a substitute for Sarsaparilla in some cutaneous affections, &c. The fruits or seeds are considered diuretic.

B. CORYMBIFERÆ. Florets of the disk all tubular, forming a level top; marginal florets often ligulate. Stigma not joined to the style.

TUSSILAGO FARFARA, *Linn.* Folia et Flores, D. Leaves and Flowers of Coltsfoot.

This is the βηχιον of the Greeks, so called from having been used in coughs, and is named *Coltsfoot* and *pas d'ane* from the form of its leaves.

Coltsfoot has a creeping root-stock. Leaves, which make their appearance after the flowers, on channeled footstalks, roundish, cordate, angular, sharply toothed, of a glaucous green above, cottony beneath. Flower-heads appear in early spring, of a bright yellow colour, solitary, drooping while in bud, erect when in flower and in seed; flower-stalk covered with smooth, scale-like, reddish bracts, many flowered, heterogamous. Florets of the ray female, narrow, ligulate, ray spreading. Florets of the disk male, tubular, with a campanulate 5-cleft limb. Receptacle naked. Involucre of one row of scales, oblong, with membranous margins. Styles of the disk inclosed, abortive; of the ray bifid, with taper arms. Achænium of the ray oblong, cylindrical, smooth, of the disk abortive. Pappus of the ray in many rows, of the disk in one row, consisting of very fine setæ.—Chalky soils in England, found in many parts of Europe, extending to Persia and the Himalayas.—St. & Ch. t. 20.

Action. Uses. This plant is mucilaginous and slightly bitter, and may be employed as a Demulcent possessed of a little tonic property. It may be prescribed in the form of an Infusion or Decoction (one or two ounces to a pint of water) *ad libitum.*

INULA HELENIUM, *Linn.* L. D. Radix, L. D. Root of Elecampane.

Elecampane has been prescribed since the time of Hippocrates.

Root perennial, thick, elongated, brownish externally, white in the inside. Stem erect, 3—4 feet high, round, leafy. Leaves large, cordate-ovate, acute, stem-clasping, unequally toothed, downy beneath; radical leaves petioled, ovate-oblong. Flower-heads few together, or solitary, large, bright yellow. Involucre imbricated in many rows, outer scales ovate, inner obovate. Florets of the ray female, ligulate, 3-toothed, subtubular; those of the disk hermaphrodite, tubular, 5-toothed. Anthers with two bristles at the base. Receptacle flat, reticulated. Achænia quadrangular, smooth. Pappus uniform, in one row, composed of roughish setæ.—Moist pastures throughout Europe, flowers in July and August. St. and Ch. 49.

The root when chewed, tastes first glutinous and then bitter, aromatic, and finally a little pungent. It is generally cut into slices, for the convenience of drying and of preserving. It contains Bitter Extractive 36·7, a peculiar principle which has been named *Inulin* 36·7, *Helenin*, or a neutral crystalline principle in some respects resembling Camphor, 0·3, Wax 0·6, acrid Resin 1·7, Gum 4·5, with Lignin, Albumen, and salts with a base of Potash, Lime, and Magnesia. *Inulin*, which has been found in many other roots, and has received different names, is a white amylaceous substance, something like Starch, but differing in a part being precipitated, on cooling from its boiling watery solution, in being rendered of a yellow colour by Iodine, and in being a little soluble in boiling Alcohol.

Action. Uses. Stimulant Tonic, Expectorant, and Diaphoretic; has been prescribed in Dyspepsia and in Chronic Catarrh.

D. Powder, ℈j—ʒj. Of the Decoc. or Inf. (℥ß. to Aq. Oj.) f℥iß.

Pharm. Prep. Conf. Piperis nigri, L. A constituent of many preparations on the Continent.

In the article Senna (p. 361) it is mentioned that a very good substitute for Senna is afforded by one of the Compositæ. This is BERTHELOTIA LANCEOLATA var. *indica*, Dec. (Prod. v. p. 376), of which the leaves, as ascertained by Dr. Falconer, are those called *ra* and *rae-Suna* by the natives of north-west India. They are mentioned in the author's Illustr. of Himal. Bot. p. 319, having been given to him as those of *Salvadora indica*, Royle, which they a good deal resemble, and are produced in the same arid tract of country extending from the banks of the Jumna towards central India. Dr. F. pronounces the leaves to be an excellent substitute for Senna, and to be remarkable for growing with their edges vertical, and for having both sides covered with stomata.

ANTHEMIS NOBILIS, *Linn.* L. E. D. Flores (D.) simplices, L. Flowers, E. *Chamæmelum.* Chamomile or Camomile.

The name ἄνθεμις occurs first in Theophrastus, and that of χαμαίμηλον in Dioscorides. Anthemis Chia is supposed to be the plant of Dioscorides. Others have been substituted, as *Matricaria suaveolens* in India; *M. Chamomila* was at one time distinguished as Common Chamomile, and another, called Noble or Roman Chamomile. The last is the present *Anthemis nobilis.*

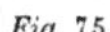

Fig. 75.

Roots perennial, with long fibres. Stems in a wild state procumbent, when cultivated, erect, about a foot long, much branched, leafy, round, furrowed, hollow. Leaves doubly pinnate, leaflets linear, subulate, slightly downy. Flower-heads terminal, solitary, with a convex yellow disk. Rays composed of hermaphrodite, tubular, 5-toothed florets. Rays white, reflexed, or spreading, formed of female florets in one row. Receptacle conical (fig. 2) with membranous scales. Involucre imbricated in a few rows, scales obtuse, hyaline at the margin. Fruit obtusely tetragonal, smooth, crowned with an obsolete margin, without pappus.

Two varieties are known. *a.* Flore simplici, fig. 75. *b.* Flore pleno, Double Chamomile, in which the florets of the disk are converted into white ligulate florets. (*v.* fig. 75, 1.)—Indigenous in gravelly places, also in other parts of Europe. Flowers in July and August. Cultivated at Mitcham in Surrey.—E. B. 980. St. and Ch. 111. 38.

The whole plant has a strong but pleasant odour, the taste is purely bitter with a little aroma. These properties are most conspicuous in the florets of the disk, and therefore the simple flowers are preferable to the double. The active principles depend on the presence of a Volatile Oil, Bitter Extractive, and a little Tannin. Both water and Alcohol take up their active properties.

Action. Uses. Stimulant, Tonic, Febrifuge, in the following forms:

INFUSUM ANTHEMIDIS, L. E. Inf. Chamæmeli, D. Infusion of Chamomile.

Prep. Macerate (infuse, E.) in a lightly covered vessel for 10 minutes (20, E. 24 hours, D.) *Chamomile* ℥v. in boiling *Aq. dest.* Oj. Strain.

Action. Uses. Tonic, especially in the form of cold Infusion, in doses of f℥jß. The hot Infusion is sometimes employed to assist the action of emetics.

DECOCTUM CHAMÆMELI COMPOSITUM, D. Decoction of Chamomile.

Prep. Boil for a little, *dried Chamomile flowers* ℥ß. *Sweet Fennel Seeds* ℨij. in *Aq.* ℔j. by measure. Strain.

Action. Uses. Similar to the Infusion, and not more useful, except for fomentations, in doses of f℥iij.—f℥xij. to assist vomiting.

EXTRACTUM ANTHEMIDIS, EXT. CHAMŒMELI, D. Extract of Chamomile.

Prep. Boil *Chamomile* ℔j. in *Aq.* Cj. down to Oiv. While hot, filter, and in the vapour-bath evaporate to the due consistence.

Action. Uses. Bitter and simply Tonic, the stimulant Oil being dissipated during the evaporation. Given in doses of gr. x.—gr. xx.

OLEUM ANTHEMIDIS, L. E. Oil of Chamomile.

By distilling the Flowers with water a Volatile Oil is obtained, which is of a yellowish-brown colour, of a strong odour, and pungent taste. It exists in the proportion of less than 1 per cent.

Action. Uses. Tonic, Stimulant, and Antispasmodic. May be added to the Extract or to pills in doses of ♏j.—♏v.

PYRETHRUM. Radix, L. D. Root, E. of ANACYCLUS (*Anthemis, Linn.* L. D.) PYRETHRUM, *Dec.* E. Pellitory of Spain.

This root was known to Dioscorides (πύρεθρον), and is still employed in Eastern medicine by the name *akurkurha.* The plant is a native of the north of Africa, whence it has been introduced into the south of Europe. Hayne believes that the root is yielded by a nearly allied species, which he calls *A. officinarum,* ix. t. 46, and which he found cultivated in Thuringia.

Anacyclus Pyrethrum has a long fusiform root. Stems numerous, procumbent, branched, pubescent. Radical leaves spreading, petiolated, rather smooth, pinnatifid; the segments pinnated, with linear tubulate lobes; stem leaves sessile. Branches one-headed. Heads many-flowered. Involucre in few rows, short, somewhat cup-shaped, scales lanceolate, pointed, brown at the edges. Receptacle convex, with oblong, obovate, obtuse paleæ. Florets of the ray female, sterile, white above and purplish beneath; of the disk, yellow, tubular, with 5 callous teeth. All the corols with an obcompressed, 2-winged tube without appendages. Style of the disk with exappendiculate branches. Achænium flat, obcompressed bordered with broad entire wings. Pappus short, irregular, tooth letted, somewhat continuous with the wings on the inner side. *Dec.*—Desf. Fl. Atl. ii. 287. Nees and Eberm. 244. St. and Ch. iii. 97.

The root, as described by Desfontaines, in its fresh state is fusiform and fleshy, about the thickness of the finger, brownish-coloured externally and white within. When handled in this state, it produces first a sensation of cold, soon followed by heat. It is without odour, but has an acrid pungent taste, and causes a copious flow of saliva. It is said to be imported from the Levant. The French obtain it from Africa. It is cultivated in Thuringia and at Magdebourg. The active principle is soluble both in Alcohol and Ether, and appears to be a Volatile Oil, which adheres with tenacity to its Resin (*Pyrethrin*) and Fixed Oil. It also contains Inulin, Gum, a little Tannin, Colouring matter, various Salts, and Ligneous fibre.

Action. Uses. Irritant, Sialogogue. Sometimes used to relieve Toothache, or as a Masticatory in Palsy of the Tongue and relaxation of the Uvula.

Artemisia, *Linn.*

Heads discoidal, homogamous, or heterogamous. Florets of the ray in one row, usually female and toothed, with a long bifid protruding style; of the disk 5-toothed, bisexual, or by the abortion of the ovary, sterile or male. Involucral scales imbricated, dry, scarious at the edge. Receptacle without paleæ, flattish or convex, naked or fringed with hairs. Achænia obovate, bald, with a minute epigynous disk. *Dec.*

The species of Artemisia extending from European to tropical countries, are most of them remarkable for their strong odour and bitter taste, and have been employed in medicine from the earliest times, as *A. Abrotanum* or Southernwood, *A. vulgaris* or Mugwort, &c. *A. Dracunculus* or *Tarragon* is employed as a condiment in Europe.

Artemisia Moxa, *Dec.* (A. chinensis, *Linn.* and A. indica, *Willd.*) D. Moxa Weed.

The Moxa of China was supposed to be formed of a soft down formed on the lower surface of the leaves of the species quoted in the D. P.; but Dr. Lindley states that it is prepared from the woolly leaves of *A. Moxa*, Dec. The inflammable cones or cylinders called Moxas are prepared in Europe from pith, cotton, &c.; these are now seldom employed, though they were at one time preferred to the actual Cautery, from acting more slowly as powerful counter-irritants, in some painful and spasmodic diseases, as well as in affections of the joints and viscera. In the Himalayas the tomentum of *Chaptalia gossypina* is used as amadou, and is applicable to the same purposes. (Himal. Bot. p. 247.)

ARTEMISIA SANTONICA, *Woodv.* D. A. maritima, *Linn.* var. β. *suaveolens*, Dec. *Semen Santonicum.* Wormseed.

This substance has long been employed as an Anthelmintic, being intended for the Αψινθιον σαρδονιον of Dioscorides, the Semen sanctum and Santonicum, *Sheeha* of the Arabs. The substance would be more correctly denominated *Cacumina*, or floral summits, than seeds; for it consists of fragments of peduncles, involucres, and half-blown flowers stripped off the tops of some little-known species of Artemisia. Esenbeck and Ebermaier have figured *A. judaica*, *A. Contra*, *A. pontica*, and *A. glomerata*, Sieb., which is now *A. Sieberi* of Besser. This, according to Batka, produces the substance called *Semen-Contra* and *Semencene*, imported from Barbary and Aleppo as a vermifuge. That from the latter is called the Levant and Alexandrian kind, distinguished by being of a strong aromatic odour and bitter taste, of a greenish-colour, and smooth, while the former is described as greyish and downy. This kind is said by some to be produced by *A. Sieberi*, and the other by *A. Contra*, a native of Asia Minor and the Oriental region.

The properties, according to Soubeiran, depend upon a Volatile Oil and a peculiar principle which has been named *Santonine*, of the nature of the concrete volatile oils, which may be separated in brilliant crystals, is without odour, and nearly insipid from its sparing solubility. The alcoholic solution is very bitter, and, like acids, forms salts with bases. It is a decided vermifuge in doses of from 4 to 6 grains, and is recommended to be given with Sugar.

Action. Uses. Anthelmintic, in doses of from gr. x. — gr. xxx. night and morning.

ABSINTHIUM, L. E. D. Summitates Florentes, D. Flowering tops. Herb. E. of ARTEMISIA ABSINTHIUM, *Linn.* *Absinthium officinale et vulgare*, Auct. Wormwood.

The Αψινθιον of the Greeks is corrupted into *Afsunteen* in the East, and species of Artemisia are substituted for the present plant.

Root ligneous, branched. Stems numerous, bushy, furrowed, leafy, the whole plant covered with close silky hairiness. Leaves alternate, silky (tripinnatisect, *Dec.*), in many deep, lanceolate, obtuse segments; lower ones on long, the upper on short and broad footstalks. Floral leaves simple. Flower heads on leafy, clustered panicles, drooping, hemispherical, heterogamous; the outer scales of the involucre linear, silky; inner roundish, scarious. Florets of a pale yellow, the outer row female. Styles deeply cloven. Receptacle convex, covered with silky hairs.—Waste ground in various parts of Europe and the north of Asia.—Eng. Bot. t. 1230. St. and Ch. ii. 58. Should be collected in July and August, when in flower.

The dried herb or the flowering top has a greyish silky look, is remarkable for its disagreeable though somewhat aromatic odour, and for its intense bitterness, whence its name has passed into a proverb.

Its properties are imparted to water, spirit, and wines. Analysed by Braconnot, it yielded, in 100 parts, Volatile Oil of a dark-green colour, upon which the odour depends, 1·5, Bitter Azotized Extract 30, very Bitter Resin 2·5, Green Resin 5, with Chlorophyll, Albumen, Salts, and among them Absinthate of Potash, which when the plant is burnt, is changed into Carbonate of Potash. This was long called Salt of Wormwood. The bitter principle has been separated by M. Righini, first by making a spirituous extract, and then by acting on it with Charcoal. It acts with great power, and without excitement. The Essential Oil ($C_{20} H_{16} O_2$) as ascertained by M. Lablanc, in its impure state, is of a dark-green colour, which begins to boil at 356° F. Its boiling point rises to 401° as it coagulates. It may be distilled off Quicklime, and when thus purified, has a fixed boiling point of 401°; is acrid, has a penetrating smell, with Sp. Gr. ·975 at 75° F.

Action. Uses. Aromatic, Bitter, and Tonic; is added to some *liqueurs;* employed in Dyspepsia; is absorbed into the system. Anthelmintic, hence its name of Wormwood.

EXTRACTUM ARTEMISIÆ ABSINTHII, D. Extract of Wormwood.

Prepare as simple Extracts.

Action. Uses. Bitter Tonic, without the stimulant properties of the Oil, in doses of gr. v.—℈j.

TANACETUM VULGARE, *Linn.* D. Folia. Leaves of Common Tansy.

Tansy is a plant of Europe, which has long been employed in medicine.

Stems about two feet high, erect, leafy, smooth. Leaves of a dark green, pinnatifid, rachis winged, and deeply serrated, as well as the leaflets, decurrent, serrated. Heads in a terminal corymb. Florets as in Artemisia, of a golden yellow, the marginal florets scarcely apparent, and often wanting. Involucre hemispherical, imbricated. Receptacle convex, naked. Achænium sessile, oblong, angular, with a large epigynous disk. Pappus formed of a membranous, quadrangular, entire, but minute crown.—Roadsides and waste places in Europe, but also cultivated.—E. B. t. 1229. St. and Ch. iii. 116.

Tansy has a strong and penetrating odour; the taste is bitter, aromatic, and camphoraceous; but the odour diminishes in drying. Its properties depend upon a Bitter Extractive, or rather Resinous matter, and a Volatile Oil of a yellow colour and a strong Tansy odour, which has an acrid and bitter taste.

Action. Uses. Stimulant, Tonic, Antispasmodic, Anthelmintic, in infusion (℥ij.—Aq. Oj.); or the Oil may be added to vermifuge powders, or its infusion used as an enema against lumbrici.

ARNICA MONTANA, *Linn.* Flores. Folia et Radix, D. Mountain Arnica. Mountain Tobacco.

Arnica has long been employed in medicine, though there is no proof that it was known to the Greeks, as inferred by Mathioli.

Several dissertations have been written upon its medical virtues, from that of Lamarche, 1719, to the present time.

Rootstock horizontal, with numerous radicles. Stem cylindric, 2 feet high, a little hairy, bearing 1—3 heads of flowers. Leaves opposite, entire, oblong, obovate, 5-nerved, the radical usually 4 in number, the cauline smaller, in one or two pairs. Flower-head large, erect, or drooping, of a golden-yellow colour. Involucre campanulate in two rows, with linear-lanceolate, equal scales, rough with glands. Receptacle fringed, hairy. Florets many, of the ray in one row, female, ligulate, of the disk bisexual, tubular, 5-toothed. Tube of the corolla shaggy, sometimes rudiments of sterile stamens remaining in the ligulæ. Style of the disk with long arms covered by down. Achænium somewhat cylindrical, tapering to each end, somewhat ribbed and hairy. Pappus in one row of close, rigid, rough hairs.—A native of the cool parts of Europe from the sea-coast to the snow-line.—Fl. Dan. t. 63. Esenb. and Eberm. t. 239.

This plant, when fresh and bruised, has rather an agreeable odour, which is apt to excite sneezing; the taste of the leaves and flowers is bitter and pungent, of the root-stock bitter and acrid. Some mountaineers smoke it like Tobacco. The properties have been supposed to depend on Volatile Oil, Acrid Resin, Bitter Extractive (*Cytisine?*) and, according to Dr. A. T. Thompson, on Igasurate of Strychnia or of Brucia.

Versmann, having been induced by Dr. Pfaff to analyse it again, finds that the aqueous infusion of the flowers of Arnica reacts as an acid, has at first a bitter, and afterwards a strong acrid flavour, depending on the presence of Gallic acid. Sol. of Gelatine renders an infusion of Arnica very turbid, and with Chloride of Iron yields a black colour, which when largely diluted becomes green. On the addition of Magnesia or its Carbonate, an intensely green colour is produced after a few hours. He was unable to ascertain the presence of Strychnia, and, indeed, considered its absence as completely proved by his experiments. (P. J. iv. 238.)

Action. Uses. Acrid Stimulant; Irritant of the Digestive Canal; Diaphoretic, and Stimulant of the Nervous system; and as such is prescribed in nervous affections requiring such treatment; also in Gout and Rheumatism; and is much esteemed for its power of discussing tumours and the effects of bruises (hence called *Panacea lapsorum* by Fehr.) when applied in the forms of Cataplasm or of the Tincture, as indicated by Scopoli.

D. Of the Powder, gr. v.—gr. x. Of the Infusion (ʒiv. in Aq. Oj.) f℥iß. Of the Tincture (℥ij. of the Root in Proof Spirit f℥xvj.), ♏x.—fʒß., or applied externally.

C. Cichoreæ.—Florets all ligulate. Style cylindrical above and pubescent as well as its long obtuse branches. Stigmatic lines prominent, narrow.—The Cichoreæ abound in milky juice, which is bitter-tasted and sometimes narcotic. By blanching, some become edible as salads.

Cichorium Intybus, or Wild Succory, which is indigenous in waste places all over Europe, is extensively cultivated on account of its root, which is much used as a substitute for, and as an addition to, Coffee. The medical properties of the plant are considered to be nearly the same as those of Taraxacum. It is used to adulterate Coffee, but is itself often much adulterated. (*v.* P. J. iv. 119.)

TARAXACUM, L. E. D. Radix, L. et Herba, D. Root, E. of TARAXACUM DENS LEONIS, *Desf.* E. (Leontodon Taraxacum, *Linn.* L. D.) Dandelion.

Dandelion, being indigenous in Europe, has long been employed in medicine.

Root spindle-shaped, milky-juiced. Leaves numerous, radical, runcinate, glabrous, of a bright shining green. Scapes 1 or more, erect, brittle, with a single head of flowers, which expand in the morning, and are of a golden-yellow colour. Involucre double, external scales spreading or reflexed, internal ones in one row, erect, without callous tips. Receptacle naked. Achænium oblong, striated, muricate at the apex, terminating in a long beak. Pappus hairy, in many rows radiating so as to form a light globe.—Fields and waste places throughout Europe, and extending even to the Himalayas.—E. B. t. 539. St. and Ch. 5.

The leaves, when young, are blanched and used as salad in some parts of Europe: their properties necessarily vary at different periods of growth. The sensible properties of the milky juice are said to be greatest just before inflorescence. The juice expressed from the bruised roots was found both by Mr. Houlton and by Mr. Squire to be of a watery nature in March, but towards the end of summer to be thick and cream-like, and bitter in taste. Mr. Squire, moreover, found that in November and December 4 lbs. of the juice yield 1 lb. of Extract; from March to May, from 6 to 9 lbs. In June, July, and August, from 6 to 7 lbs. are required to yield the same quantity of Extract. Hence it is evident, that, it is during November and December it is most abundant in solid ingredients, upon which its medical properties probably depend. Geiger pronounces the juice to be most bitter in midsummer, and that in the spring and close of autumn it is sweetish, which Mr. Squire ascribes to the effects of frost. He, moreover, says that the juice contains Gum, Albumen, Gluten, an Odorous principle, Extractive, and a peculiar crystallizable Bitter principle, soluble in Alcohol and water.

The most efficient mode of prescribing it is in the form of the inspissated juice. The root is sometimes roasted and used as a substitute for, or its dried powder mixed with ground, Coffee.

Action. Uses. Aperient, Deobstruent, Alterative, especially in affections of the Liver, in chronic cutaneous diseases, &c.

DECOCTUM TARAXACI, E. D. Decoction of Taraxacum.

Prep. Take fresh *Taraxacum* herb and root ℥vij. (℥iv. D.) *Aq.* Oij. (℔ij. D.) Boil together down to Oj. (℔j. D.) (Press out the liquor, D.) Strain.

Action. Uses. The bitter principle being exhausted by the boiling water, while the Caoutchouc is coagulated, this may be used in some cases requiring alterative treatment in doses of f℥iß.

EXTRACTUM TARAXACI, L. E. D. Extract of Taraxacum.

Prep. Prepare as Extr. Gentian, L. Take fresh root of *Taraxacum* ℔j., boiling *Aq.* Cj. Proceed as for Extr. Poppy Heads, E. (Proceed as directed for simple Extracts, employing the herb and root of *Taraxacum*, D.)

Action. Uses. This is the most commonly employed Extract, in doses of gr. x.—ʒß. It should be bitter in taste, and of a brownish colour. But the inspissated juice of the roots pressed out and evaporated spontaneously by the action of dry air, or in vacuo, is the best form for exhibition. The inspissated juice is well prepared by Mr. Squire, as well as by Mr. Hooper, and by the latter also in the form of Fluid Extract of Taraxacum, of which one or two teaspoonfuls form a dose.

LACTUCA, *Linn.*

Heads few-flowered. Involucre cylindrical, imbricated in 2—4 rows, outer row shorter, scales with a membranous margin. Receptacle naked. Achænium plano-compressed, wingless, terminating abruptly in a filiform beak.

LACTUCA SATIVA, *Linn.* Herba, D. LACTUCARIUM. Succus spissatus, L. Inspissated Juice of, (and of *Lactuca* virosa, *Linn.* E.) the Garden Lettuce.

The common Lettuce (θρίδαξ) has been used in medicine from the time of the Greeks, as it still is in the East. It is cultivated throughout Europe and in most European colonies as a salad.

The common Lettuce is an annual, with an erect, smooth stem, which is two feet high, simple below, and branched above. Leaves rounded or oval, large, erect, narrowed at the base, smooth at the keel, half embracing the stem, often much wrinkled. Flowers appearing in August, yellow, smaller than those of *L. virosa.*

The Leaves of the Lettuce when young contain a pellucid pleasant-tasted juice, containing Mucilage and Sugar; but when the flowering stem begins to appear, the juice becomes milky, bitter in taste, and of a strong odour, something like that of Opium. These characteristics increase until the flowers have blown. If slices of this stem be cut off, or incisions be made into its cortical portion, the milky juice exudes, and on drying becomes of a brownish colour, forming what is called *Lettuce Opium* or *Lactucarium*, to which Dr. Coxe of Philadelphia and Dr. Duncan, Sen. of Edinburgh first called attention. Dr. Francois subjected it to further examination, calling it *Thridace.*

Lactucarium is prepared by collecting the above exuded juice and by pressing out that of the incised stems when in flower, and then evaporating it to a proper consistence in a water-bath. This forms the best kind. It is of the consistence of a dry extract, and is sold in roundish rather hard lumps, having a brown colour, an opium smell, and a bitter, mixed with a little acid taste. It is apt to attract moisture. It yields to analysis about half its weight of a Bitter Extractive, Wax, and Resin, with a principle analogous to Caoutchouc in considerable quantities. No crystalline principle has as yet been discovered.

Action. Uses. Anodyne, Diaphoretic and slightly Diuretic in doses of gr. ij.—gr. vj. to allay Cough and Nervous Irritation, relieve the pains of Rheumatism, &c., and to induce sleep.

EXTRACTUM LACTUCÆ, L. Extract of Lettuce.

Prep. Bruise the fresh leaves of *Garden Lettuce*, sprinkled with water, in a Wedgewood or stone mortar, and then press out the juice, and evaporate to a proper consistence.

Action. Uses. Contains little of the active principle or Lactucarium, though it resembles it in colour, odour, and taste; is a little sedative in doses of gr. v.—℈j.

LACTUCA VIROSA, *Linn.* E. Folia, D. The Inspissated Juice, E. Strong-scented Lettuce.

This is supposed to be the θρίδαξ ἀγρία of Dioscorides, though *L. Scariola*, Linn. is adduced by Dr. Sibthorp as the plant. This has glaucous vertical leaves, but the same properties as the former.

The Wild Lettuce abounds in acrid milky juice, has a tap-shaped root, with round and erect, slender, glaucous stem, 2—4 feet high, a little prickly below, panicled above. Leaves horizontal, with a prickly keel, otherwise nearly smooth, finely toothed; radical ones obovate, undivided, those of the stem smaller, often lobed, auricled, and semiamplexicaul. Flower-heads numerous, panicled, with numerous small, heart-shaped, pointed bracts. Florets light yellow. Achænia striated, beak white, equalling in length the black fruit.—Dry banks and borders of fields throughout Europe.—E. B. t. 1957. Flowers about August.

This plant, distinguished by its rank smell and the blood-red spots on its stem, is preferred to the former by the E. C. Dr. Christison states from information communicated to him by Mr. Duncan of Edinburgh, that it yields a much larger quantity and a superior quality of Lactucarium, especially before the middle period of inflorescence; so Schutz has found in Germany that a single plant of L. sativa yielded only 17 grs., while one of L. virosa no less than 56 grs. of Lactucarium. That prepared near Edinburgh is "in pieces about the size of a field bean, rough and irregular, wood-brown in colour, with an ash-grey efflorescence, friable, reddish-brown in powder, of the same odour with the former, but more acrid and bitter to the taste." (*c.*) Analyzed by Walz, it was found to contain a Volatile Oil, a yellowish-red tasteless Resin, a greenish-yellow Acrid Resin, Crystallizable and Uncrystallizable Sugar, Gum, Pectic acid, Albumen, a brown Basic substance, a principle like Humus, Extractive, a concrete Oil or Wax (one part of which, insoluble in ether, is the same as the Caoutchouc of other analysts), Oxalates, and other salts, with a neutral active principle, which has been named *Lactucin.* This is in acicular crystals, colourless, without odour, very bitter, fusible, soluble in about 70 parts of water, more so in Ether, Alcohol, and diluted acids. The watery solution is very bitter, neutral, and not precipitable by any reagent. (Walz, Ann. der Pharm. xxxii.)

Action. Uses. Narcotic; suited to allay pain and induce sleep, in the same cases as Henbane, and where Opium is ineligible, in doses of gr. v.—℈j.

TINCTURA LACTUCARII, E. Tincture of Lactucarium.

Prep. Take *Lactucarium* in fine powder ℥iv. *Proof Spirit* Oij. Best prepared by percolation, but also by digestion, as Tinct. Myrrh.

Action. Uses. Anodyne, &c. as above, in doses of ♏xx. to f℈j.

TROCHISCI LACTUCARII, E. Lettuce Lozenges.

Prep. Lactucarium ʒij. To be prepared in the same proportion and manner as the Opium Lozenge.

D. May be taken to the extent of xx.—xl. daily.

M. Aubengier prepares a Spirituous Extract of Lactucarium, which contains all the active principles, and which may be prescribed in pills, or made into a syrup by adding 1 part to 500 of syrup.

LOBELIACEÆ, *Juss.* Lobeliads.

Annual or perennial herbs, usually with milky juice. Leaves alternate, without stipules. Flowers complete, sometimes from abortion diœcious, irregular. Calyx superior or half-superior, 5-parted, odd segment anterior. Corol inserted into the calyx, composed of 5 petals, usually more or less adherent, and commonly cleft longitudinally. Stamens 5, alternate with petals, and inserted into a disk crowning the ovary. Filaments united above with the anthers into a tube. Ovary 1 to 3-celled, many-ovuled. Stigma fringed. Fruit capsular or berried, 1—2-celled, rarely 3-celled, often opening at the apex, many-seeded. Embryo straight in the axis of albumen.—The Lobeliaceæ are allied to Cichoreæ and to Campanulaceæ. Found in tropical and temperate parts of the world. They secrete a milky juice, often very acrid and narcotic.

LOBELIA INFLATA, *Linn.* L. Herb, E. Indian Tobacco. *Pentand. Monog.* Linn.

This plant was first employed by the natives and then by the medical practitioners of the United States, and in this country first in 1829.

Annual or biennial, with fibrous root. Stem erect, angular, the upper part branched and smooth. Leaves irregularly serrate, dentate, hairy; the lower ones oblong, obtuse, with short petioles; those towards the middle ovate-acute, sessile. Flowers in racemes. Calyx smooth, tube ovoid, 5-lobed, segments linear-acuminate. Corol of a light blue, cleft longitudinally from above, bilabiate, the upper lip narrow, the lower broader, 3-cleft. Anthers united into an oblong curved body, the two inferior barbed at the point. Style filiform; stigma curved, inclosed by the anthers. Capsule 2-celled, ovoid, ten-angled, inflated, crowned with the calyx. Seeds numerous, small, of a brown colour.—Common weed in the United States from Canada to Carolina.—Esenb. and Eberm. 206.

The whole plant when wounded exudes a milky juice, and all parts are possessed of medicinal activity; but, according to Dr. Eberle, the root and inflated capsules are the most powerful. The dried herb is of a pale greenish-yellow colour, of a faint disagreeable smell, and a burning acrid taste, especially perceptible in the fauces. It is generally compressed into rectangular cakes by the Shaking Quakers of New Lebanon, in New York. The active properties are extracted by Proof Spirit and by Ether. It has not yet been satisfactorily analysed. Its active principle has been supposed to be of a volatile nature; but it may be long preserved with its powers unimpaired. Dr. Calhoun considers its active principle (*Lobelin*) to be of the nature of Nicotin. Dr. Pereira finds a Volatile Acrid principle (Oil?) and a peculiar Acid. Reinsch, in a later analysis, found much Gum, with Vegetable Fibre, Aromatic Resin, with other substances soluble in Alcohol, and a peculiar substance which he obtained only in an impure state.

Action. Uses. Narcotic, Acrid, Antispasmodic, acting in many respects like Tobacco; in large doses Emetic and Cathartic. Used to control attacks of Spasmodic Asthma, either by giving it in full doses so as to excite vomiting, or in small doses repeated until sickness comes on.

D. The dose of the Powder as an expectorant, gr. j.—gr. v.; as an emetic, gr. x.—℈j.

TINCTURA LOBELIÆ, E. Tincture of Lobelia.

Prep. Digest *Lobelia* dried and finely powdered ℥v. in *Proof Spirit* Oij. Or much better prepared by percolation, as Tinct. Capsicum.

Action. Uses. Expectorant in doses of ♏x.—fʒj. Antispasmodic, fʒj.—fʒij. every two or three hours. Emetic in doses of fʒiv.

TINCTURA LOBELIÆ ÆTHEREA, E. Etherial Tincture of Lobelia.

Prep. Digest in a well-closed vessel for 7 days, finely powdered dry *Lobelia* ℥v. in *Spirit of Sulphuric Ether* Oij. But it is better prepared by percolation, the materials being firmly packed in the percolator.

Action. Uses. Similar to Whitelaw's Etherial Tincture. Used chiefly as an Antispasmodic.

ERICACEÆ, *Endlicher*. Heath-worts.

Shrubs, undershrubs, or trees. Leaves evergreen, alternate, without stipules. Calyx free, or with its tube adherent to the ovary, 4 or 5-fid. Corol 4 or 5 parted, adherent to the bottom of the calyx, superior or half-superior, formed of several petals united together. Stamens 8 or 10, inserted into the corol. Anthers with the 2 cells distinct either at the base or at the apex, there opening by pores, often furnished with a bristle-like appendage. Ovary half-superior or inferior. Placentæ central. Style 1. Stigma capitate. Fruit in the genera with inferior ovary, berried or drupaceous, in those with superior ovary usually capsular. Embryo in the axis of albumen. Many-celled, many-seeded.—The Ericaceæ are found in the cold and temperate parts of the northern hemisphere, and in southern Africa. Many are astringent, some also stimulant; hence employed as substitutes for Tea, as *Gaultheria procumbens* and *Ledum latifolium*. Some have succulent edible fruit. Rhododendron Chrysanthum is much employed by the Russians as a stimulant diaphoretic in rheumatism. R. campanulatum is employed in the Himalayas as snuff.

Tribe *Ericeæ*. Fruit capsular. Anthers 2-celled. Disk hypogynous. Testa close.

Uva Ursi, L. E. Folia, L. D. Leaves, E. of Arctostaphylos (Arbutus, D.) Uva Ursi, *Spr.* Bearberry. *Decandria Monogynia*, Linn. *Trailing Arbutus.*

It is uncertain when this plant was first employed medicinally. Quer maintains that the Spaniards first discovered its antinephretic properties.

Fig. 76.

Evergreen procumbent shrub. Leaves coriaceous, obovate, obtuse, quite entire, shining, of a deep-green above, lighter colour and covered with a network of veins on the under surface, hence reticulated. Flowers in terminal clusters of 8 or 10, each supported by 3 small bracts. Calyx 5-partite, of a pale red. Corol rose-coloured, ovate-urceolate, 5-cleft, border revolute. Stamens 10, inclosed, filaments flattened. Anthers compressed, with 2 pores at the apex, and furnished laterally with 2 reflexed arms. Ovary globose, supported by 3 scales. Style short. Stigma obtuse. Berry globose, scarlet with 5 single-seeded cells.—Stony Alpine heaths of Europe, Asia, and North America. — Esenb. and Eberm. 215. St. and Ch. 91.

The leaves are the officinal parts, and usually collected in autumn. They have sometimes been adulterated with the leaves of Vaccinium Vitis Idæa, or Whortleberry; but these are *dotted* on their under surface, and have their margins revolute and somewhat crenate. The Box leaf is devoid of astringency. Uva Ursi leaves, when dried and powdered, have an odour not unlike that of hay; the taste is bitter and astringent. The active properties are extracted both by water and Spirit. They contain Gum, Resin, Extractive, and some Gallic acid, and about 36 per cent. of Tannin. The watery infusion is precipitated by Gelatine, and a bluish-black colour is produced with the Sesquichloride of Iron. The leaves are employed in tanning in some parts of Russia.

Action. Uses. Astringent Tonic and mild Diuretic. Chiefly applicable in chronic cases, where there is an increased secretion of mucus from tha bladder, in doses of the powder gr. x.—ʒß.

DECOCTUM UVÆ URSI, L. Decoction of Bearberry.

Prep. Take bruised *Uva Ursi* ℥j. *Aq. dest.* Oiß. Boil down to Oj. and strain.

Action. Uses. Tonic, mild Diuretic, in doses of f℥iß.—f℥iij. three times a day.

EXTRACTUM UVÆ URSI, L. Extract of Bearberry.

To be prepared like Extract of Gentian.

Action. Uses. Tonic in doses of grs. v.—grs. x. two or three times a day.

Tribe *Vaccinieæ*, distinguished by their baccate, fleshy, and inferior fruit, with an epigynous disk, contain Cranberries, Bilberries, and Whortleberries.

Tribe *Pyroleæ*. Fruit capsular, dry. Seeds with a loose testa. Disk 0. Anthers opening by pores.

CHIMAPHILA, L. Pyrola, E. D. Folia, L. Herba, D. Herb, E. of CHIMAPHILLA (*corymbosa Pursh*, L.) UMBELLATA, *Nuttal.* Winter Green. *Decand. Monog.* Linn.

This plant, called *Pipsissewa*, was first employed medicinally by the native Americans, and then by the European settlers. It was made known to the profession by Dr. Mitchell in 1803, and then by Mr. Carter and Dr. Somerville. (Medico-Chirurg. Trans. vol. v.)

Small evergreen shrub, with creeping root-stock. Leaves coriaceous, with short petioles, cuneate-lanceolate, coarsely serrated, smooth, and shining. Flowers drooping, in small corymbs, with linear awl-shaped bracts. Calyx 5-cleft. Petals 5, white, with a tinge of pink, spreading. Stamens 10, filaments smooth, dilated in the middle. Ovary roundish, obtusely angular, umbilicated. Style short, concealed in the umbilicus of the ovary. Stigma orbicular, 5-lobed. Cells of the capsule dehiscent at the apex, the valves unconnected by tomentum. —A native of mossy turf in the woods in the northern latitudes of America, Europe, and Asia. Flowers in June and July.—Esenb. and Eberm. 93. St. and Ch. 93.

The fresh leaves, when bruised, exhale a peculiar odour; the taste is pleasantly bitter and astringent; that of the stems and roots is said to be pungent. They contain Gum, a little Tannin, Bitter Extractive, Resin, and Saline matter with Lignin. Dr. A. T. Thomson indicates the presence of Gallic acid.

Action. Uses. Acrid and Tonic. The fresh leaves applied to the skin produce rubefaction. The infusion or decoction taken internally, acts as a Diuretic and Tonic, and has been prescribed in Dropsies accompanied with debility, and in chronic affections of the Urinary Organs, also in Scrofulous complaints.

DECOCTUM (PYROLÆ, D.) CHIMAPHILÆ, L. Decoction of Winter Green.

Prep. Take *Chimaphila* (Pyrola umbellata, D.) ℥j. *Aq. dest.* Oiß. (by measure

℔j. D.) Boil down to Oj. and strain, L. (Macerate it for 6 hours, bruise, and return it to the water. Evaporate the expressed liquor to ℔j. by measure, D.)

Action. Uses. Diuretic and Tonic. The Decoction strikes a deep green with ferruginous salts, and may be given in doses of f℥j.—f℥iij. every three or four hours.

γ. *Corolliflorœ*, Dec.

STYRACEÆ. *Rich.* Styrax Tribe.

Trees or shrubs, smooth or with stellate pubescence. Leaves alternate, without stipules. Flowers regular. Calyx free or united to the ovary, 4 or 5-cleft. Corol deeply 3 to 7, often 5-cleft, imbricate. Stamens united together at the base, inserted into the bottom of the corol, double, treble, or quadruple the number of its divisions. Ovary free or adherent, 2—3—5-celled. Ovules 4 or more in each cell, in two rows, the upper ascending, the lower pendulous. (*Lindley.*) Style simple. Stigma obscurely lobed. Drupe fleshy or dry, sometimes winged from the development of the nerves of the adherent calyx; nut 3—5, or oftener from abortion 1-celled, and in the same way 1-seeded. Seeds erect or inverse. Embryo in the axis of fleshy albumen.—Found in the tropical parts of Asia and of America, with a species in the Mediterranean region, and others in Japan. Remarkable for the secretion of Benzoic acid in Styrax and Benzoin.

STYRAX, *Linn.* *Decand. Monog.*

Calyx rather campanulate, nearly entire, or 5-toothed. Corolla campanulate at the base, deeply 3—7 cleft. Stamens 6—16, seldom 10, exserted; filaments united to the tube of the corol, sometimes adhering at the base into a ring; anthers linear, 2-celled, opening by internal longitudinal slits. Ovary inferior, Style simple. Stigma obtuse, somewhat lobed. Drupe dry, splitting imperfectly into 2 or 3 valves, with 1—2—3 stones. Seed solitary, erect, with a large, leafy, thin embryo lying in the midst of fleshy albumen, with an inferior radicle. (*Lindley.*)—Styrax officinale and S. Benzoin yield officinal products.

STYRAX, L. E. D. Balsamum, L. Resina, E. Balsamic exudation, E. of STYRAX OFFICINALE, *Linn.* Officinal Storax.

Storax (στυραξ) was well known to the Greeks. Dioscorides compares the tree producing it with the Quince tree. It is called *asteruk* in the East.

A small tree (Fig. 77), with smooth bark, and downy shoots and petioles. Leaves ovate-obtuse, of a green colour, and shining above, white and downy on the under-surface, something like those of the Quince tree. Flowers white, in terminal racemes of a few flowers, which resemble those of the Orange. Calyx downy (2), cup-shaped, 5 to 7 toothed. Corolla externally hairy, with 5 to 7 segments. Stamens (2) 10 to 16. Fruit about the size of a cherry, coriaceous, downy, with 1 or 2 nuclei.—A native of Asia Minor and Syria, common in Greece, and cultivated in the south of Europe. As this plant does not yield a balsamic exudation in all these situations, some Storax has been thought to be yielded by *Liquidambar orientale.* Du Hamel, however, states having seen it flow from a tree near the Chartreuse of Montriau.—Esenb. and Eberm. 210. Fl. Græc. t. 375. St. and Ch. 47.

Much of the Storax of commerce is yielded by this tree when incisions are made into it. It is common in Asia Minor, where Professor

Forbes was informed Storax was collected from it. Several kinds are known in commerce and described in books. (*v.* Pereira.) Of these, Storax in grains, Reed Storax (Storax calamita), called Storax amygdaloides by Guibourt, used to be most common. One kind of Liquid Storax is yielded in the islands of Cos and Rhodes by *Styrax officinale*, which is there called βουχουρι. This seems only the Arabic word *Bukhoor*, signifying incense or fumigation, put into Greek letters. Common Storax is in brown or reddish-brown masses, varying in shape (in turf-like cakes, c.), light and friable, but possessing some tenacity, softening when warmed. It has a fragrant balsamic odour, and a warm aromatic taste. It consists evidently of saw-dust united with some balsamic substance. Merat and De Lens inform us that a factitious compound is made both in the Levant and at Marseilles with the saw-dust of the wood and the juice of this tree, with a little Benzoin. The ancients also employed saw-dust, honey, wax, iris-root, &c. Storax, when exposed to the air, becomes covered with an efflorescence of Benzoic'. It burns with a white flame, leaving a carbonaceous residue; gives a slight colour and odour to water; but the greater part, with the exception of impurities, is dissolved by Rectified Spirit. Reinsch gives as the constituents of the Red Storax, a trace of Volatile Oil, Benzoic acid 1 to 2·6 per cent., Gum and Extractive 7·9 to 14, Resin 32·7 to 53·7, with much woody fibre. It hence requires to be purified.

Fig. 77.

STYRAX COLATUS, L. EXTRACTUM STYRACIS, E. Strained or Purified Storax.

Prep. Dissolve *Storax* in *Rectified Spirit,* strain with a gentle heat, distil off the Spirit to the right consistence. The E. formula is nearly the same: most of the Spirit is to be distilled off, and the liquid evaporated over the vapour-bath to the consistence of a thin Extract.

Action. Uses. Stimulant Expectorant in Chronic Coughs, in doses of gr. x.—℈j. It forms an ingredient of the Tinct. Benzoini comp.

PILULÆ STYRACIS COMPOSITÆ, L. PIL. STYRACIS, E. PIL. E STYRACE, D.

Prep. Take strained *Storax* (Resin, D.) ℥iij. (Extract of Storax 2 parts, E.), hard (Turkey, D.) *Opium* powdered, and *Saffron* āā ℥j. (1 part, E.) Rub up and beat into a uniform mass. (Divide into 4 gr. pills, E.)

Action. Uses. Narcotic, &c. gr. j. of Opium in gr. v. of the pill.

BENZOIN, L. E. D. Balsamum, L. Resina, D. Concrete Balsamic exudation, E. of STYRAX BENZOIN. *Dryander*. Benzoin. *Gum Benjamin.*

Benzoin has long been employed medicinally and as incense in the East. In Bengal it is called by a name (*looban*), which in N. W. India is applied only to Olibanum (*v.* p. 336). In Persian works on Materia Medica it is distinguished by the names of *hussee looban* and *hussee-al-jawee* (an Java?) (*v.* Himal. Bot. p. 261). The name *hussee* appears to be the original of *assa*, as mentioned at p. 414, and Benzoin we know is in old works called *asa dulcis*. Mr. Marsden ascertained it to be yielded by a tree which Mr. Dryander named Styrax Benzoin. Haynes supposed it to constitute a distinct genus, *Benzoin*, which Blume had already named Lithocarpus.

Tree of considerable size, but small when tapped, of quick growth. Branches and footstalks round, downy. Leaves oblong, acuminated, smooth above, white and tomentose underneath. Racemes compound, axillary nearly as long as the leaves, flowers on one side with short pedicels. Calyx campanulate, obscurely 5-toothed. Corol of 5 petals of a greyish white colour, which are perhaps united at the base, four times longer than the calyx. Stamens 10, inserted into the receptacle. Ovary superior, ovate, tomentose. Style filiform. Stigma simple.—Native of Java, Sumatra, Siam, and Laos, Borneo, &c.—Esenb. and Eberm. 111.

Benzoin is obtained in Sumatra by making incisions into the tree in its seventh year. The juice which flows first is the purest and most fragrant: it hardens on exposure to the air. That which flows subsequently is brownish, and some is scraped out when the tree is cut down and split open, as it is soon killed by the process of tapping. These varieties are in commerce called *head*, *belly*, and *foot* Benzoin, and have the relative values to each other of 105, 45, and 18, being esteemed according to their whiteness, semi-transparency, and freedom from admixtures. It is also produced in Siam, whence it has long been an article of commerce. The specimen given by the Author to Dr. Pereira, and named by him *Translucent Benzoin*, was obtained from Bankok. This may be produced by *Styrax Finlaysonianum;* but *S. Benzoin* may grow in the interior, as much surface moisture and umbrageous forests prevail, which will produce a climate very similar to that of a tropical island. *Benzoin in tears* is a fine kind, but seldom met with. The best kind common in commerce is in masses composed of whitish or reddish tears agglutinated together by a darker-coloured portion of the same balsam. This is *Amygdaloidal Benzoin.* An inferior kind of a dark-brown colour, is sometimes called Calcutta Benzoin. Though the history of some varieties requires investigation, this had no doubt been previously imported, as Benzoin is always mentioned as one of the imports into Calcutta, Madras, and Bombay, from countries still more to the eastward. Benzoin, though hard, is friable, presenting a resinous mottled fracture; has an agreeable fragrant odour, more perceptible if rubbed; taste somewhat sweetish and balsamic, irritating the fauces if much chewed; its powder excites sneezing. Sp. Gr. 1·092. Heated, it melts, and emits white irritating fumes of Benzoic acid, also of an empyreumatic oil, and finally burns away. It is soluble in Alcohol and also in Ether, being precipitated on the addition of water, forming a milky emulsion. Some acids dissolve it. It consists of a trace of Volatile Oil, Benzoic acid about 20 per cent., Resin from 78 to 80 per cent., some of which is soluble and the other insoluble in Ether, Woody matter, Water, and loss. Both the White and Brown Benzoins contain nearly the same quantity of Benzoic acid; but the latter contains only about 8 per cent. of the soluble, and the white Benzoin very little of the insoluble resin.

Action. Uses. Stimulant Expectorant. Formerly much employed in Chronic Catarrhs and in fumigations.

TINCTURA BENZOINI (BENZOES, D.) COMPOSITA, L. E. D. Compound Tincture of Benzoin.

Prep. Macerate for 14 (7, E. D.) days *Benzoin* ℥iijß. (℥iv. E. ℥iij. D.), Strained *Storax* ℥ijß. L. (℥ij. D.) *Balsam of Tolu* ʒx. L. (℥j. D.) *Aloes* (Indian, E. Socotrine, D.) ʒv. (℥ß. E. D. *Peru Balsam* ℥ijß. E.) in *Rectified Spirit* Oij. (by measure ℔ij. Pour off the clear liquor, D.) Filter.

Action. Uses. Stimulant, Expectorant, in doses of f℈ß.—f℈ij. sometimes made into an emulsion, or added to pectoral mixtures to improve their flavour.

ACIDUM BENZOICUM, L. E. D. Benzoic Acid.

Benzoic Acid, though named from Benzoin, is found in other substances, which are on this account called Balsams, such as Storax, and the Balsams of Peru and of Tolu. It is also produced by the action of reagents on several vegetable substances. Indeed it is supposed by Prof. Johnston to be produced in the balsams themselves by the action of heat or other reagents. It is considered to be an oxide of the hypothetical radicle *Benzoyle* or *Benzule* ($Bz = C_{14} H_5 O_2 = 125$). This has been already mentioned at p. 388. The pure Oil of Bitter Almonds being a Hydruret of Benzule. When this is exposed to the atmosphere, oxygen is absorbed, and some Benzoic acid formed. It is also formed in the Urine of some herbivorous quadrupeds, by the decomposition of Hippuric acid ; but it is usually obtained from Benzoin, either by subjecting it to the action of heat, or to that of a base, from which it is afterwards separated by a stronger Acid.

Benzoic Acid, when obtained by sublimation, is in soft, feathery, flexible crystals, which have a pearly lustre, and, when pure, are quite colourless, but as obtained by the action of heat, have a little empyreumatic oil intermixed, which increases the odour, but does not impair the medical properties. From a solution, Benzoic′ crystallizes in transparent prisms. It has a warm, acrid, slightly acid taste ; a little volatilizes at ordinary temperatures. It melts under 212°, and sublimes entirely on a little increase of temperature, and burns away ; it is soluble in about 25 parts of boiling, but not less than 200 parts of cold water, but very readily in Alcohol : it combines readily with Alkalis and Metallic Oxides. It is composed of 1 Eq. Benzule 105 + 1 Eq. Oxygen 8 = 113 + 1 Eq. of Water (Bz + Aq. = 122) when crystallized.

Tests. Colourless, sublimed entirely by heat, E. with a peculiar odour, L. Water dissolves it sparingly, but rectified Spirit readily. Solution of Potash or of Lime dissolves it entirely, and Hydrochloric acid throws it down again. L.

Prep. Take of *Benzoin* ℔j. (q. s. E.), put it into a proper vessel (a glass mattress, E.), and from a sand-bath (or by a gradually increasing heat, E.) sublime as long as anything rises. Squeeze the sublimate between folds of filtering paper to remove the oil as much as possible, and sublime the residuum again, L. E. The acid which rises contains a little empyreumatic oil intermixed, but less if the apparatus of Mohr be used. Or Benz′ may be prepared by boiling finely powdered Benzoin with Carbonate of Potash or of Soda, or, as in the D. P., with Hydrate of Lime, when a Benzoate of Lime is formed. This is decomposed by H Cl′, when Benzoic′ is precipitated, and Chloride of Calcium left in solution. This is poured off, the residual acid washed, dried, and then sublimed with a gentle heat.

Action. Uses. Stimulant, Expectorant, formerly in doses of gr. v. ℈j. but now chiefly used as an ingredient of Paragoric (*Tr.* Camphoræ Comp.) and Tinct. Opii Ammoniata, E.

Inc. Alkalis, their Carbonates and Metallic oxides.

Oleaceæ. *Lindl.* Olive Tribe.

Trees or shrubs, branches and leaves opposite. No stipules. Flowers complete, or from abortion unisexual, racemose. Calyx free, 4-toothed or divided, persistent. Corol of 4 petals, either united at their base or throughout, and these equally 4-fid, valvate, seldom wanting. Stamens 2, inserted into the tubes of the corol. No disk. Ovary free, 2-celled, ovules 2 or many, pendulous. Style short. Stigma undivided or bifid. Fruit capsular, berried or drupaceous. Seeds from abortion usually solitary, compressed, or with a membranous wing. Embryo straight in the axis of horny albumen.—The Oleaceæ are found in the warm and temperate parts chiefly of the northern hemisphere, a few in the mountainous situations in India. They are valued for the hardness of their wood and ornamental flowers ; the Olive-tree also for its fruit and oil, and the species of Fraxinus for Manna.

Divided into two groups. {
1. Oleineæ. Fruit drupaceous or baccate.
2. Fraxineæ. Fruit capsular, indehiscent samaroid.

Olivæ Oleum. L. E. D. Olive Oil. Oleum e drupis expressum L. D. Expressed oil of the Pericarp E. of Olea europæa. Linn. Europæn Olive. *Diand. Monog.* Linn.

The Olive tree, ἐλαία of the Greeks, *Zait* of the Bible, and *Zaitoon* of the Arabs, is one of the most celebrated and useful of trees.

The olive tree (Fig. 78) is usually small, evergreen, but of a dull aspect, wood hard. Leaves with short petioles, ovate-lanceolate or lanceolate, mucronate, of a greyish green colour above, hoary beneath. Flowers white, in short axillary clusters. Cal. (2) small 4-toothed. Corol with short tube and 4-cleft limb (1). Stamens 2, a little exserted. Style short. Stigma (2) bifid segments emarginate. Ovary 2-celled, 2-seeded. Drupe (3) about the size of a damson, purple coloured, containing only one sharp-pointed nut.—A native probably of Asia, early cultivated in Syria and Greece. The varieties of the olive are numerous. The var. *longifolia* is chiefly cultivated in the S. of France and Italy, and the var. *latifolia* in Spain. — Esenb. and Eberm. 212. St. and Ch. 15.

Fig. 78.

The leaves and bark of the Olive tree have been employed ; also a peculiar resinous exudation, called *Olivile* and *Olive gum* (v. p. 343), and the bark as a substitute for Cinchona.

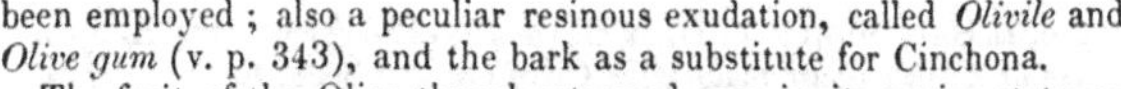

The fruit of the Olive though esteemed even in its unripe state, as an article of the dessert, having been first steeped in an alkaline ley, and then preserved in salt and water, is chiefly valued on account of the bland fixed oil which is stored up in its outer fleshy part. This is obtained by at once bruising the nearly ripe fruit with moderate pressure in a mill (*Virgin Oil*), or by the aid of boiling water and greater pressure, or when fermentation has taken place in the olives collected in heaps, *ordinary* and inferior oils are thus obtained, the worst being employed only as lamp-oils or in the manufacture of Soap. The finest oils are produced near Aix, Montpellier, Nice, Genoa, Lucca, and Florence. It is also largely produced in the kingdom of Naples, and exported from Gallipoli, on the East coast of the Gulf of Taronta, whence it is commonly called Gallipoli Oil.

Olive oil may be taken as the type of the *Fatty* or *Fixed*, called also *Expressed Oils.* It is of a pale yellow, or of a light yellowish-green colour, without smell when fresh, having a bland, somewhat sweetish, fatty taste. It is very limpid. Sp. Gr. 0·910 at 77°, insoluble in water, is readily dissolved by volatile oils, and by twice its bulk of Ether, but requires much more Alcohol. Exposed to the air it absorbs Oxygen and becomes rancid, but not drying like Linseed oil, is preferred for machinery. At 38° F. it begins to congeal, and is readily separated at 20° into two distinct bodies, of which one is fluid, called Elaine, or Oleine 72, and the other 28 per cent. solid, named Margarine from its pearly aspect. This is often deposited in jars and casks of the oil. Hyponitrous acid converts Olive Oil into a concrete mass from producing the oleaginous principle Elaidine, mentioned at p. 203. If this like other oils is heated with Alkaline solutions, or with the Oxide of Lead, great changes take place, as exemplified in the making of Soap, (v. Sapo) and of Lead Plaster, p. 161.

Tests. Olive Oil is apt to be adulterated with poppy and other oils; these are distinguished by not congealing at the same temperature as olive oil, also by retaining air, when shaken up, more readily than pure olive oil. The E. P. directs that "when carefully mixed with a twelfth of its volume of solution of Nitrate of Mercury, prepared as for the Unguentum Citrinum (v. p 202,) it becomes in three or four hours like a firm fat, without any separation of liquid oil." If 5 per cent. of any other oil be present, the consolidation is slower and less firm, but if there be 12 per cent., the foreign oil floats on the surface for several days. (*c.*) M. Gobley has invented an Elaïometer. (P. J. iii. 293.)

Action. Uses. Nutrient, Emollient, internally in irritant poisoning, externally relaxing, much employed for frictions and for embrocations, and to give consistence to Cerates, Ointments, and Plasters: good application to the hair from not drying readily. In doses of f℥j. laxative, added to enemata for its emollient effect, or to dislodge ascarides.

SAPO, L. SAPO DURUS. E. D. ex Olivæ oleo et Sodâ confectus, L. *Spanish or Castile Soap*, made with Olive Oil and Soda, E.

SAPO MOLLIS, L. E. D. Ex Olivæ oleo et Potassâ Confectus, L. Soft Soap, made with Olive Oil and Potash, E.

The manufacture of Soap was known to the Romans, and has long been practised in India. It depends upon the action of Alkalis, and of oxide of lead upon fixed oils and fatty substances. Hard, or as it is often called Castile Soap, is made by heating together olive oil, and a solution of caustic Soda. Combination gradually takes place, and a viscid homogeneous mass is formed, which is readily soluble in water. When of good quality it is "white, does not stain paper, (with oiliness,) is free of odour, and dissolves entirely in rectified Spirit," E. The mottled kind is less fit for medical use, because it is coloured by the addition of Sulphate of Iron, which becomes decomposed, and the black Protoxide is precipitated, which by the action of the oxygen of the air is converted into the red Sesquioxide of iron. Chemists conceive that at the time when soap is formed, the Elaine and the Margarine are, by a re-arrangement of their elements, converted into two acids, called Elaic and Margaric acids, and that these combine with the Soda forming Eleates and Margarates of Soda. But with the formation of the acids, a new substance, *Glycerine*, is also produced, and becomes dissolved in the water which forms one of the constituents of Soap. Some oils and animal fats contain Stearine, a substance closely allied to Margarine: in that case some Stearic acid is also formed. All Soaps are slightly alkaline, feel soft and slippery, and are detergent. The watery solution is readily decomposed by acids, also by earthy and many metallic salts, hence, when water holds any of them in solution, instead of dissolving, the soap becomes decomposed. Such waters are called *hard*, while those which are comparatively pure are called *soft* waters. Castile Soap is composed of 9· to 10·5 of Soda, 76·5 to 75·2 of Oleic and Margaric acids, and of 14·3 to 14·5 of water, (*Ure*). Common Soap made of Tallow and Soda, and Yellow Soap of Tallow, Resin, and Soda, are not so well adapted for medical use.

Soft Soap, as used in the arts, is made with Caustic Potash and Fish-Oil and Tallow; is semitransparent, of the consistence of honey, brownish-coloured, and nauseous. But that referred to by the Colleges as made with Potash and Olive Oil, Dr. Pereira was unable to meet with, and found on inquiry that common Soft Soap is usually substituted in making *Ung. Sulphuris Comp.* L.

Action. Uses. Soap is Antacid, and hence used as an Antilithic; its alkali being readily set free, it is sometimes conveniently used in poisoning by acids, and given in large quantities, without causing irritation. Its Oil being also set free, makes it useful in cases of habitual costiveness, especially when combined with Rhubarb, Aloes, or Colocynth Extract. Useful as a detergent in many cutaneous diseases, and externally, from its lubricity, it is well suited for embrocations, &c., and is hence very commonly employed in liniments.

Pilulæ Saponis Compositæ, L. Pil. Saponis cum Opio, D.

Prep. Beat together into a uniform mass *hard* (Turkey, D.) *Opium powdered* ʒiv. and *Hard Soap* ℥ij.

D. gr. v.—gr. x. as a Narcotic gr. v.—℈j. used as a Suppository.

Linimentum Saponis, L. E. D. Soap Liniment. Opodeldoc.

Prep. Dissolve *Camphor* ℥i. in *Spirit of Rosemary* f℥xvi. then add *Soap* ℥iij. and macerate till it is dissolved with a gentle heat, L. Dissolve *Soap* ℥iij. D. (Castile ℥v. E.) in *rectified Spirit* Oij. E. (Spirit of Rosemary f℥xvi. D. Digest (till dissolved, D. E.) add *Camphor* ℥i. D. (Camphor ʒiiß. and Oil of Rosemary f℥vi. Agitate briskly, E.) Soft soap is usually employed by druggists; for it was found that by following the directions of the L. P. a solid gelatinous mass is produced, owing to the Spirit of Rosemary being made with *Rectified* instead of with Proof Spirit. (*Shum.* P. J. ii. 457 and *Fisher*, p. 515.)

Action. Uses. Stimulant Embrocation. A vehicle for Opium, &c.

LINIMENTUM SAPONIS CUM OPIO, D. (*v.* Linim. Opii, L. E. p. 270.)

CERATUM SAPONIS, L. Soap Cerate.

Prep. Boil powdered *Oxide of Lead* ℥xv. in *Vinegar* cong. i. over a slow fire, constantly stirring, until they unite, then add *Soap* ℥x. and boil as before till all the moisture is evaporated, then mix with these *Wax* ℥xiiß. dissolved in *Olive Oil* Oj.—Di-Acetate of Lead is first formed. The Soda then unites with the Acetic' and the fatty acid of the Soap with the Oxide of Lead. The Oil and Wax give consistence to the compound. (*p.*) This is of a *soft* texture, but may be converted into *hard* cerate, or *Emp. Cerati Saponis,* simply by evaporating away all the vinegar. (*v.* P. J. iii. 36.)

Action. Uses. Mild application to Scrofulous and other sores. The *hard* is preferred by Dr. Houlton to keep under dressings *in situ*.

EMPLASTRUM SAPONIS, L. E. D. Soap Plaster.

Prep. Melt *Lead* (*Litharge,* E. D.) *Plaster* ℔iij. (℥iv. and *Gum Plaster* ℥ij. E.) add *Soap* (Castile) sliced ℔ß. L. D. (in shavings ℥j. E.) Boil them down to a proper consistence.—Gum Plaster is added in the E. P. to obviate the tendency to crumble possessed by the L. and D. preparations.

Action. Uses. Discutient. Gives support, and is little irritant.

EMPLASTRUM SAPONIS COMPOSITUM VEL ADHÆRENS, D. Adhesive Plaster.

Prep. Make *Soap Plaster* ℥ij. and *Plaster of Litharge and Resin* ℥iij. into a plaster. Melt and spread on linen.

Action. Uses. Less irritating than the Resin Plaster (p. 161).

Pharm. Prep. Pil. Rhei Comp. Pil. Cambogiæ Comp. Pil. Scillæ Comp.

MANNA, L. E. D. Manna. Succus concretus, L. D. Sweet concrete exudation, E. of FRAXINUS ORNUS, *Linn.* (*Ornus europœa, Persoon*), L. European Flowering Ash : probably of several species of Fraxinus and Ornus,* E. *Diandria Monog.* Linn.

The name Manna seems to be derived from the Arabic *mun*, signifying the same thing. But as there are several other sweetish exudations (*v.* Manna, *Penny Cycl.*), it is difficult to determine when Manna was first known and used. There is uncertainty also respecting the species which yields European Manna. The Flowering Ash is the Fraxinus of the ancients, while the Common Ash is the Ornus of Virgil.

FRAXINUS ORNUS (*Ornus europœa,* L.), adduced in the L. P., is a tree about 25 feet high, with leaves which are impari-pinnate, consisting of 7 to 9 stalked, oblong-acute, serrated leaflets, which are hairy at the base of the midrib on the

* The two genera are again reunited into one, v. Decand. Prod. viii. 274.

under side. Buds velvety. Panicles dense, terminal, nodding. Calyx very small, 4-cleft. Corol divided to the base into linear segments, which are white and drooping. Pericarp a narrow elongated capsule, which does not dehisce. terminated by a flat and obtuse wing.—Hilly situations in South of Europe, especially Calabria and Apulia, also in Sicily.—Fl. Græca, i. t. 4. St. and Ch. 53.

Tenore asserts that Manna is yielded by two varieties of this tree, one named *rotundifolia*, and generally cultivated on account of its Manna, and the other *O. garganica*. Prof. Gusson assured Messrs. Merat and De Lens that *Ornus rotundifolia* alone yields Manna, and that this is frequently grafted on *O. europœa*. Both this species and *O. rotundifolia* are natives of Calabria, Apulia, and Sicily ; and it is from these places that we obtain our chief supplies of Manna. But other species of Fraxinus, and even *F. excelsior* or the Common Ash, in the south of Europe, have been stated to yield Manna.

Manna is obtained chiefly by making incisions into the bark, and sticking leaves below them, in the middle of summer and in early autumn. The juice flows out as a clear liquid, and soon concretes on the stem and the leaves, as well as on straws stuck into them, forming stalactitical or *Flake Manna*. Some falls on leaves or into vessels placed for receiving it. Several kinds are known in commerce. *Manna in tears* is a pure kind, in bright and roundish white grains ; but *Flake Manna* is chiefly valued and mostly met with in this country. It is in light and porous pieces, 5 or 6 inches in length, mostly stalactitical in nature, often hollowed on one side, of a pale yellowish-white colour, easily broken. The odour is faintish, the taste mawkishly sweet, followed by acridity. Its colour changes to a yellowish-red when long kept. Inferior kinds are in smaller pieces, irregular in form, soft and sticky, of a yellowish-red or brownish colour, of an unpleasant sweetness, and often intermixed with impurities. These are called *Manna in sorts*, *Fat Manna*, *Tolfa Manna*, &c. Another set of Mannas are produced in Syria, Persia, and Arabia (*v.* Manna, *Penny Cycl.*) ; but these are never met with in European commerce. Manna melts with heat, and burns with a bluish flame. When pure, it is soluble in 3 parts of cold and in its own weight of boiling water. It is also dissolved by Alcohol. Manna consists of about 60 per cent. of a peculiar principle called *Mannite*, but which varies in different varieties of Manna, Sugar, of which some is crystallizable (*Thenard*) and some uncrystallizable, a little Gum, with some yellow nauseous Extractive, which is supposed by some to be the purgative principle. *Mannite* is in acicular 4-sided crystals, sweet, without smell, soluble in water, less so in Alcohol, incapable of undergoing fermentation.

Action. Uses. Laxative, without irritation, but less so the fresher it is ; apt, however, to create flatulence ; in doses of ℥j.—℥ij. ; but for children, to whom it is suited from its sweet taste, ʒj.—fʒij.

Pharm. Prep. Confectio Sennæ. Syrupus Sennæ, L. E. D. Enema Catharticum, D. (p. 362.)

The APOCYNEÆ contain a few plants possessing active and useful properties; but none are officinal, as the *Strychneæ* are now referred to LOGANIACEÆ. (*p.*)

The ASCLEPIADEÆ, closely allied to *Apocyneæ*, do not contain any plants officinal in the Pharmacopœias, but many which are possessed of useful medicinal virtues. *Cynanchum* (now *Solenostemma*) *Argel* has been already mentioned (p. 360) as employed to adulterate Senna. *Cynanchum Monspeliacum* and *Periploca Secamone* are said to be used to adulterate Scammony. *Secamone emetica* and *Asclepias curassavica* are emetic. *Tylophora asthmatica* has been considered an efficient substitute for Ipecacuanha, and an excellent remedy in Dysentery. *Hemidismus indicus* is considered by the medical officers of the Madras establishment to be an efficient substitute for Sarsaparilla in the treatment of Syphilis, Scrofula, and Cutaneous affections. It is there usually called Country Sarsaparilla. It has a pleasant odour, which is compared with that of Iris Root, is useful as an Alterative and general improver of the secretions. It is imported into this country and employed under the name of *Smilax aspera*. This was proved by the author to be the above *Hemidesmus indicus*. (Proc. Royal Asiat. Soc. June, 1838.) So *Calotropis procera, aka* and *mudar* of the natives of India, has long been employed as an Alterative in Cutaneous affections, and even in incipient Leprosy. The author employed a variety named *C. Hamiltonii* by Dr. Wight, which is common in North-west India, and which the natives there employ medicinally for the same purposes as the former. He has prescribed the fresh bark of the root, dried and powdered, alone and successfully in incipient cases of Leprosy and in other cutaneous affections, both in the Civil and the Military Hospital at Saharunpore.

LOGANIACEÆ, *Endlicher*. Loganiads.

Shrubs or trees, seldom herbs. Leaves opposite, entire, usually with stipules which adhere to the leafstalks or are combined in the form of interpetiolary sheaths. Flowers racemose or corymbose, rarely solitary, terminal or axillary. Calyx free, 5- rarely 4-lobed. Corol sometimes irregular, 5- rarely 4- or many-lobed. Stamens inserted into its tube, usually 5, alternate with the lobes, rarely 1 or 10 to 12, then with some opposite to the lobes. Anthers 2-celled, opening lengthwise. Pollen with 3 bands. Ovary superior, 2-celled (3- or spuriously 4-celled). Stigmas simple or 2-lobed. Fruit either capsular and 2-celled, with the valves turned inwards, bearing the placentæ, finally becoming loose, or drupaceous, with 1- or 2-seeded stones, or berried, with the seeds immersed in pulp. Seeds usually peltate, rarely erect, sometimes winged. Embryo straight, with the radicle turned towards the hilum or parallel with it, with 2 leaflike cotyledons.—Habit of Rubiaceæ, but the ovary and consequently the fruit in no way united with the calyx.—Allied on one side to Apocyneæ, and on the other to Gentianeæ, but to be distinguished by their stipules. They may be briefly defined as Rubiaceæ with free or superior ovaries.—Found in hot parts of the world.

Suborder I. SPIGELIEÆ, *Meesm.*—Are sometimes united with *Gentianeæ*, but are more allied to *Rubiaceæ*. They may be distinguished by their leaves being furnished with stipulæ. Flowers isomeric, æstivation of the corols valvate, capsules didymous, many-seeded. Seeds without wings. Embryo small; cotyledons little conspicuous.—Warm parts of the New World and in New Holland, with a few species in tropical Asia.

SPIGELIA, L. E. D. Radix, L. D. Root, E. of SPIGELIA MARYLANDICA, *Linn.* Carolina Pink. Wormseed. Perennial Wormgrass. *Pentand. Monog.* Linn.

The virtues of this plant were discovered by the Cherokee Indians, and made known in Europe about a century since.

Root perennial, branching, fibrous. Stems erect, 4-sided above. Leaves opposite, sessile, ovate-acuminate, smooth, with the margins and veins a little pubescent. Racemes terminal, 1-sided, 3-8-flowered. Calyx persistent, 5-parted, segments linear-subulate, finely serrulate, reflexed on the fruit. Corol scarlet, funnel-shaped, much longer than the calyx; the tube inflated and angular at the top, the limb in 5 acute spreading divisions, with the 5 stamens inserted between them. Anthers oblong, heart-shaped, converging. Ovary superior, ovate. Style longer than the corol, jointed near its base and bearded at the extremity. Capsule smooth, didymous, or composed of 2 cohering, 1-celled, 2-valved, globular carpels attached to a common receptacle. Seeds numerous.—Southern states of North America and Texas.—Esenb. and Eberm. 52. B. M. 80. St. and Ch. 117.

S. ANTHELMIA, *Linn.* or Pink Root of Guiana and of Demerara, has also been long employed by the natives of those countries, as well as in the West Indies; a decoction of a few fresh leaves being very efficacious against Ascaris vermicularis, or maw-worms.

The virtues of this plant reside principally in the root, which consists of numerous slender wrinkled fibres, attached to a knotty head, of a brownish colour externally; a faint smell, and of a slightly bitter, not very disagreeable taste. As sold in the shops, the stalks and leaves are usually found attached to the roots. Analyzed by M. Feneulle, they yielded a Fixed and Volatile Oil, a little Resin, a Bitter extractive matter, supposed to be the active principle, with Mucilaginous and Saccharine matter, and some salts. The leaves afforded the same principles, but a less quantity of the Bitter principle.

Action. Uses. Anthelmintic. Much used in North America. In large doses it acts as an Irritant Cathartic, and in poisonous doses as a Narcotic. It may be given in powder gr. x.—gr. xx. to a child 3 or 4 years old; ʒj.—ʒij. to an adult; or of the infusion (ʒiv. to boiling Aq. Oj.) fʒiv.—f℥j. may be given to a child. A quantity of Senna equal to the Spigelia is usually added, to ensure a Cathartic effect. (*Wood and Bache.*)

Suborder 11. STRYCHNEÆ. Flowers regular. Æstivation of the Corolla valvate. Embryo rather large. Trees or Shrubs.

Tribe 2. *Eustrychneæ.* Berry or Drupe two-celled, many-seeded, or from abortion one-celled, one-seeded, seeds peltate, apterous.

NUX VOMICA, L. E. D. Semina, L. D. Seeds E. of STRYCHNOS NUX VOMICA, Linn. Nux Vomica, or Koochla tree. *Pentand. Monog.* Linn.

Nux Vomica was early used as a medicine by the Hindoos, and it is their *Koochla* tree (Sans. Culaka, and Kataka), and being a produce of India, its properties must have been investigated long before it could be known to foreign nations. It is the *Izarakee* of Persian works on Materia Medica, but there is doubt respecting its name in Avicenna. *Khanuk-al-kulb*, dog-killer, and *Faloos mahee*, fish-scale, are other Arabic names. But under the name of *Jouz-al-Kue*, or Emetic Nut, the author obtained the fruit of a Rubiaceous shrub.

Dr. Pereira thinks that the *Nux Mechil* of Serapion is Nux Vomica; but in Persian works this name is applied to a Datura.

A moderate sized tree, with a short crooked trunk. Branches irregular, the young ones long and flexuose, with smooth dark-grey bark. Wood white, close-grained, and bitter. Leaves opposite, with short petioles, oval, smooth, and shining, 3 to 5-nerved, differing in size. Flowers small, greenish-white, in terminal corymbs. Calyx 5-toothed. Corol funnel-shaped; limb 5-cleft, valvate. Stamens 5; filaments short, inserted over the bottom of the divisions of the calyx; anthers oblong, half exserted. Ovary 2-celled, with many ovules in each cell, attached to the thickened centre of the partition. Style equal to the corol in length. Stigma capitate. Berry round, smooth, about the size of an orange, covered with a smooth somewhat hard fragile shell, of a rich orange-colour when ripe, filled with a soft white gelatinous pulp, in which are immersed the seeds attached to a central placenta. Seeds peltate or shield-like, slightly hollowed on one side, convex on the other, about ¾ of an inch in diameter, and about 2 lines in thickness, thickly covered with silky ash-coloured hairs attached to a fibrous testa, which envelopes the kernel composed of horny bitter albumen, of the form of the seed and of the embryo imbedded in a hollow in its circumference.—Roxb. Corom. i. t. 4.—A native of the Indian Archipelago and of the forests of the Peninsula of India, as well as of the Southern parts of the Bengal Presidency, as near Midnapore.—Esenb. and Eberm. 209. St. and Ch. 11. 52.

Lignum Colubrinum, supposed to be an antidote against the poison of venomous snakes, as well as a cure for intermittent fevers, is produced by other species, as *Strychnos ligustrina* and *S. Colubrina*. *S. Tieute* yields the *Upas tieute* and *Tjettek* of the Javanese, which is an aqueous extract of the bark. *S. toxifera* yields the *Woorali* or *Ourari* poison of Guyana. *S. pseudoquina* is employed in Brazil as a substitute for Cinchona Bark, and the seeds of *S. potatorum*, Roxb., *nirmulee* of the Hindoos, are employed by them to clear muddy water.

Strychnos Ignatia, usually considered as constituting a distinct genus, and called *Ignatia amava*, Linn. is however of most importance, as its seeds called St. Ignatius's beans, are frequently made to yield their Strychnia. They are ovate, triangular, of a reddish-gray colour, and about twenty of them contained in a pear-shaped fruit. They are produced in the Philippine islands, and have long been used in India, where they are called *Papeeta*, and are mentioned in the work called Taleef Sheræ. They are intensely bitter, and contain a larger quantity of Strychnia than the Nux Vomica seeds.

The wood of the Nux Vomica tree is said by Dr. Christison to be often substituted for the above *Lignum Colubrinum*, or Snake-wood. The Bark is unfortunately sold in many shops in Calcutta under the name of *Rohun*, and thus substituted for the febrifuge bark of the *Rohuna tree*, or *Soymida febrifuga* (Beng. Disp. pp. 247 and 437), which was made known by Dr. Roxburgh, written on by Dr. Duncan in 1794, and introduced into the E. P. In this way probably it came to be introduced into England, and not being found saleable, was sent to Holland, and there sold and used as Angustura-bark (v. p. 325), supposed at one time to be the bark of Brucea ferruginea. In 1804, Dr. Ronbach of Hamburgh observed that it acted as a poison, and as several fatal cases occurred, it was in consequence prohibited being used in many Continental states. From its composition Batka suspected that it was the bark of the Nux Vomica tree, or of some allied species. This Dr. Pereira confirmed by examining the specimens of

Strychnos Nux Vomica in the East Indian Herbarium. (Med. Gaz. xix. p. 492.) Dr. Christison, as he informed the author, came to the same result by examining specimens of Nux Vomica bark with French specimens of false Angustura Bark.

Dr. O'Shaughnessey in Calcutta, fully established the identity of false Angustura bark, and of the bark of the Nux Vomica tree in Journ. of Med. and Phys. Soc. of Calcutta, Jan. 1837 ; in consequence of an alkali having been obtained from what was supposed to be the bark of the *Soymida febrifuga*, but which proved to be the bark of Strychnos Nux Vomica. The whole forms a most instructive lesson on the absolute necessity of being thoroughly acquainted with the true nature of the drugs we prescribe.

Nux Vomica Bark is in flattish or slightly curved pieces, thick, hard, and compact; fracture dull and brownish; epidermis sometimes displaying a ferruginous, spongy, and friable efflorescence, at other times a yellowish-grey colour, marked with prominent greyish-white spots. Both appearances are due to alterations in the texture of the epidermis, and not to lichens, which are rare. The bark is smooth internally, its powder of a yellowish-white colour, without smell, but having an intense and permanent bitter taste. A drop of Nitric' on the external ferruginous part, turns it of a dark-greenish colour; but if applied to a transverse section or internally, a dark-red spot is produced. Analysed by Pelletier and Caventou, it yielded Brucia. Dr. Christison states that it might be employed for obtaining Strychnia. An infusion of the Bark slightly reddens Litmus, Nitric acid produces in it a red, and Sulphate of Iron a green colour, but an infusion of Galls a greyish-white precipitate.

Nux Vomica Seeds are round and flat, or rather shield-like, of a light-greyish colour, covered with a thick and tough testa, silky with fine hairs, which assist in detecting it when in powder and magnified. The seeds have little smell, but an intense and tenacious bitter taste; are so tough, that the E. C. give directions for powdering them (*v.* Extract). The powder, apt to be adulterated with substances employed to assist in pulverization, is of a yellowish-fawn colour. Water takes up some, but Proof and Rectified Spirit nearly all its active properties. Analysed by Pelletier and Caventou, the seeds were found to contain two alkalis, *Strychnia* and *Brucia* united with a peculiar acid, the *Igasuric*, called also Strychnic acid, a yellow Colouring matter, a Concrete Oil, Gum, Starch, Bassorin, and a small quantity of Wax.

Action. Uses. Cerebro-Spinant, Poisonous, producing tetanic convulsions without affecting the brain. Used as a Stimulant of the Nervous System in Paralysis, in doses of the powder gr. v.—gr. xv. or the following effective preparations may be employed.

TINCTURA NUCIS VOMICÆ, D. Tincture of Nux Vomica.

Prep. Macerate for 7 days the scraped fruit of *Strychnos nux vomica* ℥ij. in *Rectified Spirit* ℥viij. Strain.

Action. Uses. As Spirit takes up the active principles, this is an efficient preparation, sometimes applied externally to paralysed limbs.
D. ♏v.—♏x.

EXTRACTUM NUCIS VOMICÆ, E. D. Extract of Nux Vomica.

Prep. Take *Nux vomica* (scraped ℥viij. D.) *Rectified* (Proof, D.) *Spirit* (by measure ℔ij. D.) any convenient quantities, E. Expose the seeds to steam till soft, slice, dry, and grind them in a coffee-mill. Exhaust the powder by percolation or boiling with the Spirit, till it comes off free from bitterness. Distil off the greater part of the Spirit, and evaporate the residue to a proper consistence in the vapour-bath, E.

Action. Uses. A powerfully bitter-tasted extract. May be given in the form of pill in doses of gr.ß. gradually increased to gr. iij.

STRYCHNIA, L. E. Alkali of Strychnos Nux Vomica, &c. *Strychnine.*

Prep. Take bruised *Nux Vomica* ℔j. and boil it in *Rectified Spirit* Cj. in a retort fitted to a receiver, for 1 hour. Pour off the Spirit, and again and a third time boil with *Rectified Spirit,* each time Cj. and the Spirit recently distilled. Pour off the liquor. Press the *Nux Vomica,* and distil the Spirit from the mixed and strained liquors. Evaporate the residue to the consistence of an extract. Dissolve in cold water, and strain. Evaporate with a gentle heat to the consistence of syrup; while warm, add *Magnesia* gradually to saturation, shaking them together. Set aside for 2 days, then pour off the supernatant liquor. Press the residuum in cloth; boil it in *Spirit,* strain, and distil the Spirit. Add to the residue very little *dil. Sul'* diluted with *Aq.* and macerate with a gentle heat. Set aside for 24 hours, that crystals may form Press and dissolve them in water; then, frequently shaking, add *Ammonia,* to throw down the Strychnia. Dissolve this in *boiling Spirit,* and set aside to form pure crystals.—In this process, the Igasurate or Strychnate of Strychnia is dissolved in the watery solution of the alcoholic extract. On the addition of the Magnesia, decomposition ensues, Strychnate of Magnesia is formed and precipitated with the Strychnia which is set free, along with some Brucia. These are then dissolved by the Spirit, and a Sulphate of Strychnia is formed on the addition of the Sul'. This Sulphate is decomposed by the Ammonia (a soluble Sulphate of Ammonia being formed), and comparatively pure Strychnia precipitated.

In the E. P. less Spirit is required, the Nux Vomica ℔j. is powdered (*v.* Extract), decoctions are then prepared with Aq. Ov. which necessarily contain the Strychnates of Strychnia and of Brucia; and the soluble ingredients being evaporated to the consistence of Syrup, *Quicklime* ℥iß. is added in the form of Milk of Lime. The precipitate (Strychnia and Brucia, and Strychnate of Lime) is dried and powdered, and then boiled with successive portions of rectified Spirit, when the Strychnia is dissolved and afterwards obtained by distilling off the Spirit, and purified by re-crystallization, with or without animal charcoal.

By these processes a powder of a greyish or of a brownish-white colour is obtained, which is obscurely crystalline, and, though impure, amounts only to about 0·4 per cent. of the seeds. (St. Ignatius' Bean yields about 1·2 per cent.) It consists of Strychnia and Brucia, with some Colouring matter. The alkalis may be separated by Nitric', producing salts of different degrees of solubility, and then with Ammonia precipitating them from the Nitrates of Strychnia and of Brucia.

Strychnia when pure is white, crystallized in brilliant oblique octohedrons, or in elongated four-sided prisms, or it may be in a simple granular state. It is so intensely bitter, that 1 part gives a perceptible taste to 60,000 parts of water. It is very insoluble, requiring about 7000 parts of temperate and 2,500 parts of boiling water; soluble in boiling Rectified Spirit, and also in the fixed and volatile oils. It is first fused by heat, and then decomposed. It is alkaline in its reaction on Litmus, forms salts with acids, which are soluble and bitter. A white precipitate in solutions of these salts is produced by alkalis and Tinct. of Galls. Nitric acid colours it yellow; but if Brucia be present, a red colour is produced. Strychnia consists of $C_{44} H_{23} N_2 O_4$.

Tests. Strychnia is apt to be adulterated: that in common use is never pure. It consists of Strychnia, Brucia, and some colouring matter. It may contain the last in excess, as well as Lime and Magnesia. "Intensely bitter. Nit′ strongly reddens it. A solution of grs. x. in Aq. fʒiv. and *Pyroligneous acid* fʒj. when decomposed by concentrated *Sol. of Carbonate of Soda* f℥j. yields on brisk agitation a coherent mass, weighing when dry grs. x. and entirely soluble in Sol. of Oxalic acid." E. P. The precipitate should be equal to the Strychnia first employed. Lime or Magnesia will be insoluble in Oxal′. "Readily dissolves in boiling Alcohol, but not so in water. It melts by heat, and if it be more strongly urged, it is totally dissipated. This being endowed with violent power, is to be cautiously administered." L.

Brucia, which resembles Strychnia in many points, crystallizes in transparent crystals, which are usually in pearly scales. It is less bitter than Strychnia, and will dissolve in about 500 parts of water, and readily both in Alcohol and Rectified Spirit. It is alkaline, and forms crystallizable salts with acids. Nitric′ produces a deep-red colour when brought in contact with it; but the red solution becomes violet on adding solution of Protochloride of Tin. But the colour is destroyed on the addition of deoxidizing agents, as Sulphurous acid and Sulphuretted Hydrogen. It is composed of $C_{44} H_{25} N_2 O_7$.; the crystals contain 17 per cent. of water. Dr. Fuss, as quoted by Pereira, considers Brucia a compound of Strychnia and yellow Colouring matter. It may be employed for the same purposes as Strychnia, but is not above $\frac{1}{12}$ its strength; and may be given in gr. $\frac{1}{4}$—gr. v.

Action. Uses. Strychnia acts exactly as Nux Vomica, and may be employed for the same purposes, but in doses of $\frac{1}{16}$ or $\frac{1}{20}$ of a grain gradually increased to gr. j. The first effects experienced from medical doses are twitches in the muscles of the extremities, often during sleep, and frequently first in the paralysed part. Some improvement of the digestive functions is often experienced. But in larger doses, tetanic spasms ensue, and a tendency to lock-jaw, with transient intervals of relief. When the first twitches are experienced, it is necessary to intermit the use of the medicine, as the constitution does not become accustomed to its use, and some cases show a tendency to its being cumulative in its action. It is so powerful a Poison, that a girl 13 years of age was killed in about an hour by accidentally taking $\frac{3}{4}$ of a grain divided into three pills. (Edin. Med. Journal.)

GENTIANEÆ, *Juss.* Gentianads.

Herbaceous annuals or perennials, sometimes under-shrubs, sometimes twining. Leaves usually opposite, without stipules, simple, ternate in Menyantheæ, often ribbed. Flowers regular. Calyx persistent, composed of 4 to 5, seldom 6 to 8 sepals, distinct or united together, valvate or contorted. Corol hypogynous, 4—8 fid, usually withering, and twisted or plaited in æstivation. Stamens inserted into the corol, equal to and alternate with its lobes. Ovary single, 1-celled, rarely 2- or pseudo-4-celled. Styles 2, either wholly or partly united. Capsule many-seeded, with the margins of the valves bearing the seeds; or 2-celled, from the valves being turned inwards and forming a partition, when the placentæ become central, or a many-seeded berry. Embryo in the axis of fleshy albumen.

Gentianeæ veræ. Corol twisted to the right in æstivation. Leaves opposite.
Menyantheæ. Corol plaited in æstivation. Leaves alternate. Marsh plants.

The Gentianeæ are found in temperate and cold climates, often in mountainous situations. They are remarkable for the secretion of a bitter principle, which makes many of them useful as tonics.

CENTAURIUM, L. E. D. Herba, L. Folia, D. The Flowering-heads E. of ERYTHRÆA CENTAURIUM, *Pers. Chironia Centaureum.* Sm. Common Centaury. *Pentand. Monog.* Linn.

This is the κενταυριον το μικρον, or Small Centaury of Dioscorides.

Stem herbaceous, erect, about a foot high, rather quadrangular, leafy, branching above. Leaves, radical ones obovate depressed, those of the stem oval, acute, or oblong, lanceolate, 3-nerved. Flowers of a beautiful pink colour, sometimes white, arranged in a fasciculated corymb, the lateral ones with 2 opposite awl-shaped bracts. Calyx slender, 5-parted. Corol salver-shaped, with a cylindrical tube twice as long as the calyx, withering on the capsule, limb 5-fid, lobes oval, obtuse, spreading. Stamens 5, anthers yellow, rolling up spirally, after bursting. Style simple, bifurcate, with 2 stigmas. Capsule slender, imperfectly 2-celled from the much inflexed margins of the valves.—Native of heaths and pastures of most parts of Europe, flowering from June to August.—E. B. 417.

All parts of this plant possess a pure bitter taste, but the flowers in a less degree: it is therefore suited for all the purposes for which the bitter tonics are indicated. It yields its properties both to water and to Spirit.

Action. Uses. Tonic; may be given in powder ʒß. or in infusion (ʒij.—fʒiv. in Aq. Oj.) in doses of f℥jß.

GENTIANA, L. E. D. Radix. Root of GENTIANA LUTEA, *Linn.* Yellow Gentian. *Pentandria Digynia,* Linn.

Gentiana is the Γεντιανη of the Greeks, called *juntiana* in the works of the Arabs.

Root thick, perpendicular, often forked, brown externally, yellowish within. Stem straight, 2 to 3 feet in height. Radical leaves ovate-oblong, 5-nerved; stem leaves sessile, ovate-acute; those supporting the flowers cordate, amplexicaul, concave, all of a pale glaucous-green colour. Flowers (Fig. 79) in an interrupted spike of whorls, large, of a brilliant yellow. Calyx membranous, spathe-like, 3 or 4-cleft. Corol rotate, with 5 or 6 green glands at its base, 5 or 6-parted, divided usually into 5 acute veiny lobes. Stamens 5; anthers straight, subulate.

Style wanting. Stigmas 2, revolute. Ovary and capsule fusiform, 1-celled. Seeds roundish, compressed, with a membranous border.—A native of the Alps, Appenines, and Pyrenees, and other mountains of Europe.—Esenb. and Eberm. t. 199. St. and Ch. 132.

Other species also yield some of the Gentians of commerce, as the Alpine species, *G. purpurea, punctata,* and *pannonica;* while in the Himalayas *G. Kurroo* yields a similar product.

Fig. 79

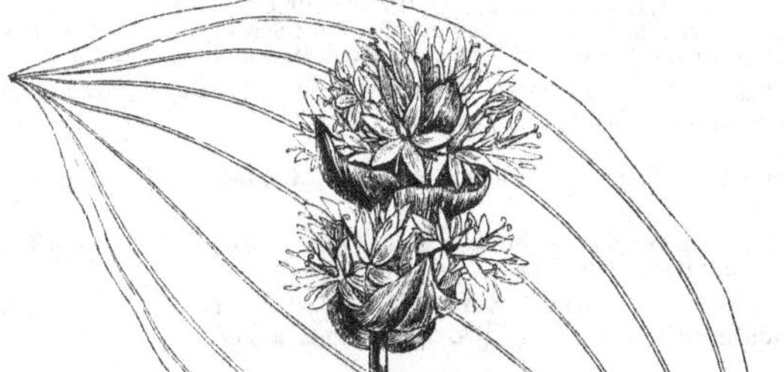

The root, which is supplied from Germany and Switzerland, is the only officinal part. France is supplied from Auvergne, &c. It varies in dimensions, but is usually about the thickness of the thumb, and several inches in length, often a little twisted, wrinkled, and of a brownish colour externally, yellowish within, rather soft, but tough ; odour feeble, but the taste at first slightly sweet, then of an intense but pure bitter. The properties are imparted readily to Water, Spirit, Wine, and Ether. The roots were found to contain Bitter Extractive matter, Gum, Uncrystallizable Sugar, Caoutchouc? Concrete Oil, Yellow Colouring matter, with a trace of Volatile Oil and an acid which has been named *Gentisic*, which in its impure state was supposed to be the active principle, but, when quite pure, is colourless, and in tasteless feebly acid crystals. Owing to the presence of Sugar, &c. Infusion of Gentian ferments with yeast, and yields a bitter distilled Spirit, prized by the Swiss and Tyrolese as a Stomachic.

Action. Uses. Bitter Tonic ; esteemed in Dyspepsia and in Convalescences. Like others of the same class, sometimes employed as an Antiperiodic and Anthelmintic.

D. Of the Powder gr. x.—gr. xxx. 3 or 4 times a day, or prescribe any of the following preparations.

INFUSUM GENTIANÆ (E.) COMPOSITUM, L. D. Compound Infusion of Gentian.

Prep. Macerate for 1 hour in a lightly covered vessel *sliced Gentian* (℥ß. E.) *dried* (and bruised bitter, E.) *Orange Peel* āā ʒij. (ʒj. E. D. Coriander bruised ʒj. E.), *fresh Lemon Peel* ʒiv. L. (ʒj. D.) in boiling *Aq. dest.* Oj. L. (℥xij. D.) Strain. (Take the solids and pour on them Proof Spirit f℥iv., after 3 hours add Cold Water f℥xvj., and in 12 hours more strain through linen or calico, E.)

Action. Uses. Aromatic Tonic. Useful in Dyspepsia, &c., and as a vehicle for acids, &c., in doses of f℥jß.

MISTURA GENTIANÆ COMPOSITA, L. Compound Gentian Mixture.

Prep. Mix *Compound Infusion of Gentian* f℥xij. *Compound Infusion of Senna* f℥vj. and *Compound Tincture of Cardamoms* f℥ij.

Action. Uses. Aperient and Tonic. Useful combination for extemporaneous use in doses of f℥jß. 2 or 3 times a day.

TINCTURA GENTIANÆ COMPOSITA, L. E. D. Compound Tincture of Gentian.

Prep. Macerate for 14 (7, E.) days *sliced* (and bruised, D.) *Gentian* ℥iiß. (℥ij D.) *dried* (bruised bitter, E.) *Orange Peel* ʒx. (℥j. D.) *bruised Cardamoms* ʒv. L. (ʒß. D.) (Canella finely powdered ʒvj. Cochineal bruised ʒß. E.) in *Proof Spirit* Oij. (by measure ℔ij. D.) Strain. (Express strongly and filter. Or, more conveniently prepare by percolation, as Comp. Tinct. Cardamom. E.)

Action. Uses. Tonic, Stomachic. Adjunct to bitter infusions in doses of ʒj.—fʒij.

EXTRACTUM GENTIANÆ, L. E. D. Extract of Gentian.

Prep. Prepare with Gentian Root and 8 times its weight of water, as other simple Extracts, D. Take *sliced* (finely powdered, q.s. E.) *Gentian* ℔ijß. and macerate it (mix thoroughly) with boiling *Aq. dest.* Cij.) half its weight of *Aq. dest.* E.) for 24 (12, E.) hours. (Put it into the percolator, and exhaust it with temperate Aq. dest. E.) Boil down to Cj. (concentrate, E.), and while hot (before it gets too thick, E.) filter. Evaporate to a due consistence (in the vapour-bath., E.)

Action. Uses. Tonic in doses of gr. v.—℈j. in pills, often with metallic salts. The Extract made from the Infusion is considered superior to that made from the Decoction; but that made according to the E. P. is still finer. (*c.*)

VINUM GENTIANÆ COMPOSITUM, E. Compound Wine of Gentian.

Prep. Digest *Gentian* ʒiv. *Yellow Cinchona Bark* ℥j. also *Bitter Orange Peel* ʒij. *Canella* ʒj., all coarsely powdered, in *Proof Spirit* f℥ivß. for 24 hours, then add *Sherry Wine* f℥xxxvj. and digest for 7 days. Strain and express the residue strongly, and filter the liquor.

Action. Uses. A good Stomachic in doses of fʒiv.—f℥j.

CHIRETTA, E. Herb and Root, E. of OPHELIA CHIRATA, *Griseb.* (*Agathotes Chirayta*, Don.) Chiretta, *Tetrand. Monog.* Linn.

The bitter called *Chiretta*, or *Chiraeta*, is as universally employed throughout the Bengal Presidency, as Gentian is in Europe. It has long been known to the Hindoos, but there is no reason to suppose it to be the *Calamus aromaticus* of the ancients. (*v.* Himal. Bot. p. 277.) The first English account is that of Dr. Fleming (in Asiat. Res. xi. p. 167), who referred it to the genus Gentiana, others to Swertia, and the late Prof. Don to Agathotes ; Grisebach now refers it to Ophelia. It is often confounded with another powerful Indian bitter, that is, Creyat, or Justicia paniculata. But there are several plants closely allied to the Chiretta, which are used for the same purposes, as stated in Him. Bot. p. 277. Thus, *Ophelia* (*Swertia*, Wall.) *angustifolia*, Don, is so in Northern India, and called *puharee*, (*i. e.* hill) *chiretta*, to distinguish it from the true or *dukhunee* (southern) *chiretta*. This is obtained from Nepal. *Exacum tetragonum* is called *ooda* (that is, purple) *chiretta*.

This is an annual, of from 2—3 feet high, with a single, straight, round, smooth stem. Branches generally decussated, nearly erect. Leaves opposite, amplexicaul, lanceolate-acute, smooth, 5—7-nerved. Flowers numerous, stalked, upper half of the plant forming elegant decussated umbel-like cymes, with 2 bracts at each division. Calyx 4-cleft, with sublanceolate persistent divisions, shorter than the corolla. Corol yellow, rotate, limb 4-parted, spreading, withering in æstivation, twisted to the right, with 2 glandular hollows protected by a fringed scale upon each segment. Stamens 4; filaments subulate, shortly connected at the base; anthers cloven at the base. Style single. Stigma large, 2-lobed. Capsules conical, rather shorter than the permanent calyx and corol, 1-celled, 2-valved, opening a little at the apex. Seeds numerous, affixed to two receptacles adhering to the sides of the valves.—Himalaya mountains, of which Nepal is one of the valleys.—Wall, Pl. As. Rar. 3. p. 33. t. 252.

Chiretta is met with in a dried state, tied up in bundles, with its long slender stems of a brownish colour, having the roots attached, and which have been taken up when the plant was in flower. The whole plant is bitter. Analysed by MM. Bousil and Lassaigne, they yielded a Resin, a yellow Bitter substance, a yellow colouring matter, Gum, Malic acid, Salts of Potash and of Lime, and traces of Oxide of Iron. Mr. Battley states that it contains—1. A free Acid. 2. A very Bitter Extractive and Resinous matter, and much Gum. 3. Muriate and Sulphate of Lime and of Potash ; also, that the Spirituous Extract is more aromatic than that of Gentiana lutea, but that the extractive and the gum are in larger proportion in the latter. Water and Spirit take up its active properties.

Action, Uses. Bitter Tonic ; Stomachic in Dyspepsia or as a Tonic in Convalescence ; either cold or hot infusion ; the former is lighter, and well suited to Dyspeptics, and not so apt to create nausea in a hot climate. Sometimes a little Orange-peel or Cardamom is added. A Tincture made as that of Gentian (or ℥ij. to Proof Spirit f℥xvj.), like other bitters, is best taken half an hour before meals.

INFUSUM CHIRETTÆ, E. Infusion of Chiretta.

Prep. Infuse *Chiretta* ℥iv. in *boiling Aq.* Oj. for 2 hours, and strain. A cold infusion, or one made with a temperature not exceeding 180°, is preferable. (*v.* Wall. *l. c.* p. 33.)

Action. Uses. Stomachic in doses of f℥iß.—f℥iij. before dinner, or twice a day.

MENYANTHES, L. E. D. Folia, L. D. Leaves, E. of MENYANTHES TRIFOLIATA, *Linn.* Buckbean or Marsh Trefoil. *Pentand. Monog.* Linn.

The Buckbean, though long employed, is now less so in European medicine.

Herbaceous, root-stock jointed, spreading horizontally, branched ascending stems, which are round and leafy. Leaves ternate, with a long alternate petiole, sheathing at base; leaflets oval or obovate, equal, wavy, a little irregular at the margin. Racemes erect, with several white, or light-lilac, beautifully fringed flowers, each opposite to a leaf. Calyx 5-parted. Corolla funnel-shaped; limb 5-parted, bearded internally with white fleshy hairs. Stamens 5. Stigma capitate, furrowed. Ovary with 5 hypogynous glands; Caps. ovoid, 1-celled, imperfectly 2-valved, with the placentæ in the axis of the valves. Seeds many, with shining testæ.—Bogs and marshes in most parts of Europe; extends to Cashmere, also to North America.—Esenb. and Eberm. 204. St. and Ch. 11. 85.

The stem and leaves are smooth, with little odour, but have a very bitter, somewhat nauseous taste. The expressed juice contains, according to Trommsdorff, a very bitter azotised Extractive (*Menyanthin*), a brown Gum, Inuline, Green Fecula, Malate and Acetate of Potash, and about 75 per cent. of water. Water and Alcohol take up its active properties.

Action. Uses. Bitter Tonic; in large doses, Cathartic and Emetic. Doses of the powdered leaves grs. xx., or of the Infusion (℥iv. to Aq. Oj.) f℥iß. two or three times a day.

CONVOLVULACEÆ, *R. Brown.* Bindweeds.

Herbs or shrubs, generally with milky, sometimes watery, juice; usually with a twining stem. Leaves alternate, without stipules. Flowers complete, pedicels axillary or terminal, commonly with two bracts. Calyx free, 5-leafed, often unequal, in one or more rows, persistent, sometimes united into a 5-fid tube. Corolla monopetalous, inserted into the receptacle, plaited and twisted. Stamens 5, often unequal, inserted into the tube, and alternate with the lobes of the corolla. Ovary free, seated on an annular disk, 2—4 celled, seldom 1-celled. Ovules solitary, or twin and collateral, erect. Style 1, often bifid. Stigmas simple or globose, in the undivided style bilobed. Fruit capsular, of 2, 3, or 4 valves, of which the margins touch the partitions projected from a central column. Seeds usually fewer than the ovules, with sparing mucilaginous albumen. Embryo curved; cotyledons crumpled.

The Convolvulaceæ are allied to Polemoniaceæ, and also to Solaneæ, and through Cordiaceæ also to Boragineæ. They abound in the plains and valleys of hot and tropical countries, some are found in the driest situations. The stems of many being annual, a few come to perfection in the summer of higher latitudes. Many are remarkable for the secretion of purgative principles, as in the Jalap, Scammony, Turpeth, *Ipomœa cærulea*, &c.

JALAPA, L. E. D. Radix, L. D. Root, E. of *Ipomœa Purga*, Wenderoth, E. (Ipomæa Jalapa, *Nuttal* and *Don*, L.) (Convolvulus Jalapa, *Willd.*) D. PURGA of the natives of Jalapa. Jalap Root.

Jalap has been known in Europe since 1609, having been introduced into England from the Mexican town of Jalapa, whence it has its name. It was at one time supposed to be produced by *Mirabilis Jalapa*, and then by *Convolvulus Jalapa*, Linn, called also *Ipomœa macrorhiza*, Mich. But all these grow in hot countries, while the Jalap, as long since stated by Humboldt (New Spain, vol. iii. p. 36) or the true "*Purga de Xalapa* delights only in a temperate climate, or rather an almost cold one, in shaded valleys, and on the slopes of mountains." The true Jalap plant seems to have been first sent from Mexico by Dr. Houston; at least, seed sent by him produced a plant which Miller has described in the 6th ed. of his Gardeners' Dict. as having smooth leaves, while the leaves of the other plant, or *Convolvulus* now *Batatas Jalapa*, are downy, especially on their under surface. In 1827, Dr. Coxe, Prof. of Mat. Med. in Pennsylvania, received from Xalapa several growing roots of the Jalap plant. Mr. Nuttal described them by the name of *Ipomœa Jalapa*, in Am. Journ. of Med. Sc. v. p. 300, Feb. 1830. Living roots were sent by Dr. Coxe to Dr. A. T. Thompson, and the description was inserted in his Dispensatory in 1831. The same plant is referred to by Mr. Don's MS. name in 1836. About the same time, or a little later, Ledanois sent the root to Paris; and Scheide, travelling in Mexico, collected at Chiconquiera on the eastern declivity of the Mexican Andes, at an elevation of 6000 feet, living plants and seeds of the true Jalap, and sent them to Germany, where they were cultivated and the plant named *Ipomœa Purga* by Wenderoth, Nees Off. Pfl. Suppl. iii. t. 13, and *I. Scheideana* by Zuccarini, Plant Nov. fasc. i. t. 12, and *I. officinalis* by G. Pelletan in France, *Exogonium Purga* by Mr. Bentham. Dr. Lindley says, "From an unpublished letter in the possession of the Horticultural Society of London, Don Juan de Orbegozo, a pupil of Cervantes, residing at Orezaba, it appears certain that this plant furnishes the Jalap of commerce." The plant is now cultivated in the open air on the Continent, at the gardens of the Horticultural Society, and that of the Society of Apothecaries. The author, by the liberality of the Hort. Soc., has been enabled to send roots to the Himalayas, where he hopes it will soon be established.

Ipomœa Purga, or *Jalapa* (Fig. 80.) The true Jalap plant, has a tuberous, fleshy rootstock, with numerous pear-shaped tubers, externally brownish coloured, internally white, with numerous long fibres. The stem, climbing to a great extent, is of a brownish colour, round and smooth, without downiness. Leaves on long foot-stalks, cordate, with a tendency to become hastate in the lower leaves, deeply sinuated at the base and acuminate at the apex, entire, very smooth. Peduncles axillary, 2-flowered, commonly only one blown at a time. Calyx without bracts; sepals 5, obtuse, mucronate, with 2 of them external. Corolla of a crimson or a light-red colour, with a long rather clavate

Fig. 80.

tube, four times longer than the calyx; limb undulated, with five plaits; lobes obtuse, subemarginate. Stamens five; filaments smooth, unequal, longer than the tube of the corol, with white, linear, exserted anthers. Stigma capitate, deeply furrowed. Capsule 2-celled; cells 2-seeded. — On the eastern declivity of the Andes of Mexico, at an elevation of about 6000 feet, where the climate is rainy but subject to frost in winter. Flowers in August and September. The tubers are gathered chiefly in the spring, when the young shoots are springing.—Nees & Eberm. Sup. 3. t. 13. Zuccarini, Plant. Nov. Fasc. 1. t. 12.

I. Mestilantica, Choisy. Dec. Prod. ix. 389, *I. orizabensis* of Pelletan, is considered to be another species closely allied to the former, and which grows in the temperate parts of the state of Oaxaca. Dr. Lindley (Fl. Med. p. 397) supposes that it may be the *Convolvulus orizabensis* of Pelletan, which Dr. Scheide had heard of under its Spanish name of *Jalapa Macho* or *Purga Macho*, or Male Jalap; but he had only seen the root, which appears very like that of *I. Purga*. Don J. de Orbegazo, as quoted by Dr. L., states that this is considered by the traders in Jalap to be extremely similar in quality, and as "it is the more abundant and larger of the two, at least in some districts," the probability is that it also forms a part of the imported samples of this drug.—Bot. Reg. 1841. t. 36.

Jalap tubers vary in size from a walnut to an orange, are usually pear-shaped or turnip-shaped, having often projecting from them smaller horn-shaped tubercles; the surface smooth, corrugated, or marked with slight furrows. The colour externally is blackish-grey. They are heavy and compact, with a brownish fracture, a very peculiar nauseous odour, and an acrid pungent taste. The larger tubers are sometimes divided into halves, quarters, or disks, and are always marked with circular or vertical incisions, made to facilitate their drying. When cut transversely, the section, if polished, appears very compact, and has the appearance of a deep-coloured wood, with still darker concentric circles, with many shining lines and points. The odour of Jalap, when cut or powdered, is strong and irritating. The powder is of a pale brownish colour.

Several adulterations are met with, especially in continental commerce, as the roots of the above *Ipomœa orizabensis*, called Stalk or Light Jalap, but by Guibourt *Jalap fusiform*, also a *False Jalap with a rose odour;* also the smaller roots of *Batatas Jalapa* of Choisy, (*Convolvulus macrorhizus* and *Jalapa* of authors,) those of a *Bryonia*, of a *Smilax*, and of *Mirabilis*, are sometimes intermixed.

Lately analysed by Guibourt, but without attempting to ascertain all the salts and principles, Jalap was found to contain of Resin 17·65 per cent., a liquid Sugar obtained, by Alcohol, containing some of the deliquescent salts, 19·00, Brown Saccharine extract, obtained by water 9·05, Gum 10·12, Starch 18·78, Woody matter, 21·60, loss 3·80 =100. This, as M. G. remarks, differs from the analyses hitherto given, but in the presence of Sugar, which he supposes to be of the nature of Cane Sugar, approximates Jalap to Batatas and other Jalap (as Rose Jalap) roots of the same family, which contain it. The Cathartic properties depend on the Resin: hence Rectified Spirit is the best solvent. Water takes up the Gum and Starch, with little of the active principle. Though Jalap is apt to be attacked by insects, its virtues are not in consequence impaired, for they leave untouched the resinous part. This Resin is of a greyish colour, opaque, brittle, acrid in taste, soluble in Alcohol, a little so in Ether, readily so in Nit′ or Ac′, and in solution of Potash. It is now often adulterated with Guaiacum, which may be detected by the blue colour produced by Nitrous gas, while Ether dissolves it, but does not dissolve pure Resin of Jalap. Dr. Kayser has named this Rhodoretine, from its producing a red colour with strong Sul′. He considers it composed of $C_{42} H_{35} O_{20}$.

PULVIS JALAPÆ COMPOSITUS, L. E. D. Comp. Jalap Powder.

Prep. Rub separately into very fine powder *Jalap* ʒiij. (℥j. E. ℔ß. D.) *Bitartrate of Potash* ℥vj. (℥ij. E. ℔j. D.) *Ginger* ʒij. Mix.

Action. Uses. Hydragogue Cathartic; useful in habitual Costiveness, &c. in doses of ℈j.—ʒj.

TINCTURA JALAPÆ, L. E. D. Tincture of Jalap.

Prep. Macerate for 14 days *powdered root of Jalap* ℥x. (℥vij. E. ℥viij. D.) in *Proof Spirit* Oij. Strain. (Prepare by digestion or percolation, *v.* Tinct. Cinchona, E.)—Contains the Resin of Jalap, with some of the principles soluble in water.

Action. Uses. Cathartic adjunct to Purgative draughts, in doses of fʒj.—fʒij.

EXTRACTUM (SIVE RESINA, E.) JALAPÆ, L. D. Extract or Resin (E.) of Jalap.

Prep. L. D Macerate *powdered Jalap root* ℔ijß. (℔j. D.) in *Rectified Spirit* Cj. (by measure ℔iv. D.) for four days. Pour off the Tincture. Boil the residue in

Aq. dest. Cij. (Cj. D.) to Cß. (℔ij. D.) Strain the liquors, evaporate the decoction, and distil the Tincture till thick; then mix the Extract and the resin, and evaporate to the proper consistence over a water-bath. The Extract must be kept *soft* for pills and hard for powder. L.

E. "Take finely powdered *Jalap,* moisten with *Rectified Spirit:* in 12 hours put it into the percolator and exhaust with Rectified Spirit. Distil off the greater part of the Spirit, and concentrate the rest to a due consistence over the vapour-bath."—As the active properties of Jalap depend upon its resinous and not upon its saccharine, gummy, or amylaceous principles, this preparation is preferable to those of the L. and D. P., where the extracts of water and of spirit are mixed together, and the produce necessarily weaker than the Resin.

Action. Uses. Cathartic in doses of gr. v.—℈j. Usually prescribed in combination. The Resin of the E. P. in doses of gr. iij.—gr. xij.

SCAMMONIUM, L. E. D. Gummi Resina, L. D. Gummy-Resinous Exudation from incisions into the root of CONVOLVULUS SCAMMONIA, *Linn.* E. Scammony.

Scammony has been employed in medicine since the time of Hippocrates. It is called *suk moonya* by the Arabs. Several varieties may be met with in commerce.

The Scammony Convolvulus has perennial tapering roots from 3 to 4 feet long and from 9 to 12 inches in circumference, fleshy, and abounding in acrid milky juice. Stems numerous, annual, round, slender, smooth, and twining over neighbouring plants, and to a great extent over the ground. Leaves petioled, quite smooth, entire, oblong, arrow-shaped, acute, truncate, and angular at the base, with acute spreading lobes. Peduncles axillary, solitary, 3-flowered, about twice the length of the leaves. Sepals rather lax, smooth, ovate-obtuse, with a reflexed point. Corol campanulate, much expanded, of a pale sulphur-yellow colour, three times as long as the calyx. Stamens 5, erect, converging, about a third of the length of the corol. Style equal to the stamens. Stigmas white, oblong, erect, parallel, distant. Ovary 2-celled, 4-seeded. Capsule 2-celled.—Common in Greece and the Levant.—Esenb. and Eberm. 195.

A Convolvulus, which is also called *C. Scammonia,* found by Capt. D'Urville in the island of Cos, having yellow flowers with reddish bands (Fl. Med. t. 317), is supposed to yield Scammony. Tournefort informs us that an inferior Scammony is obtained in Natolia, whence it is sent to Smyrna, and hence called Smyrna Scammony. Sibthorp says that Scammony is produced by two different species of Convolvulus, one the above *C. Scammonia,* and the other (perhaps *C. hirsutus*) has been supposed to be *C. farinosus,* Linn.; but, as Dr. Lindley justly observes, this is a Madeira plant, and has probably nothing to do with producing Scammony.

The root-stock of the Scammony Convolvulus was found by Dr. Russel to be a mild Cathartic. Scammony is the juice of the fresh root obtained by cutting the top obliquely off, and allowing the milky juice which exudes to be collected in shells or other vessels placed at the lowest part. The whole collected is allowed to dry in any convenient receptacle, and constitutes what is called *Virgin Scammony,* but this is very seldom to be met with in so pure a state. The greater part of that met with in English commerce is imported from Smyrna. The best accounts have been given by Drs. Pereira and Christison.

Scammony is usually in shapeless lumps, rubbed and of a dull ash-grey colour externally; the fracture is conchoidal, and when fresh, displays a glistening resinous lustre, of a pale, soon passing to a dark greenish-black colour, something like Guaiacum ; a small splinter is grey and somewhat transparent; Sp. Gr. 1·2; the whole is brittle, easily pulverized; powder of an ash-grey colour. The odour is faint, but peculiar, more perceptible if breathed upon, sometimes compared with that of old cheese; taste slight, but acrid. It should burn away without leaving much ash; form an emulsion with water, and dissolve almost entirely in boiling Alcohol, while Ether will take up from 75 to 82 parts of Resin. Dr. Christison gives as the constituents of two distinct specimens of old Scammony, Resin 81·8 and 83·0, Gum 6·0 and 8·0, Starch 1·0 and 0·0, Fibre and Sand 3·5 and 3·2, water 7·7 and 7·2. Hence it is a Gum-resin, with only a small proportion of Gum.

The Resin has a feeble Scammony odour and taste, and a dirty greenish-brown colour; but when purified, it is of a pale wine-yellow colour, and is free from both taste and smell. Its powder forms with milk a fine uniform emulsion.

The less pure kinds of Scammony, which are also the more common, and enumerated by Dr. Pereira as the *seconds* and *thirds* of commerce, are distinguished by their greater weight, less resinous, rather dull fracture, and greyish, sometimes blackish colour, sometimes with glimmering or whitish spots; also by their form, being sometimes that of the vessel in which they have been packed, sometimes in flattish cakes, at other times in amorphous spongiform masses. Some effervesce with H Cl′ from being adulterated with Chalk; others, from containing Starch, are affected by Tincture of Iodine. (*v.* Tests.)

Tests. While yet in a soft state, it is said to have mixed with it the expressed juice of the stalks and leaves, also flour, ashes, and sand. Dr. Pereira enumerates chalk, amylaceous matter, sand and guaiacum as impurities; but tragacanth is also mentioned. E. P. "Fracture glistening, almost resinous if the specimen be old and dry. Muriatic acid does not cause effervescence on its surface (if no chalk has been added). The decoction of its powder, filtered and cooled, is not rendered blue by Tincture of Iodine (showing the absence of Starch). Sulphuric Ether separates at least (75 to) 80 per cent. of Resin dried at 280°." Some of the masses appear to have been rolled in chalk, but do not contain any in their substance. Guaiacum may be detected by the action of Nitrous gas, and sand and chalk in the ashes after incineration.

Action. Uses. Drastic Cathartic. Useful from the small doses in which it can be prescribed; as for an adult, gr. x.—gr. xv.; but if pure or Virgin Scammony, gr. v.—gr. x. will suffice. It is usually given in combination with Rhubarb or Calomel, or in the following preparations; sometimes in biscuits.

Pharm. Prep. Extr. Colocynthidis Comp.

PULVIS SCAMMONII COMPOSITUS, L. D. Comp. Scammony Powder.

Prep. Rub up separately into very fine powder *Scammony* and *hard Extract of Jalap* āā ℥ij. *Ginger* ℥ß. Mix.

Action. Uses. Cathartic. May be given in doses of gr. x.—ʒß.

CONFECTIO (ELECTUARIUM, D.) SCAMMONII, L. Scammony Confection.

Prep. Rub into fine powder *powdered Scammony* ℥iß. *bruised Cloves* and *powdered Ginger*, āā ʒvj. When the Confection is to be used L. add *Syrup of Roses* q. s. and *Oil of Caraway* f℥ß. Mix well together.

Action. Uses. Stimulating Cathartic in doses of ℈j.—ʒj.

EXTRACTUM SIVE RESINA SCAMMONII, E. Scammony Resin.

Prep. Take *Scammony* q. s. in fine powder; boil it in successive portions of Proof Spirit till the Spirit ceases to dissolve anything; filter; distil the liquid till little but water passes over. Then pour away the watery solution from the resin at the bottom; agitate the resin with successive portions of boiling water till it is well washed; and, lastly, dry it at a temperature not above 240°.

Action. Uses. Active Cathartic in doses of gr. v.—gr. x. with some bland fluid, such as milk.

MISTURA SCAMMONII, E. Scammony Emulsion.

Prep. Triturate *Resin of Scammony* grs. vij. with a little, and then with the rest of, *unskimmed Milk* f℥iij. till a uniform emulsion is obtained.

Action. Uses. Cathartic Emulsion, without any disagreeable taste.

LABIATÆ, *Juss.* Labiates.

Herbaceous plants, undershrubs or shrubs; branches opposite or whorled, often 4-cornered. Leaves opposite or in whorls, simple, entire, or divided, without stipules, usually containing odorous volatile oil. Flowers perfect, irregular, often bilabiate, usually in axillary cymes, sometimes solitary. Calyx tubular, persistent, 5-toothed, or bilabiate. Corolla inserted into the receptacle, irregular, 4—5 fid, often bilabiate, imbricate. Stamens inserted into the corolla, 4 didynamous, sometimes the 2 upper ones wanting. Ovaries 4, free, seated on an hypogynous disk, 1-celled, each with one erect ovule. Style single, in the midst of the ovaries. Stigma bifid. Fruit consisting of 4 nuts, sometimes only one, included within the calyx. Embryo straight; radicle below; no albumen. —The Labiatæ are most closely allied to *Verbenaceæ* and to *Boragineæ*, more remotely to *Scrophularineæ*. They are found in most parts, but more numerously in the Old than in the New World, and most abundantly in temperate climates. They abound in volatile oil, usually containing Stearoptene, often also a little bitter and astringent principle.

Tribe *Menthoideæ.* Corolla nearly regular. Stamens distant, straight.

LAVANDULA, L. E. D. Flores, L. D. Flowering Heads and Volatile Oil of LAVANDULA VERA, *Dec.* E. (L. Spica, *Dec.* L.D.) Common Lavender.

It is unknown when Lavender was first employed in medicine.

Common Lavender forms a branched shrub, about 4 feet high. Leaves oblong, linear or lanceolate, entire, when young hoary, revolute at the edges. Spikes interrupted. Whorls of 6 to 10 flowers. Floral-leaves rhomboid-ovate, acuminate, membranous, all fertile, the uppermost shorter than the calyx. Bracts scarcely any. Flowers purplish-grey. Calyx tubular, nearly equal, shortly 5-toothed, 13- or rarely 15-ribbed. Corol, upper lip 2-lobed, lower 3-lobed; all the divisions nearly equal; the throat somewhat dilated. Stamens didynamous, declinate. Filaments smooth, distinct, not toothed. Anthers reniform, 1-celled. Ovary and fruit as in the order.—A native of barren hills in Europe, extending to the north of Africa. Cultivated in gardens; much so at Mitcham, in Surrey. Tops collected in June and July.—Esenb. and Eberm. t. 178.

LAVANDULA SPICA, *Dec.* French Lavender, sometimes called *L. latifolia*, is a distinct species from *L. vera*, but indigenous in the same countries. It may easily be distinguished by its leaves being broader and somewhat obovate or spathulate. Its odour is not so agreeable as that of Common Lavender, though more powerful.—Esenb. and Eberm. 179 as *L. latifolia.*

Lavender flowers, or rather tops, as usually dried, are well known by their spike-like appearance, greyish *lavender* colour, grateful fragrant odour, and warm bitterish taste. The properties depend chiefly on the presence of Volatile Oil.

Action. Uses. Stimulant, Carminative. Employed as an Errhine in Pulvis Asari Comp. D.

OLEUM LAVANDULÆ, L. E. D. (English) Oil of Lavender.

Prepared from Lavender Flowers distilled with water, as other volatile oils.

This Oil is of a light yellow colour, has a very grateful odour, and a pungent taste. Sp. Gr. 0·87 to 0·94. It consists of a fluid volatile oil, holding the camphor-like substance in solution, which has been called Stearoptene. It is soluble in Rectified- and in two parts of Proof Spirit. Like several other volatile oils, it will absorb Oxygen, and become acid. This Oil is apt to be adulterated with the Oil of French Lavender, commonly called *Oil of Spike*, which is a powerful but less agreeable oil.

Action. Uses. Stimulant, Carminative, in doses of ♏v.—♏x.

SPIRITUS LAVANDULÆ, L. E. D. Spirit of Lavender.

Prep. Macerate (for 24 hours, D.) *fresh Lavender flowers* ℔ijß. (℔ij. D.) in *Rectified* (Proof, D.) *Spirit* Cj. and *Aq.* Oij. L. (q. s. to prevent empyreuma, D.) with a gentle heat. Distil Cj. (Ovij. E. by measure ℔v. D.)

Action. Uses. The Volatile Oil rises with and is dissolved in the Spirit: hence this is sometimes prepared by dissolving the Oil in Rectified Spirit. It is chiefly used for making Lin. Camphoræ C. and the following preparations. It approaches in nature the so-called *Lavender Water* of the shops, which however also contains other volatile oils dissolved in Spirit.

TINCTURA (SPIRITUS, E. D.) LAVANDULÆ COMPOSITA, L. Compound Tincture or Spirit of Lavender. *Lavender Drops.*

Prep. Macerate for 14 (7, E. 10, D.) days bruised *Cinnamon* ʒijß (℥j. E. ℥ß. D.) bruised *Nutmeg* ʒijß. (℥ß. E. D.) *Red Sandal Wood Shavings* ʒv. (ʒiij. E. ℥j. D.

Cloves bruised ʒij. E. D.) in *Spirit of Lavender* Ojß. (Oij. E. by measure ℔iij. D.) and *Spirit of Rosemary* Oß. (f℥xij. E. by measure ℔j. D.) Strain.

Action. Uses. This compound Tincture contains the Volatile Oil of Lavender, and that of the other aromatics used, dissolved in Spirit and coloured by the Red Sandal Wood. It is Stimulant and Cordial; is used in Hysterical cases and in Flatulent Colic in doses of ♏xv.—fʒij.

MENTHA, *Linn.* Mint.

Calyx nearly equal, 5-toothed. Corol with the tube inclosed; limb nearly equal, 4-cleft, the upper segment broader. Stamens 4, equal; anthers with 2 parallel cells. Stigmas at the points of the bifid style. Fruit dry, smooth.

Several of the Mints, remarkable for their odour and taste, have long been used in medicine (μινθα, 'Ηδυοσμος and Καλα μινθα of the Greeks, *nana* of the Arabs), and some as sweet herbs; but it is difficult to distinguish one species from another by the short descriptions given.

MENTHA VIRIDIS, L. E. D. The whole Herb. Spearmint.

Spearmint has long been employed in medicine.

Root creeping. Stem smooth, erect. Leaves sessile, lanceolate, acute, unequally serrated, glabrous, glandular below, those under the flowers bract-like, these and the calyxes hairy or smooth. Spikes linear-cylindrical; bracts subulate. Whorls approximated, or the lowest or all of them distant. Corol glabrous. Stamens rather long.—Marshy places in the milder parts of Europe, introduced into many parts of the world. Collected when about to flower.—E. B. 2424. Esenb. and Eberm. 166.

This plant has an agreeable odour, and a pleasant aromatic taste, with some bitterness.

Action. Uses. Stimulant, Carminative, in some of the following forms:

OLEUM MENTHÆ VIRIDIS, L. E. D. Oil of Spearmint.

Prep. Distil the fresh herb with Aq. as for other volatile oils.

Action. Uses. Pale yellow in colour, becoming reddish by age, of a strong, rather grateful odour, and pungent taste, followed by a sensation of coolness, giving its properties to the plant (of which it forms about 1-500th part), and also to the preparations. Stimulant, Carminative, in doses of ♏ij.—♏x.

INFUSUM MENTHÆ SIMPLEX, D. Infusion of Spearmint.

Prep. Take dried leaves of *Mentha viridis* ʒij. pour on boiling *Aq.* q. s. and strain off by measure ℥vj.

INFUSUM MENTHÆ COMPOSITUM, D. Comp. Inf. of Spearmint.

Prep. Digest for half an hour in a covered vessel dried leaves of *Mentha viridis* ʒij. in boiling *Aq.* q. s. Strain off by measure ℥vj. When cold, add *purified Sugar* ʒij. *Oil of Mint* ♏iij. dissolved in *Comp. Tinct. Cardamoms*, ℥ß. Mix.

Action. Uses. Stomachic and Carminative Infusions: the latter more stimulant: given in doses of f℥iß. every 2 or 3 hours.

AQUA MENTHÆ VIRIDIS, L. E. D. Spearmint Water.

Prep. To be prepared as Aq. Menth. Pip. If the fresh herb be employed, take double the weight of the dried, L. Mix *Spearmint*, fresh, ℔iv. if dry, ℔ij. (℔jß. D.) *Rectified Spirit* f℥iij. *Aq.* Cij. (q. s. to avoid empyreuma, D.) Distil off Cj. E. D. Or add *Essential Oil of Spearmint* ℥iij. to each gallon of *Aq.* and distil. D.

Action. Uses. Carminative, and used as a vehicle in doses of f℥iß.

SPIRITUS MENTHÆ VIRIDIS, L. D. Spirit of Spearmint.

Prep. Prepare as Spir. Menth. Pip. L. Add *Oil of Spearmint* by weight ℥ß. to *Rectified Spirit* Cj. *Aq.* q. s. to prevent empyreuma. With gentle heat distil Cj.

Action. Uses. Stimulant adjunct in doses of fʒß.—fʒij.

MENTHA PIPERITA, L. E. D. *Linn.* Herba, D. Herb and Volatile Oil, E. of Peppermint.

Peppermint seems to have been introduced into practice in this country in the last century.

Root creeping. Stem procumbent, ascending, smooth, or with a very few spreading hairs. Leaves stalked, ovate, lanceolate, acute, rounded at the base, smooth, serrated, floral leaves smaller, lanceolate. Spikes lax, the uppermost whorls collected into a short obtuse spike, the lower ones removed from each other. Calyx tubular, glabrous below, with lanceolate subulate teeth.—Watery places in England, and also in other parts of Europe.—Cultivated at Mitcham, and collected when the flowers begin to blow. Flowers from July to September. —Esenb. and Eberm. 165. E. B. 687. St. and Ch. 45.

Peppermint is remarkable for its diffusive aromatic odour, and its warm but agreeable taste, feeling at first warm, but afterwards cool. Its properties depend on a Volatile Oil, a Bitter principle, and some Tannin, and these are taken up by Spirit, and to some extent by water. It may be prescribed in the form of its Oil or Spirit, or in its distilled Water, or in an Infusion.

Action. Uses. Stimulant, Carminative. Much used in Flatulent Colic, &c., or where a diffusible Stimulant is indicated, or a medium required to counteract nausea or griping, or to cover the taste of other Medicines.

OLEUM MENTHÆ (PIPERITIDIS, D.) PIPERITÆ, L. E. Oil of Peppermint.

Prep. Distil the fresh or dry herb with water, as above.

Peppermint Oil, obtained in the proportion of about a 200th part, is at first colourless, but soon becomes of a pale greenish-yellow colour, and of a deeper colour with age, has a fragrant penetrating odour, and a pungent but cooling taste. Sp. Gr. 0·902 (0·899, *Per.*). Boils at 365°. At a temperature of —12°, or by spontaneous evaporation, or pressure, white needle-like crystals of Stearoptene are obtained. But from some kinds of Oil from North America and also from Canton, this Stearoptene separates spontaneously. This Oil is composed of $C_{12}H_{10}O$, and its Stearoptene of $C_{10}H_{10}O$. According to Walter, these numbers should be doubled.

Action. Uses. Stimulant, Carminative, in doses of ♏ij.—♏v. on a piece of Sugar.

AQUA MENTHÆ (PIPERITIDIS, D.) PIPERITÆ, L. E. Peppermint Water.

Prep. Take dried *Mentha Piperita* ℔ij. or fresh, ℔iv. (℔jß. D.) or *Oil of Peppermint* ʒij. (ʒiij. D.) *Proof Spirit* f℥vij. L. *Aq.* Cij. (q.s. to prevent empyreuma, D.) Distil Cj. Prepare as Aq. Menth. Vir. E.

Action. Uses. Carminative. Much used as a vehicle for other medicines in doses of f℥j.—f℥iij.

SPIRITUS MENTHÆ (E.) PIPERITÆ, L. D. Spirit of Peppermint.

Prep. Mix *Oil of Peppermint* ʒiij. L. (by weight ℥ß. D.) with *Proof Spirit* Cj. and *Aq.* Oj. L. (Rectified Spirit Cj. Aq. q. s. to prevent empyreuma, D.) With heat slowly distil Cj. (Take of Peppermint fresh ℔jß. Proceed as for Spirit of Caraway, E.)

Action. Uses. Stimulant in doses of fʒß.—fʒij. Essence of Peppermint consists of Oil of Peppermint fʒj. dissolved in Rectified Spirit f℥j.

MENTHA PULEGIUM, *Linn.* Herba, D. Herb, E. of Pennyroyal.

Supposed to have been the Γλήχων of the Greeks and the Pulegium of Pliny.

Creeping root. Stem much branched, prostrate, rooting. Leaves about half an inch long, stalked, ovate, or elliptical, crenate, upper ones smaller, all with pellucid dots, a little hairy. Whorls sessile, all remote, globose, many-flowered. Calyx hispid, tubular, bilabiate, villous in the inside of the throat. Corols of a light purple.—Wet places in many parts of Europe. Collected when beginning to flower.—E. B. 1026. Esenb. and Eberm. 167. St. and Ch. i. 45.

The whole herb has a powerful fragrant odour, and warm, aromatic, as well as bitter taste. Its properties depend on Volatile Oil and Tannin, and are very similar to those of other species of Mint.

OLEUM MENTHÆ PULEGII, L. E. D. Oil of Pennyroyal.

Obtained by distilling the herb with water.

Action. Uses. Stimulant, Carminative, in doses of ♏ij.—♏v.

AQUA MENTHÆ PULEGII, L. E. D.

Prep. Employing the fresh herb (if dried, half the weight L.), distil as for Aq. Menth. Pip. L. D., Aq. Menth. Vir. E. Or, by adding to *Aq.* Cj. *Essential Oil of Pennyroyal* ʒiij. D.

SPIRITUS MENTHÆ PULEGII, L. D. Spirit of Pennyroyal.

Prep. Add *Oil of Pennyroyal* ʒiij. (℈vj. D.) to *Proof Spirit* Cj. and *Aq. dest.* Oj. L. With gentle heat distil (Cj. L.)

Action. Uses. These preparations are applicable to the same purposes and in the same doses as the preparations of Mint and of Peppermint.

Tribe *Monardeæ*. Corolla 2-lipped. Stamens 2, fertile, parallel under the upper lip.

The tribe *Monardeæ* contains SALVIA OFFICINALIS, *Linn.*, or Garden Sage, which has been employed in medicine from the times of the Greeks, and is no doubt as useful as any of the other Labiatæ for many of the same purposes. It is pungent and aromatic, and its Oil contains Stearoptene, while the plant abounds also in Bitter principle.

ROSMARINUS OFFICINALIS, *Linn.* L. E. D. Cacumina, L. D. Tops E. of Common Rosemary. *Diandria Monog.* Linn.

Rosemary was called *Libanotis coronaria*, which the Arabs translated *akleel-al-jibbul*, or the Mountain Crown.

A very leafy shrub, 5—6 feet high. Leaves sessile, elongated, narrow, revolute at the margin, hoary beneath. Flowers few, in short, axillary, subsessile, opposite racemes, forming altogether a kind of spike. Floral leaves shorter than the purplish calyx, which is 2-lipped, the upper entire, the lower bifid. Corol of a grayish-blue or lavender-colour, not ringed in the inside, somewhat inflated in the throat, upper lip emarginate, the lower trifid, with the middle lobe larger, concave, and hanging down. Filaments shortly toothed near the base: anthers linear, with two divaricating confluent cells. Upper lobe of style very short.—Rocky hills of the south of Europe, Asia Minor, and Syria.—Flora Græca, t. 14. St. and Ch. i. 24.

Rosemary-tops should be collected when coming into flower. They have a powerful odour, a warm and bitter, slightly astringent taste. Their properties depend on Volatile Oil, Bitter principle, and Tannin.

Action. Uses. Stimulant, Carminative. Supposed to be useful in preserving the hair. Much employed as an ingredient in some perfumes, as Hungary Water, Eau de Cologne. "The admired flavour of Narbonne Honey is ascribed to the bees feeding on the flowers of this plant." *Lindl.*

OLEUM ROSMARINI, L. E. RORISMARINI, D. Oil of Rosemary.

Distil Rosemary-tops with water.

The Oil of Rosemary is obtained in the proportion of 4 or 5 ounces from a cwt. of the herb: sometimes scarcely any is yielded. (P. J. ii. 516.) It is colourless, having all the properties of the plant. Sp. Gr. 0·88. The imported Oil is usually very impure.

Action. Uses. Stimulant, chiefly applied externally, and used as an ingredient of perfumes.

SPIRITUS ROSMARINI, L. E. RORISMARINI, D. Spirit of Rosemary.

Prep. Mix *Oil of Rosemary* ʒij. L. (℈vj. D.) with *Rectified Spirit* Cj. and *Aq.* Oj. (Proof Spirit Cj. D.), and with a gentle heat distil Cj. L. Or the D. and E. C. direct as follows: Take (fresh, D.) *Rosemary-tops* ℔ijß. (℔jß. D.), (Rectified, E.) *Proof Spirit* Cj. and with a gentle heat distil ℔v. D. Proceed as for

Spirit of Lavender, E.—Mr. Fisher prefers the L. formula of 1815, of *fresh tops* ℔ij. to *Rectified Spirit* Cij. If only this quantity is distilled, some portion of the water which is put into the still to prevent burning, will necessarily rise and dilute the Spirit. (*v.* Linim. Saponis, p. 477.)

Action. Uses. Stimulant Spirit. Often employed to impart an agreeable odour to Lotions. An ingredient of Tinct. Lavandulæ Comp. and of Linimentum Saponis.

Tribe *Satureineœ.* Cor. 2-lipped. Stamens 4, distant. Anther-cells separate, divergent.

ORIGANUM VULGARE, L. E. D. Herb, E. Oleum, D. Common and Wild Marjoram.

The ὀρίγανος of the Greeks and *satar* of the Arabs is supposed to be this plant.

Root creeping. Stem erect, 1—2 feet high. Leaves stalked, broad, ovate, obtuse, often slightly serrate. Spikes oblong, 4-sided, imbricated, with bracts, clustered in corymbose panicles. Bracts ovate, obtuse, coloured, longer than the calyx, which has 5 equal teeth, and is 10—13-nerved, throat hairy. Corol upper lip straight, nearly flat; lower spreading, 3-fid. Stamens divergent, connective subtriangular. Achænia rather smooth.—Europe, the Mediterranean region, and extending to the Himalayas.—E. B. 1143. St. and Ch. 131.

ORIGANUM MAJORANA, *Linn.* now MAJORANA HORTENSIS, *Mœnch.*, or Sweet Marjoram, a native of the South of Europe and of Syria, is officinal in the D. P. on account of its agreeable odour and pleasant aromatic taste, on which account it is cultivated in gardens and much used as a Sweet Herb.

Wild Marjoram has a strong rather agreeable odour, and a bitter aromatic taste, which it retains in its dry state. Its properties depend chiefly on its Volatile Oil.

Action. Uses. Stimulant, Carminative. May be used in Infusion.

OLEUM ORIGANI, L. D. Oil of Marjoram: called *Oil of Thyme.*

Distil the herb with water. A reddish oil is obtained, which becomes colourless on redistillation.

Action. Uses. Stimulant, in doses of ♏v. ♏x. Chiefly used externally, with Olive Oil, &c., as a remedy for toothache.

Tribe *Melissineœ.* Corol 2-lipped. Stamens distant. Anther-cells connected above.

MELISSA OFFICINALIS, *Linn.* E. D. Herba, D. Herb, E. Common Balm. *Didyn. Gymnospermia*, Linn.

This plant is supposed to be the Μελισσοφυλλον of Dioscorides.

Stem branched, 1—2 feet high. Leaves ovate, acute, cordate at base, crenate. Flowers white, in axillary unilateral racemes. Calyx 13-nerved, subcampanulate, slightly ventricose in front, 2-lipped, upper lip flat, truncate, with 3 short broad teeth, lower with 2 lanceolate teeth. Corol, upper lip concave, lower spreading, trifid, with apices of stamens connivent under the upper lip of the corol. Anther-cells divergent.—South of Europe; cultivated in English gardens. —Esenb. and Eberm. 180.

Balm has an agreeable odour, like that of the Citron, and a mild aromatic taste, with a little astringency, its properties depending, as in the other Labiatæ, on volatile Oil, Bitter principle, and Tannin.

Action. Uses. Mild Stimulant. Much used on the Continent in the slighter Nervous affections, generally in the form of Infusion (℥iv. —Aq. Oj.) or *Balm Tea.*

Tribe *Stachydeæ.* Stamens approximating, parallel under the upper lip of the corol, 2 inferior largest. Calyx tubular or bell-shaped. spreading in front.

MARRUBIUM VULGARE, *Linn.* L. D. Herba, D. White Horehound.

This is considered to be the πρασιον of the Greeks, and Marrubium of Pliny.

Stem bushy, erect, hoary. Leaves ovate and attenuated into a petiole, or roundish-cordate, crenate, surfaces wrinkled and veiny, more or less woolly. Flowers many, white, in dense whorls. Calyx woolly, with 10 subulate, recurved, spreading teeth. Corol upper lip erect, cloven, lower 3-lobed, middle lobe the largest. Stamens included within the tube of the corol. Anther-cells divaricating, bursting longitudinally. Style with short obtuse lobes. Achænia flatly truncate.—Europe and northern parts of Asia.—E. B. 410. Nees von E. 174.

Horehound has an aromatic, somewhat musky odour, and a warm and bitter taste. It contains Volatile Oil, a Bitter principle, and Tannin (Gallic acid, *e. & v.*)

Action. Uses. Stimulant and Tonic. Much used in popular medicine, in Chronic Catarrhs, and supposed also to possess some Emmenagogue properties, in the form of Infusion (℥iv.—Aq. Oj.)

SCROPHULARINEÆ, *R. Brown.* Figworts.

Herbs or undershrubs, with roundish, or four-cornered, jointed stems. Leaves alternate or opposite, sometimes whorled, entire or cut, sometimes pinnately so, decurrent, without stipules. Flowers complete, irregular, sometimes regular, inflorescence various. Calyx persistent, 5- or 4-parted. Corolla inserted into the receptacle, irregular or unequal, 5-lobed, imbricate. Stamens inserted into the corolla, and usually fewer than its lobes, often 4, didynamous, sometimes 2; anthers without appendages. Ovary free, 2-celled, many-ovuled. Style 1 Stigma entire or 2-lobed. Fruit 2-celled, or a berry; capsules opening by valves or opercula. Embryo in the axis of a fleshy albumen, straight, with the radicle towards the hilum.—The Scrophularineæ are closely allied on one side to *Labiatæ*, &c. and on the other to *Solaneæ;* in fact, Verbascum is sometimes placed in Solaneæ, sometimes in Scrophularineæ. These are found in all parts of the world, but chiefly in temperate climates, though some are to be found within the tropics, and a few in cold countries. They are chiefly mucilaginous, some slightly acrid and bitter; but Digitalis is possessed of active properties.

SCROPHULARIA NODOSA, *Linn.* Folia, D. Leaves of Knotted Figwort. *Didynamia Angiospermia*, Linn.

Scrophularia has long been an article of domestic medicine.

Root thick and knotty, whence the specific name of this species. Stems 2—3 feet high, acutely 4-angled. Leaves ovate, acute, subcordate, smooth, deeply serrate, lower serratures largest, all acute. Inflorescence a lax cyme. Calyx 5-lobed, divisions roundish-ovate, with a narrow membranous margin. Corol globose, of a greenish-purple colour, sometimes white; limb minute, of 2 short lips, upper 2-lobed, lower 3-lobed. The rudiment of the fifth stamen transversely oblong, slightly emarginate. Capsules ovate, opening by 2 valves, with their margins inflexed, 2-celled.—Indigenous in moist situations; flowers in July.—E. B. 1544.

The leaves have a rather disagreeable odour, and a bitter, slightly acrid taste.

Action. Uses. Possessed of a little irritant property; hence its fomentation was formerly applied, as well as its ointment, to cutaneous affections, tumours, &c.

UNGUENTUM SCROPHULARIÆ, D. Ointment of Scrophularia.

Prep. Boil fresh leaves of *Scrophularia nodosa* ℔ij. in *prepared Hogs' Lard* ℔ij. and *prepared Mutton Suet* ℔ij. till crisp. Strain by expression.

DIGITALIS FOLIA, L. D. Leaves, E. and DIGITALIS SEMINA, L. Seeds of DIGITALIS PURPUREA, *Linn.* Foxglove.

Foxglove does not appear to have been known to the ancients. Fuchsius was the first to describe it, and to name it Digitalis, from the resemblance of its flowers to the finger of a glove. It was admitted into the L. P. of 1668 and 1721, rejected in that of 1745. Withering brought it into permanent notice in 1775.

Biennial. Root fibrous. In the first year a tuft of radical leaves is thrown up, from the midst of which rises, in the second year, a Stem 1—5 feet high, which is erect, wand-like, and leafy, slightly angled and downy, in some varieties with a purple tinge, as well as on the lower surface of the leaves. Leaves alternate, ovate-lanceolate, or oblong, crenate, and rugose, downy, especially on the under surface, tapering at the base into winged footstalks. Racemes terminal, long, and lax, on which the pendulous flowers appear on one side in slow succession. Flowers crimson, purple, marked with eye-like spots, and hairy within, sometimes white. Calyx 5-parted, segments ovate, or oblong-acute. Corol declinate, much longer than the calyx, contracted at the base, campanulate and ventricose above, with an oblique limb; upper limb emarginate, lower 3-fid, with the middle lobe the largest, all short, obtuse. Stamens 4, didynamous, ascending; anthers smooth. Stigma bilamellate. Capsule ovate-acute, with a septicidal dehiscence. Seeds very small, of a pale brownish colour, and pitted.—Indigenous and also common chiefly in the western parts of the Continent: found on pastures and exposed hill sides, as also in plantations: begins to flower in June and July, and ripens its seed in August and September.—Nees von E. 154. St. and Ch. i. 18.

Both the leaves and seeds are officinal, but the latter are seldom employed. The roots, collected in the autumn or winter of their first year, are possessed of active properties. The leaves of this plant, like the leaves of all biennial plants, Dr. Houlton says, should be gathered in the second year of their duration, and as soon as possible after the first flowers have expanded: he also prefers those of the plants having a purplish stem. Dr. Christison, however, thinks this a needless—it is at least a safe—restriction. He has observed that their bitterness, which probably measures their activity, is very intense both in February and September, and that their extract is highly energetic as a poison in the middle of April, before any appearance of the flowering stem. Full-grown and perfect leaves should be chosen, especially of such plants as grow spontaneously in open situations. They should be carefully dried in a dark airy room, and the midrib separated, and kept so that the light be excluded. They should be renewed annually, have a dull, but when powdered, a fine green colour, a slight odour, with the strong bitterness of the recent plant. The juice of the fresh plant may be expressed and evaporated to the consistence of an extract, or its active properties imparted to water or Spirit. The leaves of Digitalis have been found to contain traces of Volatile Oil, Fixed Fatty matter, a red extractiform Colouring matter, Chlorophylle, Albumen, Starch, Sugar, Gum, salts of Potash, of Lime, and of Magnesia, but also an acid, partly free and partly combined, probably with a Bitter principle (*Digitaline*), on which the activity of the plant seems to depend. (*Homolle.*). This Bitter matter is soluble in Alcohol, a little so in Ether, and dissolves in water with the aid of the substances with which it is combined and mixed. It has been known that Sesquichloride of Iron produces a greenish-black and Tincture of Gall-nuts a greyish precipitate, and it was by means of Tannin and Oxide of Lead that M. Homolle and subsequently M. Henry have succeeded in isolating Digitaline, which is excessively bitter, a little irritant, scarcely soluble in water, very soluble in Spirit; melts with heat, and may be drawn into long threads of a pearly appearance; cooled, it is easily reduced to a yellowish-white powder, which must be kept from air and light. From its solution in Spirit it may be separated in beautiful white scales. Dr. Morries Sterling, by the destructive distillation of the dried leaves, obtained an empyreumatic oil, containing a crystalline principle possessed of narcotic properties.

The leaves of Foxglove are apt to be intermixed with those of *Verbascum Thapsus*, also with those of *Symphytum officinale*, and sometimes with those of *Conyza squarrosa*, but they may be distinguished by attending to the description, or by comparison with genuine leaves.

Action. Uses. Indirectly Sedative, that is, first exciting and then greatly diminishing the force and frequency of the heart's action. The intestinal canal is apt to be disordered by large doses, as well as the brain and organs of the senses affected by vertigo, &c. The kidneys are often acted on, and the secretion of urine increased. It is cumulative in its effects; therefore when nausea or intermission of the pulse occurs, its use should be discontinued for a time, and the patient should not rise from the recumbent position when under its influence. It has been used to control the circulation, in diseases of the Heart, in Fever, in Inflammations, and in Pulmonary affections after the acute symptoms have subsided, and is useful in excitement from nervous irritability. It is much prescribed as a Diuretic in Dropsies of all kinds, but is most useful in those associated with a debilitated and generally diseased state of the constitution.

D. Of the powder to act as a Sedative, gr. j.–gr. jss. should be given 5 or 6 times a day, carefully watching its effects. As a Diuretic, gr. j. —gr. iij. 3 times a day, usually with some aromatic; but those containing Tannin may precipitate its active principle.

Antidotes. In cases of poisoning, or of excessive doses, evacuate the stomach, and assist the vomiting with diluents; prescribe astringents containing Tannin, as Infusion of Nutgalls, of Oak-bark, of Green Tea; preserve the recumbent position; administer Ammonia, Wine, Brandy, Aromatics.

INFUSUM DIGITALIS, L. E. D. Infusion of Foxglove.

Prep. In a vessel lightly covered infuse for 4 hours *dried leaves of Digitalis* ʒj. (ʒij. E.) in boiling *Aq. dest.* Oj. (f℥xviij. E. by measure ℔ss. D.) Strain (through linen or calico, E.) Then add *Spirit of Cinnamon* f℥j. (f℥ij. E. ℥ss. D.)

Action. Uses. Effective preparation in doses of fʒiv.—f℥j. every 3 or 6 hours.

TINCTURA DIGITALIS, L. E. D. Tincture of Foxglove.

Prep. Macerate for 14, L. (7, D.) days *dried leaves* (rejecting the larger, D. and in moderately fine powder, E. D.) *of Digitalis* ℥iv. (℥ij. D.) in *Proof Spirit* Oij. (by measure ℔j. D.) Strain. (Much better prepared by percolation, as Tinct. Capsicum. If f℥xv. of Spirit be passed through, the density is 944, and f℥j. contains gr. xxiv. of solid contents, E.)

Action. Uses. Sedative, Diuretic, in doses of ♏x.—♏xl. gradually increased. Much larger doses have been given without detriment; but much depends upon the nature of the preparation.

EXTRACTUM DIGITALIS, L. E. Extract of Foxglove.

Prep. From fresh leaves of *Digitalis*, E. prepare as Extractum Aconiti, L., or by any of the processes given for Extr. Conium, E.

Action. Uses. Effective, if carefully prepared, in doses of gr. ss. —gr. j.

PILULAE DIGITALIS ET SCILLAE, E. Foxglove and Squill Pills.

Prep. Beat into a proper mass, with *Conserve of Red Roses, Digitalis* and *Squill* āā 1 part, *Aromatic Electuary* 2 parts. Divide into 4-gr. pills.

Action. Uses. Diuretic in doses of gr. iv.—gr. viij., the certainty of action being increased by combination with the Squill.

LINIMENTUM DIGITALIS. The Diuretic effects of Digitalis may be often secured by rubbing the Tincture with Soap Liniment on the abdomen. Or make a Liniment with *Inf. Digitalis* f℥ij. *Liq. Ammoniae* ʒij. *Ol. Papaverum* ʒiv. to be used 2 or 3 times a day, diluting it if necessary.

VERBASCUM, D. Folia, D. Leaves of VERBASCUM THAPSUS, *Linn.* Great Mullein. *Pentand. Monog.* Linn.

The genus Verbascum is placed by some botanists among Solaneæ, and by others in this family. This species is supposed to be the *φλομος* of Dioscorides.

Biennial. Stem single, 4—5 feet high, woolly, thrown up in its second year. Leaves ovate-oblong, crenate, densely woolly on both sides, all decurrent. Spike long, terminal, very dense, pedicels shorter than the deeply 5-parted calyx. Corol yellow, about twice as long as the calyx, rotate; limb spreading, 5-cleft, unequal, segments oblong, obtuse. Stamens 5, unequal; filaments woolly, two longer nearly glabrous, and about 4 times longer than their slightly decurrent anthers, which are all nearly equal, adnate, by confluence 1-celled. Capsules 2-celled, 2-valved, the valves slightly bifid.—Indigenous throughout Europe in waste ground.—E. B. 549.

The leaves, which are alone officinal, are woolly on both sides, and have a mucilaginous and slightly bitter taste.

Action. Uses. Demulcent. The Infusion or Decoction may be taken internally or used externally as a fomentation, or the boiled leaves applied as a cataplasm.

SOLANEÆ, *Jussieu.* Nightshades. *Pentand. Monog.* Linn.

Annual or perennial herbs or shrubs, with watery juice. Leaves alternate, the upper ones often in pairs, unequal, simple, often lobed. Stipules wanting or spurious. Flowers regular, axillary, or often out of the axils; pedicels without bracts. Calyx 5-fid or in 5 divisions, persistent, or the limb deciduous and base persistent Corolla inserted into the receptacle, usually regular, and 5-fid; æstivation plaited, rarely valvate. Stamens 5, inserted into the corolla, and alternate with its lobes. Anthers at the apex of an acute filament. Ovary free, 2 celled, sometimes 4- or 5-celled; placentæ attached to the partition or projected from the central angle. Style 1. Stigma simple. Fruit capsular or berried, with 2 or 4 cells. Seeds numerous, compressed laterally or from the back. Embryo in the former straight, in the axis of fleshy albumen, in the latter curved, peripheral, or spiral.—The Solaneæ are allied to Convolvulaceæ, &c., especially to Hydroleaceæ. They are with difficulty distinguished from some of the Scrophularineæ. (*v.* Verbascum.) The Solaneæ chiefly inhabit tropical regions, where many are shrubby and even arboreous; a few extend into the temperate and even cold climates of higher latitudes. Several of the species are remarkable for their narcotic properties.

DULCAMARA, L. E. D. Caulis, L. D. Twigs, E. of SOLANUM DULCAMARA, *Linn.* Bitter Sweet. Woody Nightshade.

Dulcamara is supposed to have been employed by the ancients, but has been dsitinctly known only since the time of Tragus.

Root woody. Stem shrubby, flexible, twining in hedges and over shrubs to the height of 12 or 15 feet. Leaves cordate-ovate, the upper ones more or less auriculate, halberd-shaped, all generally smooth, acute, and entire at the margin. Racemes spreading, cyme-like, opposite to the leaves, or terminal. Flowers drooping. Bracts minute. Calyx permanent, 5-parted. Corol rotate, 5-parted, purple—coloured with 2 green spots at the base of each segment. Anthers 5, yellow, erect, connivent, opening by 2 pores at the apex. Berry scarlet, ovoid, juicy, many-seeded.—Indigenous in woods and hedges throughout Europe; found also in Asia and America.—Nees von E. 188. St. and Ch. 17.

Solanum nigrum, a small leafy plant, with obtusely angled acute leaves, with white rotate flowers, and berries about the size of peas, is said to have the same properties as the above; but it is also narcotic. Its leaves are sometimes sold for those of Belladonna. The twigs of the Potato (*Solanum tuberosum*), of which the tubers are so important on account of their starch, are also said to possess some of the same properties.

The officinal part is the stem and twigs, which should be collected in autumn. They are about the thickness of a pen, usually cut into short pieces, sometimes split down the middle; and when dry, they are light, wrinkled, containing much pith, and of a greyish colour. In this state they are scentless, but have a bitter taste, followed by a slight degree of sweetness. It is probable that the root, leaves, and berries have the same properties, which are taken up both by water and Spirit. Analysed, the twigs have been found to contain an alkali, *Solanine* or *Solania*, Gum, Gluten, with Potash and Lime salts. Pfaff indicates the presence of a Bitter principle with a sweet after-taste, which he names *Dulcamarine. Solania*, when purified, is white, pearly, imperfectly crystalline. It restores the colour of Litmus reddened by an acid. Iodine and Iodide of Potassium produce a permanently dark and turbid brown colour with the solutions of Solanine and its salts. It has a faint bitter taste; its salts scarcely crystallize; it does not dilate the pupil, but is said to be a powerful narcotic.

Action. Uses. Alterative in Cutaneous diseases, &c., having a slight determination to the skin and kidneys, also slightly Narcotic.

DECOCTUM DULCAMARÆ, L. E. D. Decoction of Dulcamara.

Prep. Take *sliced Dulcamara* ʒx. (℥j. E. D.) *Aq. dest.* Ojß. (f℥xxiv. E. by measure ℔jß. D.) Boil down to Oj. (f℥xvj. E. ℔j. D.) Strain.

Action. Uses. Alterative, &c. in doses of f℥iß. 2 or 3 times a day with some aromatic water.

BELLADONNA, L. E. Folia, L. Leaves, E. Belladonnæ Folia et Belladonnæ Radix, D. Leaves and Root, D. of ATROPA BELLADONNA, *Linn.* Deadly Nightshade, or Dwale.

This plant has been supposed to be the Mandragora of Theophrastus, and the *strykhnos manicos* of Dioscorides; but it has been distinctly known only since the time of Tragus, and is said to have been first used in Germany as a cure for cancer.

This plant (fig. 81) has a lurid hue, and, when bruised, a fœtid odour. Root perennial, branched, but fleshy, white internally. Stems annual, herbaceous, 3—5 feet high, branched, round, slightly downy or velvety, with a tinge of red. Leaves with short footstalks, lateral, often in pairs of unequal size, broadly ovate-acute, entire, smooth and soft, 4 or 5 inches in length, often with hairs on under-surface. Flowers solitary, imperfectly axillary, stalked, about an inch in length, rather drooping. Calyx campanulate, 5-cleft. Corol (1) campanulate, an inch long, or twice the length of the calyx, greenish towards the base, but of a dark purple towards its 5-lobed equal border. Stamens 5, distant above. Style (2) as long as the corol. Stigma (3) capitate. Berry (4) seated in the enlarged calyx, globose, 2-celled, of a shining violet-black colour, about the size of a small cherry, with a longitudinal furrow on each side, 2-celled, containing numerous reniform seeds in a mawkishly sweet but neither agreeable nor nauseous pulp.—Indigenous in waste, often shady places, in many parts of Europe. Flowers in June and July, and its berries are ripe in September.—E. B. 592. Nees von E. 191. St. and Ch. 1.

Fig. 81.

The root of Belladonna, which is branched, thick, fleshy, and often a foot or more in length, is white internally when fresh, becomes of a grayish colour when dried. The taste is slight, but bitter; the odour feeble, but its properties energetic. It should be collected in autumn or spring, and the leaves about the time of flowering. These, when stripped from their stems and carefully dried, have a dull-green colour, very little odour, with a slight bitter taste. The leaves of *Solanum nigrum*, as well as of *S. Dulcamara* (*v.* p. 508), are sometimes actually sold by herbalists for those of Belladonna, and consequently must be frequently employed medicinally by those who look for the powerful effects of this medicine, and being disappointed, will afterwards pronounce upon the inefficiency of the drug.

The leaves of Belladonna, analysed by Brandes, yielded Gum, Starch, Albumen, Chlorophylle, a little Wax, several Salts, Lignine, and water, with two nitrogenous substances, and an acid Malate of Atropia. This alkali, upon which he considered the medical properties to depend, has since been obtained by other chemists in white crystals, which are without odour; are fused by heat in closed vessels, volatilized above 212°, but burn in air; soluble in Alcohol, sparingly so in Ether, more so in boiling water. It combines with acids, forming salts, which are bitter, and have the poisonous properties of Atropia. This is bitter in taste, also a little acrid, dilates the pupils, and is very poisonous, but forms only a small proportion of the plant. The proportion of its constituents is uncertain. ($C_{22}\,H_{15}\,O_3\,N$?)

Action. Uses. Anodyne, Antispasmodic; externally Anodyne, and used by surgeons for dilating the pupil. The roots possess the same properties as the leaves, and the berries have frequently proved poisonous to children. Dryness and stricture in the throat, difficulty of swallowing, nausea, &c., dimness of vision, dilatation of the pupil, vertigo, mirthful or extravagant delirium, followed by coma, are experienced. It sometimes induces sleep from relieving pain. Anodyne in Neuralgic and other pains; more applicable to those which are external than to internal pains. Antispasmodic in Hooping and other coughs. Thought by some to be Prophylactic against Scarlatina.

D. Of the powder gr. j. gradually increased to gr. v. or until dryness of the throat is experienced. Atropia $\frac{1}{10}$ of a grain produces all the same symptoms, and has been used for dilating the pupil.

EXTRACTUM (SUCCUS SPISSATUS, D.) BELLADONNÆ, L. E. Extract of Belladonna.

Prep. To be prepared (from fresh leaves of *Atropa Belladonna*, like Succus spissatus Aconiti, p. 241, D.) like Extr. Aconiti, L. Bruise into a uniform pulp in a marble mortar fresh *Belladonna* q. s. Express, moisten with water, and again express. Unite the expressed fluids, filter, and evaporate the filtered liquids in the vapour-bath to the consistence of firm extract, stirring constantly towards the close, E.

Action. Uses. The Extract of Belladonna is an uncertain preparation, because it is not always prepared with care. The E. P. directions are suited to insure a good preparation, and the Extract prepared in vacuo is an energetic one. Dr. Christison suggests the preparation of an Alcoholic Extract, like that of Aconite of the E. P. (*v.* p. 241.)

D. gr.ß. or gr. j. 2 or 3 times a day, gradually increased to gr. v. until the peculiar effects of Belladonna are observed. It is often diluted with water and applied on the eyebrow, to dilate the pupil, or a solution dropped into the eye; or it may be applied externally as a liniment, or applied endermically to relieve severe pains. It has also been applied to the os uteri in protracted first labours, and in stricture of the urethra, and spasm of the sphincter ani, &c.

EMPLASTRUM BELLADONNÆ, L. E. D. Belladonna Plaster.

Prep. In a water-bath (with gentle heat, E.) melt *Resin Plaster* (Soap Plaster ℥ij. D.) ℥iij. to this add *Extract of Belladonna* ℥iss. Agitate briskly and mix. (Make a plaster, D.)

Action. Uses. Anodyne in Neuralgic and other pains. Belladonna may also be applied externally in the form of its Infusion, as a Lotion, or as an Ointment, with some of the Extract rubbed up with water, or with simple Ointment.

SUCCUS BELLADONNÆ, as prepared by Mr. Bentley, is an effective preparation.

TINCTURE OF BELLADONNA. Macerate *Belladonna leaves* dried ℨij. in *Proof Spirit* for 14 days. May be given in doses of ♏xv.—♏xxx.

Antidotes. Emetics and Purgatives, Astringent Infusions? application of cold to the head, and the use of the ordinary external stimuli; Ammonia internally in the Comatose state, as in Digitalis, p. 506.

CAPSICUM, L. E. D. Baccæ, L. Capsulæ cum seminibus, D. Fruit, E. of CAPSICUM ANNUUM, *Linn.* Capsicum. Chillies.

The several species of Capsicum are natives of South America, whence they have been introduced into the Old World, and become universally diffused, from the fondness of Asiatics for warm condiments. The Hindoos, though cultivating the Capsicum extensively, have no specific name for it, but call it *Red Pepper.* "Chilli, either simply or in composition, being the Mexican name for all the varieties and species of this genus" (*R. Brown*), indicates that the genus is American. Many varieties have no doubt been raised to the rank of species. The genus is distinguished by its berry-like but dry fruit.

The officinal Capsicum is annual, smooth, dark-green in colour, from 1—2 feet high, with branched, furrowed, angular stems. Leaves ovate, acuminate, sometimes lanceolate, entire, shining, sometimes hairy beneath on the veins. Flowers small, white, axillary, solitary, drooping. Calyx 5-cleft. Corol rotate, equal. Stamens 5; filaments short; anthers dark-coloured, connivent, opening longitudinally. Fruit firm, succulent, 2-celled, containing numerous dry flat seeds. The fruit varies much in form, being round, oblong, cordate, or horned, and either scarlet or yellow, and more or less pungent in taste. The horn-shaped variety is most common, from 2—3 inches in length, and from ½—1 inch diameter at the base, and usually called Capsicums, and the plant C. annuum. One variety, called Cockspur-pepper, has the fruit long and slender. Sometimes the fruit is globose or lobed: the variety is then called *C. baccatum.* When the fruit is small, elongated, and pointed, the variety is called Bird-Pepper, and botanically *C. minimum.* When the plants are allowed to grow beyond the year, they become shrubby, and form the species or variety called *C. frutes-*

cens.—Cultivated in all hot countries, but also under glass in this country.—Nees von E. 190. St. and Ch. 44.

The Berry or fruit of the Capsicum, in its dried state, is the only officinal part. These, when powdered, form Cayenne Pepper, but are often preserved in vinegar as a pickle, and the fluid likewise employed under the name of Chilly Vinegar. The active properties are taken up also by water, Spirit, Ether, and fixed oils. Analysed by Forchhammer, a red Colouring matter, a nitrogenous substance, Mucilage, and some salts; among these Nitrate of Potash and an alkaline body, *Capsicine*, white, brilliant, pearly, and very acrid. But Braconnot describes the acrid principle as of an oleaginous nature, very acrid in taste, readily volatilizing, and diffusing a very acrid vapour.

Action. Uses. Rubefacient, Acrid Stimulant. Much used as a Condiment in hot countries. Sometimes used as a Counter-Irritant, with salt as a Stimulant in Scarlatina maligna, as a Gargle in relaxed sore throat, or in the form of Cayenne Lozenges.

TINCTURA CAPSICI, L. E. D. Tincture of Capsicum.

Prep. Macerate for 14 (7, E.) days *bruised Capsicum* ʒx. (℥j. D.) in *Proof Spirit* Oij. (℔ij. D.) Strain. (Squeeze and filter; or it is better prepared by percolation, to be commenced as soon as the Capsicum in fine powder is made into a pulp with a little of the Spirit, E.)

Action. Uses. Irrritant. Stimulant in doses of ♏v.—fʒß. or as a Gargle (fʒiv.—Inf. Rosæ f℥viij.) Dr. Turnbull uses a concentrated Tincture (℥iv. to Rect. Sp. f℥xij.) as a Counter-Irritant.

STRAMONII FOLIA, L. D. (Stramonium, E.) Leaves (Herb, E.) and STRAMONII SEMINA, L. D. Seeds of DATURA STRAMONIUM, *Linn.* Thornapple.

Species of Datura (Sans. *Dhatoora*) have long been employed medicinally by the Hindoos, and were thus made known to the Arabs, who curiously give Stramonia as a synonyme of Datura. It is their *jouzmasil*, that is, *masil* or *methel*, which has long been referred to Datura. *D. Stramonium* occurs in the Himalayas (*v.* Himal. Bot. p. 279), and is probably indigenous in the Hindoo Khoosh, whence most likely it was taken to Constantinople, having been obtained by Gerard from that city, and by Fuchsius from Italy.

The Thornapple (fig. 82) is an annual of vigorous growth, about 3—5 feet high. Stem much branched, dichotomous above, bushy, fœtid, smooth. Root large, white, and fibrous. Leaves from the forks of the stem, large, unequal at the base, ovate, unequally sinuate-dentate, smooth, variously and acutely sinuated and toothed, simply veined, of a light dull-green colour. Flowers axillary, erect, white, sweet-scented, especially at night, about 3 inches long. Calyx oblong, tubular, ventricose, 5-angled, 5-toothed, dropping off and leaving a circular mark round the base of the ovary. Corolla funnel-shaped, regular, angular, plaited with mucronate lobes. Stamens 5 Stigma thick, obtuse, 2-lobed. Ovary 4-celled. Capsule as large as a walnut, dry, very prickly, 4-valved, with 2 partially bipartite cells, containing many brownish or black flattened reniform seeds.—Waste places and dung-heaps in all parts of Europe, also in North America.

No doubt introduced from Asia. Flowers in July.—Nees von E. 193. St. and Ch. 6.

Fig. 82.

The whole plant has a rank odour, which may be detected at a distance. All parts possess medicinal properties; but the leaves and seeds are alone officinal. The *seeds* are brownish or black, flattened, kidney-shaped, without odour, except when bruised, but with a bitter weakish taste, often employed for poisoning in India, where pulses form so large an article of diet, or for stupifying only, given in sweetmeats—(O. S.) The *leaves* should be gathered when the flower-buds begin to blow. They have a fœtid odour, especially when bruised; this they lose in drying. Their taste is rather bitter and nauseous.

Analysed by Brandes, the seeds yielded Fixed Oil, Wax, Resin, Extractive, Gum, Albumen, &c., with salts, and a Malate of *Daturia.* This alkali has been obtained by Geiger and Hesse, who describe it as occurring in brilliant crystals, without odour, and colourless, having a bitterish, tobacco-like taste, alkaline, easily soluble in Alcohol, and forming salts with acids. The fresh leaves of Stramonium did not yield Promnitz anything except the ordinary vegetable constituents, though they must also contain the Daturia. Mr. Morries Stirling, by the destructive distillation of Stramonium, obtained an empyreumatic oil, which contains an active poisonous principle.

Action. Uses. Anodyne. Antispasmodic, and as such may be combined with Valerian. By relieving pain, it will induce sleep; and affects the constitution much in the same way as Belladonna, in doses of the powder, gr. j.—gr. v. In Neuralgic and in Rheumatic pains it has given relief both when taken internally and applied externally. It has also been considered calmative in Mania. In Spasmodic Asthma smoking the leaf (gr. x.—ʒß.) often gives instantaneous relief; but it must be exhibited with care. M. Trousseau recommends its being smoked with an equal quantity of Sage leaves, in a roll of paper. In India I used to order it to be added to the ordinary chillum of Tobacco; or the inhaling warm water in which Datura leaves had been infused. Mr. Skipton found f℥ij. of an infusion of the root (℥j. to aq. Ojß.) of Datura fastuosa give great relief. But the most convenient practice is that of smoking the Stramonium cigars, which are prepared by some chemists. Daturia is an energetic poison, and very small quantities cause dilatation of the pupil.

EXTRACTUM STRAMONII, L. E. D. Extract of Stramonium Seeds.

Prep. L. D. Macerate for 4 hours in a lightly covered vessel (near the fire, L.) in boiling *Aq. dest.* Cj. *Seeds of Stramonium* ℥xv. Take the seeds out, and bruise them in a stone mortar, return them to the liquor, and boil down to Oiv. (℔iv, D. while hot, L.) filter, and evaporate to the proper consistence

E. Grind in a coffee-mill *Seeds of Stramonium* q. s. Rub the powder into a thick mass with *Proof Spirit,* which transmit through the pulp in a percolator, till it passes colourless. Distil off the Spirit, and in the vapour-bath evaporate the residuum to a proper consistence. This alcoholic extract is the best form.

Action. Uses. Anodyne. Antispasmodic in doses of gr. $\frac{1}{4}$—gr. iij. or it may be made moist and applied over a pained part.

A tincture of Stramonium (*Seeds* ℥iv.—*Proof Spirit* ℥xxxij.) is officinal in the United States Ph., and may be given in doses of ♏x.—fʒß. 2 or 3 times a day, or it may be rubbed along the course of a pained nerve.

Antidotes. Stimulant emetics, cold affusion, with blisters to nape of neck, in cases of poisoning with Belladonna.

HYOSCYAMI FOLIA, L. D. (Hyoscyamus, E.) Leaves. HYOSCYAMI SEMINA, L. and, Seeds of HYOSCYAMUS NIGER, Linn. Henbane.

Henbane has been employed in medicine from the earliest times; is the *ὑοσκυαμος* of the Greeks, and the *bunj* of the Arabs. The seeds are known by the name of *Khorassani Ujwain* in India.

Henbane (fig. 83) is annual or biennial, that is, plants grown from the seed of the biennial variety will, in favourable conditions of the soil and climate, come to full perfection in the first year. Roots spindle-shaped, those of biennial plants having considerable resemblance to small parsnip-roots in the winter and spring. (*Houlton.*) The plants in the first year throw up a tuft of radical leaves which are petiolated, woolly, and possess little of that clamminess and odour which are peculiar to the mature plant. In the second spring, another set of leaves make their appearance with, and attached to the flowering stem. This is from a foot to 3 feet high, seldom branched, hairy, hairs glandular, viscid. Leaves sessile, subamplexicaul, occasionally decurrent, lower ones sometimes stalked, oblong-acute, coarsely and unequally cut or sinuate, appearing pinnatifid, clammy, and fœtid, of a pale dull green colour, slightly pubescent, with long glandular hairs, like those of the stem, upon the midrib. Flowers nearly sessile, axillary, subsolitary, unilateral, erect, much shorter than the leaves. Calyx funnel-shaped, 5-lobed, villous. Corol (2) funnel-shaped, limb spreading, 5-lobed, not quite equal, of a dull straw-colour, marked with dark purple veins. Stamens 5, declinate; filaments pubescent. Ovary ovoid, shining, 2-celled, with numerous ovules attached to the placentæ. Style filiform. Stigma (1) capitate. Capsule opening transversely by a convex lid, 2-celled, many-seeded. Seeds small, roundish, finely dotted, of a light grey colour.—Indigenous in waste grounds throughout Europe, also in the Persian region of Botanists.—Nees von E. 192. St. and Ch. 9.

Fig 83.

Henbane plants come into flower about the beginning of June, but the annual plants a little later: the seeds ripen from August to October. Mr. Houlton is of opinion that the biennial plant should alone be employed medicinally, and that the leaves should be collected when the first flowers begin to appear. But it has not been proved that annual plants, when properly grown, are devoid of active properties. The author was in the habit of largely cultivating Henbane in the Botanic garden at Saharunpore, where, from the nature of the climate, the whole process of cultivation, including the ripening of the seed, was completed between the months of October and March. The Extract made from these plants was highly approved of by several medical officers, and pronounced by Mr. Twining, after trial in the General Hospital at Calcutta, to be of "most excellent quality" (Himal. Bot. p. 281). But the secretions of plants growing in a colder and moister climate, or in seasons having these characteristics, may not come to as great perfection in the first year. Dr. Christison states, from experiments made in the Royal Infirmary at Edinburgh, "that inferiority of cultivated plants, if it exists at all, seems not appreciable in practice;" and, with respect to the period at which the leaves acquire their activity, he says, "I have found them sufficiently active even in the spring, before the appearance of the flowering stem." When collected, they should as soon as possible be separated from the stem, spread out, and dried in a warm airy room. They ought to have a mucilaginous, slightly bitter taste, and should retain some of the peculiar odour of the plant.

Analysed by Brandes, the seeds yielded Gum, Starch, Albumen, a large proportion of Fixed Oil, with a variety of salts, Ligneous fibre, and an oily-like alkali, resembling Conia, which was highly poisonous.

Geiger and Hesse, however, obtained groups of radiated needle-like crystals, fusible and volatile, but readily decomposed when distilled, alkaline in nature, neutralizing acids and forming crystallizable salts, soluble in Alcohol and Ether, less so in water. These have been considered pure *Hyoscyamia*, which is very poisonous, dilates the pupils, and when moistened, smells strongly of Tobacco. The oily-like liquid of Brandes is thought to hold a little of this principle in solution. A highly poisonous empyreumatic oil is obtained by destructive distillation, as from the other Solaneæ and from Foxglove.

Action. Uses. Narcotic, Anodyne, and Soporific. Available for a variety of cases where we wish to relieve pain, allay irritability, and procure sleep, having the advantage of not constipating the bowels like Opium. Hence it is frequently prescribed with Calomel, Purgatives, or with Antispasmodics. It may be given internally in powder in doses of gr. v.—gr. x. or in Extract or Tincture; or it may be applied externally in the form of fomentation, or in cataplasms of its leaves, or its Extract or Tincture used as those of Belladonna, p. 511.

EXTRACTUM (SUCCUS SPISSATUS, D.) HYOSCYAMI, L. E. Extract of Henbane.

Prep. To be prepared from the (fresh herb of, D.) *Hyoscyamus niger* as Extractum (Succus Spissatus, D.) Aconiti, L. (p. 241) (or by any of the processes directed for Extract of Conium, E.)

Action. Uses. The expressed juice evaporated spontaneously in a dry current of air, or in vacuo, forms an excellent preparation. A still more powerful preparation may be obtained by the action of Alcohol. Adapted for all the purposes of Henbane in doses of gr. v.—℈j. The author has also found the *Succus Hyoscyami* of Mr. Bentley a very good form of preparation.

TINCTURA HYOSCYAMI, L. E. D. Tincture of Henbane.

Prep. Macerate (digest, E. D.) for 14, L. (7, D.) days *dried leaves of Hyoscya-*

mus niger (in fine powder, E.) ℥v. in *Proof Spirit* Oij. (℔ij. D.) Strain. (Much better prepared by percolation, as the Tinct. of Capsicum, E.)

Action. Uses. Narcotic, &c. in doses of ♏x. as soothing,—fℨj. or fℨij. as a Hypnotic.

TABACUM, L. E. Folia exsiccata, L. D. Leaves of NICOTIANA TABACUM, *Linn.* Tobacco.

Tobacco was introduced from the New World about the middle of the 16th century, and is now extensively cultivated in most parts of the world.

Root fibrous. Stem erect, branched, and viscid, from 2 to 6 feet high. Leaves sessile, oblong, lanceolate, the lower ones decurrent, very large, a little hairy, viscid. Flowers in terminal panicles. Bracts linear-acute. Calyx tubular, swelling, 5-cleft, hairy, glutinous. Corol rose-coloured, funnel-shaped, throat inflated, ventricose, limb spreading, plicate, with 5-cleft acuminate segments. Stamens 5, declinate. Ovary ovate. Style long. Stigma emarginate. Capsule usually 2-celled, 2-valved, opening crosswise at top, valves finally bifid. Seeds numerous, small, kidney-shaped, attached to fleshy placentæ.—Warm parts of America, but now cultivated in most parts of the world.—Nees von E. 194. St. and Ch. 37.

Most of the Tobacco of commerce, as that of Virginia, is yielded by this species, as is that of India. Small Havannah cigars are said to be formed of the leaves of *N. repanda;* the Syrian and Turkish Tobaccos by *N. rustica* and the fine Shiraz Tobacco by *N. persica,* Lind.

Tobacco, as it occurs in commerce, is of a yellowish-brown colour, soft and pliable, a little clammy, with something of a honey, mixed with a narcotic odour; the latter, however, is not obvious in the fresh leaves. The taste is bitter, acrid, and nauseous. Virginian Tobacco, though the strongest, is best adapted for medical use, in order to observe uniformity of strength. Its active properties are taken up by water, Spirit, and Wine, but are destroyed by heat. Tobacco was elaborately analysed by Vauquelin. *Nicotianin* was discovered by Hermbstadt in 1821. The analysis of Posselt and Reimann displayed the presence of *Nicotina* ·06, of *Nicotianin* 0·01, Extractive 2·87, Gum 1·74, Chlorophylle 0·26, Vegetable Albumen and Gluten 1·30, Malic acid 0·51, Lignin and Starch 4·65, Salts 0·73, Silica 0·08, Water 88·28 = 100 nearly. *Nicotina* has since been studied by Boutron and Henry. It is obtained much in the same way as Conia, and in the form of a limpid, oily, volatile liquid, devoid of colour, having an acrid taste, and a weak smell of Tobacco, unless when heated. Its vapours are extremely acrid, with an overpowering odour of Tobacco. It has a Sp. Gr. of 1·048, is alkaline, forms salts with acids, and is soluble in Alcohol, Ether, and water, and in fixed and volatile Oils. It is the active principle of Tobacco, and is extremely poisonous. When heated, it is decomposed, becomes resinoid, and disengages Ammonia. Nicotina exists in combination in Tobacco, and is found varying in proportion from 4 to 12 parts in 1000. It is composed of $C_{10}\,H_8\,N$. *Nicotianin* is a camphoraceous volatile oil, bitterish in taste, having the odour of Tobacco, and seeming to owe its properties to a little Nicotina intermixed with it. By the destructive distillation of Tobacco, an *empyreumatic oil* is formed, which is better known as produced in tobacco-pipes, and as being highly poisonous. It seems to be a volatile oil holding some Nicotina in solution.

Action. Uses. Local Stimulant, hence used as an Errhine and Sialogogue: secondarily Sedative, Antispasmodic, also Emetic, Laxative, and Diuretic; and acts upon the system, to whatever surface it is supplied. Chiefly employed to produce relaxation in Spasmodic affections, as in strangulated Hernia, obstinate constipation from spasm of the bowels, or retention of urine from that of the urethra.

ENEMA (INFUSUM, D.) TABACI, L. E. Tobacco Enema.

Prep. Macerate for 1 (½, E.) hour (in a covered vessel, D.) *Tobacco* ʒj. (gr. xv. to ʒß. E.) in boiling *Aq.* Oj. (f℥viij. E. by measure ℔j. D.) Strain.

Action. Uses. Sedative, Antispasmodic. Used only in the above cases. ℈j. is sufficient for trial at first.

VINUM TABACI, E. Tobacco Wine.

Prep. Digest *Tobacco* ℥iij. in *Sherry* Oij. for 7 days. Strain, express the residue strongly. Filter.

Action. Uses. Sedative and Diuretic. Capable of producing the full effects of Tobacco in doses of ♏x.—♏xl.

d. *Monochlamydeæ* or *Apetalæ.*

POLYGONEÆ, *Juss.* Buckwheats.

Herbs, seldom shrubs, stem and branches jointed. Leaves alternate, simple, sometimes undulate, or cut; petioles sheathing at the base, or united into a tube called ochrea. Flowers complete, or by abortion unisexual, inflorescence various. Perianth herbaceous or subcorolline, inferior, 3- 5- or 6-partite, imbricate in æstivation, often permanent, and growing with and covering the fruit. Stamens definite, but varying in number, inserted into the narrow margin of the receptacle and adhering to the perianth. Ovary single, 1-celled, with one erect ovule. Styles 2 to 3. Fruit indehiscent, nut-like, or fleshy, often triangular, naked or covered by the interior segments of the perianth. Embryo inverted, straight, and central, or curved and unilateral or peripheral. Radicle superior, remote from the hilum. Albumen farinaceous.—They are found in the greatest numbers in the temperate regions of the northern hemisphere, but some in almost all parts of the world. The young shoots of many are acid (chiefly Oxalic); when older, an astringent together with a purgative principle is secreted by species of Rheum and of Rumex. The seeds of many, as the Buck-wheats, or Fagopyrum, afford nutritious flour. *Coccoloba uvifera,* or Seaside Grape is said to yield Jamaica Kino.

POLYGONUM BISTORTA, *Linn.* Radix, D. Root of Bistort. *Octand. Monog.* Linn.

Bistort has long been employed in European medicine.

Rootstock creeping, often twice bent on itself, of a dark-brown colour externally, and rugose with annular rings. The stem is annual, simple, erect, 1—2 feet high. Leaves ovate, bluntish-pointed, wavy. Footstalks tubular and sheathing, with jagged stipules. Radical leaves heart-shaped, and yet decurrent, so as to form a wing to the petioles. Cluster of flowers spike-like, terminal, with membranous brown bracts. Flowers pink, with short pedicels. Perianth 5-parted, spreading. Stamens 8, half as long again as the perianth. Styles 3, distinct, stigmas obtuse. Nut triquetrous, its faces ovate, smooth. Embryo in the centre of farinaceous albumen. Cotyledons large, foliaceous, twisted, and contorto-plicate.—Moist meadows. Flowers in June.—E. B. 509. St. and Ch. 47.

The rootstock of Bistort contains a large proportion of Tannin, some Gallic acid, and Starch, with woody fibre, and has a rough astringent taste

Action. Uses. Astringent, either internally in doses of gr. xv.–ʒß. or made into a decoction, in doses of f℥iß.; or externally as a lotion; has also been prescribed in Intermittents.

Rumex, *Linn.* *Hexandria Trigyn.* Linn.

Perianth 6-parted, the 3 external segments spreading, permanent, more or less united at the bottom; the 3 interior petal-like, large, connivent; in some species bearing a dorsal grain or tubercle. Stamens 6, disposed in pairs. Ovary, triangular, rather turbinate. Styles 3. Stigmas large, in tufted segments. Nut triangular, polished, with 3 sharp edges, covered by the enlarged inner sepals, single-seeded; embryo lateral.

Rumex aquaticus, *Linn.* Radix, D. Root of Water Dock.

Water Dock was formerly much employed by the name of *Herba Britannica.*

Plant 3 to 4 feet high. Leaves very large, lanceolate, lower somewhat cordate, with channeled petioles; enlarged sepals broadly cordate, membranous, entire, or wavy, without tubercles.—Indigenous in ditches and damp places in the North.—E. B. S. 2698.

R. Hydrolapathum, or Great Water-Dock, is sometimes used instead of the foregoing.

R. obtusifolius was found by Dr. A. T. Thomson useful in obstinate Icthyosis.

The root of Water Dock is large, without odour, but has an austere bitter taste. It yields its virtues readily to water.

Action. Uses. Astringent, Alterative. This was formerly much employed in Skin diseases. *Extract of Water Dock* is sold by chemists, being prescribed by many practitioners as an Alterative in Cutaneous diseases.

Rumex Acetosa, *Linn.* Folia, L. D. Leaves of Common Sorrel.

Common Sorrel is supposed to be the ὀξυλαπαθον of the Greeks.

Plant of 1—2 feet. Leaves oblong, arrow-shaped, acid. Whorls leafless. Flowers diœcious; enlarged interior sepals roundish, cordate, entire, membranous, with a very minute tubercle at the base; exterior sepals reflexed.—Pastures throughout Europe. 2 B. t. 127.

The leaves and herbaceous parts of this plant, so well known by the name of Sorrel, are pleasantly acid, with a slight degree of astringency, owing to the presence of Binoxalate of Potash, some Tartaric acid, and Tannin. There is also Mucilage, Woody Fibre, and Starch.

R. Acetosella, or Field Sorrel, may be substituted for it.

Action. Uses. Refrigerant. Eaten in salads. Acid drinks may be made with its leaves.

RHEUM, L. E. D. Radix, L. D. Root, E. of an undetermined species of RHEUM, *Linn.*, of R. palmatum, L. D. and of R. undulatum, D. Rhubarb.

The name Rheum is derived from the *ρεον* of Dioscorides; but his description does not well apply to modern Rhubarb. This was, however, known to Paulus Ægineta, &c. The Arabs were acquainted with several kinds, as Indian, Khorassanee, Chinese (their *rewund sini*). The Persians give *reon* as the Greek synonyme of their *rawund*, which is Rhubarb, and of which the plant they say is called *ribas*. Rhubarb is no doubt the rootstock of a species of Rheum, but the species is still unknown. The author, after giving in another work (Him. Bot. p. 314—318) an account of the commerce of Rhubarb, stated that "This would bring the Rhubarb country within 95° of E. long. and 35° of N. latitude, that is, into the heart of Tibet. As no naturalist has visited this part, and neither seeds nor plants have been obtained thence, it is as yet unknown what species yields the Rhubarb." This seems now the general opinion. Sievers, an apothecary sent in 1790 by the Russians to investigate the subject, had previously said, that "his travels had satisfied him that as yet nobody, that is, no scientific person, has yet seen the true Rhubarb plant." Dr. Fischer when in London, subsequent to the above publication, informed the author that all the information obtained of late years in Russia, only confirmed what was previously known, that *Rheum palmatum* is not the species, but that the genuine plant is a small one with roundish denticulate leaves. So more recently, Calau, apothecary in the Rhubarb factory at Kiachta, says: "All that we yet know of the Rhubarb plant or its origin is defective and wrong; every sacrifice to obtain a true plant, or the seed, has been in vain; nor has the author been enabled to obtain it." Dr. Falconer entered Tibet from the side of Cashmere, and proceeded as far as the Muztagh range, or about long. 77° E. and lat. 36°, a region where Rhubarb is sent as a present to the Chief Ahmed Shah from the true Rhubarb country, but was unable to learn anything respecting commercial Rhubarb. He discovered new species of Rheum, and obtained specimens of genuine Extract of Rhubarb *ossareh-rewund*, or Rhubarb-juice, a name which he as well as the author found applied in north-west India to Gamboge. He also found Rhubarb-root employed there as a yellow dye. Some information might probably be procured respecting Rhubarb from the traders to Upper Assam.

RHEUM, *Linn.* *Enneandria Monog.* Linn.

Flowers complete; perianth petaloid, 6-parted, with equal segments. Stamens usually 9, inserted in pairs into the base of the 3 outer segments, and singly into the 3 interior; filaments subulate; anthers versatile. Ovary triangular, 1-celled. Ovule single, basilary, orthotropous. Styles 3, short, reflexed. Stigmas 3, entire, subdiscoid, spreading. Achænium 3-cornered, broadly winged, supported by the withered perianth at the base. Seed erect, triangular. Embryo straight, antitropous, in the axis of farinaceous albumen. Cotyledons flat, radicle short, superior.

Herbaceous plants, with perennial and branching rootstocks, which are thick and succulent. Stem of most 4 to 10 feet high, except in No. 1 and 2. Leaves large, more or less cordate, wavy at the margin, sheathing at the base, either all radical, or where cauline, alternate. Inflorescence paniculate, or spicato-racemose. The species are valuable not only on account of their rootstocks or Rhubarb, but also from the agreeable acidity of their leaf-stalks, employed for making sherbets, tarts, &c. They inhabit cold parts of the world, as the southern part of Russia, Siberia, Tibet, the north of China and the Himalayan mountains, also Affghanistan and Persia. Hence all may be grown in the open air in Europe, and several are so cultivated. As no species seems more entitled than another to be considered as yielding either the Russian or Chinese Rhubarb of commerce, we shall briefly enumerate all, without describing any.

With Spike-like Racemes.

1. RHEUM SPICCIFORME, *Royle.* (Illustr. Himal. Bot. p. 318. t. 78.) Kherang Pass and other places in Kunawar. Found by Dr. Falconer in Tibet.

2. R. MOORCROFTIANUM, *Royle.* (*l. c.* p. 318. Lindl. Med. Bot. p. 356.) Niti Pass in the Himalayas. Found by Dr. Falconer in Tibet.

These two species differ in their inflorescence from the other described species. Their roots are more dense in texture and of a more yellow colour than those of *R. Emodi* and *R. Webbianum.* The powder of both is of a light and bright yellow colour. Dr. Falconer met with both in Tibet, and discovered another species of this group. It is probable that the commercial species will be found to resemble these in habit, from being indigenous, like them, in the elevated, arid, and cold regions of Tatary.

With Compound Racemes.

3. R. EMODI, *Wall.* Bot. Mag. t. 3508. *R Australe,* Don. Sweet Fl. Gard. t. 269. Nees von E. Suppl. t. 31. A. and B. Lindl. Fl. Med. p. 354.

4. R. WEBBIANUM, *Royle, l. c.* p. 318. t. 78, a. Choor Mountain. Niti Pass. This yielded the Rhubarb submitted to experiment by Mr. Twining. Trans. Med. Soc. Calc. iii. p. 439.

5. R. RIBES, *Linn.* Dill. Elth. t. 158. f. 192. An. Mus. 2. t. 49. *Ribas* and *rivash* of the Affghans and Persians, much celebrated among them, and much esteemed on account of the agreeable acid of its leafstalks. The root is said to be *rawund.* It is the *Riwas* of Serapion, who mentions it as making a good sherbet. It is said also to be found on the mountains of Syria.

6. R. RHAPONTICUM, *Linn.* Alpin Rhapont. i. t. 1. Nees von E. 113-14-15. Borders of the Euxine, and on the north of the Caspian Sea, Deserts near the Volga, and in Siberia, as it is known to yield Siberian Rhubarb. Supposed to have yielded the Rhabarbarum of the ancients. Cultivated in this country on account of its stalks, and extensively at Banbury on account of its roots, and also at Rheumpole near Lorient, in the department of Morbihan in France.

7. R. CRASSINERVIUM, *Fischer.* Sent from St. Petersburgh to the Apothecaries' Garden at Chelsea. Roots large, and said to have the colour and odour of Turkey Rhubarb.

8. R. LEUCORRHIZUM, *Pallas.* R. NANUM, *Sievers.* Ledebour, Il. Pl. Ross. t. 492. Found in the deserts of the Kirghis, and South of Siberia, and Altai Mountains. Said to yield *White* or *Imperial Rhubarb.*

9. R. UNDULATUM, *Linn.* Amœn. Acad. iii. t. 4. Nees von E. 116, 117. St. and Ch. 177. Lindl. Fl. Med. p. 357. A native of Siberia, Tatary, and China? Seeds of what was called the genuine Rhubarb plant were given to Kauw Boerhaave by a Tatarian merchant, and these seeds produced this species and *R. palmatum*, both of which are admitted as officinal in the D. P. This is said to be cultivated in France as one of those yielding French Rhubarb.

10. R. CASPICUM, *Fischer.* Caspian Shores and Altai Mountains. Lindl. p. 557.

11. R. COMPACTUM, *Linn.* Mill. Dict. 218. Nees von E. 121. A native of Chinese Tatary, said to yield some of the Rhubarb cultivated in France, which forms a fair imitation. Valued in this country on account of its stalks.

12. R. PALMATUM, *Linn.* L.D. This species is easily distinguished from the others by its roundish, cordate, half palmate leaves, with the lobes also deeply cut. Root large, branched, brown externally, of a deep yellow internally. Supposed to be a native of the mountains of Mongolia near the great wall of China. The seeds of this were received by K. Boerhaave with those of *R. undulatum* as those of the genuine Rhubarb; and it is thought, when cultivated in Europe, to resemble Chinese-Russian Rhubarb more closely than any other kind, in taste, odour, internal structure, and the action of some chemical reagents.

R. HYBRIDUM, *Murray*, is a doubtful species, but is remarkable for the great size of its roots as cultivated both on the Continent and in this country. This plant, with R. Rhaponticum, compactum, and emodi, with hybrids from them, are those most commonly cultivated in this country on account of their stalks.

The greater part of the Rhubarb of commerce grows in Chinese Tatary, on the mountains and plains surrounding Lake Kokonor (*Pallas and Rehman*), especially in the province of Gansun (an *Kansu? Calau*) and is gathered in summer from plants of six years of age. When dug up, it is cleansed, peeled, cut into pieces, bored through the centre, strung on a string, and dried in the sun. In the autumn it is brought to Sinin, where the Bucharian traders reside, and from thence sent to the Russian frontier town of Kiachta, and to Pekin, Canton, Macao. Considerable care is bestowed at the Russian factory in examining and separating the good from the inferior pieces, also from impurities; and in paring the Rhubarb to remove remaining portions of the bark, also the upper part of the root, and in perforating all pieces so as to examine their interior; because many, though sound externally, soon decay internally, from the rapid drying it is thought. The author, however, found most old roots of R. Webbianum more or less decayed. It is collected in quantities of 40,000 pounds before it is imported into the European parts of Russia, and is packed in bags, and placed where there is a free current of air, afterwards in chests.

Russian, called also *Turkey Rhubarb*, but which in Russia is called *Chinese Rhubarb*, is imported into the frontier town of Kiachta, and thence sent to Moscow and St. Petersburgh, whence it is distributed to the rest of Europe. It varies in shape, being irregularly roundish, and angular, from the bark having been shaved off with a knife; some pieces are cylindrical, a few flattish, many of them pierced with holes. Externally smooth, of a yellow colour; internally, the texture is rather dense; fracture uneven, irregularly marbled with white and red veins, having a strong and peculiar, slightly aromatic odour, a bitter, rather astringent taste, feels gritty when chewed, tinges the saliva yellow, and produces a powder of a bright yellow colour. Mr. Quekett obtained from 35 to 40 per cent. of raphides, or conglomerated crystals of oxalate of lime, which are situated within the cells, and which are the cause of the grittiness experienced when Rhubarb is chewed.

Bucharian Rhubarb, which makes its way to Vienna by Brody and Nischny, seems, from specimens and information afforded by Mr. Faber to Dr. Pereira, to be the inferior kinds of the above Rhubarb, and which, as inferred by the latter, would be burnt, if presented to the Russian authorities; it therefore finds its way into Europe by other channels. It is intermediate between the Russian and the Chinese Rhubarb, and generally of inferior quality.

Chinese, sometimes called *East India Rhubarb*, consists of two or three varieties. 1. One called by Dr. Pereira *Dutch-trimmed*, or *Batavian Rhubarb*, and, according to the shape, called *Flats* or *Rounds* in the trade, is closely allied to, and is derived, with very little doubt, from the same sources as the Russian Rhubarb; some of which, as mentioned above, finds its way to Canton, and is thence imported into Europe, or first into one of the Indian ports, and thence re-exported to Europe. It resembles the above in appearance, as the cortical portion appears to have been sliced off, and not scraped. The holes with which the pieces are perforated often contain within them pieces of the string by which they had been strung together. 2. Another variety, which is more particularly called *Chinese Rhubarb*, also *half-trimmed*, is distinguished from that called Russian, as being irregular in shape, never angular, but the edges rounded, as if the bark, instead of being sliced, had been scraped off, often some of it still remains adherent; the roots are, besides, of less uniform good quality than the Russian. Externally of a duller yellow; many of the pieces heavier from being more compact, the reticulation less regular, and of a yellowish-brown colour. 3. A third variety has lately been described by Dr. P. under the name of *Canton Stick Rhubarb*. This is in *cylindrical* pieces, about two inches long, and from half to three-quarters of an inch in diameter. These are probably produced in the mountains which bound China, as those of the province of Sechuen, and perhaps of Kansu.

Siberian Rhubarb. Small quantities of this kind have been imported by Mr. Faber, and have been proved to be those called by Grassman and others *Siberian Rhopontic Root*. This occurs in long, thin, almost cylindrical or spindle-shaped pieces, decorticated and perforated by a hole. Colour, externally pale yellow, internally brownish yellow, or reddish white. Odour and taste of good Rhubarb, but weaker; does not feel gritty. Dr. Pereira compares it with English *Stick Rhubarb* (*v.* P. J. iv. 448 and 500). It has since been proved (P. J. vi. p. 74) that the Rhubarb cultivated at Banbury is yielded by the same species, that is *Rheum Rhaponticum*.

Himalayan Rhubarb. This is produced by different species, is of very different quality. That yielded probably by *R. Moorcroftianum*, given to the author by Major Hearsey, the companion of Mr. Moorcroft, was of a bright but light yellow colour, and, as stated by the author in 1827, "appeared both in sensible qualities and medical virtues to equal the best Rhubarb that he had ever seen" (*Trans. Med. Soc. of Calcutta*, iii. p. 439.) The Rhubarb of *R. Webbianum* was tried by the author in both the Military and the Civil Hospital at Saharunpore, and found to be of very good quality. Some of it was subsequently submitted by the Medical Board to the late Mr. Twining, for experiment in the General Hospital at Calcutta. After trial in 43 cases, he reported that in doses of ℈j. or ʒß. it has a good purgative effect, operating nearly as freely as the best Turkey Rhubarb; and, further, that the effects of small doses of the remedy, as a tonic and astringent, are highly satisfactory; also, that it "is very efficacious in moderate doses for such cases as Rhubarb is generally used to purge." After a further trial in 4 other cases of Diarrhœa, he reports, "If further experiments should confirm the efficacy of the Himalaya Rhubarb in such cases, the acquisition of this remedy to the Materia Medica of this country will be of the utmost importance" (l. c. p. 445).

This Rhubarb differs much in appearance from that of commerce. The bark, of a brownish colour, has not been taken off. The texture is radiated, rather spongy, the colour a yellowish brown, the powder of a dull brownish-yellow colour, with little aroma (but when fresh dried, the root was described as aromatic by Mr. Twining), with a bitter and rather astringent taste. The author selected the branches of the roots, for he found the root-stock generally decayed in the centre. They were cut into short pieces, and slung upon string for the facility of drying. Some of the Himalayan Rhubarb is probably yielded by R. Emodi, perhaps by other undiscovered species of Nepal, and of the passes towards Bootan. Their value must be decided by their medical effects in the cases for which they are suited, and not by their differing in appearance from the roots of other species.

English Rhubarb. This is cultivated at Banbury, in Oxfordshire, to the extent of twenty tons annually, and is the produce of *Rheum Rhaponticum* (*v.* P. J. vi. p. 75). It is the kind frequently sold by men dressed up as Turks as Turkey Rhubarb. The pieces vary in shape, some being ovoid, others cylindrical (*English Stick Rhubarb*), smoothed externally, and rubbed with a yellow powder; light, rather spongy, with a reddish hue. It is rather mucilaginous in taste, and a little astringent. Its odour feeble, but unpleasant. It is supposed that much of this is employed for adulterating the Asiatic Rhubarb when in a powdered state (*v.* P. J. vi. p. 74 and 76).

A variety of analyses of Rhubarb have been made by different chemists, but with such varying results, that Drs. Schlosberger and Dæpping say that "not one satisfactory analysis is to be met with." They present as the result of their own labours that

1. Rhubarb appears, as regards its chemical and therapeutic properties, to be a mixture of Resin, Extractive Matter, and Chrysophanic Acid.

2. The Chrysophanic', of Parmelia parietina, is identical with the pure yellow Chrystalline obtained from Rhubarb, which has been described in its impure state as *Yellow Principle of Rhubarb, Rhein* or *Rhababarinas*, by Geiger; the *Rhabarberic Acid*, by Brandes.

3. Resins are among the chief constituents of Rhubarb, although their presence is denied by Dulk; they are, by the intermedium of other substances, as the so called Extractive matter, &c. partially soluble in water.

The three chief Resins are Aporetine, Phæoretine, and Erythroretine ; the two former appear isomeric; but all three are chiefly characterized by their different degrees of solubility.

4. The taste, odour, the relation to chemical re-agents, and the therapeutic action of Rhubarb, appear to be modified essentially by the joint co-operation of the *Resins*, the *Colouring Matter*, and the *Extractive Matter;* and probably, also, in a less degree by the Tannin, Gallic Acid, Sugar, Pectine, and the copious Salts of Lime which it contains. (Ann. der Chemie und Pharmacie, May 1844. P. J. iv. 322.)

The active principles of Rhubarb are taken up by water, either cold or hot, as also by Proof Spirit. The Alkalis produce a red-coloured solution with Rhubarb. The Acids cause a precipitate in its infusion, as does Gelatine. Sesquichloride of Iron produces a green-coloured precipitate. Tincture of Iodine produces a tawny muddiness.

Action. Uses. Cathartic, also mildly Astringent and Tonic. Acts chiefly upon the muscular fibre, and thus produces fæcal rather than watery evacuations. The cathartic is followed by its astringent effect, making it particularly valuable in cases of Diarrhœa, where it first evacuates and then strengthens the intestinal canal. In small doses it acts as a Stomachic and Tonic. Its colouring matter is readily absorbed, and may soon be detected in the urine. It is much used as a Laxative for children, especially in combination with Magnesia, sometimes with Calomel, and is equally suitable as a Purgative in cases of Diarrhœa, with an antacid and aromatic, or in cases where a mild Cathartic is required.

D. Of the powder gr. x.—℈j.

PULVIS RHEI COMPOSITUS, E. Comp. Rhubarb Powder.

Prep. Mix thoroughly *Magnesia* ℔j. *finely powdered Ginger* ℥ij. and *finely powdered Rhubarb* ℥iv. Preserve in well-closed bottles.

Action. Uses. Laxative and Antacid; well known as *Gregory's Powder.* In doses of ℈j.—ʒj. For children, gr. v.—gr. x.

PILULÆ RHEI, E. Rhubarb Pills.

Prep. Beat into a proper mass *finely powdered Rhubarb* 9 parts, *Acet. Potash* 1 part, *Conserve of Red Roses* 5 parts. Divide into 5 gr. pills.

Action. Uses. Aperient in doses of gr. x.—gr. xv.

PILULÆ RHEI COMPOSITÆ, L. E. Comp. Rhubarb Pills.

Prep. Mix *powdered Rhubarb* ℥j. (12 parts, E.) *powdered Aloes* ʒvj. (9 parts) *powdered Myrrh* ℥ß. (6 parts, E.) Then rub into a proper mass with *Soap* ʒj. (Castile 6 parts, E.), *Oil of Caraway* fʒß. L. (Oil of Peppermint 1 part, E.) *Syrup* q. s. L. (Conserve of Red Roses 5 parts, E.) till thoroughly mixed. (Divide into 5 gr. pills. Or, if preferred, omit the Oil of Peppermint, E.)

Action. Uses. Cathartic in doses of gr. x.—℈j. Well suited to a sluggish state of the bowels.

PILULÆ RHEI ET FERRI, E. Rhubarb and Iron Pills.

Prep. Beat into a proper mass *dried Sulph. Iron* 4 parts, *Extr. Rhubarb* 10 parts, *Conserve of Red Roses* 5 parts. Divide into 5 gr. pills.

Action. Uses. Tonic and Aperient in doses of gr. x.—gr. xv.

EXTRACTUM RHEI, L. E. D. Extract of Rhubarb.

Prep. L. D. Macerate for 4 days (with a gentle heat, L.) *powdered Rhubarb* ℥xv. (℔j. D,) in *Proof Spirit* Oj. (℔j. D.) and *Aq. dest.* Ovij. (℔vij. D.) Strain. Set by for the dregs to subside. Pour off the liquor and evaporate to the proper consistence.

E. Cut Rhubarb ℔j. into small pieces, macerate in Aq. Oiij. for 24 hours, filter through cloth, express moderately, macerate the residue with Aq. Oij. for at least 12 hours, filter through the same cloth, and express strongly. Filter again, if necessary, and evaporate to the due consistence in the vapour-bath. The Extract may be obtained of fine quality by evaporation in vacuo with a gentle heat.

Action. Uses. Cathartic in doses of gr. x.—ʒß. A good preparation may be obtained with cold water and percolation, when Spirit is unnecessary (*c*), and still better if evaporated in vacuo, as recommended in the E. P.

INFUSUM RHEI, L. E. D. Infusion of Rhubarb.

Prep. Infuse in a lightly covered vessel for 2 (12, E.) hours in boiling *Aq. dest.* Oj. (f℥xviij. E. by measure ℔ß. D.), *Rhubarb* in coarse powder ʒiij. (℥j. E. ʒj. D. Add Spirit of Cinnamon f℥ij. E.) Strain (through linen or calico, E.)

Action. Uses. Aperient and Stomachic in doses of f℥jß. repeated. The boiling water is ineligible, as a precipitate takes place on cooling: this is intended to be prevented by the addition of the Spirit. A good preparation may be made with cold water and percolation.

VINUM RHEI, E. Rhubarb Wine.

Prep. Digest for 7 days coarsely powdered *Rhubarb* ℥v., coarsely powdered *Canella* ʒij. in *Proof Spirit* f℥v. and *Sherry* Oj. and f℥xv. Strain; express strongly the residue. Filter.

Action. Uses. Stomachic in doses of fʒij. Purgative f℥ß.—℥j.

TINCTURA RHEI, E. Tincture of Rhubarb.

Prep. Mix powdered *Rhubarb* ℥iijß. and bruised *Cardamoms* ℥ß. Proceed by percolation with *Proof Spirit* Oij. as in Tinct. Cinchona Or prepare by digestion.

Action. Uses. Stomachic in doses of fʒj. Purgative f℥ß. A good preparation, especially if prepared by percolation, as Proof Spirit is an excellent solvent.

TINCTURA RHEI COMPOSITA, L. D. Comp. Tinct. of Rhubarb.

Prep. Macerate for 14 (7, D.) days *cut Rhubarb* ℥ijß. (℥ij. D.) *bruised Liquorice* ʒvj. (℥ß. D.) *cut Ginger* and *Saffron* āā ʒiij. (ʒij. D. bruised Cardamoms ℥ß. D.) in *Proof Spirit* Oij. (by measure ℔ij. D.) Strain.

Action. Uses. Cordial, Stomachic in doses of fʒj. Purgative f℥ß.—f℥j.

TINCTURA RHEI ET ALOES, E. Tincture of Rhubarb and Aloes.

Prep. Mix powdered *Rhubarb* ℥jß. *Socotrine* or East Indian *Aloes* powdered ʒvj., bruised *Cardamoms* ʒv., and with *Proof Spirit* Oij. proceed as for Tinct. Cinchona.

Action. Uses. Warm Cathartic in doses of f℥ß.—f℥j.

TINCTURA RHEI ET GENTIANÆ, E. Tinct. of Rhubarb and Gentian.

Prep Mix powdered *Rhubarb* ℥ij., powdered or finely cut *Gentian* ℥ß. and with *Proof Spirit* Oij. proceed as for Tinct. Cinchona.

Action. Uses. Stomachic in doses of fʒj. and Aperient in f℥ß.—f℥j.

THYMELÆÆ, *Juss.* (*Daphnoideæ.*) Daphnads.

Shrubs or undershrubs, with tenacious bark. Leaves scattered or opposite, simple, without stipules. Flowers complete, or by abortion unisexual. Perianth inferior, coloured, tubular; limb 4, seldom 5-fid, imbricate. Stamens equal to, or double the number of the divisions of the perianth, and inserted into its tube; anthers 2-celled, opening by 2 longitudinal chinks, sometimes abortive and scale-like. Ovary free, 1-celled, usually with 1 pendulous ovule. Style lateral or subterminal, usually very short. Fruit drupaceous or a nut, 1-seeded, seldom 2 to 3-seeded. Seed inverted; albumen wanting, or very thin. Embryo straight; radicle superior.—They are found in the central parts of the temperate zones, and in mountainous situations, and are remarkable for the tenacity and lace-like appearance of their bark, as well as for its acridity.

MEZEREUM, L. E. D. Radicis Cortex, L. Cortex, D. Root-Bark, E. of DAPHNE MEZEREUM, *Linn.* Mezereon. *Octandria Trigynia*, Linn.

This plant is supposed to be included with *Daphne oleoides* under the *χαμελαια* of Dioscorides. It is called *Mazrioon* in Persian works on Materia Medica and *Khamela* assigned as its Greek name.

Mezereon (fig. 84) is a small shrub. Leaves lanceolate, tapering below, smooth, evergreen. Flowers subternate, lateral, arranged in a spike-like manner, appearing before the leaves, rose-coloured. Perianth 4-fid, segments ovate acute, tube hairy. Stamens 8, short, inserted (1) in the tube of the perianth in two rows. Ovary (2) oval, oblong, with a short style and peltate stigma. Berry bright-red, fleshy, 1-seeded (*v.* 3, where some of the sarcocarp has been removed, to show the seed).—Woods of central Europe, less common in Great Britain; cultivated in gardens as an ornamental shrub.—Nees von E. 125. St. and Ch. 65.

D. Gnidium, (Fr. *Garou*), *D. Laureola,* Spurge Laurel, and other species, are also employed on the continent. The bark of the latter forms much of what is used, even in this country, for Mezereon.—Squire, P. J. i. 395.

Fig. 84.

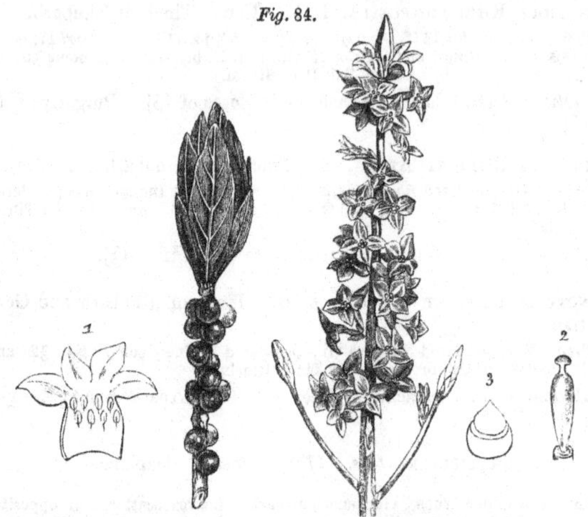

The bark of the root is officinal in the L. and E. P. and is the most efficacious; that of the stem and branches in the D. P.; but all these parts as well as the berries, are acrid. The bark is tough and fibrous, as in all Daphnes: it is met with in strips which are of a light greyish colour externally, whitish and shining within; when fresh dried, it has a slight but peculiar odour. The taste, especially of the inner part of the bark, is hot, acrid, and durable, though at first a little sweetish. These properties are imparted to water, Alcohol, Oils, and Vinegar. Analysed, it has been found to contain Sugar, Wax, Colouring matter, a neutral principle (*Daphnine*), together with an acrid Resin; but much of its active principle is volatilized by heat, and M. Vauquelin infers it is analogous in nature to Conia. The berries, according to Pallas, are employed as Cathartics, but in large doses will prove poisonous.

Action. Uses. Epipastic. A piece of the bark moistened in vinegar, and applied to the skin, and renewed, will produce a blister. Guibourt recommends an Ointment as a substitute for Savine Ointment. Also, Stimulating Diaphoretic and Diuretic, but chiefly used as an ingredient of the Decoction of Sarsaparilla.

Decoctum Mezerei, D. Decoction of Mezereon.

Prep. Mix *Mezereon* (bark of Daphne Mezereum, D.) in chips ℥ij. *bruised Liquorice Root* ℥ß. in *Aq.* Oij. (℔iij. D. with gentle heat, E.) Boil down to Ojß. (℔ij. D.) Strain.

Action. Uses. Diaphoretic in doses of f℥ij. 3 or 4 times a day; but most of the active principle is dissipated during the boiling.

Pharm. Prep. Decoctum Sarzæ Compositum, L.

Myristiceæ, *R. Brown.* Nutmegs.

Trees, often lofty, or shrubs, with acrid juice, which becomes of a reddish colour in the air. Leaves alternate, in two rows, without stipules, simple, entire. Flowers diœcious, usually inconspicuous, white, or yellowish. Perianth simple, trifid, rarely bi- or 4-fid, valvate. Stamens united into a column, or separate;

Fig. 85.

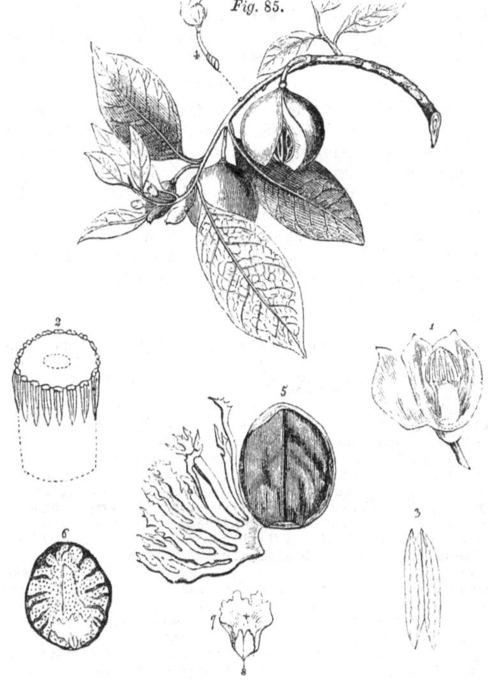

anthers 3 to 15, 2-celled, turned outwards and opening longitudinally, connate or distinct. Ovary free, single with 1 erect ovule, seldom 2 ovules. Style very short. Stigma somewhat lobed. Fruit drupaceous, dehiscent, 2-valved. Seed nut-like, enveloped in a many-parted fleshy aril. Embryo very small, at the base of a fleshy, fat-containing ruminate albumen. Cotyledons foliaceous; radicle inferior.

The Myristiceæ resemble *Laurineæ* in properties. By Dr. Lindley they were placed near *Anonaceæ*, which they resemble in the structure of their flower and seed, ruminate albumen, and position of embryo. He now places them in his Alliance Menispermales, in connection with which may be mentioned an interesting fact, first noticed by Dr. Falconer, that is, that the seeds of *Spærostemma grandiflorum* have the albumen remarkably aromatic, nearly as much so as Nutmegs, which in aroma and taste they closely resemble when bruised or chewed. Myristiceæ are found within the tropics of Asia and America.

MYRISTICA, L. E. Nux Moschata, D. Nuclei, L. D. Kernel of the Fruit, E. of MYRISTICA OFFICINALIS, *Linn.* E. (*M. Moschata*, Thunb.) L. D. Nutmegs. MACIS, Involucrum, D. (arillus) of the Nut. Mace. Adeps et Oleum, *v.* infra. *Diœcia Monadelph.* Linn.

Nutmegs, being the produce of the distant Spice Islands, were probably first known to the Hindoos (Sans *Jae-phul*, Java-fruit?), and through them to the Arabs, being the *jouz-al-teeb* or fragrant nut of Avicenna. The Dutch long endeavoured to confine the Nutmeg to three of the small Banda Isles. But when these were in the possession of the English from 1796 to 1802, Dr. Roxburgh brought away and introduced numerous plants into the English settlements of Bencoolen and Penang and into the Calcutta Botanic Garden. The Nutmeg has also been introduced into the Mauritius, French Guiana, and West India Islands. Nutmeg and Clove-trees have flowered this year at Syon House.

The Nutmeg-tree (fig. 85) is about 25 to 30 feet high, with some resemblance to a Pear-tree. Leaves faintly aromatic, alternate, sub-bifarious, with short petioles, oblong, somewhat obtuse at the base, acuminate, glabrous, above dark-green, paler beneath. *Male.* Racemes axillary. Flowers small, yellowish, the pedicels of each supported by a minute bract. Calyx (1) urceolate, 3-toothed, thick and fleshy, with short reddish pubescence. Filaments (1 and 2) united into a thick, oblong, and obtuse column. Anthers (2) about 9 (9 pairs, *Roxb.*), linear-oblong, attached round the upper part of the filamentous column, 2-celled, free at their base, opening longitudinally (3). *Female* (4). Peduncles usually solitary, axillary. Perianth much as in the male. Ovary ovate. Style short. Stigma 2-lobed, persistent. Fruit pyriform or nearly spherical, about the size of a peach. Pericarp fleshy, splitting from the apex into two equal, thick, fleshy, astringent valves, and displaying the deep-orange or scarlet-coloured *arillus* or *Mace*, which, cut into many irregular denticulate stripes, embraces the nut so tightly as to impress it with superficial furrows. Nut (5) ovoid, attached by a large umbilicus to the bottom of the cell; its shell is hard, of a dark brownish-black colour, and glossy, with its inner coat of a light-brown colour, thin, but spongy, closely investing the seed. *Seeds* or *Nutmeg* conform to the shell, and consisting chiefly of albumen, into which the inner coat of the shell dips deeply, giving it a variegated, brownish-veined, or *ruminated* appearance (6); while fresh rather soft, juicy, and more fragrant than after being dried. Embryo at base of albumen (6), erect, patelliform. Cotyledons 2, thick, fan-shaped, margins irregularly cut (7) Plumule of 2 unequal lobes. Radicle inferior, hemispherical.—Roxburgh chiefly, Fl. Ind. iii. p. 844. Corom. Plant. iii. t. 267. Nees von. E. 133. St. and Ch. 104.

Nutmegs are imported chiefly from the Spice Islands. A few other species yield aromatic nuts, as *M. tomentosa*, sometimes called Wild or Male Nutmegs: so *M. Otoba* in South America. *Virola sebifera* yields a large quantity of oil. Nutmegs are imported generally without their shells, and the *Mace* separated from them. Both are carefully dried in the sun, and the nuts then dipped in milk of lime to protect them from the depredations of insects. They are roundish or ellipsoidal, the finest rather small and heavy, marked externally with a network of furrows, internally of a light reddish-grey colour, marked with darker-coloured veins. The odour is agreeably aromatic; the taste warm, a little bitter, but gratefully aromatic. 500 parts analysed by Bonastre, were found to contain of Volatile Oil 6 per cent., Stearine or solid fatty matter 24, Elaine or liquid oil, coloured, 7·6, Acid 0·8, Starch 2·4, Gum 1, and Lignin 54 per cent. The dark veins especially contain the oily matter. The properties are taken up both by Alcohol and Ether.

MACE (*Macis*), as seen in the fresh Nutmeg, or in wet preserved specimens, is of a deep-orange or crimson colour; in a dry state, it is of a yellow or dull-orange colour; in flat, irregularly cut, somewhat horny, but also brittle pieces. It has the odour and taste of Nutmegs, and when analysed yields the same principles, that is, a volatile oil by distillation and a fixed oil by pressure; so that its active properties are soluble in Alcohol and Ether.

Action. Uses. Nutmegs and Mace are both Aromatic and Stimulant; in large doses Narcotic. Both are employed as Condiments, but Nutmegs as an adjunct to many officinal preparations.

MYRISTICÆ ADEPS, E. MYRISTICÆ OLEUM EXPRESSUM, L. (*v.* Emp. Picis.) Butter of Nutmegs, commonly but erroneously called *Expressed Oil of Mace*.

Butter of Nutmegs is an ingredient of the Emp. Picis, L. (*p.*) E., otherwise it is little used. It is imported from the Moluccas, being prepared there by partially heating and subjecting Nutmegs to pressure. It is solid, in brick-shaped cakes, of an orange-colour, with the odour of Nutmegs. It contains a little Volatile Oil, a fluid Elaine, and another fatty principle, which is solid and crystallizable; melts at 118°, is composed of $C_{28} H_{27} O_3$, and, when saponified, yields Myristic acid. (*Gregory.*)

Action. Uses. Emollient and slightly Stimulant as an Embrocation.

OLEUM MYRISTICÆ, L. E. Volatile or Essential Oil of Nutmegs.

Obtained by distilling *Nutmegs* reduced to powder with water. Usually imported of a pale yellow colour, with the odour and taste of Nutmegs. Sp. Gr. 920—948. This after a time deposits crystals of Stearoptene, the Myristicine of some authors. The truo Volatile Oil of Mace is similar in nature and properties.

Action. Uses. Stimulant in doses of ♏j.—♏iij.

SPIRITUS (NUCIS MOSCHATÆ, D.) MYRISTICÆ, L. E. Spirit of Nutmeg.

Prep. Mix bruised *Nutmegs* ℥ijß. (℥ij. D.) in *Proof Spirit* Cj. and *Aq.* Oj.

(q. s. to prevent empyreuma. Macerate for 24 hours, D.) with gentle heat, L. Distil Cj.

Action. Uses. Aromatic Stimulant. Used as an adjunct in doses of f℥j,—f℥iv.

LAURINEÆ, *R. Brown.* Laurels.

Trees, seldom shrubs, generally with handsome foliage. Leaves exstipulate, alternate, sometimes approximated, so as to appear opposite, simple, entire, coriaceous, and evergreen, sometimes glandular and dotted below. Flowers complete, or by abortion unisexual, regular, racemose, or paniculate, sometimes umbelliform; pedicels tribracteate. Perianth calyx-like, inferior, 4 to 6 cleft or divided, imbricated. Disk fleshy, attached to the bottom of the perianth. Stamens inserted into the base of the perianth, either 6 in a single row, or 12 in a double or treble row, the fertile alternate with the barren ones. Anthers adnate, 2—4-celled, opening by valves, recurved from the base to the apex. Glands often present at the base of the inner filaments. Ovary free, 1-celled, with 1 to 3 pendulous ovules. Style 1. Stigma 1, obscurely 2—3 lobed. Fruit a berry or rather drupe. Seed without albumen; embryo straight; cotyledons large; radicle superior. — The Laurineæ, from the structure of their anthers, are allied to *Atherospermeæ* and to *Gyrocarpeæ:* from the *Thymelææ* they are easily distinguished by their structure. They have been well elucidated by Nees von Esenbeck. In their red-coloured juice, and *aromatic* properties, they resemble *Myristiceæ.*—Tropical regions of Asia and America, with two species extending to the north of Africa and south of Europe.—*Enneand. Monog.* Linn.

LAURI FOLIA, L. D. et LAURI BACCÆ, L. D. Leaves and Berries of LAURUS NOBILIS, *Linn.* Laurel or Sweet Bay. *Enneandria Monog.* Linn.

This is the Δαφνη of the Greeks. From its leaves having been employed in making chaplets for their gods and crowns for their heroes, it was called *Laurus nobilis* by Linnæus. It is the *ghar* of the Arabs, and probably the *Ezrach* of the Bible. (*v.* Bibl. Cycl. i. p. 692.)

Evergreen (fig. 86) from 15 to 25 feet high, with dense leafy branches. Leaves oblong, lanceolate, acute, wavy at the margin, hairless, with the exception of a fine beard and small pore at the axils of the lower veins. Umbels 4—6 flowered, (1.) axillary, supported by scarious concave scales. Flowers diœcious, yellowish, dotted with fine glands. Perianth 4-parted (2). Fertile Stamens 12 in 3 rows, the external alternating with the segments of the perianth. Filaments each with 2 glands at, or above, the middle. Anthers oblong, 2-celled, all looking inwards, opening with 2 turned-up valves (3). Female flowers with 2 to 4 castrated stamens. Stigma capitate. Berry ovoid, about the size of a field-bean, bluish-black, single-seeded. Cotyledons large, oleaginous, convex on the back.—North of Asia Minor; common in Mediterranean region; shrubberies in England.—Nees von E. t. 132.

Lauri Folia, L. D. Laurel leaves, which must be carefully distinguished from the poisonous leaves of the Cherry Laurel, p. 390, have a fragrant odour, and an aromatic, rather bitter taste. These properties they owe chiefly to the presence of a yellow-coloured Volatile Oil, which may be separated by distillation with water. The leaves of a species of Cinnamomum are still employed in India for the same purposes as these, and were, no doubt, the Malabathrum of the Ancients. They are called *tej-pat*, and the bark, *tej* and *putruj*.

Lauri Baccæ, L. D. Laurel Berries (*hab al ghar* of the Arabs) are oblong, ellipsoid; when dry of a dark-brown colour, with a wrinkled, friable sarcocarp, covering the two oval fatty cotyledons. They contain a warm, fragrant, Volatile Oil, which may be obtained by distillation with water, and about $\frac{1}{4}$ of a greenish-coloured fat, which may be separated by expression.

Fig. 86.

Oleum Lauri expressum, or Oil of Bays, obtained by pressure, and the aid of heat, both from the fresh and the dry drupes of the Bay tree. It is imported from the S. of Europe. Like the Butter of Nutmegs, it contains a Volatile Oil, Elaine, and Stearine.

Action. Uses. Leaves and Berries, Aromatic Stimulants, but are not much used now. The infusion of the leaves is Diaphoretic, or they may be used in a bath. Oil of Bay, is a stimulant Embrocation.

SASSAFRAS. Radix, L. D. Root, E. Lignum et Oleum Volatile, D. Wood and Volatile Oil of SASSAFRAS OFFICINALE, *Nees von Esenbeck* E. (*Laurus Sassafras*, Linn.) L. D.

Sassafras was discovered by the Spaniards in Florida in 1528, but is common throughout the United States. The name is considered to be a corruption of Saxifrage (*De Theis*). Sassafras nuts have nothing to do with this plant, but are probably those of a Nectandra.

A small diœcious tree, but growing to a great height in favourable situations, with a trunk about a foot in diameter. Bark rough, furrowed, greyish-coloured, but the twigs smooth, and bright green. Leaves alternate, petiolate, downy when young, membranous, varying much in form and size, some being oval and entire, others with a lobe on one side only, the generality 3-lobed, but all tapering into the petiole. Flowers slightly fragrant, of a pale yellowish-green colour, racemose, with deciduous subulate bracts. Perianth 6-parted. Stamens 9, the 3 inner with a thick stipitate gland on each side. Anthers linear, 4-celled, all looking inwards. Female flowers usually with fewer sterile stamens, the inner often all united together. Drupe oval, about the size of a large pea, of a deep-blue colour, placed on the thickened apex of the reddish-coloured peduncle, surrounded by the cup formed by the remains of the perianth.—North American woods from Canada to Florida. Flowers in May in the north, but earlier in the south.—Nees von E. 131. St. and Ch. 126.

Sassafras Wood is light, porous, and fragile, whitish in the young, and reddish in the old tree, but feebly aromatic, and seldom employed in America. The Bark is sometimes separated.

The *Root* is much more efficacious, and usually seen in irregular and branched fragments; wood, of a brownish-white, light and porous; bark brittle, spongy, in layers of a rusty cinnamon hue; fresh exposed surfaces of a lighter hue, sometimes covered with a brownish epidermis. Dr. Reinach has lately analysed the *Bark of the Root* of the Sassafras, in which he found, as was to be expected, a much larger proportion of the active principle than in the Wood. He found of Heavy and Volatile Oils 8, Fatty matter 8, Resin and Wax 50, *Sassafrid* (a peculiar principle which may be arranged with Tannin) 92, Tannin 58, Sassafrid with Tannin and Gum 68, Albumen 6, Gum, Colouring matters, &c. 30, Starch with reddish-brown Colouring matter 54, Starch with Tannin, &c. 289, Lignin 247=1000 parts.

The odour is pleasantly fragrant, its taste sweetish and gratefully aromatic, dependent on the presence of its Volatile Oil. These yield their properties to hot water as well as to Alcohol; but the Volatile Oil will necessarily be dissipated, if much heat is employed.

Sassafras Pith is described by American writers as in slender, very light, and spongy pieces; mucilaginous in taste, with a slight flavour of the Sassafras; forming a limpid mucilage with water, which is much used as a demulcent. The leaves even are said by Dr. Lindley to be used in Louisiana for thickening soup, from containing much mucilage.

Action. Uses. Stimulant; Diaphoretic in infusion; but chiefly used in combination.

Pharm. Prep. Decoct. Guaiaci Comp. Aqua Calcis Comp. Decoct. Sarzæ Comp.

OLEUM SASSAFRAS, L. E. D. Volatile Oil of Sassafras.

Prep. Distil bruised Sassafras Root, wood, and bark, with water.

A light yellow-coloured Oil; Sp. Gr. 1·094; with Sassafras-odour and pungent taste. It appears to be composed of two oils, one of which swims, the other sinks in water. After a time it deposits a Stearoptene.

Action. Uses. Warm Stimulant in doses of ♏ij.—♏v.: It is one of the Oils recommended to be added to the Fluid Extract of Sarsaparilla, E.

CAMPHORA, L. E. D. Camphor. Concretum sui generis sublimatione purificatum, L. Produce of CAMPHORA OFFICINARUM, *Nees v. Esenbeck*, E. (*Laurus Camphora*, Linn.) L. D.; also of Dryobalanops Camphora, *Colebr.* D.

Camphor is a principle found in many plants, but only in two in any great abundance. One of these, erroneously referred to as yielding some of the Camphor of European commerce, has been mentioned at p. 290. The other is a Chinese and Japanese plant. Camphor, like several substances the produce of countries to the southward or eastward of India, was unknown to the ancients. (Hindoo Med. p. 93.) It was known to the Arabs, and called by them *Kaphoor*.

The Camphor tree is an evergreen, grows to a considerable size, is straight below and branched. All parts emit a camphoraceous odour when bruised. Wood white, fragrant, much used in China for making trunks, boxes, &c. Branches somewhat lax, smooth, with a greenish bark. Leaves alternate, with long petioles, ovate-lanceolate, rather coriaceous, smooth, shining, and bright-green above, paler beneath, triple-nerved, with a sunken gland opening by a pore beneath, at the axils of the principal lateral veins. Leaf-buds scaly. Flowers small, hermaphrodite, smooth externally, in naked, axillary, and terminal corymbose panicles. Perianth 6-cleft, with a deciduous limb. Fertile stamens 9, in 3 rows, the 3 inner supported at the base with 2 stipitate compressed glands. Anthers 4-celled, opening by as many ascending valves, the 3 interior looking outwards, the others opening inwards. Three sterile stamens subalternate with those of the second row, three others stipitate, each with an ovate head. Drupe situated in the truncate cup-like base of the perianth.—Native of China, principally near Chinchew in the province of Fokien; also of Formosa and Japan.—Nees von Esenb. 130. St. and Ch. 129.

Camphor is diffused through all parts of the plant, and is separated from the root, trunk, and branches, which when cut into chips, are boiled in water and then sublimed into inverted straw cones contained within earthen capitals. It is thus obtained in the form of *Crude Camphor*, chiefly from the province of Fokien and the opposite island of Formosa, but some of good quality is also procured from Japan. The Dutch exported from thence into Europe 310,520 lbs. in seven years. It is sometimes imported into this country from Batavia. The ordinary Crude Camphor is in small greyish-coloured, slightly sparkling grains, which by aggregation form greyish crumbling cakes, with all the properties of purified Camphor. This is separated from impurities by being mixed with lime and sublimed in thin glass vessels, which being afterwards cracked, the Camphor is obtained in a concavo-convex cake about three inches thick with a hole in its middle.

Camphor is solid, colourless, and translucent, with a crystalline texture, has a strong, penetrating, aromatic odour, and a bitter, rather pungent taste, followed by a sensation of coolness; though brittle, it is not, from its toughness, easily pulverized. Sp. Gr. ·98 to ·99; so that it floats on water, and, evaporating, produces a circulatory movement. From its volatility, it volatilizes at ordinary temperatures, and crystallizes on the inside of bottles. It melts at 288° and boils at 400°, burns with a bright flame; is little soluble in water, but easily so in Alcohol, Ether, the volatile and fixed Oils, also in Acetic acid (*v.* Acid. Aceticum Camphoratum), and the diluted acids, and in water charged with Carbonic acid, and with the aid of trituration in about 8 times its weight of Milk. Nitric Ether f℥ij. will retain Camphor ℈j. dissolved in f℥iv. of water. (*c.*) Nitric acid, by yielding Oxygen, converts it into Camphoric acid. Camphor is considered an oxide of *Camphogen* ($C_{10}H_8$), or as a solid volatile oil composed of $C_{10}H_8O$. The E. C. give as a test of its purity: "Its powder evaporates entirely when gently heated."

Borneo Camphor is in white crystalline fragments, as found in the wood of the Dryobalanops Camphora. Sp. Gr. 1·009. Its odour is not of so diffusible a nature, otherwise it closely resembles the above. The Liquid Camphor of the same tree seems to be of the nature of Camphogen. Dr. A. T. Thomson, by passing a current of Oxygen gas through it, converted it into Camphor.

Action. Uses. The action of Camphor is variously described. Applied locally for a time to any delicate surface, it will act as an irritant. Its action when taken internally, is chiefly on the nervous system. In moderate doses it will exhilarate, and also allay nervous irritation, and produce quietude and placidity of feeling. Being absorbed into the system, in large doses, or in particular constitutions, the circulation may be affected. It afterwards passes off by the skin and bronchial membranes, but not by the urine; the pulse is increased in fulness, and diaphoresis produced, especially if the patient be covered over; consequent relief of febrile symptoms ensue. But other cases occur, in which sedative effects are perceived. In large doses it acts as a Narcotic, and might be poisonous. It is chiefly useful as a Calmative and Anodyne in various Nervous, especially Hysterical, affections, in doses of gr. v.—gr. x., and in some Nervous and Typhoid fevers.

Mistura Camphoræ, L. E. D. Camphor Mixture.

Prep. L. D. Rub up *Camphor* ʒß. (℈j. D.) with *Rectified Spirit* ♏x. (add purified Sugar ℥ß. D.) stirring, gradually add *Aquæ* Oj. (hot, by measure ℔j. D.) Filter through linen (bibulous paper, D.)

E. Rub together *Camphor* ℈j. and pure *Sugar* ʒß., add *Almonds* ʒß. previously steeped in hot water and peeled; beat the whole into a smooth pulp; constantly stirring, add *Aq.* Oj. Strain.

Action. Uses. The Camphor Mixture of the L. P. is feeble, but has the odour of Camphor, and is a grateful vehicle for more powerful medicines, in many nervous affections, in doses of fʒij.—fʒiij. With the aid of Sugar or the emulsion, a larger quantity is taken up in the other preparations. Camphor may be given dissolved in Milk.

MIST. CAMPHORÆ CUM MAGNESIA, E. D. Camphor Mixture with Magnesia.

Prep. Rub up together *Camphor* gr. x. (gr. xii. D.) *Carb. Magnesia* gr. xxv. (ʒß. D.) gradually add *Aq.* fʒvj.

Action. Uses. Camphor by the aid of the Magnesia is dissolved in larger quantity, but will separate from the water. This as an antacid may be more useful in some cases than the above, in doses of fʒß.—fʒiij.

TINCTURA CAMPHORÆ, L. E. D. SIVE SPIRITUS CAMPHORATUS, D.

Prep. Dissolve *Camphor* (in small fragments, E.) ʒv. (ʒijß. E, ʒj. D.) in *Rectified Spirit* Oij. (by measure ʒviij. D.)

Action. Uses. For external use, Stimulant and Anodyne (*v.* Linim. Ammoniæ Comp. p. 59). With the aid of Sugar to suspend the Camphor in water, it may be given internally in doses of ♏x.—fʒß.

TINCT. CAMPHORÆ COMP., or Tinct. Opii Camphorata. Paregoric. (*v.* p. 269.)

Paregoric contains Opium gr. i. in fʒß. D. f3ß.—f3iv.

LINIMENTUM CAMPHORÆ, L. E. OLEUM CAMPHORATUM, D.

Prep. Rub together in a mortar, and dissolve *Camphor* ʒj. (ʒj. D.) in *Olive Oil* fʒiv. (ʒj. Strain, D.)

Action. Uses. Externally as a Stimulant and Anodyne. Commonly called Camphorated Oil. Camphor is also an ingredient of Soap Liniment. (p. 477.)

LINIMENTUM CAMPHORÆ COMPOSITUM, L. D. Compound Camphor Liniment.

Prep. Mix *Liquor Ammoniæ* (Aqua caustica, D.) fʒvijß. (by measure ʒvj. D.) with *Spirit of Lavender* Oj. (by measure. ℔j. D.) With a slow fire distil from a glass retort Oj. (by measure ℔j. D.) in which dissolve *Camphor* ʒijß (ʒij. D.)

Action. Uses. Rubefacient and Stimulant for external use.

CINNAMOMUM, L. E. D. Cortex, et Oleum e cortice destillatum, L. D. Bark and Volatile Oil of the Bark of CINNAMOMUM ZEYLANICUM, *Nees von Esenbeck*, E. (*Laurus Cinnamonum*, Linn.), L. D. Cinnamon. *True Cinnamon* of the shops.

Cinnamon is the *Kinnemon* of Exod. xxx. 23 (see Bibl. Cycl. ii. p. 210), and the κιννάμωμον of Herodotus, a name which he states the

Greeks learned from the Phœnicians. The name seems derived from the Cingalese *Cacynnama* (dulce lignum), or the Malay *Kaimanis*, which Mr. Marshall says is sometimes pronounced *Kainamanis*. (*v. Antiq. of Hind. Med.* 84 and 141.)

Fig. 87.

The Cinnamon tree of Ceylon (fig. 87) is about 30 feet high. The root has the odour of Cinnamon as well as that of Camphor, and yields this principle upon distillation. The twigs are somewhat 4-cornered, smooth, shining, and free from any downiness. The leaves are liable to variation, ovate, or ovate-oblong, terminating in an obtuse point, triple or three-nerved, that is, there are three principal nerves, which sometimes remain separate to the very base, but usually approach each other a little above the base, but without uniting; there are, moreover, in many cases, two shorter nerves external to these. Leaves reticulated on the under side, smooth, shining, the uppermost the smallest, with a good deal of the taste of cloves. The leaf-buds are naked. Panicles terminal and axillary. Flowers usually bisexual, rather silky. Perianth 6-cleft (2), segments oblong, the upper part deciduous. Fertile stamens 9, in 3 rows, the 3 inner with 2 sessile glands at the base (6). Anthers ovate, 4-celled (4—6), the 3 interior opening outwards. 3 abortive capitate stamens (staminodia) in the interior of all. Ovary 1-celled, with a single ovule. Stigma disk-like. Drupe (or berry) 1-seeded, seated in the cup-like six-lobed base of the perianth (7). Seed large, with large oily cotyledons (8—10); embryo above. —Native of Ceylon, now cultivated elsewhere, as on the Malabar coast, in Java, Cayenne, &c.—Nees von E. as Laurus Cinnamomum 128. St. and Ch. 121.

Cinnamon is cultivated in plantations situated on the south-west of the island of Ceylon, between Negombo and Matura, where the soil is nearly a pure quartzose sand, the climate damp, showers frequent, and the temperature high and equable. (*Dr. Davy.*) Trees may be cut when six or seven years old. Branches three years old, or which are from half an inch to three inches in diameter, are selected and lopped off, commencing in May and continuing till October. The bark is divided by longitudinal incisions, of which two opposite to each other other are made in the smaller shoots, several in the larger, and then peeled off in strips. After twenty-four hours, the epidermis and the green matter under it are scraped off, after the strips of bark have been placed on a convex piece of wood. The bark soon contracts into the form of quills, which are about forty inches in length, of which the smaller are introduced within the larger ones, and form the ordinary rolls of Cinnamon. They are dried first in the shade and then in the sun, and sorted into Cinnamon of different qualities, known in commerce as first, second, and third Cinnamon. It is imported chiefly from Ceylon; some also from Tellicherry on the Malabar coast, probably grown at Anjarakandy. (*v.* p. 371, and Buchanan's Mysore, p. 546.) Some is also exported from other parts of the Madras presidency, where it was long since introduced by Dr. Anderson, and grown at Tinnivelly, &c. *v.* Roxb. Fl. Ind. ii. p. 296. It has been exported from Quilon. *v.* p. 544.

Besides Cinnamon, Oil of Cinnamon is produced; and from the ripe fruits a fatty substance called *Cinnamon Suet* is expressed, which the author supposed to be the *Comacum* of Theophrastus. (*Antiq. of Hind. Med.* p. 546.) He finds it noted in Rheede, *Hort. Mal.* i. p. 110.

Ceylon Cinnamon of the best quality is in long and slender cylindrical bundles, about forty inches in length, composed of numerous quills rolled up within one another, each about the thickness of cartridge paper, smooth, pliable, breaking readily with a splintery fracture, and easily powdered. The colour of the bark is a dull, yellowish-brown, now usually called a Cinnamon-colour. The surface is intersected by pale glistening fibrils. It has a pleasant, grateful odour, a warm, sweetish, and very agreeable taste. Besides the three qualities of Ceylon Cinnamon, it is also imported of different qualities from the Malabar Coast; and Dr. Wight has ascertained that the Cassia of the Indian Peninsula is sometimes exported as Cinnamon. *v.* p. 544. But the Cinnamon plant itself has been introduced into so many places, that small quantities are occasionally imported from them, as for instance from Cayenne. Some of these are employed for adulterating the superior and more expensive Ceylon Cinnamon. The inferior kinds are thicker and less grateful, and more resembling Cassia. Analysed, Cinnamon is found to contain Volatile Oil (about 6 parts in 1000), Tannin (of the nature of Catechu Tannin), Mucilage, Resin, Colouring matter, Cinnamic acid, and Ligneous fibre.

Action. Uses. Aromatic and Stomachic, slightly Astringent. The most grateful of condiments, and much used as an ingredient of chocolate. The powder in doses of gr. v.—℈j. will check Nausea, relieve Flatulence, and some Cramps. Much employed in combination in Diarrhœas; in low states of the constitution; and as a constituent of various preparations.

Pharm. Prep. Infusum Catechu. Decoct. Hæmatoxyli. Elect. Catechu. Confectio Aromatica. Pulv. Cretæ Comp. Pulv. Kino Comp. Tinctura Catechu. Tinct. Cardamomi Comp. Sp. Lavand. Comp. Sp. Ammoniæ Arom. Acid. Sulphuric. Arom. Vinum Opii.

PULVIS CINNAMOMI COMPOSITUS, L. PULV. AROMATICUS, E. D. Compound Cinnamon or Aromatic Powder.

Prep. Rub up into very fine powder *Cinnamon* ℥ij. *Cardamoms* ℥jß. *Ginger* ℥j. (Cardamoms and Ginger āā equal parts, D., of all three equal parts, E.) *Long Pepper* ℥ß. L. (℥j. D. Keep in well-closed glass vessels, E.)

Action. Uses. Aromatic Stimulant in doses of gr. v.—℈j.

ELECTUARIUM AROMATICUM, E. Aromatic Electuary.

Prep. Of the above *Aromatic Powder* 1 part to be triturated with *Syrup of Orange-peel* 2 parts. This, though similar to the *Aromatic Confection*, L. D. (p. 111), in the nature of the spices, differs in not containing any Chalk.

Action. Uses. Carminative in doses of gr. x.—ʒß.

OLEUM CINNAMOMI, L. E. D. Oil of Cinnamon.

This Oil is imported from Ceylon, being obtained by macerating the powdered rejected bark in a saturated solution of salt, and then distilling. The water which passes over is milky, from holding the Oil in suspension; but this soon separates. About 8 ounces are obtained from 80 ℔. of recently prepared Cinnamon. Some of it is heavier and some lighter than water.

The E. P. gives as its characteristics: "Cherry-red when old, wine-yellow when recent; odour purely cinnamonic. Nitric′ (concentrated, added drop by drop) converts it nearly into a uniform crystalline mass (a compound of the acid and of the oil)." The colour varies according to age. Oil of Cassia is often substituted for it; other adulterations are used, which the Nitric′ detects by producing a less distinct effect. Oil of Cinnamon is remarkable for its grateful aroma and spicy cinnamonic taste, less pungent and acrid than Oil of Cassia. It is composed of $C_{18} H_7 O_2 + H$, and is considered by Chemists a Hydruret of Cinnamyle. On exposure to the air it absorbs Oxygen and some Cinnamic acid is formed; also two resins. This acid may also be obtained by distilling the Balsams of Tolu and of Peru. By powerful oxidizing agents this acid may be converted first into Bitter Almond Oil, and then into Benzoic acid. (*v.* Fownes, P. J. iv. 264.)

Action. Uses. Grateful but powerful Stimulant in doses of ♏j.—♏iij. Much used for flavouring, by cooks and confectioners.

Aqua Cinnamomi, L. E. D. Cinnamon Water.

Prep. Take *bruised Cinnamon* ℔jß. (℥xviij. E. ℔j. D.) or *Oil of Cinnamon* ʒij. L. (ʒiij. D.) *Proof Spirit* f℥vij. L. (Rectified f℥iij. E.) *Aq.* Cij. (q. s. to prevent empyreuma. Macerate for 1 day, D.) Distil off Cj.

Action. Uses. Carminative, but chiefly used as a vehicle in doses of f℥jß.—f℥iij. It is made with greater facility with the Oil, and is sometimes prepared by diffusing the Oil through water by means of Sugar or Magnesia. But Cinnamon water made from the Oil is much more apt to spoil, from the formation of Cinnamic acid: distilling from the bark is therefore the preferable method.

Spiritus Cinnamomi, L. E. D. Spirit of Cinnamon.

Prep. Mix *Oil of Cinnamon* ʒij. *Proof Spirit* Cj. *Aq.* Oj. With slow heat distil Cj. L. Take *Cinnamon* in coarse powder ℔j. (Proceed as for Spir. Caraway, E.) *Proof Spirit* Cj. *Aq.* q. s. to prevent empyreuma. Macerate for 24 hours. Distil Cj. D.

Action. Uses. Stimulant adjunct in doses of fʒj.–fʒiij. to draughts.

Tinctura Cinnamomi, L. E. D. Tincture of Cinnamon.

Prep. Take *Cinnamon* powdered ℥iijß., *Proof Spirit* Oij. (by measure ℔ij.) Macerate for 14 days, and strain. (Proceed by percolation or digestion, as directed for Tinct. Cassia, E.)

Action. Uses. Grateful adjunct (fʒj.—fʒiv.) to draughts of different kinds.

Tinctura Cinnamomi composita, L. E. Compound Tincture of Cinnamon.

Prep. Macerate for 14 (7, E.) days *bruised Cinnamon* ℥j., *bruised Cardamoms* ℥ß. (℥j. E.) *Long Pepper* ground ℥ijß. (ʒiij. E.) *Rasped Ginger* ʒijß. in *Proof Spirit* Oij. (Strain express, E.) Filter. (Best prepared by percolation, as Compound Tincture of Cardamoms, E.)

Action. Uses. Aromatic adjunct to Astringent and other draughts, in doses of fʒj.—fʒij.

Emplastrum Aromaticum, D. Aromatic Plaster.

Prep. Melt together *Thus* (Abietis Resina) ℥iij. and *Yellow Wax* ℥ß. Strain. When thickening as it cools, mix in powdered *Cinnamon* ʒvj. rubbed up with *Essential Oil of Allspice* and *Essential Oil of Lemons* āā ʒij. Make a plaster.

Action. Uses. Applied over the Stomach in Nausea, &c.

Cassiæ Cortex, E. D. Bark of Cinnamomum (Cassia, *Blume*) aromaticum, *Nees v. Esenbeck*, E. (Laurus Cassia, *Linn.*), D. and of other species.

Cassia is mentioned by early Greek writers; in the Bible by the name *Kiddah* (κιττώ is one kind of Cassia in Diosc. i. c. 12). It is translated Cassia in Exod. xxx. 24 (*v.* Kiddah, and Kinnamon, *Bibl. Cycl.* and *Antiq. of Hind. Med.* p. 84). Now, there are several distinct sources of Cassia, though it is often described as produced by the *Laurus Cassia* of Linnæus, a plant, said to be of Ceylon and the Peninsula of India, but which it is difficult to determine. Dr. Wight has shown that no less than three species were included by Linnæus under one name: one plant is *Litsæa zeylanica*, Dawalkurunda of the Cingalese; another, apparently *Cinnamomum sulphuratum* of Nees, is the *C. perpetuo-florens* of Burmann; and the third, the *Carua* of Rheede (Hort. Mal. i. 57), Dr. Wight considers to be the *Cinnamomum iners* of Nees. Hermann's own plant of Cassia lignea, in the British Museum, is named *Walkurunda*, and is a true Cinnamomum.

Cassia bark, called *Cassia lignea* (and by the Chinese *Kwei Pe*, or Cassia-skin), we learn from the Chinese Repository, ii. 455, is exported from China to all parts of the world, sewed up in mats, usually two or more rolls in each mat, and a pound in each roll. From China it is imported into Singapore, Calcutta, and Bombay. It is imported into Great Britain from all these places, and also from Manilla, as well as from Quilon and Madras, and is kept in all shops and sold as Cinnamon, this being distinguished by the name of *true* Cinnamon, and must be specially asked for by this name.

Chinese Cassia, or *Cinnamon*, as it is also called, is, according to Mr. Reeves, produced in the province of Kwangsi, whose principal city Kwei-ling derives its name from the forests of Cassia by which it is surrounded. It is also said to be produced in Cochin-China in the dry sandy districts lying north-west of the town of Faifoe, between 15° and 16° of N. lat., and imported into Canton, &c. to the extent of 250,000 lbs.; also, that it is preferred to the Cinnamon of Ceylon. Mr. Crawford (Embassy to Siam, p. 470) says that the epidermis is not freed from the bark, as it is in that of Ceylon, that the superior kinds are retained for consumption in China, and the inferior re-exported to Singapore, &c. and to Europe.

Cassia, though it bears a considerable resemblance to Cinnamon, is usually in single quills, seldom more than double, from ¼ to ½ or even 1 inch in diameter. It is thicker, rougher, more dense, and breaks with a shorter fracture, and is of a darker red colour than Cinnamon, and its powder of a reddish-brown tint. Its taste is more pungent and stronger, but not so sweet and grateful as that of the *true* Cinnamon, and its odour less agreeable. That from China is now always stripped of its epidermis, and is probably produced by the plant described below under the name of *Cinnamomum aromaticum.*

Cassia Buds (called by the Chinese *Kwei-tsze*, or Cassia-seeds), formerly officinal under the name of Flos Lauri Cassiæ. They are exported from Canton in considerable quantities, have some resemblance to cloves or to nails with round heads, or are cup-shaped when the perianth bud falls off. Those that are fresh and plump are preferred. They are imported here chiefly for use in confectionery, forming one of the ingredients of many old receipts. From the Chinese Repository, the opinion of Mr. Reeves, of Nees, &c., it is more than probable that both *Cassia buds* and *Cassia lignea* are obtained from the same tree. The buds have the flavour and pungency of taste of Cassia, and yield an essential oil upon distillation.

CINNAMOMUM AROMATICUM, *Nees v. Esenb.* A tree of considerable size. There is one 18 feet high in the Edinb. Bot. Garden. Branches angular, twigs and petioles covered with down. Leaves often nearly opposite, though usually alternate, oblong-lanceolate, acute at each end, triple-nerved, or with three nerves which unite into a single nerve above the insertion of the leaf-stalk, and disappear towards the apex of the leaf; the nerves are, like the twigs, covered with broken (strigulose) downiness, with curved veinlets on the under surface. Panicles narrow, silky.—This is the *C. Cassia* of Blume, introduced from China, and the *Laurus Cassia*, t. 3. of the brothers Nees in their paper on the Cinnamon, also of Hort. Kewensis, ii. p. 427, and the Laurus Cinnamomum of Andrew's Repos. t. 595, often quoted 596. The leaves taste mucilaginous and Cinnamonic (*c.*) Dr. Lindley says that as grown in stoves they are almost insipid, mucilaginous, and somewhat astringent. I have received a specimen of this plant (named *Cinnamomum chinense*) from the Messrs. Loddiges, who inform me that it was imported by their house from China in 1790. This is probably the source of the various plants in hot-houses. The leaves taste mucilaginous, but also aromatic.

Cassia, on the authority of Mr. Marshall, has been usually thought to be only the coarser kinds of the Cinnamon of Ceylon, or that separated from the larger branches or thick roots, and that though intended for consumption in Asia, has been imported into England and sold as *Cassia lignea.* But this, as has been observed, can have seldom taken place, for it cannot be traced among the exports of Ceylon, where, moreover, all Cinnamon, whether coarse or fine, pays an export duty of 3*s.* a pound, and the Cassia, even in England, is not worth more than 1*s.* a pound.

MALABAR CASSIA. Cassia has long been known to be a product of the forests of the Indian Peninsula. It is stated to have been a cause of jealousy between the Dutch and Portuguese that the latter sold some of the *Wild Cinnamon* growing in Cochin, &c. Buchanan, in his Travels in Mysore (ii. 336), mentions it as common on all the hills of the Malabar coast, also that Mr. Brown had planted it at Anjarakandy, as well as the Ceylon Cinnamon (ii. 545—6). He states that at Mangalore it was called *Dhal-China* (that is, Cinnamon), and exported to Muscat, Cutch, Scind, and Bombay; also that the buds of the tree were called *Cabob-China,* and likewise exported to the same places. This export still continues, as it appears by the reports of the commerce of Madras and of Bombay that 107,856 lbs. of Cassia were exported from Malabar and Canara, and that in 1844—45, there were imported into Bombay from Malabar and Canara, of Cassia, 52,686 lbs., and of Cassia Buds, 69,860 lbs. Some of these probably find their way into the English market.

This plant Dr. Wight considers to be the famous *Carua* of Rheede, Hort. Mal. i. t. 57, referring it, doubtfully however, to *C. iners*, Reinwardt, a plant of Java and Penang, but which seems to be sufficiently distinguished by being devoid of the aromatic qualities of the Malabar plant. *Cinnamomum iners* is, however, closely allied to another species, *C. eucalyptoides*, which is intermediate between *C. iners* and *C. nitidum* (Nees). Dr. Wight's drawing of the Malabar Cassia Plant sent to the India House (*v.* his Icones), appears to me to agree very closely with that of *C. eucalyptoides*, given by Nees and Ebern., in their Pflanz. Med., and which Nees v. Esenb. describes as

CINNAMOMUM EUCALYPTOIDES, *Nees von Esenb.* Branches roundish, smooth; lower leaves elliptico-oblong, upper ones subovate, acute at the base, subattenuate, obtuse at the apex, 3-nerved, nearly veinless. Panicles sessile, subterminal and axillary. Flowers with silvery silkiness; segments of the perianth obovato-cuneiform, deciduous from the middle.—*Laurus Malabrathrica*, Roxb. Hort. Calc .p. 30. Wall. Cat. n. 2583, B. "Sapor et odor foliorum fortis et acris caryophyllorum cum levi Camphoræ tinctura." The berry not seen.

C. INERS of *Reinw.* is distinguished chiefly by the leaves being more lanceolate, 3-nerved, nearly veinless. Panicles pedunculate. Segments of the perianth deciduous below the apex.—But both require careful re-examination with good specimens, and comparison with the Malabar plant, before any one can determine whether this be identical with either or different from both. In the latter case, it ought to be called *C. Carua.* The Chinese *C. aromaticum* might, no doubt, be successfully cultivated where this thrives so well.

Dr. Wight says, "a set of specimens (submitted officially to his examination) of the trees furnishing Cassia on the Malabar coast, presented no fewer than four distinct species, including among them the genuine Cinnamon plant, the bark of the older branches of which it would appear are exported from that coast as Cassia." And, besides, he infers that all sorts of Cinnamon-like plants, yielding bark of a quality unfit to bear the designation of Cinnamon, are passed off as Cassia. Mr. Huxham, of Quilon, also states, that the only difference between Malabar Cassia and Malabar Cinnamon is, that the former is coarser and thicker than the latter. Both are obtained from the same tree, the Cassia being the bark of the larger parts of the tree, and the Cinnamon being peeled from the younger shoots and small branches.

Cassia lignea, analysed by Bucholz, yielded of Volatile Oil 0·8, Resin 4·0, Gummy Extractive 14·6, Bassorin with Ligneous fibre 64·3, water and loss 16·3=100. But as Sesquichloride of Iron and Gelatine both produce precipitates, that of the former of a dark-green colour (*p.*), it must also contain Tannin. Its active properties, dependent chiefly on the volatile oil, are taken up by Spirit, and partially by water, as in the preparations.

Action. Uses. Aromatic Stimulant, in doses of gr. x.—ʒß.

OLEUM CASSIÆ, E. Oil of Cassia.

The Volatile Oil of Cassia-bark, obtained by distillation with water, is imported from Singapore, and is still probably all produced by the Chinese Cassia or Cinnamon, C. aromaticum.

Cassia Oil, when pure, has a pale wine-yellow colour, which does not deepen with age. (*c.*) Sp. Gr. 1·095. It has in a remarkable degree the Cassia odour and taste. Like Oil of Cinnamon, it absorbs Oxygen, and is converted into Cinnamic acid. Nit′ also converts it into a mass of crystals, in which the oil appears to be combined with the acid. It appears in all essentials to resemble the Oil of Cinnamon, and though not so grateful, is often sold for Oil of Cinnamon.

Action. Uses. Stimulant Carminative in doses of ♏j.—♏v.

AQUA CASSIÆ, E. Cassia Water.

Prep. Mix bruised *Cassia* ℥viij. *Rectified Spirit* f℥iij. *Aq.* Cij. Distil off Cj.

Action. Uses. Carminative vehicle, and used as Cinnamon-water.

SPIRITUS CASSIÆ, E. Spirit of Cassia.

Prep. With powdered *Cassia* ℔j. proceed as for Spir. Caraway.

TINCTURA CASSIÆ, E. Tincture of Cassia.

Prep. Digest powdered *Cassia* ℥iijß. in *proof Spirit* Oij. for 7 days; strain, express strongly, and filter. Or more conveniently prepare by percolation, previously macerating the Cassia in the Spirit for 12 hours.

Action. Uses. Stimulant adjuncts to draughts in doses of f℈j.—f℈iv.

BEBEERINE. Alkali of NECTANDRA RODIEI, *Schomburgk.* Greenheart Tree.

A considerable quantity of a wood called Greenheart* is imported into this country for ship-building. It is large in size, heavy, hard, durable, takes a polish, but is apt to split, and is of different tints of olive-green, varying from pale to dark.

Sir R. Schomburgk, Hooker's Journ. of Bot. Dec. 1844 (British Assoc. 1845), has described the tree which yields the *Greenheart* timber of Guiana (called *Bebeera* by the Indians of Demerara, and *Sipeeri* by the Dutch colonists). It is a new species of the Laurels, belonging to the genus *Nectandra,* and which has been named *N. Rodiei,* in compliment to Mr. Rodie, late a surgeon in the R. N., who first, in 1834, directed attention to its valuable febrifuge properties and indicated the presence of an alkali in the bark of this tree. Dr. Warburg also prepared what he called "Vegetable Fever Drops" from some part of this tree, which he distributed extensively, and which were favourably reported on by various medical officers. Dr. Maclagan in April 1843 read before the Royal Society of Edinburgh an able paper on the Bebeera Tree, its chemical composition, and its medical uses; and the nature of the alkali Bebeerine has been further elucidated by himself and T. Tilley, Esq., Professor of Chemistry in Birmingham, in a paper

* This the author in the Catalogue of woods published by Mr. Holtzapffel in 1843 referred to *Laurus Chloroxylon,* a tree not well known. The Greenheart of Browne's *Jamaica* resembles a laurel, and the leaves with their three-arched nerves are compared by Browne with those of the Camphire tree, both in shape, size, and texture. The fruit is like that of a Laurel, and the specimens of *Laurus Chloroxylon,* Lin., in the British Museum, from Jamaica, are very like Browne's figure, and are those of a true Laurel. But in this case, the flowers which he describes could not have belonged to it; neither does it follow that the Greenheart of Jamaica is identical with the Greenheart of Guiana.

read before the Chemical Society. The medical virtues of this alkali, or rather, of its Sulphate, have been detailed by Dr. Maclagan, &c. in the Lond. and Ed. J. of Med. Science, July, 1843 and April 1845.

The bark of the Bebeera tree occurs in large flat pieces, is about four lines in thickness, heavy, and with a rough fibrous fracture, of a dark cinnamon-brown colour, rather smooth within, but covered externally by a splintering greyish-brown epidermis. It has little or no aroma, but a strong, persistent, bitter taste, with considerable astringency. These properties depend on the presence of an alkali, which has been called Bebeerine. Dr. M. at first thought that there were two alkalis; but this, from his second paper, does not appear to be the case. It is contained also in the seeds, as is evident from Dr. M.'s analysis of both the bark and seed.

	Bark.	*Seeds.*
Alkalis (not quite pure)	2·56	2·20
Tannin and Resinous matter	2·53	4·04
Soluble matter (Gum, Lignin, Salts)	4·34	9·40
Starch	0·	53·51
Fibre and Albumen	62·92	11·24
Ashes (chiefly Calcareous)	7·13	0·31
Moisture	14·07	18·13
Loss	6·45	1·17
	100·00	100·00

The Tannin resembles that which has been found in the Cinchona Bark. The author has received fruits of the Bebeera tree from Guiana, which were stated to be those of the tree employed in making Warburg's Fever Drops. They are 2¼ inches in length, and 1½ inch in breadth, and correspond with Dr. Maclagan's description.

The alkali is separated from the rest of the bark by being boiled in water acidulated with Sul′, as in the ordinary process for obtaining Sulph. of Quinine, and is then precipitated, from the impure Sulphate obtained, by Ammonia. The alkaline matter thus separated is, after washing with water, triturated with about an equal weight of freshly precipitated and moist Hydrated Oxide of Lead. The magma thus formed is dried over the water-bath, and the alkali is then taken up by absolute Alcohol. On distilling off the Spirit, the organic base is left in the form of a transparent, orange-yellow, resinous mass. This, on being dried, pulverized, and treated with successive portions of pure Ether, is in great part dissolved in this fluid. The dissolved portion, on distilling off the Ether, is obtained in the form of a translucent, amorphous, but homogeneous, resinous-looking substance, of a pale-yellow colour, and possessed of all the properties of an organic alkali. This is pure *Bebeerine*, which does not crystallize. It is very soluble in Alcohol, less so in Ether, and very sparingly in water. Heated, it fuses, and the heat being continued, it swells up, giving off vapours of a strong peculiar odour, and burning without residue. Subjected to the action of oxidizing agents, it gives with Bichromate of

Potash and Sul′ a black, and with Nit′ a yellow Resin. It forms with acids, salts which are all uncrystallizable; with Perchloride of Gold, Mercury, Copper, Iron, and Platinum, it gives precipitates which are soluble to a certain extent in hot water and Alcohol, but which, on the solution cooling, are not deposited from it in a crystalline form. Messrs. Maclagan and Tilley found it to be composed of Oxygen 71·92, Hydrogen 6·49, Nitrogen 4·75, Oxygen 16·84 = 100, and that its formula is $C_{35} H_{40} N_2 O_5$. Dr. M. obtained some Bebeerine from Warburg's Fever Drops.

The authors observe it as remarkable that it should be isomeric with Morphia, which acts as a pure narcotic. The atomic constitution of Morphia, calculated from the formulæ deduced by Liebig and Regnault from their analyses, agrees perfectly with that given above for Bebeerine. The composition of the two bodies is, in fact, identical. (That of Morphia, according to Turner and Gregory, is given at p. 258.) From this the authors conclude, "that similarity of physiological properties does not depend upon similarity in the properties of their constituents. It seems probable that the mode in which their atoms are grouped has an important share in modifying their physiological actions. The difference in their physical properties, in fact, proves that their elements are differently arranged."

Action. Uses. Tonic, Antiperiodic, Febrifuge. From the original experiments of Mr. Rodie, and those made with Warburg's Fever Drops, there was little doubt of the Bebeera bark being a powerful Antiperiodic. These have been confirmed by the experiments of Dr. Maclagan, and of Dr. Watt of George Town, Demerara, with the Sulphate of Bebeerine, and of Dr. Anderson and others at Kamptee, &c., in the Ague and Remittent Fever of India, by Drs. Bennett and Simpson, in Periodic Neuralgia. Dr. Christison has stated to the author that the Sulphate of Bebeerine has come into general use in Edinburgh as a Tonic and Stomachic, and also as an Antiperiodic, in the very same diseases and for the very same purposes, as Sulphate of Quinine, and that it appears not so apt to occasion headache. He had employed it in a very severe case of periodic Tic douloureux, and with complete success, exactly as if Sulphate of Quinine had been used. It is given in 2 or 3 grain pills every hour, or three or four times a day, according to the case, so that ℈j. or so, may be given before the accession of a paroxysm, or it may be given in gr. x. doses, morning and evening. Considerable improvement in the manufacture has been made by Mr. M'Farlane of Edinburgh, who now prepares it in considerable quantities for medical use in the form of the Sulphate of Bebeerine.

ARISTOLOCHIEÆ, *Juss.* Birthworts.

Herbs or shrubs, with creeping rootstocks, often with twining stems, without rings of wood. Leaves alternate, sometimes without, often with, leafy stipules opposite to the leaves. Flowers hermaphrodite, usually of a dull colour, axillary, solitary, or clustered. Perianth with the tube adherent to the ovary; limb undivided and obliquely truncate, or trifid, with the segments valvate in æstivation. Stamens 6 or 12 or 9, definite, free, and inserted into the apex of the ovary, or united with the style and stigma. Ovary inferior, 3 to 6-celled. Placenta central, with many ovules. Fruit 3 to 6-celled, many-seeded. Embryo small, at the base of fleshy or subcartilaginous albumen. Radicle near the hilum, centripetal or below.—The Aristolochieæ are most closely allied to Nepentheæ, more remotely to Cucurbitaceæ, and perhaps also to Dioscoreæ and Taccaceæ, &c. They are chiefly found in hot countries, though a few species extend to northern latitudes. They secrete a bitter principle and volatile oil.

ASARUM, L. D. Folia. Leaves of ASARUM EUROPÆUM, *Linn.* Asarabacca. *Dodecandria Monogynia*, Linn.

This is the ἄσαρον of Dioscorides, the *asaroon* of the Arabs.

Rootstock creeping, with numerous branched root-fibres. Stems very short, round, each bearing two kidney-shaped leaves, which are of a dark green colour. shining above, but a little hairy, with long downy footstalks. In the axil of the two leaves there is a single drooping flower, about an inch long, fleshy, lurid in aspect. Perianth coriaceous, campanulate, 3-lobed; segments incurved. Stamens 12, inserted on the ovary; anthers attached to the inner side of the filaments, below the summit, each of two round separated cells. Ovary turbinate; style short. Stigma stellate, 6-lobed. Capsule coriaceous, 6-celled. Seeds ovate, with horny albumen.—Hilly woods, mountains of England, &c. Flowers in May; said to be collected near Kirkby Lonsdale, Westmoreland.

The root-fibres, when bruised, have a spicy odour and an acrid taste. The leaves are scentless, but have a bitter and acrid, slightly aromatic taste. Both are employed on the Continent, but the leaves only are officinal, though the whole plant is usually sold. Analysed, the root was found to contain a Volatile Oil, Bitter Extractive, which will itself excite nausea, a Camphor-like body, named *Asarine*, which is volatile, has an acrid taste, excites nausea and vomiting. The leaves yield a volatile oil.

Action. Uses. Acrid, formerly employed as a Purgative and Emetic, but now seldom used except as an Errhine (*v.* the officinal Powder), for which it is very effectual, producing sneezing and a copious secretion from the nostrils, and is therefore employed as a counter-irritant in some head affections.

PULVIS ASARI COMPOSITUS, D. Comp. Asarabacca Powder.

Prep. Rub together into powder *dried leaves of Asarum* ℥j. and *dried Lavender flowers* ʒj. Sometimes Lavender and Marjoram āā ʒß. are added.

SERPENTARIA, L. E. D. Radix, L. D. Root of ARISTOLOCHIA SERPENTARIA, *Linn.* Virginia Snake-root. *Gynandria Hexandria*, Linn.

Several species of Aristolochia were employed by the ancients, and and still are so on the Continent, as well as in Asia. The officinal species was probably first brought to notice as a *Snake-root* to settlers in America. It is first mentioned in Johnson's edition of Gerard's Herbal.

Rootstock perennial, roundish, with numerous root-fibres; throwing up several herbaceous stems 8 to 10 inches high, slender, flexuose, jointed at irregular distances, often of a reddish colour at the base. Leaves alternate, shortly petioled, cordate, acuminate, smooth, and of a pale yellowish-green colour, a little downy beneath. The peduncles are produced on the stem, but near the root, nearly unifloral, with one or more bracts. The perianth is tubular, contorted like the letter S, inflated at its two extremities, its throat surrounded by an elevated ridge, and its border expanded into a broad irregular margin, forming an upper and under lip. The anthers 6, attached to the sides of the fleshy style, which is situated in the bottom of the perianth, covered by the spreading 6-lobed stigma. Capsule obovate, 6-angled, 6-celled, with many flat seeds.—A native of the middle, Southern, and Western States of North America. The root is collected in Western Pennsylvania, and Virginia, in Ohio, Indiana, and Kentucky. Nees von E. 143. St. and Ch. 180.

American writers state that the roots of two other species, *A. tomentosa* and *A. hastata*, are also collected. The latter is closely allied to *A. Serpentaria*, the other climbs to the tops of tall trees; but their roots are said scarcely to differ from those of *A. serpentaria*. The roots of *Spigelia marylandica* are also sometimes found intermixed. Nees von Esenbeck, moreover, mentions a species, *A. officinalis*, Med. Pfl. t. 144, but which may be only a variety of *A. Serpentaria*.

Virginia Snake-root is in tufts of long, slender, frequently interlaced, and brittle fibres, attached to a short, contorted, knotty head or caudex. The colour, which in the recent state is yellowish, becomes brown by time. The smell is strong, aromatic, and like Camphor and Valerian; the taste warm, very bitter, and camphorous. The root yields its virtues to water, to Alcohol, and to Proof Spirit. The active ingredients are probably the Volatile Oil and a bitter Extractive, which is also acrid, and was detected by Chevallier and by Bucholz.

Action. Uses. Stimulant Tonic, Diaphoretic, and Emmenagogue in doses of gr.x.—ʒß. In large doses it causes nausea and purging. Used to be employed in cases of atonic fevers, &c., and in Exanthemata, where the eruption is tardy in appearing, or has receded.

Pharm. Prep. Tinct. Cinchonæ Comp. L. E. D. Electuarium Opii, E.

INFUSUM SERPENTARIÆ, L. E. D. Infusion of Snake-root.

Prep. Macerate *Serpentaria* ℥ß. in boiling *Aq. dest.* Oj. for 4 hours in a lightly covered vessel. L. Strain (through linen or calico, E.)

Action. Uses. Diaphoretic in doses of f℥jß. every two or three hours.

TINCTURA SERPENTARIÆ, L. E. D. Tincture of Snakeroot.

Prep. Take bruised *Serpentaria* ℥iijß. (℥iij. D.) and *Proof Spirit* Oij. (by measure ℔ij. D. Cochineal bruised ʒj. E.); macerate for 14 (7, D.) days, and strain. (Proceed by percolation or digestion, as for Tinct. Cinchona, E.)

Action. Uses. Stimulant adjunct to Tonics and Diaphoretics in doses of from fʒj.—fʒij.

EUPHORBIACEÆ, *Adr. de Jussieu.* Spurgeworts.

Herbs, shrubs, or trees, with watery, often milky juice, some with a fleshy Cactus-like stem. Leaves usually alternate. Stipules small or wanting. Flowers monœcious or diœcious, frequently incomplete, often enclosed within an involucre. Calyx free, 4, 5, or 6-fid, or divided, sometimes wanting. Corol often wanting, or consisting of scales, or of petals, usually twisted in æstivation, often alternating with scales or glands. *Male.* Stamens definite or indefinite, distinct or united, inserted into the centre of the flower below the abortive ovary. *Female.* Ovary free, 2—but usually 3—seldom many-celled. Ovules solitary or in pairs, suspended from the inner angle. Styles equal in number to the cells, at first united into one, and then dividing into as many stigmas. Fruit usually capsular, seldom berried, 2 to 3-celled, often of cocci or carpels separating from the central persistent column. Seeds often furnished with caruncula or arillus. Embryo straight in the axis of a fleshy albumen.

The Euphorbiaceæ, multiform in habit, are usually placed among apetalous orders; but as many of the genera are furnished with both calyx and corolla, and as they are allied to Celastrineæ and Rhamneæ, and also to Burseraceæ, through Juglandeæ, it would perhaps be preferable to place them near these families. They abound in tropical countries; many species, however, are found in the southern parts of the temperate zone, but in the northern only herbaceous species, with Buxus as a shrub. The Euphorbiaceæ are remarkable for acridity, which is contained in the milky juice; hence some are used as local, and some as general stimulants; others are poisonous in nature. A few secrete volatile oil in the bark, as Cascarilla, and others, fatty oil united with acrid principle in their seeds, as in Castor and Croton oils. The roots of Janipha Manihot secrete fecula and acrid principle, which is dissipated by heat.

Tribe *Ricineæ, Adr. de Jussieu.*

CROTON, *Linn.* *Monœcia Monadelphia.*

Flowers monœcious, or very rarely diœcious. Calyx 5-parted. *Males.* Petals 5. Stamens 10 or more, distinct. *Females* Petals none. Styles 3, divided into two or more partitions. Capsule tricoccous. *Adr. de J.*

CASCARILLA, L. E. D. Croton, L. D. Bark, probably, of CROTON ELEUTERIA, *Swartz*, &c., and possibly of other species of the same genus, E.; of Croton Cascarilla, *Linn.* L. D. Cascarilla.

Cascarilla was first made known by V. Garcias Sabat in 1692. The name, signifying a little bark, is applied by the Spaniards to a variety of barks. It is intended in the Pharmacopœia to indicate a bark known for 150 years, and which Dr. Pereira has shown comes principally from the Bahamas, and which Dr. Lindley has proved (Fl. Med. p. 179) to be the produce of *Croton Eleuteria* of Swartz, and not of *Croton Cascarilla* of Don (the *C. pseudo-China* of Schlechtendal); for this yields Copalchi bark. Nees von Esenbeck has also shown that *C. micans* of Swartz, a Jamaica plant, yields some of the Cascarilla of the Continent. Both these plants are figured in the Pfl. Med. of Nees Suppl. t. 22.

The *C. Cascarilla* of Linnæus (*C. lineare* of Jacquin) the *Wild Rosemary bush* of Jamaica, does not appear to yield any bark like Cascarilla, or to have the sensible properties of that bark.

Croton Eleuteria forms a small tree. Branches and twigs angular, rather compressed, striated, downy, ferruginous. Leaves stalked, alternate, ovate, with a short but obtuse point, green on the upper surface, silvery and densely downy beneath. Flowers monœcious. Racemes axillary and terminal, branched. Males uppermost and smallest; females below, few, and on short stalks. Filaments 10 to 12. Ovary roundish. Styles 3, bifid. Stigmas obtuse. Capsule roundish, minutely warted, about the size of a pea, with 3 furrows, 3 cells, and 6 valves. *Lindley.*—Thickets of Jamaica and other West India Islands.—This species, having the name of Eleutheria, from one of the Bahamas, has been proved by Dr. Lindley, from information and authentic specimens from the Bahamas, to yield the true Cascarilla bark, as had been before stated by Drs. Wright and Woodville.—Nees von Esenb. t. 139. St. and Ch. 150.

Cascarilla may be confounded not only with Copalchi, but also with that kind of Cinchona called Grey or Huanuco bark. It consists, however, of irregular fragments, which are thin, two to three, sometimes four inches in length; these are moderately quilled, a little twisted, or flat, about the thickness of a pencil or that of the little finger, of a grayish colour externally, much fissured, covered in many parts with a whitish lichen, the substance of the bark of a brownish colour, and its internal face smooth. It is compact, fracture short, brittle, the powder of a light-brown colour. It has a weak though aromatic odour; the taste is bitter, a little acrid, but also spicy. M. Duval in recently analysing it, found it to contain Albumen, Tannin, a Bitter crystallizable substance (*Cascarilline*), Red Colouring matter, Fatty matter with a nauseous smell, Wax, Gum, Volatile Oil with an agreeable smell, Resin, Starch, Pectic acid, Chloride of Potassium, Salts of Lime, Woody fibre. *Cascarilline* when pure is white and crystalline, without odour, has a bitter taste, which, however, is not at first perceptible from its sparing solubility; but it communicates its bitterness to a large quantity of water. It is very soluble in Spirit and Ether, and appears to be a non-azotised neutral substance of the nature of Salicine. The properties of the bark no doubt depend chiefly on the Volatile Oil and the Cascarilline. These are taken up by Spirit, partially by water.

Action. Uses. Stimulant Tonic; has been considered Febrifuge, and may be advantageously prescribed with Cinchona. It is chiefly employed in Dyspepsia and in other complaints requiring a warm Tonic.

D. Of the powder gr. x.—℈jß. well given with Soda in Milk.

INFUSUM CASCARILLÆ, L. E. D. Infusion of Cascarilla.

Prep. Macerate for 2 hours in a lightly (L.) covered vessel powdered *Cascarilla* ℥jß. (℥ß. D.) in boiling *Aq. dest.* Oj. (℔ß. D.) Strain (through linen or calico, E.)

Action. Uses. A light warm Tonic in doses of f℥jß.

MISTURA CASCARILLÆ COMPOSITA, L. Comp. Cascarilla Mixture.

Prep. Mix *Infusion of Cascarilla* f℥xvij. *Vinegar of Squills* f℥j. and *Tinct. Camphoræ Comp.* f℥ij.

Action. Uses. Warm Tonic and Expectorant in Chronic affections of the Lungs in doses of f℥jß. two or three times a day.

Tinctura Cascarillæ, L. E. D.

Prep. Take *powdered Cascarilla* ℥v. (℥iv. D.) *Proof Spirit* Oij. (by measure ℔ij. D.); macerate for 14 (7, D.) days and strain. (Proceed by percolation or digestion, as for Tinct. Cinchonæ, E.)

Action. Uses. Stimulant and Tonic adjunct to draughts in doses of fℨj.—fℨij.

Tiglii Oleum, L. Crotonis Oleum, E. Crotonis Tiglii Oleum, D. Oleum e seminibus expressum, L. Expressed Oil of the seeds of Croton Tiglium, Linn. Croton Oil.

The seeds called *Jamalgota* were given to the author when in India as the *dund* of the Arabs. They were, no doubt, employed at those times, as well as subsequently in Europe, under various names, as *Grana Tilli*, also *Tiglia*, &c., and its wood *Lignum Moluccense.* Having passed out of practice, they were re-introduced by the notices of Dr. White and of Mr. Marshall, published in Ainslie's Materia Medica of India, and by a publication of Mr. Conwell.

Croton Tiglium forms a small tree, of 15 to 20 feet in height, with the young branches smooth and roundish. Leaves oval-oblong, acuminate, 3—5-nerved, with shallow glandular serratures, thin, membranous, with 2 glands at their base, the younger leaves covered with minute stellate scattered hairs. Petioles short, somewhat angular, with a few stellate hairs when young. Stipules 2, subulate, minute. Racemes terminal, erect, simple, male at apex, female single, below. Flowers downy. *Male,* Calyx 5-cleft. Petals 5, lanceolate, and woolly. Stamens 15 (15 to 20, woolly at base, *Roxb.*), distinct. *Female,* Calyx 5-cleft, permanent. Styles long, bifid. Capsules oblong, obtusely triangular, the size of a hazel nut, closely covered with minute stellate hairs; the cells completely filled with the solitary seeds. Skin of the seeds pale dull-brown, overlying a harder, dark, and smooth integument.—*Hamilton* and *Roxburgh*. *C. Jamalgota,* Ham. Linn. Trans. xiv. 258.—Indigenous everywhere in Bengal; found also in the Indian Peninsula and in Ceylon; *Lindley,* Fl. Med. p. 181.

Croton Pavana (Ham. *l. c.* p. 258), having ten stamens, and the seeds much smaller than their cells, is another species, which yields what Dr. Hamilton considers the original Tiglium seeds. It is a native of Burma, Assam, and Silhet, perhaps also of Amboyna.

Croton Roxburghii, Wall. (C. polyandrum, *Roxb.*) is a native of the Circar mountains; bears seeds to which the name *Jamalgota* is also applied.

The *Croton Tiglium* has a disagreeable smell, and the taste of the leaves is exceedingly nauseous and permanent. (*Roxb.*) All parts of the plant seem provided with an acrid purgative principle, but the seeds are alone now officinal. These are about the size of a grain of Coffee, oblong, rounded at the extremities, with two faces, the external more convex than the internal, separated from each other by longitudinal ridges, and each divided by a similar longitudinal ridge, forming altogether an irregular quadrangular figure. Sometimes when there are only two seeds in the capsule, the internal surface is flat, with a groove formed by the central axis. The shelly covering of the seeds being sometimes partially removed, they have a mottled appearance; but if entirely so, they are blackish. The kernel is oily, of a yellowish-white colour when fresh, but becoming brownish by age. It has a large embryo, with leafy cotyledons. In India the seeds with their coverings are subjected to torrefaction, and the embryo usually separated before they are prescribed medicinally. In Europe their effects are obtained by prescribing the Oil, commonly called *Croton Oil.* This is obtained by bruising the kernels, and subjecting them to pressure, when about 50 per cent. of Oil may be obtained. But some Oil is also imported from Ceylon. According to Dr. Nimmo, the seed consists of 64 parts of kernel and 26 of covering, and the kernels yield 60 per cent. of Oil.

The analyses of Pelletier and Caventou, and of Brandes, display a trace of Volatile Oil, *Crotonic acid,* which is acrid and volatile, Fixed Oil, *Crotonine,* an alkaline and crystalline body; Resin, Gum, Albumen and Gluten, Salts, Lignin, &c. The active properties depend on the Crotonic acid which passes off with the fixed oil, whether this is separated by expression or by being dissolved in Ether.

Action. Uses. Croton seeds are powerfully Cathartic, and in very common use in India as Purgative Pills, commonly called *Jamalgota Pills.* The natives usually separate the embryo, and combine the albumen of the seed (which, however, is sufficiently active) with Catechu or Pepper. About a grain, or half a seed, is sufficient for a dose, though they do take larger doses.

TIGLII OLEUM, L. CROTONIS OLEUM, E. Croton Oil.

Expressed from the seeds, and usually imported from India, chiefly from Madras and Bombay, but also from Ceylon.

Croton Oil, when quite pure and fresh, is nearly colourless, but as usually met with, it is rather viscid, yellowish, and even of an orange-colour, from over-roasting of the seed. It has a faint but peculiar smell, and a permanent acrid taste, which is most felt in the throat. The Oil is soluble in Ether, as well as in the volatile and fixed oils, partially so in Alcohol—that is, the acrid portion of Croton Oil, which is composed of Crotonic acid and Resin, is dissolved, while the Oil itself is insoluble. Hence the E. P. gives as a test of its purity: "when agitated with its own volume of pure Alcohol, and gently heated, it separates on standing, without having undergone any apparent diminution." Castor Oil, which is the only oil likely to be used in adulterating it, is soluble in Alcohol.

Action. Uses. Drastic Purgative in obstinate Costiveness and torpid state of the Intestinal canal, or when a Hydragogue, or the speedy action of a Cathartic, is required. Sometimes is very useful in nervous disorders, as in Tic Douloureux. Rubbed on the skin, it acts as a Rubefacient, and is frequently employed as a Counter-Irritant to relieve internal affections, either in its pure state, or dissolved in twice its bulk of Olive Oil, &c. Bouchardat recommends 20 parts being added to 8 parts of diachylon plaster as a revulsive.

D. As a Purgative ♏j.—♏ij. or ♏iij. made up into pills, to which Opium is sometimes added. The liquid form is objectionable on account of the acrid sensation produced in the throat. A *Croton Soap* is prepared by Mr. Morson, of which gr. i.—gr. iij. forms a dose.

RICINI OLEUM, L. E. D. Oleum e seminibus expressum, L. D. Expressed Oil of the Seeds, E. of RICINUS COMMUNIS, *Linn.* Castor Oil Plant. *Monœcia Monadelphia,* Linn.

This plant appears to be the Gourd, or the plant so translated, in Jonah iv. 6, 7, 9, 10. (*v. Kikayon,* in Bibl. Cycl. ii. p. 203.) It is also the κικι or κρότων of Dioscorides, and its Oil has been employed in medicine from the earliest times by Hindoos, Egyptians, Greeks, and Arabs. The Greek names are taken from the insect called the tick (in Latin *ricinus*) which the seeds resemble.

Different opinions are held respecting the number of species belonging to this genus ; but several varieties have no doubt been raised to the rank of species. These are found in Java, and throughout India. One has been named *R. europæa,* but it must have been introduced from the East, and is annual because unable to withstand the cold of winter. The common species may be seen in India, especially at the borders of fields, with stems of considerable thickness, and attaining a height of sixteen to twenty feet, and surviving for many years. The Oil is valued as a medicine, and for burning, and the leaves for feeding the Arendy silk-worm.

Root perennial or annual, long, thick, and fibrous. Stems (Fig. 88) round, thick, jointed, channelled, hollow, glaucous, of a purplish-red colour upwards. Leaves large, palmato-peltate, deeply divided into 7 lanceolate, serrated segments, on long, tapering, purplish petioles, with glands at the apex of the petiole. Flowers monœcious, in terminal panicles, the lower male, the upper female, all articulated with their peduncles, and sometimes supported by biglandular bracts. Calyx 3—5 cleft, valvate. Petals wanting. *Male.* Stamens numerous, with the filaments branched (*a*) and united below, with distinct globose cells of the anthers (*b*). *Female.* Style 1. Stigmas 3, bipartite (*c*), plumose, coloured red. Capsule tricoccous, covered with spines, 3-celled (*d*); cells 1-seeded. Seeds pendulous, elongated, ovate, convex externally, somewhat flattened on the inside, of a pale gray colour, but marbled with darker colours (*e*). The seed is covered by a thin, coriaceous, smooth seed-coat, composed of two layers ; at its upper end is observed the fleshy swelling which has been termed *Strophiole,* with a delicate white membrane investing the nucleus, which is large, oleaginous, and consists of albumen containing in the middle a large leafy embryo (*f*). Native of India ; cultivated in many countries.

Two varieties of Castor Oil seed are known, one large, the other small. The latter is thought to yield more oil, and of a superior quality. Geiger found in 100 parts of these seeds, exclusive of moisture,

Fig. 88.*

23·82 parts of seed-covering, and 69·09 of kernel. These 69·09 parts contained 46·19 parts of fixed Oil, 2·40 of Gum, 20·00 of Starch and Lignin, and 0·50 of Albumen. The kernel, when fresh, is of a white colour, and sweetish almond-like taste, followed by some acrimony. The Oil may be extracted from the seeds by decoction in water, or expression, with or without the aid of heat, and for experiment by the agency of Alcohol. Sometimes the Oil is boiled with water to dissolve out the Mucilage and to coagulate the Albumen. Dr. Christison sums up the results of various papers by stating that "by simple expression a mild oil of excellent quality may be extracted alike from the small and large varieties of the seed: that when so prepared, it is apt to become sometimes rancid (Wright), but may be prevented from doing so if heated to about 200°, so that its Albumen is coagulated and detached: that the embryo is scarcely more active than the Albumen of the nucleus, and that the husk and perispermal membrane are inert (Boudron and Henry): that if the seeds be boiled in the Eastern way, without first roasting them, or driving off the residual water from the Oil by heat, an Oil of fine quality is obtained, which keeps well (Guibourt), but is probably not quite so active: that the active part of the Oil is probably volatilizable during decoction with water (Guibourt), so that long ebullition may materially impair its energy: and that if the seeds be roasted before being expressed, or the Oil be exposed to a considerable heat, as in the American process, peculiar acids are engendered (called the Ricinic, the Elaïodic, and Margaritic), which greatly increase the acridity (Bussy and Lecanu).

Castor Oil is imported in the largest quantities from the East Indies, and this is commonly called *cold-drawn* Castor Oil. Some is also imported from North America and the West Indies. It is of a pale straw-colour, a faint but unpleasant smell, a mild oily taste, sometimes accompanied with a little acrimony. Though heavier than most fixed oils, it is lighter than water, and viscid. Sp. Gr. 0·969 at 55° F. If exposed to a cold of 32°, it deposits a few grains of Margarin. Exposed to the air, it becomes rancid, and dries up. It is soluble in all proportions in both Alcohol and Ether. By the action of Hyponitrous acid, a solid fatty matter is produced, which has been called *Palmine.* The alkalis saponify it, but produce acids apparently identical with those generated during its distillation. Differing in many respects from other fixed oils, chemists are inclined to consider it as consisting "of a single and peculiar oleaginous principle," others as composed of three fatty acids combined respectively with Glycerine. Its purity may be ascertained by its being "entirely dissolved by its own volume of Alcohol." E.

Action. Uses. Purgative. Castor Oil seeds, though mild tasted, even in their fresh state, are acrid.* The Oil is a mild but certain laxative, acting quickly, and is particularly eligible whenever it is wished to produce as little irritation as possible along the intestinal canal. With Oil of Turpentine f℥ij. a particularly efficient purgative may be formed. (*c.*)

* The author was once called in a great hurry to the hospital-boat when proceeding with a battalion of Artillery up the Ganges, with the statement that several men had been poisoned or seized with cholera. The seeds immediately revealed what the "Indian filberts" were which the men had been picking and eating. The majority recovered rapidly; but three of them suffered severely, and were not discharged from hospital for some time.

D. Of the Oil f℥ß.—f℥jß. swimming on weak Spirit and water, or on hot Milk, Coffee, shaken up with Vinegar, &c. For children, f℈j. or f℈ij. made into an emulsion.

TAPIOCA, E. Fecula of the root of JANIPHA MANIHOT, *Humbl. and Bonpl. Jatropha Manihot* of Linnæus.

Tapioca, first mentioned by Piso in his Nat. Hist. of Brazil, p. 52, is a starch-like substance yielded by the above plant.

A shrub, 4—6 feet high. Root large, tuberous, fleshy, and white, with a milky, acrid, poisonous juice. Leaves palmate, 5—7 parted, smooth, glaucous beneath; segments lanceolate, quite entire. Flowers axillary, racemose, monœcious. Calyx campanulate, 5-parted. Petals none. Stamens 10. Filaments unequal, distinct, arranged around a disk. Style 1. Stigmas 3, consolidated into a rugose mass.—Adr. de Juss. and Hooker.—Cultivated in the West Indies and in many parts of South America.—B. M. 3071.

Of this plant there are two distinct varieties, one known as the Bitter the other as the Sweet Cassava. The former is about six feet high, the leaves of a darker green, and the stem of a dark-brown colour; the roots are longer in coming to maturity, much larger, about twenty inches in length, and ten in circumference. The juice is acrid and poisonous, owing, it is said, to the presence of Hydrocyanic acid (*Henry* and *c.*), and not always entirely dissipated by heat. It is cultivated for making the Tapioca of commerce, and Cassava Bread. This is made by grating the fresh roots, squeezing out the juice, and then baking into cakes on an iron plate. The Tapioca is also prepared by beating the root into a pulp, washing it with cold water, and then allowing the fæcula to subside from the milky fluid which flows from it. Being then dried on heated plates, it becomes of a granular form.

The Sweet Cassava,* is about four feet high, the root about a foot in length, and seven or eight inches in circumference, of a light-brown colour. It is very juicy, something resembling chesnuts in taste, and is used as a vegetable either boiled or roasted. But much of it is employed in making a fermented liquor from the root scraped into a pulp, and from which the liquor is squeezed. This is called *Piwarry,* and drank by the Indians as an intoxicating liquor. (Mr. Gill.)

The irregular grains of which Tapioca consists are about the size of large shot, whitish, and like other kinds of Starch, without odour or taste. The grains are very minute, but regular in form, and most resembling those of wheat-starch. Tapioca has the general characteristics of Starch, of which it is a pure form.

Action. Uses. Dietetical, Demulcent. Much approved of as a diet for the sick-room and for infants at the time of weaning.

* This is sometimes considered a distinct species, and called *Janipha Lœflingii. v.* Hamilton. Pharm. J. v. p. 27. In the Synop. Plant. of Humboldt and Bonpland by Kunth (I. 417), *J. Lœflingii,* the *Yuca* of the natives, is described with 5-partite cordate leaves; segments acuminate, very entire, the middle one panduriform, while *J. Manihot,* their *Yuca dulce,* has leaves from 5 to 7-partite, glaucous on the under surface; segments acuminate, very entire.

Tribe *Euphorbieæ*.

EUPHORBIUM, L. E. D. Concrete resinous juice of undetermined species of Euphorbia, E. Gummi-Resina, of Euphorbia officinarum. L. of E. canariensis. D.

EUPHORBIA, *Linn.* *Monœcia Monadelphia*, Linn.

Flowers incomplete, collected into monœcious heads composed of one female and numerous male flowers. Involucre campanulate, with 5 divisions and 5 alternate glands. *Male.* Naked, consisting of a single stamen upon a pedicel, intermixed with scales, surrounding the female. *Female*, A single pistil. Styles 3. Stigmas bifid. Capsule 3-celled, bursting at the back. Seeds solitary, pendulous.—This genus is multiform in habit, some being cactus-like, among which must be the officinal species, having jointed angular stems, with branches of a similar structure, and double prickles at their angles. When wounded, they exude an acrid milky juice, which concretes upon the surface, usually upon these prickles, and constitutes the Euphorbium of commerce.

Euphorbium was employed by the early Greek physicians, and is noticed by the Arabs by the name *Furfioon*. The species yielding it is still uncertain. The Euphorbium of the ancients was obtained from Mauritania, that of modern commerce apparently entirely from Mogadore. The D. P. assigns *E. canariensis*, a plant of the Canary islands, and the L. P. *E. officinarum*, which is said to be found in Arabia and the hotter parts of Africa, while *E. antiquorum*, common in Arabia and all over India, which is also adduced, the author found comparatively inert. The only positive information is that of Jackson (*Account of Morocco*), who describes the inhabitants of the lower Atlas range making incisions in the branches of the plant found there, from which a milky juice exudes, which is very acrid, hardens on the plant, and drops off in September. The people who collect it are obliged to tie a cloth over their mouth and nostrils, to prevent the small dusty particles from annoying them, as they produce incessant sneezing. Bruce also describes the violent sneezing produced on his party on wounding some Euphorbia plants in a dry state. But this species (which he calls *Dergmouse*) is not well ascertained; it seems to approach *E. officinarum* in some of its characters. Dr. Pereira, from examining the branches found mixed with the Euphorbium of commerce, considers that *E. tetragona* agrees most closely with it in the size of the stems, the number of angles, and the number and direction of the spines.

Euphorbium is in irregular shaped tears, usually pierced with one or with two diverging holes, made by the double prickles of the plant on which it had dried. These sometimes remain in the holes. The colour is of a dull yellowish-white, something like that of inferior Tragacanth. It is friable, with little odour, but the dust causes violent sneezing, and irritation to the eyes, requiring the face to be well protected; it is very irritant to whatever part it is brought in contact with, and the taste is after a short period acrid and burning. It is composed of an Acrid Resin about 60 per cent., of Wax 14, Malate of

Lime 12, Malate of Potash 1, Bassorine, probably a little Caoutchouc, Lignin, water, and loss. It is, therefore, a Cereo-Resin, and not a Gum-Resin. It burns with a pale flame and rather an agreeable odour. Water takes up but little of it; Alcohol and Ether are its best solvents. Its active principle is identical with or associated with the Resin.

Action. Uses. Powerful Irritant; will produce incessant sneezing, and even bloody discharges, and ophthalmia if blown into the eyes. Acts as an Emetic or Cathartic; apt to produce inflammation of the intestinal canal. Largely diluted with starch, it is sometimes used as an Errhine in obstinate affections of the head; and occasionally as a Rubefacient and a Counter-Irritant, as in *Acetum Cantharidis*, E., and added to a Burgundy Pitch plaster in chronic affections of the joints.

Antidotes. Oil, emollient drinks, oleaginous enemata. Obviate inflammation by blood-letting, baths.

PIPERACEÆ, *Richard.* Pepper-worts.

Shrubs or herbs. Leaves without stipules. Flowers usually hermaphrodite and sessile, in spikes. Stamens definite (usually 2) or indefinite, arranged on one side or round the ovary, to which they adhere more or less. Anthers 1- or 2-celled, with or without a fleshy connective. Ovary superior, 1-celled, containing a single erect ovule. Stigma sessile, simple, rather oblique. Fruit somewhat fleshy, indehiscent. Seed erect, with the embryo lying in a fleshy sac or vitellum placed at that end of the seed which is opposite the hilum, on the outside of the albumen.

The Piperaceæ are allied in some respects to Polygoneæ, also to Urticeæ, &c. They are sometimes placed among Endogens, at other times among Exogens. They inhabit the tropical parts of Asia and America, with a few species extending to higher latitudes, and are characterized by pungent and aromatic properties. Besides the officinal species, the Piper Betle is much cultivated and famed for its moderately pungent and aromatic properties, its leaf being employed to envelope the fragments, of Areca Catechu, of pale Catechu, and of Lime, which form the famed masticatory of the East, known as Pan or Betle.

PIPER, *Linn.* *Diandria Trigynia*, Linn.

Spike covered with flowers on all sides. Flowers hermaphrodite, each supported by a scale. Stamens indeterminate in number, often two. Anthers 2-celled. Ovary 1-celled. Ovule solitary, erect. Stigma trifid or multifid. Berry 1-seeded.—Shrubs, rarely trees, aromatic, with knotted, jointed branches. Leaves alternate, very entire, often nerved. Spike supported by a spathe at the base, opposite to the leaves, rarely terminal, cylindrical, sometimes subglobular. H. B. and K. This genus has been subdivided by Miquel.

PIPER LONGUM, *Linn.* L. E. D. Fructus immaturus exsiccatus, L. D. Dried Spikes, E. of Long Pepper. Chavica Roxburghii, *Miq.*

Long Pepper has been employed by the Hindoos in medicine from the earliest times. Its Sanscrit name *pippula* seems to have been the original of the Greek πεπερι, and the πεπεριως ριζα would appear to be its roots, which, called *pippula moul*, are still extensively employed throughout the East. (*v.* Hindoo Med. p. 86.)

Root woody. Stems shrubby, climbing, jointed. Lower leaves ovate-cordate, 3—5-nerved; upper ones on short petioles, oblong, acuminate, oblique, and somewhat cordate at the base, obsoletely 4 to 5-nerved and veined, coriaceous, smooth. Peduncles erect, longer than the petioles. Spikes almost cylindrical. A native of the woody hills of the Circars, as well as along the foot of the Himalayas; cultivated in Bengal.—Nees von E. 26. St. and Ch. 174.

This plant is cultivated both on account of its roots, which, as well as the thickest part of the stems, are cut into small pieces and dried, and form an article of commerce all over the East. The spike of berries forms a long nearly cylindrical body, varying from an inch to an inch and a half in length. The berries are most pungent in their immature state, and are therefore dried, and the whole become of a greyish colour. They have a faint aromatic odour when bruised, but a powerfully pungent taste. Analysed by Dulong, its composition was found to be analogous to that of Black Pepper, as it contains Piperin, a concrete Oil, upon which its acrimony depends, and a Volatile Oil, to which it probably owes its odour.

Action. Uses. Stimulant, and a substitute for Black Pepper. It is probably retained in the Pharmacopœia as being a constituent of several old-established preparations, as Pulv. Aromaticus, Pulv. Cretæ Comp., Tinct. Cinnamomi C. Confect. Opii.

PIPER NIGRUM, *Linn.* L. E. D. Baccæ, L. Semina, D. Dried unripe berries, E. of the Black Pepper, which, decorticated, form *Piper album*, or White Pepper.

The πεπερι of Hippocrates and Dioscorides is no doubt our Pepper, the name being derived from the Persian *Pilpil.* (*v.* P. longum.) The Hindoos were no doubt the first to investigate the properties of Pepper. It grows in abundance on the Malabar coast, &c., whence it is now imported, as well as from the Malay Peninsula, Sumatra, and other islands.

The Pepper-vine is a perennial, with trailing or climbing, round, flexuose stem, from 8 to 12 feet in length, dichotomously branched, articulated, swelling near the joints, and often radiating. The leaves are distichous, broadly ovate, acuminate, occasionally somewhat oblique, 5 to 7 nerved, the nerves prominent beneath, connected by lesser transverse ones, of a dark green colour and glossy above, pale glaucous green beneath. Petioles rounded, nearly an inch in length. Spikes opposite the leaves, stalked, from 3 to 6 inches long, slender, drooping; apparently some male, others female, while sometimes the flowers are furnished with both stamens and pistil. (*Lindl.*) Stamens 3. Fruits distinct, round, sessile, about the size of a pea, at first green, then red, afterwards black, covered by pulp. Native of India and the Indian Islands.—Nees von E. 21. St. and Ch. 174.

Dr. Roxburgh's *Piper trioicum* yields the Pepper of the Rajahmundry Circars, and which he described before he had seen the true *P. nigrum.* But after he had done so, he observes that the leaves of *P. trioicum* have a glaucous appearance, which readily distinguishes it from *P. nigrum*, which has shining dark leaves. Dr. Heyne, who succeeded him in the superintendence of the Pepper cultivation, says the want of success in culture at Rajahmundry was owing to defects in cultivation, where they had *starved* these plants into celibacy. (*v.* Royle, Product. Resources of India, pp. 53 and 67.)

Black Pepper is formed by the above berries, gathered before they are quite ripe, and dried in the sun. They then become black and wrinkled from the drying up of the pulpy part, which covers a round grayish-white coloured seed.

White Pepper is the same berry allowed to ripen, when its pulpy part is easily removed by soaking in water and subsequent rubbing. The dried pulpy covering of the Black Pepper has in this country been removed by mechanical means, to form a white Pepper.

Pepper in both these states has, when bruised, an aromatic smell, and a hot, spicy, pungent taste, which is milder in the White Pepper. These properties are taken up partially by water, completely so by Ether, Alcohol, or Proof Spirit. Analysed by Œrsted and by Pelletier, they were found to contain a peculiar neutral principle, which has been called *Piperin*, an Acrid Resin or Concrete Volatile Oil, a little Balsamic Volatile Oil, Gum, Starch, Bassorine, Extractive, Malic and Tartaric acids, Salts, and Lignin.

Piperin, when perfectly pure, is in colourless rhombic crystals, neutral and not alkaline, insoluble in water, soluble in Alcohol and Acetic acid, less so in Ether; fusible at 212°, and volatile. Pelletier says, that, when quite pure, it is tasteless, and ascribes any active properties to a portion of the acrid resin. Dr. Christison, however, states "the very whitest crystals I have been able to obtain were as acrid as those which were brownest, and also that it exists in as large quantity in white as in black pepper, and is more easily separated, because combined with less resin." It is composed of $C_{40}\ H_{22}\ O_8\ N$. (*Liebig.*)

The *Acrid Resin* is soft, becomes solid at 32° F., is soluble in Alcohol and Ether, and unites readily with all fatty bodies. Its taste is extremely pungent and acrid, and it is very abundant in Black Pepper. Some conceive that the properties of Pepper depend chiefly on this Resin.

Action. Uses. Hot Stimulant; pungent, grateful Condiment, and as such universally employed: thought to be Febrifuge. Chiefly used to correct the effects of other medicines in causing nausea, &c., in doses of gr. v.

UNGUENTUM PIPERIS, D. Pepper Ointment.

Prep. Make into an ointment *Black Pepper* in powder ℔iv. with prepared *Axunge* ℔j.

Action. Uses. Stimulant application to Tinea Capitis.

CONFECTIO PIPERIS NIGRI, L. D. ELECTUARIUM PIPERIS, E.

Prep. Rub to a very fine powder *Black Pepper* ℔j. *Inula* ℔j. L. D. *Fennel*

Seeds ℔iij. Powdered (Liquorice Root ℔j. E.) *Sugar* ℔ij. Keep in a close vessel, and when required, rub up with *Honey* ℔ij. The E. and D. colleges direct this to be done at first.

Action. Uses. Moderate Stimulant; has been introduced as a substitute for *Ward's Paste*, which obtained celebrity as a cure for Hæmorrhoids. Sir B. Brodie conceives that it acts on them as a gentle stimulus in consequence of some of it passing along the colon.

D. ʒj.—ʒij twice or thrice a day.

CUBEBÆ, L. E. D. Baccæ, L. Fructus, D. Fruit, E. of PIPER CUBEBA, *Linn.* Suppl. The Cubeb Pepper. *Diandria Trigynia*, Linn.

Cubebs were probably first made known through the Hindoos to the Arabs, being the *kubabeh* of the latter, and the *kubob-chini* of the former. It is not probable that they were known to the Greeks. (*v.* Hindoo Med. p. 85.) Dr. Pereira has adduced evidence that they were employed in England 500 years ago.

Stem climbing; branches round, the thickness of a goose-quill, smooth, rooting at the joints; when young, petioles minutely downy. Leaves 4 to 6½ inches long, 1½ to 2 inches broad, stalked, oblong, or ovate-oblong, acuminate, rounded, or obliquely cordate at base, strongly veined, netted, coriaceous, very smooth. Spikes at the end of the branches, opposite the leaves, diœcious, on peduncles the length of the petioles. Fruit rather larger than Black Pepper, globose, on pedicels about half an inch long. (*Lindley.*) A native of Java and Prince of Wales' Island.—Nees von E. 22. St. and Ch. 175.

Dr. Lindley has ascertained that this is the *P. Cubeba* of the Linnean Herbarium. Blume says that the fruits of this, although of good quality, are not sent to Europe, but those that are furnished by *P. caninum*, Rumph. v. t. 28. p. 2; of this the fruit is smaller and shorter-stalked, having a distinct Anise flavour, and less pungent than the fruit of *P. Cubeba*. Dr. L., however, observes, "I cannot perceive any difference in the flavour of the dried fruit of this species and of the Cubebs sold in the London shops." Fl. Med. p. 314.

Cubeb berries, when dried, resemble Black Pepper, but are of a brownish colour, with raised veins forming a network over their surface, and are, moreover, distinguished by having a short stalk; hence Cubebs were called *Piper caudatum* by old writers. The sarcocarp is thin, the shell hard, seed spherical, white, oleaginous. The odour of Cubebs when bruised is aromatic and rather agreeable; the taste warm, peppery, and camphoraceous. Analysed by Vauquelin, and subsequently by M. Monheim, Volatile Oil (v. *Oleum Cubebæ*), was recognised, also *Cubebin*, which is a neutral substance apparently of the nature of Stearoptene; a soft and acrid Balsamic Resin, Extractive. The Volatile Oil, upon which the active principles chiefly depend, will evaporate with age, and therefore Cubebs should be powdered only as required. The powder is of a dark colour, and somewhat oily in appearance. It is said to be sometimes adulterated with Allspice powder.

Action. Uses. Stimulant; used as a Stomachic in the East. Having also the power of arresting excessive discharges from the Urethra, it is much employed in Gonorrhœa, for which it is in many cases an effectual cure. It is, perhaps, best prescribed immediately the first inflammatory symptoms have subsided; but requires caution, as it is apt to create irritation in the Urinary passages, and to cause swelled Testicle.

D. Of the powder, ℈j.—ʒij. three or four times a day.

OLEUM CUBEBÆ, E. Oil of Cubebs.

Obtained from pounded Cubebs by distillation with water, in the proportion of about 10 per cent.

Oil of Cubebs is colourless, or nearly so, lighter than water, thick, with the odour of Cubebs, and their pungent spicy taste. If rectified with water, it leaves a soft and resinous mass. It cannot be distilled by itself without undergoing decomposition, and some water being given off. It is composed of C_{15} H_{12}. By standing for some time, it deposits a Stearoptene, which has been called Camphor of Cubebs by Mr. Winkle.

Action. Uses. Appears to possess all the virtues of Cubebs in doses of ♏x.—fʒß. It may be given with Sugar in water.

TINCTURA CUBEBÆ, L. TINCT. PIPERIS CUBEBÆ, D. Tincture of Cubebs.

Prep. Macerate for 14 days *bruised Cubebs* ℥v. (℥iv. D.) in *Proof Spirit* Oij. (℔ij.) Strain.

Action. Uses. Stimulant. Used as Cubebs in curing Gonorrhœa, in doses of fʒj. two or three times a day.

MATICO is a name applied in South America and Mexico apparently to the leaves of several very different plants. Martius, in the *Phar. Central Blatt.*, considered it to belong to the genus Phlomis. Mr. Hartweg informs Dr. Lindley that "Matico is the vernacular name applied by the inhabitants of Quito to *Eupatorium glutinosum*, or the Chussalonga." He adds, "That it is the true Matico of the inhabitants of Quito and Riobamba, I have not the smallest doubt. I have also a small quantity of powdered leaves of some shrub possessing the same virtue as the Matico, collected in Bolivia, where it is known under the name of Moxo-Moxo. From bits of square stems which I find in the parcel, I suspect this to belong to some Labiatæ." (Lindley, Veg. Kingd. p. 707.) But it is equally certain that what has been of late years imported here, and of which specimens were distributed by, and for some of which the author is indebted to, Dr. Jeffreys of Liverpool, are the leaves, with portions of the stem and flowering-spikes, of a species of Piper, supposed to be *Piper angustifolia*, but which is now named *Artanthe elongata.* (*v.* P. J. iii. 472 and 525, and Lindl. *l. c.* p. 517.)

The Matico was first brought into notice by Dr. Jeffreys as a Styptic (*Lancet*, Jan. 7, 1839) in leech-bites and wounds of arteries, and has been found efficacious in many obstinate cases of bleeding, as from the nostrils, and even from the tongue. Its *under* surface, which is reticulated with veins, and covered with hairs, should be applied, as it is probably on this structure that its utility chiefly depends. Its Infusion and Tincture have also been recommended internally in affections of the Urinary organs, on which, by its stimulant action combined with a little astringency, it would appear to produce a salutary effect, as Cubebs are frequently known to do. Its properties, by the analysis of Mr. Morson, appear to depend chiefly on its Resin and Volatile Oil, its aqueous extract having only a slightly bitter and astringent taste. It has also been prescribed in discharges of blood from the urethra and rectum, as well as in uterine hæmorrhage, and has been used as an injection in Leucorrhœa, and as an external application to hæmorrhoidal affections, both as an ointment and as a lotion, by Mr. Young and Dr. O. Ferral, &c. The Infusion may be prepared with *Matico* ʒiv. increased—℥j. to *Aqua* Oj., and given in doses of f℥iß.; and the Tincture (*Matico* ℥iij. to *Proof Spirit* Oj.) to be given in doses of fʒß.—fʒj. two or three times a day.

URTICEÆ, *Juss.* Nettleworts.

Diclinous herbs, shrubs, or trees, with watery or milky juice. Leaves opposite or alternate, usually rough, as well as other parts of the plant, and covered with (often stinging) hairs. Stipules entire or lobed, usually persistent. Flowers small, polygamous, spiked, capitate or paniculate, sometimes placed on a fleshy receptacle. Perianth calyx-like, 4 to 5-parted, imbricate, in the female, often reduced to a single spathe-like scale or sepal. Stamens 4 to 5, inserted into the bottom of the perianth. Ovary free, 1-celled or 2-celled, 1 or 2 styles, with a single ovule in each cell. Fruit indehiscent. Embryo straight or spiral. Radicle superior.

The Urticeæ contain a great many plants very unlike each other, except in the structure of their inconspicuous flowers and small fruit. They are, however, divided into several tribes, which are now as often considered distinct families. They are widely diffused in tropical and temperate climates; the shrubby and arboreous species in the former, the herbaceous ones in the latter. Many secrete an acrid principle. Of the true Urticeæ none are officinal.

Tribe *Cannabineæ.* Annual or perennial, with watery juice. Flowers diœcious; male paniculate. Perianth calyx-like, 5-parted, imbricate in æstivation. Stamens 5, inserted into the bottom of the perianth. Female flowers in a spike or catkin with bracts. Perianth urceolate or spathe-like. Ovary free, 1-celled, 2-styled, with a single pendulous ovule. Nut bivalved, 1-seeded. Seed pendulous. Embryo without albumen, hooked or spiral, with the radicle superior. The genera Humulus and Cannabis are officinal; each contains only a single species.

LUPULUS, L. E. HUMULUS, D. Strobili exsiccati, L. D. Catkin, E. of HUMULUS LUPULUS, *Linn.* The Common Hop. *Diœcia Pentandria,* Linn.

The Hop plant was known to the Romans, being considered the *Lupus salictarius* of Pliny. It is found wild in many parts of Europe, and by Bieberstein among the bushes and hedges of the Caucasus. It is found in China, said to be wild in North America, and to be a native of this country. Humalineæ, or Hop-grounds, are mentioned in the ninth century in Germany. In the thirteenth century, Hops were introduced into the breweries of the Netherlands. Its culture is supposed to have been introduced into this country from Flanders in the reign of Henry VIII. Both Hops as well as Newcastle Coals were petitioned against by the city of London, the former

Fig. 89.

"in regard they would spoyl the taste of drink and endanger the people;" whence Henry VIII. issued an injunction "not to put any hops or brimstone into the ale." As in the history of many other prejudices, we observe a complete reversal of opinion, as Hops are now considered indispensable in the brewing of all malt liquors.

The Hop (fig. 89) is a perennial rooted plant, with annual pliable stems, which on poles or in hedges climb to a great extent, twining from right to left, slender, somewhat angular, rough, with little asperities and minute reflexed hairs. The leaves are opposite, the upper alternate, on long, often winding petioles, the smaller heart-shaped, the larger 3 to 5-lobed, serrated, veiny, and extremely rough, with prickle-like pubescence. Stipules 2, bifid, between the petioles, reflexed. Flowering branches axillary. Flowers numerous, of a yellowish green colour. *Males* (*a*) on a separate plant (a few on the female), in axillary panicles. Perianth 5-parted; segments oblong, spreading. Stamens 5; filaments short; anthers with a projecting apex, oblong, 2-celled, opening by longitudinal lateral slits Pollen globose. *Females* (*b*), like the males, on a separate plant, in dense catkins or strobiles, with membranous concave bracts (*d*), each supporting a flower. In place of perianth there is a membranous scale or sepal, which embraces the ovary and grows with it (*c*). Ovary ovate, subcompressed, 1-celled, with a single ovule. Stigmas 2, elongated. Fruit a strobile or catkin, formed by the enlarged bracts and scales or sepals, which are glandular and embrace the nuts. These are small, subglobular, erect, 1-seeded. Pericarp hard, but fragile, covered with yellow, cellular, superficial, aromatic glands (*lupuline*) (*f*). Seed pendulous. Testa membranous. Embryo (*e*) without albumen, spiral, with long cotyledons. Radicle roundish, turned towards the hilum.—Nees von E. 101. St. and Ch. 41. Fig. 89, where a bit of the male plant is shewn on the right, and the female on the left and above the male.

Hop plants grown from root-sets come to perfection in the third year from planting. They spring out of the ground about the end of April, and come into flower about the end of August. The catkins are fit to gather from the beginning of September to the middle of October, according to the sort cultivated, but chiefly owing to differences in the seasons. They are then picked, dried by artificial heat in kilns, and packed in large long bags, the finer in pockets. Hops consist of the leaf-like bract and of the scale-like sepal which invests the seed-nut. This, or rather the scale and the base of the bract, are covered with numerous superficial glands, which have been called *Lupulinic glands*, and simply *Lupulin*, though this name is objectionable, as also indicating the peculiar or Bitter principle. Dr. Ives of New York, by thrashing, rubbing, and sifting, procured from 6 lb. of Hops about 6 oz. of these *grains;* but there is always intermixed some fragments of the bracts and scales. The *glands* are yellow, shining, roundish, or kidney-shaped, cellular, somewhat transparent, and sessile; the point of attachment is called the hilum (*v.* a magnified view in fig. *f* from Raspail). Hops are remarkable for their bitter taste combined with a very agreeable odour, especially when being picked or collected in kilns or in breweries. The bitterness resides partly in the bracts, but also in the glands, to which the aromatic qualities are especially due. The medicinal properties also depending on them, the E. C. directs that these glands or grains are to be used in making the Tincture of Hops. Analysed by MM. Payen, Chevalier, and Pelletier, these *Hop* glands were found to consist of Volatile Oil 2 parts, Bitter Extract (which has been called *Lupuline* and *Lupulite*) 10 parts, Resin 50 to 55, with Gum, Extractive, Ozmazome, Fatty matter,

Malic acid, Malate of Lime, and other salts. The bracts, analysed by the same chemists, yielded only a trace of the Volatile Oil, Bitter Extract, and Resin, but some Tannin and Colouring matter, Chlorophylle, Gum, Lignin, with some free acid and different salts. A portion of the active properties both of Hops and of the Hop glands are taken up by water, but completely so by Spirit.

Two varieties of the Hop plant are particularly distinguished: one cultivated near Canterbury and in East Kent, of which both the plants and catkins are smaller; the latter ovoid, about an inch and a half in length, of a pale but lively yellowish-green colour, and of a fine aromatic fragrance. The West Kent or Sussex Hop grows to a much larger size, is considered hardier, and its catkins are about two and a half, sometimes four inches in length, but do not bring so high a price in the market as the East Kent Hops. As root-sets from the female plants are alone planted by cultivators, the author was led to inquire how the seed was perfected? and if not, whether the Hop glands were produced in as great abundance and perfection as they might be if some *male* were set along with the *female* plants. He is informed by Mr. Alderman Masters of Canterbury that some *male* blossoms are always produced on the female plants, and suffice for the purpose of fertilizing them. The author has been unable to learn whether the female ever changes into a male plant, or *vice versa*, as has been observed with the Nutmeg plant. He may mention, that, owing to the kindness of the above gentleman, as well as of Joseph Royle, Esq., of Stuppington, where the finest Hops are grown, he has been enabled to introduce the Hop plant into the Himalayas, where it is now flourishing alongside of the China Tea plant. The root-sets, with the ends dipped in wax, wrapped in cotton, and enveloped in caoutchouc-cloth, were sent by the overland mail to the East India Company's Botanic Garden at Saharunpore. They arrived there in a living state, as reported by Dr. Jameson, and have produced fine plants, as well as the seeds sent with them. The successful cultivation of Hops would make malt liquors more within the reach of European soldiers, and assist in detaching them from the pernicious spirituous compounds of the bazaars, which now destroy the health and shorten the lives of thousands. Well *hopped* ale, moreover, is well known to be one of the best Stomachics and Tonics for convalescents from many Indian diseases.

Action. Uses. Stomachic and Tonic, slightly Narcotic. The property of Hops of giving the bitter to Beer, and, by preventing acetous fermentation, of enabling it to be kept much longer, is well known. To it no doubt is owing a portion of the stomachic properties of malt liquors, as we see exemplified in the bitter, often called Indian, ales. Hops are Hypnotic, especially when stuffed into a pillow, but they should be first moistened with Spirits, to prevent the rustling noise. Fomentations also have been used. Hops are thought to be Diuretic (as is also the root), and to be useful in correcting Lithic acid deposits. The Lupulinic or Hop glands may be given in doses of from gr. vj.—gr. xij. made up into pills.

INFUSUM LUPULI, L. Infusion of Hops.

Prep. Macerate for 4 hours in a lightly covered vessel *Hops* ʒvj. in boiling *Aq. dest.* Oj. Strain.

Action. Uses. Tonic, slightly Narcotic in doses of f℥jß.

TINCTURA LUPULI, L. E. TINCT. HUMULI, D. Tincture of Hops.

Prep. Macerate *Hops* (dried, D.) ℥vj. (℥v. D.) in *Proof Spirit* Oij. (℔ij. D.) for 14 days (7 days, continually stirring, D.) Strain.

Tincture of Hop Glands. From freshly dried *Hops* q. s. separate by friction and sifting the yellowish-brown powder attached to their scales, and of this take ℥v., *Rectified Spirit* Oij. Prepare by percolation, as Tinct. Capsicum, E.

Action. Uses. The E. Tincture of the glands, though called by the same name, is superior in efficacy to the others. The E. Kent Hops contain a larger proportion of glands than the Sussex Hops. Rectified Spirit is also the best solvent of the Hop glands. Doses of f3ß.—f3ij.

EXTRACTUM LUPULI, L. E. EXT. HUMULI, D. Extract of Hops.

Prep. Prepare from *Hops* as Extr. Gentian, L., as Extr. Logwood, E., as Extracts generally, D.

Action. Uses. Tonic ; being bitter, without aroma : in doses of gr. v.—Эj.

CANNABIS SATIVA and its variety *C. indica.* The Leaves and Resin of Hemp.

The Hemp appears to be a plant of the Persian region, where it is subjected to great cold in winter, and to considerable heat in summer. It has thus been able to travel on one hand into Europe, and on the other into India ; so that the varieties produced by climate have by some been thought to be distinct species, the European being called *C. sativa*, and the Indian *C. indica.* The name κανναβις, by which it was known to the Greeks, seems to be derived from the Arabic *kinnub*, the *canape* of the middle ages, Dutch *kinnup* and *hinnup*, German *hanf*, whence the English *hemp.* Herodotus mentions it as Scythian. Bieberstein met with it in Tauria and the Caucasian region. It is well known in Bokhara, Persia, and abundant in the Himalayas. It seems to have been employed as an intoxicating substance in Asia and Egypt from very early times, and even in medicine in Europe in former times, as we find it noticed in Dale (*Pharmacologia*, i. 133) and Murray (*Apparat. Medicaminum*, iv. p. 608—620), where it is arranged, as in this work, next to the Humulus. It has of late years been brought into European notice by Dr. O'Shaughnessy.

The Hemp is a diœcious (occasionally monœcious) annual, from 3 to 10 feet high, according to soil and climate. Root white, fusiform, furnished with fibres. The stem erect ; when crowded, simple ; but when growing apart, branched even from the bottom, angular, and, like the whole plant, covered with fine but rough pubescence. The leaves are opposite or alternate, on long petioles, scabrous, digitate, composed of from 5 to 7 narrow, lanceolate, sharply serrated leaflets, of which the lower are the smallest, all tapering at the apex into a long entire point. Stipules subulate. *Males* on a separate plant. Flowers in drooping, axillary, or racemose panicles, with subulate bracts. Perianth 5-parted; segments not quite equal, downy. Stamens 5; filaments short; anthers large, pendulous, 2-celled; cells united by their backs, opening by a longitudinal slit. *Females* in a crowded spike-like raceme, with leafy bracts. The perianth consists of a single, small, spathe-like sepal, which is persistent, acuminate, ventricose at the base, embraces the ovary, and is covered with short brownish glands. Ovary subglobular, 1-celled, with one pendulous ovule. Style short. Stigmas 2, elongated, glandular. Nut ovate, greyish-coloured, smooth, covered by the calycine sepal, bivalved but not dehiscing, and inclosing a single oily seed. Seed pendulous. Testa thin, membranous, marked at the apex with a coloured hilum. Embryo without albumen, doubled upon itself. Radicle elongated, turned towards the hilum, and the apex of the nut separated from the incumbent plano-convex cotyledons (by a small quantity of albumen, *Lindley*).

The Indian plant has by some been thought to be a species distinct from the European one; but, like Dr. Roxburgh and others, the author was unable when in India to observe any difference between the plant of the plains and that of the hills of India, nor between these and the European plant. The Indian secretes a much larger proportion of resin than is observable in the European plant, but a difference is observed in this point in India between plants grown in the plains, and those of the mountains, and also when grown thickly together. The natives plant them wide apart, to enable them to secrete their full powers. In Europe, the thick sowing, and moister, often dull, climate will prevent the due secretion of the peculiar principles of a plant of the Persian region. But the plants grown in the past season, from the great heat and light, ought to be more resinous than usual. It is not without interest to observe that both the Hop and Hemp, belonging to the group *Cannabineæ*, owe their properties to glandular resinous secretions. The author, in calling attention to the uses of this plant, in his *Illustr. of Himalayan Botany*, stated that "the leaves are sometimes smoked in India, and occasionally added to Tobacco, but are chiefly employed for making *bhang* and *subzee*, of which the intoxicating powers are so well known. But a peculiar substance is yielded by the plants on the hills, in the form of a glandular secretion, which is collected by the natives pressing the upper part of the young plant between the palms of their hands, and then scraping off the secretion which adheres. This is well known in India by the name of *cherrus*, and is considered more intoxicating than any other preparation of the plant; which is so highly esteemed by many Asiatics, and serves them both for wine and opium: it has in consequence a variety of names applied to it in Arabic, some of which were translated to me, as "grass of faqueers," "leaf of delusion," "increaser of pleasure," "exciter of desire," "cementer of friendship," &c. Linnæus was well acquainted with its "vis narcotica, phantastica, dementens" (anodyna et repellens). It is as likely as any other to have been the *Nepenthes* of Homer." (*l. c.* p. 334.)*

Dr. O'Shaughnessy has described in detail the different preparations, as—

1. *Churrus*, the concreted resinous exudation from the leaves, slender stems, and flowers. This is collected in various ways; that of the Himalayas is much esteemed, that of Herat and of Yarkund still more so. For a specimen of the last the author is indebted to Dr. Falconer.
2. *Ganjah.* Dr. O'S. describes it to be the dried hemp plant which has flowered, and from which the resin has not been removed. The bundles are about two feet long, and contain twenty-four plants. In N. W. India the name *Ganjah* is applied to the whole growing plant.
3. *Bang*, *Subjee*, or *Sidhee*, is formed of the larger leaves and capsules without the stalks.

The leaves of common Hemp have been analysed, but the analysis requires to be repeated and carefully compared with that of the Indian plant. The properties seem to depend on a Volatile Oil, which is as yet but little known, and upon the Resin. This is very soluble in Alcohol and Ether, as well as in the fixed and volatile Oils, partially soluble in alkaline, insoluble in acid solutions; when pure, of a blackish-grey colour. (The Yarkund specimen is of a dark blackish-green, another kind is of a dirty olive.) Its odour is fragrant and narcotic; taste slightly warm, bitterish, and acrid. The *Ganjah*, which is sold for smoking chiefly, yields to Alcohol 20 per cent. of resinous extract, composed of *churrus* and Chlorophylle. Dr. Farre found that already a substitute (*Apocynum cannabinum*, called Indian Hemp in America) is sold for this, though having no resemblance to it, and possessing only emetic and cathartic properties.

Action. Uses. All these preparations are capable of producing intoxication, whether the *churrus* be taken in the form of a pill, or with conserve, or the dried leaf be rubbed up in milk and water with a little sugar and spice, or smoked. As a medicine, it was tried by Dr. O'S. in Rheumatism, Hydrophobia, Cholera, and Tetanus. In the last such marked benefit and cures were produced, that the Hemp was pronounced an Anticonvulsive remedy of the greatest value. Its general effects are, alleviation of pain (generally), remarkable increase of appetite, unequivocal Aphrodisia, and great mental cheerfulness. Its more violent effects were, delirium of a peculiar kind, and a cataleptic state. Dr. Pereira was among the first to submit it to experiment, but failed in obtaining any results, probably from changes having

* Dr. O'S. states that "no information as to the medicinal effects of Hemp exists in the standard writers on Materia Medica to which we have access. It is only in the later writers that it is omitted. Linnæus was acquainted with them, as the author quoted in the above briefly, as being a botanical work.

taken place in the drug. Dr. Laurie pronounced it uncertain, and not to be trusted to as a narcotic. Mr. Ley, however, found it useful in relaxing spasm, producing sleep, and during its action abatement of pain. Mr. Donovan found its power great in temporarily destroying sensation, and subduing the most intense neuralgic pain. Professor Miller of Edinburgh considers its virtue to consist in a power of controlling inordinate muscular spasm. Dr. Clendinning says that in his hands its exhibition has been followed by manifest effects as a soporific or hypnotic in conciliating sleep, as an anodyne in lulling irritation, as an antispasmodic in checking cough and cramp, and as a nervous stimulant in removing languor and anxiety. The Hemp may be used in the following preparations and doses; but Dr. O'S., when in England, found that he was obliged to give as much as 10 or 12 grs. and even more; though in India he considered gr. ½ a sufficient, and 1½ gr. of the Extract a large dose.

EXTRACTUM CANNABIS. Resinous Extract of Indian Hemp.

Prep. Boil the rich adhesive tops of the dried *Ganjah* in *Rectified Spirit* until all the Resin is dissolved out. Distil off the Spirit with a gentle heat.

D. This Extract is effectual in gr. ß. and gr. j. doses; but 10 and 20 grs. have been given in Hydrophobia and Tetanus.

TINCTURA CANNABIS. Resinous Tincture of Indian Hemp.

Prep. Dissolve *Extract. Cannabis* gr. iij. in *Proof Spirit.* fʒj. A weaker Tincture may also be made with the dried herb or Ganjah.

D. ♏x.—fʒj. with the dried herb or *Ganjah.* A drachm or so may be given in Tetanus every half-hour, until the paroxysms cease, or Catalepsy is induced.

Mr. Donovan states the only preparation to be relied on is the Tincture of the Resin prepared from properly collected Hemp. He advises of the Resinous Tincture ♏xv. to be added to Rectified Spirit ♏xlv. and taken as a draught; or, if added to water, it should instantly be swallowed, or the Resin would precipitate and adhere to the vessel.

Tribe *Artocarpeæ.* Shrubs or trees, with white or yellowish milky juice. Leaves alternate, large, convolute. Flowers unisexual, in a consolidated, fleshy receptacle or head, seldom spiked. Ovary 1 or 2-celled, with 1 to 2 styles. Ovule 1, erect, straight. Fruit berried, 1-seeded, often growing together, or in a fleshy receptacle. Embryo without albumen. Radicle superior.—Tropical family, with a few species in higher latitudes. Among them are many secreting acrid principles, some very poisonous, as the *Antiaris toxicaria,* or Upas-tree of Java; but there are also some which yield edible fruits.

MORA, L. Fructus, L. Baccæ, D. Mulberries. Fruit of MORUS NIGRA, *Linn.* The Common Mulberry. *Monœcia Tetrandria,* Linn.

The Mulberry is the Μορέα ἢ Συκαμινέα of Dioscorides and of other Greeks, and is mentioned in Luke xvii. 6, as συκάμινος. It has no doubt been known from the earliest times.

Tree of 25 to 30 feet in height. It is often described as watery in juice; but Mr. Sievier at the author's request examined and found it to contain Caoutchouc. (*Antiq. of Hind. Med.* p. 10.) Leaves alternate, roundish, often lobed, cordate, rather acuminate, coarsely serrated, pubescent. Stipules oblong, deciduous. Flowers monœcious, thickly set, or distinct. Unisexual catkins. Perianth 4-lobed; in each the lobes concave. *Male* flowers in a spike. Stamens 4, alternate with the segments of the perianth. *Female* flowers clustered in ovoid catkins. Sepals 4, scale-like, overlapping each other, becoming fleshy. Stigmas 2, linear, glandular. Fruit formed by the accretion of the sepals of the perianth become fleshy, each inclosing a lenticular nucule. Seed pendulous. Embryo curved, in fleshy albumen.—Native of Persia, early introduced into the south of Europe.—Nees von E. 100. St. and Ch. 39.

Mulberries, formed by the lateral aggregation of the several female flowers, constitute an ovoid spurious berry; they are at first reddish, but become of a deep purple colour when ripe, and contain an agreeable subacid juice. They are refrigerant and slightly laxative.

Syrupus Mori, L. (*Mororum.*) Mulberry Syrup.

Prep. With the aid of gentle heat dissolve *Sugar* ℔ijß. in filtered juice of *Mulberries* Oj. and proceed as for Syrup of Lemons.

Action. Uses. Refrigerant. Used also for colouring draughts.

Fici, L. E. D. Fructus siccatus, L. D. The dried fruit E. of Ficus Carica, *Linn.* The Common Fig. *Diœcia Triandria*, Linn.

The Fig (σῦκον) has been employed in diet and in medicine from very early times.

A small tree. Leaves cordate, often palmately lobed, scabrous above, pubescent beneath. Flowers monœcious, numerous, stalked, and inclosed within a pear-shaped fleshy receptacle, which converges so as to leave only a small orifice at the apex, forming what is commonly called the fruit or Fig, with a few bracteal scales at its base. *Male.* Perianth 3-lobed. Stamens 3. *Female.* Perianth 5-parted. Ovary semi-adnate. Style single. Stigmas 2. Utricle single, covered with the persistent, somewhat fleshy, perianth, and sunk into the fleshy receptacle. Achœnium lenticular, hard. Embryo curved, within fleshy albumen.—Native of Asia, long introduced into Europe. Nees von E. 97. St. and Ch. 154.

Formerly, as in the present day, the process of caprification was practised, to assist the ripening of the fruit. This consists in puncturing the fruit with a sharp instrument covered with oil.

The trunk and branches of the common as of other Fig trees abound in milky, usually acrid juice. This is found also in unripe Figs; but as they ripen, mucilaginous and saccharine matter is produced, the fig becomes soft, juicy, and of a delicate flavour in all favourable climates. When nearly ripe, they are dried in large quantities in the south of Europe, and are exported to this country. They form also an article of commerce in Asia, imported into India from Affghanistan and Persia.

Action. Uses. Figs are dietetical, slightly laxative with those unaccustomed to their use. Chiefly employed as a Demulcent; or heated and split open, applied as Cataplasms; or used as additions to such preparations as Decoct. Hordei Comp. and Confectio Sennæ.

DORSTENIA, *Linn.*

Dwarf herbaceous plants, with scaly rhizomata, monœcious; flowers arranged upon a fleshy receptacle, usually flat, and expanded (basket-shaped), but extremely variable in form. *Male*, on the surface of the receptacle, 2-lobed, fleshy, diandrous. *Female*, immersed in the receptacle, also 2-lobed in most species. Ovary 1—2-celled, with a single suspended ovule in each cell. Style 1. Stigma 2-lobed. Achænia lenticular, imbedded in the fleshy receptacle, from which they are projected with elasticity when ripe. (*Lindley.*)

CONTRAJERVA, L. Radix. Root of DORSTENIA CONTRAJERVA, *Linn.* L. and probably of other species. Contrajerva. *Monœcia Tetrandria*, Linn.

This root is supposed to have been first made known by Monardes; others say that it was first sent by Sir F. Drake to Lecluse, who named it *Drakena radix.* The name signifies *counter-poison.*

Though the L. P. mentions only one species of Dorstenia as yielding Contrajerva root, there is reason to believe, from the statement of Martius and of others, that several species, as *D. braziliensis, Houstoni, Drakena,* all yield it. Dr. Pereira states that none of the roots of *D. Contrajerva* are met with in commerce.

D. CONTRAJERVA, *Linn.* L. Caulescent; stem (rhizoma?) covered with spreading green scaly stipules. Leaves palmate, the lobes lanceolate-acuminate, coarsely serrated and gashed, occasionally almost pinnatifid. Receptacle on a very long stalk, quadrangular, wavy, or plaited.—Native of Mexico and the West Indies.

D. braziliensis, Linn. Rootstock oblong, woody, præmorse, powerfully aromatic. Stemless. Leaves cordate, oblong, obtuse, crenulated, serrated, or toothletted, cucullate at the base. Scape as long as the petioles. Receptacle orbicular, somewhat cup-shaped, crenated at the margin.—Native of mountains of San Paulo and Minas in Brazil, Jamaica, Trinidad.—Bot. Mag. t. 2804.

The Contrajerva root of commerce is imported from Brazil, and probably yielded by *D. braziliensis*, especially as it resembles it in character. The part which is officinal is the rootstock, which is præmorse, an inch or two in length, scaly or wrinkled, of a greyish colour externally, paler within, with numerous slender radicles from its sides, as well as one or two long tapering ones from its base. The odour is somewhat aromatic; the taste slightly bitterish, warm, and aromatic. The radicles have less of these sensible properties, which are readily extracted by Spirit, and partially by boiling water. They depend chiefly on a Volatile Oil, Resin, Bitter Extractive, and Starch.

Action. Uses. Stimulant, Tonic, and Diaphoretic; but little used in the present day, though formerly employed (in the form of *Pulv. Contrajervæ Comp.*) in low states of Fever and malignant states of the Exanthemata.

D. Of the powder ℈j.—℈ij.; or it may be given in infusion.

Tr. *Ulmeæ.* Ovary 2-celled. Seed pendulous. Embryo straight.

ULMUS, L. Cortex, L. Cortex interior, D. Bark of ULMUS CAMPESTRIS, *Linn.* The Elm.

The Elm is supposed to be the πτελέα of Dioscorides.

A tree of 60—80 feet, with rugged bark. Leaves rhomboid-ovate, acuminate, wedge-shaped, and oblique at the base, always scabrous above, downy beneath, doubly and irregularly serrated, sometimes incurved. Branches wiry, slightly corky, when young, light brown, and pubescent. Flowers perfect. Perianth bell-shaped, 5-cleft, persistent. Stamens 5. Styles 2. Capsule compressed, oblong, with a broad membranous wing all round, deeply cloven, naked. (*Lindl.*) European forests, &c.—Nees von E. t. 104.

The inner bark, which is officinal, should be stripped from the tree in spring, and its epidermis and outer layer of bark afterwards removed. The pieces are broad, thin, tough ; taste mucilaginous and slightly bitter, from containing Gummy matter and a little Tannin.

Action. Uses. Demulcent, Tonic; thought also to be Alterative in Cutaneous affections. Used in decoction in doses of f℥iij.

DECOCTUM ULMI, L. D. Decoction of Elm Bark.

Prep. Take recently bruised *Elm Bark* ℥ijß. (℥ij. D.) *Aq. dest.* Oij. by measure ℔ij. Boil down to Oj. (℔j. D.) Strain.

AMENTACEÆ, *Juss.*

Flowers monœcious or diœcious, rarely perfect. Barren flowers capitate, or in catkins (amentum), sometimes with a membranous perianth. Female flowers clustered, solitary, or in catkins. Ovary usually simple. Stigmas 1 or more. Fruit as many as the ovaries, bony, or membranaceous. Albumen usually wanting. Embryo straight or curved, plain. Radicle mostly superior. Young leaves with stipules. (*Babington.*) The Amentaceæ are found chiefly in temperate climates, with the exception of Salix, which is more widely diffused. They yield valuable timber and some, hardly less valuable bark, which, on account of its astringency, is used as a medicine, for tanning, and as a dye. The acorns of some of them are employed as articles of diet.

Tribe SALICINEÆ. Flowers all in catkins. Fruit naked, 2-valved, 1-celled, many-seeded. Seeds erect, comose.

SALIX, *Linn.* Willow. *Diœcia Diandria*, Linn.

Catkins consisting of imbricated scales. Stamens 1—5. Fruit a single-celled follicle, with 1—2 glands at its base. No perianth.

The bark of different species of Willow (ἰτέα) has been long employed medicinally, and its use has been revived in modern times. The species are numerous, and no less than three of them are officinal. They are all difficult to distinguish from each other. But the best practical rule is, " Select those whose barks possess great bitterness combined with astringency." (*Pereira.*)

I. Catkins on a leafy stalk, lateral, coetaneous. Scales of the catkins deciduous.

Sect. *Fragiles.* Trees with glabrous leaves. Stamens 2.

SALICIS FRAGILIS Cortex, D. Bark of SALIX FRAGILIS, *Linn.* The Crack Willow.

A large tree, with round, very smooth branches, brown, brittle in the spring. Leaves lanceolate, pointed, serrate ; floral leaves somewhat obovate, recurved, often blunted. Ovary tapering, stalked, glabrous. Style short. Stigma bifid. —Marshy ground.—E. B. 1807.

Salix Russeliana, *Smith*, found in marshy wood, is very similar to *S. alba*, and is said by Sir J. C. Smith to be much the most valuable species, from its bitterness and astringency. St. and Ch. 139.

Sect. *Albæ*. Trees with their leaves, when young, hairy with adpressed silky hairs. Catkins lax.

SALICIS ALBÆ Cortex, D. Bark of SALIX ALBA, *Linn.* White Willow.

Tree of 50 to 80 feet in height, with silky branches. Leaves elliptic-lanceolate, glandular, serrate, acute, silky on both sides when young. Ovary nearly sessile, ovate-acuminate, glabrous. Style short. Stigmas thick, recurved, bifid. —Moist situations.—E. B. 2430. Nees von Esenb. Suppl. 17.

II. Catkins lateral, sessile, without, or nearly without, leaves.

Sect. *Capreæ*. Trees or shrubs. Stamens 2. Anthers yellow. Catkins bracteated. Stalks of the capsule at least twice as long as the gland.

SALICIS CAPREÆ Cortex, D. Salicis Cortex, E. Bark of SALIX CAPREA, *Linn.* Great Sallow, or Round-leaved Willow.

A small tree, 15 to 20 feet high. Leaves large, ovate, or elliptical, flat, acute, crenate, serrate, wavy at the margins, deep green, with a downy midrib, whitish above, and cottony beneath. Stipules subreniform. Ovary lanceolate, subulate. Style very short. Buds glabrous. Catkins very thick, blunt.—Woods and hedges in dry places.—E. B. 1488.

Willow bark will of course vary somewhat according to the species from which it is obtained ; but it is thin, flexible, rolling up into a quill, or like shavings, with a brown epidermis, white in the inside ; reduced with difficulty to powder, having a slight odour, but a powerfully bitter and astringent taste. Analysed by Pelletier and Caventou, Willow bark was found to contain Green Fatty matter, a Bitter Yellow Colouring matter, Tannin (which is not precipitated by Tartar Emetic), Resinous Extract, Gum, a Magnesian salt, and an organic acid. Buchner discovered the peculiar neutral principle *Salicine*, which is no doubt separated with the bitter and yellow substance. From the presence of Tannin, the bark is sometimes used in tanning, and a greenish colour is produced by sesqui-salts of Iron. Water and Alcohol take up its active properties.

Salicine is very bitter, crystallizes in white silky needles or laminæ, and has no alkaline reaction. It differs also from the vegetable alkalies in not containing Nitrogen, and not forming salts with acids. It is soluble in 5·6 parts of cold, and in much less of boiling water; soluble in Alcohol, but insoluble in Ether. Sulphuric acid decomposes it, producing a bright red colour. It is composed of $C_{21}\,H_{12}\,O_{11}$. C. Gerhardt has since stated its composition to be $C_{42}\,H_{28}\,O_{22}$. It is found in several species of *Salix* and of *Populus ;* of the former, in *S. Helix, alba*, &c. It may be obtained by acting on a saturated decoction with Acetate or Oxide of Lead, getting rid of the Lead by means of Sul′ or a current of Sulphuretted Hydrogen gas, then evaporating the solution until the Salicine crystallizes, and purifying it with animal charcoal and recrystallization.

Action. Uses. Astringent Tonic. Useful as a Stomachic, and even for arresting Agues. It may be given in Infusion (dried bark ℥j.

—Aqua Oj.), or in Decoction, in doses of f℥jß. every two or three hours.

Salicine may be used as a febrifuge in doses of 2—8 or even 20 grs., like the Sulphate of Quinine.

Tribe CUPULIFERÆ. Male flowers in a catkin. Female solitary, or aggregated, or spiked. Perianth adnate to the ovary, with a denticulated limb, sometimes evanescent, surrounded by a coriaceous involucre.

QUERCUS, *Linn.* Oak. *Monœcia Polyandria*, Linn.

Monœcious. Male catkins long, pendulous, lax (Fig. 90, *a*). Stamens 5 to 10 (*b*). Perianth (*b*) 5 to 7-cleft. Female flower solitary, with a cup-shaped scaly involucre (*c*, magnified). Stigmas 3 (*c*). Ovary 3-celled, 2 of which are abortive. Nut or acorn 1-celled, 1-seeded, surrounded at the base by the enlarged cup-shaped involucre. (*d*, the young fruit; *e*, the same magnified and cut vertically, that the perianth, ovary, and ovules may be seen. *f*, a cotyledon with the radicle.)

QUERCUS CORTEX, E. D. Quercus, L. Bark of QUERCUS PEDUNCULATA, *Willd.* L. E. Q. Robur, *Linn.* D. The Common Oak.

Species of the Oak (δρυς of the Greeks, and *allon* of the Bible) have been esteemed for their strength and astringency from the earliest times.

The common English Oak, which by some botanists is named *Q. Robur*, Linn., and by others *Q. pedunculata*, Willd. (Fig. 90), has its acorns borne on long peduncles, and is thus distinguished from *Q. sessiliflora*, Salisb. (*Q. Robur*, Willd.), which has its acorns clustered upon a very short stalk, or sessile, with leaves on elongated stalks. E. B. t. 1845. Nees von E. t. 92. Dr. Lindley states that the timber of this kind is very superior to that of the former; but opinions differ respecting the timber of these species: for medical purposes one is probably as good as the other. Dr. Greville states that the characters of the different kinds pass insensibly and completely into each other.

QUERCUS ROBUR, *Linn.* (*Q. pedunculata*, Willd.) Young branches glabrous. Leaves on short footstalks, cuneately oblong, pinnatifid, slightly pubescent beneath. Lobes oblong, rounded, with deep, narrow, somewhat acute sinuses; bases biarticulate, equal. Female catkins on long footstalks. Acorns oblong. —Woods.—E. B. t. 1842. Nees von E. t. 93.

The Oak is stripped of its bark in spring and in the beginning of summer. It is usually in long strips, of a coarse fibrous texture, and not easily reduced to powder. When deprived of its epidermis, it is of a light brown colour externally. The odour is faint, but the taste bitter and roughly astringent. Its properties are readily extracted by water and by Proof Spirit. Its constituents are Tannin (about 15 per cent.), Gallic acid, Uncrystallizable Sugar, Pectin, Tannates of Lime, of Magnesia, and of Potash, &c. The inner part of the bark contains the largest portion of Tannin, and in the spring of the year. From the presence of this principle, a precipitate necessarily takes place with Gelatine, and a blackish-coloured one on the addition of a sesqui-salt of Iron.

Fig. 90.

Action. Uses. Powerful Astringent, in Gargles, Lotions; and Baths for children; sometimes doses of ʒß.—ʒij. of the powder given as a Febrifuge. Applied externally, made into a poultice, in flabby ulcers and external gangrene.

DECOCTUM QUERCUS, L. E. D. Decoction of Oak Bark.

Prep. Take bruised *Oak Bark* ʒx. (℥j. D.) *Aq. dest.* Oij. (℔ij. D.) Boil down to Oj. (℔j. D.) Strain.

Action. Uses. Astringent; internally in chronic Diarrhœa; as a Gargle in relaxed Uvula; or as an Injection in Leucorrhœa. It has been recommended for the injection into the cyst of Hydrocele, &c.

EXTRACTUM QUERCUS, D. Extract of Oak Bark.

Astringent Extract, prepared by evaporating the Decoction.

GALLÆ, L. E. D. Galls. Gemmæ morbidæ, L. Diseased Buds. Excrescences, E. formed by Diplolepis (or Cynips) Gallæ Tinctorum, on QUERCUS INFECTORIA, *Oliv.* The Gall-Oak.

Galls were known to Hippocrates, and are described by Dioscorides (i. c. 147) under the name κηκις, which the Indo-Persian writers have converted into *fikees.* They are the *afus* of the Arabs, and well known in India by the name of *majoo-phul.* Galls are imported into England from Smyrna, being produced in Asia Minor; also from Aleppo, the produce of the vicinity of Mosul in Kurdistan. They are also imported into England from Bombay (sometimes to the extent of 1000 cwt.), having been first imported there from the Persian Gulf. Mr. Wilkinson, of the house of Wilkinson and Jewsbury, informs me that formerly, when he paid much attention to this trade, he observed that whenever the prices were low at Smyrna, the Galls came from Bombay, and *vice versa;* but the supply was never abundant from both sources in the same year. They are imported into Bombay from Basra (Bussorah), which is not a great deal farther from Mosul than is Aleppo. They are therefore most probably the produce, like Aleppo Galls, of Kurdistan and of other Persian provinces. Dr. Falconer, when travelling in the Punjab, was informed that Galls were produced on the Balloot Oak, *Quercus Ballota.*

Galls are produced on different species of Oak, as well as on some other plants, as the Tamarisk; Aleppo Galls, by the female of the above Diplolepis piercing the buds of *Q. infectoria* with its ovipositor, and there depositing its eggs. These producing irritation, cause the juices of the plant to flow towards the wound, and the subsequent enlargement of the part into the form of galls round the larva. This, when fully developed, escapes by a hole which it perforates in the gall.

Quercus infectoria, now generally acknowledged to be the species producing the Galls of commerce, is a small tree or shrub, with a crooked stem, not above 6 to 8 feet high. Leaves on short stalks, 1—1½ inch long, ovate-oblong, with a few coarse mucronated teeth on each side; apex bluntly mucronate, rounded, and rather unequal at the base, smooth, shining on the upper side. Acorn solitary, obtuse, 2 or 3 times longer than its hemispherical scaly cup.—A native of Asia Minor; found by Capt. Kinnier in Armenia and Kurdistan.—Nees von E. t. 94. St. and Ch. 152.

Besides the names applied from the places whence they are obtained, Galls are distinguished by their physical characters, as into Blue and White Galls. The Blue Galls vary in size, and are of a bluish-grey colour. They are gathered before the insect has become perfect, or worked its way out. Some of these are larger, and are called Green Galls from being of a greenish colour. They display on their otherwise smooth surface a number of bluntly-pointed tubercles, which would appear to be the apices of leaves stimulated into unnatural growth. The best are heavy, hard, shining, and break with a short flinty fracture. White Galls are so called from being of a lighter colour than the others, but still of a greyish or yellowish hue. They are distinguished by being perforated with a small round hole, that by which the insect had escaped. They are usually less heavy than th others, have a larger internal cavity, and are not so astringent. Both are easily reduced to powder, which is without odour, but with a simple powerful astringent taste. They yield their properties to water, which is the best solvent; also to Proof Spirit, and slightly to Alcohol and Ether. From 500 parts Sir H. Davy obtained 185 parts of matter soluble in water, of which he states 130 were Tannin, 31 Gallic acid with a little Extractive, 12 of Mucilage, &c., and 12 of saline and calcareous salts, the insoluble matter consisting chiefly of Lignin. But a larger proportion of Tannin has been obtained by other chemists, as from 30 or 40 to 60, instead of the above 26 per cent. The little colouring matter in Galls makes them particularly valuable to tanners.

TANNIN, or TANNIC ACID, is usually described with Galls, as existing in them in large quantity, being generally obtained from them, though a constituent of many other astringents, as Oak bark, Catechu, &c. It is the type of astringents. Ordinary Tannin is amorphous, brownish-coloured, and consists of impurities united to the Tannic acid. Pure Tannin, or Tannic′ (being so named because its solution reddens Litmus and effervesces with Carbonates), is sometimes white, but usually with a yellow tinge, spongy, shining, without odour, but extremely astringent. It is most easily obtained from Nutgalls by the action of Ether. It is very soluble in water and in weak Spirit. When heated, it swells up, is decomposed, leaving a bulky charcoal. It precipitates Gelatine from its solutions, and combines with the Gelatinous part of skin, and thus forms leather. It forms precipitates (Tannates), most of which are nearly insoluble, with most metallic oxides, and likewise with alkalis and their Carbonates, including vegetable alkalies. The mineral acids, combining with the Tannic′, also form precipitates in concentrated solutions. With Sesqui-salts of Iron it is well known to form a black precipitate (ink); the Tannin of Sumach, Catechu, &c., as has been frequently mentioned, forms a very dark green precipitate with the same salts: no effect is produced on the Proto-salts. Tannin is composed of $C_{18}H_5O_9+3HO$.

GALLIC ACID. Though Galls are stated by Sir H. Davy to contain about 6 per cent. of this acid, a much larger quantity may be obtained from them, because it is formed by the conversion of the Tannic′ into Gallic acid by the absorbing of Oxygen from the atmosphere, Carbonic′ being given off. It is colourless, with an acid and astringent taste, and is usually seen in the form of a grey crystalline powder. It has an astringent taste, but is of no use in tanning.

Action. Uses. Galls are powerfully Astringent; seldom given internally; the author frequently prescribed from 10 to 20 grs. of the powder several times a day, or in Infusion, in the obstinate chronic Diarrhœas of the natives of India. The natives themselves prescribe them in Intermittents. Its Tincture is much used as a test or the salts of Iron. An Infusion may be employed as a Gargle, Wash, or Injection, or as an antidote to poisoning by vegeto-alkalis; but the diluted Tincture affords a more ready antidote.

TINCTURA (GALLARUM, E. D.) GALLÆ, L. Tincture of Galls.

Prep. Macerate for 14, L. (7, D.) days powdered *Galls* ℥v. (℥iv. D.) in *Proof Spirit* Oij. (℔ij. D.) Strain. (Or prepare by percolation, as Tinct. Capsicums, E.)

Action. Uses. Astringent in doses of f℈ß.—f℈ij. May be diluted with water as a Lotion, or for exhibition in cases of poisoning with vegeto-alkalis.

UNGUENTUM GALLARUM, D. Ointment of Galls.

Prep. Mix finely powdered *Galls* ℥j. with prepared *Hog's Lard* ℥viij. Make an Ointment.

Action. Uses. Astringent application to external Hæmorrhoids.

UNGUENTUM GALLÆ (ET OPII, E.) COMPOSITUM, L. Compound Ointment of Galls.

Prep. Triturate into a uniform mass very finely powdered *Galls* ʒij. *hard Opium* powdered ʒß. (ʒj. E.) *Hog's Lard* ℥ij. (℥j. E.)

Action. Uses. Astringent and Anodyne application to Hæmorrhoids. The E. preparation is much stronger than that of the L. P., as it contains more Opium and less Lard. Dr. Paris suggests dissolving Morphia in Olive Oil, and adding the Ointment of Galls.

BALSAMACEÆ, *Lindl.*, contains the genus *Liquidambar*, of which one species, *L. styraciflua*, is indigenous in North America. This yields in Mexico and Louisiana a liquid balsam of an aromatic odour and taste, containing Styracin and Benzoic acid. Dr. Pocock found *L. orientale* in Cyprus, where it was called *Xylon Effendi*, the tree of our Lord. It produces an excellent Turpentine. It is probable that this yields some of the liquid Storax of commerce; as some liquid Balsam, under the names of *Rose Maloes* and *Rosa Mallas*, makes its appearance in the accounts of the commerce of the Red Sea and Persian Gulf; and Petiver, as quoted by Dr. Lindley, states that the tree which yields it is the *Rosa Mallas*, and grows in Cobross, an island at the upper end of the Red Sea, near Cadess, which is three days' journey from Suez. It is sent in barrels by way of Jidda to Mocha. This is supposed by some to be yielded by *Liquidambar Altingia* of Blume, a native of Java, which is there called *Ras-sa-mala*, and undoubtedly yields the fine liquid Storax or *Rosamala* of the Malayan Archipelago. (*Lindley.*) But Dr. Pereira has ascertained that all the liquid Storax imported for the last seven years comes from Trieste. He also states that the strained Storax (*Styrax colatus*) sold to the perfumers is prepared from this variety of liquid Storax. This is a subject of inquiry for those visiting the shores of the Red Sea or its islands.

GYMNOSPERMÆ, *Lindl.* Gymnosperms.

This division has been made of some Exogens, in consequence of their ligneous tissue being dotted with disk-like marks, and their ovules being truly naked, so as to be fertilized directly through the foramen of the ovule.

The CYCADEÆ form a small family somewhat resembling Palm trees in appearance, and were at one time thought to be allied to them and to Ferns. By Mr. Brown they have been shown to be most closely allied to Coniferæ.

A kind of Sago is said to be procured from the cellular substance occupying the interior of the stem of *Cycas revoluta*, a native of Japan, and also of *C. circinalis*. Both exude a clear insipid mucilage, which hardens into a firm transparent gum, like Tragacanth, but clearer. Dr. Roxburgh was unable to ascertain that any of the species yielded Sago, or a substitute for it, though species of Cycas are quoted as yielding Sago in the E. P.

Dr. Lindley states that one of the best kinds of Arrow-root is prepared in the Bahamas from the trunk of some species of Zamia which is a native of the West India Islands.

CONIFERÆ, *Juss.* Conifers.

Trees or shrubs, with a branched trunk, abounding in Turpentine. Leaves simple. Flowers monœcious or diœcious. *Male* flowers of 1 or more monadelphous stamens, collected in a deciduous catkin about a common axis. Anthers of 2 or more lobes, bursting outwards, often terminated by a scale-like crest. *Female* flowers usually in cones, sometimes solitary. Ovary spread open in the shape of a scale, and placed in the axil of a membranous bract: in the solitary flowers apparently wanting. Ovules naked, in pairs on the face of the ovary, and inverted, or in the solitary flowers erect. Fruit a cone, or solitary naked seed. Testa hard, crustaceous. Embryo in the axis of fleshy albumen. Radicle next the apex.—Yield valuable timber, as Deal, Cedar, &c., and most of the species Turpentine, which is a compound of resin and of volatile oil.

The products of Coniferous plants officinal in the Pharmacopœias are so numerous and obtained from so great a variety of sources, and are yet so similar to each other, that it is hardly possible to refer them with correctness to their respective plants. It is preferable, therefore, as has been done by Dr. Pereira and in Duncan's *Edinburgh Dispensatory*, to enumerate the several Pine-trees which are supposed to yield these products, and then to treat of the products themselves,—that is, of Turpentine, Resins of different kinds, and then of those obtained with the aid of heat, as Oil of Turpentine, Tar, and Pitch.

PINUS, *Linn.* Pine. *Monœcia Monadelphia*, Linn.

Flowers monœcious. *Males.* Catkins racemose. Filaments short. Anthers crested, 2-celled, bursting longitudinally (or Stamens 2, Anthers 1-celled). *Females.* Catkins solitary, or from 2 to 3. Scales imbricated, with membranous bractlets. Ovules 2, at the base of the scales, collateral, inverted, their points lacerated and directed downwards. Scales of the cone hard, woody, and truncated, hollowed at the base for the reception of the seeds. Seeds prolonged at the base into a membranous wing. Leaves evergreen, usually acicular, in fascicles, surrounded at the base by a membranous tubular sheath.

PINUS SYLVESTRIS, *Linn.* L. D. Scotch Fir. Red Deal. Leaves in pairs. Young cones stalked, recurved, ovate-conical. Wing thrice as long as the seed.—Lamb. Pin. t. 1. Nees von E. t. 79.—Scotland, Norway, woods of Europe, north of the Alps.—This species yields much Turpentine, Pitch, and Tar, though at present little of it is imported into this country.

P. MARITIMA, *Dec.* (P. Pinaster of Lambert), Nees von E. t. 76, 77, is abundant on the southern coasts of Europe, as well as of England, and in the south of France in the department of the Landes. It yields Bourdeaux Turpentine, Galipot, Pitch, and Tar.

P. PALUSTRIS, *Lambert.* The Swamp Pine and Long-leaved Pine. A large tree, spreading from the southern States of Virginia to the Gulf of Mexico. "This tree furnishes by far the greater proportion of Turpentine, Tar, &c. consumed in the United States, or sent from them to other countries." *Wood and Bache.*

P. PINEA, *Lamb.* and *P. Cembra,* the Siberian Stone Pine, are interesting, as the seeds of both, sometimes called *Pine-nuts,* are eaten, as are those of *P. Geradiana,* in Affghanistan and Tibet. *P. longifolia,* Lamb. is an Himalayan species; which yields a very fine Turpentine, resembling pure white granular honey; much used by the natives of India in medicine, and called *bireeja,* &c.

ABIES, *Tourn.* Fir.

Monœcious. *Males.* Catkins solitary. Anthers bursting transversely. *Females.* Catkins simple. Scales (or carpels) imbricated, thin at the apex, rounded, flat, instead of being hollowed for the seeds; when ripe, falling from the axis. Leaves solitary in each sheath, never fascicled. In other respects agreeing with Pinus.

ABIES EXCELSA, *Dec.* E. (*Pinus Abies,* Linn.) L. D. Norway Spruce Fir.—Leaves scattered, tetragonal. Cones cylindrical, pendulous; the scales rhomboidal, flattened, jagged, and bent backwards at the margin.—Northern parts of eastern Europe, Alps, northern parts of Asia.—Nees von E. t. 80. Yields Abietis Resina by spontaneous exudation.

A. Picea, Lindl. The Silver Fir, with distichous leaves and erect cones. A native of the mountains of central Europe. Yields Strasburgh Turpentine.

A. BALSAMEA, *Marsh.* E. (*Pinus balsamea,* Linn. L. D.) Canadian Balsam and Balm of Gilead Fir. Leaves solitary, flat, subpectinate, suberect above. Acuminate apex of the scales of the cone when in flower reflexed.—Northern parts of North America.—Lamb. Pin. t. 41. Nees von E. t. 82.

A. canadensis, Lindl. Hemlock Spruce Fir is said to exude a Turpentine similar to that of the foregoing. *A. nigra,* the Black Spruce Fir, is interesting as yielding the Essence of Spruce.

LARIX, *Tourn.* Larch.

Monœcious. Catkins and cones lateral. *Males.* Catkins simple, ovate. Anthers numerous, with their filaments united into a thick column. Anthers crested, bursting longitudinally. Leaves, when first expanding, in tufted fascicles, becoming somewhat solitary by the elongation of the new branch.

LARIX EUROPÆA, *Dec.* (*Abies Larix,* Lam. E., *Pinus Larix,* Linn. D.) The Larch is a lofty tree, with wide-spreading branches; when well grown, the extremities droop gracefully. The Leaves deciduous. Flowers reddish. Cones ovate-oblong. Edges of scales reflexed, lacerated. Bracts panduriform. *Lambert.*—Nees von E. 83. St. and Ch. 75.—A native of the Alps, much cultivated in this country. Yields Venice Turpentine, and a kind of Manna called "Manna de Briançon."

Larix (or Cedrus) *Deodara* (Deodar and Kelon), or Himalayan Cedar, is an elegant and lofty tree, hardy as the Larch, and yielding valuable timber. It has been extensively introduced into this country by the East India Company, and is interesting as having been long employed in medicine by the Hindoos, and known even to Avicenna. (*Hindoo Med.* 36.) Its Turpentine, known by the name *kelon-ke-tel*, is in great repute in the North-west of India, from its stimulant properties and power of healing deep-seated ulcers, as in elephants and camels.

TEREBINTHINA VULGARIS, L. D. Resina liquida, L. Fluid Resin of PINUS SYLVESTRIS, *Linn.*, of various species of Pinus and of Abies, E.

Common Turpentine either exudes naturally or from incisions from most trees of the Pine tribe, as also from *Pistacia Terebinthus* (p. 333). It consists of Resin intimately mixed with a Volatile Oil, known in its separated or distilled state as Oil of Turpentine. In time, all Turpentines become converted into Resins, from the evaporation of the Oil and by its oxidation. They all soften by heat, burn readily, are soluble in Alcohol and Ether, unite with the fixed Oils, and resemble each other very closely in taste and smell; but differ in being more or less white or dark-coloured, and in the odour and taste being more or less agreeable. Water acquires only a little of their properties, but they may be made into an emulsion with eggs or vegetable Mucilage.

Common Turpentine used to be procured from *Pinus sylvestris*, as it still is in many parts of Europe, and also from *P. maritima*, which yields the Bourdeaux Turpentine, and in winter the *galipot* of the French. But Dr. Pereira has shewn that almost the whole quantity of Turpentine imported here is from America. This is procured chiefly from *P. palustris*, partly also from the *P Tæda.* This Turpentine is viscid, semifluid, of a dull light yellowish colour, with a warm, acrid, rather bitter taste, and a moderate terebinthinate odour. When fresh, it yields 17 per cent. of Oil of Turpentine. (*W.* and *B.*) The Bourdeaux Turpentine is whitish, turbid, separates upon standing into a transparent liquid and into a granular honey-like semifluid. It is acrid and nauseous in taste, and of a disagreeable smell; yields about 20 per cent. of Oil. M. Faure discovered that it might be solidified by the aid of a 32nd part of Magnesia. Common Turpentine yields Oil of Turpentine and Resin, q. v., and is a constituent of Ung. Elemi (p. 345) and of Emp. Galbani (p. 424).

TEREBINTHINA VENETA, E. D. Fluid Resinous exudation of *Larix europæa.* Venice Turpentine. This, when genuine, is a thick tenacious fluid, usually of a cloudy appearance, of a yellowish-green tint, acrid and bitter in taste, of a strong peculiar odour. It is sold in Paris as Strasburgh Turpentine, and is distinguished by being less liable than others to solidify. (*Guibourt* and *Pereira.*) Dr. Thomson stated long since that the Venice Turpentine of the shops was imported from America. This is therefore only a substitute. What is usually sold is a mixture of Oil of Turpentine with common Resin. Venice Turpentine is intended to be a constituent of Emp. Cantharidis, E. and of Ung. Infusi Cantharidis, E.

TEREBINTHINA (BALSAMUM, E. D.) CANADENSIS, L. Fluid Resin of ABIES (Pinus, L. D.) BALSAMEA. Canada Balsam is procured by breaking the vesicles which naturally form upon the trunks and branches, and then collecting their fluid contents. It is often called Balm of Gilead. When fresh, it is nearly colourless, of a light yellow colour, transparent like thin honey; solidifies slowly; is of a strong, rather agreeable odour, and a bitterish, rather acrid taste. It is also obtained by making incisions into the tree. Strasburgh Turpentine is sometimes substituted for it.

ABIETIS RESINA, L. THUS, D. Resin of ABIES EXCELSA, E. (*Pinus abies*, Linn.), L. D. The Resin of the Norway Spruce Fir may be arranged with the Turpentines as being a spontaneous exudation, and with the Resins as having lost by evaporation most of its Volatile Oil. It used to be called *Thus*, or Frankincense, as it still is in the D. P. It is collected in the form of concrete tears, which are hard and brittle, but soften readily at the temperature of the body. It is of a light yellowish or brownish-yellow colour externally, lighter within; slight terebinthinate odour and acrid bitter taste. The substance which the French call *galipot* or *barras* is the concretion produced on the Pine of the Landes, late in the year or in winter, when the collection of Bourdeaux Turpentine has ceased. A very fine Resin is spontaneously yielded by the Himalayan *Pinus Morinda*, Royle.

This Resin is intended to be employed in making Pix Burgundica, Emp. Aromaticum, D. p. 541, Emp. Opii, L. E. D., Emp. Thuris, D. p. 129, Emp. Galbani, L. D., and Emp. Picis, L. E.

PIX ABIETINA, L. PIX BURGUNDICA, E. D. Burgundy Pitch is the above Resin melted in water immediately after being scraped from the tree, and strained through a cloth. It is thus freed from mechanical impurities, with a loss of a little of its Volatile Oil. In all other respects it corresponds with it. But most of that which is sold is a factitious compound of Resin rendered opaque by the incorporation of water, and coloured by Palm Oil, or made from concrete American Turpentine. (*Pereira.*) This is used for making the following Plaster as well as the Emp. Calefaciens (*v.* Cantharidis), Emp. Opii, E. D., Ung. Resinæ albæ, D., and Emp. Cantharidis Comp. E.

Action. Uses. The above Resin and Pitch are both Rubefacient.

EMPLASTRUM PICIS, L. E. Burgundy Pitch Plaster. Warm Plaster.

Prep. Take *Burgundy Pitch* ℔ij. (℔jß. E.), *Resin* ℔j. (℥ij. E.) and *Bees' Wax*

℥iv. (℥ij. E.); melt them together with a gentle heat, then add *Resin of Spruce Fir* ℔j. L., *Expressed Oil of Nutmegs* ℥j. L. (Oil of Mace ℥ß. E.), *Olive Oil* f℥ij. (f℥j. E.), *Aq.* f℥j. (f℥ij. E.); mix well, and boil till the mixture acquires the proper consistence.

Action. Uses. Warm Rubefacient Plaster to the chest and joints, &c.

RESINA, L. E. D. Residue of the distillation of the Volatile Oil from Turpentines of various species of *Pinus* and of *Abies*, E.

When any of the Pinic Turpentines are subjected to distillation with or without water, the Volatile Oil rising when much heated, leaves behind a solid Resin, which is often called Colophony (Fr. *Colophane*), from the Greek κολοφωνια, but usually Black Rosin, though it is only of a brownish-yellow colour, transparent, and a little empyreumatic. When the distillation is not carried quite so far, or if more water is added during the process, and agitated with it while in fusion, some of it becomes incorporated with the Resin, which becomes opaque or of a whitish colour. This is the *Resina flava* of the Pharmacopœia, or *Yellow*, sometimes called *White Resin*. The incorporated water escapes by evaporation or it may be expelled: the Resin then becomes of a pale yellow colour and transparent.

Resin is solid and transparent, very brittle, with a glassy fracture, is a little heavier than water, differs in colour according to its purity, with a weak terebinthinate odour and taste, melts at a moderate heat, becomes decomposed at a higher, producing both an oil and a gas, and burns with a smoky flame. It unites when in fusion with Wax, fats, and fatty oils, also Spermaceti; is readily dissolved by Alcohol, Ether, and many volatile oils, and is insoluble in water. The strong acids decompose it: the alkalis unite with it, and form soaps. Instead of being simple, it is found to be a compound of two acid bodies, one called Sylvic, the other Pinic acid, and of a neutral resinous principle. The Sylvic' is more soluble in cold and diluted Alcohol, and may thus be separated from the other. It crystallizes in small, quadrangular, rhombic prisms, is colourless, insoluble in water, soluble in Ether, strong hot Alcohol, and in volatile oils. Pinic acid ($C_{20} H_{15} O_2$) is considered isomeric with the Sylvic, and has many of the same properties. A third acid, the Pimaric, has been detected in the Bourdeaux Turpentine. The acid of Colophony, called the Colophonic, is considered somewhat different, being of a brown colour, and sparingly soluble in Alcohol.

Action. Uses. Mild Stimulant; used externally, but chiefly on account of its adhesive properties, in various Cerates, Unguents, and Emplastra. For Emp. Resinæ, *v.* p. 161.

CERATUM RESINÆ, L. UNG. (RESINOS. E.) RESINÆ ALBÆ, D. Resin Cerate, or Basilicon Ointment.

Prep. With a gentle heat melt together *Resin* ℔j. (℥v. E., white Resin ℔ij. D.)

Wax ℔j. (℥ij. E. Lard ℥viij. E. ℔iv. D.); then add *Olive Oil* f℥xvj. and press the Cerate while hot through linen, L. D. (Stir the mixture briskly while it cools and concretes, E.)

Action. Uses. A mild Stimulant, applied to foul or indolent ulcers.

TEREBINTHINÆ OLEUM, L. E. D. Volatile Oil of the Turpentine of various species of *Pinus* and of *Abies*, E. Oil of Turpentine.

Oil of Turpentine, separated from the Resin by the process of distillation, is found swimming on the surface of the water with which it is distilled, and is not observed to differ materially when obtained from different Pine trees. The author, when in India, distilled the Oil from the Turpentine of *Pinus longifolia*, which, when sent to the General Hospital at Calcutta, was pronounced to be "of very superior quality." Dr. Pereira states that American Turpentine is now chiefly employed for obtaining the Oil, at the rate of about 14 to 16 per cent., and that the Bourdeaux Turpentine yields an inferior Oil and Resin.

Prep. Distil in a copper alembic *Common Turpentine* ℔v. with *Aqua* Oiv. Yellow Resin will remain after the distillation.

OLEUM TEREBINTHINÆ PURIFICATUM, L. E., RECTIFICATUM, D. Purified Oil of Turpentine.

Prep. Cautiously distil *Oil of Turpentine* Oj. (℔ij. by measure, D.) with *Aqua* Oiv. (As long as oil comes over with the water, E., till Ojß. of oil is obtained, D.) or agitate it with ⅛ of Alcohol. Mr. Flocton redistils from a solution of caustic Potash, to get rid of all traces of resinous and acid matters (*Per.*)

This purified Oil is limpid, colourless, with a powerful penetrating odour, and pungent bitterish taste. Sp. Gr. 0·865; boils at about 312°, but, as volatilization proceeds, at 350°. Sp. Gr. of its vapour 4·764. It is very inflammable, producing much black smoke. It is slightly soluble in water, more readily in Ether and in Alcohol; miscible in all proportions in the fixed oils; dissolves resins and fats, and is one of the few solvents of Caoutchouc. When moist and exposed to great cold, it deposits crystals, which are a Hydrate of the Oil. Sul′ chars it, Nit′ and Chlorine set it on fire. It absorbs H Cl′ acid gas, and a substance called *artificial Camphor* (C_{20} H_{16} H Cl) is produced. The composition of Oil of Turpentine is C_5 H_4. When exposed to the air, it absorbs Oxygen: therefore Oil which has been long kept usually contains some. Oil of Turpentine is now considered to be composed of two different but isomeric *oils*, as the changes in the boiling point seem to indicate. One of these, or that which combines with the H Cl′ gas, has been called Radical Oil of Turpentine, and also *Camphene.*

Action. Uses. Rubefacient and Counter-Irritant when applied externally. Stimulant when taken internally, acting as a Diuretic and Diaphoretic; in large doses, as a Cathartic; useful as an Anthelmintic. For this purpose, and to act as a Cathartic, it is sometimes conjoined (f℥ij) with a little Castor Oil (f℥vj.) Its action is, however, somewhat uncertain. Occasionally, intoxicating effects are produced by large doses. In doses of ♏viij.—f℈ß. frequently repeated, it acts as a Stimulant, becomes absorbed, and is exhaled both by the skin and lungs, while the urine acquires a violet odour. In larger doses (f℈iv.—f℥ij.) it acts as a Cathartic, and its irritant effects, as those of Strangury, are not perceived, from absorption not having taken place. It should be made into an emulsion with yolk of one egg for every f℈ij. and diluted to the patient's taste, with water, plain or aromatized. (*c.*)

LINIMENTUM (TEREBINTHINATUM, E.) TEREBINTHINÆ, L. D. Turpentine Liniment.

Prep. Shake together till mixed *Oil of Turpentine* f℥xvj. (f℥v. E. ℔ß. D.), *Camphor* ℥j. L. (℥ß. E.), *Soft Soap* ℥ij. L. (Resin Ointment ℥iv. E. ℔j. D.) (Melt the Ointment, and mix with it gradually the (Camphor, E.) Oil of Turpentine, E.D.) (till a uniform Liniment be obtained, E.)

Action. Uses. Stimulant Liniment, but chiefly used by applying lint soaked in it, to burns and scalds. (*Dr. Kentish.*)

ENEMA TEREBINTHINÆ, L. E. D. Turpentine Enema.

Prep. Mix together *Oil of Turpentine* f℥j. (℥ß. D.) with *Yolk of Egg* q. s., and gradually add *Barley Water* f℥xix. L. (*Water* (not higher than 100° F.) ℥x. D. f℥xix. E.)

Action. Uses. Antispasmodic; Anthelmintic in cases of Ascarides.

PIX LIQUIDA, L. E. D. Tar, from various species of Pinus and Abies, E.

Tar has been employed in medicine from very early times. It is imported into this country both from the north of Europe and from North America. It is prepared by submitting the roots and branches of different Pine trees to a smothered combustion. The resinous matter is melted and also somewhat altered by the heat, and the Tar flows out as a viscid and tenacious semifluid, of a brownish-black colour, having a bitter, resinous, and a little acid taste, and an empyreumatic odour. It is a very complex mixture of Resin and Oil of Turpentine, both somewhat modified, and some Empyreumatic Oil, Charcoal, and Pyroligneous acid, with various products of the destructive distillation of the wood. By subjecting it to distillation, Oil of Tar and Pyroligneous acid are obtained, and Tar-water by agitating it with water. Tar is soluble in Ether, Alcohol, and the fixed and volatile oils. Several kinds of it yield Creosote, Paraffin, Eupion, &c., when the whole of the liquid parts are evaporated, and Pitch is left.

Action. Uses. Tar, taken internally, is an Alterative Stimulant; applied externally it promotes a healthy action in indolent ulcers, and in some cutaneous diseases. The vapour is sometimes inhaled in chronic bronchial affections.

AQUA PICIS LIQUIDÆ, D. Tar Water.

Prep. Mix *Tar* by measure ℔ij. with *Aq.* Cj., agitating with a wooden stick for ½ hour. When the Pitch has subsided, filter, and keep in well-stoppered vessels.

Action. Uses. Slightly Stimulant and Alterative. Recommended by Bishop Berkeley as almost a panacea: it is now rarely used.

UNGUENTUM PICIS LIQUIDÆ, L. E. D. Tar Ointment.

Prep. Melt together *Tar* ℔j. (℥v. E. ℔ß. D.), *Suet* ℔j. L. (℔ß. D. Bees' Wax ℥ij. E.) Express through linen, L. (a sieve, D. Stir briskly while it concretes in cooling, E.)

Action. Uses. Stimulant. Useful in Ringworm and some Ulcers.

PIX (ARIDA, E.) NIGRA, L. Pitch.

Pitch is left after the distillation of the liquid parts of the Tar. It is well known for its black colour and firm texture, and consists of many of the same constituents as Tar.

Action. Uses. Stimulant and Alterative. Used in Ichthyosis in doses of gr. x.—ʒj. in pills.

UNGUENTUM PICIS NIGRÆ, L. Pitch or Black Basilicon Ointment.

Prep. Melt together *Black Pitch, Wax, Resin,* āā ℥ix. *Olive Oil.* f℥xvj. Express through linen.

Action. Uses. Stimulant application to Porrigo or to indolent ulcers.

Tribe *Cupressineæ.*

Flowers diœcious, rarely monœcious, upon different branches. *Males.* Catkins axillary or subterminal, ovate, small. Anthers 4—7, 1-celled, inserted on the lower edge of the subpeltate scales. *Females.* Flowers few, in an axillary ovate catkin, imbricated, with bracts at the base, lower ones barren. Scales 3—6, united at the base, and containing usually 3 ovules, which are erect, perforated at the apex. Fruit a galbulus, consisting of the scales become succulent, and consolidated into a drupe-like body. Seeds osseous, triquetrous.

JUNIPERUS, *Linn.* *Diœcia Monadelphia,* Linn.

JUNIPERI CACUMINA et FRUCTUS (Baccæ, D.), L. E. D. Tops and Fruits (*Berries*) of JUNIPERUS COMMUNIS, *Linn.*

The Juniper (ἀρκευθος) was employed by the Greeks, and subsequently by the Arabs, being their *abhool.* Species are mentioned in the Bible. (*v.* Bibl. Cycl.)

The Juniper forms a bushy shrub. Branches smooth and angular towards their extremities. Leaves evergreen, 3 in each whorl, crowded, linear, subulate, channeled, stiff and sharp-pointed, longer than the galbulus, of a shining green colour on their lower surface, but having a broad glaucous line along the centre of the upper, which is resupinate. Flowers axillary, sessile, the males discharging much yellow pollen. Females on a separate shrub, green, on scaly stalks. The fruit ripens in the autumn of the second year.—A native of the northern parts of Europe, Asia, and America.—Nees von E. 86. St. and Ch. 141.

All parts of this plant when bruised exhale a more or less agreeable terebinthinate odour. The wood is officinal on the Continent, but the tops and fruits in this country. The latter are imported from the north, but the best come from the south of Europe. They are globular, marked with three radiating furrows at the summit and below by the bracts; are of a purple-black colour with a glaucous bloom, and contain a brownish-yellow pulp. Their taste is sweetish, followed by bitterness, slightly terebinthinate, as is the odour, and somewhat aromatic. These properties are imparted partly to water and readily to Alcohol, depending on the presence of Volatile Oil (q. v.) 1 per cent., Wax 4, Resin 10, Gum 7, Grape Sugar with salts of Lime, 33·8, the remainder being Lignin and water = 100.

Action. Uses. Berries Stimulant, Diuretic; but seldom used. Largely employed in the manufacture of Hollands.

OLEUM JUNIPERI, L. E. D. Oil of Juniper.

Oil obtained from the fruits (and other parts of the plant) by distillation with water. It is colourless, or of a light green tinge, lighter than water, corresponding very closely (comp. $C_{10} H_8$) with the Oil of Turpentine, like it little soluble in Alcohol. It has the odour of the fruit, and a warm aromatic taste.

Action. Uses. Stimulant, Diuretic. Considered very certain in its effects in doses of ℳiv.—ℳvj. Its effects may be assisted by combination with Spirit of Nitre and with Digitalis. Hollands Gin owes its Diuretic properties to the presence of this Oil.

SPIRITUS JUNIPERI COMPOSITUS, L. E. D. Comp. Spirit of Juniper.

Prep. Macerate *bruised Juniper Berries* ℥xv. (℔j. E. D.), bruised *Caraways* and bruised *Fennel Seed* āā ℥ij. (℥jß. E. D.) in *Proof Spirit* Cj. (Ovij. E.) *Aq.* Oij. (q. s. to prevent empyreuma, D.) for 2 days (24 hours, D.), (then add the water, E. D.,) and with a gentle heat, L. distil Cj. (Ovij. E.)

Action. Uses. Stimulant adjunct in doses of fʒij.—fʒiv. to Diuretic draughts. May be substituted for Hollands in prescriptions.

SABINA, L. E. D. Cacumina recentia et exsiccata, L. Tops E. both fresh and dried, L. Folia, D. Leaves of JUNIPERUS SABINA, *Linn.* Savin.

Savin is the βραθυς of Dioscorides, converted by the Arabs into *buratee.*

A small, bushy, very compact shrub, disposed to spread. Branches slender, completely invested by the short imbricating leaves. Leaves small, ovate, convex, opposite, decussate, deeply imbricated. Fruit round, of a bluish purple, about the size of a currant.—A native of the midland parts of Europe, of the mountains of the south of Europe, and of Russia in Asia.—Nees von E. 87.

The whole plant exhales a strong fœtid odour, and has an acrid, bitter, and disagreeable taste. The officinal parts are the young branches, which are completely enveloped in the small imbricated leaves, and retain a portion only of the properties of the fresh plant. These are taken up by Spirit, fixed oils, and fats, partially by water. They depend on the presence of a Volatile Oil, Resin, Gallic acid, &c. A deep green colour is formed on the addition of a Sesqui-salt of Iron to its watery infusion.

Action. Uses. Irritant, so as even to be poisonous. Sometimes used to destroy warts, and its ointment to keep open issues. In small doses, Stimulant, Diuretic, and Emmenagogue. Often taken to cause abortion: it can only do so by producing inflammation, and thus destroy the mother, sometimes without causing the expulsion of the child. Dr. Pereira recommends it as an Emmenagogue, in the form of an infusion in 64 parts of water, in doses of f℥iv.—f℥j., or the Volatile Oil in some mucilaginous vehicle.

Oleum Sabinæ, L. E. D. Savin Oil.

Obtained, in the proportion of about 3 per cent., by distilling with water the fresh tops of the plant; is light in colour; in composition resembles Oils of Juniper and of Turpentine; has the strong odour and the disagreeable acrid taste of the plant.

Action. Uses. Acrid Stimulant. Emmenagogue in doses of ♏ij.—♏v. with Sugar, gr. x.—℈j. or with some mucilaginous substance.

Ceratum (Unguentum, D.) Sabinæ, L. E. Savine Cerate or Ointment.

Prep. Take bruised *Savin* ℔j. (2 parts, E. ℔ß. D.), *Hog's Lard* ℔ij. (4 parts, E.), *Wax* (yellow, D.) ℔ß. (1 part, E.) Melt the Wax and Lard together, and boil with the leaves till crisp. Strain; express through linen (after which the D. C. directs the Wax to be added, and the whole melted).

Action. Uses. Acrid application to keep open blistered surfaces, and the discharge from setons. But much of the oil is dissipated. When made in a porcelain vessel on a water-bath, it is of a yellowish green colour, efficient, and active, and will keep good for a long time. It ought not to be made in a copper vessel. P. J.

II. MONOCOTYLEDONES *vel* ENDOGENÆ.

Monocotyledons are so called from having a single cotyledon, or, if more than one, arranged alternately. They are also called Endogens in consequence of growing by additions to their centre. They are sometimes divided into groups, according as they have the ovary free or superior, as in the families from *Palmæ* to *Smilaceæ;* or adherent to the perianth, or inferior, as in the families from *Orchideæ* to *Irideæ;* and into the Glumales, or those with a chaffy perianth.

Palmæ, *Juss.* Palms.

Stem simple, rarely forked. Leaves terminal, very large, pinnate or flabelliform, plaited in vernation. Spadix inclosed in a valved spathe. Flowers small, hermaphrodite, or polygamous. Perianth 6-parted, persistent. Stamens inserted into the base of the perianth, definite or indefinite. Ovary 3-celled, or deeply 3-lobed, with an erect ovule. Fruit baccate or drupaceous. Albumen cartilaginous or fleshy. Embryo in a cavity at a distance from the hilum. (*Lindley.*)

Palms by Linnæus were styled the Princes of the vegetable kingdom. It has been said that Flour, Sugar, Oil, Wax, and Wine are all yielded by the family of Palms, as well as thread, utensils, timber for habitations, and leaves for thatching.—They inhabit chiefly the tropical parts of the world, though a few extend to higher latitudes.

Phœnix dactylifera is the Date tree, affording Dates as the chief article of diet to thousands of Arabs and Africans, and which are imported here as an article for the dessert.

Phœnix sylvestris in India and *Arenga saccharifera* in Java yield large quantities of Sugar by the simple evaporation of their sap.

Cocos nucifera, the Cocoa Nut Palm, one of the most useful trees, is valuable on account of the kernel of its fruit, (mentioned by Avicenna as the Indian Nut,) which is not only edible, but yields large quantities of Oil.

Elæis guineensis and *melanococca* yield the immense quantities of Palm Oil which is imported from the west coast of Africa. The Oil is obtained by bruising the fleshy part of the fruit. It has a solid consistence, is of an orange-yellow colour, and rather a grateful odour. It consists of about 30 parts of Stearine and 70 of Elaine, besides the principles which give it odour and colour. It is emollient, and sometimes used as an embrocation to spasms and bruises; but its chief employment is for making soap. The E. C. in their last Pharmacopœia erroneously indicated the *Cocos butyracea*, a South American Palm, as the source of Palm Oil.

Areca Catechu, *Linn.* E. The Catechu or Betle Nut Palm, *foful* of the Arabs, receives its specific name from a kind of Catechu being made from its seeds, which, being boiled down, yield an astringent Extract, from the large quantity of Tannin which they contain. On this account the tree is mentioned in the E. P. as one of the sources of Catechu; but it does not seem probable that any of this Catechu reaches Europe (*v.* p. 349), for Catechu is one of the regular imports into the ports of the Madras Presidency. The seeds or nuts are, however, always in demand, and are an extensive article of commerce, because they form one of the ingredients of the celebrated masticatory of the East, called Pan or Betle. (*v.* p. 559.) They are sometimes sold cut into transverse slices.

Calamus Draco, Willd., is one of the species which yields the reddish resinous substance known as Dragon's-blood (the *dum-al-akhwain* of the Arabs). It is only used for colouring.

Ceroxylon, now *Iriartea andicola*, yields wax, as does another Palm, called Carnauba in Brazil.

Sagus Rumphii, *Willd.* L. (the *Sagus spinosus* of Rumph. Herb. Amb. i. p. 75) is made officinal in the L. P. as the plant yielding

Sago, which is described as *Medullæ Fæcula;* but Endogens have no proper pith. In the E. P. it is said to be the Farina from the interior of the trunk of various Palms and species of Cycas.

The Malays are said to prefer the Sago of this tree (*Sagus Rumphii*). Dr. Roxburgh states that the granulated Sago is made from *Sagus* (*inermis*, Roxb.) *lævis*, Jack. Blume states it to be produced by *Sagus lævis* and *S. genuina.* The *Saguerus saccharifer* also, (called likewise *S. Rumphii*, and *Arenga saccharifera*,) when exhausted of its saccharine sap, yields Sago of good quality. *Caryota urens* and *Phœnix farinifera* also yield a Sago-like farina in India. It is, therefore, difficult to select one as the sole officinal species. Sago is also said to be produced by species of Cycas, but the statement has never been properly confirmed. Some *Zamias* are said to yield one kind of Arrow-root in the West Indies.

Sago is imported chiefly from the Moluccas and from Singapore; but it is first imported into the latter place in very large quantities for granulation and re-exportation. It is first obtained by cutting and splitting the stem, washing and stirring it up with water, when the fecula becomes suspended, and is passed through a sieve. The fecula then subsides, and forms a powder like Arrow-root, but of a dirty-white colour, or *Meal Sago*, of which 500 or 600 lbs. may be yielded by a single tree. It is in this state imported in large quantities into Singapore, as has been related by Mr. Crawford and Mr. Bennett. It seems to be first made into a paste with water, and then, granulating in drying, is rubbed into Sago of different sizes; but the processes are not well ascertained. Common Sago was formerly of the size of Coriander Seeds, but unequal, and of a reddish or brownish-white colour, and mixed with some of the meal. It is formed of the Starch-globules of the meal aggregated together, but unbroken. The Chinese settled at Singapore, some time since, introduced methods of refining and producing it in much smaller grains, which was first distributed by Mr. W. Johnson. This is now called Pearl Sago, and is in small grains about the size of a pin's head, hard, whitish, with a pearly lustre, sometimes even translucent, without odour, and with very little taste. This kind is sometimes bleached, and has been subjected to heat, as the starch-globules are ruptured, so that even in cold water Iodine produces a blue colour with this Sago.

M. Planche has arranged the Sagos of commerce under six different heads, according to the places whence they are procured. M. Guibourt arranges them all under three heads. 1. Uncut Sago. 2. Seed Sago. 3. Tapioca Sago, which is the same as Pearl Sago. The unbroken grains of Sago are more or less of an ovoid form, but most appear as if truncated or muller-shaped. An imitation of Sago is made with Potato-Starch near Paris.

Sago is insoluble in cold water, but by long boiling becomes soft, and then transparent, and ultimately forms a gelatinous solution, which in all essentials corresponds with Starch.

Action. Uses. Sago is Nutrient and Demulcent, and well suited for invalids.

LILIACEÆ, *Dec.* Lilyworts.

Plants with fibrous or fasciculate roots. Stem either wanting or formed of sheaths of leaves (a bulb), or tuberous, creeping, or arborescent; simple or branched at the apex, leafy or leafless, scape-like. Leaves simple, very entire, sheathing at the base or sessile, often linear, with parallel veins. Flowers complete, usually terminal, with scarious or spathe-like bracts, regular, occasionally sub-irregular. Perianth inferior, petaloid, 6-leafed, in 2 rows, distinct or united below into a tube, and 6-fid above. Stamens 6, inserted into the receptacle or into base of perianth, seldom 3; anthers opening inwards. Ovary 3-celled; ovules in 2 rows, affixed to the central angle. Stigmas 3-sided or 3-lobed. Fruit 3-celled, sometimes a berry. Embryo straight, within a fleshy albumen.

The Liliaceæ as at present constituted include several groups, which are often treated of as distinct families, as *Tulipeæ*, including *Lilium candidum*, at one time officinal chiefly as a demulcent; *Hemerocallideæ; Asphodeleæ*, including Allium, Scilla; *Aloineæ*, the Aloe plants, *Asparageæ*, and others (*v.* Lindl. Veg. King. p. 200.)—The true Lilies are allied to Palms, also to *Melanthaceæ*, and to *Amaryllideæ*. They are widely diffused, but are chiefly found in temperate climates.

ALLIUM, *Linn.* *Hexandria Monogynia*, Linn.

Flowers umbellate, with a spathe of 1 or 2 leaves. Perianth 6-parted, spreading, or campanulate. Stamens 6, inserted into the base of the perianth; filaments simple or tricuspidate; anthers incumbent. Ovary 3-celled. Style filiform. Stigma simple. Capsule commonly bluntly 3-cornered or 3-lobed, depressed at the apex, 3-celled, bursting into 3 valves through the dissepiments. Seeds solitary or 2 in each cell, affixed to the base, black, compressed, angular. Embryo falcate, placed out of the centre.

ALLIUM SATIVUM, *Linn.* Bulbus, L. D. Bulb, E. of the Common Garlic.

Garlic is the *σκοροδον* of the Greeks, *Som* of the Arabs, and *Shumim* of Numbers, xi. 12. It has been used as an article of diet and likewise in medicine from very early times.

Bulbs clustered, several enveloped in the same silvery skin. Stem about 2 feet high, leafy below the middle. Leaves glaucous, distichous, channelled above, obscurely keeled, acute. Spathe single, calyptriform, horned. Umbels bulbiferous. Flowers, if any, pink, red, or whitish, rather larger than the stamens. (*Lind.*)—Cultivated from early times in the East, and now in all gardens. Its native country unknown.—St. and Ch. 111.

ALLIUM CEPA, *Linn.* Bulbus, D. The Bulb of the Common Onion.

The Onion is the *κρομμυον* of the Greeks, *Busl* of the Arabs, an the *Betzulim* of Numbers, xi. 12.

Biennial. Bulb simple, roundish, invested with shining, thin, dry membrane. Stem 1—3 feet high, fistular, leafy, and often ventricose at the base. Leaves glaucous, distichous, fistular, terete-acute, shorter than the stem. Spathe reflexed. Umbels large, regular, globose, many-flowered, not bulbiferous. Flowers whitish or greenish, the segments with a green keel, linear-elliptic, obtuse, shorter than the stamens.—Cultivated in gardens in Europe; flowers in July. Probably a native of the Persian region.—Fl. Græca, t. 326.

PORRUM, L. Bulbus, L. Bulb of ALLIUM PORRUM, *Linn.* The Leek.

The Leek is the πρασον κηπαιον of the Greeks, *Korros* of the Arabs, and the *Chatzir* of Numbers, xi. 12.

Bulb oblong, simple. Stem 2—4 feet, round, leafy below the middle. Leaves subopposite, keeled, linear-lanceolate, glaucous, 1 to 2 feet long, 1 to 2 inches broad at the base, tapering and acute at the apex, with the keels and margins subserrulate. Spathe single, calyptriform, horned. Umbel large, globose. Pedicels filiform, 1—2 inches long. Perianth slightly purplish, or greenish-white, half-spreading; segments oblong; keel green. Stamens a little longer than the perianth, the alternate ones tricuspidate; the lateral appendages filamentous.—Cultivated in the East from the earliest ages, and introduced into Europe.—Blackwell, Ic. 421.

The Garlic, Onion, and Leek are so well known as articles of diet, or rather as condiments, that it is unnecessary to dwell long on their properties, which indeed are so similar that it does not seem necessary that more than one of them should be retained in the Pharmacopœias. The Leek has been considered Diuretic and Expectorant: as an article of diet, it is rather indigestible, and apt to cause flatulence. The Onion is possessed of similar properties, but is somewhat more powerful, though less so than the Garlic. In its roasted or boiled state it is occasionally applied as a poultice to suppurating tumours or to sluggish ulcers. The Garlic, when applied externally, will act as a Rubefacient; taken internally, it acts as a Stimulant, and being absorbed, as an Expectorant and Diuretic, producing at the same time a disagreeable odour in the breath. It is sometimes employed as an Anthelmintic in Ascarides.

SCILLA, L. E. D. Bulbus recens, L. D. Bulb, E. of SQUILLA MARITIMA, *Steinhill*, E. (*Scilla maritima*, Linn.), L. D. The officinal Squill.

The Squill (Σκιλλα) was employed by the ancient Greeks. The Asiatics substitute for it an allied species, *Squilla indica*, to which they apply the name of *iskeel*. From the difference in character of this from the ordinary species of Scilla, M. Steinhill formed it into the genus Squilla.

Bulb roundish-ovate, very large, half above ground; integuments greenish or reddish. Leaves all radical, appearing after the flowers, spreading, large, rather fleshy, broad, lanceolate, channelled, recurved. Scape from 2 to 4 feet high, rising from the centre of the leaves, simple, cylindrical, terminated by a long, dense, ovate raceme of flowers, with long bracts. Flowers of a pale yellowish-green colour. Sepals 3, coloured, spreading. Petals very like them, and scarcely broader. Stamens 6, shorter than the perianth; filaments smooth, somewhat dilated at the base, acuminate, entire; anthers yellow. Ovary 3-parted, with 3 nectariferous glands at the apex. Style smooth, simple. Stigma obscurely 3-lobed, papillose. Capsule rounded, 3-cornered, 3-celled. Seeds numerous, in 2 rows, flattened, winged, with a membranous testa. (*Lindley.*)—Native of both the north and of the south sides of the Mediterranean and of the Levant. Flowers about August.—B. M. t. 918. Nees von E. 55. St. and Ch. 153.

The Squill bulb is formed of scales applied over each other, of which the external ones are dry, membranous, and often coloured; the inner are colourless, thicker, and full of acrid viscid juice. It is sometimes imported in its entire state, packed in sand; the average weight is from half a pound to four, but occasionally ten pounds in weight, and as large as a child's head. But being in this state very retentive of moisture and of life, it is directed in the L. P. to be, like Colchicum, preserved in dry sand, and before drying, the dry outer coats are to be removed, and the others cut into transverse narrow strips, and dried with moderate heat. Squill is, however, usually imported in this state, the pieces being white or yellowish-white, often contorted, translucent, of a mucilaginous and bitter taste, at first rather tough, but when quite dry, brittle enough to be powdered. Many spiral vessels may be detected in these scales, as well as numerous acicular raphides, the *Pulvis Scillæ* containing 9 or 10 per cent. of these crystals. (*p.*) Analysed, Squill contains of moisture $\frac{4}{5}$ths, Gum, uncrystallizable Sugar, traces of Tannin, Phosphate of Lime, Lignin, and a bitter resinous Extractive, from which a peculiar principle called *Scillitine* has been extracted. This is crystalline, capable of neutralizing alkalis, moderately soluble in Alcohol, but insoluble in water; bitter, but not acrid. The properties apparently depend upon this, perhaps partly upon an acrid volatile matter. They may be extracted by Alcohol, Spirits, and by Vinegar.

Action. Uses. Irritant Poison. In small doses, Expectorant, Diuretic; and in larger, Emetic and Cathartic. Used as an Expectorant in chronic Catarrh; as a Diuretic in general Dropsy, especially when there is a deficiency of tone. Its effects may be increased by combination with other Diuretics and Expectorants.

PULVIS SCILLÆ, D. Squill Powder.

Prep. Take off the membranous integument from the bulb of the *Squill*, cut it into transverse slices, and dry it with a moderate heat, powder it, and keep it in stoppered glass phials.

Action. Uses. Expectorant and Diuretic in doses of gr. j.—gr. iij. Emetic gr. x.—gr. xv.

PILULÆ SCILLÆ (E.) COMPOSITÆ, L. D. Compound Squill Pill.

Prep. Mix *fresh dried Squill* in powder ℨj. (5 parts, E.) with powdered *Ginger* (ℨiij. D.) and *Ammoniacum* (ℨij. D.) āā ℨij. (āā 4 parts, E.) Then rub up with *Soap* ℨiij. (4 parts, and *Conserve of Red Roses* 2 parts, beat into a uniform mass and divide into 5 gr. pills, E.) Add *Syrup* (Molasses, D.) q. s. to obtain the proper consistence, L. D.

Action. Uses. Expectorant in doses of gr. v.—gr. xv.

ACETUM SCILLÆ, L. E. D. Vinegar of Squill.

Prep. Macerate fresh dried and cut *Squill* ℥xv. (℥v. E. ℔ß. D.) in distilled *Vinegar* Ovj. (Oij. E. by measure ℔iij. D.) in a covered glass vessel with the aid of gentle heat (continually agitating, D.) for 24 hours (7 days, E. D.) Express strongly, strain, to separate the dregs, and add *Proof Spirit* Oß. (f℥iij. E.) (*Recti-*

fied Spirit by measure ℥iv. D.) This might, like the Tincture, be prepared by percolation.

Action. Uses. Expectorant and Diuretic in doses of f℥ß.—f℥j.

Oxymel (Syrupus, E.) Scillæ, L. D. Oxymel of Squill.

Prep. Take *Honey* (Sugar ℔vij. E.) ℔iij. (by weight, D.) and *Vinegar of Squill* Ojß. (Oiij. E. by measure ℔ij. D.) Boil in a glass vessel, over a slow fire (agitating, E.), to the proper consistence.

Action. Uses. These preparations, though differently named, are very similar to each other, and the Honey of the one has probably little advantage over the Sugar of the other. Expectorant in doses of f℥j.—f℥ij. Sometimes used as an Emetic for children in doses of f℥j.

Tinctura Scillæ, L. E. D. Tincture of Squill.

Prep. Macerate for 14, L. (7, D.) days dried slices of the fresh bulb of *Scilla maritima* ℥v. (℥iv. D.) in *Proof Spirit* Oij. (by measure ℔ij. D.) Strain. (Or prepare by percolation, but without packing the pulp firmly in the percolator, E.)

Action. Uses. Expectorant and Diuretic in doses of ♏x.—f℥ß.

Aloe, L. Inspissated Juice of the Leaves of Aloe spicata, *Dec.* L. Aloe socotorina, E. D., of Aloe spicata, *Persoon*, D. Aloe Barbadensis, E. Aloe (indica, E.) hepatica, D. Inspissated Juice of *A. vulgaris*, Dec. D. The E. C. refers these different Aloes to one or more undetermined species of Aloe.

The word Aloe, in our translation of the Bible, is confounded with *ahila*, or Eagle-wood. (*v.* Ahalim, *Bibl. Cycl.*) Aloes were known to Dioscorides, to Galen, and to Celsus. The Arabs describe three kinds, Socotrine, Arabic, and Semegenic. The Indo-Persian writers give *Sibr*, or *Sibbur*, as its Arabic, and *bol-seah* (black Myrrh) as its Persian name. The Hindoo *elwa* is very similar to the word *Aloe*, and the Greek *fekra* which the Arabs quote seems to be derived from πικρος.

It is still imported from Socotra (*Wellsted*) and from the east coast of Africa, as Melinda, &c. Dr. Malcolmson writes from Aden, that, "besides the island of Socotra, it is produced in almost every part of Yemen. Small quantities are also brought from Abyssinia. That supplied by Arabia is principally exported from Macculla. It is much superior to the other Arabian Aloes, but greatly inferior to that of Socotra." Some is prepared in India. Dr. Falconer has ascertained, as the author had also heard, that it is manufactured from the Aloe indica, in the country between the Jumna and Sutlej. A little is produced in Guzerat; but it is largely imported into Bombay. The late Dr. Malcolmson informed the author that in Madras they obtained their Aloes from Bombay, and found an extract of it of excellent quality for hospital use, and inferior kinds from Salem and Trichinopoly, both to the southward of Madras. Aloes are, moreover, obtained from the West Indies and from the Cape of Good Hope.

ALOE, *Linn.* Aloe.

Succulent plants, with spiked inflorescence. Perianth tubular, 6-cleft, sometimes so deeply divided as to appear 6-petaled, converging below into a tube, with the limb regular, spreading, or recurved, somewhat fleshy, nectariferous at the base; segments ligulate, the interior equal to or larger than the exterior, and imbricate. Stamens hypogynous, ascending, as long as the tube, or projecting beyond it. Style as long or almost wanting, 3-furrowed. Stigma simple or triple, minute, and replicate. Capsule membranous, scarious, obtusely or acutely triangular, 3-celled, 3-valved; valves bearing the septa in the middle. Seeds numerous, in two rows, roundish, flattened, or 3-cornered, winged or angled.

Aloes is the bitter proper juice inspissated of several species of Aloe. It is contained in the vessels lying under the epidermis of their fleshy leaves, the interior being filled with much watery colourless sap. In many cases the leaves are cut into pieces and boiled with water, so as to form an extract rather than an inspissated juice; but both kinds are sold under the same name. That of commerce being derived from a variety of sources, is necessarily produced by different species. The Asiatic species, with those of Abyssinia and Socotra, are enumerated in the author's *Illustr. of Himal. Bot.* p. 389. The following may be referred to, until more accurate information is available respecting the African and Arabian species.

ALOE RUBESCENS, *Dec.* Stem suffruticose. Leaves amplexicaul, spreading, thorny at the margin. Peduncle compressed, branched. Branches sub-bracteate.—*Pl. grass.* t. 15. A native of Arabia.

A. BARBADENSIS, *Mill.* Stem somewhat shrubby, offshoots from the root. Leaves sword-shaped, sinuato-serrate. Corol yellow. Var. of *A. vulgaris*, Nees von E. 50.—South of Europe, perhaps in the Peninsula of India. (Rheede, ii. t. 3.) Introduced into the West Indies.

A. ABYSSINICA, *Lam.* Subcaulescent. Leaves long and lanceolate, rather erect, hard, of a deep green colour, rather concave above; margin sinuato-dentate, reddish, flowers of a greenish yellow. Var. of *A. vulgaris*, Linn.—Abyssinia.

A. SOCOTRINA, *Lam.* Stem shrubby, thick, dichotomous. Leaves ensiform, greenish, incurved towards the apex, with the marginal serratures small, white, and numerous. Flowers scarlet at the base, pale in the middle, green at the point.—Nees von E. 50. St. and Ch. 110.

A. ARABICA, *Lam.* *A. variegata*, Forsk. With spotted leaves.

A. SPICATA, *Thunb.* Caulescent. Leaves flat, ensiform, dentate. Flowers spiked, campanulate, horizontal.—Interior of the Cape of Good Hope; probably yields some Cape Aloes.

"Ex hujus succo optima gummi resina aloes paratur: ex reliquis speciebus vilior." *Linn. fil.* But, as Dr. Pereira has been informed, all species of Aloe at the Cape are indiscriminately used.

Dr. Christison remarks it as probable that *A. linguæformis* of Thunberg, and *A. Commelini* of Willdenow likewise yield some Aloes.

ALOE INDICA, *Royle.* A low plant, with spikes of red flowers, which grows in dry barren places in N. W. India. This, if known to Roxburgh, was probably included by him in *A. perfoliata.* Col. Sykes has a species from the Deccan also with red flowers.

The Aloes in common use are the Socotrine, Hepatic, (under which term seems to be included the Aloes exported from Bombay, and which is no doubt the produce of Arabia and of Africa;) also Barbadoes Aloes, and that of the Cape of Good Hope.

Socotrine Aloes is distinguished by being of a redder colour when compared with the other kinds. This in the E. P. is described as "being of a garnet-red, in thin and translucent pieces;" when quite dry, is of a golden-red (*p.*); but by exposure the colour is changed, and this Aloes becomes of a brownish-red. The fracture is conchoidal, usually smooth and shining, but sometimes a little rough. The odour is rather fragrant, especially when the Aloes is fresh and heated. It is easily reduced to a golden-yellow powder, and is nearly all soluble in Spirit of the Sp. Gr. 0·950. There can be no doubt that some of this Aloes is produced in the island of Socotra, as Lieut. Wellsted describes the plants as growing in parched and barren places at 500 to 3000 feet above the sea, and the leaves as being plucked and the juice allowed to exude into a skin. Specimens were procured for the author by a medical friend at Socotra, but unfortunately after his ship had been supplied, when two adulterated skins (those mentioned by Dr. Pereira) were supplied as genuine Socotrine Aloes. About two tons are exported from Socotra. Some used to be carried up the Red Sea, and thence into the Mediterranean. The semifluid pieces are dried in thin layers, and the inferior portions are strained in this country.

Hepatic Aloes, so called from its usual liver-brown colour, is referred to in the E. P. as *Aloe indica*, but is certainly not produced in that country. This is evident from the specimens, all inferior in quality, in the author's collection, and those procured at his request by the late Dr. Malcolmson, and noticed by Dr. Pereira at p. 971 of his *Elements.* The Hepatic, therefore, is the Aloes imported into Bombay from Arabia and Africa, and which is known in India by the name of Bombay Aloes. Some of it is probably obtained from the same sources as the Socotrine, which it resembles in odour; and, as Dr. Pereira states, "the two are sometimes brought over intermixed, the Socotrine occasionally forming a vein in a cask of the Hepatic Aloes." It is of a liver-brown colour, has a dull, somewhat waxy fracture, is less fragrant. The taste is nauseous and intensely bitter; the powder of a golden-yellow colour. This kind is inferior in quality to the fine Socotrine.

Barbadoes Aloes is prepared in the West Indies, chiefly from the variety of *A. vulgaris*, but also probably from *A. socotrina* and *A. purpurascens*, which are said to be cultivated there. Browne, in his *Nat. Hist. of Jamaica*, states that the largest and most succulent leaves are placed upright in tubs, that the juice may dribble out. This, evaporated, forms what is sold as Socotrine Aloes; but the common Aloes is obtained by expressing the juice out of the leaves, boiling it with water, evaporating and pouring into gourds, whence this kind is often called Gourd Aloes. It sells for a high price, as being much in demand for veterinary medicine, and may be distinguished by its dark brown or blackish, sometimes liver-brown colour; by its usually dull appearance, and its disagreeable odour, especially when breathed upon. Being more gummy, it is tougher, and difficult to pulverize.

Cape Aloes, procured no doubt from a variety of species of Aloe, is especially distinguished by its vitreous lustre, and is hence called *Aloe lucida* by some authors. The finer qualities are of a deep brown colour externally, with a tinge of olive-green; thin laminæ are translucent, with something of a yellowish-red colour. It is very brittle, easily pulverized, its odour strong, rather disagreeable, and its powder of a yellow colour. Some of the inferior kinds are black in colour, vesicular, and with a rough fracture.

All the kinds of Aloes have an extremely disagreeable taste, which is very permanent in the fauces. The odour is peculiar, and is more perceptible when the Aloes is breathed upon. A great portion of Aloes is dissolved by water, and much of what at first appears insoluble is dissolved by boiling water, but is again deposited on the water cooling. This is what some chemists call a resin, but it is by Braconnot considered extractive matter which has become oxidized. Diluted Spirit is the best solvent of Aloes. Aloes was at one time considered to be a Gum-Resin. But the portion which was thought to be of the nature of Gum is now considered to be a variety of Extractive, yet sufficiently peculiar to have the name of *Aloisin* applied. Its proportion varies from 50 to 80 per cent. in different Aloes. Aloisin is soluble in water and in weak Spirit, and may be obtained from the former solution in thin translucent layers, which are of a reddish-brown colour, but of a fine yellow when powdered, very bitter, and extremely active as a Cathartic. Besides this, there is an oxygenated Extractive matter, of which the quantity is increased when Aloes are boiled. The quantity of this varies from 6 to 42 per cent., and in some kinds there is Vegetable Albumen, as might be expected where the juices of leaves are pressed out and then boiled. The presence of an acid in a solution of Aloes is also indicated by Litmus-paper. Trommsdorf considered this to be the Gallic, but Dr. Pereira has named it Aloetic acid. But Aloes requires a fresh and carefully conducted analysis.

Action. Uses. Aloes, in small doses, is Tonic; in larger, Cathartic. It is considered by some to stimulate the liver, and also to supply the place of the deficient bile in torpidity of the intestinal canal. Its action seems directed to the larger intestines, especially the rectum, and thus to be useful in evacuating them, but detrimental when there is already irritation or Hæmorrhoids. By this determination it no doubt stimulates neighbouring organs, and is thus useful as an Emmenagogue, and is hence often ranged with them.

EXTRACTUM ALOES PURIF. L. Extr. Aloes Hepaticæ, D. Extract of Aloes.

Prep. Macerate with a gentle heat for 3 days *bruised Aloes* ℥xv. (Hepatic 8 parts, D.) in boiling *Aq.* Cj. (8 parts, D.) Strain, and set aside for the dregs to subside. Pour off the clear liquor, and evaporate to a proper consistence.

Action. Uses. Cathartic, in doses of gr. v.—gr. xv. Useful preparation, as freed from all mechanical impurities.

PULVIS ALOES CUM CANELLA, D. Aloes and Canella Powder.

Prep. Rub separately into powder *Hepatic Aloes* ℔j., *White Canella* ℥iij., and mix.

Action. Uses. Cathartic and Stomachic in doses of gr. x.—℈j.

PULVIS ALOES COMPOSITUS, L. D. Compound Aloes Powder.

Prep. Rub separately into powder *Aloes* (Hepatic, D.) ℥jß. *Resin Guaiacum* ℥j. and add *Comp. Cinnamon* (Aromatic, D.) *Powder* ℥ß.; mix.

Action. Uses. Warm Cathartic and Diaphoretic in doses of gr. x. —℈j.

PILULÆ ALOES (E.) COMPOSITÆ, L. D. Compound Aloes Pill.

Prep. Beat into a pill mass bruised *Aloes* (Socotrine and Castile Soap equal parts, E. Hepatic, D.) ℥j. L. D., *Extract Gentian* ℥ß. L. D. *Oil of Caraway* ♏xl. L. D. *Syrup.* q. s. L. D. (Conserve of Red Roses q. s. E.)

Action. Uses. Cathartic and Tonic, in doses of gr. x.—ʒß. The bitter and the Soap are both thought to promote the action of Aloes.

PILULÆ ALOES (CUM MYRRHA, L. D.) ET MYRRHÆ, E. Aloes and Myrrh Pills.

Prep. Rub separately into powder *Aloes* (Socotrine or E. I. 4 parts, E. Hepatic, D.) ℥ij. and *Myrrh* ℥j. (2 parts, E.), then rub them together till incorporated with *Saffron* ℥j. (1 part, E.), *Syrup* q. s. L. D. (Conserve of Red Roses, E.)

Action. Uses. Cathartic and Emmenagogue in doses of gr. x.—℈j. every night.

PILULÆ ALOES ET ASSAFŒTIDÆ, E. Aloes and Assafœtida Pills.

Prep. Beat into a proper pill mass Socotrine or E. Indian *Aloes*, *Assafœtida*, and *Castile Soap* āā. equal parts with *Conserve of Red Roses*, q. s.

Action. Uses. Cathartic and Antispasmodic in doses of gr. x.—gr. xv. thrice a day.

DECOCT. ALOES, (E.) COMP. L. D. Compound Decoction of Aloes.

Prep. Boil *Extract of Liquorice* ʒvij. (℥ß. E. D.) *Carb. Potash* ʒj. (℈ij. E. D.) bruised (Socotrine, E. or Hepatic, E. D.) *Aloes*, powdered *Myrrh* and *Saffron* āā ʒß. (ʒj. E. D.) in *Aq. dest.* Ojß. (f℥. xvj. E. by measure ℔j. D.) till only Oj. (℥xij. E. D.) remains. Filter, and add *Compound Tincture of Cardamoms* f℥vij. (f℥iv. E. D.)

Action. Uses. Cathartic; Emmenagogue in doses of f℥ß.—f℥ij. The boiling must not be carried to any extent, as some of the Aloes becomes insoluble.

VINUM ALOES, L. E. D. Wine of Aloes.

Prep. Take (separately, D.) powdered *Aloes* (Socotrine, E. D. or E. Indian, E.) ℥ij. (℥jß. E. ℥iv. D.) and *Canella* ℥iv. L. (℥j. D. Grind Cardamom Seeds and Ginger āā ʒjß. E.), pour upon them *Sherry Wine* Oij. (by measure ℔iij. mixed with Proof Spirit by measure ℔j. D.) Macerate for 14 (7, E.) days, continually agitating; then strain (through linen or calico, E.)

Action. Uses. Warm Cathartic in doses of f℥ß.—f℥j.

TINCTURA ALOES, L. E. D. Tincture of Aloes.

Prep. Macerate for 14 (7, E) days bruised (Socotrine, E. D. or Indian, E.) *Aloes* ℥j. (℥ß. D.), *Extract of Liquorice* ℥iij. (℥jß. dissolved in boiling Aq. ℥viij. D.) in *Aq. dest.* Ojß. L. (Oj. and f℥viij. E.) and *Rectified* (Proof, D.) *Spirit* Oß. (f℥xij. E. by measure ℥viij. D.) (Agitate occasionally, E.) Strain. (Not conveniently prepared by percolation, E.)

Action. Uses. Cathartic adjunct to Purgative or Emmenagogue draughts in doses of fʒß.—fʒij. The weak Spirit is an excellent solvent of the active properties.

TINCTURA ALOES (COMPOSITA, L. D.) ET MYRRHÆ, E.

Prep. Macerate for 14 (7, E.) days bruised (Socotrine E. D. or Indian, E.) *Aloes* ℥iv. (℥iij. D.), *Saffron*, L. E. ℥ij. in *Tincture of Myrrh* Oij. (by measure ℔ij.) Strain. (Not well prepared by percolation, E.)

Action. Uses. Emmenagogue. Stimulant Cathartic and adjunct to draughts and mixtures in doses of fʒß.—fʒij.

PILULÆ ALOES ET FERRI, E. Aloes and Iron Pill.

Prep. Pulverize separately *Barbadoes Aloes* 2, and *Sulphate of Iron* 3 parts; add *Aromatic Powder* 6, and *Conserve of Red Roses* 8 parts. Mix and beat into a proper mass; divide into 5 grain pills.

Action. Uses. Chalybeate Tonic and Laxative in doses of gr. x.—gr. xv. The conjunction of the Iron with the Aloes assists the action of the latter, besides being itself useful in some cases of Amenorrhœa.

Aloes is an ingredient of the Pilulæ Cambogiæ. Pil. Cambogiæ Comp. Extr. Colocynthidis Comp. Pil. Colocynthidis et Hyoscyami. Pil. Sagapeni Comp. Tinct. Benzoini Comp. Pil. Rhei Comp.; and Tinct. Rhei et Aloes.

MELANTHACEÆ. *R. Brown.* (*Colchicaceæ.*)

Herbaceous plants. Root fibrous or fasciculate. Rhizoma sometimes a fleshy corm, sometimes creeping horizontally. Stem simple, seldom branched. Flowers complete, or from abortion unisexual, often radical. Perianth free, corol-like, 6-leaved, all distinct, or united at the base into a tube. Stamens 6, inserted into the receptacle, or into the perianth. Anthers opening outwards. Ovary 3-celled. Style 3-parted (fig. 3). Ovules affixed to the inner angle of the cells. Capsule divisible into 3. Seeds with a membranous testa. Embryo straight within a fleshy or cartilaginous albumen. (*R. Brown* chiefly.) The Melanthaceæ are allied to Junceæ and to Liliaceæ, are found in temperate parts of the world, and are remarkable for secreting Veratria.

COLCHICI CORMUS, L. E. (Bulbus, D.) COLCHICI SEMINA, L. E. D. Cormus and Seeds of COLCHICUM AUTUMNALE, *Linn.* Colchicum, or Meadow Saffron.

Colchicum is well described by Dioscorides. It was used by the Arabs, and is their *sorinjan;* they give *kuljikoon* as its Greek name. The Hermodactyls (Sweet and Bitter *sorinjan* of the Arabs) of the later Greeks and Arabs were no doubt species of this genus. The author's specimens have been described by Dr. Pereira.

The true root is fibrous, and below the underground stem or cormus (the bulbo-tuber of some authors), which is ovate, about the size of a chesnut, solid, fleshy, enveloped in a brown-coloured tegument. It is rather convex on one side, flattened on the other, or with a longitudinal furrow made by the growing plant. It is largest and in full perfection in June or early in July, when a new but minute corm is seen at its lower end close to the radicles. This new corm flowers in autumn, is then small, but enlarges before spring, when the young seed-vessel rises with the leaves, having remained underground from the time of flowering in autumn. The seeds ripen about midsummer. The parent corm becomes more spongy and watery as the new flower rises, but retains its size and form till next April, the second spring of its own existence. But by the end of May it has become shrivelled and leathery, and is attached to the lower part of its progeny, now the perfect and full-sized corm. In Scotland, where the seeds do not ripen, the plant propagates itself by little corms being thrown off from the large corm during the second or last spring of its existence. The leaves are broadly lanceolate, flat, somewhat keeled, about a foot in length, dark green, smooth, and appearing in the spring with the capsules. Flowers several, leafless, rising from the corm with a long white tube; limb of a pale purple or rose-colour. Perianth funnel-shaped, with a very long tube; limb 6-parted, petaloid. Stamens 6, inserted into the throat of the perianth. Capsules 3, connected throughout, 1-celled (fig. 1 and 4), opening at the inner edge (2), many-seeded. Seeds (fig. 5) roundish, with a brown shrivelled skin, and large strophiole, which gives them a rough appearance.—Native of moist meadows throughout Europe.—Nees von E. 49. St. and Ch. 70.

Fig. 91.

Dr. Christison has given the fullest and best account of the growth of the corms of the Colchicum, which the author has above very much abridged. Its activity is considered to be greatest in July and August, that is, when the leaves have withered and the flowers of the new corm have not appeared. Dr. Lindley says he has seen many *cwt.* sent to town of those which had flowered; and the flowers broken off, so as to prevent the circumstance from being observed.

Dr. C. observes that the corm is whitest, firmest, and largest at the end of June and beginning of July, and then abounds in Starch, and that there is no other corm connected with it. But if taken up in April, two are found united, one spongy, the other plump and firm. Though they are generally collected when the corm is single, large, and plump, Dr. C. doubts whether this is essential, as the corm in April, though more watery, is at least as bitter; and he quotes Stolze as showing that in autumn the corm contains of water 80 per cent., Starch 10, Bitter Extract 2, and Sugar 4, with a little Gum, Resin, and Lignin; but in April the proportion of Bitter Extract is greater. In the L. P. it is directed to be kept in sand, and to have its brownish integuments removed. It is then to be cut into transverse slices, and dried with a moderate temperature. The slices should be dry, firm in texture, and of a greyish-white colour, and readily change their colour to blue when moistened with vinegar and touched with Tincture of Guaiacum, as pointed out by Dr. A. T. Thomson. The taste is disagreeable, bitter, and somewhat acrid. The seeds have the same properties, and, if collected ripe, are necessarily more uniform.

According to Pelletier and Caventou, the corm contains the alkali Veratria united with an excess of Gallic acid, Fatty matter united with a volatile acid, Yellow Colouring matter, Gum, Starch, Inulin in abundance, and Lignin. Geiger and Hesse announced the presence of a peculiar alkali, *Colchicina* and *Colchiceà;* but it is probable this may be only Veratria in a modified form. The active principles are partially taken up by water, but readily by Alcohol, diluted Spirit, and Vinegar, which are used as the officinal solvents.

Action. Uses. Colchicum is Irritant, and in large doses a Narcotic Acrid Poison. In small doses repeated, it stimulates some of the secretions, and acts as a Nauseant and Cathartic, Diuretic and Diaphoretic, producing at the same time Sedative and Anodyne effects; thus relieving the pain of Gout and of Rheumatism, and controuling the action of the heart in Inflammatory diseases. It is doubtful whether its full effect as a medicine is experienced, unless some of the inconveniences attending its use are felt, as slight colic and diarrhœa, or headache and giddiness, and then the doses should be diminished.

D. The corm (bulb) or the powdered seeds may be given in doses of gr. j.—gr. v. 3 times a day, with aromatics and sometimes a little opiate. Mr. Wigan recommends it in 8 grain doses every hour, until either vomiting, purging, or sweating are experienced.

EXTRACTUM COLCHICI CORMI, D. Extract of Colchicum.

Prepared from the fresh corms of the Colchicum, like Extr. Aconite.

Action. Uses. Efficient preparation in doses of gr. j. every three or four hours.

EXTRACTUM COLCHICI ACETICUM, L. Acetic Extract of Colchicum.

Prep. Rub fresh *Corms* (bulb, E.) of *Colchicum* ℔j. to a pulp, and gradually add *Acetic acid* (Pyroligneous acid, E.) f℥iij. Express the liquid, and evaporate it in an earthenware vessel not glazed with lead (over the vapour-bath, E.) to the due consistence.

Action. Uses. Anodyne, &c. in Gout and irregular Rheumatism, in doses of gr. j.—gr. iij. three or four times a day.

ACETUM COLCHICI, L. E. D. Vinegar of Colchicum.

Prep. Macerate fresh sliced *Colchicum Corms* ℥j. in *distilled Vinegar* f℥xvj. (℔j. D.) in a covered glass vessel for 3 days. Strain, express strongly, and filter for the impurities to subside; then add *Proof Spirit* f℥j.

Action. Uses. Mild but efficient preparation in doses of f℈ß.—f℈ij. every three or four hours.

OXYMEL COLCHICI, D. Oxymel of Colchicum.

Prep. Macerate fresh bulb of *Colchicum* in thin slices ℥j. in *distilled Vinegar* by measure ℔j. in a glass vessel for 2 days. To the strongly expressed and filtered liquor add *purified Honey* by weight ℔ij. Then stirring with a wooden stick, boil to the thickness of Syrup.

VINUM COLCHICI, L. E. Wine of Colchicum.

Prep. Macerate for 14 (7, E.) days dried and sliced *Colchicum Corms* (bulb, E.) ℥viij. in *Sherry Wine* Oij. Strain (express strongly the residuum, and filter, E.)

Action. Uses. Irritant, &c. Sedative in doses of ♏x.—f℈j. three times a day. The Wine of the Seeds is preferred by some (Seeds ℥ij. to Sherry Wine Oj.) in the same doses.

TINCTURA COLCHICI COMPOSITA, L. Compound Tincture of Colchicum.

Prep. Macerate for 14 days *bruised Colchicum Seeds* ℥v. in *Aromatic Spirit of Ammonia* Oij. Strain.

Action. Uses. Recommended by Dr. Williams in cases where Acidity prevails, in doses of ♏x.—f℈j.

TINCTURA (SEMINUM, D.) COLCHICI, L. E. D. Tincture of the Seeds of Colchicum.

Prep. Macerate (digest, E.) for 14 days bruised (ground, E.) *Colchicum Seeds* ℥v. (℥ij. D.) in *Proof Spirit* Oij. (℔j. D.) Strain. (Or better by percolation, like Tinct. Cinchonæ, E.)

Action. Uses. As the seeds are more uniform in strength than the corm, and as Proof Spirit is a good solvent, this and the wine of the seeds are preferred by many practitioners in doses of ♏xv.—f℈j.

VERATRUM, L. Rhizoma, L. Radix, D. Rootstock of VERATRUM ALBUM, *Linn.* White Hellebore.

White Hellebore is believed to be the Ἑλλέβορος λευκος of Dioscorides, the *Khirbuk abiuz* of the Arabs.

Rootstock rugose, oblong, præmorse, rather horizontal, which when dry is of a brownish colour on the outside, but internally of a greyish colour, with long cylindrical radicles. The stem is $1\frac{1}{2}$ to 4 feet high. Leaves plicate, elliptic, or elliptico-lanceolate, pubescent below, passing obliquely into the sheath. Racemes paniculate, terminal, pubescent. Flowers polygamous, yellowish-white, green at the back. Segments of the 6-parted perianth oblong-lanceolate, denticulate, without glands at the base, spreading, much longer than the pedicel. Stamens 6, inserted into the base of the segments; anthers reniform, opening transversely. Ovary with 3 spreading stigmas. Capsules 3, united below, horned above, separating into 3 many-seeded follicles. Seeds compressed, or winged at the apex. —Meadows in the south of Europe, and also in central Europe. Nees von E. 46. St. and Ch. 136. *Veratrum Lobelianum* may be used for the same purposes.

All parts of this plant are acrid and poisonous; but the rootstock with its radicles is alone officinal. It seems when first tasted a little sweetish, but its bitter and acrid disagreeable taste soon overwhelms every other. They are usually imported from Germany, and are rough with the remains of the detached radicles. The properties depend chiefly on the presence of Veratria, but another alkali has been detected by E. Simon, which he has called *Jervine*, from the Spanish word signifying poison. The other constituents resemble those of the Cormus of Colchicum.

Action. Uses. Irritant Poison. Applied to the nose, it produces sneezing and coryza, and when taken internally, vomiting and purging. It used to be employed as a Hydrogogue Cathartic, and before Colchicum as an Anodyne in Gout. It is now chiefly employed to destroy vermin infesting the skin or hair; sometimes as an Errhine, one or two grains being mixed with Starch or Florentine Iris.

DECOCTUM VERATRI, L. E. Decoction of Veratrum.

Prep. Add sliced *Veratrum* ʒx. (℥j. D.) to *Aq. dest.* Oj. (by measure ℔ij. D.) Boil down to Oj. (℔j. by measure, D.), and when cold add *Rectified Spirit* f℥ij. (by measure ℥ij. D.) Express and strain, L.

Action. Uses. This is chiefly employed as a Wash to destroy Pediculi.

UNGUENTUM VERATRI, L. D. Veratrum Ointment.

Prep. Mix powdered *Veratrum* ℥ij. (℥iij. D.) *prepared Hog's Lard* ℥viij. (℔j. D.), *Oil of Lemons* ♏xx. L. (Make an Ointment, D.)

Action. Uses. Used as a substitute for the Sulphur Ointment in Scabies. Veratrum is a constituent of the Compound Sulphur Ointment. (p. 30.)

VINUM VERATRI, L. Veratrum Wine.

Prep. Macerate *sliced white Hellebore* ℥viij. in *Sherry Wine* Oij. for 14 days. Strain.

Action. Uses. Emetic, Cathartic, Anodyne. Sometimes used in Gout and Rheumatism in doses of ♏x. thrice a day, with a little Laudanum.

SABADILLA, L. E. (Cebadilla). Semina, L. Seeds of ASAGRÆA OFFICINALIS, *Lind.* (*Helonias officinalis*, Don), L. Fruit of Veratrum Sabadilla, *Retz.*, of Helonias officinalis, *Don*, and probably of other Melanthaceæ, E. Cevadilla.

Sabadilla, Cevadilla, or Cebadilla (from *Cebada*, the Spanish for Barley, from the resemblance of its flowering spike to an ear of barley) seeds, or rather fruits, were known to Monardes in 1573. Occasionally employed for the destruction of vermin; they are now recognized as the source for obtaining the alkali Veratria. Cevadilla was long supposed to be the fruit of *Veratrum Sabadilla*, Retz., but Schiede discovered that it was produced in Mexico by a different plant, which has been referred to Veratrum, to Helonias, and is now formed into a new genus, Asagræa.

ASAGRÆA OFFICINALIS, *Lindl.* Bulbous. Plants cæspitose. Leaves linear grass-like, tapering, smooth, channeled above, carinate below, 4 feet long, lax. Scape naked, 6 feet high, simple. Raceme very dense, 1½ foot long. Flowers polygamous, racemose, naked, yellowish-white. Perianth 6-partite; segments linear. Stamens alternately shorter; anthers cordate, after dehiscence shield-shaped. Ovaries 3, simple. Stigma obscure. Follicles 3, papery. Seeds scimetar-shaped, winged.—Eastern side of the Mexican Andes.—Nees von E. Suppl. 4.

VERATRUM SABADILLA, *Retz.* Is chiefly distinguishable from Asagræa by the fruit in the latter being crowded round the stem, while in this plant they are few in number and attached to one side only.—A native of Mexico and of the West Indies.—Nees von E. 48.

Cebadilla or Sabadilla seeds, as they are called, consist of the loose seeds and the 3-celled, thin, dry follicles, of a reddish-grey colour, which are either empty, or each contains two brownish-black seeds, without odour, but having a bitter, acrid, and persistent taste. Analysed by Pelletier and Caventou, they were found closely to resemble the corm of Colchicum in composition, as they contain Fatty matter composed of Stearine, Elaine, and Cevadic acid, Wax, Veratria in combination with an excess of Gallic acid, Yellow Colouring matter, Starch, Lignin, Gum, with some saline substances. (*v.* Merat and De Lens.) A more elaborate analysis has been given by Meisner.

Action. Uses. Anthelmintic.

VERATRIA, L. E. (*Sabadillin*, Meisner) was first discovered in the Sabadilla seeds, and subsequently in the rootstock of *Veratrum album*, and in both in combination with Gallic acid in excess.

Prep. Boil *Cebadilla Seeds bruised* ℔ij. with *portions of Spirit* Cj. three times (Gallate of Veratria, Colouring matter, and some other compounds are dissolved). Press, and distil the Spirit, so that the residue be brought to the consistence of an extract. To this, boiled in water three or more times, *dil. Sul'* is added (Sulph. Veratria is formed). The mixed liquors are evaporated to the consistence of a syrup, when *Magnesia* is added to saturation, frequently shaking (the Sulph. of Veratria is decomposed, and the Veratria precipitated, but mixed with other substances). Digest the precipitate in *Spirit* (the Veratria is dissolved). Distil off the Spirit: the residue is boiled in water for a ¼ hour, and a little *Sul'* and *animal Charcoal* added, boiled and strained (the Sulph. Veratria is formed, and freed from colour). The liquid is then evaporated to the consistence of syrup, *Sol. of Ammonia* q. s. is added (the Veratria is precipitated, and Sulph. Ammonia remains in solution). The Edinburgh process is essentially the same, but differs slightly in details.

Veratria is pulverulent and not crystallizable. When melted, it becomes on cooling a transparent yellowish mass; burns entirely away when ignited in the air. In cold water nearly insoluble; boiling water takes up only 1-1000th of its weight, and the solution is acrid; Alcohol dissolves it readily, and Ether but sparingly. It restores the blue colour of Litmus, and forms neutral salts with acids, which crystallize with great difficulty. According to Couerbe, the Veratria of commerce contains, besides Veratria, also Veratrin and Sabadilline, as well as the Gum-Resin of Sabadilline. Veratria is composed of $C_{34}H_{22}O_6N = 288$. "Dissolves but slightly in water, is more soluble in Alcohol, but most so in Sulphuric Ether. It has no smell, and a bitter taste. It is to be cautiously administered." L.

Action. Uses. Irritant Poison. In small doses, applied externally, acts as a Rubefacient; also as an Errhine, excites violent sneezing; taken internally, excites nausea, vomiting, and purging; has been applied externally in Neuralgia, in Rheumatism, and Gout, after the acute symptoms have been mastered. In doses of $\frac{1}{12}$ of a grain, it has been given internally in Gout, Rheumatism, and painful nervous affections, as a substitute for Colchicum.

SMILACEÆ, *R. Brown.* Sarsaparillas.

The *Smilaceæ* are so closely allied to the berry-bearing *Asphodeleæ*, i. e. *Asparageæ*, as only to be distinguished from them by their seeds being neither black nor crustaceous, by a thicker albumen, and somewhat different habit. They are also allied (though these have the ovary inferior) to *Dioscoreæ*, in which is included *Tamus communis*, or Black Bryony, of which the scraped roots, applied as a poultice, promote the absorption of effused blood. From some peculiarities of structure, as the broad, reticulated, and deciduous leaves, and the wood of the root in a solid concentric circle, Dr. Lindley has formed these, with a few other small orders, into his class of Dictyogens. (*v.* Veg. King. p. 211.) The Smilaceæ have bisexual or hexapetaloid flowers, several consolidated carpels, and axile placentæ. They are found in the temperate and tropical parts of Asia and America, but also beyond these limits. They are chiefly remarkable for yielding Sarsaparilla.

SMILAX, *Linn.* *Diœcia Hexandria*, Linn.

Diœcious. Perianth 6-partite, nearly equal, spreading (female persistent). Stamens 6, inserted into its base. Anthers linear, erect. Ovary 3-celled. Cells 1-seeded. Style very short. Stigmas 3, spreading. Berry 1—3-seeded. Seeds sub-globular. Albumen cartilaginous. Embryo very small, remote from the umbilicus.—Evergreen climbing shrubs. Root fibrous or tuberous. Stems often prickly. Leaves alternate, petiolate, cordate or hastate, reticulate, venose; cirrhiferous stipules between the petioles. Flowers sessile on a globular receptacle, subcapitate, pedicellate, or umbellate,—a few in temperate, but the majority in warm and tropical regions of both hemispheres.

SARZA, L. E. SARSAPARILLA, D. Radix, L. D. Root of SMILAX OFFICINALIS, *Humb.* and *Bonpl.* L. E., and probably of other species E. of Smilax Sarsaparilla, *Linn.* D. Sarsaparilla.

The word *Smilax* occurs in Greek authors, and the plant, *Smilax aspera*, continues to be employed in medicine. The name Sarsaparilla (from the Spanish *Sarsa*, a bramble, and *Parilla*, a vine) is applied to species of the same genus, and to their roots, first introduced into Europe from the New World in the 16th century. Several kinds, as Jamaica, Honduras, Brazilian, &c., are known in commerce; but it is extremely difficult, if not impossible, at present to determine the species of *Smilax* which yield the several varieties of drug, because much of it is brought over by the Indians from the little known Mosquito coast to Jamaica, and thence imported into this country: the greater portion of the remainder is imported from Mexico, Guatimala, Brazil, and Peru.

SMILAX OFFICINALIS, *H. B.* and *K.*, was discovered by Humboldt and Bonpland in New Granada on the banks of the Magdalena, in Columbia. As its roots are collected by the natives, called *Zarzaparilla*, and taken to Cartagena, and thence exported to Jamaica, it is more than probable that this yields some of the Sarsaparilla of commerce; hence it is adopted in the L. and the E. P. Dr. Pereira concludes it to be probably the source of Jamaica Sarsaparilla. Martius states that, according to Pohl, the roots are collected in the western parts of the province of Minas Geraes.

S. SARSAPARILLA, *Linn.* D., a native of the United States of America; but it does not, according to the evidence of American authors (*v.* Wood and Bache), yield any of the Sarsaparilla of commerce.

S. PAPYRACEA, *Poir.* (S. syphilitica, *Mart.* non Humb.) is a native of Brazil. Martius ascertained that its roots were collected by the Indians on the Rio Negro and other places in the vicinity of the Amazon river, and that they form the *Salsa*, *Salsaparilha*, *Sarza*, or *Zarza*, which is named the Sarsa of Maranhao, of Para, and of Lisbon, and that it abounds more than the others in Parigline.*

S. SYPHILITICA, *Willd.*, is a distinct species, found by H. and B. on the Rio Cassiquiare, in Brazilian Guiana.

S. MEDICA, *Schlecht*, was found by Schiede on the eastern slope of the Mexican Andes. The roots are dried and exported from Vera Cruz, but are little known here.

Dr. Hancock states that there is but one species that yields genuine Sarsaparilla which manifests to the taste any of the sensible properties of Sarsaparilla; and this grows chiefly on the elevated lands of the Rio Imiquem, at Unturana and Caraburi; also that the Sarsa of the Rio Negro, which comes by way of Angostura or Para, is the best. This is probably yielded by the above *S. papyracea*. Several other species are, however, enumerated by Martius, as *S. Japicanga*, *brasiliensis*, and also *Herreria Salsaparilha* as employed for the same purposes, and says "Recentes multo efficaciores, quam exsiccatæ, vetustæ." So *S. Cumanensis*, the *Azacoreto* of the natives, *S. cordato-ovata* of Pöppig, *S. Purhampuy*, referred doubtfully to *S. officinalis* by Dr. Lindley, *S. China*, yielding the *China-root* of the shops, long famed in the East. It is probable that some of the Indian species are possessed of similar virtues. (*v. Himal. Bot.* p. 383.)

---

* "Ejus radices præ cæterarum Smilacum pollent materia illa extractiva, saporis amaricantis, fauces vellicantis, *Parillinum* dicta, cui efficacia medicaminis præcipue debetur."

Sarsaparilla roots are usually imported in bundles, formed of the roots folded up, but sometimes unfolded, as in the Brazilian variety, frequently still adhering to the *rhizoma* (chump of druggists). These roots are flexible, several feet in length, about the thickness of a quill, cylindrical, but wrinkled longitudinally, with radicles attached along their length. The colour varies, probably owing to adventitious circumstances. They are composed of a thick cellular cortex, covered by a thin epidermis, and of the meditullium or duramen, an inner layer of ligneous intermixed with cellular tissue, having a central pith, often containing Starch, in its interior. Hence a transverse section resembles one of an exogenous stem but without medullary rays. Sarsaparilla is without odour, often with little else than a mucilaginous taste; but when good and fresh it is a little bitter, nauseous, and acrid, which affords, according to Dr. Hancock, the best criterion of its goodness. The roots are often split up the middle and cut into short pieces, for the facility of making preparations. In this, it is more difficult than in the entire state, to distinguish the different kinds.

Jamaica Sarsaparilla, which is usually the most esteemed, is in bundles of from 12 to 20 inches in length, and from 4 to 5 in breadth, distinguished from other kinds by its reddish colour, and by having more rootlets attached to it, hence it is sometimes called *red-bearded* Sarsaparilla. It is less mealy, but yields more extract than the other kinds, and the bark five times more than the woody part. (Pope.) Its powder is of a reddish colour, and does not produce so blue a colour as the Honduras, and similar kinds, when tested with Iodine. *Brazilian*, called also *Lisbon* and *Rio Negro* Sarsaparilla, ought to be of as good quality as any other, if yielded by the above *S. papyracea*, which, according to Martius, has the sensible properties more marked than other species, and may be the plant alluded to by Dr. Hancock. Some is yielded perhaps also by the *S. cordato-ovata*. The roots are in bundles of from 3 to 5 feet in length, and not folded up, often with the chump attached, are less wrinkled longitudinally, have fewer radicles, a reddish-brown colour, and are amylaceous. The *Honduras* Sarsaparilla is greyish-brown in colour, and, like the last, has but few radicles attached, is very amylaceous, so as to appear mealy when broken, and becomes blue when either its powder or decoction are tested with Iodine. *Lima Sarsaparilla*, though originally brought from Lima, is also imported from Valparaiso and Costa Rica (*Per.*) On account of its resemblance to Jamaica Sarsaparilla, it is often sold for it. It is folded in bundles 3 feet long and 9 inches across, having the chump still attached, and contained in the interior.

Sarsaparilla roots are sometimes adulterated, the inferior being often passed off for superior kinds. The roots of *Agave*, and of *Furcræa*, also of *Herreria stellata*, and *Aralia nudicaulis*, and even the stems of Dulcamara and of the Hop, are substituted. Those of *Phormium tenax* have been likewise used for Sarsaparilla.

Sarsaparilla contains much Lignin, Starch, and Mucilage, a little

Acrid Bitter Resin, a trace of Volatile Oil, (which has the odour and acrid taste of Sarsaparilla, p. 608,) and a peculiar principle, which has been variously named, because thought to be different by different chemists, as, first, *Pariglin* by Pallota, Smilacin, Salsiparin, and Parallinic Acid, all which have been proved to be identical by Poggioli. *Pariglin* (*Smilacin*) is white, crystallizable, without odour, but having a bitterish taste, very slightly soluble in cold, more so in boiling water, and in hot spirit, also in Ether and Oils. Strong Sul′ turns it red, and finally yellow. HCl′ dissolves it, and becomes red. It is composed of Carbon 62·53, Oxygen 28·8, Hydrogen 8·67 = 100 (Poggioli). The active properties of Sarsaparilla are taken up both by hot and cold water, but are impaired by long boiling. They are also extracted by diluted Spirit; a little addition of this, therefore, in making its preparations, is probably useful, while long boiling is positively injurious.

Action. Uses. Alterative, Diaphoretic; will sometimes create nausea and vomiting. Improves the state of the constitution, slightly strengthens and induces plumpness in Cachectic cases, and in depraved states of the general health; useful in secondary Syphilis and Mercurio-Syphilitic cases. Many ascribe its good effects to the care bestowed on the comfort and diet of the patient.

INF. SARSAPARILLÆ COMP., D. Comp. Infusion of Sarsaparilla.

Prep. Macerate *Sarsaparilla root,* first washed with cold water and then sliced, ℥j. and *Lime-water* by measure ℔j. in a covered vessel for 12 hours, occasionally agitating. Strain. The Lime-water is now considered unnecessary.

Action. Uses. Alterative, in doses of f℥iij. two or three times a day.

DECOCTUM (SARSAPARILLÆ, D.) SARZÆ, L.E. Decoction of Sarsaparilla.

Prep. Macerate sliced *Sarza* (washed in cold water, D.) ℥v. (℥iv. D.) in boiling *Aq. dest.* Oiv. (℔iv. D.) for 4 (2, E.) hours in a lightly covered vessel near the fire, L. (at a temperature somewhat below ebullition, E.) Then take out the Sarza, and bruise it. Replace it and in the same way macerate for 2 hours Then boil down to Oij. (℔ij. by measure, D. Squeeze out the decoction, E.) and Strain.—The long boiling, L. D., is injurious; as good a preparation may probably be made by the maceration without the boiling down. The Sarza ought to be well divided, perhaps bruised. The Extract is often prescribed with it, and the Syrup also added.

Action. Uses. Alterative. Much used in doses of f℥iij—f℥vj. two or three times a day.

DECOCTUM (SARSAPARILLÆ, D.) SARZÆ COMPOSITUM, L. E. Comp. Dec. of Sarsaparilla.

Prep. Boil together for ¼ of an hour *Decoction of Sarza* Oiv. (by measure ℔iv. D.), bruised *Sassafras,* rasped *Guaiacum Wood,* and bruised (fresh, E.) *Liquorice root* āā ʒx. (℥j. D.), *Mezereon* ʒiij. (℥ß. E.) Strain.—The boiling will necessarily dissipate the Volatile Oil of the Sassafras.

Action. Uses. Alterative. A substitute for the *Lisbon diet drink,* in doses of f℥iij.—f℥vj. two or three times a day.

EXTRACTUM (SARSAPARILLÆ, D.) SARZÆ, L. Extract of Sarsaparilla.

Prep. Prepare as Extr. Gentian, L. Take *cut Sarsaparilla Root* ℔j. and *boiling Aq.* Cj. Macerate for 24 hours, then boil down to ℔iv.; while hot, strain, and with heat evaporate to the proper consistence, D.

Action. Uses. Alterative. Often given with the Decoction in doses of ℈ß.—℈ij.

EXTRACTUM SARSAPARILLÆ (SARZÆ, E.) FLUIDUM, E. D. Fluid Extract of Sarsaparilla.

Prep. Take *Sarza* sliced ℔j., boiling, E. *Aq.* Ovj. (by measure ℔xij. D.) Digest (boil, D.) the root for 2 (1, D.) hours in the (Oiv. E.) water. (Pour off the liquor, D.) Bruise the Sarsaparilla, replace it, and boil for 2 hours, E. (Add Oxij. of Aq. and repeat the boiling and pouring off, D.) Filter, squeeze out the liquid, set aside for the dregs to subside. (Boil the residuum in the remaining Aq. Oij. filter, squeeze out this liquor also; mix the liquors, E.) Evaporate by continual boiling to the consistence of thin syrup (℥xxx. D.) When cool, E. add *Rectified Spirit* (℥ij. D.) as much as will make in all f℥xvj. Filter. This fluid Extract may be aromatized with Volatile Oils or warm Aromatics, E.—The evaporation may be carried on without boiling, and the long roots and rootlets of Jamaica Sarsaparilla alone employed. This preparation is made of good quality by several chemists, as by Mr. Battley, Messrs. Herring, &c.

Action. Uses. Alterative in doses of f℈ß—f℈ij. with water, flavoured as agreeable.

SYRUPUS (SARSAPARILLÆ, D.) SARZÆ, L. E. Syrup of Sarsaparilla.

Prep. Macerate cut *Sarza* ℥xv. (℔j. D.) in boiling *Aq.* Cj. for 24 hours. Then boil down to Oiv. (℔iv. D.), and while hot, strain. Then add *Sugar* ℥xv. L. E. and evaporate to the consistence of Syrup. (Proceed as for making Syrup, D.)

Action. Uses. Alterative in doses of f℈iv. with water, or added to the Decoction.

ORCHIDEÆ, Juss. Orchids.

The *Orchideæ,* though so interesting to horticulturists, are of little importance for their uses. The Vanilla, however, is remarkable for its aromatic fragrance, and some of the family yield the highly nutritious tubers known by the name of Salep, Salop, and Saloop. A name which seems to be derived from the Arabic *Salib.* These radical tubers are hard and horny in appearance, whitish, semi-transparent, with little odour, but a mild mucilaginous taste. They are composed chiefly of Bassorin, some soluble Gum, with a little Starch, and are considered by many as containing the largest portion of nutritious matter in the smallest space. The plant yielding the finest kind, a produce of the Persian region, is imperfectly, if at all known. A plant brought to the author from near Cashmere he named *Eulophia vera,* and he himself prepared some very good Salep from the tubers of *E. campestris*, found at the foot of the Himalayas in North West India (*v.* Himal. Bot. p. 370). Dr. Falconer informs the author that he was told by Dost Mahomed that the finest Salep is produced near Candahar. Sprengel considers *Orchis papilionacea* to be the Orchis or Salep of the ancients. M. Beissinhirtz says that *Orchis Morio*, *mascula*, and *militaris*, give the best Salep in Europe. Dr. Cullen says, "I have seen it prepared in this country from *Orchis bifolia*, as pure and as perfect as any that comes from Turkey."

Action. Uses. Nutritious and unirritant diet for the sick, convalescents or children, boiled with water or milk, and flavoured as Sago and other farinaceous foods.

CANNEÆ, *R. Brown.* *Marantaceæ*, Lindl.

Herbs, with fibrous roots, often with creeping rhizomes. Stem simple or branched at apex, formed of sheathing petioles of leaves surrounding the scape. Leaves alternate, simple, broad, with parallel veins diverging from the midrib. Flowers perfect, irregular, racemose or paniculate, supported by bracts. Perianth double, both superior. Calyx of three sepals. Corolla of six divisions, in two whorls, with the segments unequal, variously united, or abortive. Stamen 1, in consequence of the two lateral ones being abortive; filament petaloid, bearing the anthers on its edge; anthers simple or with 1 cell. Ovary 3-celled, or 1-celled. Stigma hooded and incurved. Capsule 1-celled, sometimes berried or 3-celled, 3-valved. Seeds solitary or numerous, hard, without arillus (vitellus, *Br.*) Albumen mealy or horny. Embryo straight, with its radicle touching the hilum.

The Canneæ resemble Musaceæ in habit, and are very closely allied to Scitamineæ. They are found in tropical parts of Asia and America, and are destitute of aroma, and remarkable for the secretion of Starch.

MARANTA, L. E. Fecula of the Rhizoma and Tubers of MARANTA ARUNDINACEA, Linn. L. E. and of M. INDICA, *Tussac.* E. Arrow Root. *Monandria Monogynia*, Linn.

The fecula, root-stocks, and tubers, have long been familiar to the inhabitants of S. America; but West Indian Arrow Root has only been used in England during this century.

Rootstock white, horizontal, annulated, from which proceed root-fibres, some of which swell into tubers, and become jointed stocks, similar to the rhizoma, but covered with scales. These often elongate, curve upwards, and rising out of the ground, become new plants. (*v.* Nees and Ebermaier, Pfl. Med. 69 and 70.) Stem 2—3 feet high, much branched, slender, finely hairy, tumid at the joints. Leaves alternate, with long, leafy, hairy sheaths, ovate, lanceolate. Panicles terminal, lax, spreading, with long, linear, sheathing bracts at the ramifications. Calyx green, smooth. Corolla white, small, unequal, one of the inner segments in the form of a lip. Anther attached to the petal-like filament. Style hooded, petal-shaped. Ovary 3-celled, smooth. Stigma 3-sided. Fruit even, dry, 1-seeded.

The Starch, or Arrow Root, is obtained by beating into a pulp one-year old tubers (the Rhizoma does not seem to be used, neither in this nor in the Curcumas, *q. v.*), then throwing them into water, agitating and straining it so as to separate the amylaceous from the fibrous portion. The Starch suspended in the water gives it a milky appearance. When allowed to stand the fecula subsides, is washed with a fresh portion of water, and afterwards dried in the sun. It is then snow-white, and is composed of minute granules, usually elliptical in form, often like a muller in shape, rarely quite spherical or ovate. Many are only a 2000th, others as much as a 750th of an inch in length. Some Arrow Root is also produced in the West Indies by a plant there called "Tous les Mois," referred to in the E. P. as "An imperfectly determined species of Canna." This by some, as Dr. Waterson of St. Kitt's, is stated to be *C. coccinea.* Dr. Hamilton describes it as cultivated in Barbadoes, St. Kitt's, and the French islands, as attaining in rich soils a stature of fourteen feet, having tuberous roots, equal in size to the human head. He suspects it may be identical with the Achira of Choco. This is the *Canna edulis* of the Bot. Reg. ix. 775. Dr. Pereira has described the Starch particles as longer than those of any other commercial fecula (many a 300th, some nearly a 200th of an inch long). Their shape oval or oblong, generally more or less ovate. The circular hilum is placed at the narrow extremity; the rings are numerous and close.

In India, *Maranta arundinacea* has been successfully cultivated near Calcutta. Some Arrow Root is yielded by another species, *M. ramosissima.* Besides these, much is yielded by *Curcuma angustifolia, leucorhiza, rubescens*, &c.; but the greater portion of the Arrow Root which is exported is produced in Travancore, and we have no information respecting the species which yields it, as *Curcuma angustifolia* has not been proved to be indigenous there. Some is also made in India from the tubers of *Ipomea Batatas.*

In Europe Potato Starch is often substituted for it. The granules are mostly elliptical, but irregular in form, from a 400th to a 300th of an inch in length.

Action. Uses. Arrow Root participates in all the properties of Starch (*q. v.*), and is Nutritious and Demulcent. Well adapted for the diet of the sick, convalescents, and for infants when weaned. As a Demulcent in Urinary and Bowel complaints. Arrow Root makes a firmer jelly than most of the feculas, with the exception, perhaps, of that of Tous les Mois.

SCITAMINEÆ, *R. Brown.* Zingiberaceæ, *Lindl.*

Herbaceous plants, remarkable for aromatic properties, with creeping or tuberous rootstocks, rarely perennial, with fibrous roots, stemless, or with a simple stem formed of the sheaths of the leaves investing the scape. Leaves radical, or alternate, sheathing, with parallel veins diverging on either side of midrib. Flowers complete, irregular, terminal or radical, spiked, racemose or panicled, each in the axils of sheath-like bracts. Perianth double, both superior; the exterior (calyx) tubular, usually coloured, 3-dentate or trifid; the interior (corolla) corol-like; tube more or less lengthened; limb 6-partite, in 2 rows. Stamen single, in consequence of the two lateral ones being abortive, inserted into the throat within the anterior divisions of the outer series of the corolla. Filament not petaloid, often extended beyond the anther, which is 2-celled. Ovary 3-celled. Cells many-ovuled, attached to the inner angle. Style filiform, received in a furrow of the filament. Stigma dilated, hollow. Capsule 3-celled, 3-valved, many-seeded, sometimes berried. Seeds roundish, with or without an aril. Albumen farinaceous. A fleshy vitellus (sacculus amnioticus) sheaths the apex of the embryo.

The Scitamineæ, peculiar in habit, are most closely allied to Canneæ. They are remarkable for yielding a variety of spices, as Ginger, the various Cardamoms, and others now less known in Europe, as Zedoary, Zerumbet, Galangal, and several of the Curcumas, which, besides Turmeric, &c., yield Starch, or Arrow-root.

ZINGIBER, L. E. D. Rhizoma, L. E. Radix, D. Rootstock of ZINGIBER OFFICINALE, *Roscoe.* (*Amomum Zingiber*, Linn.) D. *Monandria Monog.* Lin.

Ginger, the ζιγγιβερις of Dioscorides, and the *Zinjàbil* of the Arabs, seems to derive its name from the Sanscrit Shringaveram. Pliny says it was thought to be the root of Pepper, and called Zimpiperi,—" quanquam *sapori simile.*"

Rootstock biennial, creeping. Stem annual erect, 3 or 4 feet high, enclosed in the sheaths of distichous leaves. Leaves subsessile, linear-lanceolate, smooth. Spikes radical but elevated, oblong-obovate, strobiliform, formed of single-flowered, imbricated, acute bracts. Corolla with outer limb 3-parted, the interior unilabiate. Lip 3-lobed. Anthers double, crowned with a single incurved beak. Capsule 3-celled, 3-valved. Seeds many, arilled.—Cultivated in Asia and also in America.—Nees von E. t. 61.

Ginger is propagated by planting cuttings of the rootstock of the plant. When the rhizome is young* it may be preserved in syrup, having been first scalded and scraped; it then forms the much-esteemed *Preserved Ginger.* In the autumn the rhizomes are taken up, and scalded in hot water, to stop the vegetative principle: they are then dried, without scraping, when they form what is called *Black Ginger*, or being scraped, they become *White Ginger.* But it is said that there are also differences in the plants; as there are in all which have been long cultivated. We require further information on the subject. Most of the *Black* Ginger of commerce is brought from the East Indies, where it is cultivated both in the plains and mountains; the *White* Ginger comes principally from the West Indies. Much is further whitened by being bleached with Chloride of Lime, &c. Both kinds are remarkable for their warm and fiery, but grateful aromatic taste, and yield their principles readily both to water and to spirit. Ginger consists of Lignin, Starch, Gum, Bassorin, Acidulous Extractive, a yellow acrid Volatile Oil, and some soft, very acrid Resin.

* The late Hon. F. Shore, when in the Deyra Doon, in conjunction with the Author, succeeded in making very good preserved Ginger, by taking up the rhizomes when very young, scalding them in hot water, and then preserving in Syrup. That prepared in India is usually very stringy. The Chinese Ginger, though in large pieces, is yet comparatively tender throughout.

Action. Uses. Grateful and Warm Aromatic. Rubefacient. Errhine, Sialagogue. Stimulant Stomachic, much used to give a tone to the Stomach, and in Flatulence. Used also as a condiment and as an addition to various officinal preparations, chiefly aromatic and purgative compounds. Besides the Syrup and Tincture, the effects may be obtained from its powder, in doses of gr. x.—℈j. or the Essence, which is a concentrated Tincture, or from Lozenges made with it.

TINCTURA ZINGIBERIS, L. E. D. Tincture of Ginger.

Prep. Macerate for 14 days *powdered Ginger* ℥ijß in *Proof Spirit,* L. D. (Rectified Spirit, E.) Oij. (by measure ℔ij. D.) Strain. (Proceed as for Tinct. Cinchonæ, E.)

Action. Uses. Warm Carminative Adjunct, in doses of ♏ x.—f℥i. The Tincture is best made with Rectified Spirit, as it is not then so apt to become turbid.

SYRUPUS ZINGIBERIS, L. E. D. Syrup of Ginger.

Prep. Infuse *bruised Ginger* ℥ijß. (℥iv. D.) in *boiling Aq.* Oj. (by measure ℔iij. D.) for 4 (24, D.) hours. To the filtered liquor add *Sugar* (pure, E.) ℔ijß. (q. s. D.) Dissolve (with the aid of heat, E. and make a Syrup, D.)

Action. Uses. A pleasant adjunct to draughts, &c., in doses of f℥ij.—f℥iv.

CURCUMA, L. E. D. Rhizoma, L. E. Radix, D. of CURCUMA LONGA, *Linn.* Turmeric.

Curcuma appears to be the κυπειρος ινδικος of Dioscorides; but the name is no doubt derived from the Persian name *kurkoom.* Turmeric is extensively cultivated in almost every part of India, being employed as a condiment by the natives.

Rhizoma perennial, having many elongated ramifications; like it, yellow in colour, with numerous root-fibres proceeding from the rhizoma, many of which, as in the Maranta (*v.* p. 612), swell into white tubers. The leaves are all radical, bifarious, with long sheathing petioles, broad, lanceolar, of a uniform green. The scape rises from the midst of the leaves, is short, and formed into a spike by numerous imbricated and united bracts, in the lower only of which are from 3 to 5 flowers, supported by bracteoles. Corolla with a tube gradually enlarged upwards; limb double, each 3-parted. Anther double, incumbent, bicalcarate at the base. Style capillary. Capsule 3-celled. Seeds numerous, arillate.—Cultivated in India and China. Bot. Reg. t. 1825. Nees von E. 59 as *Amomum Curcuma.*

The ramifications of the rootstock form Turmeric, while Arrowroot is procured from the white tubers, as in Maranta, q. v. The Turmeric is sometimes divided into *round* and into *long,* but there are a great many varieties. The latter is most common, about the size of the little finger, curved, pointed, and marked externally with transverse annular wrinkles, of a yellow colour, but somewhat of a reddish-brown internally. The powder is of a bright yellow colour. The odour of Turmeric is peculiar, usually very conspicuous in Currie

Powder. The taste is warm and bitterish, but spice-like. It contains a yellow Starch, a yellow Colouring matter, and an odorous acrid Volatile Oil (*Curcumin*).

Action. Uses. Turmeric is a mild aromatic, and much employed as a condiment. It is also used in dyeing. It is officinal chiefly for making Turmeric-paper, this being turned from yellow to a reddish brown, and thus being a ready method of detecting any alkaline excess.

CARDAMOMUM, L. E. D. Semina, L. D. The Fruit, E. of ELETTARIA CARDAMOMUM, *Maton.* (referred to Alpinia, L. Renealmia, E., and to the genus Amomum in the D. P.) The Lesser or Officinal Cardamom. *Monand. Monog.* Linn.

Cardamoms were probably the καρδαμωμον of the Greeks, as they are produced in the same tract as Pepper, though it is difficult to prove the point. A great variety are known, and have been ably examined by Dr. Pereira in his Elements ; but it is equally difficult to refer them to their respective plants. There is no doubt, however, that the officinal Cardamom is produced in Wynaad and Coorg, on the Malabar coast, and by the plant so fully described and figured by Mr. White, and communicated by the Directors of the East India Company to the Linnean Society (v. Trans. x. p. 229), as well as by Dr. Roxburgh. (Fl. Indica, ed. Wall. i. p. 68.) It was formed into a new genus, *Elettaria*, by Dr. Maton, where it is best retained until a re-examination of the family is made by a competent botanist.

Fig. 92.

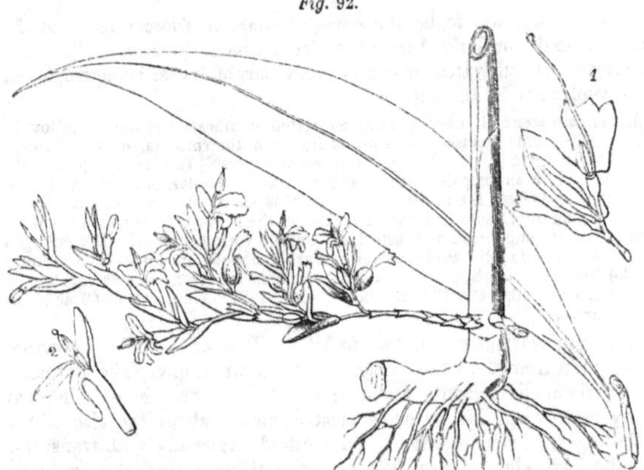

Rhizoma with numerous fleshy fibres. Stems from 6 to 9 feet high. Leaves lanceolate, acuminate, pubescent above, silky beneath. Scapes or flowering racemes from the base of the stem compoundly flexuose, procumbent. Outer limb of the corolla in three oblong lobes, inner a single lip. Anther of two distinct lobes. Filament with two transverse lobes at the base, emarginate, and simple at the summit. Capsule of 3 cells and 3 valves, with a central receptacle. Seeds rough tunicated.—Trans. Linn. Soc. x. t. 4 and 5. Nees von E. 66. St. and Ch. 106.

The Cardamom plant springs up spontaneously after the felling of large trees and the clearing away of the undergrowth. In the February (or April, *White*) of the fourth year, four or five flowering suckers are seen to spring from the stem near the root. Of these the fruit is ripe by the following November, when it is collected, and requires only drying in the sun.

Cardamoms in the capsule are from 4 to 7 lines long, from 3 to 4 thick, 3-sided, with rounded angles, obscurely pointed at both ends, longitudinally wrinkled, and of a yellowish-white colour. The seeds are small, angular, irregular, dotted on their surface, of a brown colour, easily reduced to powder. Varieties are distinguished by the names of Shorts, Short-longs, and Longs, probably all produced by the same plant. Mr. White describes the fruits as collected being separated into three or four sorts, head, middle, and abortive fruits. The odour of Cardamoms is fragrant, the taste warm, slightly pungent, and highly aromatic. These properties are extracted by water and Alcohol, but more readily by the latter. They depend on a Volatile Oil (about 4 per cent); which rises with water in distillation. The seeds also contain Fixed Oil 10·4. Alcohol extracts from them 12·5 per cent. of an acrid burning Resin and Extractive matter. The Lignin amounts to about 77 per cent.

TINCTURA CARDAMOMI, L. E. Tincture of Cardamoms.

Prep. Digest for 14 (7, E.) days *bruised Cardamoms* ℥iijß. (℥ivß. E.) in *Proof Spirit* Oij. Strain. (Squeeze the residuum and filter the liquor; or better prepare by percolation, *v.* Tinct. Caps., first grinding the seeds in a coffee-mill, E.)

TINCTURA CARDAMOMI COMPOSITA, L. E. D. Comp. Tincture of Cardamoms.

Prep. Macerate for 14 (7, E.) days *Cardamoms* and *Caraways* āā bruised ʒijß. (ʒij. D.), *Cochineal* bruised ʒj. L. E. bruised *Cinnamon* ʒv. (℥ß. D.), *Raisins*, ℥v. L. E., *Proof Spirit* Oij. (by measure ℔ij. D.) Strain. Express strongly, and filter. Or prepare by percolation, beating together the solid materials, and leaving them moistened with a little Spirit for 12 hours before they are put in the percolator, E.

Action. Uses. Aromatic adjuncts to draughts, &c. in doses of fʒj.—fʒij.

IRIDEÆ, *Juss.* Corn Flags.

Herbs or undershrubs, with tuberous rootstocks. Leaves usually radical, distichous. Flowers regular or irregular, each with 2 spathe-like bracts, and a common 2-leaved spathe. Perianth petaloid, 6-fid, or 6-partite, divisions in 2 rows, but confounded together. Stamens 3, opposite to and inserted into the base of the exterior segments of the perianth; anthers opening outwards.

Ovary inferior, 3-celled. Style single, with 3 often petaloid stigmas. Capsule 3-celled, 3-valved, bearing the dissepiments in the middle. Seeds many, attached to the central angle. Albumen horny or densely fleshy, enclosing the cylindrical embryo. Radicle turned towards the hilum.—*R. Brown* chiefly.

The Irideæ are naturally allied to Orchideæ. They are found in temperate parts of the world. The rootstocks of different species of Iris have long been employed in medicine. *Iris florentina*, yields Iris (orrice) root, which is collected chiefly near Florence and Leghorn and sent to other parts of the world and finds its way to India, where it is called *Bekh-bunufsha* (violet-root). It has a pleasant odour resembling that of the violet, a bitterish acrid taste, much fecula with an acrid volatile oil. It is now used chiefly to impart an agreeable odour to the breath, and its powder as perfumery; much used by the French for making small round balls to keep open issues. The Costus of the ancients (*koot* and *puchuk* of the natives) is often called Orrice-root in North-west India.

Crocus, L. E. D. Stigmata exsiccata, L. Stigmata, E. D. of Crocus sativus. *Allioni.* Saffron. *Triandria Monog.* Linn.

Crocus is mentioned by Homer and Hippocrates: and as *Carcom* in the Old Testament. It is *kurkoom* of the Persians and *zafran* of the Arabs.

Fig. 93.

The cormus (fig. 93) is roundish, and from its lower surface proceed numerous radicles. The leaves are 7 or 8 inches long, very narrow, traversed by a white midrib, and having long membranous sheaths at the base. They appear just as the flowers begin to fade. These are of a purplish colour, and make their appearance in autumn. They are axillary, with a 2-valved membranous spath, funnel-shaped, with a long tube and 6-parted limb, the throat bearded. Stamens 3, inserted into the tube. Anthers sagittate. Style filiform, with 3 long linear stigmas, wedge-shaped and notched at apex, drooping on one side, of a deep orange-colour. Capsule oval, acuminate, 3-celled, many-seeded. Seeds roundish. — Saffron was early cultivated in Egypt and Persia. The author obtained it from Cashmere. (*v.* Himal. Bot. p. 2.) It has long been introdnced into Europe. Saffron Walden was so named from its being cultivated there.

The stigmata are the only officinal parts of the plant. These with a portion of the style are separated and carefully dried on paper by artificial heat. When dried, they form narrow shreds about an inch in length, of a brownish-red colour, and are called *Hay Saffron;* the produce of about 60,000 flowers are required to make up a pound. At other times the Saffron is pressed together, and forms what is called *Cake Saffron*, as is done with that of Persia, which is highly esteemed when imported into India: the Hay Saffron being the produce of Cashmere. But the Cake Saffron procurable in the shops here is made up of Safflower (*Carthamus tinctorum*) and gum-water. (*Per.*) Saffron is now imported both from France and Spain, a little from other parts of Europe, and some from Bombay, which must be the produce of Cashmere or of Persia. Saffron must necessarily be dear, from the space and labour required to produce even a small quantity; and therefore it is frequently adulterated with Safflower, Marygold, occasionally with shreds of dried beef; and old Saffron is sometimes oiled to make it look fresh, or that of which the colour has been extracted is sold as good and fresh Saffron.

Saffron has a strong aromatic odour, and warm bitter taste, and is of a deep orange-colour, which it imparts readily to water or to Spirit, and tinges the saliva yellow when chewed. Analysed, it yielded of Volatile Oil 7·5, Gum 6.5, a yellow Colouring matter (*Polychroite*) 65 per cent., the remainder consisting of Wax, Albumen, a little Saline matter, Lignin, and moisture. The properties depend probably on the Volatile Oil and also on the Colouring matter, as this has been separated into a Volatile Oil and a bitter-tasted red substance.

Action. Uses. Slightly Stimulant, and highly esteemed in Eastern countries, as it formerly was in Europe. Much used as an ingredient in the cookery of the East, as in that of the Continent; but is chiefly employed here as a colouring ingredient; sometimes in nervous affections, in doses of gr. x.—ʒß. Forms an ingredient of several officinal compounds, as Confectio Aromatica.

Syrupus Croci, L. E. Syrup of Saffron.

Prep. Macerate *Saffron* ʒx. in *boiling Aq.* Oj. in a lightly covered vessel for 12 hours, then strain, and add *Sugar* ℔iij. (Proceed as for Syrup of Orange-Peel, E.)

Action. Uses. Useful as a colouring addition to draughts.

Tinctura Croci, E. Tincture of Saffron.

Prep. Digest *Saffron* chopped fine ℥ij. in *Proof Spirit* Oij. Or prepare by percolation.

Action. Uses. Emmenagogue, but used also as a Colouring ingredient.

Acoraceæ, *Lindl.* (Now a tribe of *Orontiaceæ.*)

A spadix naked and closely covered with flowers. Flowers surrounded with 6 scales. Ovaries 3-celled, about 6 suspended ovules in each cell. Stigmas 3-lobed. Berries 1-seeded. (*Lindl.*) The *Acoraceæ* are now made a tribe of *Orontiaceæ* by Dr. Lindley (Veg. King. p. 194), and are like these allied to *Araceæ*, many of which secrete much Fecula often united with acrid principle; but as the Starch may be separated by washing, as in the case of what is called Portland Sago or Arrowroot, yielded by *Arum maculatum*, so the rootstocks of several form articles of diet in different countries.

Acorus Calamus, *Linn.* L. E. Rhizoma, L. E. or Rootstock of Common Sweet Flag. Calamus aromaticus, E.

The ἄκορον of the Greeks, the *wuj* of the Arabs, and *buch* of the Hindoos. It has no claim to the name assigned it in the E. P.

Fig. 94.

Rhizoma thick, rather spongy, aromatic like every other part of the plant. Leaves erect, 2—3 feet high, about an inch broad, of a bright green colour. Stalk 2-edged or leaflike, but thicker below the spadix, which issues from one of the edges, about a foot above the root, 2 or 3 inches long, tapering, covered with numerous thickly set, pale-green flowers, characterised as in the family. — Native of Europe and of moist and cool parts of India, also of North America.—St. and Ch. 32. Nees von E. 24.

The rhizoma, or creeping procumbent stem, which throws up leaves from its upper and roots from its lower surface, is flattened, jointed, or marked with the semicircular impressions of the leaves, of a light brownish colour externally, with a reddish tinge in the inside. The odour is strong and aromatic, but not very agreeable; the taste warm, bitterish, aromatic, and a little acrid. It contains Volatile Oil, Resin, Extractive, Salts, woody fibre, and water.

Action. Uses. Aromatic Stimulant. The author has frequently prescribed it in conjunction with bitters, as the Chiretta and Bonduc Nut, and with success as an Antiperiodic in Agues. In powder in doses of gr. x.—℈j.; or in infusion (℥jß.–℥ij.—Oj. Aq.) in doses of f℥jß.

GRAMINEÆ, *Juss.* Grasses.

Stem cylindrical, usually fistular. Leaves alternate, with split sheaths. Flowers perfect or unisexual, in spikelets consisting of a common rachis with imbricated bracts, of which the exterior are called *glumes*, the interior, or those immediately enclosing the stamens *paleæ*, and the innermost, at the base of the ovary, *scales*. These are 2 or 3 in number, sometimes wanting. Stamens hypogynous, 1, 2, 3, 4, 6, or more; anthers versatile, notched at both ends. Ovary simple; styles 2, very rarely 1 or 3; stigmas feathery. Pericarp membranous. Albumen farinaceous; embryo on one side of the albumen, lenticular.—*Lindley* chiefly.

Grasses constitute the most important of families, being found in all parts of the world. Their herbaceous parts afford fodder for cattle, and secrete fecula in their seeds, which forms the chief food of mankind. Sugar is secreted by some, but especially by the Sugar-Cane, and a Volatile Oil by *Andropogon Calamus aromaticus*, Royle, and several other species. This oil, often called *Oil of Spikenard*, is extremely grateful for its fragrance, powerful as a stimulant, and especially useful as an embrocation, with one-half or two-thirds of Olive Oil, in rheumatism of the joints, &c. Silex is deposited on the surface of most grasses, as well as in the joints of the Bamboo, forming *Tabasheer*.

Tribe *Aveneæ*.

AVENA, L. E. D. Semina integumentis nudata, L. Seeds, E. of AVENA SATIVA, *Linn.* The Common Oat. Farina ex seminibus, D. Oatmeal.

The Oat (*βρῶμος* of Dioscorides) was known to the Greeks.

The Oat is distinguished among cereal grains by its loose panicle. Spikelets 3-2 flowered. Florets smaller than the glumes, naked at the base, alternately awned. Outer palea with lateral nerves, awned, ending in 2 points. Awn dorsal, kneed and twisted. Stamens 3. Ovary hairy at the top. Stigmas 2. Scales 2. Grain long, crested, and furrowed.—A native probably of the Persian region. Several varieties are cultivated in Europe.—Nees von E. 28.

The grains of Oat when deprived of their integuments form *Groats*, when these are crushed, *Embden* and *Prepared Groats*. When the grain is kiln-dried, stripped of its husk and delicate outer skin, and then coarsely ground, it constitutes the *oatmeal* of Scotland. "The husk, with some adhering starch from the seed, is sold under the inconsistent name of *Seeds*." (*c.*) Oats, according to Vogel, consist of 34 of husk and 66 per cent. of meal, and Oatmeal, in 100 parts, of 59 of Starch, 4·3 of Albuminous matter, Bitter Extractive and Sugar 8·25, Gum 2·5, with 23·95 of Lignin and moisture. Dr. Christison finds as much as 72 per cent. of Starch, and that it consists therefore of nearly five-sixths of real nutriment.

Action. Uses. Groats and Oatmeal are nutrient and demulcent. When boiled with water (℥j. to Aq. Oj. boiled to Oß.) Gruel is formed, which is so useful as diet for the sick. Oatmeal, when of thicker consistence, forms Porridge, and may be employed for making poultices.

PULVIS PRO CATAPLASMATE. Poultice or Cataplasm Powder.

Prep. Linseed after the Oil has been expressed 1 part, *Oatmeal* 2 parts. Mix

CATAPLASMA SIMPLEX, D. Simple Poultice.

Prep. Take *Cataplasm powder* and boiling *Aq.* q. s. to make a poultice, which should be smeared over with *Olive Oil* while warm.

Tribe *Hordeæ.*

HORDEUM, L. E. D. Semina (decortica, D.) integumentis nudata, L. Pearl Barley. Decorticated Seeds, E. of HORDEUM DISTICHON, *Linn.* Common or Long-eared Barley. *Triandria Digynia*, Linn.

Barley formed one of the ancient articles of diet (Exod. ix. 31, v. Bibl. Cycl.) The Hebrew name *shoreh* is very similar to the Arabic *shair*. It is the *κριθη* of Dioscorides.

Several species of Barley are cultivated. *H. vulgare* or Spring Barley, having its grains arranged in four rows, and *H. hexastichon*, or Winter Barley, having the same in 6 rows, and the officinal species, *H. distichon*, or Common Barley. Spikelets 3 together. Glumes 2, terminating in long awns, with 1 perfect flower, which is distichous, close pressed to the stem, awned, the lateral florets male, awnless, with the upper flower a subulate rudiment placed next the rachis. Paleæ 2, the inferior one ending in an awn. Stamens 3. Ovary hairy at the apex. Segments 2, feathery. Scales 2. Grain oblong internally, with a longitudinal furrow, adherent to the ovary.—Probably a native of Tatary, (Reideul.)

The grains of Barley, deprived of their husks, which according to Einhof amount to 18·75 per cent., form the *Hulled* or *Scotch Barley*, and when ground, *Barley Meal.* When the process of decortication is carried further, and the grains become rounded or ovoid, but still retain the mark of the longitudinal furrow, they form the officinal article, *Pearl Barley.* This ground to powder, forms *Patent Barley.* (*p.*) It abounds in Starch, with a little Gluten, Sugar, and Gum. According to the analysis of Einhof, Barley consists of Meal 70·05, Water 11·20, and of Husk 18·75 in 100 parts, while the Meal consists of 67·18 of Starch, 5·21 of uncrystallizable Sugar, 4·62 of Gum, 3·52 of Gluten, 1·15 of Albumen, 6·24 of Phosphate of Lime, and 7·29 of Vegetable fibre, the remainder being water and loss. Proust, however, considers some of the Starch to be peculiar, and intermediate in its nature between Starch and Lignin, and he calls it *Hordein.* But its nature is nct well understood. When Barley is malted, Sugar and Gum are produced at the expense of the Starch (of the Hordeine).

DECOCTUM HORDEI, L. D. Barley Water.

Prep. Take *Barley* ℥iiß. (℥ij. D.) wash off extraneous matters with water, pour on it *Aq.* Oß. (by measure ℔ß. D.) and slightly boil, throw this away and pour on it boiling *Aq.* Oiijß. (by measure ℔v. D.) and boil down to Oij. (half, D.) Strain.

Action. Uses. Mucilaginous Demulcent, containing the soluble parts of the Barley.

DECOCTUM HORDEI COMPOSITUM, L. D. MISTURA HORDEI, E. Compound Decoction of Barley, or Barley Mixture.

Prep. Take (Pearl Barley ℥iiß., if necessary clean it by washing, and with Aq. Oivß. boil down to Oij. E.) *Decoction of Barley* Oij. (by measure ℔iv. while boiling, D.) add stoned *Raisins, Figs* sliced āā ℥iiß. sliced and bruised *Liquorice Root* ℥v. (℥ß D.) *Aq.* Oj. L. E. Boil down to Oij. (℔ij.) and strain.

Action. Uses. Demulcent, and useful as a pleasant diet drink.

SECALE CEREALE, *Linn.*, or Common Rye, is sometimes made officinal, and mentioned as *Secale cornutum*, Spurred Rye, or Ergot, ERGOTA, L. E.; but as the properties of this substance seem to depend entirely on the presence of a fungus, it is preferable to treat of it with the Fungi. The Rye cultivated in Europe is considered to be a native of the Caucasico-Caspian Desert; Dr. Falconer met with it in Tibet and Toorkistan, where it is called *Deo gundum*, or Devil's Wheat. The meaning of *Kussemeth*, translated Rye and Fitches in our Bible, is uncertain.

TRITICUM VULGARE, E. TRITICUM HYBERNUM, L. D. Wheat.

Wheat is very similar in sound to the Hebrew *khittah*, Arabic *hinteh*, and there is no doubt it was cultivated by and formed the food of the earliest civilized nations. It was the πυροι of the Greeks.

T. vulgare, var. *hybernum*, the kind most commonly cultivated, is sown in Autumn and reaped in the following summer. Spike four-cornered, imbricated, with a tough articulated rachis. Spikelets solitary, generally 4-flowered. Flowers distichous. Glumes 2, nearly opposite, equal, the upper one bicarinate; the keels more or less aculeate, ciliate, ventricose, ovate, truncate, mucronate, compressed below the apex, round and convex at the back, with a prominent nerve, awned or awnless. Stamens 3. Ovary pyriform, hairy at the apex. Stigmata 2, feathery. Scales 2. Grain loose, externally convex and internally marked with a deep furrow.—Cultivated everywhere; said to be a native of Tatary.

Besides this, *T. vulgare*, var. *æstivum* or Spring Wheat is cultivated. *T. compositum* or Egyptian Wheat is distinguished by its compound spikes. *T. Spelta*, Bere or Spelt, much cultivated in France, and *T. monococcum*, remarkable for having only a single row of grains.

The grain of Wheat differs from that of both Barley and Oats in not adhering to its perianth, so that this is easily separated in the process of thrashing. It is reduced into Flour, *Farina* by grinding. The *Bran*, which constitutes from 25 to 32 per cent., according to the variety of wheat, is separated by sifting.

FARINA. Farina Seminum, L. E. D. Flour.

Flour, according to analysis of Vauquelin, consists of Starch 68·08, Gluten 10·80, Sugar 5·61, Gum 4·11, Water 10·25; but the proportion of these constituents necessarily varies. The ashes of Wheat, which amount only to 0·15 per cent., consist, according to Henry, of Biphosphates of Soda, of Lime, and Magnesia.

Flour, though officinal, is seldom applied to any medicinal purpose. Its nutritious properties, and its superiority to all other grains for making bread, are well known. Both are dependent on the presence of *Gluten*, which was at one time thought to be a simple substance, but is now known to be compound. The Starch and Gluten may easily be separated by kneading Wheat Flour in water, when the particles of Starch are washed out, suspended, and afterwards deposited, in the same way as Sago, Arrow Root, and Tapioca. There remains behind a greyish-white adhesive mass, which is also ductile and elastic. This is Gluten. Its properties are fully described in chemical works. It is remarkable for containing a large proportion of Nitrogen, and approximating in nature to Albumen, Gelatine, and Fibrine. Owing to the presence of Gluten, the paste made with Wheat Flour is very tenacious, and bread made with it is light, porous, and well raised.

AMYLUM, L. E. D. Seminum Fæcula, L. Fecula of the Seeds, E, of Triticum vulgare, &c. Starch.

Starch may be procured by the above process; but it is obtained on a large scale by steeping the Wheat Flour for some time in water, when Sugar, Gum, and Salts are dissolved the liquor becomes sour, from the production of Lactic Acid. The Gluten, which adheres to the Starch with great tenacity, is in a great measure then dissolved by the acid, and the Starch more easily separated. When the Starch has been separated, it is allowed to drain, and then subjected to pressure. In drying it assumes the form of irregular prisms, so characteristic of manufactured Starch. It has of late been obtained of fine quality from Rice, by the action of a weak solution of Caustic Soda. Good Wheat Starch is white, and, without odour or taste, appears a soft homogeneous powder. But, when examined under the microscope, it is found to be composed of granules smaller than those of most other kinds of Starch, which are unequal in size, mostly globular, each displaying a series of concentric rings, surrounding a central point, which has been named the hilum. These granules are now considered to be composed of an external integument, named *Amylin,* which contains matter of the nature of pure Starch, now called *Amidine.* Starch globules are insoluble in cold water, but boiling water bursts the *Amylin,* or membranous tegument, and then the Amidine is dissolved, though not completely. If the quantity be sufficient, a gelatinous mass is produced on cooling. If brought when cold into contact with free Iodine, a deep blue colour (*v.* p. 41) is produced. This colour is destroyed by heat. Starch is insoluble in Alcohol, but this removes a little Volatile Oil which is attached to the Amylin, also in Ether, as well as in Fixed and Volatile Oils. By the action of Dil. Sul' Starch is converted into Sugar, and by Nit' into Oxalic Acid. In the process of germination, as in that of malting, it is likewise converted into Sugar by the action of a principle called Diastase. The composition of Starch is variously given by chemists, some stating it to be $C_7H_6O_6$, others as $C_{12}H_{10}O_{10}$, and also doubling the numbers of the last.

Action. Uses. Starch is Nutritious and Demulcent, extensively employed as an article of diet; and for the sick, in the form of Sago, Arrow Root, and Tapioca. As hair powder, it is employed for powdering the irritated skin.

Pharm. Prep. Trochisi Acaciæ, E. Pulvi Tragacanthæ comp. L.

DECOCTUM (MUCILAGO, E. D.) AMYLI, L. Mucilage of Starch.

Prep. Rub up *Starch* ʒiv. (ʒvj. D.) with *Aq.* Oj. (℔j. by measure, D.) Boil.

Action. Uses. Demulcent. Useful in Dysenteric and Urinary Complaints as an injection; and also for suspending powders.

Other Grasses abound in Fecula and afford nutritious diet. Rice, *Oryza sativa* contains at least 85 per cent. of Starch (*v.* p. 624), about 3·5 Gluten, and a little Gum, Sugar, Oil, Water, Lignine, and Phosphate of Lime. It forms a good substitute for Potatoes, &c.; if carefully boiled and steamed, the grains then dry, remain soft and separate, instead of forming a pulpy mass. Maize or Indian Corn *Lea Mays*, which, like Rice, forms the chief food of millions, and is highly esteemed by the Americans, is nutritious, containing Starch 80 per cent., *Zeine*, Hordeine, (*Bizzio*) a little Sugar, Gum, Oil, and Salts. The flour is sold by the name Polenta. Coarsely ground it makes excellent Gruel, and may be used for Poultices. *Sorghum vulgare*, the *joar* of India *durra* of the Arabs, is well suited to the same purposes, also many of the smaller grains, especially of the Tribe *Paniceæ*.

Tribe *Saccharineæ*.

SACCHARUM. Succus præparatus, L. of SACCHARUM OFFICINARUM, L. E. D. Saccharum purum, E. Succus concretus purificatus, D. Sugar. Purified or White Sugar.

Saccharum commune, E. Succus concret. non purific. D. Brown Sugar.

Sacchari Fæx, L. E. Syrupus empyreumaticus, anglice *Molasses*, D. Treacle. (But this is sometimes distinguished from Molasses as draining from Sugar in the process of refining.)

Sugar is a principle very generally diffused in the vegetable kingdom. In the East and West Indies it is obtained chiefly from the Sugar Cane, but in the East also from Palms; in France, from the Beet-root and Mangel-wurzel; in America, from the Maple; but it is also found in many fruits, roots, &c. It is probable that it was first discovered by evaporating the juice of Palms in India, of which the Sugar is called *jaggary*. But the Sugar of the Cane has been known both in India and Egypt from very early times, and the ancients were acquainted with it. (v. Essay on the Antiq. of Hindoo Med. p. 83.)

The Sugar Cane grows from 6 to 12 feet high, with a jointed stem, hard and dense externally, but juicy in the inside. Leaves long, linear, strap-shaped, enveloping the stem with their sheaths. Panicles 1 to 3 feet long, elegantly diffuse and waving, silvery from the quantity of long hairs attached to each floret. Spikelets all fertile, in pairs, the one sessile, the other stalked, articulated at the base, 2-flowered, the lower floret neuter, with 1 palea, the upper hermaphrodite, with 2 paleæ. Glumes 2, membranous, obscurely 1-nerved, with very long hairs on the back. Paleæ transparent, awnless, those of the hermaphrodite flowers minute, unequal. Stamens 3. Ovary smooth. Styles 2, long. Stigmas feathered. Scales 2, obscurely 2 or 3-lobed at the point, distinct. Grain little known.—A native probably of India, the Indian Islands, or of China. —Nees von E. 33. 34·35. St. and Ch. 148.

The Sugar Cane is cultivated from cuttings, and takes about a year to come to maturity. It is then cut down close to the earth, topped, and stripped of its leaves, and crushed between iron rollers, or in a wooden mill. The juice is first mixed with Lime to saturate the acid which is present, and then heated. The clear liquor is separated and evaporated till it becomes granular. It is then put into casks, and the uncrystallizable parts (the *Molasses* or *Treacle*) allowed to drain off, and the Sugar left in the state of the Raw or Muscovado Sugar of commerce. The quantity of this is diminished, and that of the Sugar increased by a less degree of heat and by boiling in vacuo. It undergoes purification in various ways, by solution in water, fining with albuminous matter, &c., filtration through a stratum of animal charcoal, evaporation, and recrystallization, and by passing pure syrup through it. It then forms pure, refined, or loaf Sugar.

Cane Sugar ($C_{12} H_9 O_9$, *Peligot*, $C_{12} H_{10} O_{10}$, *Thomson*, or $C_{24} H_{18} O_{18} + 4 H O$ when crystallized), allowed to crystallize slowly from its solution, forms large crystals of hydrated Sugar, or Sugar Candy, in oblique rhombic prisms. Sugar is well known for its pure and sweet taste. Sp. Gr. 1·6. It is white, and without odour, soluble in water, forming Syrup, less freely so in Alcohol. It is unchanged in the air, but when heated melts, and again solidifies on cooling in the form of glassy clear *Barley Sugar*. When heated to a greater degree, it becomes decomposed, swells, emits a peculiar odour, becomes of a deep brown colour, and is called Burnt Sugar or *Caramel*, which is much used for colouring Spirits. It burns away at a higher heat. Sugar combines with the alkalis: after a time the alkaline character disappears, especially of Lime and Baryta, and an acid (the Glucic) is formed ; and also with some metallic oxides, as that of Lead. When pure, no precipitate takes place with Diacetate of Lead. It prevents both the Iodide and Carbonate of Iron being readily decomposed, and it renders the fixed and volatile oils to a certain extent miscible with water. Nit′ converts it into Oxalic acid. Sul′ chars it; but long boiled with diluted Sul′, it is converted into Grape Sugar. A weak watery solution exposed with yeast to a temperature between 50° and 80°, undergoes fermentation (v. p. 633).

Grape Sugar ($C_{24} H_{26} O_{21} + 5 H O$ when crystallized), or Sugar of Fruits is found in the Grape as well as other fruits, and differs in several particulars from Cane Sugar, first in containing more Oxygen and Hydrogen. It is also less sweet and less soluble in water, crystallizes in warty granular masses, and combines with difficulty with Lime, Baryta, and Oxide of Lead. It undergoes fermentation, and is converted entirely into Alcohol and Carb. acid, while Cane Sugar requires an equivalent of water to be decomposed.

Action. Uses. Dietetic, Nutrient, Demulcent. Sugar is much used to cover the taste of medicines, also in Syrups, Conserves, Confections, Electuaries, and Lozenges; to suspend oily in aqueous liquids. Treacle, remaining soft, is well adapted for making pills.

SYRUPUS (L.) SIMPLEX, E. D. Syrup.

Prep. Dissolve *Sugar* ℔x. (℥xxix. D.) (boiling, E.) *Aq.* Oiij. (℔j. by measure gradually mixed, D.) with aid of gentle heat, E. (in a covered vessel, D.)

Action. Uses. Syrup is applied to all the pharmaceutical uses of Sugar. It is preserved at a temperature of 50°. It sometimes requires to be purified by boiling with the white of egg.

The *Cyperaceæ* may be called the Grasses of moist situations, and very closely resemble them in appearance; but they may readily be distinguished by their stems being solid, often triangular, and their leaves with entire, not split sheaths. They are much less useful than the plants of that family. The famed Papyrus belongs to the Cyperaceæ. Though a few do secrete fecula in their tuberous rootstocks, as the Water Chesnut of the Chinese, &c., others secrete a little Volatile Oil, as *Cyperus longus* and *rotundus.* They are mentioned here because the creeping rhizomes of *Carex arenaria*, and of a few allied species, are sometimes used medicinally, under the name of German Sarsaparilla.

III. ACOTYLEDONES, *vel* CRYPTOGAMIÆ.

Substance of the plant composed of cellular tissue chiefly (except in the Acrogens). No woody fibre. No true flowers with stamens and pistils. No distinct embryo or cotyledons. Reproduction taking place by spores, or by a mere dissolution of the utricles of tissue. (*Lindl.*) These Cryptogamic, or Flowerless Plants, are divided into Acrogens and Thallogens.

Acrogens grow by an extension of the stem point, do not increase in thickness when once formed, and contain some ducts among the cellular tissue.

FILICES, *Juss.* Ferns.

Leafy plants, with a rhizoma, in some creeping, or rising into a palm-like trunk (Tree Ferns). Leaves (or Fronds) coiled up in a circinate manner in vernation, simple or divided, with dichotomous veins of equal thickness, with scalariform vessels in the interior. Reproductive organs arising from the veins on the under side or edge of the leaf, and consisting of 1-celled *Thecæ* or Sporangia, which contain the spores, and are either stalked, with an elastic ring, or are sessile, and without a ring, collected in *Sori* either naked or with an indusium. The rhizomes of many Ferns are astringent, some contain a volatile oil, and some an acrid principle. The fronds are mucilaginous when young, and are used as food in some countries.

ASPIDIUM, L. FILIX, E. Rhizoma, E. Radix, L. D., of LASTREA FILIX MAS, *Presl.*, referred to Aspidium, by *Smith*, &c. L. D. to Nephrodium by *Richard*, E. Male Shield Fern. *Fern Root.*

This is supposed to have been the πτερις of Dioscorides. Several Ferns were no doubt employed medicinally by the ancients.

The Rhizoma horizontal, thick, with numerous tufts (the bases of the fronds) ranged along the common axis, separated from one another by brownish-yellow silky scales. The true roots emerge from between these tubercles, and descend downwards. The fronds or leaves ascend upwards in tufts 1 to 4 feet high. Fronds bipinnate, rising in a circle from the tufted rhizoma; pinnules obtuse and serrated, only slightly narrowed downwards, and the lowest leaflet of considerable size, lobes usually a little combined at the base. Veins distinct, after leaving the midrib, not uniting with those of the adjoining pinnule. Stipes, or footstalk and midrib, either glabrous, yellow, or densely clothed with purple scales. Sori roundish, scattered, covered by an indusium, which is reniform, attached by the sinus. Sori placed in two rows near the central nerve, and below its lower half.—Indigenous in woods, but found in other parts of Europe, &c.—Nees von E. 27.

The rhizoma, according to M. Peschier of Geneva, should be collected in summer. M. Geiger directs that the inner parts of the fresh root-stock and of the portions of leaf-stalk attached to it, which are fleshy and of a light greenish colour, should alone be preserved; the black and discoloured parts with the fibres and scales should be separated, and the other parts carefully dried, powdered, and kept in small well-stopped bottles, and renewed annually. The powder should be of a light greenish-yellow colour, of rather a disagreeable odour, a bitter and astringent taste. Analysed by Geiger, it was found to contain of a Fat Oil 6·9, Resin 4·1, with Tannin, Starch, Gum, Uncrystallizable Sugar. Morin of Rouen indicates a Volatile Oil. M. Peschier of Geneva found its active principle soluble in Ether, an aromatic and strong smelling Fixed Oil, Adipocire, &c. Ether extracts the Adipocire along with the active ingredient, but deposits the former on standing.

Action. Uses. Anthelmintic: has been so used from early times; formed the basis of Madame Nouffer's remedy for expelling tapeworms. Dr. Peschier, a brother of the above, Brera, Ebers, have all borne testimony to its efficacy, in the form of the Etherial Extract, of which from 12 to 24 grs. form a dose (at night and again in the morning) or from 1 to 3 drachms of the powder. A Decoction also (℥j.—Aq. Oj.) has also been employed in divided doses. A dose of Castor Oil is exhibited after the second dose of the Etherial Extract in cases of Tape-worm, especially of that more common on the Continent, the Tænia lata.

Thallogens. Grow by development in all directions from one common point.

Lichenes, *Juss.* Lichens.

Perennial plants, growing in the air, and spreading in the form of a leafy expansion or lobed thallus, which is formed of a cortical and of a medullary layer, the former being simply cellular, the latter both cellular and filamentous. Reproductive matter of two kinds: 1. Sporules lying in membranous tubes, immersed in shields or disks (*apotheciæ*), which burst through the cortical layer, and colour and harden by exposure to the air: 2. Separated cellules of the medullary layer of the thallus. (*Lindley.*)—Crawl upon the earth, or on rocks, or on the bark of trees, sometimes burrowing into its substance. Some are mucilaginous and nutritious, others bitter and astringent, and a few remarkable for yielding colouring matter.

CETRARIA, L.E. Planta D. of CETRARIA ISLANDICA, *Achar.* (*Lichen islandicus*, Linn.) D. *Iceland Lichen* or *Moss.*

This Lichen was first employed by the natives of Iceland.

Plant, erect, 2 to 4 inches high, formed of a dry, leathery, smooth, laciniated, foliaceous thallus, the lobes of which are irregularly subdivided, channeled, and fringed at their edges. Those divisions upon which the reproductive matter is produced are more dilated, smooth, of a light brownish colour, paler on the under surface, rather reddish towards the base. The fructifications or *apotheciæ* are shield-like, or like shallow saucers, with a harder elevated rim, of a deeper brown colour, and project from the surface of the thallus near its border.—Mountains of both the Old and New World.—Nees von E. 10. St. and Ch. 69.

This Lichen in its dried state varies in colour from greyish-white to reddish-brown, is without smell, but has a mucilaginous bitter taste. When moist, it is a little leathery, but when dry, may be powdered. Cold water takes up only a small portion, but boiling water about 65 per cent. of its substance, forming a slimy and nearly colourless liquid, which, if the decoction be strong, forms a jelly on cooling. Alcohol dissolves the bitter principle, which has been called *Cetrarin*, and has been obtained in white crystals, which are extremely bitter, and have been used as a substitute for Cinchona. Alkalis readily combine with it, and form soluble compounds, and this forms the best method of freeing this and other similar Lichens from their bitter principle; as by macerating them in 24 times their weight of a solution formed of 1 part of an alkaline carbonate and 375 parts of water. The inhabitants of Iceland and Lapland, however, free it of the bitter principle by repeated maceration, and make use of it as an article of diet, either made into bread or boiled with milk. They find it nutritious from the principle called *Lichenin*, or Lichen-Starch. This does not dissolve, but swells up in cold water; the solution is not affected, although its jelly is rendered blue, by Iodine. It may be converted into Grape Sugar by dil. Sul′, and most nearly resembles the jelly of the sea-weeds. Iceland Lichen contains 44·6 parts in 100 of Lichenin or this Starch-like principle, 3·0 of Cetrarin, 7·5 of Gum and uncrystallizable Sugar, 36·2 of Lignin, with a little Wax, Colouring matter, and Salts.

DECOCTUM (LICHENIS ISLANDICI, D.) CETRARIÆ, L. Decoction of Iceland Lichen.

Prep. Take *Cetraria* ʒv. (℥ß. D.) *boiling Aq.* Ojß. (by measure ℔j. Digest for 2 hours in a covered vessel, D.) Boil down to Oj. (for ¼ hour, and while hot, D.) strain.

Action. Uses. Demulcent: Tonic, for cases where more stimulant remedies are unsuitable, as Phthisis and other chronic Pulmonary affections, in doses of f℥jß.—f℥iij. every 3 or 4 hours.

LACMUS, L. E. LITMUS, D. Thallus præparatus, L. Litmus. A peculiar Colouring Matter from ROCCELLA TINCTORIA, *Achar.* Dyers' Orchil or Weed.

Orchil (written also Archill) is the name of a dye, as well as of the plant yielding it. But several distinct kinds are employed for the same purpose, distinguished by different names according to the country from whence they are imported. Also, by manufacturers into *weed* and *moss*, the former term being applied to the filiform Lichens of botanists, belonging to the genus *Roccella*, while the term *moss* and Rock Moss is applied to the crustaceous Lichens belonging to the genus *Lecanora* and others like it.

The most valuable of these dye-lichens is imported from the Canaries, and consists of *Rocella tinctoria;* but some *R. fuciformis* is also imported under the name of Madeira Weed. It is doubtful, however, whether either is employed in the preparation of *Litmus*, as this is said to be prepared from *Lecanora Parella* and *tartarea*, the first called by the French *Parelle d'Auvergne;* the latter is the *Cudbear* of English commerce. The colouring matter is developed by the action of Ammonia, though the exact method of preparing Litmus is unknown, which, besides colouring matter, contains Chalk and Ligneous matter. (*v.* Thomson, Org. Chem. p. 399.) Guibourt states that the colour of Litmus is given by *Crozophora tinctoria*, or Turnsol, one of the Euphorbiaceæ.

Uses. Litmus is officinal only as a test for ascertaining the presence or excess of acids and alkalis; blue Litmus-paper being changed to a red by acids, and reddened Litmus has its colour restored by alkalis.

ALGÆ, *Juss.* Sea Weeds.

Leafless plants, with no distinct axis, growing in water, consisting either of simple vesicles, or of articulated filaments, or of lobed fronds, which are formed of uniform cellular, with some filamentous, tissue interspersed. Reproductive matter either apparently wanting, or contained in the joints of the filaments, or deposited in *thecæ* or peculiar receptacles varying in form, size, and position. Spores in germination elongating in two opposite directions. (*Lindl.*)

Several of these Algæ abound in gelatinous matter, as the Ceylon Moss prepared by Mr. Previté, and respecting which several favourable testimonies have been given, and which is stated to contain from 54 to 63 per cent. of jelly. This appears to be, as stated by Dr. Lindley, a species of *Gracillaria*, as the edible Birds'-nests are composed of one collected by swallows. A *Gelidia* from Ceylon, is in Rottler's Herbarium as yielding jelly; so *Chondrus crispus*, or the Carrageen Moss of Ireland, has of late years been introduced as a Nutrient and Demulcent, and used both in the form of Decoction and of Jelly. *Gigartina Helminthocorton*, or Corsican Moss, is officinal on the Continent as an Anthelmintic. It probably acts, like Cowhage, as a mechanical irritant, from the fragments of numerous corallines mixed with it. (*v.* Fee and Merat and De Lens.) Some species of *Porphyra* and *Ulva* yield the Laver, which is used as an article of diet, as Laminaria is in Lapland. Some Sea-weeds are employed as manure for land. Some are burnt for Kelp (*v.* p. 90), or impure Carbonate of Soda, but which is now chiefly valuable on account of the

Iodine it yields (*v.* p. 40), on which account seemingly a Laminaria from the China seas finds its way to the foot of the Himalayas, where it is employed as a cure for Goitre. (*v.* p. 40.)

FUCUS VESICULOSUS, *Linn.* Herba cum fructu, D. Sea or Bladder Wrack.

Frond plane, compressed, linear, dichotomous, entire at the margin, coriaceous. Air-vessels large, roundish-oval, in pairs, innate in the frond. Recepta cles in pairs terminating the branches, mostly elliptical, turgid, containing tubercles imbedded in mucus, and discharging their spores (sporangia) by conspicuous pores. (*Greville.*)

The ashes of this plant, burned in a covered crucible, were formerly highly esteemed under the name of Vegetable Æthiops for the cure of lymphatic tumours and Goitre. Its properties are usually ascribed to the presence of Soda, but are no doubt owing to the small quantity of Iodine the ashes contain; to be given in doses of gr. x.—ʒij.

FUNGI, *Juss.* Mushrooms.

Plants conspicuous for great diversity of form and structure, sometimes consisting of simple cells or chains of cells, among which filaments are occasionally intermixed; increasing in size by additions to their inside, their outside undergoing no change after its first formation. They are often ephemeral, and variously coloured. Spores lying either loose among the tissue, or enclosed in membranous cases, called *sporidia.*—Hence Fungi are distinguished from Lichens by having their thecæ concealed by a covering of some kind. Fungi generally abound in moist situations, are generated on leaves and stems, are sometimes subterranean, but are most frequently found on organized bodies in a state of decomposition. A few of them are edible, as the common Mushroom (*Agaricus campestris*), and Truffle (*Tuber Cibarium*), but all are suspicious. *Ammanita muscaria* is remarkable for its intoxicating properties: many are poisonous. The Agarics of the old Materia Medica are now referred to the genus *Polyporus*, and *Amadou*, used as tinder, is made from *P. igniarius.* They are remarkable among plants for consuming much Oxygen, and giving out Hydrogen and Carbonic acid gas. Analysed by Vauquelin and Braconnot, they have been found to contain some peculiar principles, as Fungin and Boletic and Fungic acids. They are very destructive to plants and property in the forms of Mildew, &c. and of Dry Rot, and of late, if not in causing, at least greatly accelerating the destruction of the Potato crop.

ERGOTA, L. E. An undetermined Fungus, with degenerated seed of *Secale Cereale*, Linn. The Fungus is erroneously named Acinula Clavus in the L. P. Ergot of Rye.

Ergot seems to have been first used as a medicine by the profession in France and the United States towards the end of the 18th and the beginning of the present century, but in this country not before the year 1824. Its effects seem to have been long popularly known in Germany, and pestilential diseases have long been ascribed to eating ergotised grain as food. (*v.* Burnett's Outlines of Botany, p. 207.) Various opinions have been entertained respecting the nature of Ergot, some considering the Ergot as a Fungus, which has been named Spermoedia

Clavus by Fries. The opinion which has always appeared to the author as the most satisfactory, is that which considers the Ergot as the grain of Rye stimulated into diseased action by the presence of the spores or sporidia of a Fungus. This opinion, promulgated by Leveillé (*v.* Merat and De Lens), has been fully confirmed by others, more recently by Mr. Quekett, who has shown that the Ergot is the altered grain, from its articulation to the receptacle, the scales at its base, the hairy crown of the grain, and frequent remains of the stigma on its top. Some beautiful drawings, now in the British Museum, have been made of it in its different states by Mr. Bauer, (*v.* Trans. Linn. Soc. vol. xviii., and the Penny Magazine, where some of them have been published). The first appearance of the Fungus, which Mr. Quekett calls *Ergotætia abortifaciens*, is indicated by the young grain and its appendages becoming covered with a white coating, which is formed by a multitude of sporidia mixed with cobweb-like filaments. A sweet fluid, which by degrees becomes viscid, and is found to contain the sporidia, oozes from the Ergot or parts around it. When half-grown, it shows itself above the floral envelopes, and is of a dark purplish colour, and the production of sporidia then nearly ceases, and the upper part of the grain is observed to be of an undulated vermiform appearance, which Leveillé considers to be the Fungus, and calls *Sphacelia segetum*, but which, according to Mr. Quekett, consists of myriads of sporidia. The Ergot, come to its full size, is of a violet-black colour, and projects much above the paleæ. Many other Grasses and some Cyperaceæ are affected by Ergot, which is most prevalent in damp situations and in moist seasons. *v.* Linn. Trans. vol. xviii. t. 32. 33.

The Sporidia are described by Mr. Q. as elliptical, moniliform, finally separating, transparent, and containing seldom more than one, two, or three well-defined (greenish) granules.

The Ergot of Rye is sometimes called Spurred Rye, from its elongated and curved form resembling the spur of a cock. It is either cylindrical or somewhat angular, tapering towards both extremities, from half, to an inch and a half in length, and two or three lines in diameter, with two furrows along its length, often terminated at the apex by a greyish projection. It is on the outside of a purplish colour, internally of a greyish-white with a tinge of red. The smell is peculiar, nauseous, and musty; the taste is slight, bitterish, a little acrid. Ergot is brittle, easily pulverised when dry. The surface, where glaucous, is found to be composed of sporidia, and the interior of the cellular tissue, is characteristic of the Albumen of the grain, within which are globules of oil, according to Mr. Quekett.

Most of the Ergot used is imported from the Continent and from America, and requires to be renewed every year or two, as it is apt to be destroyed by an Acarus, which produces much excrementitious matter.

Various analyses have been made of Ergot. Wiggers found a fixed Oil 35·00, Fungin 46, a peculiar principle called Ergotin 1·25, which has a heavy odour and a disagreeable acrid taste, and was supposed to possess all the active properties of the drug; besides these, a little Phosphoric acid combined with Lime, Potash, and Iron, a little Gum, Sugar, Albumen, Vegetable Ozmazome, and Wax. Dr. Wright found it to contain of Oil 31, Fungin 11·4, modified Starch 26, Mucilage 9, Gluten 7, Ozmazome 5·5, Colouring matter 3·5, and Salts 3·1, with free Phosphoric acid. He states that the activity of the drug resides in the Fixed Oil, which may be separated from its powder by Ether, and this afterwards evaporated. M. Bonjean states that there are two active principles: 1st. the Oil, which is of a yellow colour, acrid, and poisonous in nature: 2nd. the aqueous Extract, obtained from its powder either deprived of its Oil or not; brown, of a thick consistence, and musty smell, and possessing very decided anti-hæmorrhagic properties. Boiling water takes up the active properties, forming a reddish-coloured acid Infusion or Decoction. Alcohol and Ether also in like manner remove its active principles. Hence both aqueous and spirituous preparations are possessed of useful properties.

Action. Uses. The effects of Ergot were first observed in the diseases produced by it when taken for some time with the ordinary food, that is, in Convulsive Ergotism and in Gangrenous Ergotism, both accompanied with formication. In single doses of ʒij. Dr. Wright and others have observed that it created nausea, vomiting, colic pains, and headache, sometimes stupor and delirium. In many cases it has also been observed to depress the pulse. Given to women when in labour, Ergot has been found so constantly to excite labour pains and to cause the speedy expulsion of the child, that it has now become established as a safe and effectual remedy in cases where slowness of labour is dependent only on insufficiency of uterine contraction. It may be prescribed also for expelling the placenta, clots of blood, or hydatids, or to produce contractions, and restrain hæmorrhage. It has also been prescribed as an Emmenagogue, likewise in Leucorrhœa, Chronic Dysentery, Colica Pictonum, &c.

D. ℈j.—ʒß. in fine powder, or with Syrup and some aromatic, or in flavoured infusion (℈j.—Aq. ferv. f℥iij.) repeated if necessary at intervals of 15 or 30 minutes, for two or three times. A Tincture is sometimes prescribed in fʒj. doses. Dr. Wright recommends the Oil of Ergot obtained by evaporating the Etherial Tincture, which he finds produces uterine contractions in doses of ♏xx.—♏l. which may be given in any convenient vehicle.

PRODUCTS OF FERMENTATION.

Organic substances are known to undergo spontaneous decomposition, and to form new compounds, from the affinity which exists between their constituents. Some are, however, very permanent in nature, as the vegetable acids and alkalis, also the resins; others are ready to undergo a transposition of their elements when under the influence of an external agent, or, in other words, prone to pass into a state of fermentation. These substances belong to that group of organic products which contain Carbon with Hydrogen and Oxygen in the proportion in which these exist in water, such as Starch, Sugar, and Mucilaginous substances; while the *ferments* belong to the *Albuminous* group, or such as contain much Nitrogen in their composition, such as Gluten. The conversion of Starch or Fecula into Sugar, as exemplified in the ripening of fruit, or in the process of germination, and seen on a great scale in the operation of malting, is by some called *Saccharine fermentation.* But the term Fermentation is rather applied to the production of Alcohol and Carbonic′ gas at the expense of Sugar, as seen in Vinous Fermentation, while the further change which under peculiar circumstances takes place of Alcohol into Vinegar, is called Acetous Fermentation.

ALCOHOL. Sp. Gr. 0·815, L. 0·810, D. 0·794—6, E.

SPIRITUS RECTIFICATUS, L. E. D. Rectified Spirit of commerce. Sp. Gr. 0·838.
Spiritus Tenuior, L. E. Spiritus Vinosus tenuior, D. Proof Spirit. Sp. Gr. 0·920, L. D. Sp. Gr. 0·912 (7 over Proof). E.
Spiritus Vini Gallici, L. Spirit of French Wine. Brandy.

The process of distillation has been long familiar to the natives of India, as exemplified in their several *araks* or Spirits, and their Rosewater and *attar* of Roses. From them it was no doubt made known to the Arabs.

When Sugar is dissolved in water, and some ferment, such as *yeast* is added, in a temperature of between 60° and 80°, brisk motion is observed to take place, the liquid becomes turbid, froth collects upon its surface, Carbonic acid gas is copiously evolved, the impurities finally subside, and the liquid becomes clear. The Sugar has disappeared, and Alcohol has been produced, which may be separated by distillation. The Sugar which has disappeared has been considered equivalent to the united weight of the Alcohol and Carbonic acid gas which has been produced; but a little water is also decomposed when Cane Sugar undergoes fermentation, that it may be first converted into Grape Sugar. Gay-Lussac calculated that 90·72 parts of Sugar are capable of supplying 46·68 parts of Alcohol, and 44·24 of Carb′ gas, making together 90·72, or an equal weight. Alcohol is considered to be composed of C_2H_3O, but according to other chemists, of $C_4H_6O_2$, and an equivalent of Grape Sugar as forming 4 Eq. of Alcohol, 8 of Carb′, and 4 of water.

Spirit, however, is not usually obtained from Sugar in this country, though it is so in the East and West Indies in the form of Rum. On the Continent it is obtained chiefly from the juice of the Grape, which contains all the elements for due fermentation. Spirit may also be distilled from feculent roots, as Potatoes; or from grain, as Rice in

India, or from malted Barley, as in this country. In these cases the Fecula or Starch has been first converted into Grape Sugar, before the vinous fermentation takes place. The Spirit first obtained is comparatively weak, being mixed with some water and a trace of Essential Oil, which is known by the name of *Grain Oil.* In this state it is usually called *Raw Spirit.* By a second distillation it is freed from much of this water and Oil, and may be procured of the Sp. Gr. of ·835, which is the strongest *Rectified Spirit,* or *Spirit of Wine* of commerce, but which still contains about 13 or 14 per cent. of water. The following are given as the characteristics of the different officinal Spirits.

SPIRITUS RECTIFICATUS. Sp. Gr. ·838 at 62°. "Colourless, not rendered turbid by water, tastes and smells vinous. This may be reduced to the state of Proof Spirit by diluting 5 pints of it with 3 pints of distilled water." L. "f℥iv. treated with *Sol.* of *Nitrate of Silver* ♏ij. exposed to bright light for 24 hours, and then passed through a filter purified by weak Nit′, so as to separate the black powder which forms, undergo no further change when exposed to light with more of the test." E.

This is owing to the decomposition, by means of the Oxide of Silver, of the Oil which Rectified Spirit still contains, and from which it is with difficulty freed. Its presence may also be detected by adding an equal vol. of pure Sul′. The properties of Rectified Spirit are essentially the same as those of Alcohol; but though necessarily weaker as a Spirit, it is an excellent solvent for many of the officinal Resins, of some of which it dissolves more than Alcohol.

ALCOHOL. Sp. Gr. ·815 (Sp. Gr. ·810, D.) Colourless; entirely vaporizable; unites with water or ether; smells and tastes vinous. L. Sp. Gr. ·794—6; when mixed with a little solution of Nitrate of Silver, and exposed to bright light, it remains unchanged, or only a very scanty dark precipitate forms. E.

As Spirit even when rectified still contains water to the extent of $\frac{1}{10}$ or $\frac{1}{11}$, or 18 water to 82 Alcohol in 100 parts; and, as a stronger Spirit is required for some purposes, processes are given for getting rid of much of this water. A common method, as recommended in the D. P., is to add *Pearlash,* dried, powdered, and still warm, lbiijß. to *Rectified Spirit,* Cong. 1, and digest in a close vessel for seven days, with frequent agitation. The dry Pearlash having a great affinity for water, detaches it from the Alcohol, and being itself insoluble in this menstruum, forms a semifluid mass, with the purer Spirit floating above, which can then be decanted off. Other substances have as great an affinity for the water, but being soluble in the Spirit, require this to be separated by distillation. Thus, as recommended by Mr. Brande, agitate together equal weights of Spirit and Quicklime, or, as adopted in the E. P., add *Rectified Spirit* O. j., *Lime,* broken into small fragments, ℥xviij., allow this to slake with a gentle heat, keep cool the upper part of the vessel; then attach a proper refrigeratory, and with a gradually increasing heat, distil off f℥xvij. The density of the Alcohol obtained should be ·796. The L. C. prefer the Chloride of Calcium for its superior deliquescing power, and direct *Chloride of Calcium,* lbj. to be added to *Rectified Spirit*, Cong. 1., and when this is dissolved, distil off Ovij. and f℥v. The D. C. direct *Chloride of Calcium*, lbj., to be added to the Spirit which has been already freed of water by Carb. of Potash, and then distil with a moderate heat, till the residuum grows thick. The Sp. Gr. of the Alcohol should be ·810.

Alcohol still contains a small proportion of Water, of some of which it may be freed by repeated distillation off Chloride of Calcium. By which means it is brought to ·796 at 60°, which is called Absolute Alcohol in the E. P.; but some of lower density has been obtained. The Sp. Gr. of ·825 is adopted as that of Alcohol by the Excise regulations. It may be obtained of considerable strength by enclosing Chloride of Calcium in a vessel, either in vacuo or not, with Rectified Spirit. The water is absorbed by the Chloride, and the Alcohol left comparatively pure. This, when pure, is light, limpid, and colourless, of a peculiar, rather agreeable odour, and a warm, burning taste. It is very volatile, and produces considerable cold during its evaporation. Its boiling point is from 173° to 175°, when Sp. Gr. ·820 at 60°. The stronger the Alcohol, the lower is the boiling-point. Sp. Gr. of its vapour 1·613. Burns readily, without smoke, water and Carb′ being produced. It has never been frozen; is hence well adapted for making thermometers for ascertaining cold. It unites with water in all proportions, some condensation and evolution of heat taking place, and will abstract it from the air. It is a powerful solvent of many substances, as the Vegeto-Alkalis and the Fixed Alkalis, but not their carbonates, many crystalline neutral Resins, Volatile and Fixed Oils; also some elementary substances, as Iodine, and many salts.

SPIRITUS TENUIOR. Proof Spirit. Sp. Gr. ·920, as defined by the laws of Excise. L. Sp. Gr. ·912 (7 over proof). E. The other tests as for Rectified Spirit. Dr. Christison states that the E. C. adopted the standard of Proof Spirit ·920 in its Pharmacopœia of 1839, but had been led to alter the density to ·912, because a Spirit of this strength is produced by mixing *one measure of water* and *two of commercial Rectified Spirit*, and has been long adopted in practice for preparing Tinctures by all the leading druggists of Edinburgh.

The properties of Proof Spirit are necessarily of the same nature as Rectified Spirit, and though weaker as a solvent of some things, is more useful for such as are of the nature of Gum-Resins. But as it consists of nearly one-half Alcohol, it is sufficiently powerful as a stimulant.

SPIRITUS VINI GALLICI, L. Spirit of French Wine. Brandy.

Spirits distilled from various fermented substances form our several varieties of Ardent Spirit, which may be considered as Alcohol diluted to the strength of Proof Spirit, and mixed with some volatile ingredients. Brandy, besides being made from Wine, is distinguished as being free from Grain Oil. Rum is obtained from fermented Molasses. Whiskey from malted Barley or Rye. Holland Gin from malted Barley and Rye, and rectified from Juniper Berries. Common Gin from malted Barley, Rye, or Potatoes, rectified with common Turpentine. The *Arrak* of the East is described as being made from Rice; but the word signifies Spirit, and is made from a variety of substances. Mr. Brande has ascertained that these several spirits contain from 51 to 54 per cent. of Alcohol.

Action. Uses. All are Diffusible Stimulants, and well known for their intoxicating properties. Even in moderate quantities they produce temporary excitement of all parts of the system, followed by corresponding depression. In small quantities they are sometimes useful to health; but in general their use can be abstained from with benefit. Dr. Paris has particularly distinguished Brandy as being Cordial and Stomachic; Rum as Heating and Sudorific; Gin and Whiskey both as Diuretic. Spirit diluted is often used as a cooling lotion, but it must be allowed to evaporate; covered up, it will act as a Rubefacient. Both Alcohol and Proof Spirit are used as the solvents for numerous officinal preparations.

MISTURA SPIRITUS VINI GALLICI. Brandy Mixture.

Prep. Mix together *Brandy* and *Cinnamon-water* āā f℥iv. *Yolks of 2 eggs, pure Sugar* ℥ß. *Oil of Cinnamon* ♏ij.

Action. Uses. Cordial Stimulant. Useful in cases of depression, when the rapid action of a diffusible stimulant is required, as in sinking stages of typhus.

VINUM XERICUM, L. VINUM ALBUM, E. VINUM ALBUM HISPANICUM, D. Sherry Wine.

Sherry, commonly called White Wine, is officinal as a solvent for some active medicinal substances. But Wine is also employed as an important Stimulant and Tonic; and, though all fermented liquors obtained from the juice of fruits are called Wines, good Wine is prepared only from the Grape, because its juice (v. p. 311), besides Sugar dissolved in a large portion of water, and a glutinoid substance, or vegetable albumen in its husk, contains, as its acid principle, Bitartrate of Potash, which being insoluble in Alcohol (v. p. 87), is deposited as the fermentation proceeds, and thus removes a great portion of the acid out of the Wine. The albuminous matter of the Grape juice, absorbing Oxygen from the atmosphere, is considered by chemists to be passing into decomposition, and thus to act as a ferment to the Sugar, and to cause it to be converted into Alcohol, when the same changes take place which have already been described. Besides this, there is also developed a little Volatile Oil, and, according to Liebig and Pelouze, a small quantity of an aromatic substance, which they have called Œnanthic Ether, to which, and especially to the oil, wines owe their flavour. Differences are observed in the Wines of every locality; but they are sometimes divided into *dry* and *sweet* Wines, also into *still* and *sparkling*. Thus, in cases where the proportion of Sugar is small, and that of the albuminous matter large, the Sugar becomes entirely converted into Alcohol, and the Wine is said to be dry, and having become *still*, may be kept for some time to ripen. In other cases, where the proportion of ferment is small, and that of the Sugar is large, and remains unconsumed, the Wine is sweet. *Sparkling* Wines are those which have been bottled before the fermentation, though advanced, has entirely ceased, the Carb' gas naturally escapes as soon as the pressure is removed, causing the appearance of sparkling. The acidity of Wine may be caused by the Bitartrate of Potash, or by the formation of Acetic Acid. The colour of Wine is of different degrees of straw-colour when the juice of the Grape is alone used; but it is red when the skins or husks are left in the liquor when in a state of fermentation. As ascertained by Mr. Brande, the stronger Wines, such as Lissa, Raisin Wine, Marsala, Port, Madeira, Sherry, Teneriffe, Constantia, Malaga, contain from 18 or 19 to 25 per cent. of Alcohol. The lighter Wines, such as Claret, Sauterne, Burgundy, Hock, Champagne, Hermitage, and Gooseberry Wines, from 12 to 17 per cent. of Alcohol. Wines are considered less intoxicating than Spirit and water of the same strength, because the Alcohol is supposed to be combined with the Mucilaginous, Extractive, Colouring, and Astringent principles of the Wine, and to be in this state less diffusible in its action. Sherry is preferred as a solvent, in consequence of being more free from colouring matter, and containing less acid. As a Cordial and Tonic, Wine must be selected according to the nature of the case.

CEREVISIÆ FERMENTUM, L. D. Yeast.

Though neither Ale nor Beer are officinal, they may be noticed as differing from Wine, in containing a larger proportion of mucilaginous and extractive matters, derived from the Malt with which they are made. They often contain a free acid, and are ready to enter into the acetous fermentation. The bitter principle of the Hop assists in preserving Malt liquors, as well as adds to their tonic effects. According to Mr. Brande, Ale and Porter contain from 4 to near 10 per cent. of Alcohol. The Yeast, which makes its appearance mostly on the surface of fermenting Wort, and is produced from the Glutinoid substance in the brewing of Malt liquors, is, however, officinal. It is a light, soft substance, of a greyish-yellow colour, which readily putrifies if kept moist; if dried, it becomes brownish, and may be kept for some time. When magnified, it appears composed of vesicles containing globules, and is by some supposed to be an infusory plant.

Action. Uses. Stimulant when applied externally, and used in the form of a poultice.

CATAPLASMA (CEREVISIÆ, D.) FERMENTI, L. D. Yeast Poultice.

Prep. Mix *Wheat Flour* ℔j. with *Yeast* Oß. (℔j. D.), and heat them gently till they swell up.

ÆTHER SULPHURICUS, L. E. D. Sulphuric Ether, or simply Ether.

If Alcohol, or, better, if Rectified Spirit, be mixed and distilled with Sulphuric Acid, a light, very inflammable liquid is produced, which is well known by the name of Ether, which is sometimes called Rectified, but by the Colleges Sulphuric Ether. The Ethers produced by the action of the other acids contain a portion of the acid or its elements, and differ essentially from each other, and are distinguished by the name of the particular acid.

Sulphuric Ether is colourless and transparent, very light and limpid, of a powerful and peculiar, but rather pleasant odour, and of a warm pungent taste, afterwards feeling cool. Sp. Gr. ·712—·720, or .715 at 60°, extremely volatile. Sp. Gr. of its vapour is high, being 2·586. It evaporates even in pouring from one vessel to another, feeling and producing cold if evaporated from the surface of the hand; and, being very inflammable, it is apt to take fire on the near approach of a light. Its vapour inhaled is exhilarating, and will produce a kind of intoxication. It boils at 96°, under the ordinary pressure of the atmosphere, and has been frozen at —47°, becoming a white, crystalline mass. It burns with a bright flame, producing Carbonic Acid and water. When much exposed to the atmosphere it becomes by degrees converted into Acetic Acid and Water. Ether may be mixed in all proportions with Alcohol, and one part with nine of water. It readily dissolves Resins, Caoutchouc, Volatile and Fixed Oils, a little Sulphur and Phosphorus in a smaller proportion, several Vegeto-Alkalis, and some neutral crystalline principles. Ether is composed of C_4H_5O; its elements are represented to be variously combined by different chemists, some conceiving it to be an Hydrate of Etherine ($C_4H_4 + HO$), or an Oxide of the hypothetical base Ethyle ($C_4\ H_5O$).

Prep. It is directed to be prepared in the L. P. by pouring *Rectified Spirit* ℔ij. into a glass retort; add *Sulphuric acid* ℔ij. and on a sandbath raise the heat so that the liquor may quickly boil. Ether is produced by the action of the acid on the Alcohol, and passes into the receiving vessel, which should be kept cooled with ice or water. The process is to be continued until a heavier liquid portion begins to pass over. Then add to the liquor in the retort previously cooled *Rectified Spirit* ℔j. and continue to distil Ether as before. The distilled liquors are to be mixed, and the supernatant portion poured off, when *Carbonate of Potash* ℥j. previously ignited, is to be added, shaking them occasionally together. (This is to abstract any free acid and any water dissolved by Ether: if it contains water, this may be separated by agitation with Lime.) The Ether is purified by a second distillation.

The D. P. directs a *Sulphuric Etherial liquor* to be first prepared by mixing together *Rectified Spirit* and *Sulphuric acid* ℥xxxij. and distilling with a sudden and sufficiently strong heat f℥xx.; then adding *Rectified Spirit* ℥xvj. and distilling off more of the same Sulphuric Etherial liquor.

D. To prepare Sulphuric Ether, take of the above liquor f℥xx. and mix with it *Carbonate of Potash*, dried and powdered, ʒij. and from a very high retort distil with a gentle heat f℥xij. into a cooled receiver. Sp. Gr. of the Ether should be ·765.

An improvement in the process consists in the regulation of the temperature and also in the allowing an extra proportion of Alcohol to dribble into the mixture as the process proceeds to make up for the loss of that consumed. By this means the same quantity of Sulphuric acid converts a much larger proportion of Alcohol into Ether; while by the regulation of temperature it is stated (*v.* Brande's *Chemistry*) that any quantity of Alcohol may be etherized by the same portion of acid, which is no further altered than by foreign matters which may be accidentally present, or by the volatilization of a minute portion along with the etherial vapour. Some of these improvements have been introduced into the E. P. and are fully described by Dr. Christison, who however observes that he does not find it requisite to carry the conversion beyond double.

Prep. Take of *Rectified Spirit* f℥l. *Sulphuric Acid* f℥x. Pour f℥xij. of the Spirit gently over the Acid contained in an open vessel, and then stir them together briskly and thoroughly. Transfer the mixture immediately into a glass mattrass connected with a refrigeratory, and raise the heat quickly to about 280°. As soon as the etherial fluid begins to distil over, supply fresh Spirit through a tube into the mattrass in a continuous stream, and in such quantity as to equal that of the fluid which distils over. This is best accomplished by connecting one end of the tube with a graduated vessel containing the spirit,—passing the other end through a cork fitted into the mattrass,—and having a stop-cock on the tube to regulate the discharge. When f℥xlij. have distilled over and the whole spirit has been added, the process may be stopped. Agitate the impure Ether with f℥xvj. of a saturated solution of Muriate of Lime, containing about ℥ß. of Lime recently slaked. When all odour of sulphurous acid has been thus removed, pour off the supernatant liquor, and distil it with a very gentle heat so long as the liquid which passes over has a density not above ·735. More Ether of the same strength is then to be obtained from the solution of Muriate of Lime. From the residuum of both distillations a weaker Ether may be obtained in small quantity, which must be rectified by distilling it gently again.

SPIRITUS AETHERIS SULPHURICI. E. Spirit of Ether.

Prep. Mix *Sulphuric Ether* Oj. with *Rectified Spirit* Oij. The density of this preparation ought be ·809.

The changes which take place in the conversion of Alcohol into Ether are supposed to consist essentially in the former ($C_4H_6O_2$) being deprived of 1 eq. of water, HO, and thus converted into C_4H_5 O, the eq. of Ether. But the mere abstraction of water is not all, in the case of the action of Sul′ on Rectified Spirit. Mr. Hennel proved a new acid was produced, the Sulpho-Vinic, which contains Sul′ and Quadri Hydro Carbon, or Etherine, also that this is first formed, and then so evolved as to unite with water to form Ether. By Liebig, Ether is regarded as an oxide of an hypothetical base, which has been called Ethule or Ethyle, C_4H_5+O Ether, and that Alcohol is the hydrated oxide of the same radicle, $C_4H_5O + HO$. Liebig also considers Sulpho-Vinic Acid an essential step in the process of etherification, and that it is no sooner formed (as he conceives, of 2 eq. S′, 2 aq. and 1 Alcohol) than it is resolved into Sul′, Ether, and water. Ether being formed by substracting 2 eq. of Anhydrous Sulph′ and 1 eq. aq. from 1 eq. of Sulpho-Vinic Acid.

Tests.—Ether is apt to be adulterated with Rectified Spirit and a little water. Sometimes it contains Etherial Oil, " L.," Sp. Gr. ·750; but that of commercial Ether varies from ·733 to ·765 ; evaporates away entirely in the air; slightly reddens litmus; unites sparingly with water, that is, at the rate of f℥j. to Oß., and remains limpid, (if

Etherial Oil is present it will cause turbidity,) "E., Sp. Gr. ·735, or under:" when agitated in a minim measure, with half its volume of concentrated solution of Chloride of Calcium its volume is not lessened (if Spirit be present, it will be absorbed, and the Ether will float at the surface, diminished in bulk). The L. and D. preparations, as evident by the Sp. Gr., are weaker than the E. Sulphuric Ether.

Action. Uses. Diffusible Stimulant and Antispasmodic, Carminative. Frequently prescribed in Spasmodic affections, and often to relieve Flatulence. In Hysterical and in Nervous complaints in general, in Dyspnœa, Nervous Colic, and similar affections, it is very effective, especially if prescribed with Laudanum, or a solution of the Salts of Morphia.

D. ♏xv.—f℥j. and repeated in a short time if necessary.

OLEUM ÆTHEREUM, L. Liquor Æthereus Oleosus, D. Etherial Oil. *Oil of Wine.*

This substance is formed towards the end of the distillation of Sulphuric Ether. It is an oily-looking liquid, which when washed has a bitter, somewhat aromatic taste, and a peculiar odour. It is insoluble in water, but soluble in Rectified Spirit and in Ether.

Prep. Mix *Rectified Spirit* ℔ij. cautiously with *Sul'* ℔iv. and distil until a black froth appears, when the retort is to be immediately removed from the fire. (Ether, water, Sulphurous acid, and an oily liquid which floats upon the water, are produced.) The light fluid is to be separated from the heavier, and to be exposed for a day to the air. (The Ether present evaporates.) *Sol. of Potash* f℥j. or q. s. mixed with an equal quantity of water, is to be well shaken with the oily liquid. (The free Sulphurous acid is removed.) Separate the Etherial Oil which subsides, and wash it well. D. P. "Take what remains in the retort after the distillation of Sulph. Ether, and distil down to one-half with a moderate heat," D.

The composition of this body is unsettled. Mr. Hennel considered it to be a Sulphate of the Hydrocarbon Etherine. A compound probably of Sulphuric' and Sulph. Ether. Liebig thinks it is composed of 2 Eq. Sul', $C_8H_8 + HO$.

This Oil, though officinal, is not used medicinally by itself, but forms an ingredient in the following Spirit, which is intended as a substitute for the Anodyne liquor of Hoffman.

SPIR. ÆTHERIS SULPHURICI COMP. L. Compound Spirit of Sulphuric Ether.

Prep. Mix together *Sulph. Ether* f℥viij. *Rectified Spirit* f℥xvj. *Etherial Oil* f℥iij.

Action. Uses. Stimulant, Antispasmodic, Anodyne; hence useful in cases of Nervous Irritation, and want of sleep, often prescribed with Opiates, in doses of f℥ß.—f℥ij.

ÆTHER NITROSUS, D. Hyponitrous or Nitrous Ether.

Nitrous, or, more correctly, Hyponitrous Ether, is of a pale yellow colour, very volatile, has an agreeable, fragrant odour, like that of apples, and a sweetish, cooling, slightly acid taste. Its boiling-point is 62°, and its Sp. Gr. about ·900. It is very inflammable. Miscible with Alcohol and Ether in all proportions, but requires 48 parts of water to dissolve it. It is officinal in the D. P., of the Sp. Gr. ·900, and directed to be made without the aid of heat, by presenting Rectified Spirit to Nitric Acid, while this is forming from the action of Sul′ on Nitr′, or Nitrate of Potash. It is decomposed by the alkalis. But, as it is prepared with difficulty, and apt to undergo change, Nitrous Acid being at length formed, it is now seldom used. It can be made by the following process of the E. P., Sp. Gr. ·899 at 60°, and, according to Dr. Christison, with safety and dispatch, if all the directions are carefully attended to. If the ebullition should become tumultuous, it may at once be stopped by blowing cool air across the mattrass. The Ether being accompanied by some water, undecomposed Alcohol, and a little acid : this is removed by Milk of Lime, and the water and Alcohol by the concentrated solution of Chloride of Calcium; upon which a very pure Ether separates and floats on the saline solution. It is supposed to be composed of 1 Eq. of Ether (AeO or C_4H_5O) with 1 Eq. of Hyponitrous (NO_3), and the changes which take place in its manufacture consist in both the Nit′ and the Alcohol becoming decomposed; the latter, losing its Eq. of water, becomes Ether, while the Nit′, losing 2 Eq. of Oxygen, becomes Hyponitrous Acid. The disengaged Oxygen, Hydrogen, and Nitrogen, give origin to various other compounds, some of which are disengaged as gas, and others remain in the retort.

Spiritus Ætheris Nitrici, L. E. Spir. Æthereus Nitrosus, D. Spirit of Nitric, or rather of Nitrous Ether. Hyponitrous Ether, with (four volumes of, E.) Rectified Spirit. Sweet Spirits of Nitre.

Prep. Take of *Rectified Spirit* Oij. f℥vj. *pure Nitric acid* (Sp. Gr. 1·500) f℥vij. Put f℥xv. of the Spirit, with a little clean sand, into a two-pint mattrass, fitted with a cork, through which are passed a safety-tube terminating an inch above the Spirit, and another tube leading to a refrigeratory. The safety-tube being filled with pure Nitric acid, add through it gradually f℥iijß. of the acid. When the ebullition which slowly rises is nearly over, add the rest of the acid gradually, f℥ß. at a time, waiting till the ebullition caused by each portion is nearly over before adding more, and cooling the refrigeratory with a stream of water, iced in summer. The Ether thus distilled over, being received in a bottle, is to be agitated first with a little milk of lime, till it ceases to redden litmus-paper, and then with half its volume of concentrated solution of Muriate of Lime. The pure Hyponitrous Ether thus obtained, which should have a density of ·899, is then to be mixed with the remainder of the Rectified Spirit, or exactly four times its volume.

Spirit of Nitric Ether ought not to be kept long, as it always undergoes decomposition, and becomes at length strongly acid. Its density by this process is ·847. E.

L. P. Add gradually *Nitric acid* f℥iv. to *Rectified Spirit* ℔iij. and distil f℥xxxij. The D. C. use the residuum of the distillation of Nitrous Ether and the Spirit employed in that process for condensing the elastic vapours. The distilled liquor is afterwards mixed with alkaline matter until it ceases to redden Litmus. The Sp. Gr. of the liquid obtained afterwards by again distilling is ·850.

The Spirit of Nitrous Ether is thus obtained in the E. process, by diluting Nitrous Ether with four times the quantity of Rectified Spirit, and in the L. formula by distilling Nit′ with more spirit than is consumed in the production of Nitrous Ether; but the E. preparation is at least twice, perhaps thrice as strong as the L. one (*c*). The E. Spirit is of a light straw colour; that of the L. P. colourless, having the peculiar but agreeable odour and flavour of the Nitrous Ether, though of course less strong. It is mobile, volatile, and inflammable, almost always a little acid, especially if it has been kept any time. It mixes in all proportions with water, and with Alcohol. The Sp. Gr. varies much; but the strength may also be ascertained by the E. method.

Tests.—It is apt to be adulterated by the addition of spirit or water, and to spoil by keeping, from the formation of Nitrous Acid. "Sp.Gr. ·834. It slightly reddens litmus, and does not effervesce with Carbonate of Soda, and is recognised by its peculiar odour." L. "Sp. Gr. ·847. It effervesces feebly, or not at all, with solution of Bicarbonate of Potash. When agitated with twice its volume of concentrated solution of Chloride of Calcium, 12 per cent. of Ether slowly separates." E.

Action. Uses. Stimulant and Antispasmodic. Diuretic, and, by management, Diaphoretic in doses of f℥ß.—f℥ij. As a Diuretic it is best combined with others, as Squills, Acetate of Potash, &c.

Acetous Fermentation and Destructive Distillation.

Acetic Acid exists in plants either in a free state, or combined with Potassa, Soda, Lime, &c. Vinegar, being producible from simply exposing to a warm temperature the liquors which have undergone, or are susceptible of, the vinous fermentation, has been known from the earliest times. But in all cases some ferment requires to be present, as these substances must first undergo the vinous fermentation: for it is the Alcohol which is the subject of Acetous fermentation, and it may in other ways be made to yield the same products. Thus, though Alcohol when burnt in the open air produces Carbonic Acid and water, if, when diluted with a little water, it be dropped by degrees upon finely-divided Platinum, the Oxygen of the air, attached to the extended surface of the metal, coming in contact with the thin film of spirit, by combining with, changes its nature, and converts it into Acetic Acid, of which the vapours may easily be perceived by their pungent odour. So the same spirit, with a little yeast, exposed to the action of the air, will speedily become converted into Acetic Acid. Vinegar on the Continent is made by exposing Wine to the action of the air in partially filled vessels. In this country a less pure Vinegar is made from an inferior kind of Beer, to which $\frac{1}{1000}$ part of Sul′ is allowed to be added, to prevent further change taking place. Several of these forms of Vinegar are officinal. Besides these, Acetic Acid may also be obtained by the destructive distillation in close vessels of some hard dry woods (*v.* Pyroligneous Acid). All these are, however, diluted forms of

Acetic Acid, which may, however, be obtained from them in a concentrated form.

ACETUM, L. ACETUM GALLICUM, E. ACETUM VINI, D. Vinegar, L. French or Wine Vinegar, E. D.

Though the L. C. does not distinguish Foreign from British Vinegar, yet as by their tests they require only pure Vinegar to be used, it may be treated of with the Foreign Vinegar.

L. Yellowish; of a peculiar odour; f℥j saturates *Carbonate of Soda* gr. lx. and *Sol. Chloride of Barium* produces in this quantity not above 1·14 gr. of Sulph. of Baryta. Its colour is not altered by Sulphuretted Hydrogen. The Sulphate of Baryta is equivalent to the quantity of Sul′ (1 part in 1000) which is allowed by the excise laws to be added. The Sulph. Hydrogen will indicate the presence of Lead. The colour and odour are indications of purity, and the Soda will prove its strength. Wine Vinegar, according to the E. P., has a Sp. Gr. of 1·014 to 1·022. Ammonia in slight excess causes a purplish muddiness, and slowly a purplish precipitate. In f℥iv. complete precipitation takes place of all the Sul′ present with *Sol. of Nitrate of Baryta* ♏xxx. E.

ACETUM BRITTANICUM, E. British Vinegar.

British Vinegars are usually much paler, but sometimes much darker, from the presence of Extractive and other matters, than the White Wine Vinegar, and are deficient in aroma. They often contain a much larger than the recognised proportion of Sul′, besides some Lime.

The E. P. gives as tests of its being sufficiently pure, that it has a Sp. Gr. of 1·006 to 1·0019. Sulphuretted Hydrogen does not colour it; complete precipitation takes place in f℥iv. with *Sol. of Nitrate of Baryta* ♏xxx.

ACETUM DESTILLATUM, L. E. D. Distilled Vinegar.

Prep. Take of *Vinegar* (French by preference, E.) Cj. (10 parts, D.), distil in a glass retort, from a sand-bath, and into a glass receiver, Ovij. (Dilute the product if necessary with *Aq. dest.* till the Sp. Gr. is 1·005, E.) Reject the first tenth, and preserve the next seven-tenths, of which the Sp. Gr. is 1·005. D.

By distillation, the Colouring Matter, Sul′, and other impurities, being left behind, the Vinegar becomes a colourless diluted Acetic Acid, and used to be called Acetous Acid. With the acid and water rises a little of the Mucilage, and also the Ethero-Spirituous substance (or Acetic Ether?), which characterizes all good Vinegar, and is the source of its peculiar aroma, that is, if the first part has not been unnecessarily rejected. The odour and taste of the distilled are less agreeable than that of good Vinegar.

L. "It should be entirely dissipated by heat. Nothing should be precipitated from it on the addition of Acetate of Lead, or Nitrate of Silver, or Iodide of Potassium (showing the absence of Sul′ and H Cl′, as well as of Lead). Neither Sulphuretted Hydrogen nor Ammonia alter its colour (showing absence both of Lead and of Copper). Silver being digested with it, nothing is precipitated on the addition of H Cl′ (showing absence of Nit′). 100 grs. saturate Carb. of Soda grs. xiij. (100 minims saturate grs. viij. Sp. Gr. 1·005, E.) The Sp. Gr. of the L. distilled Vinegar is 1·007 or upwards, according to Dr. Christison. The E. C. order dilution because distilled Vinegar is often above 1·005.

ACIDUM PYROLIGNEUM, E. "Diluted Acetic Acid, obtained by the destructive distillation of Wood." Pyroligneous Acid is considered to be a discovery of Glauber; but it is probable, as stated by Berzelius, that it was known to the Egyptians, as may be inferred from a passage of Pliny, in which Pine wood, heated in a furnace, gives out "sudore, aquæ modo, fluit canali: hoc in Syria *cedrum* vocatur, ac tanta est vis, ut in Ægypto, corpora hominum defunctorum, eo perfusa, serventur." In the present day the distillation is usually conducted in iron cylinders, with condensers attached. The woody matter being decomposed by heat, its elements unite to form fresh compounds, and by distillation an acid liquor passes over with water, tarry matter, Empyreumatic Oil, and much inflammable gas, while a large proportion of excellent Charcoal is left in the retort. The Pyroligneous Acid is a brown, transparent liquid, consists essentially of Acetic Acid, diluted with water, holding in solution tar, with some Empyreumatic Oil, and has a smoky smell. It is distilled, and further purified, by the addition of Carb. of Soda.* This is retained as Acetate of Soda (*v.* p. 100), or used for making a purer Acetic or Pyroligneous Acid. When the acetate of the above Alkali is formed, it is purified by crystallization and re-solution, afterwards decomposed with Sul′: the Acetic′ set free is again distilled, and the processes repeated until a nearly colourless acid, with the odour of the Acetic, is produced, but which is often Empyreumatic.

E. Nearly or entirely colourless; Sp. Gr. at least 1·034; ♏c. neutralize at least grs. liij. of Carb. of Soda; unaffected by Sulphuretted Hydrogen or sol. of Nitrate of Baryta.

ACIDUM ACETICUM, L. E. D. Acetic Acid, E., a little diluted, L. and D.

Acetic acid, which is the basis of the foregoing acids, may be obtained in a concentrated state by decomposing an anhydrous Acetate, as that of Soda, with Sulphuric acid. When set free, being volatile, it rises and is then condensed, and is readily recognized by its peculiar and grateful odour. It is limpid and colourless, acrid in taste, and will blister the skin unless it is moderately diluted. It is exceedingly volatile, even at ordinary temperatures. Its vapour is inflammable. It may be crystallized at 60°, forming large colourless crystals, when it is called Glacial Acetic Acid. Sp. Gr. ·1063. It is composed of $C_4H_3O_3$=Ac′ with 1 Eq. of water. It may be mixed in all proportions with water, Alcohol, and Ether. It dissolves Camphor, several Resins and Volatile Oils, which therefore are frequently employed for aromatising it. It forms numerous important salts with metallic oxides, alkalis, and vegeto-alkalis. But the Acetates are decomposed by most of the acids, except the Carb′. An important fact, first investigated by Mr. Clark, which is referred to in the E. P., is, that the

* Sometimes Chalk is first added, and Acetate of Lime is formed, which is decomposed by digestion with Sulphate of Soda.

Sp. Gr. does not always bear an exact relation to the strength of the acid. The strength and the density go on increasing in a pretty uniform ratio till the latter arrives at 1·077; but as the strength increases still further, the density gradually sinks again to 1·063. Dr. Christison observes that the density is a tolerably correct measure of strength up to 1·062; and above this point it becomes equally so, on observing whether the addition of a small per centage of water raises or lowers it. It may be here stated that 15 parts by weight of the London acid with 85 of water are equal in strength to distilled Vinegar. But the London acid is much weaker than that prepared by the E. formula. The Dublin acid is composed of about 3 Eq. of water to 1 of acid.

Prep. To *Acetate of Soda* ℔ij. (Ac. Potash 100 parts, D.) add *Sulphuric acid* f℥ix. (52 parts, D.) previously diluted with *Aq. dest.* f℥ix.; distil from a sand-bath, carefully regulating the heat towards the end. (Pour the acid into a tubulated retort: add gradually the acetate, waiting after each addition till the mixture cools: distil with a moderate heat to dryness. Sp. Gr. 1·074, D.)

Take *Acetate of Lead* q. s. heat it gradually in a porcelain basin by means of a bath of oil or fusible metal (8 tin, 4 lead, 3 bismuth) to 320° F.: and stir till the fused mass concretes again. Pulverise this when cold, and heat the powder again to 320° with frequent stirring, till the particles cease to accrete. Add ℥vj. of the powder to f℥ix. and a half of pure Sul′ contained in a glass mattrass; attach a proper tube and refrigeratory, and distil from a fusible metal bath with a heat of 320° to complete dryness. Agitate the distilled liquid with a few grs. of red Oxide of Lead to remove a little Sulphurous acid (apt to rise with the Ac′). Allow the vessel to rest for a few minutes, pour off the clear liquid, and redistil it. The density is commonly 1·063 to 1·065, but not above 1068·5. E.

Tests. L. "Sp. Gr. 1048 at 62. Grs. c. saturate *Carb. of Soda* grs. lxxxvij., and the fluid gives by evaporation crystals of Acetate of Soda, vaporizable, &c., as in Acetum Destillatum, L. Density not above 1·0685, and increased by 20 per cent. of water; colourless; unaltered by Sulphuretted Hydrogen or Nitrate of Baryta (showing absence of Lead, Copper, and Sul′); ♏c. neutralize at least grs. ccxvj. of Carb. of Soda. E.

Action. Uses. Though so many forms are officinal, Ac′ is not proportionally useful. The strong acid is one of the quickest Vesicants, and an excellent escharotic for warts and corns. Acetum or Vinegar is used as a solvent for several vegetable principles, and is an excellent and grateful Refrigerant when applied externally as a lotion, or for sponging the body: its vapour when inhaled is useful in various affections of the throat and larynx. It is sometimes given internally, and the *Syrupus Aceti*, E. *Good Vinegar* (French by preference) f℥xi. to *pure Sugar* f℥xiv. diluted is a grateful form.

Acid. Aceticum Camphoratum, E. D. Camphorated Acetic Acid.

Prep. Pulverise *Camphor* ℥ß. with a little Rectified Spirit, and dissolve in *Acetic acid* f℥vjß using the acid of the respective colleges.

Action. Uses. Stimulant. Grateful in faintness, &c., but still more so in the form of the aromatised preparations of Acetic acid to which various volatile oils have been added.

CREASOTUM, L. CREAZOTON, E. An Oxy-Hydro-Carburet, prepared from Pyroxylic Oil, L.

This substance was discovered by Reichenbach in 1830 with several other compounds of Carbon, Hydrogen, and Oxygen, in the products of the destructive distillation of wood. It is found in Tar, in Pyroligneous acid, in wood-smoke, and in other substances which no doubt owe to it some of their properties. When pure, it is colourless and transparent, fluid, limpid like a volatile oil, of a powerful smoky odour, and a pungent burning taste. Its Sp. Gr. is ordinarily stated to be 1·037. Dr. Christison has ascertained it to be 1·067, of high refractive power; when pure, unchanged by exposure to light; greatly expanded by heat; boils at 397°; burns with a sooty flame. It forms two compounds with water, one a Hydrate of 1 part in 10 of water, and the other a solution of 1·25 part in 100 (*Phillips*). It is soluble in Alcohol, Ether, and Naphtha, but more readily in Acetic acid, also in the alkaline solutions; but is neither acid nor alkaline. It is decomposed by Potassium, Nit', and Sul'. Its most characteristic property is that of coagulating Albumen, and of preserving meat indefinitely, whence its name (from κρέα, *flesh*, and σώζω, *I preserve*).

The process for obtaining it pure is very tedious (*v.* Brande's *Chem.*) It consists in forming a compound with solution of Potash and Oil of Tar, which contains Creasote, then separating the alkali by means of Sul', redistilling and separating the Creasote, and repeating the process.

Tests. L. "Oleaginous, colourless, transparent, of a peculiar odour. It boils at 397°, and does not freeze at —50°, soluble in Acetic acid." E. "Colourless, and remains so under sunshine; Sp. Gr. 1·066; entirely soluble in its own volume of Acetic acid: a drop on filtering paper heated for 10 minutes about 212°, leaves no translucent stain.

Action. Uses. Creasote applied to the tongue, causes pain, and on the skin a burning sensation; and is fatal to small animals. It is the active ingredient of Tar and Tar-water; useful in healing ulcers, and inducing a healthy action in cutaneous affections, and in gangrene, scrofulous ulcers, &c. It often gives great relief in toothache, and is extremely useful in allaying vomiting. (*Elliotson.*) The medicine should be well diluted, at least ½ oz. of water to 1 drop.

MISTURA CREAZOTI, E. Mixture of Creasote.

Prep. Mix *Creasote* ♏xvj. with *Acetic acid* ♏xvj.; gradually add *Aq. dest.* f℥xiv., then *Comp. Spirit of Juniper* and *Syrup*, of each f℥j.

D. f℥j.—f℥ij.; each f℥j. containing ♏j. of Creasote.

UNGUENTUM CREASOTI, L. E. Creasote Ointment.

Prep. Melt *Lard* ℥ij. (Axunge ℥iij. E.). add *Creasote* ʒß. (ʒj. E.) Stir them briskly, and continue to do so as the mixture concretes on cooling.

PETROLEUM (E.) BARBADENSE, L. D. Petroleum. Barbadoes Tar.

Petroleum, as its name indicates, is an oil-like exudation from rocks, which has been employed in medicine from the earliest times, though little used now. It is very abundantly diffused, and in other forms, as of Asphalte, Naphtha, &c.

Besides in Barbadoes and Trinidad, Petroleum is found floating on some springs of water in this country, as at Colebrooke Dale, &c., in many parts of Europe, also at Baku on the shores of the Caspian, and very abundantly at Ranan-goong, or Earth-oil Creek, on the banks of the Irrawaddy. Col. Symes describes the wells as about 500 in number, and that upwards of 400,000 hogsheads are taken away in boats. It may readily be obtained by digging into the sand in warm weather. This was examined by Drs. Christison and Gregory, who found in it Paraffin, Eupion, which Reichenbach met with among the products of the destructive distillation of wood; whence they infer that "Rangoon Petroleum is the product of the destructive distillation of vegetable matter," probably from subterraneous causes now in operation.

Petroleum has the consistence of treacle; reddish-brown or black colour, with a bituminous taste; floats on water, in which it is insoluble; burns with a dense black smoke, leaving a carbonaceous residuum. Acids, alkalis, and Rectified Spirit have little effect on it; Ether and both Volatile and Fixed Oils dissolve it. Exposed to the air, it hardens into Asphalte: if exposed to heat, a yellowish coloured liquid distils over, which resembles the Naphtha obtained in making Coal-gas. Both are free from Oxygen, and therefore used for preserving Potassium and dissolving Caoutchouc.

Action. Uses. Stimulant; has been recommended externally in Rheumatism and in Cutaneous affections; internally as a Vermifuge. The Rangoon is probably as good, if not better, than other kinds.

SUCCINUM, L. D. Amber.

Amber (ηλεκτρον) and its property of attracting light bodies was known to the Greeks, also to the Arabs, being their *kah roba* (grass-attractor). The term Electricity has been derived from its Greek name. It is no doubt a fossil product, usually washed up by the sea in different parts of the world. This country is supplied chiefly from the Baltic, it being cast on shore between Konigsberg and Memel. In India it is obtained both in Cutch and Assam. It is probably the resin of some Coniferous tree, as such wood is found in a fossil state. Insects and parts of plants are inclosed in amber found associated with Lignite beds. It is met with in irregular-shaped brittle pieces, of a yellowish or yellowish-red resinous appearance, translucent, devoid of taste and smell. It is not acted on by water or Alcohol. Subjected to distillation, it yields first a yellow liquid which contains Acetic acid, and afterwards a thin yellowish oil, with a yellow crystalline sublimate, which is the Succinic acid, D. Continuing the heat, the Oil gradually deepens in colour, finally becomes black and of the consistence of pitch, requiring to be redistilled before it is fit for use.

ACIDUM SUCCINICUM ET OLEUM SUCCINICUM, D. OLEUM SUCCINI, L. Succinic acid and Oil.

Prep. Distil with a gradually increasing heat from *Amber in coarse powder* and *pure Sand*, of each 1 part, an acid liquor, an oil, and an acid in crystals. Press these in bibulous paper to get rid of moisture, and sublime them again. The oil may be detached from the acid liquor by filtering through bibulous paper. In the L. P. Amber is directed to be similarly distilled, and the oil again and a third time, to obtain *Oleum Succini*, L.

Action. Uses. Succinic acid has been supposed to be Expectorant. The Oil is Stimulant and Antispasmodic in doses of ♏v. It is an ingredient of Tinctura Ammoniæ Comp. (p. 59.)

ANIMAL MATERIA MEDICA.

The animal creation, so interesting and essential an object of study for acquiring a scientific knowledge of the anatomy and physiology of the human frame, is less important as connected with Materia Medica, because the progress of medicine has caused the disappearance from our books and practice of a crowd of inert and at the same time disgusting remedies, which could only have operated through the imagination; while modern chemistry has shown that others can be obtained more easily and as pure from the mineral or vegetable kingdom. Thus Corals, burnt Oyster-shells (p. 658), and Crabs-claws were valuable on account of their being formed of Carbonate of Lime; burnt Bones on account of Animal Charcoal, and their ashes for Phosphate of Lime, yielded also by Cornu ustum (p. 662); burnt Sponge (p. 650) for the salts of Iodine (p. 650). Oils and Fats in the animal kingdom are obtained of nearly the same nature essentially as vegetable oils and fats, as in the form of Lard (p. 665) or Suet (p. 663). So Bees' Wax (p. 657) is like that of Palms and Myricas. Spermaceti (p. 660) is considered a peculiar principle. These, however, continue to be retained but chiefly as external applications. Saccharine matter is contained in Honey (p. 657). Gelatine is obtained from Isinglass (p. 659) and also from Hartshorn (p. 662), and even from Bones (p. 660); Albumen from the White of the Egg (p. 659); and Caseine from Milk, which also contains Butter and Whey. The peculiar secretions called Musk and Castor (p. 662) are, however, still considered to possess some power in controlling a few nervous affections. The only animals which are officinal in their entire state are the Cochineal (p. 656) for its colouring matter, Cantharides (p. 652) for their vesicating properties, and the Leech (p. 651) to draw blood.

I. CYCLO-NEURA *v.* RADIATA.

Class PORIFERA, *Grant.*

SPONGIA OFFICINALIS, *Linn.*, E. D. The Officinal Sponge.

Sponge is so well known for its economic uses, that it does not re-uire a detailed description. Numerous species are known, with soft porous bodies, traversed by tortuous canals; but the officinal Sponge is imported from the Mediterranean and Red Seas. Some of a coarser kind from the West Indies. Those of the British Seas would probably answer equally well for burning. When collected, Sponge contains numerous small fragments of corals and minute shells, from these it must be freed before it can be used for surgical purposes; for which it is well suited, from its soft and porous nature. Sponge *tents* are sometimes used for dilating sinuses. These are prepared by dipping strips of Sponge into wax, and as this melts by the heat of the body, the Sponge absorbs moisture, and swells; or pieces of Sponge may be tightly wound with thread, which can afterwards be cut. Sponge is composed of Gelatine and Coagulated Albumen. (*Hatchett.*) When burnt, its ashes give Carbon, and some Silex, Carbonate and Phosphate of Lime, Carbonate of Soda, Chloride and Iodide of Sodium, Bromide of Magnesia, with a little Oxide of Iron,

PULVIS SPONGIÆ USTÆ, D. Burnt Sponge.

Prep. Having cut the *Sponge* into pieces, beat it to free it from stones, then burn it in a covered iron vessel till it is black and friable. Reduce this to powder.

Action. Uses. Alterative. Formerly much used in Goître, Scrofulous and obstinate Cutaneous Affections. On the Continent it is still preferred by many to the preparations of Iodine, to the presence of which Burnt Sponge no doubt owes its efficacy, in doses of about ʒj. with Honey and Aromatics.

Class POLYPIFERA, *Grant.* The skeletons of Polypiferous animals, so well known by the name of Corals were long officinal, as they still are on the Continent, and in the East. They owe their medical properties to Carbonate of Lime, with the addition of a little Oxide of Iron.

II. DIPLO-NEURA *v.* ARTICULATA.

Class ANNELIDA.

HIRUDO, L. HIRUDO MEDICINALIS, *Linn.* L. D. SANGUISUGA, *Savigny.*

Leeches and their effects must have been known from the earliest times. They were early employed therapeutically by the Hindoos and the Arabs adopted their practice. (Royle, *Hindoo Med.* p. 38, and Wise on *Hindu Medicine*, p. 177.) Themison mentions the employment of Leeches by the ancients. Herodotus alludes to one kind, *Bdella nilotica.*) Dr. Pereira infers that *Sanguisuga ægyptiaca*, the species from which the French soldiers in Egypt suffered, is that referred to in the Bible (Prov. xxx. 15) by the name of *Olukeh* or *Aluka.* The latter, or *Aluk*, is also the Arabic name for Leech.

Fig. 95.

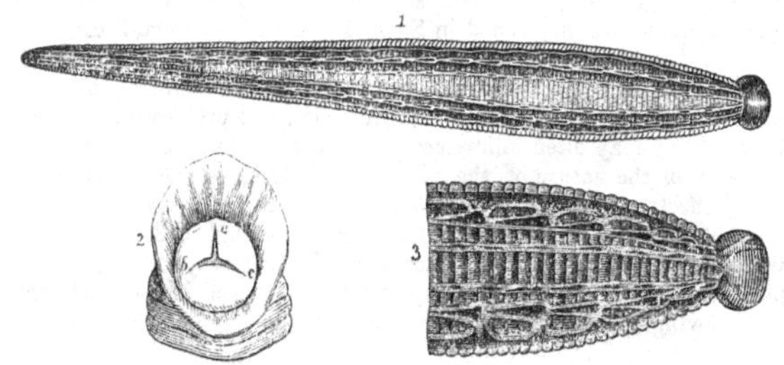

Leeches are included by Cuvier in the genus *Hirudo.* This has since been subdivided into several genera. Savigny calls that which includes the leeches used in medicine SANGUISUGA, the *Iatrabdella* of Blainville. Leeches are characterised by having an elongated, plano-convex body, tapering towards both extremities, wrinkled transversely, and composed of from 90 to 100 soft rings. The mouth furnished with a lip, and the posterior extremity provided with a flattened disk, both adapted to fix upon bodies by suction, and to serve the leech as the principal organs of locomotion. Underneath the body two series of pores are observed, which lead to as many interior pouches, which are regarded as organs of respiration. The intestinal canal is straight, inflated from space to space, as far as two-thirds of its length, where there are two cæca. The blood swallowed is preserved there red and unchanged for many weeks. The subgenus *Sanguisuga* has the upper lip divided into several segments, the aperture of its mouth is tri-radiate, and it contains three jaws, each armed on its edge with two ranges of very fine teeth. Ten blackish points are observed on the head, which are taken for eyes. The anus is small, and placed on the dorsal surface of the last ring.

The two species most commonly used, and which by some are considered varieties of one another, are *Hirudo medicinalis* and *H. officinalis.*

H. OFFICINALIS is distinguished by its unspotted olive green belly and by the dark green back, along which and the flanks are observed six longitudinal, often interrupted rusty-red stripes. Six of the eyes are said by Savigny to be very prominent. Teeth about 70 in number A native of the south of Europe, as of France and Germany. It is usually called the Green Leech, sometimes the Hungary Leech, from being a native of that country.

H. MEDICINALIS (fig. 95, 1 to 3), the kind usually employed here, and readily distinguished from the foregoing by its belly, which is of a yellowish green colour, but covered with black spots, which vary in number and size, forming almost the prevailing tint of the belly, the intervening spaces appearing like yellow spots. On the back are six longitudinal reddish or yellowish red bands, spotted with black and placed on an olive green or greenish brown ground. The number of rings varies from 93 to 108. Teeth 79 to 90 in number. A native of almost all parts of Europe, often called the English, the speckled, the true, the brown Leech, &c.

Other species are figured by Brandt, as *Hirudo provincialis*, *H. Verbana*, *H. obscura*, and *H. interrupta.* In the United States they use *H. decora.* In India Leeches are extremely abundant, procurable both in the tanks of Bengal and in the north-west provinces, as well as along the foot of the Himalayas. Six kinds of useful and six venomous Leeches are mentioned in Susruta and by Avicenna, *l. c.*

Uses. Leeches are effectual for the local abstraction of blood, affording indeed the best method in many cases, as in inflammation of the abdomen, scrotum, in hæmorrhoidal tumours, and prolapsus of the rectum. They may often follow general depletion; but, according to the quantity or the nature of the case, will themselves produce constitutional effects, especially in the cases of children. They act by a sawing motion, and draw about ʒjß. of blood, though f℥ß. may be obtained by fomentation, &c. Excess of bleeding may be stopped by pressure, application of Matico; sometimes caustic is required, and even sewing up the wounds with a fine needle.

Class Insecta. Order *Coleoptera.*

Cantharis, L. E. The whole Fly, E., or Cantharis vesicatoria, *Latreille*, D. The Blister Beetle.

The name κανθαρις was applied by the Greeks to a species of Coleopterous Insect which possessed the properties of the officinal Blistering Beetle, but it was distinguished by *yellow* transverse bands. This is the characteristic of species of Mylabris, one of which, *M. Fusseleni*, occurs in the south of Europe, and another, *M. Cichorii* (fig. 96, 4), in Syria, and apparently throughout the East. In India it is called *telee* and *telee mukhee*, or the Oily Fly, no doubt from the oil-like exudation which the insects of this genus give out from the articulations of their legs when seized. Another species, *M. Trianthemæ*, is mentioned by Dr. Fleming, and the *Lytta gigas*, Fab. is found there as well as in Senegal. One is mentioned by the Arabs under the name of *zurareh.* It is not known when the officinal Blistering Fly came to be used, but it has had a variety of names. It was called *Meloe vesicatorius* by Linnæus, *Lytta vesicatoria* by Fabricius, and *Cantharis vesicatoria* by Geoffroy, and now by the Pharmacopœias. Geoffroy grouped the Vesicatory Beetles in a small tribe corresponding nearly with the Linnean genus Meloe, and distinguished it by the title Cantharideæ. This he divides into eleven genera, among which are *Cantharis*, *Mylabris*, and *Meloe*, all of which species have been employed as vesicatories. *Meloe majalis*, or Mayworm, is figured at 96 fig. 3, as a specimen of the genus.

Fig 96.

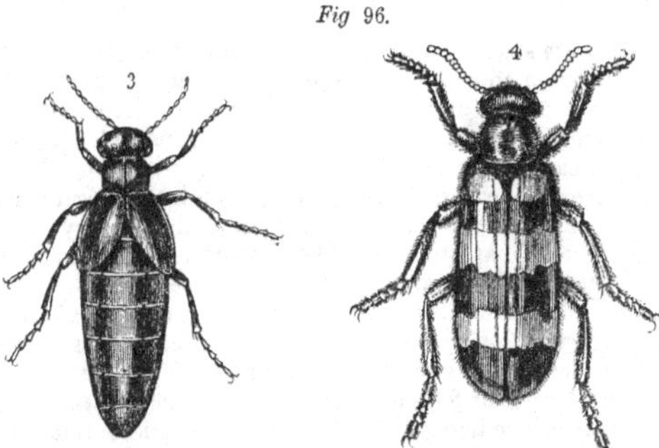

Cantharis vesicatoria (fig. 97, 1 and 2) is of an elongated, almost cylindrica form, from 6 to 10 lines in length by about 2 in breadth, the male somewhat smaller than the female. It is easily distinguished by its two beautiful elytra or wing-cases, which are long and flexible and of a shining golden green colour, and cover two thin brownish membranous wings. The head is large and sub-cordate, with a longitudinal furrow along the head and the thorax: this is not larger than the head, rather quadrate. Its thorax chiefly, but also the rest of the body covered with whitish gray hairs; antennæ black, long, simple filiform. The maxillæ support the jointed palpæ, of which the terminal joint is some-

Fig. 97.

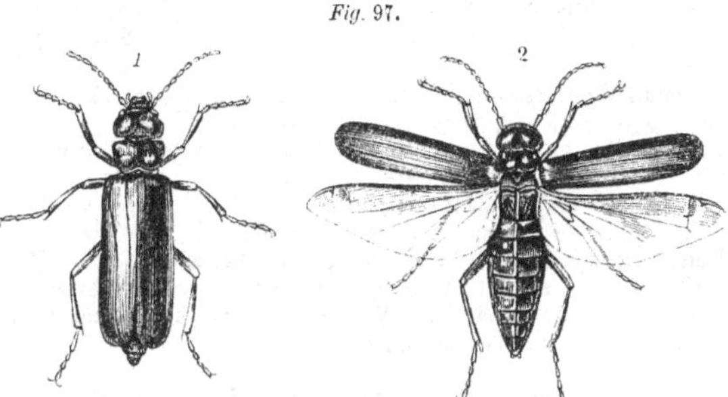

what ovate. The legs are from 4 to 6 lines long, smooth, 5 joints to the first pair of the tarsi, and 4 only to the last, all violaceous. Single spine on leg and notch in tarsus. The last joint of tarsus with a pair of claws, each of which is bifid. Near the anus of the female are two articulated caudal appendages, somewhat similar to palpi.—It abounds in the S. of France, Spain and Italy; and it has spread into Germany and the S. of Russia. It is found upon the Ash, Lilac and Privet especially, but also upon the Elder and Honeysuckle, more rarely on the Plum-tree, Rose, Willow, and Elm. M. Farines states, that the insects produced in warm places, and such as are exposed to the sun, are most energetic.

Cantharides are stated not to live above eight or ten days. When alive, they exhale a strong, fœtid, and penetrating odour, by which their presence is readily detected, and so offensive, that public walks sometimes become deserted until they have disappeared. They are usually caught early in the morning, when persons with covered hands and faces shake them off the trees, plunge them into vinegar, or expose them in sieves to the vapour of vinegar, and then dry them in the sun or in warmed apartments. They should be preserved in well stoppered bottles, as they are subject to be destroyed by other insects, introducing with them a little Alcohol, Petroleum, Camphor, Chloride of Lime (*Derheims*), or Pyroligneous acid (*M. Farines*), as it is preferable to employ fresh, well-dried, and smooth, not dusty, insects. They long preserve their form and colour, also to some extent the disagreeable odour of the living insect, and have an acrid burning taste. The powder is of a greyish-brown colour, interspersed with shining particles, which are the fragments of the elytra, head, and feet; and, though comparatively inert, these parts are very indestructible, and thus often serve to detect their presence in fatal cases.

Action. Uses. Narcotic, Acrid Poison; Rubefacient and Irritant; Stimulant Diuretic; sometimes producing inflammation of the Urinary organs; but useful in smaller doses in inducing a healthy state of the mucous membrane in Chronic Gonorrhœa, Leucorrhœa, &c. Best known and most extensively employed for raising Blisters when applied to the skin, in a space of from six to twelve hours, sometimes accompanied with the disadvantage of producing painful Strangury.

TINCTURA CANTHARIDIS, L. E. D. Tincture of Cantharides.

Prep. Macerate for 14 (7, E. D.) bruised *Cantharides* ℥ij. (℥ß. E.; ʒij. D.) in *Proof Spirit* Oij. (by measure ℔iß, D.) Strain. (Express strongly and filter, or more conveniently by percolation, E.)

Action. Uses. Stimulant Diuretic.

D. ♏x., but cautiously increased—f℥j. with some demulcent. Used internally in Gleet, Leucorrhœa, incontinence of Urine. Proof Spirit is a good solvent; it is to be remarked that the three Tinctures are unequal in strength. With Soap or Camphor Liniment used as a Rubefacient in Rheumatism, &c.

ACETUM CANTHARIDIS, E. (EPISPASTICUM), L. Blistering Vinegar of Cantharides.

Prep. Macerate for 8 (7, E.) days powdered *Cantharides* ℥ij. (℥iij. also Euphorbium coarsely powdered, ʒiv. E.) in *Acetic acid* Oj. (f℥v. mixed with Pyroligneous acid f℥xv. E.) occasionally agitating. Express strongly and filter.

Action. Uses. Rubefacient and Epipastic. Useful as a powerful method of raising a blister; too strong and irritant for internal exhibition.

CERATUM (UNGUENTUM, E.) CANTHARIDIS, L. Cantharides Cerate or Ointment.

Prep. With heat melt *Spermaceti Cerate* ℥vj. (Resinous Ointment ℥vij. E.); add very finely powdered *Cantharides* ℥j.; mix. (Stir briskly as it concretes on cooling, E.)

Action. Uses. Irritant; employed to keep open blisters and issues. The active principles are dissolved in the fatty matter aided by the heat, though too great heat will diminish the activity.

UNGUENTUM (UNG. INFUSI, E.) CANTHARIDIS, L. D.

Prep. L. D. Boil very finely powdered *Cantharides* ℥j. (℥ij. D.) in *Aq. dest.* f℥iv. (by measure ℥viij. D.) down to one-half. Strain. To the filtered liquor add *Resin Cerate* ℥iv. (℥viij. D.) Evaporate to the proper consistence.

E. Infuse for one night powdered *Cantharides* ℥j. in boiling *Aq.* f℥v. Squeeze strongly, and filter the expressed liquid; add *Axunge* ℥ij.; boil till the water is dispersed. Then add *Bees-wax* and *Resin* āā ℥j. When liquid, remove the vessel from the fire; add *Venice Turpentine* ℥ij. Mix thoroughly.

Action. Uses. Irritant. Used as the Cerate; but is milder in its action.

EMPLASTRUM CANTHARIDIS, L. E. D. Blister Plaster.

Prep. Melt together *Lard* ℔ß. L. D. (*Suet* ℥ij. E. ℔ß. D.) and *Wax plaster* ℔jß. (Resin āā ℥ij. E. Yellow Wax ℔j. and Yellow Resin ℥iv. D.) Remove from the fire, and when near concreting from cooling, sprinkle in very finely powdered *Cantharides* ℔j. (℥ij. E.) Mix, and make a plaster, stirring briskly, E.

Action. Uses. Vesicant: the most commonly employed application for raising blisters. In some cases the skin requires to be previously stimulated with an Embrocation or Sinapism, but generally the action is certain and complete.

EMPLASTRUM CANTHARIDIS COMPOSITUM, E. Compound Blister Plaster.

Prep. Melt *Bees-wax* ℥j. and *Burgundy Pitch* ℥iij. add *Venice Turpentine* ℥ivß. and while hot sprinkle into it finely powdered and mixed *Cantharides* ℥iij. *White Mustard Seed* and *Black Pepper* āā ʒij. and *Verdigrise* ʒß. Stir the whole briskly as it concretes on cooling.

Action. Uses. Powerful and Irritant Blistering Plaster; but seldom used.

Several substances have of late years been introduced as substitutes for Blistering Plaster, which are both elegant and efficient, under the names of *Tela vesicatoria*, *Blistering Tissue*, &c., and of which other forms are the Parisian *Taffetas vesicant*, *Papier et Taffetas epispastique.* (*v.* Soubeiran, ii. p. 210.) They are made with an etherial or alcoholic Extract of Cantharides or of Cantharidin, mixed with wax, and spread in a very fine layer upon these tissues, previously oiled or waxed.

EMPLASTRUM CALEFACIENS, D.

Prep. With a medium heat melt together *Cantharides Plaster* 1 part and *Burgundy Pitch* 7 parts; mix, and make a plaster.

Action. Uses. Rubefacient. Useful for relieving internal pains, as in the joints, &c.

Order *Hemiptera*,

COCCUS CACTI, *Linn.* L. E. D. Cocci, L. The entire Insects, E. Cochineal.

Cochineal, so valuable as a dye, is of little importance in medicine. Several of the genus are used as dyes, as the Kermes insect, or *Coccus Ilicis*, found on the Ilex Oak; the *Coccus Lacca*, or Lac insect, found in various trees and shrubs in India, much used as a substitute for Cochineal and its Resin (Shell-Lac) for various purposes; the *Coccus polonicus*, found on the roots of *Scleranthus perennis.* Something similar is found on the roots of a plant in the marshes of Herat. (*Burnes.*) The true Cochineal or *grana fina* of the Spaniards, is found in the cool parts of Mexico, as near Oaxaca, whence one kind of Elemi is obtained. The *grana sylvestra*, an inferior species, is found in hot parts, as near Vera Cruz, and in Brazil. The female of the Cochineal insect is alone collected from off the Cactus plant, where the impregnated females have previously been placed by the natives of Mexico to produce their numerous young. These, when matured, are brushed off, and killed by artificial heat. Three harvests are annually collected. The *Silver* and *Black* varieties of the *Grana fina* are known. The insects are oblong, roundish, plano-convex, from one to two lines in length, wrinkled, the former of a colour resembling silver-paper, owing to fine down with which they are covered, and by which the genuine may be distinguished from any imitation of the colour with powdered Talc, &c. They are of a purplish-grey colour, while the *black* variety is without bloom, and of a dark-reddish colour. An inferior kind (granilla) made up of smaller insects, and broken fragments is also sold. Cochineal, when powdered, is of a carmine colour, without odour, but having a slightly-bitter taste. It contains some fatty matters, with a brilliant colouring principle, which has been called *Cochenilline*, making it valuable as a dye. It is used for giving colour to some tinctures.

Order *Hymenoptera*, *Linn.*

APIS MELLIFICA, Linn., L. E. D. The Honey-Bee is officinal only on account of the Honey and Wax which it secretes or stores up.

MEL, L. E. D. Humor e floribus decerptus et ab Ape præparatus, L. Saccharine secretion, E. Honey.

Honey is secreted by the nectaries of flowers, sucked by the Bee into its crop, where it undergoes some slight changes, and is then stored up in the comb. The finest Honey is that which is allowed to drain from thence; and, if obtained from hives which have never swarmed, it is called Virgin Honey. It partakes of the properties of the plants from which the bees have collected it. When of fine quality it is liquid and viscid, but translucent, having a fine, though peculiar odour, and a very sweet taste, but the best appears to some people slightly acrid, from the uneasiness experienced in the fauces. After a time honey becomes thick, white, and granular in texture. Inferior qualities are of a reddish-brown colour, granular, and intermixed with impurities, and are usually obtained by pressure of the comb. Honey is soluble in water, and a great part is taken up by boiling Alcohol. It is composed of crystallizable, with some uncrystallizable Sugar, a small proportion of Mannite, and a little aromatic principle. The inferior qualities contain Wax, some acid matter, and impurities. Diluted with water, it undergoes the Vinous fermentation, and Hydromel, or Mead, is produced, "Honey should be clarified by despumation before being used. A watery solution is not rendered blue by Iodide of Potassium added along with an acid," L. proving that it is not adulterated with Starch.

MEL DESPUMATUM, D. Clarified Honey. Melt Honey in the vapour-bath, and then remove the scum.

Action. Uses. Demulcent, and slightly Laxative. Chiefly used in gargles, &c. It is a constituent of Oxymels, and of the Mel Boracis and Mel Rosæ.

CERA, L. E. D. (CERA FLAVA, E. D.) Concretum ab Ape paratum; Waxy Secretion, E. CERA ALBA L. E. D. idem dealbatum. Bleached Bees' Wax.

Wax, like Honey, has been known and employed from very ancient times. It is secreted in considerable quantities by various plants, as the Palms, p. 591, by species of *Myrica*, especially at the Cape of Good Hope, and by less known plants in China and Japan. It used to be supposed that the Bee merely conveyed it from the plant; but it is now known that it is secreted by the Bee in glands situated on its abdominal scales, and thence collected by it for the construction of its cells.

CERA FLAVA. Yellow or Common Wax is obtained by subjecting the comb to division and expression, melting the residue in boiling water, keeping it hot for some time, to allow the impurities to separate or be dissolved. On cooling the Wax concretes. It is melted again, and then strained, and sold in cakes. It is of a dull yellow colour, and has a peculiar, somewhat agreeable odour, and is bleached by agitation with water, &c. Thus, by making it fall, previously melted, in small streams, upon a revolving, wetted cylinder, it concretes in thin ribbon-like layers, which being exposed for some time to the united influence of light, air, and moisture, become bleached, and of a yellowish-white colour. Spermaceti is often mixed with it, to improve its colour; and Starch, Tallow, &c., as adulterations. Wax is firm in consistence, but melts at 155°; burns with a bright light. It may be dissolved in boiling Alcohol and in Ether, but very readily so in the Fixed and Volatile Oils. Resin and fat unite with it, and imperfect soaps are formed by its union with alkaline solutions. Dr. John conceived it to be composed of two principles *Cerin* and *Myricin*. It is now considered to be a simple principle.

Action. Uses. Demulcent; made into an emulsion, it is sometimes useful in sheathing abraded and irritable surfaces, as in Catarrh and Chronic Dysentery; but its chief value is for external application in the various forms of Cerates, Ointments, and Plasters.

CERATUM, L. UNGUENTUM (SIMPLEX, E.) CERÆ ALBÆ, D. Simple Cerate.

Prep. Melt *White Wax* ℥iv. (℥iij. E. ℔j. D.) add it to *Olive Oil* f℥iv. (f℥vß. E. prepared Hogs-lard ℔iv. D.) Mix, and stir briskly while it concretes in cooling, E.

LINIMENTUM SIMPLEX, E. Simple Liniment.

Prep. With gentle heat dissolve *White Wax* 1 part in *Olive Oil* 4 parts; agitate well, as the fused mass cools and concretes.

Action. Uses. For Emollient dressings. The Liniment is used for softening the skin, &c.

EMPLASTRUM (SIMPLEX, E.) CERÆ, L. Wax Plaster.

Prep. Melt together (with a moderate heat, E.) *Wax* ℔iij. (℥iij. E.) *Suet* ℔iij. (℥ij. E.) and *Resin* ℔j. (℥ij. E.) Strain. (Stir briskly till the mixture concretes on cooling, E.)

Action. Uses. Chiefly employed as a basis for other plasters.

UNGUENTUM CERÆ FLAVÆ, D. Ointment of Yellow Wax.

Prep. As *Ung. Ceræ albæ*, D., substituting *Yellow* for *White Wax.*

Action. Uses. Soft dressing; but more stimulant than the White Cerate.

Class CONCHIFERA, *Lamarck.*

OSTREA EDULIS, *Linn.* TESTÆ, L. The Shells of Oysters.

Oysters are well known as articles of diet. Immense quantities are dredged up on the coast of Kent, near Whitstable, and also on the opposite coast of Essex. The Shells are alone officinal on account of the Carbonate of Lime of which they are chiefly composed, with a small portion of the Phosphate, and a trace of animal matter.

TESTÆ PRÆPARATÆ, L. Prepared Oyster Shells.

Prep. Wash the *shells*, first freed from dirt, with boiling *Aq*, then prepare as directed with Chalk. (*v.* p. 110.)

Action. Uses. Antacid as Chalk, sometimes preferred on account of the Phosphate of Lime.

IV. SPINI-CEREBRATA, *v.* VERTEBRATA.

CLASS PISCES.

ICTHYOCOLLA, a name derived from ιχθυς a fish, and κολλα glue, is translated Isinglass, a word derived from the German *Hausenblase*, from *hausen*, the great sturgeon, and *blase*, a bladder, being one of the coats of the swimming-bladder of Fishes, chiefly of the genus Acipenser, or Sturgeon, and of which the best qualities are imported from the rivers of Russia, flowing into the Black and Caspian Seas, but also from the Sea of Aral, and the Lake Baikal. Isinglass is also imported from Brazil, and likewise from India. Of late the quality of this has been much improved. Isinglass is the purest known form of Animal Jelly; and it is therefore as Gelatine that it is valuable. Gelatine, when pure, is transparent, and nearly colourless, devoid of both taste and smell, easily preserved when in a dry state, but soon putrefying when moist. It is soluble in the different dilute acids, as well as in the fixed alkalis. Its solution forms a copious precipitate on the addition of Tannin, and which smells like tanned leather. As Corrosive Sublimate does not precipitate Gelatine, it serves to distinguish it when in solution from Albumen. Gelatine unites with a large proportion of water, and on cooling becomes a solid tremulous mass. By boiling with Sul', it may be converted into a kind of Sugar. Isinglass in its purest form is white, semitransparent, devoid of smell or taste, softening in cold water, and dissolving in boiling water, with the exception of a minute proportion of earthy impurities. Bengal Isinglass, analysed by Mr. E. Solly, yielded, in three specimens, respectively 86·5, 90·9, and 92·8 per cent. of Gelatine. For the sources of Isinglass *v.* Brandt and Ratzeburg Medicinische Zoologie, Pereira's Elements of Materia Medica, and the Author's Pamphlet on the Production of Isinglass along the Coasts of India, 1842.

Action. Uses. Demulcent. Used chiefly as a nutritious and little irritating article of diet for the sick.

OVUM, L. The Egg, E. of the Hen of GALLUS DOMESTICUS, *Temm.* (*Phasianus Gallus*, Linn.), L. E.

The common Fowl, domesticated everywhere, is probably derived from the Jungle Fowl of India. Its Egg is well known as a highly esteemed article of diet. The shell, consisting almost entirely of Carbonate of Lime, is seldom now employed. Within the shell there is a white semi-opaque albuminous membrane, which contains the White or *Albumen Ovi*, a glairy viscid fluid, contained in very delicate membranous cells. The liquid may be considered a solution of Albumen, as it consists of 12 per cent. of this principle, and 85 of water, 2·7 of Mucus or uncoagulable matter, and 0·3 of saline substances, including Soda and traces of Sulphur. The glairy liquid is miscible with water, coagulated by heat under 212°, as in boiling an egg, also by acids. It then becomes white, opaque, and insoluble. The white of Egg is precipitated by Corrosive Sublimate, Diacetate of Lead, Muriate of Tin, Tannin, &c.

The Yolk or Yelk, *Vitellus Ovi*, is also fluid, opaque, and yellow in colour, without odour, but with a bland and pleasant taste, forming a milky emulsion when agitated with water, and assisting to suspend in it many insoluble substances. It consists of Oil (which contains much Elaine with a little Stearine) 28·75, Albumen 17·47, and water 53·8, with a little free Sulphur and Phosphorus in combination.

Action. Uses. The White of Egg is used chiefly for the clarification of watery liquids with the aid of heat, and of spirituous ones without heat. It may be used as an antidote in poisoning by Corrosive Sublimate and the salts of Copper. It is sometimes employed as an emollient application. Agitated with Alum, it forms an astringent poultice. (*v.* p. 125.) The Yolk, besides being nutritive, is extremely useful in pharmacy in making Emulsions, and suspending Oils, Oleoresins, and Resins.

Class MAMMALIA, *Linn.*

OSSA, D. Bones.

The bones of Mammals are no doubt intended and chiefly employed. They consist of Gelatine and of about 60 per cent. of earthy matter, the greater portion of which is Phosphate of Lime with about 1-5th of Carbonate of Lime, and small portions of other salts. This is obtained by burning away the gelatinous part in an open fire, and then powdering the earthy remains. These are used for making the Phosphas Calcis præcipitatum, D. and Sodæ Phosphas. L. E. D.

If bones are burnt in close vessels, *Carbo animalis*, L. E. (p. 50) or Animal Charcoal, is obtained, which consists of Charcoal and Phosphate of Lime. This is much used for purifying Sugar, the vegeto-alkalis, &c.

The Gelatine of Bones is not officinal in the P., but it is often used dietetically in making nutritious soups. The earthy parts, being soluble in dilute Hydrochloric or Muriatic acid, are removed by bones being digested in such an acid solution, having first been carefully cleaned by boiling and scraping, and subsequently washed; the Gelatine is then boiled out, and flavoured, or with vegetable additions converted into soup. The diet of man, to be properly nutritious, requires to be of a mixed nature, and Bone-Gelatine is as well suited as anything else to be one of its constituents, though, like other proximate principles, it will not answer by itself.

Class CETACEA, *Linn.*

CETACEUM, L. E. D. Concretum in propriis capitis cellulis repertum, L. Cetine, nearly pure, E. of PHYSETER MACROCEPHALUS, *Linn.* Spermaceti of the Sperm Whale or White Cachalot.

Spermaceti, as defined in the L. P., is found in peculiar cells situated in the great head (but a little is also found in other parts, as the blubber) of the above Whale, which inhabits the Pacific Ocean and China Sea. It is itself of the nature of a concrete fat, and occurs as an oily substance disposed in numerous cells situated in a great cavity along the upper jaw of the animal, and separated from its blubber by a ligamentous covering, and supported by ligamentous partitions projected across, as fully described by John Hunter. The liquid contents of the head being taken out and boiled, on cooling, the Spermaceti concretes, and the valuable Oil swims, and is further separated by draining and pressure. Subsequent fusion and the action of a weak alkaline solution purifies the Spermaceti, which is then seen in beautiful white, pearly, crystalline masses, soft and slightly unctuous, with little odour or taste ; Sp. Gr. 0·94 ; capable of being powdered with the addition of a little Spirit ; melts at 112° ; burns readily ; is insoluble in water, slightly soluble in Alcohol, more so in Ether, and readily in fixed and volatile oils. Spermaceti may be separated from any oil it contains by boiling Alcohol, which dissolves its pure principle.

CETIN. This is considered a neutral fatty body intermediate in nature between Wax and Fats. It is like Spermaceti in all respects, being only separated from oil, and melts at 120°, is saponified with difficulty, when two substances are formed, one a neutral crystallizable fat, which can be sublimed, and has been called *Ethal*, and the other *Ethalic* acid, which approaches Margaric acid in nature. Cetin is composed of Carbon 81·66, Hydrogen 12·85, Oxygen 5·48 = 100. (*Chevreul*).

Action. Uses. Emollient; formerly much used internally with Mucilage or yolk of Egg; now chiefly externally in cerates and ointments.

CERATUM (UNGUENTUM, D.) CETACEI, L. CER. SIMPLEX, E. Spermaceti Cerate.

Prep. Melt together *Spermaceti* ℥ij. (1 part, E. ℔j. D.) and *White Wax* ℥viij. (3 parts, E. ℔ß. D.), and add them to *Olive Oil* Oj. (6 parts, E. prepared Hogslard ℔iij. D.) Stir briskly while cooling.

Action. Uses. The Spermaceti Cerate and Ointment are both emollient dressings, the former of softer consistence than the latter.

UNGUENTUM CETACEI, L. Spermaceti Ointment.

Prep. With gentle heat melt together *Spermaceti* ʒvj. and *White Wax* ʒij. in *Olive Oil* f℥iij. Stir well till cold.

CORNU, L. E. CORNUA CERVINA, *Ramenta*, D. Horn and Horn-shavings of the Stag. CERVUS ELAPHUS, *Linn.*

The antlers of the Stag, or rather of the male called Hart, are known officinally as Hartshorn and Hartshorn-shavings. They are selected because they differ in chemical composition from common horn, as of oxen, sheep, &c., which is analogous to coagulated Albumen. The antlers of the Stag, which are shed annually in spring, have the composition of Bone, that is, Gelatine 27 and Phosphate of Lime 57·5, with 1 part of Carb. of Lime. The former, however, is more soluble in boiling water than the Gelatine of bones. Subjected to destructive distillation, an impure Carbonate of Ammonia is obtained. This is so well known by the name of Spirit of Hartshorn that other preparations not so obtained are called by the same name. When burnt, Hartshorn yields ashes consisting almost entirely of Phosphate of Lime.

Action. Uses. Hartshorn shavings, boiled in water, yield a nutritious and colourless jelly, which may be used for the same purposes as that obtained from Isinglass or other sources.

CORNU USTUM, L. PULVIS CORNU CERVINI USTI, D. Burnt Hartshorn.

Prep. Burn fragments of Bone in an open vessel till white, then rub into very fine powder (as directed for Chalk, L. 110.)

Action. Uses. Consisting chiefly of Phosphate of Lime it may be used for the same purposes as Bone-ashes.

MOSCHUS, L. E. D. Humor in folliculo præputii secretus, L. Concretum Moschus dictum, D. Inspissated secretion in the follicles of the prepuce, E. of MOSCHUS MOSCHIFERUS, *Linn.* Musk.

The Musk animal differs from common Ruminants in the absence of horns, and in having long canine teeth on each side of the upper jaw. It inhabits the mountainous regions of central Asia, extending from the Himalayas to the Altai mountains, and from these to China. Hence, as in the case of Rhubarb from the same regions, we have Russian, China and Indian Musk. It is singular that the common Hindoo name of the Musk, and in the Himalayas that of the Musk animal, is *kustooree*, a name similar to Castoreum, a substance which Musk so closely resembles in nature. The name *musk* is no doubt derived from the Arabic *mishk* or *mooshk*, which is evidently the same word as the Sanscrit *mooshka*. This has been used as a perfume and as a medicine by the Hindoos from very early times. It seems to have been adopted from the Hindoos by Serapion, but it was previously mentioned by Ætius.

The animal bears a close resemblance to the Deer tribe in shape and size. It is usually less than three feet in length, with the haunches somewhat more elevated than the shoulders. The want of horns and the projecting canines have already been mentioned. There are altogether 32 teeth : namely, 8 incisors in the lower jaw, 2 canines in the upper, and 24 molars. The canines are not met with in the female. The ears are long and narrow, and the tail very short. The fleece, which consists of strong, elastic, undulated hairs, varies in colour with the season, the age of the animal, and perhaps the place which it inhabits. The general colour is a deep iron-grey. The individual hairs are whitish near the root, and fawn-coloured or blackish towards the tip. The gestation of the female was quite unknown until Mr. Hodgson in Nepal ascertained that it was about 170 days. They are extremely timid, mild, and gentle in their nature. Found on the tops of difficultly-accessible and generally open mountains, usually in the neighbourhood of the snow, but coming nearer to the plains according to the inclemency of the seasons, springing from rock to rock with great agility.

The Musk animal is particularly distinguished by the males secreting the remarkably strong-smelling secretion called Musk, in a plano-convex, oval, hairy bag, of which the orifice is situated just before the præputial orifice. The sac is flat, smooth, and naked above where it is applied against the abdomen, convex below, and hairy, composed of several coats. The Musk is secreted by small gland-like bodies situated in little pits on the most internal of these coats. The quantity in each sac varies from 1½ to 3 drachms. It is most abundant in the rutting season, and when fresh is soft, and of a reddish-brown colour. When dried, and contained in its native sacs, it forms the Musk of commerce. The kinds known are the Chinese and Siberian; the last is inferior in its fragrance to the other. The Chinese is probably from the same kind of cold and lofty region as the Himalayan. Some of this is imported into and apparently consumed in India.

Musk is in grains or lumps, soft and unctuous to the touch, of a reddish-brown colour, with a powerful, penetrating, and diffusive smell. It is usually adduced as an instance of the subtlety of the particles of matter. The taste is bitter, disagreeable, and somewhat acrid; readily inflammable. Rectified Spirit and Ether are the best solvents of Musk. Analysed by different chemists, it has been found to contain a variety of principles, as Stearine, Elaine, Cholesterine, an Oily Acid combined with Ammonia, free Ammonia, various salts, and animal principles, as Albumen, &c., an odoriferous principle which seems particularly attached to the Ammonia. But the quantity and proportion of the constituents vary considerably, perhaps from the difficulty of obtaining specimens which have not been subjected either to abstractions or to additions. The hunters even are said to adulterate it, and if they do not the Chinese merchants do so.

MISTURA MOSCHI, L. Musk Mixture.

Prep. Rub up *Musk* and *Sugar* āā ʒiij., then add powdered *Acacia* ʒiij., gradually adding *Rose Water* Oj,

TINCTURA MOSCHI, D. Tincture of Musk.

Prep. Digest for 7 days powdered *Musk* ʒij. in *Rectified Spirit* by measure ℔j. Strain.

Action. Uses. Stimulant, Antispasmodic, slightly Hypnotic. M. Trousseau considers it useful in some nervous affections, and that it is Aphrodisiac. It may be given in 5-grain pills every three or four hours, and increased to ℈j.; or in doses of f℥j. to f℥jß. of the foregoing Mixture. The quantity of Musk in the Tincture is very small in proportion to the Spirit.

SEVUM, L. E. ADEPS OVILLUS, D. Fat of Ovis Aries. The Sheep. Suet. Adeps Ovillus præparatus, D.

The Sheep domesticated from the earliest times is supposed to be derived from *Ovis Argali.* Its flesh in the form of Mutton and Lamb is well known as a highly nutritious article of diet. The Suet or Fat taken chiefly from near the kidneys is alone officinal. It is prepared by being cut into pieces, melted with a moderate heat, sometimes with a little water, and strained, D. Suet is white, firm in consistence, brittle, with little smell, and of a bland taste. According to Chevreul, it consists of Stearine and Elaine, with Hircine and a little Margarine. The two first are abundant also in Lard. Hircine is a liquid like Elaine, but differs in being much more soluble in Alcohol. Suet requires for its fusion a higher temperature (103°) than any other animal fat, owing to the very little Margarine it contains. It is composed of Carbon, Hydrogen, and Oxygen.

Action. Uses. Emollient. Useful in giving greater consistence to plaster, &c.

LAC. Milk : a peculiar liquid secreted by the mammæ of Mammiferous animals for the support of their young. That of the Cow is most commonly employed in this country, though that of Goats and Asses is also employed, and that of Buffaloes and Camels in Eastern countries. It is remarkable that the milk of Carnivora is of the same general nature as that of the vegetable-feeders. Cows' Milk, however, being that usually made use of as a demulcent, or as an antidote in cases of poisoning, and for making the *Mistura Scammonii*, E., it will be sufficient to notice its constituents.

Milk is white, and appears homogeneous, but is actually an emulsion composed of a transparent serous fluid, with numerous globules of fatty matter floating in it. When allowed to rest, these separate in the form of *Cream*, which by agitation, and pressing out the whole of the liquid (*Butter-milk*), and washing with water, is converted into *Butter*, which retains some of the serous fluid. It is sometimes further purified by melting it, as in the case of the *clarified Butter* or *ghee* of India. Butter is composed of Stearine and Elaine, the proportions of which vary at different times. There is also a volatile odoriferous principle, which has not been isolated, but has been called *Butyrine*, and a yellow Colouring matter. Milk is said to be *skimmed* when the cream has been removed, and will of itself become acid, when the clots termed *curds* separate from it. But if an acid or rennet be added, an albuminous substance separates, which is the basis of Cheese, and has been called *Caseum* and *Caseine*. This differs from the Albumen of Egg (p. 659) by not coagulating with heat. The residual serous fluid, or *Whey*, will on evaporation yield *Sugar of Milk*, Lactic acid, and other substances. The composition of Milk, according to M. Haidlen, is water 873, solid residue 127, Butter 30, Caseine and insoluble salts 51, Sugar and soluble salts 46. The salts are combinations of Phosphoric with Lime, Magnesia, and Peroxide of Iron, Chloride of Sodium and Potassium, and Soda in combination with Caseine. The Sp. Gr. of Milk varies from 1·030 to 1·035. When fresh, it has always an alkaline reaction, but soon becomes acid from the speedy conversion of its Sugar into Lactic acid. (*v.* Simon, *Animal Chemistry*, ii. p. 62.)

Action. Uses. Milk is nutritious as an article of diet, and useful as a Demulcent and often in cases of acrid poisoning, as in that of Corrosive Sublimate, Sulphate of Copper, Nitrate of Silver, &c.: whey as a Diluent.

Order PACHYDERMATA.

ADEPS, L. Adeps præparatus. ADEPS SUILLUS, D. AXUNGIA, E. Fat of SUS SCROFA, *Linn.* The Hog. Hogs-lard.

The Hog is well known both in its wild and domesticated state, and that its flesh was prohibited as an article of diet both to the Jews and Mahomedans. The fat about the loins having greater consistence than that of other parts, is preferred for medical use; but that of the omentum and mesentery is also employed, as well as the subcutaneous fat. It is prepared much in the same way as Suet, but it should be kept stirred to prevent the separation of its constituents. That commonly sold contains salt, which unfits it for medical use, until it has been melted in boiling water to remove the salt, as directed in the D.P.

Lard is white, with little odour and taste, of a soft consistence, though granular in appearance; fusible at a temperature of from 80° to 90°; partially soluble in Alcohol, readily so in Ether and the volatile Oils; when fused, melts Wax and the Resins; is converted into Soap by union with the alkalis. Exposed to the air, it becomes rancid, that is, acrid in taste and disagreeable in odour. It is composed of Carbon 79, Hydrogen 11, and Oxygen 9 parts in the hundred, and of three proximate principles, Stearine and Margarine, amounting to 38, and of Elaine 62 per cent. These may be separated by the action of Alcohol, which deposits Stearine upon cooling as a white, concrete, and crystalline body (*v.* p. 476), and the Elaine when evaporated. The Margarine may be separated from Stearine by being more soluble in cold Ether; or they may be separated from Elaine by pressure or congelation.

Action. Uses. Lard, like other fats, is Emollient, and used as a basis for various Cerates and Ointments. It is occasionally employed as an ingredient in laxative Enemata.

Order RODENTIA.

CASTOREUM, L. E. D. Concretum in folliculis præputii repertum, L. A peculiar secretion from the præputial follicles E. of CASTOR FIBER, *Linn.* Castor from the Beaver.

Castor, a substance analogous in nature to Musk, has been employed in medicine from the time of Hippocrates. It is described by the Arabs under the head of *joond bedustur.* The description by Dioscorides leaves no doubt about the animal, which is so interesting and remarkable for its *building* habits in North America, though those of Northern Europe, from *burrowing*, are supposed to constitute distinct species; but Cuvier states, that after the most scrupulous comparison of the Beavers which burrow along the Rhone, the Danube, and the Weser, he has been unable to find any characters to distinguish them from those of North America. The Beaver is moreover distinguished from other Rodentia by its nearly oval tail (*a*), which is flattened horizontally, and covered with scales: it is peculiar also in the Castor-sacs, which are found both in the male and female, and of which a detailed account is given by Brandt and Ratzeburg. From them the accompanying illustration has been copied, as the Castor sacs (*h.h.*) are often confounded with the testicles (*w. w*), and their position is so difficult to understand. They can be distinctly seen only on the removal of the skin of the abdomen. Besides these there are two others (*e. e.*), which are Oil-sacs. All are situated between the pubic arch and the cloaca, a common hollow which is covered by a wrinkled hairy protuberance, into which open the Oil and Castor sacs, and the rectum (*b*,) and prepuce (*i.*) The Castor-sacs are somewhat pear-shaped and compressed, communicate by the same opening at their narrow extremities, but their fundi diverge. Like the musk-bags, these sacs have several coats; within all there is a convoluted mucous membrane, covered with scales, with a small brownish body, supposed to be a gland. The secreted matter, or Castor, in these sacs is at first of a yellow-orange-colour, but becomes of a brownish colour as it becomes exposed to the air.

Fig. 98.

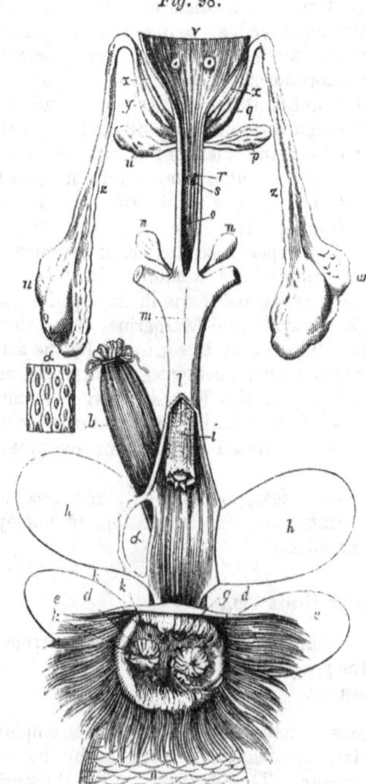

Two kinds of Castor are known in commerce: one American, imported by the Hudson's Bay Company, and the other, Russian, but which is very rare. This may be distinguished by a Tincture of 1-16th part in Alcohol being of the colour of deep Sherry; while that with the American Castor is of the colour of London Porter. The American, moreover, effervesces when dropped into an acid, which the Russian does not do. (*Per.*) The sacs are usually united together by a part of the above excretory duct, and sometimes the oil-sacs may be seen with them. Internally they are divided into numerous cells, of which the membranes may be seen when the Castor is dissolved out; or when torn they may be seen intermixed with the Castor, which often breaks with a resinous fracture, and is of a reddish-brown colour. The odour is strong, fœtid, heavy, and the taste bitter, rather disagreeable. The chemical constituents are a volatlie Oil, Resin, Ozmazome, Albumen, Mucus, Urate, Carbonate, Benzoate, Phosphate, and Sulphate of Lime, with salts of Soda and Potash, some Carbonate of Ammonia, and a peculiar non-saponifiable substance, which crystallizes, and has been called *Castorin;* but there does not appear any proof of its being the active principle.

Action. Uses. Moderate Stimulant and Antispasmodic. Mr. Alexander, as also M. Joerg and his pupils having taken full doses of the Castor, experienced only eructations; but M. Trousseau has justly observed that we are not justified in inferring that because a medicine does not affect those in health, that therefore it will have no effect on those labouring under disease, especially when this is of a nervous nature. He found it decidedly useful in many nervous and spasmodic affections, and in all its actions to resemble Valerian and Assafœtida rather than Musk. He recommends its union with Tincture of Assafœtida or of Aloes. It may be exhibited in powder or in pills in doses of ʒß.—ʒij.

TINCTURA CASTOREI (ROSSICI, D.), L. E. Tincture of Castor.

Prep. Macerate for 14 (7. D.) days bruised (Russian, D.) *Castor* ℥ijß. (℥ij. D.) in *Rectified* (Proof, D.) *Spirit* Oij. (by measure ℔ij. D.) Strain. (Prepare by digestion or percolation, as Tinct. Cassia, E.)

Action. Uses. Intended to be Antispasmodic, but is too weak a preparation.

TINCTURA CASTOREI AMMONIATA, E. Ammoniated Tincture of Castor.

Prep. Digest for 7 days in a well-closed vessel *bruised Castor* ℥ijß. and *Assafœtida* in fragments ʒx. in *Spirit of Ammonia* Oij. Strain and express strongly the residuum. Filter the liquor. Not conveniently prepared by percolation.

Action. Uses. Stimulant Antispasmodic in doses of fʒj.—fʒij. The Spirit of Ammonia is a good solvent, and both itself and the Assafœtida are useful in the same class of cases as the Castor.

THERAPEUTICAL ARRANGEMENT OF THE MATERIA MEDICA.

Therapeutical arrangements of Medicines are nearly as numerous as the authors who have written on this subject. That of Dr. Murray is one of the most clear and simple, and has been adopted as the basis of their arrangement by Dr. Paris and by Dr. A. T. Thomson. It is also sufficiently comprehensive, with the additions made by Dr. Paris, to fit it for practical purposes. We have in the following table retained together the Remedies which are called Chemical, though only a portion of their effects are due to their agency as such, and a great part, like that of all other medicines, to the agency of the vital functions. The Evacuants are necessarily all grouped together, and are followed by the Depressents or Contra-Stimulants, as being generally employed as parts of the Antiphlogistic treatment, and, like many of the Evacuants, to diminish action generally; while the General Stimulants are employed to rouse the flagging powers, and to give permanent tone and strength to the system. The author has reversed the order of Dr. Murray's arrangement, in order to treat first all Remedies most simple in their action, and of those which are employed to depress, before those which are employed to excite and to strengthen.

MECHANICAL REMEDIES.

Diluents.	Emollients.
Demulcents.	

CHEMICAL REMEDIES.

Acids.	Disinfectants.
Alkalis.	Escharotics.
Antilithics.	Antidotes.

EVACUANTS OR SPECIAL STIMULANTS.

Rubefacients.	Expectorants.
Counter-Irritants.	Diaphoretics.
Vesicants.	Diuretics.
Errhines.	Emmenagogues.
Sialogogues.	Anthelmintic.
Emetics.	Alteratives.
Cathartics.	

DEPRESSENTS OR CONTRA-STIMULANTS.

Refrigerants.	Narcotics.
Sedatives.	Antispasmodics.

GENERAL STIMULANTS.

Diffusible Stimulants.	Astringents.
Aromatics.	Tonics.

CHEMICAL REMEDIES.

ACIDA. Acidulæ. Antalkalis.

Acids are ranked among chemical remedies only when employed to counteract an alkaline state of the secretions, as in the Phosphatic diathesis (*v.* ANTILITHICS), and perhaps when employed as Antiscorbutics. Acids in a concentrated state are well known to act as Caustics; but if moderately diluted, and applied to the skin or other mucous membrane, a pungent sensation with a little astringency is produced, followed by stimulant reaction. So, taken internally they will act as poisons; but if much diluted, a sensation of coolness is experienced; less diluted, a tonic effect is produced, and in large doses considerable irritation. They may therefore be employed as Refrigerants or Astringents, also as Tonics. From their irritant effects, they are employed as Rubefacients and Caustics, and the Vegetable acids often act as Laxatives.

Acidum Sulphuricum Dil. 33.
—— —— Aromaticum, 33.
Potassæ Bisulphas, 82.
—— Phosphoricum Dil. 36.
—— Nitricum Dil. 40.
Spiritus Ætheris Nitrici, 641.
—— Hydrochloricum Dil. 48.
—— Nitro-Hydrochloricum Dil. 49.
—— Carbonicum, 53.
in Carbonic acid water.
—— Oxalicum, 315.
Binoxalate of Potash, 316.
Oxalis Acetosella, 314; and Rumex Acetosa, 519.

Acidum Citricum, 300.
Limonum Succus, 299; Lemonade, 299.
—— Tartaricum, 312.
Potassæ Bitartras, 87.
—— Aceticum, 645.
—— Pyroligneum, 645.
Acetum Pyroligneum, 644.
—— Destillatum, 644.
Syrupus Aceti, 645.
Acid Fruits.
Lemons and Limes, 296, 297.
Vine Juice, 311.
Tamarinds, 354.

ALKALIS. ANTACIDS.

Alkalis or alkaline Earths introduced into the stomach, or making their way into the intestinal canal, will neutralize any acid present, in the same way that they would do out of the body. They are therefore frequently prescribed in cases of Heartburn and Dyspepsia to counteract acidity, whatever may be the cause, and the treatment is consequently only palliative. But in poisoning by acids, they will, by neutralizing, put a stop to their corroding effects. It must be remembered, however, that some of them are as powerfully corrosive as the acids themselves: excess therefore must be carefully avoided. In diminishing acidity they at the same time allay irritation. When a little in excess, they produce some stimulant effect on the stomach; but their continued use is injurious in neutralizing the healthy acid of the gastric juice. Absorbed into the system, they may be detected in the excretions, and will thus diminish any acid state of the secretions generally. By their stimulant effect when thus diffused, they are useful in some glandular affections, and are hence employed as ALTERATIVE-

STIMULANTS. Some of them, as Chalk, &c., are occasionally called *Absorbents*. Rubbed on the surface in a diluted state, they will act as RUBEFACIENTS, and when concentrated, as CAUSTICS. Ammonia is in some cases preferred for its *Stimulant* and Potash for its *Alterative* effects, Magnesia as a Laxative, though apt to form concretions, and Chalk as an apparent Astringent, while Soap with some Purgative is preferred as an Antacid in sluggish states of the intestinal canal.

Ammoniæ Liquor, 57.
  Spir. Ammoniæ, E.
  Tinct. Opii Ammoniata, 269.
Carbonate of Ammonia.
  Spir. Ammoniæ, 60.
  —— —— arom. 61.
  —— —— fœtidus, 61.
Ammoniæ Sesquicarbonas, 61.
  Liq. Ammoniæ Sesquicarb. 61.
Ammoniæ Bicarbonas, 64.
Potassæ Liquor, 71.
—— Carbonas, 77.
  Liq. Potassæ Carbonatis, 77.
  —— —— effervescens, 79.
Sodæ Carbonas, 91.
—— —— ex siccata, 92.
Sodæ Bicarbonas, 93.
Liq. Sodæ effervescens, 94.
  Sapo durus, 476.
Calcis Liquor, 108.
Carbonate of Lime, 108.
Creta præparata, 110.
  Testæ præparatæ, 639.
  Mistura Cretæ, 111.
  Trochiscus Cretæ, 111.
Pulv. Cretæ compositus, 111. Tormentil and Saffron.
—— —— c Opio, 111, 267, with Opium.
  Confectio Aromatica, 111.
  Hydrargyrum cum Creta, 182.
Magnesia, 116.
  Pulv. Rhei comp. E. 525.
Magnesiæ Carbonas, 117.
  Mist. Camphoræ c Magnesia, 587.
Bicarbonate of Magnesia,
  or Soluble Magnesia, 119.
  Hydrargyrum cum Magnesia, 183.
Alkaline Mineral Waters, as of Malvern, Vichy, &c.
Some Oxides of Metals, as of Zinc, 149.
Oils also act as Antacids.

ANTILITHICS. *Lithontriptics.*

Antilithics (from *αντι*, *against*, and *λιθος*, *a stone*) is a preferable name to *Lithontriptics*, and is applied to remedies which counteract the tendency to the deposition of Urinary Calculi. The Urine, compound in nature and very variable, is, in a healthy state, a little acid, from the presence of Super-Lithate of Ammonia. But there may be an excess of this, as in the Lithic acid diathesis, from irregularities of the digestive organs, check to the functions of the skin, &c. A deposit takes place of reddish powder, or rather crystals, consisting chiefly of the Super-Lithate of Ammonia, sometimes with some pure Lithic acid. But when there is a deficiency of acid in the Urine, a white sandy deposit takes place, consisting chiefly of an Ammonio-Magnesian Phosphate with some Phosphate of Lime. Sometimes there is a deposition of Oxalate of Lime, as in the *Mulberry* Calculus.

Attention to diet. Vegetable food in some cases, meat in others. Exercise. Baths. Attention to the skin. Diluents. Water, distilled or mineral, but pure. Alteratives. Diaphoretics. Tonics.

1. In the Lithic acid Diathesis Antacids (q. v.) are indicated.

Potassæ Liquor. Potassæ Carbonas et Bicarb. Liq. Potassæ Carb. and Liq. efferves. 71—79.

Effervescing Saline Draughts produce an alkaline reaction in the Urine.

Potash and its Carbonates are more eligible than Soda, because the Lithate of Potash is soluble, that of Soda insoluble.

Soda and its Carbonates. Soda siccata and Liq. Sodæ effervescens, p. 91—94. Sapo durus. Waters of Vichy, and other alkaline mineral waters.

Ammonia and its Carbonates, 57—64: act as stimulants and as antacids in the stomach.

Calcis Aqua. Creta præparata. Testæ præparatæ, 108—111.

Magnesia or its Carb. 116, 117. Magnesia-water, the Bicarb. with excess of Car' gas, 119.

Colchicum and Mercury both diminish the acidity of the Urine.

2. In the Phosphatic Diathesis an acid is indicated.

Nitric and especially Muriatic, also Dil Sul', and Phosphoric acids. Mr. Ure has particularly recommended Benzoic acid and soluble Benzoates. Carbonic acid. (*v.* ACIDS.) Vegetable acids, as Vinegar, may sometimes be used as articles of diet; but Tonics and Vegetable Bitters are required, with generous diet, Wine, and Opium and the avoidance of everything depressing.

3. In the Oxalic' Diathesis, mineral acids with tonics, the Muriatic or Nitro-muriatic. Meat, and nourishing farinaceous diet.

Local Lithontriptics, as injecting very weak Nit' into the bladder, or weak alkaline solutions. Electro-chemical action. Lithotrity. Dr. E. Hoskins (Phil. Trans. 1843) recommends the introduction of weak solutions of chemical decomponents (as the Nitro-Sacchorate of Lead) instead of solvents, into the living bladder.

DISINFECTANTS.

These are substances calculated to free the air of buildings and rooms, as well as infected bodies in general, of the invisible, usually imperceptible particles which constitute infection and propagate disease. Some of the means employed are purely mechanical, others chemical, in their action. Fumigations and Pastiles only conceal the smell, without removing the causes, and are therefore often worse than useless.

Ventilation. Caloric. Diffusive Gases which act chemically, as Chlorine, 45. Liquor Chlorini, 45. Liq. Sodæ Chlorinatæ, 96. Calx Chlorinata, 112. Acid fumes, as Sulphurous acid gas, Hydrochloric' gas, 47. Nitrous acid fumes and Acetic and Pyroligneous acids are less effectual.

Destruction of infected matter by application of heat, of Quicklime, 106; of Charcoal, 50. Fumigating Pastiles, Balsamic Resins, and Aromatic Vinegar, only diffuse an agreeable odour.

ESCHAROTICS with some Local Stimulants.

Escharotics (from Eschar) are often called *Caustics*, occasionally *Potential Cauteries*, to distinguish them from the Actual Cautery or great Heat used for the same purposes. Concentrated acids and alkalis destroy the vitality of a part by forming a chemical union with one or more of the constituents of the animal body. Their action is afterwards followed by a stimulant reaction. Some of those enumerated below may be considered rather as Local Stimulants, or such as excite the parts to which they are applied, but which by continued action will also cause the erosion of a part. The Liquid Caustics have the disadvantage of spreading, but they are useful in cases of the bites of rabid animals, or of snakes, in following the sinuosities of a wound.

Sulphuric, 33, Nitric, 40, and Acetic acids, 645. Arsenious acid, 205, in some cases of Cancer.

Potassæ Liquor, 70. Potassæ Hydras, 67. Potassæ Carbonas, 77. Potassa cum Calce, 70,

Calx recens usta, 106. Calcis Hydras, 107. Ammoniæ Liq. fort. 57. Actual Cautery.

Local Stimulants.

Alumen exsiccatum, 124. Tinct. Ferri Sesquichloridi, 134. Zinci Chloridum, 150. Antimonii Sesquichlorid. 174.

Argenti Nitras, 216. Cupri Sulphas, 154. Cupri Diacetas, 157. Linim. Æruginis, 158. Ung. Æruginis, 158. Cupri Ammonio-Sulphas, 155.

Hydrargyri Oxydum, 186 (Black Wash). Hydrargyri Binoxydum, 187. (Yellow Wash, 189.) Hyd. Nitrico-Oxydum et Ung. 188.

Hydrargyri Bichloridum, 195. Hydr. Biniodidum, 190. Ung. Hydrarg. Nitratis, 202.

EVACUANTS OR SPECIAL STIMULANTS.

These are remedies which cause an increased secretion (hence called Evacuants) from different organs, by first exciting them to increased action. They are therefore called Local Stimulants by Dr. Murray, but as the term is objectionable, Dr. Paris proposes that of Special Stimulants, as producing "an effect which is supposed to be confined to one particular organ, though remote from the seat of application." They may be prescribed for the purpose of restoring natural secretion, but as increased secretion is followed by more or less of exhaustion, they are also very frequently employed to lessen the mass of circulating fluid, or to relieve one organ by exciting another. Hence some of them always form a part of Antiphlogistic treatment. The Rubifacients however, except when used as Vesicants, do not produce an increased secretion, but they relieve the interior by causing a determination to the surface, and thus act on the same general principle.

RUBEFACIENTS. IRRITANTS. VESICANTS.

Rubefacients, as their name indicates, produce redness of the skin, with warmth, &c. If long applied, or more concentrated, vesication will ensue ; and on continuance of the application, a suppurative discharge, whence Epispastic, from ἐπισπάω, *I draw.* Tartar Emetic produces a small pustular eruption. These local effects sometimes react upon the constitution, so as to induce a state of general excitement. The local external effect produced by the Rubefacient or Epispastic very frequently has the effect of relieving some internal irritation or deep-seated, even distant, pain ; and therefore it is for their COUNTER-IRRITANT or Revulsive effects that these remedies are applied : sometimes only to relieve slight internal inflammation, as Hartshorn and Oil on the neck to relieve sore throat, or a blister behind the ear to relieve toothache. But it is usually in chronic affections of the chest, of the abdominal cavity, or of the joints, that they are most employed, or in spasmodic attacks. Sometimes the head is relieved by hot pediluvia, or Sinapisms to the feet. Issues, Setons, and Acupuncture, are employed on the same general principles. Occasionally stimulant frictions and Sinapisms are applied to rouse the system in great depression of the vital powers. Some produce healthy granulations.

Friction. Heat; in form of Hot water, Steam, Heated Sand, Metals, and Hip and Foot-Baths.

Gases: as hot dry Air, Chlorine, Carbonic acid, 52, and Sulphurous gases, 53.

Acid Solutions, as Nitric, Acetic, &c. Ung. Acidi Sulph. 34. Ung. Acidi Nitrici, D., made with Olive Oil Oj. Lard. ℥iv. Nit′ f℥vß. Acetic′ is epispastic.

Alkaline Solutions, as of Ammonia, Potash, and their Carbonates. Liquor Ammoniæ fortior, 57. Lin. Ammoniæ, 59. Lin. Ammoniæ comp. 59. Ammoniacal Ointment, 60. Lin. Ammoniæ Sesquicarb. 64.

Antimonii Potassio-Tartras, 177: Sol. et Ung. 180. Argenti Nitras, 215, or in solution.

Ammoniæ Hydrochloras, 67. Potassii Sulphuretum et Aqua, 79. Sodii Chloridum, 95. Borax, 97. Mel Boracis, 98.

Ung. Hydrarg. Nitr. 202. Ung. Hydr. Chloridi; Hydr. Nitrico-Oxydi; Ung. Hydr. Ammonio-Chloridi, &c. as Local Stimulants.

Vegetable Irritants employed as Rubefacients, Local Stimulants, and Epispastics.

Creasote, 647, pure or diluted. Ung. Creasoti, 647. Crotonis Oleum, 554. Toxicodendron, 335.

Ranunculus acris, 235. R. Flammula, 236. Staphisagria, 238. Delphinia, Tinct. et Ung. 239. Cocculus indicus, 249. Ung. Cocculi, 250. Armoracia, 722. Sinapis alba et nigra, 274, 276. Cataplasma Sinapis, 277. Volatile Oil of Mustard, 276. Pyrethrum, 458. Capsicum et Tinct. 511. Mezereon, 527. Euphorbium, 558. Sabina et Ceratum, 599. Veratrum Dec. et Ung. 605. Piper nigrum, 560. Allium, 593. Zingiber, 614.

Volatile Oils (*v.* STIMULANTS) may be used as Rubefacients; Oil of Calamus Aromaticus (*Spikenard*), also others less agreeable, as Oleum Rutæ, &c.

Turpentines, as Terebinthina Chia, 333. T. vulgaris, Veneta, et Canadensis, 583, 584. Terebinthinæ Oleum, 586, et Linimentum, 587.

Resins, as Resina, 585. Abietis Resina et Pix Burgundica, 584. Emp. Picis. Elemi et Ung. comp. with Turpentine, 345. Galbanum et Emp. Galban comp. 424. Pix liquida. Pix arida et Ung. 587, 588. Cerevisiæ Fermentum et Cataplasma, 638. Emp. Aromaticum, 541.

Cantharides, 654. Tinct. Acetum epispast. Ceratum. Ung. Emp. et Emp. comp. 655. Emp. Calefaciens, 656.

ERRHINES.

Errhines (from εν and ῥιν, *the nose*) include all medicines which are applied to the mucous membrane of the nostrils; among them Sternutatories which cause sneezing. They may be applied in a dry, soft, liquid, or gaseous state. They may be Demulcent, Astringent, or Stimulant. Of the latter the different preparations of Ammonia and Acetic acid are frequently employed. Though generally local in their effects, they often afford relief by revulsion. Sternutatories are obtained both from the mineral and from the vegetable kingdom.

Powdered leaves of Labiatæ, as Melissa, Lavandula, Rosmarinus, Origanum, 496—502. Teucrium Marum is often called the Headache Plant. Powdered Iris-root, 618. Asarum, and Pulv. Asari comp. 548, with Lavender-flowers. Tabacum, 517, as Snuff.—Ammonia and its Carb. Acetic acid.

Some *acrid* substances, as Veratrum album, 605. Veratria, 606. Euphorbium, 558, or Hydrargyri Subsulphas flavus (*Disulphas*), 201, all with mild powders. The fumes of Biniodide of Mercury (*c*).

SIALOGOGUES.

Sialogogues (from σιαλον, *saliva*, and αγω, *to drive*) are medicines which increase the flow of saliva. This may be effected by chewing a nearly inert substance, like Mastic, or an Astringent, like Catechu; an Astringent and Purgative, as Rhubarb; or such as are acrid; or some of the warm Spices; or by the application of warm Stimulants. Salivation may also be produced by the action of Mercurial preparations (see Alterative Stimulants). They may be useful from their local or their derivative effects.

Acrids.—Armoracia, 272. Pyrethrum, 458. Mezereon, 527. Iris root, 618. Angelica, 411.
Spices.—Zingiber, 614. Pepper of different kinds, 559. Capsicum, 511.
Astringent, &c.—Catechu, 349. *Astringent and Purgative.*—Rhubarb, 520. Mastiche. 334.
Stimulants.—Application of Oil of Cloves, 349. Of Creasote, 647, &c.
Mercurial Preparations, 182—203, see *Alterative Stimulants.*

EMETICS.

Medicines which evacuate the stomach by vomiting: an act produced partly by the influence produced on the stomach, and partly by that induced by the brain and nervous system. The latter we see in Sea-sickness, and the want of it in the difficulty with which Emetics act in narcotic poisoning, when the brain is in a comatose state. Emesis is also produced by tickling the fauces with a feather. Emetics differ much among themselves, some acting only when introduced into the stomach; others, as Tartar Emetic, if applied to any other part of the body, so as to be absorbed into the system. The effect is not altogether dependent upon the nature of the substance, for Ammonia and Mustard, which in small doses act as Stimulants, and Sulphates of Zinc and Copper as Tonics, will in large doses evert the action of the stomach, and produce an emetic effect, generally quickly, and without debilitating the system. Others act more slowly, and produce long-continued nausea, with the depressing symptoms which accompany such a state, and which are known to favour absorption. These are therefore, as well as from their slow action, not suited to cases of poisoning. With both, the act is accompanied by a series of concussions which favour the excretion and secretion of the biliary, pancreatic, and intestinal fluids, causing a determination to the skin. But this very concussion makes them dangerous when there is a determination to the head, or in advanced stages of pregnancy, in hernia, &c. But it makes them useful before the accession of an Intermittent, also in Bilious Fever, likewise in Asthma, Hooping-cough; or they may be used for merely evacuating the stomach.

Direct Emetics, and acting quickly.

Ammoniæ Liq. 58. Ammoniæ Sesquicarb. Liq. 64 (f℥ß.—f℥j. of either taken in a glass of cold, followed immediately by some warm, water). Sodii Chloridum, 96, or common Salt is usually readily available.

Zinci Sulphas, 151. Cupri Sulph. 154. Cupri Ammonio-Sulph. 155. Ærugo, 157.
Sinapis nigra, 274. S. alba, 276.

Indirect Emetics.

Antimonii et Potassæ Tartr. (Tartarum Emeticum, D.), 177. Vinum, 180.
Antimonii Oxidum, E. 172. Sesquisulphuret. et Oxysulphuretum, 175, 176.
Ipecacuanha, 433. Pulv. Vin. et Syr. 436. Emetine, 435. Viola odorata, 278.
Scilla, Pulv. Tinct. et Syr. 594, 595. Asarum, 548. Euphorbium, 558, but is too acrid.
Anthemis, Inf. et Dec. comp. 458 : assists vomiting.
Tabacum, 518. Lobelia inflata, 467; but both are unsafe as Emetics.
Ipecacuanha and Tartar Emetic are often combined together, or the latter may be prescribed with a Cathartic, forming an Emeto-Cathartic.

Cathartics.

Cathartics (from καθαιρω, *to purge*) are medicines which increase the peristaltic movements of the intestinal canal, evacuate its contents, and usually augment its mucous secretions. They were formerly distinguished into *Hydrogogues*, as causing watery evacuations, *Chologogues*, as favouring the evacuation of bile, &c. They are now distinguished chiefly according to their energy of action, as into Laxatives, which merely evacuate the intestinal contents, and Purgatives, which stimulate secretion as well as evacuation. But among the latter the more violent are distinguished by the name of Drastics and of Hydrogogue Cathartics. They differ likewise according to the part of the intestinal canal to which their action is more particularly directed. Saline Purgatives, often called Cooling, act on the whole intestinal canal producing copious watery evacuations. Castor Oil stimulates superficially the mucous surface of the small intestines. Senna also acts upon them, but with much activity. Rhubarb evacuates, and exerts a tonic effect. Aloes stimulates the colon and rectum. Colocynth, Scammony, and Jalap act with energy upon the whole of the intestines. Hellebore, Elaterium, and Croton Oil, still more so, producing watery motions. Mercurials stimulate the intestinal glands and the secretion of bile from the liver. The influence of Cathartics from proximity is propagated to the uterus, as also to the kidneys and bladder; but the secretion of urine is generally diminished by the action of purgatives. As the copious watery evacuations must diminish the quantity of fluid in the body, they necessarily favour absorption and diminish excitement: they thus form a part of Antiphlogistic treatment. Hence they are useful in a variety of complaints, as in Fevers and Inflammations, avoiding in some cases those that are irritant ; also in Nervous and Spasmodic affections. They are employed to clear out in Diarrhœa, and to favour absorption in Jaundice, Dropsy, &c.; in many cases to act as Counter-irritants.

Laxatives from the Mineral Kingdom.

Sulphur sublimatum et lotum, 27. Magnesia 116. Magn. Carb. 117.

Saline Purgatives.

Magnesiæ Sulphas, 120, purum, D. (made by adding Sul' to Carb. Magnesia). Potassæ Sulph. 80. Bisulph. 81. This may be given in effervescence with Carb. Soda.

Potassæ Tartras, 87 : Bitartras, 88, also in Pulv. Jalapæ comp. 493.

Potassæ Sulphas cum Sulphure, 82. Potassæ Acetas (seldom used).

Sodæ Sulphas, 98. Sodæ Phosphas, 99 : mild. Sodæ et Potassæ Tartras (*Soda Tartarizata*), 100. Sodæ Acetas, 101.

Sodii Chloridum, 95. Pulv. Salinus comp. 82. This is formed of Sulphates of Potash and of Magnesia with Common Salt : is also Diuretic.

Mercurial Purgatives.

Pil. Hydrarg. 183. Hydrarg. c. Creta, 182. c. Magnesia, 183. Hydrarg. Oxydum, 187. The last is uncertain in its action. Hydrargyri Chloridum (Calomel), 191, usually combined with, or followed by some other Cathartic, to insure its purgative effect.

Laxatives from the Vegetable Kingdom.

Manna, 478, in Conf. and Syr. Sennæ. Cassiæ Pulpa et Conf. 355, 356, with Manna and Tamarinds in Conf. Sennæ. Tamarindus, 354, in Conf. Sennæ, Conf. Cassiæ, and Inf. Sennæ comp. Pruna, 389, in Elect. Sennæ. Uvæ passæ, 311. Fici, 572.

Viola odorata et Syr. 278. Rosa centifolia et Syr. 384.

Fixed Oils. Amygdalæ Ol. 388. Olivæ Ol. 475. Lini Oleum, 285.

Purgatives.

Ricini Oleum, 554. Senna, 356, Syr. Inf. comp. 362, with Ginger, Inf. c. Tamarindis, 362. Conf. 363, with Pulp of Cassia, Senna, Tamarinds, Prunes, Figs, and Coriander, 363. Tinct. comp. with Raisins, Caraway, and Coriander, 363.

Rheum, 520. Pil. Extr. Inf. Vinum, 526. Pulv. comp. with Magnesia and Ginger, 525. Pil. comp. with Aloes, Myrrh, and Caraway, 525. Pil. Rhei et Ferri, 526. Tinct. E. with Cardamoms, 526. Tinct. comp. with Ginger and Saffron (Cardamoms, D.), 527. Tinct. Rhei et Aloes, E. Tinct. Rhei et Gentianæ, 527.

Colocynthis et Extr. 403. Extr. comp. Spirituous Extract with Aloes, Scammony, and Cardamoms, 404. Pil. comp. E. D. with Scammony, Sulphate of Potash, and Oil of Cloves, 405. Enema, 405.

Ecbalium Elaterium, 405, and Extr. or Elaterium, 406. Tiglii, *vel* Crotonis Oleum, 552.

Jalapa, 491. Tinct. et Resina Jalapæ, 493. Pulv. comp. with Cream of Tartar and Ginger. 493.

Scammonium, 494. Resina. Mistura 496, with Milk. Conf. with Cloves and Ginger, 496. Pulv. comp. with Resin of Jalap, and Ginger, 496.

Cambogia, 304, 307. Pil. comp. with Aloes and Ginger, 308.

Aloes, 596. Extr. Tinct. with Liquorice, 601. Tinct. comp. with Saffron in Tinct. Myrrhæ, 601. Pulv. c. Canella, 600. Pulv. comp. with Guaiacum and Aromatics, 600. Pil. comp. with Soap, Gentian Extr. and Caraway, 600. Pil. Aloes c. Myrrha. et Pil. Aloes et Assafœtidæ, 600. Aloes et Ferri, 601. Dec. Aloes comp. with Myrrh, Saffron, Carb. of Potash, and Tinct. of Cardamoms, 600. Vinum Aloes with Canella, Cardamoms, Ginger, 600.

Helleborus, 236. Veratrum, 604. Colchicum, 603. Rhamnus et Syr. 331.

Linum catharticum, E omitted at p. 283, is seldom used though a good indigenous cathartic, about six inches high, with small smooth leaves, very bitter, ʒj. may be given, or ʒij. in infusion.

Terebinthinæ Oleum, 583, 586. Euphorbium, 559, but this is too irritant.

Enema Catharticum, E. D. 362; Senna, Sulph. Magnesia, and Olive Oil, E. Manna and Sulph Magnesia in Comp. Dec. Chamomile, D.

EXPECTORANTS.

Expectorants are medicines which are supposed to have the power of favouring the expulsion and excretion of mucus (*ex pectore*) from the chest, that is, "from the trachea and cells and passages of the lungs." These, like several following groups, are *relative* agents, that is, their action bears a relation to the nature of the case, the state of the patient, and the period of the disease. Thus in a state of excitement, with dryness of the skin, &c., Venesection, Warm-bath, Nauseants, and Demulcents may be useful. While in other cases, where there is deficiency of action, or, with sufficient secretion, a deficiency of power to expel the mucus, &c. secreted, Stimulants are necessary, either such as, taken in the form of lozenges, may come in contact with the upper part of the trachea, and thus have their effects propagated by sympathy, or others which may be inhaled in the form of gases: while some, taken internally, are absorbed, and have their particles carried to the mucous surface of the lungs, where they act as Stimulants. Tonics may be useful in improving the state of the constitution, and thus restore its secretions to a healthy condition, and that of the bronchial passages with power of expectoration among the rest.

Emetics acting mechanically favour expectoration. Sulph. of Zinc, 153, or Sulph. of Copper, or Carb. of Ammonia.

Demulcents and *Refrigerants*, by allaying irritation, and by obviating a dry state of the skin, favour expectoration.

Inhalations of warm Water, and of Demulcent Decoctions. Warm Baths and Pediluvia, are useful by relaxing the skin.

Gummy Substances taken slowly. Isinglass. Jujubes. Pate de Guimauve. Liquorice. Quince and Linseed, &c. (*see* DEMULCENTS.)

Nauseating Expectorants.—Antimonials. Vinum Antim. 180. Tartar Emetic, 177. Antimonii Oxydum, E. 172. Pulv. Antimonialis, 173.

Ipecacuanha. Pulv. Syr. et Vinum, 436. With *Narcotics, &c.* in Pulv. comp. with Opium, and in Pil. comp. 437, with Opium, Squill, and Ammoniacum.

Pil. Conii comp. with Ipecac. 430. Anod. Expect. so Ipecac. and Henbane. Syr. Violæ. 430.

Tinct. Camphoræ comp. 269, with Opium and Benzoin. Smoking of Stramonium, 510, and of Belladonna, 514. Also of Tobacco in some cases.

Stimulant Expectorants.

Sulphur, 27, and Alkaline Sulphurets, 79. Senega, Dec. et Inf. 279.

Balsams. B. Peruvianum, et B. Tolutanum, Syr. et Tinct. 368. Styrax, 471. Benzoin, 472. Tinct. comp. with Myrrha, 338; in Pil. Galbani comp. 424. Benzoic Acid in Paregoric, 269.

Fœtid Gum-Resins. Assafœtida, 416. Mist. and in Pil. Scillæ comp. 423. Ammoniacum, 423. Galbanum, 424.

Copaiba, 364, in Emulsion or in pills with Magnesia. Scilla, Pulv. Acetum, Tinct. et Oxymel, 594. Pil. comp. with Ammoniacum and Ginger, 595. Mist. Cascarillæ comp. 551, with Acetum Scillæ et Inf. Cascarillæ, 551. Allium sativum, &c. 593.

Succinic acid and Oil, 649. Petroleum, 648. Naphtha,* 54, 648.

---

* *Vide* a paper on the distinction between genuine and spurious Barbadoes Petroleum in P. J. iv. 73. The author understands the term Naphtha as applicable to the pure Hydrocarbon which is obtained by distillation from Petro-

Stimulant Lozenges, as of Capsicum, or of Astringents, as of Catechu, Tinct. Catechu on a lump of sugar will also answer in cases of relaxation.

Inhaling Stimulant vapours, as of Benzoin and Benzoic′ placed in warm water, of Acetic acid, of much diluted Chlorine, Ammonia, &c.

Demulcent Tonics. Cetraria, 629. Tussilago Farfara, 455. Inula Helenium, 456. Marrubium vulgare, 503. Archangelica, 411, and other Tonics.

Diaphoretics.

Diaphoretics are medicines which increase the natural function of perspiration. To these, when acting so as to produce sweating, the term of *Sudorifics* is applied. The function of perspiration is useful in keeping down temperature, and in carrying off much Carbon, Hydrogen, and even Oxygen from the blood. Like the secretion of urine, this function varies in the same individual at different times, because under the influence of different circumstances, as the state of the constitution, the nature of the food, the temperature, dryness, or rarefaction, moisture, or coldness of the atmosphere. The action of a Diaphoretic is influenced by many of the same circumstances. Hence it frequently depends entirely upon our keeping the patient in bed that it promotes Diaphoresis; while if the patient is up, and has the skin exposed to the cool air, it will act as a Diuretic. These two functions are very often antagonistic to each other: therefore when Diaphoretics act freely, much aqueous matter will be carried off by the skin, and the quantity of urine diminished, as well as the secretions of the intestinal canal, by a determination being thus caused to the surface. Some act by relaxing the surface, others act at first as stimulants, and then produce sweating. The patient requires to be kept *in bed*, the *skin* should be *clean* and *warm*, hence the double utility of warm *water* and *vapour* baths, and whatever determines to the surface, as *friction* with brushes, application of *heated bodies*, of *Rubefacients*, of *dry air*, of some *gases*, as *Carbonic′ gas*, *Chlorine*, and *Sulphuretted Hydrogen*. In other cases *exercise*, or the sympathetic influence of a glass of *cold water* will produce diaphoresis. From their mode of action and effects, it is evident that Diaphoretics are of extensive application, as those which are relaxing, in febrile and inflammatory affections, others, in rheumatic and some chronic diseases. According to the nature of the case or the period of the disease, either the relaxing or stimulant Diaphoretics will be eligible in Pulmonary affections, in Bowel complaints, in Cutaneous diseases, or in Dropsy.

leum. He is best acquainted with that obtained from Rangoon, which is a dark greenish fluid: from some, which the author received direct from thence, the late Professor Daniell obtained about 80 per cent. of Naphtha, which he employed for preserving Potassium. The Naphtha used by Dr. Hastings is stated in the P. J. iii. p. 33, to have been examined by Dr. Ure, and to be what is called by chemists Pyroacetic Spirit or Acetone. This is distinct from Wood or Pyroxylic Spirit, called also Hydrated Oxide of Methyle, though both are sold for and used as Naphtha, but both differ from it in being miscible with water, and in containing Oxygen. Dr. Ure has pointed out that Nit′ Sp. Gr. 1·45 produces a red colour, but no effervescence with Pyroxylic Spirit; but with Pyroacetic acid no change of colour, but an effervescence from copious evolution of gas.

Antimonials. Antimonii Oxidum, E. 172. Pulv. Antim. comp. 173. Jacobi verus, 172. Antimonii Sesquisulphuretum præp, 175. Ant. Oxysulphuretum, 176. Antimonii et Potassæ Tartras, 177 (Tartar Emetic). Vinum, 188.

Mercurials. Pil. Hydr. &c. 183. Pil. Hydr. Chloridi comp. (Sulph. Antimony et Guaiac.), 194. Pil. Calomelanos et Opii, 267. Hydr. Sulphuret. c. Sulph. 200.

Ipecacuanha. Emetine, 435. Pulv. Syr. et Vinum, 436. Pulv. Ipecac. comp. with Opium, 437. Pil. Ipecac. c. with Opium, 436.

Stimulant Diaphoretics. Sulphur, 27. S. lotum. et S. præcipitatum, 29. Potassii Sulphuretum et Aqua, 79, 80.

Ammoniæ Liq. 58. Sp. Ammon. arom. Liq. Ammoniæ Acet. 66. Effervescing Draughts, 301.

Alcoholic and Etherial Draughts. Sp. Ætheris Nitrici, 642. Petroleum. Naphtha, 648, and note, 677.

Opium (*see* Narcotics.) Pulv. Ipecac. comp. 437. Pil. Ipecac. c. Opio, 267. Pil. Calomel. et Opii, 267. Morphiæ Hydrochloras et Sol. 261. Morphiæ Sulphas, 261. Morphiæ Acetas, 262.

Senega, Inf. et Dec. 279. Guaiacum, Mist. Dec. Tinct. et Tinct. Ammon. 319. Toxicodendron, 335. Arnica montana, 462. Inula, 456. Sassafras, 534.

Serpentaria, Inf. et Tinct. 549. Contrajerva, 573. Mezereon, 527.

Infusions of Vegetable Excitants, as of Sage, Rosemary, &c.

The milder Diaphoretics and Alteratives are Sarza, 608, Hemidesmus indicus, 480, Calotropis, 480, Dulcamara, 507.

Diuretics.

Diuretics are medicines which are supposed to have the power of augmenting the secretion or excretion of urine. As in the case of the function of perspiration, so in that of the secretion of urine, many external circumstances control its due performance. We have also seen that these two great functions mutually supply the place of and alternate with each other, and that frequently the causes which favour the one secretion will interfere with the due performance of the other. It follows, therefore, that an opposite course requires to be followed with regard to the treatment of the patient. The skin must be kept cool, the patient rather kept out of, than in bed, and hence the daytime is frequently the best time for prescribing Diuretics. As the operation of some is incompatible with that of others, it is absolutely necessary to pay attention to their mode of action. This has been best explained by Dr. Paris, whose arrangement we have adopted, but reversed the series, in order better to compare the list with that of the Diaphoretics, as well as of other Therapeutic agents.

Whatever acts upon the system so as to promote absorption will appear to act as a Diuretic, as Tonics and Stimulants in cases of debility: but when diminution of urine is caused by an inflammatory and febrile state of the system, then Venæsection, Warm-baths, Acidulous drinks, even Diaphoretics, by producing relaxation, will restore the secretion of urine to a natural state. Diuretics are useful in Fevers and in Dropsies, and usually require Diluents to be prescribed with them. Some of the Stimulant Diuretics, as the Oleo-resins, stimulate the mucous membrane of the Urethra, and are hence employed in restoring it to a healthy state in cases of Gonorrhœa, &c.

1. *Medicines which act primarily on the Stomach or System and secondarily on the Urinary Organs.*

a. **By diminishing arterial action, and increasing that of absorption.**

Venesection and some parts of the Antiphlogistic treatment may be considered to act in this way, as well as Digitalis. Pulv. Inf. Tinct. et Extr. 506. Pil. Dig. et Scillæ, 506. Linim. with Ammonia and the Inf.; or the Tinct. and Soap Liniment. Tabacum, Vinum, 518. Lactucarium, 465, and other Narcotics.

b. **By increasing the tone of the Body in general, and that of the Absorbent System in particular.**

Bitter Tonics, q. v.; the effects of some Diuretics, as Chimaphila, Uva Ursi, Diosma, may be ascribed partly to their Tonic effects.

c. **By producing Catharsis, and thereby increasing the action of the Exhalants directly, and that of the absorbents indirectly.**

Elaterium. Gambogia. Jalapa. Pulv. Jalapæ comp. (*See* CATHARTICS.)

2. *Medicines which act primarily on the Absorbents, and secondarily on the Kidneys.*

Mercurials, &c. 181. Hydr. Chloridum, 191. Bichlorid. 195. Iodineum. Potassii Iodidum, 71. Liq. et Tinct. comp. 73.

3. *Medicines which act primarily on the Urinary Organs.*

Potassæ Liquor, 70. Potassæ Carb. et Liq. 76, 77. Bicarb. 77. Potassæ Aqua efferves. 79.

Potassæ Nitras 84, et Potassæ Chloras, 85.

Potassæ Acetas, 89. Citras, 88. Bitartras, 87, also in Pulv. Jalapæ comp.

Nitre in gr. x. doses sometimes prevents Incontinence of Urine, as does Tinct. Ferri Sesquichloridi. Dr. Bennett states that Gendriu gives Nitre in doses of ʒvj.—ʒxij. in 24 hours without injurious effects.

Sodæ Acetas, 101, and Citras. Sodæ Potassio-Tartras, 100.

Sodæ Carb. et Bicarb. 91—94. Sapo durus, 476. Sodæ Biboras, 96. Magnesiæ Sulphas, 120. Diluted Mineral acids.

Stimulant Diuretics.

Ammoniæ Liq. 58, et Sesquicarb. 64. Spiritus Ætheris Nitrici 642. Rhine Wines, especially with Squill and Bitter Tonics.

Armoracia, 272. Inf. comp. with Mustard Seed and the Comp. Sp. 273, which contains Orange-peel and Nutmegs. Cochlearia, 272.

Scoparium, Inf. et Dec. comp. with Juniper-berries and Bitartrate of Potash. Carotæ Fructus, 425. Parsley and other Umbelifers. Asparagus.

Juniperi Baccæ et Cacumina. Ol. et Spir. comp. 589, with Caraway and Fennel Fruits. Terebinthina et Oleum, 586.

Squilla, 594. Pulv. Acetum, Tinct. 596. Allii species, 593, 594. Colchicum, 603. Veratrum, 604.

Senega, Inf. et Dec. 279. Diosma (Buchu), Inf. and Tinct. 323, 324. Chimaphila (Pyrola), 467. Uva Ursi, Dec. 468.

***Demulcent, &c.* Arctium minus (Lappa), 455. Pareira brava, Inf. et Extr. 245. Sarza, 608. Dulcamara, Dec. 508. Ulmus Cortex, 573.**

Cantharides, Tinct. 654.

Copaiba, 364, et Oleum, E. 367. Cubebæ, 562. Ol. Cubebæ, stimulate the mucous membrane of the Urethra.

Emmenagogues.

Medicines which are considered to have the power of promoting the menstrual discharge when either retained or suspended. As this is sometimes the primary, and at other times the consequence of other diseases, the treatment must necessarily differ ; especially as Amenorrhœa is as often dependant on a want of constitutional energy in a leucophlegmatic habit, as on a plethoric state of the constitution, with irritation of the uterine system, brought on perhaps by an application of cold, &c. In either case attempts must first be made to restore the constitution to a natural state, and then to prescribe those remedies, all more or less stimulant, which are considered to have a specific effect as Emmenagogues, though it is doubtful whether there are many such, most seeming to act by contiguous sympathy. Dr. Paris has observed, "that as the uterus is not an organ intended for the elimination of foreign matter, it is necessarily less under the control of medicines."

In cases of *Plethora*, Venesection may be necessary, or Cupping on the Loins with Leeches to the Loins or Groins, and Legs or Feet. Purgatives. Hip-baths. Pediluvia of Hot Water, or made more stimulating with Mustard-flour. —Exercise, especially on horseback, in a salubrious air.

In cases of *depression*, Warm Purgatives are equally necessary, accompanied with the alterative action of Blue Pill; at first mild then stronger Tonics, followed by the preparations of Iron. Frictions in the Lumbar region. Electricity. Nourishing Diet. Exercise. Fresh air. Sea-bathing. Shower-baths. Alteratives often necessary, or Mercurials, as Pil. Hydrargyri, &c. 183, and Pil. Hydrargyri Chloridi comp. (*Plummer's Pill*), 194. Iodine, in form of Iodide of Potassium, 71, or the Iodide of Iron.

Purgatives, employed as Emmenagogues.—Aloes, 599. Pil. Aloes c. Myrrha, et Dec. Aloes, 600, et Tinct. Aloes, 601, with Assafœtida in Pil. Aloes et Assafœtidæ, E. 600. With Iron, in Pil. Aloes et Ferri, E. 601.

Colocynthis, 403. Senna, 356. Cambogia, 302. Helleborus, 236.

Antispasmodics—as Assafœtida, 414. Moschus, 662. Castoreum, 665.

Galbanum, 423, in Pil. Galbani comp. 424, with Assafœtida, Sagapenum, and Myrrh.

Myrrha, 338. Tinct. 342, with Iron, in Pil. Ferri et Mistura Ferri, 140, 141, also in Pil. Galbani comp. 424; and with Aloes. (*v.* supra.)

Mineral Tonics.—Ferri Sulphas, 137. Pil. et Mist. Ferri comp. 140, 141. Ferri Carbonas Saccharatum et Pil. 140. Ferri Iodidum et Syrupus, 131, 133.

Emmenagogues.—Rubia Tinctorum, 432, and Meum Athaminticum, now little used. Senega, Inf. et Dec. 179. Serpentaria, Inf. et Tinct. 548. Rutæ, Ol. et Conf. 321. Tanacetum, 461.

Juniperus, 588. Sabina et Ol. 589. Ergota, 631, has been found useful by Dr. Locock (Cycl. of Prac. Med. i. p. 70), who also states the same of a combination of Myrrh, Aloes, Sulphate of Iron, and the Essential Oil of Savine.

Anthelmintics.

As some of the Anthelmintics are of a mechanical nature, they may be treated of with the other Mechanical Remedies, though some of the most important are adopted from other classes.

Alteratives, or Alterative Stimulants.

The term *alterative*, so commonly employed, is differently interpreted by different authors. Müller includes under it all such Remedies as are neither Stimulant nor Sedative, and have the power of effecting changes in the state of the living solids, and consequently in the functions which they perform. The term is, however, usually applied to such as, taken in comparatively small doses, and continued for some time, by degrees and almost without any perceptible effect, *alter any disordered actions* (*Conolly*) or secretions. Under this head Trousseau and Pidoux include Mercury, Iodine, Gold, and the Alkalis; so also Edwards and Vavasseur, calling them "medicamens excitans qui agissent specialement sur certaines glandes et sur l'absorption en general." In Alterative treatment is usually included the taking of various decoctions of the *woods*, or substitutes for them in the form of the Decoct. Sarzæ, &c., which, taken with large quantities of water, must operate by its diluting and solvent properties, and partly by the stimulant effect, though small in quantity, of the active principles of the several ingredients in these diet drinks conveyed into the capillaries.

Mercury, in some of its preparations, acts as an Irritant (*v.* Local and Special Stimulants, Escharotics, Errhines, Cathartics). But when some of its suitable preparations, as Blue or Plummer's Pill, or Corrosive Sublimate, are prescribed in small doses, with considerable intervals, as every night or so, there is by degrees perceived an improvement in the function of digestion, as well as in the evacuations, with a softer state of the skin. If larger doses are given, or more frequently, some excitement in the circulation may be observed, as well as in the absorbent system, and in the several secretions, as is instanced in the stimulation of the liver, the kidneys, and in the healthful perspirable state of the skin. The less observable effects, dependent on smaller doses, spread also over a greater space of time, will produce *alteration* in *disordered* actions, so as to cause an improvement in the digestive and nutrient functions, the disappearance of eruptions, and the removal of thickening of the skin or of other tissues. A greater degree of the same action will promote the absorption of glandular enlargements, or of indurated structure, and thus the *Deobstruent* effects of Mercury may be obtained. All this may be short of its constitutional effects, indicated by fœtor of the breath, redness of the gums, followed by salivation, often very profuse, and even by ulceration; but often the beneficial effects are only observed when some of these phenomena display themselves, though in a very slight degree. These effects of Mercurialism, as the state is called, may be produced either by repeated small doses, by one or two large ones, or by rubbing Blue Ointment into the skin, or using the Mercury in the form of fumigations. But many deleterious consequences follow from the unadvised use of Mercury and its preparations.

Iodine, like Mercury, will in concentrated doses act as an Irritant on the surface to which it is applied. Hence some of its preparations are poisonous. But if taken for some time in small doses, the stimulant effects are observed in the increased perspiration, often in the improved secretion of bile as well as of urine; while the mucous membrane of the nostrils becomes inflamed, as in Catarrh. But the characteristic phenomenon in the action of Iodine is the disappearance of glandular enlargements, as in the case of Goitre, or even of glands in a healthy state, as of the mammæ and testicles, under its long continued use. Other symptoms are also observed, included under the term Iodism, which in addition to some of the foregoing, is especially characterized by giddiness and headache, nausea, want of appetite, restlessness, weakness, and emaciation, with a weak but frequent pulse. The medicine ought to be intermitted for a time on the first appearance of any of these symptoms. In the Iodide of Iron, the alterative effects of the Iodine are combined with the tonic effects of the Iron. Bromine and the Bromide of Potassium may be used for many of the same purposes as Iodine.

The preparations of Gold, as the Powder and the Chloride of Gold and Sodium, are likewise stimulant of the absorbents, and may with great benefit be used in Scrofula or Secondary Syphilis.

Arsenious acid, mentioned as an Antiperiodic under the head of Tonics, is a powerful Alterative in many skin diseases, as Lepra and Psoriasis. It is generally discontinued when the symptoms of accellerated pulse, weakness and itching of the eyes, griping, restlessness at night, or a great feeling of weakness and lowness of spirits, are experienced. Mr. Hunt (*Lancet*, 1846) recommends diminishing the dose, and continuing the medicine so as not to lose its effect.

Before the discovery of Iodine, the Chlorides of Calcium and of Barium were frequently employed as stimulants of the glandular and lymphatic systems. They were formerly much employed in scrofulous diseases, in Bronchocele, and other glandular complaints, as well as in chronic skin diseases. The *alkalis* also, as Liq. Potassæ and the Carbonates, when taken for some time, besides the effects described under the head of Antacids, produce many of the same effects as Iodine, &c. in removing glandular swellings. They are supposed to diminish the consistence of the blood, rendering it more watery, and reducing the habit to a state resembling Scurvy. Dr. Pereira proposes the term of Liquefacients as synonymous with the *verflussigende mittil* of Sundelin for medicinal agents which augment the secretions, check the solidifying, but promote the liquefying, processes of the animal economy, and which by continued use create great disorders in the functions of assimilation.

Mercurials.—Hydrargyri Pil. 183. Hyd. c. Creta, 182. Hyd. c. Magnesia, 183. Ung. fort. et mitius, 184.

Hydrargyri Cerat. comp. with Soap and Camphor, 185. Lin. comp. with Camphor and Liq. Ammoniæ, 185, 186. Emp. et Emp. Ammoniaci c. Hydr. 187.

Hydrarg. Oxydum, 187. Iodidum, Pil. et Ung. 189. Biniodid. et Ung. 191.
Hydrarg. Chloridum, 191. (Calomelanos). Pil. comp. 194, with Oxysulph. of Antimony and Guaiacum, 194. Ung. 196.
Hydrarg. Bichloridum, 195, et Liq. 197, with Sal Ammoniac. Hydrargyri Acetas, 203.
Hydrarg. Bisulphuretum, 200, and Sulphuretum c. Sulphure, 200, for fumigations.
Hydrarg. Nitratis Ung. 202. Lin. 203. Ung. Hydrarg. Ammonio-Chlorid. 198.
Iodineum, 40. Iodide of Starch, 42. Tinct. 42. Tinct. et Ung. comp. 42, with Iodide of Potassium. Ung. Iodinei, D. (made with Iodine ℈j. to Lard ℥j.)
Potassii Iodidum, 71. Ung. 73. Ung. et Tinct. comp. with Iodine, 42, 73. Liq. comp. with Iodine.
Ferri Iodidum et Syr. 131, 133. Hydrargyri Iodid. 189, et Biniodidum, 191. Plumbi Iod. 163, et Ung.
Burnt Fuci, 631. Some Sea-weeds, 40. Burnt Sponge, 650.
Bromineum, 43. Potassii Bromidum, 74. Bromide of Iron, 133.
Acidum Arseniosum, 206. Liq. Potassæ Arsenitis, 213. Iodide of Arsenic and Mercury, 213.
Alkalis, &c.—Liq. Potassæ, 71. Potassæ Carb. 76. Bicarb. 79.
Calcis Aqua, 107. Calcii Chloridum, 113. Liquor, 114.
Barii Chloridum, 104. Liq. 105.
Auri Pulvis, 219. Chloride of Gold and Sodium, 219.
Acid. Nitro-Muriaticum, 49. Antimonii Oxysulph. 176. Tartar Emetic, 177.
Mild Vegetable Alteratives. Sarza, 608. Dulcamara, et Dec. 508.
Taraxacum, 463. Dec. et Extr. Rumex aquaticus et Hydrolapathum, 519. Ulmus, Dec, 574.

DEPRESSENTS OR CONTRA-STIMULANTS.

The author has grouped together the Refrigerants, Sedatives, Narcotics, and Antispasmodics, not that they can be considered as resembling each other physiologically in action (though the Sedatives are often united with Narcotics), but because all are employed to subdue inordinate action, the Refrigerants and Sedatives when occurring in the circulation; Narcotics, at first excitant, are followed by collapse, and are employed to assuage pain, control restlessness and spasm, and to procure sleep, while Antispasmodics have a quieting effect on the disordered nervous system.

Refrigerants.

Refrigerants are placed by Dr. Murray among chemical remedies; but the ingenuity of his views has been more admired than their correctness. Dr. Paris has contrasted them with those of Liebig. Refrigerants are employed to diminish the heat of the body, and to reduce the force of the circulation. They are either applied externally or given internally.

External Cool Air. Cold Water. Continued Sponging.* Evaporating Lotions. Freezing mixtures. Ice.

* This local application may be made to produce the most powerful constitutional effects. The author has kept patients labouring under the most severe attacks of Jungle Remittent continually sponged, so that the circulation was never

Vegetable Acids, q. v. Trochisci Acidi Tartarici, E.* Limonum Succus et Syr. 299. Lemonade. Aurantii Fructus, 296. Acetum, 644. Syrupus Aceti, 645. Tamarindus, 354. Acetosella, 314. Rumex Acetosa, 519.

Potassæ Bitartras, 87. Potassæ Nitras, 83. Potassæ Chloras, 85.

Mild Diaphoretics. Liq. Ammoniæ Acet. 65. Effervescing Draughts.

Sedatives.

Sedatives, when strictly defined, are medicines which directly depress the energy of the nervous system, without causing any previous excitement. Though there are few medicines which can be rigorously brought under this definition, there are several which may be prescribed in diseases of increased action, to repress any undue excitement of the nervous or circulatory systems.

Sedatives form a class of remedies respecting which there is considerable difference of opinion, some denying that there are any medicines which can be strictly so called, others uniting them with Narcotics, q. v.; while Dr. Paris, Dr. A. T. Thomson, and Dr. Billing, admit Sedatives as a distinct class, distinguishing them from Narcotics as directly and primarily depressing the powers of life, without previously exciting. They form a group of medicines well suited to control inordinate action, especially as displayed in the circulation. However much writers may differ respecting the proper position of Sedatives in a classification, practitioners know that there are medicines which may be prescribed beneficially in cases of excitement, when they would avoid Narcotics. They should not unite these with true Sedatives in a prescription. There are, however, a few remedies, such as Digitalis, which, though exciting the circulation at first, depress it in so much greater a degree, that they may frequently be prescribed even in diseases of the Heart. Dil. Hydrocyanic', so useful in allaying irritation and cough, also gives great relief in certain painful affections of the Stomach.

Nitrogen, 24. Sulphuretted Hydrogen, 53 (so Carburet. Hydrogen), acts as a Sedative. Also Carbonic acid gas, when inhaled, but all are poisonous,

Aconitum, 239. Extr. Alcoholicum et Tinct. 241. Aconitina, Tinct. and Ung. 242.

Acid. Hydrocyanic. Dil. 391. Lauro-Cerasus et Aq. 389. Amygdalæ amaræ. 388. Mist. D. 387.

Ferrocyanide of Potassium, 143. Ferri Percyanidum, 144.

Antimonii et Potassæ Tartras, 177. Vinum, 180. Ipecacuanha, Pulv. et Vin. both in nauseating doses.

Plumbi Acetas et Pil. Opiatæ, 167. Plumbi Diacet. Liq. et Dil. 168, 169.

Digitalis, Inf. Tinct. Extr. 504, 506. Linim. with Soap and Ammonia. Pil. Dig. et Scillæ, 506.

---

allowed to rise: on one occasion, while the patient lay in a comatose state for six days and nights. Drying of the skin was immediately revealed by the moaning of the patient, when his pulse would immediately become hard, full, and bounding. Blisters to the nape of the neck and Sinapisms to the feet were also made use of, while Calomel and Purgatives were producing a change in the system.

* These are the Lemon or Acidulated Drops, made with Sugar ℥viij. Tar' ʒij. Oil of Lemons ♏x. and Mucilage q. s.

Tabacum. 517. Enema et Vinum, 518, or the Smoke.

Venesection. Leeches, 651. Application of Cold, as by continued sponging.

Antiphlogistic treatment generally, which includes Blood-letting, with low diet. Purgatives. Refrigerants. Demulcents.

Hydrargyri Chloridum, 194, in gr. x. and gr. xx. doses in Dysentery, &c. Pil. Calomelanos et Opii, E. 194.

NARCOTICS.

Narcotics are medicines which have been named from ναρκη, *torpedo* (which stupefies any other animal that it touches), because stupefaction is the most striking symptom of some, though not of all the medicines usually included under the head of Narcotics, Hence Dr. Pereira has suggested the name of Cerebro-Spinants. He includes under this class several which are here placed in other groups ; but it will be found advantageous to pay attention to his subdivisions, because the several Narcotics differ very remarkably from each other in their mode of action. Narcotics are distinguished from true Sedatives by producing when prescribed in moderate doses "an increase of the actions of the Nervous and Vascular systems, but which is followed by a greater depression of the vital powers than is commensurate with the degree of previous excitement, and which is generally followed by sleep." *Paris.* From the varied effects which they produce, some are called Anodynes, from relieving pain, others Hypnotics and Soporifics, from inducing sleep. Their tendency is to weaken and even to destroy more or less completely the functions of the nervous system. They are had recourse to chiefly to assuage pain and to procure sleep; hence they may be useful in a great variety of affections, chiefly in the treatment of nervous and spasmodic complaints, painful diseases, as Neuralgia, Rheumatism, &c., and in the last stages of other painful complaints.

Papaver somniferum, 252, Capsulæ, 253. Dec. Syr. et Extr. 254

Opium, 254. Extr. 266. Pil. Trochisci, 267. Conf. vel. Elect. Tinct. 268 (Laudanum). Vinum. Acetum, 269. Liq. Opii sedativus and Black Drop, 270. Enema, Lin. et Emp. 270.

Pills with Opium. Pil. Saponis comp. and Pil. Styracis comp. 267. Pil. Calomelanos et Opii, 194, 267. Pil. Plumbi Opiatæ, 167, 267.

Tinct. Camphoræ comp. with Opium, Camphor, and Benzoic', 269. Tinct. Opii Ammoniata, E. with Sp. Ammon. E.

Opium with Ipecacuanha. Pil. et Pulv. Ipecacuanhæ comp. 436, 437.

Opium with Astringents. Pulv. Kino comp. 267, 373. Elect. Catechu, 268, 352. Pulv. Cretæ comp. c. Opio, 267. Ung. Gallæ comp. 580.

Morphia, 258. Hydrochloras (Murias, E.), 259. Muriatis Sol. 261. Trochisci, 261. Troch. Morphiæ et Ipecac. 261.

Morphiæ Sulphas, 261. Morphiæ Acetas, 262. Citrate and Bimeconate, 262.

Lactuca sativa, 464, virosa, 465. Lactucarium, 464. Extr. 465. Tinct. et Trochisci, 466.

Hyoscyamus niger, 514. Extr. et Tinct. 516.

Anodynes, which dilate the pupil.—Belladonna, 508. Extr. Succus. Tinct. et Emp. 511. Stramonium. Extr. 512.

Benumbent and Sedative.—Aconitum Napellus, 239. Extr. et Succus Spissatus, L. D. 241. Extr. E. 241. Extr. Alcoholicum Aconiti et Tinct. 241. Aconitina, 242. Solutio et Ung. 242.

Paralysers.—Conium maculatum, 429. Extr. (Succus spissatus), Tinct. 429. Pil. comp. with Ipecac. 430. Ung. et Cataplasma, 430.

Humulus Lupulus, 564. Inf. et Tinct. and Tincture of Hop-Glands, 568.

Anticonvulsive and Anodyne.—Cannabis sativa, 568. Extr. et Tinct. 571. Resinous Tincture (*Donovan*), 571.

Acro-Narcotics and Cathartics.—Staphisagria, 238. Colchicum, 601. Veratrum, 604. Sabadilla, 606.

Cocculus indicus, 249. Lobelia inflata, 466. Spigelia, 481.

Nux Vomica, Toxicodendron, and Arnica montana, *see* EXCITANTS.

ANTISPASMODICS.

Antispasmodics are medicines prescribed for the purpose of allaying the irregular muscular contractions denominated Spasms. As these may arise from various causes, so whatever removes the cause will in many cases assuage the spasm and the pain which accompanies it. This may at one time be an Antacid or a Purgative which removes a source of irritation from the intestinal canal ; at another time, an Anodyne, which, by lulling pain, stops the irregular movements to which this gives origin ; and if these should depend on debility, then strengthening diet and Tonics will prove Antispasmodic. The name is however usually applied to a group so peculiar in their action, as to be accounted excitant by some and sedative by others. Though exciting the circulation, they have a sedative effect upon the nervous system in disease, apparently by strengthening and thus restoring it to a healthy state. They are remarkable for rapidity of action, as well as for their effects being temporary, and thus requiring a repetition and increased doses of the medicine. They are chiefly prescribed in nervous complaints, especially Hysteria, and in Spasmodic and Convulsive affections ; also in Asthma, Spasm in the stomach or intestinal canal ; sometimes in the advanced stages of Typhoid disease.

Valeriana, Inf. Tinct. 453, 454, and Tinct. Ammon. *Fœtid Gum-Resins.* Assafœtida, 414. Mist. Tinct. et Enem. Emp. E. with Galbanum, 421. Galbanum, Tinct. et Pil. comp. with Myrrh, Assafœtida, and Sagapenum, 424. Emp. 425, with Resin and Turpentine. Sagapenum, 413. Pil. comp. with Aloes and Ginger. Opopanax, 413.

Ruta, 321. Ol. et Conf. comp. with Bay-berries, Sagapenum, and Black Pepper. Tanacetum, 461. Anthemis et Ol. 458.

Camphora, 536. Mist. et Mist. c. Magnesia. Tinct. et Tinct. comp. with Opium. Linim. with Oil, and Lin. comp. with Ammonia and Sp. Lavender, 537,

Æther Sulph. 639. Sp. Ætheris Sulph. c. 641. Sp. Ætheris Nitrici, 642.

Ammoniæ Sp. arom. 61. Sp. Ammoniæ fœtidus, 61. Ammoniæ Sesquicarb. 61.

Petroleum, 54. Naphtha, 538, 677. Succini Ol. 649.

Moschus, Mist. et Tinct. 662, 663. Castoreum, 665. Tinct. and Tinct. Ammon. 667.

Tonics.—Argenti Oxydum, 215. Chloridum, 215. Nitras, 218. Zinci Oxydum et Sulphas, 150. Cupri Ammonio-Sulph. 155. Pil. 156. Bismuthi Trisnitras, 169. Cinchona, and other Tonics.

Narcotics.—Opium, 254. Belladonna, 508. Stramonium, 512. Hyoscyamus, 516, &c. Tabacum, Enema et Vinum, 517. Lobelia, 465. Tinct. and Tinct. Ætherea.

GENERAL STIMULANTS.

These are distinguished from the Special Stimulants in not having their influence confined to one or two organs, but in accelerating all the principal functions of life, as the sanguineous, muscular, and secreting systems by directly influencing the nervous system. Temporary increase of action is, however, not to be considered increased strength, for all such excitement is followed by exhaustion, and therefore the true Diffusible Stimulants can only be used for temporary purposes. They were judiciously divided by Dr. Murray into Diffusible and into Permanent Stimulants, the latter including Astringents and Tonics. With the former may be considered the Aromatics.

STIMULANTS. *Excitants. Exhilarants.*

Stimulants, as stated under the head of Rubefacients, when applied externally, produce redness, a sensation of warmth quickly communicated to surrounding parts, often followed by pain, according to the more or less susceptibility of the organ. If taken internally, the sensation of warmth is experienced in the stomach and intestinal canal, followed by a marked increase in the vital energy and contractility, with activity of digestion, often accompanied by thirst and dryness of the mouth. In large doses, those which are more acrid will prove irritant to the intestinal canal, and act as Cathartics. These effects are very temporary with many Stimulants; with others, they are quickly communicated to the Heart and Circulation, which is increased in force and frequency ; more animal heat is developed, and transpiration promoted both from the cutaneous and pulmonary surfaces, as well as many of the secretions increased; and with some of the Stimulants the organs of generation participate in the general excitement. In all these cases it is the nervous or true cerebro-spinal system which is primarily affected, and through it, by reflex action, all the cerebro-spinous organs. "The nervous energy becomes more equable and rapid, and the muscular contractility more energetic ; the senses more delicate and perfect, and the intellectual faculties even seem to acquire more activity and development." (*Guersent.*) All excitement is, however, followed by exhaustion; and though the collapse is proportionally much greater with Narcotics, it may be produced to as great an extent by the use of a large quantity of a Diffusible Stimulant; but in such a case it ceases to be employed as a Stimulant only.

Stimulants are useful in cases of debility, where this is real and unaccompanied by inflammation, and not merely apparent, whether the debility be the consequence of profuse Hæmorrhage or of other inordinate discharges, or a consequence of Asphyxia or of Syncope : in great general debility, or in Anœmia and Cachexia without any local inflammation, or in the last stages of many grave diseases, when the powers flag and life appears about to be extinguished. Their use, however, requires the greatest caution and discrimination; and though they may be prescribed beneficially in some cases even of chronic Inflammation, as is done with stimulant lotions to the eye in a state both of incipient and of chronic Inflammation, their employment is usually limited to prescribing them with other classes of remedies.

Heat. Electricity. The Vital Stimuli, as Heat, Air, Food, Drink, will all act as Stimulants in cases of Debility, or where the patient has been deprived of them; otherwise their use is not followed by exhaustion, but strengthens.

Ammoniæ Liq. 57. Sp. Ammoniæ et Sp. Ammoniæ arom. Ammoniæ Sesquicarb. et Liq. 61, 64.

Aromata or Spices, grateful in odour and taste; are hence used as Condiments. They stimulate the stomach; are useful as Carminatives, and as Adjuncts to remedies of different kinds, as to Tonics, Antispasmodics, and Cathartics.

Canella alba, 309. Of *Myrtaceæ*, Caryophyllus, 397, Ol. et Inf. Pimenta, 399, Ol. Sp. et Aqua. Cayaputi Ol. 395. Of *Laurineæ*, Cinnamomum, 537. Ol. Sp. Tinct. Aq. 541. Tinct. comp. Pulv. comp. et Elect. with other Aromatics. Confect. Aromat. 111, with Chalk also. Cassia, E. 541, Ol. Sp. Tinct. et Aqua, 545. Laurus nobilis. Folia, 532. Sassafras Ol. 535. Of *Myristiceæ*, Myristica, 530. Ol. et Sp. 531. Of *Piperaceæ*, Piper nigrum, 560. Conf. et. Ung. 561. P. longum, 559, in Conf. Opii, and with other aromatics. Of *Scitamineæ*, Curcuma, 618. Cardamomum, 616. Tinct et Tinct. comp. 617. Zingiber, 614. Tinct. et Syr. 615. Of *Irideæ*, Crocus, 618.

Capsicum, 511. Tinct. 512. Sinapis nigra et alba, 274, 276.

Others valuable chiefly on account of their *Volatile Oil*, as Ol. Calami Aromat. (*Spikenard*), 621. Of *Aurantieæ*, Limonis Ol. et Aurantii Ol. 298. Of *Umbelliferæ*. Anisum. Carui Ol. Sp. et Aqua, 409. Fœniculum dulce. Ol. et Aqua, 411. Anethum. Ol. et Aq. 412. Cuminum, 426. Coriandrum, 430. Of *Labiatæ*. Melissa, 502. Mentha viridis, Piperita, and Pulegium. Ol. Sp. et Aq. 498, 500. Lavandula, Ol. et Sp. Tinct. comp. 497. Rosmarinus, Ol. et Sp. 501. Origanum vulgare et Majorana et Ol. 502.

Others less agreeable, as Ol. Rutæ et Conf. 321. Ol. Anthemidis, 458, et Tanaceti, 461. Ol. Juniperi, 589. Ol. Sabinæ, 589. Ol. Jecoris Aselli, or Cod-liver Oil; but this is rather alterative than stimulant.

Turpentines—as Terebinthina Chia, 333, vulgaris, &c. 583. Oleum Terebinthinæ, 586.

Resins—as Elemi, 343. Mastiche, 334. Olibanum, 336. Myrrha, 338, and Balsam of Peru, &c. 367.

Ordinary Excitants—as Green Tea, 292. Coffee, 437.

Petroleum, 648. Naphtha, 648. Creasotum et Mist. 647. Camphora, 536.

Nux Vomica, 481, Tinct. et Extr. Strychnia, 484. Brucia, 485. Toxicodendron, 335. Arnica montana, 461.

See also STIMULANT TONICS, DIAPHORETICS, EXPECTORANTS, AND DIURETICS.

DIFFUSIBLE STIMULANTS.—These, which include Alcohol and Ether, are usually distinguished from other Stimulants from the rapidity with which they excite all the tissues through the medium of the brain and nervous system. They are quickly followed by exhaustion and collapse. The Alcoholic are seldom employed alone. The Etherial are useful in nervous and hysterical affections. The Wines, in moderate quantities, are in many cases useful Tonics.

Diffusible Stimulants.—Spir. Rectificatus, 634. Sp. Tenuior, 636. Sp. Vini Gallici et Mist. 637. Vinum Xericum, and other Wines, 637.

Æther Sulphuricus, 639. Sp. Ætheris Sulph. comp. 640. Sp. Ætheris nitrici.

Astringents.

Astringents are such substances as when brought in contact with the living body produce a corrugation and contraction of its fibres, and at the same time exercise a tonic influence through the medium of the Vital agency. The first effect is visible when an Astringent (then called a Styptic) is brought in contact with a bleeding wound, in the contraction which stops the bleeding from small vessels. The second effect may be observed in their occasionally curing intermittent fever, and other states of the system connected with debility. The possession of astringency in a body may be readily recognized by its corrugating the tongue when merely tasting it, otherwise there is no principle common to the whole. Some of the Astringents are mineral acids, or metallic salts, as those of Zinc, Lead, and Copper, or an earthy salt like Alum. But the greatest number of valuable Astringents are yielded by the vegetable kingdom, owing to the presence of Tannin, as in Catechu, but usually associated with Gallic acid, which produces an inky blackness with the sesqui-salts of Iron. Their effects must be ascribed partly to the mechanical effect of corrugation which they produce in the intestinal canal, but in a considerable degree also to the sympathetic influence, of which the effects are felt at a distance. They thus diminish inordinate secretion in cases of Diarrhœa, and also in distant organs, as in cases of Leucorrhœa. They are sometimes useful in cases of incipient inflammation, as in that of the eye and fauces, though it is usually only after the acute symptoms have subsided that they are admissible.

Cold. Cold Water. Freezing Mixtures. Acids, mineral as medicines: Acid. Sulph. dil.; vegetable, diluted as drinks. Refrigerants. Acetum.

Alumen, 122. Pulv. comp. with Kino, 124. Liq. comp. with Sulph. Zinc. 125. Cataplasma Aluminis, 125.

Plumbi Acetas, 166. Ung. 167. Pil. Plumbi Opiatæ, 167, 267. Plumbi Diacet. Liq. et dilutus, et Ceratum, 168, 169.

Zinci Sulphas, 156. (*v.* Liq. Aluminis comp. 125.) Zinci Acetat. Tinct. 153.

Ferri Sulphas, 137. Tinctura Ferri Sesquichloridi, 133.

Cupri Sulphas, 154. C. Ammonio-Sulphas, 155. Cupri Diacetas, 157.

Calcis Liquor, 108. Creta præparata, 110, by neutralizing acids.

Tannic and Gallic acids, 579. Hæmatoxylon, Dec. et Extr. 352, 354. Krameria (*Ratanhia*), Inf. et Extr. 281, 282.

Catechu, 349. Inf. Tinct. et Elect. with Opium, 352. Kino, 370, 374. Tinct. et Pulv. comp. with Aromatics and Opium, 373,

Granatum. Fructûs Cortex, 401. Dec. (℥ij.—Aq. Ojß. boiled to Oj.), L. Prunus spinosa, 389. Tormentilla et Dec. 381. Bistorta, 518.

Quercûs Cortex, Dec. et Extr. 576, 577. Gallæ, 578 : Tinct. Ung. and Ung. comp. with Opium, 580.

Rumex aquaticus, 519. Geum urbanum, 380. Lythrum Salicaria, 402.

Rosa gallica, Conf. Syr. Mel. et Inf. comp. with Dil. Sulphuric acid, 383

In Hæmorrhages: Styptics, Compresses, Plugs. Matico. Creasote. Quietude. Ergot in Uterine Hæmorrhage. Opium, 254, 266, allays irritability, and restrains inordinate discharges; hence united with Astringents as above. (*v.* Narcotics.)

TONICS.

Tonics are medicines possessing the power of gradually increasing the tone of the muscular fibre when relaxed, and the vigour of the body when weakened by disease. Though resembling Astringents in some of their effects, they do not produce corrugation, unless when combined with Astringent principle, as is the case with some true Tonics. Acting like Excitants on the vital principle through the medium of the nervous system, they differ from them in the slowness with which they produce, as well as in the permanence of their effects, and in their use not being followed by exhaustion or perceptible collapse. They are hence defined by Dr. Murray as "stimulants of considerable power, permanent in their operation." If carried to excess, or too long continued, they may act as Irritants, or be productive of debility; for "if given when the powers of the system are at their maximum, Tonics will assume the characters of excitants, and their administration be followed by collapse." (*Paris.*) When a Tonic is fitly prescribed, as in a case of debility, its effects are gradually perceived; the energy of the stomach and the appetite are increased, digestion is facilitated, the force of the circulation augmented without corresponding quickness, and respiration becomes fuller and more vigorous. In consequence of the more healthful performance of these functions, nutrition becomes necessarily more perfect. Absorption is performed with more vigour, as is first evident in the constipation which usually follows the successful exhibition of Tonics, but is soon perceptible in other parts, whence the œdematous swellings of invalids disappear. Secretions become more natural, the urine more scanty and high-coloured; inordinate and partial sweats disappear, and the skin returns to a natural state of softness, and the countenance resumes the natural glow of health. The senses and all the faculties become more active with the strength increased; and thus the patient labouring under diseases of real not apparent debility, or recovering from acute disease, or the effects of depressing and evacuating remedies, is restored to pristine health and energy.

Some of the Tonics, as Cinchona and its alkali Quinine, with the Arsenious acid, especially in the form of the Arsenical Solution, are prescribed as Antiperiodics, that is, as remedies which, taken in the intervals of paroxysmal diseases, as Ague and Remittent Fever, or attacks of Neuralgia, and even of Rheumatism, which observe some periodicity in their accession, are very frequently controlled by small doses frequently repeated in the intervals of, or a larger one immediately before a paroxysmal attack. The most violent Remittent will often be affected by a few drops of Arsenical solution, if prescribed immediately after the acute symptoms have been controlled by other means, and anything like a remission is observed in the febrile attack.

Tonics are prescribed either in substance or in Infusion or Decoction, with the addition frequently of a Tincture of the same or some other Tonic, or of one of the aromatic Stimulants. They require to be prescribed in moderate doses frequently repeated every two or three hours, sometimes changing them. It is usually preferable to begin with the milder before proceeding to the more powerful metallic Tonics, as the preparations of Iron.

Nutritious Diet. Cold. Exercise in the open air. Cold and Sea-Bathing.

Demulcent Tonics.—Cetraria et Dec. 629. Ulmus et Dec. 573. Pareira, Inf. et Extr. 245, 247. Calumba, Pulv. Inf. et Tinct. 247, 249: a mild tonic, thought also to be a little sedative; and being, like Quassia, without Tannin, may be prescribed with the salts of Iron.

Bitter Tonics.—Calumba. Quassia, Inf. et Tinct. 327. Simaruba et Inf. 329. Gentiana, Inf. Tinct. et Vinum comp. 487, 488. Chiretta, Inf. 489. Centaurium, 486. Menyanthes, 490. Centaurea benedicta, 455. Nux Vomica, Extr. 483: a powerful Bitter and Stomachic.

Stimulant Tonics.—Drimys Winteri, 244. Canella alba, 309. Aurantii Cortex, Conf. Tinct. Syr. et Inf. comp. 293, 294. Limonum Cortex, 298. Cusparia, Inf. et Tinct. 326.

Ruta et Extr. 321. Absinthium, Extr. 460. Tanacetum, 461. Archangelica, 411. Marrubium, 503. Cascarilla, Inf. et Tinct. 550. Mist. comp. with Squill, 552. Lupulus, Inf. Tinct. et Ext. 564, 568. Acorus, 620.

Antiperiodic and Astringent Tonics.—Cinchona Coronæ, 441. cinerea, 442, flava, 443, rubra, 444; Inf. Dec. Extr. et Tinct. 451; Tinct. comp. with Orange-peel, Saffron, and Serpentaria, 451. Quinia and Quinæ Disulphas, 446, 447. Amorphous Quinine, 450. Bebeerine, 545. Salicis Cortex, Dec. Salicine, 574. Piper nigrum? Uva Ursi. Chimaphila.

Mineral Antiperiodics.—Acid. Arseniosum, 202. Liq. Potassæ Arsenitis, 213.

Mineral Tonics.—Dil. Sul'. Dil. Nit'. Dil. Mur'.

Ferrum, 127. Ferri Sesquioxydum, 128. Emp. Ferri et Emp. Thuris, 199. Ferrugo, E. (Hydrated Sesquioxide), 129. Ferri Oxydum Nigrum, E. D. 130 (a compound of Protoxide and Sesquioxide). Chalybeate Springs.

Tinct. Ferri Sesquichloridi, 134. Tinct. Ferri Ammonio-Chloridi, 135. Ferri Sulphas, 137, et Pil. E. 138. Liq. Oxysulphatis, 139. Fer. Carbon. Saccharat. E. 140. Pil. Fer. comp. 140, with Carb. of Iron, Myrrh, and Sulph. Soda, 140. Mistura, with Myrrh, Nutmeg, Sulph. Potash, 141.

Ferri et Potassæ Tartras, 145. Vinum Ferri, 146. Ferri Citr. et Ammonio-Citras, 147. Aqua Chalybeata, 141. Ferri Acetas, D. 148. Lactate and Malate, 148. With Iodine, Ferri Iodidum et Syr. 131, 133. With Aromatics, Ferri Mist. Arom. 145. Pil. Rhei et Ferri, 526; Aloes et Ferri, 601.

Zinci Oxyd. et Sulph. 150, 151. Cupri Sulph. et Ammonio-Sulph. 154, 155. Bismuthi Trisnitras, 169.

MECHANICAL REMEDIES.

These, acting only as ordinary physical agents, or by their simple mechanical properties, are necessarily of less importance than the other groups of remedies, and might have commenced the series at p. 669. There occur, however, numerous cases in which we are required to protect an abraded surface, sheathe an irritated canal, or dilute an acrid state of the secretions, or increase the solvent powers of an excretion. For such purposes the more powerful remedies are as unsuited as these milder agents would be unfitted for controlling the more urgent symptoms of disease.

EMOLLIENTS.

Emollients, as their name indicates (from *Emollire*, to soften), are medicines calculated to soften the tissues with which they are brought in contact, and may thus include some of the Expressed Oils, Liniments, and Embrocations, with many Cerates and Ointments, such as the Cataplasms and Fomentations, of which the effects must be ascribed "to the relaxing effects of warmth and moisture upon the extreme vessels of the surface, propagated by contiguous sympathy to the deeper seated organs." (*Paris.*) It seems desirable to retain the term of Emollients for external applications, and that of Demulcents for those intended for internal exhibition.

Moist Heat. Fomentations. Papaveris Dec. 254. Malvæ Dec. comp. with Chamomile, 286. Anthemidis Inf. et Dec. 458. Cataplasm, with Malva, Verbascum, or Bread and Milk. Dauci Radix, 436. Cataplasma Lini, 285. C. simplex, 622, with Figs, 572.

Papaveris Oleum, 253. Lini Ol. 285. Amygdalæ Ol. 388. Olei Ol. 475. Cacao Butyrum, 289. Myristicæ Oleum expressum, 531. Palm Oil, 591.

Sambuci Ung. 432. Sevum vel Adeps Ovillus, 663. Adeps Suillus, 665. Cetaceum, Cerat. et Ung. 661. Cera alba et flava, Ceratum, Lin. et Emp. 657, 658.

Sapo, 478. Linim. Saponis with Camphor and Sp. of Rosemary, 476. Lin. c. Opio, 478. Emp. et Emp. comp. 478.

Application to Burns. Cotton, p. 288. Linim. Calcis, 108.

DEMULCENTS.

Demulcents in signification and in nature are the same as the Emollients, with which indeed they are usually united. But it is convenient to retain in a separate group the mucilaginous, starchy, saccharine, and gelatinous substances which are so frequently found useful in softening an irritated surface, and diminishing its sensibility to pain, either when applied externally, or taken internally, as in coughs, inflammation of the intestinal canal, or irritation of the urinary passages. But in these cases it is more than probable that their utility is chiefly due to the large quantity of water in which the Demulcent is dissolved.

Lini Inf. comp. 284, with Liquorice. Malva, 286. Althæa, Syr. et Mist. E. 287, with Raisins. Pate de Guimauve. Cydonia et Dec, 385.

Acaciæ Gummi, 346. Mucil. 348. Mist. E. with Sugar and Almonds, 349. Trochisci, 349. Tragacantha, 377. Mucil. et Pulv. comp, 379, with Starch and Sugar. Amygdalæ dulces, Conf. et Mist. 386, with Gum and Sugar. Emulsio Arabica, D. is nearly the same as the Mistura Acaciæ, E. 349, or the Mistura Amygdalæ, L. 387. Glycyrrhiza, Dec. Extr. et Troch. 375, 376. Verbascum, Inf. et Dec. 507.

Uvæ passæ, 311. Jujubes, 331. Fici, 672.

Amylum et Dec. 624. Tapioca, 557. Arrowroot, 612. Tous les mois, 613. Sago, 592. Salep, 611. Tritici Farina, 623. Avena, 621. Hordeum et Dec. 622. Rice and Maize, 625.

Saccharum et Syr. 626, 627. Mel, 656.

Icthyocolla, 659. Cetaceum, 660. Cornu, 661. Hartshorn for Jelly. Ovum 659. Lac, 664.

Diluents.

Diluents are very closely allied in nature to the Demulcents; indeed the same substances dissolved in a larger quantity of water form the group of Diluents, though it is the water here which is the powerful agent, as it will dilute acridity, and diminish viscidity. It requires to be of the purest kind when it is intended to make use of its solvent powers in Urinary complaints.

Aqua, 26. Distilled, rain, or pure spring Water. Toast and Water. Barley-Water. Rice-Water. Thin Gruel. Whey. Weak Demulcent Decoctions.

Anthelmintics. *Vermifuge.*

The word Anthelmintic is sometimes employed to indicate not only the medicines prescribed to prevent the production of worms, but also those which destroy or expel them, and the term Vermifuge is then applied to the latter only. As in other classes, very different medicines may be employed to produce the same effects, because worms may exist in different states of the constitution; therefore, whatever rectifies this, makes the intestinal canal less suitable to the residence of these parasites. Purgatives are frequently required to clear the intestinal canal, but Tonics are often as necessary to give it a healthy tone. Some Anthelmintics act mechanically by irritating the worms, as they press their bodies against the sides of the intestinal canal. Some again are specifically injurious to them, and others act chiefly as acrid and drastic purgatives. The worms commonly occurring in the intestinal canal in this country are the Tænia solium, or common Tape-worm. (Bothriocephalus latus, or Broad Tape-worm, occurs in Switzerland, &c.) Trichocephalus dispar, or Trichurus, Long Thread-worm, Oxyuris vermicularis, Maw-worm, or Ascarides, and Ascaris lumbricoides, common or long Round-worm. (See Steph. *Med. Zool.* Pl. 29.)

Mechanical Anthelmintics.—Stanni Pulvis, 171. Ferri Limatura, 126.

Mucuna pruriens, 379. The strigose pubescence of Rottlera tinctoria is also used as an Anthelmintic in India. Gigartina Helminthochorton, 630. from fine spiculæ of Corols, &c.

Specific Anthelmintics.—Granatum. Radicis Cortex. Dec. 401 (omitted in the L. P.) Filix Mas. Pulv. and Etherial Extract, 627.

Andira vel Geoffroyæ Dec. 364. Spigelia marylandica Pulv. et Inf. with Senna, 481. S. Anthelmia, 481. Persicæ Folia, 386.

Terebinthinæ Oleum, 586. Rutæ Ol. 321. Tanacetum, 461. Absinthium, 460. Santonicum, 460.

Purgatives, &c. as Anthelmintics.—Calomel. Gamboge, 307. Jalap. Scammony. (*See* Cathartics.)

Bitters generally injurious to worms, but useful also in giving tone.

Enemata against Ascarides. Sol. of Salt in Inf. Quassiæ. (*Paris*). Enema Aloes, L. (made with *Aloes* ℈ij. *Carb. Potash* gr. xv *Barley-water* Oß.)

Injections of cold Water or of Bitter Infusions, of Camphor in Oil.

ANTIDOTES, ETC., TO POISONS.

POISONS *as arranged by Dr. Christison, with the* ANTIDOTES, *&c., mentioned in the foregoing pages.*

In most cases the stomach requires to be quickly evacuated, either by emetics or by the stomach-pump; but with corrosive poisons this is not always safe. The suitable antidotes are to be prescribed, sometimes viscid substances to involve the poison. In many cases irritation is to be allayed, and inflammation subdued; but in others, moderate stimulants are necessary. With the poisonous gases, fresh air is essential, and cold affusion useful.

IRRITANT POISONS.

Sul′, 33. Nit′, 40. Phosphorus and acid, 35.

Acids.—Muriatic′ and Nitro-Muriatic acid, 48. Tinct. of Muriate of Iron. Chalk to be avoided.

Oxalic′, Tartaric′, Citric′, or Acetic acid, 316.

Alkalis.—Ammonia, 58. Potash, 71, its Carb. 77. Soda, &c. 92. So Lime. Nitre, 84.—Alkaline Sulphurets, 54.

Sol. of Chlorine, 40. Iodine, 42, and Iodide of Potassium, 73 : so also Bromine and the Bromide of Potassium.

Baryta, salts of. Chloride of Barium, treated with Sulphates of Magnesia and Soda, also Carbonates. Stomach-pump or Emetics.

Arsenic. Arsenious acid. Liq. Potassæ Arsenitis. Sulphurets, 212. (See Hydrated Sesquioxide of Iron. 129.) Magnesia in a gelatinous state, or very light Magnesia, will remove about 1-25th of its weight of Arsenic from its solution in water. (*v.* P. J. vi. 137.)

Mercury, Bichloride, Bicyanide, and its irritant salts, 198. Dr. Paris recommends Tartar Emetic as an Emetic in poisoning by Corrosive Sublimate.

Copper, salts of, 154. Hydrated Oxide of Iron has been recommended.

Antimony, salts of, 180. Zinc, as for Antimony. Lead, salts of, 167, with Milk and Albumen, Sulphate of Soda and Magnesia. Silver, Nitrate of, &c.—Administer Common Salt and some of the Incompatibles at p. 217.

Vegetable Acrids.—Euphorbium, 559. So Croton. Colocynth. Elaterium. Mezereon. Gamboge. Jalap. Savine.

Animal Acrids.—Cantharides.—Evacuate Stomach. Demulcents. Allay irritation with Camphor, Dover's Powder, &c.

NARCOTIC POISONS.

Opium and its preparations, 271. So Henbane. Lactuca.

Hydrocyanic′. Laurel-water. Oil of Bitter Almonds, 394.

Poisonous Gases.—Chlorine, 40. Ammonia, 58. Hydrosulphuric′, 54. Carbonic acid, 53. Carburetted Hydrogen. With all, exposure to pure air, artificial respiration, and affusion of cold water.

NARCOTIC ACRID POISONS.

Belladonna, 511. Stramonium, 514. Tobacco.

Conium, 429, or Hemlock and poisonous Umbelliferæ.

Nux Vomica, 483. Strychnia, 485. Remove poison from stomach.

Aconite, 243 : so Black and White Hellebore. Colchicum. Sabadilla.

Cocculus indicus. Digitalis, 506.

Alcohol, 637. Ether.—Evacuate stomach; cold affusion over head, and evaporating lotions; Leeches. Ammonia as a Stimulant.

COLOURING INGREDIENTS.—Rhœas, 251. Dianthus Caryophyllus, 282. Pterocarpus santalinus, 369. Syr. et Tinct. Croci, 619. Coccus, 656.

GAUBIUS' TABLE,

Regulating the ordinary Proportion of Doses according to the Age of the Patient.

For an adult, suppose the dose to be	1	or	1	drachm,
Under 1 year will require	$\frac{1}{12}$	,,	5	grains.
,, 2 ,, ,, ,,	$\frac{1}{8}$	,,	8	,,
,, 3 ,, ,, ,,	$\frac{1}{6}$	,,	10	,,
,, 4 ,, ,, ,,	$\frac{1}{4}$	,,	15	,,
,, 7 ,, ,, ,,	$\frac{1}{3}$	,,	1	scruple.
,, 14 ,, ,, ,,	$\frac{1}{2}$	,,	$\frac{1}{2}$	drachm.
,, 20 ,, ,, ,,	$\frac{2}{3}$	,,	2	scruples.
,, 21 to 60, the full dose, or	1	,,	1	drachm.

Above this age, an inverse gradation must be observed.

INDEX.

LATIN AND ENGLISH.

THE END.

London:
Printed by S. & J. Bentley, Wilson, and Fley,
Bangor House, Shoe Lane.

ADDENDA.

Page 450, line 19, *add*:—Mr. Redwood (P. J. vi. 163) states that Quinoidine and the Amorphous Quinine are both mixed products, containing several proximate principles, and that one part of these principles is soluble, the other insoluble in Ether, and that in fact amorphous Quinine is the same substance which has long been known in commerce under the name of Quinoidine or Chinoidine.

To Calcis Murias	page 113	line 22	*add*	E.
„ Potassii Ferrocyanidum	„ 143	„ 1	„	E.
„ Plumbi Diacet. Sol.	„ 168	„ 6	„	Aqua.*
„ Plumbi Subacetatis Liquor comp.	169	„ 2	„	D.
„ Names of Ant. Pot. Tart.	„ 177	„ 33	„	Tartarum Emeticum, D.
„ Hydr. e Creta	„ 182	„ 7	„	Grey Powder.
„ Tinct. Opii Camph.	„ 269	„ 25	„	Elixir Paregoricum, D.
„ Sinapis alba	„ 276	„ 36	„	D.
„ Krameria	„ 281	„ 16	„	Ratanhia, D.
„ Limonum Oleum	„ 298	„ 3	„	L. E.
„ Oxalis Acetosella	„ 314	„ 25	„	Acetosella, L.
„ Pil. Colocynthidis, E.	„ 405	„ 6	„	Comp. D.
„ Cuminum Cyminum	„ 426	„ 2	„	Cyminum, L.
„ Pulv. Ipecac. comp.	„ 436	„ 35	„	L. E. D.
„ Decoct. Cinchonæ	„ 451	„ 18	„	D.
„ Cnicus benedictus	„ 455	„ 1	„	Centaurea benedicta, D. *syn.*
„ Extr. Anthemidis	„ 458	„ 25	„	E.
„ Mentha Pulegium	„ 500	„ 19	„	L.
„ Origanum Majorana	„ 502	„ 21	„	D.
„ Ung. Piperis	„ 561	„ 41	„	Nigri.

CORRIGENDA.

Page 45	line 44	*dele*	Clk.
„ 125	„ 4	*for*	Lime *read* Zinc.
„ 138	„ 32	„	exsiccatus *read* exsiccatum, E.
„ 193	„ 31	„	Calomelus *read* Calomelas.
„ 195	„ 12	„	Corrosivus Sublimatus *read* Sublimatus Corrosivus, E.
„ 201	„ 28	„	Hydrarg. Sulph. Oxydum *read* Hydrarg. Oxyd. Sulphuric. D.
„ 252	„	„	Fig. 39 *read* 41.
„ 253	„ 20	„	Fig. 39 *read* 41.
„ 302	„ 17	„	Cambogia *read* Gambogia, D.
„ 336	„ 20	„	Boswellia serrata *read* B. THURIFERA, Roxb. *B. serrata*, L.
„ 343	„ 8	*dele*	Tinct. Ammon. comp.
„ 406	„ 3	*for*	Rech *read* Rich.
„ 429	„ 22	„	inspissatus *read* spissatus, D.
„ 432	„ 3	„	Leucorrhœas *read* Leucorrhœa.
„ 474	„ 33	„	Paragoric *read* Paregoric.
„ 544	„ 6	„	Ebern. *read* Eberm.

---

* Called Aqua, p. 30, and Solutio, p. 111, E. P.

Printed in the United States
By Bookmasters